STUDY GUIDE & PROBLEMS BOOK

ORGANIC CHEMISTRY

SECOND EDITION

BROWN & FOOTE

Brent L. Iverson
University of Texas, Austin

Sheila A. Iverson
Austin, Texas

William H. Brown
Beloit College

Saunders College Publishing

Harcourt Brace College Publishers

Fort Worth Philadelphia San Diego New York Orlando Austin
San Antonio Toronto Montreal London Sydney Tokyo

Brown & Foote: Student Study Guide and Problems Book to accompany *Organic Chemistry* Second Edition.Iverson, Iverson & Brown.

0-03-020453-4

567 202 7654321

To Carina, Alexandra, Alanna, and Juliana
with love

This Study Guide and Problems Book is a companion to the second edition of *Organic Chemistry* by William Brown. This volume provides a detailed section-by-section *overview* of the major points covered in the text. Reading these overviews before and after reading a text chapter should help identify and summarize important elements of the material. The overviews are written in a very compact *outline format*. *Key terms* are printed in boldface the first time they appear. Bracketed sentences in italic print provide *hints for studying and pitfalls to avoid*. Especially important ideas are identified with the symbol "☆".

A *summary of reactions* is presented in tabular form. The starting materials are listed on the vertical axis, while the products of the reaction are listed across the top. Where the two intersect in the table is the section number where a particular reaction can be found in the text. Below the table are generalized descriptions and explanations of the different reactions in the table.

All of the *problems* from the text have been reprinted in this guide, so there is no need to flip back and forth between the text and the guide. Detailed, stepwise *solutions* to all of the problems are provided. This guide was reviewed for accuracy by Brent and Sheila Iverson, Richard Luibrand at the California State University at Hayward, and William Brown at Beloit College. If you have any comments or questions, please direct them to Professor Brent Iverson, Department of Chemistry and Biochemistry, the University of Texas at Austin, Austin, Texas 78712.
E-mail: biverson@utxvms.cc.utexas.edu.

Brent and Sheila Iverson
Austin, Texas
June, 1997

CONTENTS:

1   Covalent Bonds and Shapes of Molecules   1
2   Alkanes and Cycloalkanes   37
3   Acids and Bases   72
4   Stereochemistry   87
5   Alkenes I   111
6   Alkenes II   133
7   Alkyl Halides and Radical Reactions   174
8   Nucleophilic Substitution and β-Elimination   201
9   Alcohols and Thiols   231
10   Alkynes   269
11   Ethers and Epoxides   291
12   Mass Spectrometry   329
13   Nuclear Magnetic Resonance Spectroscopy   346
14   Infrared and Ultraviolet-Visible Spectroscopy   372
15   Aldehydes and Ketones   388
16   Carboxylic Acids   453
17   Functional Derivatives of Carboxylic Acids   478
18   Enolate Anions and Enamines   518
19   Aromatics I: Benzene and its Derivatives   571
20   Aromatics II: Reactions of Benzene and its Derivatives   619
21   Amines   656
22   Conjugated Dienes   690
23   Organic Polymer Chemistry   714
24   Carbohydrates   733
25   Lipids   764
26   The Organic Chemistry of Metabolism   782
27   Amino Acids and Proteins   794
28   Nucleic Acids   825

CHAPTER 1: COVALENT BONDS AND SHAPES OF MOLECULES

1.0 OVERVIEW
- Organic Chemistry is the study of compounds that contain carbon atoms in combination with other types of atoms such as hydrogen, nitrogen, oxygen, and chlorine.
- The Lewis model of bonding qualitatively describes coordination numbers of atoms and molecular geometries.
- Valence bond theory and molecular orbital theory comprise a more accurate theoretical framework with which to understand relationships between molecular structure and reactivity.
- Organic chemists are concerned primarily with where electrons are located in an atom, molecule, or ion, because then they can understand or predict structure, bonding, and reactivity. ✶

1.1 ELECTRONIC STRUCTURE OF ATOMS
- Electrons are found around atoms in defined regions of space called **atomic orbitals.**
    - An atomic orbital can hold up to two electrons (the **Pauli Exclusion Principle**), one with spin quantum number of +1/2, and the other with spin quantum number -1/2.
    - Atomic orbitals are classified as s, p, d, or f.
    - For most of organic chemistry, s and p atomic orbitals are the only types of orbitals that we need to consider.
    - The d atomic orbitals are important for third row elements such as sulfur (S) and phosphorus (P).
- Orbitals with the same **principal quantum number** (1, 2, 3, etc.) form what is called a **shell of electrons.** For example, the 2s and 2p orbitals are in the same shell (the 2nd shell) while 1s and 2s orbitals are in different shells.
    - There is only one s orbital for a given shell
    - There are three p orbitals for shells with principal quantum number 2 and higher. The three 2p orbitals are orthogonal to each other and are designated $2p_x$, $2p_y$, and $2p_z$.
    - There are five d orbitals for shells with a principal quantum number of 3 or higher.
    - All the orbitals of the same type in a given shell have the same energy; that is, they are said to be **degenerate.** For example, all three 2p orbitals are degenerate.
- Different elements have different numbers of electrons, and these are placed in orbitals beginning with the lowest energy orbital (the **Aufbau Principle**).
    - Orbitals with the smallest principal quantum number are lowest in energy and are filled first.
    - For orbitals with the same principal quantum number, s orbitals are filled before p orbitals which are filled before d orbitals.
    - According to **Hund's rule**, one electron is added to each degenerate orbital before two electrons are added to any one of them.
- Chemists are primarily interested in the **valence electrons** of an atom. **Valence electrons** are the electrons in the outermost shell of an atom. ✶
    - For H, the valence electron is in the 1st shell, namely the 1s orbital.
    - For C, N, O, and F the valence electrons are in the 2nd shell, namely the 2s and the three 2p orbitals.
    - For Si, P, S, and Cl the valence electrons are in the 3rd shell, namely the 3s, and the three 3p orbitals. For P and S, the 3d orbitals are important.

1.2 THE LEWIS MODEL OF BONDING
- Atoms can gain or lose electrons to fill their valence shells. An atom that gains electrons has an overall negative charge and is called an **anion**, while an atom that loses electrons has an overall positive charge and is called a **cation**. ✶
- Atoms can also take part in **chemical bonds** to fill their valence shells.
- **Chemical bonds** in molecules are made from electrons in the valence shells of atoms. ✶
    - Bonds do not involve electrons from shells below the valence shell.
    - Atoms overwhelmingly prefer to be surrounded by a filled valence shell of electrons (**noble gas configuration**). ✶

- A filled valence shell for H is 2 electrons, and a filled valence shell for C, N, O, and F is 8 electrons; the **"octet rule."**
- P and S can have 8 electrons in their valence shells, but the valence shell for S may contain as many as 10 electrons, and the valence shell of P may contain as many as 12 electrons. This is due to the presence of 3d orbitals in these third row elements.

• In an **ionic bond**, an atom transfers one or more electrons to a different atom creating negatively and positively charged ions that then attract each other.
- Ionic bonding is only observed when the electron transfer creates filled valence shells for both of the ions.

• In a **covalent bond**, atoms share pairs of electrons, thus increasing the number of electrons around each atom. In this way the valence shell of each atom is filled.
- Sharing a pair of electrons holds the two atoms together.
- Organic chemistry is concerned primarily with covalent bonds.
- Noble gases do not normally take part in bonding because their valence shells are already filled.

• Electron pairs taking part in covalent bonds are not necessarily shared evenly between the atoms. The more **electronegative element** taking part in a bond attracts the majority of the electron density of the shared electrons. *
- The unequal sharing of electrons of a covalent bond can be analyzed using a table of electronegativities of the elements.
- On the Periodic Table of the elements, **electronegativity** increases from the bottom left hand corner to the upper right hand corner (See Table 1.3 in book).

• Knowing how to identify the electron-rich and electron-poor regions of molecules is the key to understanding and learning organic chemistry. *
- The unequal sharing of electrons in covalent bonds forms the basis for reactivity in a molecule, so understanding how electrons are distributed in a given molecule allows chemical reactions to be predicted accurately. *[By the time you have finished studying organic chemistry you should be able to look at the structure of a molecule, and then, based on your understanding of the electron distribution in the molecule, predict chemical reactions. This prediction approach is much more successful than trying to memorize reactions without understanding the reasons they take place.]*

• A **nonpolar covalent bond** has a relatively small difference in electronegativities, less than 0.5, between the two atoms of the bond. A **polar covalent bond** has a relatively large difference in electronegativities, (between 0.5 and 1.9) between the two atoms of the bond.

• The polarity of a covalent bond can be measured as the **bond dipole,** μ, the product of one of the charges times the distance separating the charges. *

• **Lewis structures** are used to represent molecules. *[Lewis structures are not as hard as they look, but you need a lot of practice to get the hang of them.]*
- In a Lewis structure, a line between two atoms represents a pair of electrons taking part in a bond and a pair of dots represents an unshared or lone pair of electrons.

• **To draw a Lewis structure:**
- **First**, count all the valence electrons of each atom in the structure. Remember that in the neutral states, H has 1 valence electron, C has 4, N has 5, O has 6 and F has 7. Add a valence electron for each unit of negative charge on an atom or ion, and subtract a valence electron for each unit of positive charge.
- **Second**, determine how the atoms are connected to each other. Usually the connectivity information must be provided for you in the form of a condensed structural formula.
- **Third**, draw single bonds (a line) between all of the atoms known to be connected to each other. For example, for the condensed structural formula CH_3CHO write down:

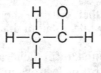

- **Fourth**, draw any remaining bonds (double or triple bonds) and add any lone pairs of electrons that may be necessary so that each atom in the molecule is surrounded by a **filled valence shell** of electrons. *

For H, a filled valence shell is 2 electrons (1 bond), and for C, N, O, and F; a filled valence shell is 8 electrons distributed as described in the following table:

Atom	# Of Bonds (Single bonds count as 1 bond, double bonds as 2, and triple bonds as 3)	# Of Lone Pairs Of Electrons
C	4	0
N	3	1
O	2	2
F	1	3

Finishing the example, two lone pairs of electrons are added to the oxygen atom and a double bond is added between the carbon and oxygen atoms to fill all valence shells and complete the structure:

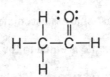

In most organic molecules, the halogens Cl, Br, and I are treated the same as F. On the other hand, the neutral atoms B, P, and S have some unusual bonding patterns as described in the following table.

Atom	# Of Bonds (Single bonds count as 1 bond, double bonds as 2, and triple bonds as 3)	# Of Lone Pairs Of Electrons
B	3	0
P	3	1
P *	5	0
S	2	2
S	4	1
S *	6	0

(* 3d orbitals are involved, so more than 8 valence electrons can be accommodated in the valence shell.)

If there is a negative formal charge (see below) on an atom, add a lone pair of electrons and use one less bond.

If there is a positive formal charge on an atom other than carbon, add a bond and use one less lone pair of electrons.

For carbon with a positive formal charge, you cannot add a bond because carbon already has four. In this case, you should use one less bond and no lone pairs (the carbon is surrounded by only six valence electrons).

• **Some Helpful Hints** for drawing Lewis structures:

- After drawing the single bonds between all connected atoms, draw the lone pairs of electrons. For neutral atoms, the number of lone pairs on a given element does not change from molecule to molecule.

In the neutral, uncharged state, C has zero lone pairs, N has one lone pair, O has two lone pairs, and F has three lone pairs. For example, in the molecule CO_2, draw the single bonds and lone pairs of electrons around oxygen as follows:

$$\ddot{O}-C-\ddot{O}$$

- After drawing the lone pairs, fill in multiple bonds as necessary. If there are not enough single bonds to other atoms surrounding an atom to give a filled valence shell, then use multiple bonds.

In these cases you should draw the multiple bond(s) to the adjacent atom(s) that also need multiple bond(s) to fill their valence shell. The CO_2 example is completed by adding a double bond to each oxygen atom, thereby filling the valence shell of carbon as well:

$$\ddot{O}{=}C{=}\ddot{O}$$

• **Formal charges** are useful as a bookkeeping method for keeping track of charges on a molecule. ✳
   - For the computation of formal charge, the electrons in bonds are counted as being distributed evenly between the bonded atoms. One electron is counted for each atom taking part in a single bond, two electrons for each atom taking part in a double bond, and three electrons for each atom taking part in a triple bond.
   - Lone pairs of electrons and electrons in shells lower than the valence shell (1s electrons for C, N, O, F, etc.) are counted as belonging entirely to the atoms to which they are attached.
   - Total formal charge derives from comparing the number of electrons counted as above to the number of protons in the nucleus of the atom (an extra electron results in a formal charge of -1, an extra proton results in a formal charge of +1, etc.).
   - The following tables show the formal charges associated with certain atoms commonly found in organic molecules and reaction intermediates.
   - The sum of formal charges in a molecule is equal to the net charge.

Table Of Atoms With +1 Formal Charge

Atom	# Of Bonds (Single bonds count as 1 bond, double bonds as 2, and triple bonds as 3)	# Of Lone Pairs Of Electrons
H	0	0
C	3	0
N	4	0
O	3	1
S	3	1
P	4	0

Table Of Atoms With -1 Formal Charge

Atom	# Of Bonds (Single bonds count as 1 bond, double bonds as 2, and triple bonds as 3)	# Of Lone Pairs Of Electrons
C	3	1
N	2	2
O	1	3
S	1	3
F	0	4

1.3 FUNCTIONAL GROUPS
• **Condensed structural formulas** are highly abbreviated versions of Lewis structures that are used to describe molecules.
   - The number of hydrogen atoms attached to a given atom is denoted with a subscript. For example -CH_3 means there are three hydrogen atoms attached to the carbon atom.
   - Lines are used to denote bonds. A single line (-) between two atoms represents a single bond, a double line (=) denotes a double bond, etc.
     Lone pairs of electrons are not drawn.
   - Parentheses are used to denote branching in a molecule. In other words, whole groups of atoms attached to a given atom are placed in parentheses. For example, $(CH_3)_3CH$ indicates that there are three -CH_3 groups attached to the remaining carbon atom.

• Carbon combines with other atoms to form characteristic structural units called **functional groups** such as hydroxyl groups (-OH), carboxyl groups (-CO$_2$H), and carbonyl groups (C=O) that are important for three reasons: ✳
 - **First**, they are sites of chemical reactions, and a particular functional group, in whatever compound it is found, undergoes the same types of chemical reactions.
 - **Second**, they are used to divide organic molecules into classes in terms of their physical properties.
 - **Third**, they provide a basis for naming compounds.

1.4 BOND ANGLES AND SHAPES OF MOLECULES

• The **Valence-Shell Electron-Pair Repulsion (VSEPR)** model of molecular structure assumes that *areas* of valence electron density around an atom are distributed to be as far apart as possible in three-dimensional space. ✳
 - When using the VSEPR model, lone pairs of electrons, the two electrons in a single bond, the four electrons in a double bond, and the six electrons in a triple bond are each counted as only a single area of electron density.
 - Four areas of electron density around an atom adopt a tetrahedral shape with bond angles near 109.5°, such as in methane, CH$_4$.

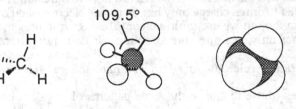

 - Three areas of electron density around an atom adopt a trigonal planar shape with bond angles near 120°, such as in formaldehyde, H$_2$C=O.

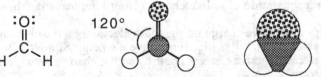

 - Two areas of electron density around an atom adopt a linear shape with bond angles near 180°, such as in acetylene, HC≡CH.

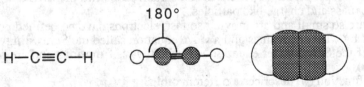

 - The VSEPR model predicts shape but does not explain why the electrons are located where they are. Thus, it is only a useful model and not a theory. The theory of electronic structure is presented in Sections 1.7 and 1.8.

1.5 POLAR AND NONPOLAR MOLECULES

• **The dipole moment, μ,** of a molecule is the vector sum of all the individual bond dipoles. Thus, to understand the dipole moment for a molecule, an understanding of the bond dipoles must be combined with an understanding of molecular structure. For example, all the bond dipoles cancel in symmetrical molecules such BF$_3$ and CCl$_4$, but not in molecules such as H$_2$O and NH$_3$. ✳

1.6 RESONANCE

• **Resonance theory** is used to depict and understand those special chemical species for which no single Lewis structure provides an adequate description. ✳ *[This also requires a lot of practice.]*

- Resonance theory is particularly good at helping to understand cases of partial bonding (for example 1.5 bonds between two atoms, etc.) or when a formal charge is distributed between more than one atom. In these situations, the true structure (referred to as a **resonance hybrid**) is thought of as a composite of two or more **contributing structures**.

  In drawing resonance structures, a straight, double headed arrow (↔) is placed between contributing structures. Curved arrows (⌒) are used to indicate how electrons can be redistributed to make one contributing structure from another. Always draw the curved arrows to indicate where a pair of electrons started (tail of arrow) to where the electrons end up (head of arrow).
- No atoms are moved between contributing structures, and only *certain* kinds of electrons are moved. ✳
  Lone pairs of electrons or a pair of electrons taking part in a multiple bond are moved.
  A lone pair of electrons from an atom can only move to an adjacent bond to make a multiple bond. A pair of electrons in a multiple bond can only move to an adjacent atom to create a new lone pair of electrons or to an adjacent bond to make a multiple bond.
  - Contributing structures cannot have atoms with more than a filled valence shell of electrons.✳
    There can be no more than 2 electrons around H or more than 8 electrons around C, N, O, or F.
    You cannot have more than a filled valence shell because there are no more orbitals in which to place the extra electrons. On the other hand, atoms can have less than the filled valence shell, for example a carbon atom with +1 formal charge only has 6 valence electrons and one empty 2p orbital.
- Keep in mind that even though we use multiple contributing structures to describe a molecule or ion, the molecule or ion in fact only has <u>one</u> true structure. It does <u>not</u> alternate between the contributing structures.✳
- The following qualitative rules are used to estimate the **relative importance** of **contributing structures**.
  - Equivalent structures (those that have the same patterns of covalent bonding) contribute equally. For example, carbonate ion (CO_3^{2-}) has three equivalent contributing structures.
  - Structures in which all atoms have filled valence shells (complete octets) contribute more than those in which one or more valence shells are not filled.
  - Structures involving separation of unlike charges contribute less than those that do not involve charge separation.
  - Structures that carry a negative charge on the more electronegative element contribute more than those with the negative charge on the less electronegative element. Similarly, structures that carry a positive charge on the less electronegative element contribute more than those with the positive charge on the less electronegative element.

1.7 QUANTUM OR WAVE MECHANICS
- Electrons have certain properties of **particles** and certain properties of **waves**. ✳
  - Electrons have mass and charge like particles.
  - Because they are so small and are moving so fast, electrons have no defined position. Their location is best described by wave mechanics and a wave equation called the **Schrödinger equation**.
    Solutions of the Schrödinger equation are called **wave functions** and are represented by the Greek letter ψ (psi).
    Each wave function (ψ) describes a different orbital.
    There are many solutions to the Schrödinger equation for a given atom.
  - The **sign of the wave function** ψ can change from positive (+) to negative (-) in different parts of the same orbital. This is analogous to the way that waves can have positive or negative amplitudes. ✳
    The sign of the wave function does not indicate anything about charge. *[This can be confusing. Make sure that you understand it before you go on.]*
  - The value of ψ^2 is proportional to the probability of finding electron density at a given point in an orbital.
    Note that the sign of ψ^2 is always positive, because the square of even a negative value is still positive.
    In a **2p** orbital, it is just as probable to find electron density in the negative lobe as it is to find electron density in the positive lobe. *[Make sure you understand this statement.]*
  - A **node** is any place in an orbital at which the value of ψ and thus ψ^2 is zero.
    A nodal surface or nodal plane are surfaces or planes where ψ and ψ^2 is zero. There is absolutely no electron density at a node, a nodal surface, or a nodal plane.

- The Schrödinger equation can in principle describe covalent bonding, but, even with powerful computers the equation is too complicated to be solved exactly for large molecules.

1.8 MOLECULAR ORBITAL THEORY TO COVALENT BONDING

• **Molecular orbital theory** assumes that individual electron pairs are found in **molecular orbitals** that are distributed over the *entire* molecule. ✳

• Molecular orbitals are analogous to atomic orbitals and are described by the following four rules:
   - **First**, combination of n atomic orbitals in a molecule or ion forms n molecular orbitals, each of which extends over the entire molecule or ion.
      The number of molecular orbitals is equal to the number of atomic orbitals combined, because atomic orbitals can be combined by both addition and subtraction.
   - **Second**, molecular orbitals, just like atomic orbitals, are arranged in order of increasing energy.
   - **Third**, filling of molecular orbitals is governed by the same principles as the filling of atomic orbitals. (See Section 1.1)
      Electrons are placed in molecular orbitals starting with the lowest energy orbitals first.
      A molecular orbital cannot hold more than two electrons.
      Two electrons in the same molecular orbital have opposite spins.
      When two or more **degenerate** (same energy) molecular orbitals are available, one electron is placed in each before any one of them gets two electrons.

• When two atomic orbitals combine to form a molecular orbital, the wave functions are both added and subtracted to create one **bonding molecular orbital** and one **antibonding molecular orbital**.
   - A bonding molecular orbital occurs when the electron density of the orbital is concentrated between the atomic nuclei.
      Electrons in bonding molecular orbitals stabilize covalent bonds because they serve to offset the repulsive forces of the positively-charged atomic nuclei. Both nuclei are attracted to the electrons between them.
      The energy of a bonding molecular orbital is lower than the energy of the uncombined atomic orbitals.
   - An antibonding molecular orbital (designated with an *) occurs when the electron density of the orbital is concentrated in regions of space outside the area between the atomic nuclei.
      Electrons in antibonding molecular orbitals do not stabilize covalent bonds because the electrons are *not* positioned to offset the repulsive forces of the positively charged atomic nuclei.
      The energy of an antibonding molecular orbital is higher than the energy of the uncombined atomic orbitals.

• A σ **(sigma) bond** occurs when the majority of the electron density is found on the bond axis.
   - For example, a σ bond results from the overlap between two 1s orbitals.
   - Because rotating a σ bond does not decrease the overlap of the orbitals involved (σ bonds have cylindrical symmetry), a σ bond can rotate freely about the bond axis.

• A π **(pi) bond** occurs when the majority of the electron density is found above and below the bond axis.
   - For example, a π bond results from the overlap of two 2p orbitals that are parallel to each other, and orthogonal to the σ bond that exists between the two atoms.
   - Because rotating a π bond by 90° destroys the orbital overlap, π bonds cannot rotate around the bond axis. *[Understand this before going on.]*

• An **electronic ground state** occurs when all of the electrons are in the molecular orbitals of lowest possible energy. An **electronic excited state** occurs when an electron in a lower lying orbital is promoted to an orbital that is higher in energy. This can occur when light is absorbed by a molecule, for example.

• For elements more complicated than hydrogen, it is helpful to combine (**hybridize**) the *valence* atomic orbitals on a given atom before looking for overlap with orbitals from other atoms. ✳
   - For C, N, and O hybridization means the 2s atomic orbital is combined with one, two, or all three 2p atomic orbitals.

• The results of the orbital combinations are called **hybrid orbitals**, the number of hybrid orbitals are equal to the number of atomic orbitals combined.
   - An **sp³ hybrid orbital** is the combination of one 2s orbital with three 2p orbitals.
      Four sp³ orbitals of equivalent energy are created.

Each sp^3 orbital has one large lobe and a smaller one of opposite sign pointing in the opposite direction (with a node at the nucleus). The large lobes point to different corners of a tetrahedron (109.5° bond angle). This explains the tetrahedral structure of molecules like methane, CH_4.

- An sp^2 **hybrid orbital** is the result of combining the 2s orbital with two 2p orbitals.

Three sp^2 orbitals of equivalent energy are created.

Each sp^2 orbital has one large lobe and a smaller one of opposite sign pointing in the opposite direction (with a node at the nucleus). The large lobes point to a different corner of a triangle (120° bond angle). This explains the trigonal planar structure of molecules like formaldehyde, $CH_2=O$.

The left over 2p orbital lies perpendicular of the plane formed by the three sp^2 orbitals.

• **An sp hybrid orbital is the combination of one 2s orbital with one 2p orbital.**

Two sp orbitals of equivalent energy are created.

Each sp orbital has two lobes of opposite sign pointing in opposite directions (with a node at the nucleus).

The lobes with like sign point in exactly opposite directions (180° bond angle). This explains the linear structure of molecules like acetylene, $HC\equiv CH$.

The two left over 2p orbitals are orthogonal to each other, and orthogonal to the two sp hybrid orbitals as well.

- Carbon atoms in molecules are either sp^3, sp^2, or sp hybridized. 1s Orbitals are not considered for hybridization with C, N, or O because the 1s orbitals do not participate in covalent bonding.

- The hybridization of a given atom (sp^3, sp^2, or sp) determines the geometry and type of bonds made by that atom. The important parameters associated with each hybridization state are listed in the following table.

Carbon Atom Hybridization State Parameters

Hybridiztion State	# Of Hybrid Orbitals	# Of 2p Orbitals Left Over	# Of Groups Bonded To Carbon	# Of σ Bonds	# Of π Bonds	Geometry Around Carbon
sp^3	4	0	4	4	0	Tetrahedral
sp^2	3	1	3	3	1	Trigonal Planar
sp	2	2	2	2	2	Linear

• **Bonding in complex molecules** can be *qualitatively* understood as **overlap of hybrid orbitals**. ✳
• Organic chemistry is primarily concerned with **two types of covalent bonds**, namely **sigma (σ) bonds and pi (π) bonds.**

- A **σ (sigma) bond** can be formed in a variety of ways. ✳

A σ bond results from the overlap between an s orbital and any other atomic orbital.

A σ bond also results from the overlap of an sp^3, sp^2, or sp hybrid orbital and any s, sp^3, sp^2, or sp hybrid orbital **along the bond axis.** *[You should be able to picture these different types of orbital overlap that all lead to σ bonds.]*

Because rotating a σ bond along the bond axis does not decrease any orbital overlap, there is only a small barrier to rotation. Thus, **single bonds rotate extremely rapidly around the bond axis.** This explains why molecules with only σ bonds are highly flexible, able to adopt an almost infinite number of rapidly interconverting conformations in solution. ✳

- A **double bond** in molecules such as $H_2C=CH_2$ can be understood in terms of:

One σ bond formed by sp^2 hybridized orbitals on each carbon atom.

One π bond formed between the 2p orbitals on each carbon atom.

Because rotating a π bond by 90° destroys the orbital overlap, there is a large barrier to rotation and for all practical purposes **double bonds cannot rotate around the bond axis.** *[Make sure this makes sense to you before moving on. The fact that π bonds cannot rotate adds rigidity to molecules* *that contain them. This rigidity imparted by π bonds has a major influence on the conformations of complex molecules in solution.]* ✳

- A **triple bond** in molecules such as HC≡CH can be understood in terms of:

One σ bond formed by sp hybridized orbitals on each carbon atom.

Two π bonds formed between the two 2p orbitals on each carbon atom. Because rotating a π bond by 90° destroys the orbital overlap, triple bonds also cannot rotate around the bond axis. *[It is absolutely essential that students understand bonding in complex molecules in terms of the overlap of hybrid orbitals. This orbital picture of molecules forms the theoretical foundation for understanding the reactions of the different functional groups discussed in the rest of the book. Do not go any further until these concepts are thoroughly understood.]* ✳

CHAPTER 1
Solutions to the Problems

Problem 1.1 Write and compare the ground-state electron configurations for the following:
(a) Carbon and silicon

$$C \text{ (6 electrons) } 1s^2 2s^2 2p^2$$
$$Si \text{ (14 electrons) } 1s^2 2s^2 2p^6 3s^2 3p^2$$

Both carbon and silicon have four electrons in their highest (valence) shells

(b) Oxygen and sulfur

$$O \text{ (8 electrons) } 1s^2 2s^2 2p^4$$
$$S \text{ (16 electrons) } 1s^2 2s^2 2p^6 3s^2 3p^4$$

Both oxygen and sulfur have six electrons in their highest (valence) shells.

(c) Nitrogen and phosphorus

$$N \text{ (7 electrons) } 1s^2 2s^2 2p^3$$
$$P \text{ (15 electrons) } 1s^2 2s^2 2p^6 3s^2 3p^3$$

Both nitrogen and phosphorus have five electrons in their highest (valence) shells.

Problem 1.2 Show that the following obey the octet rule.
(a) Sulfur (atomic number 16) forms sulfide ion, S^{2-}.

$$S \text{ (16 electrons): } 1s^2 2s^2 2p^6 3s^2 3p^4$$

$$S^{2-} \text{ (18 electrons): } 1s^2 2s^2 2p^6 3s^2 3p^6$$

(b) Magnesium (atomic number 12) forms Mg^{2+}.

$$Mg \text{ (12 electrons): } 1s^2 2s^2 2p^6 3s^2$$

$$Mg^{2+} \text{ (10 electrons): } 1s^2 2s^2 2p^6$$

Problem 1.3 Judging from their relative positions in the Periodic Table, which element is more electronegative?
(a) Lithium or potassium

In general, electronegativity increases from left to right across a row and from bottom to top of a column in the Periodic Table. This is because electronegativity increases with increasing positive charge on the nucleus and with decreasing distance of the valence electrons from the nucleus. Lithium is higher up on the Periodic Table and thus more electronegative than potassium.

(b) Nitrogen or phosphorus

Nitrogen is higher up on the Periodic Table and thus more electronegative than phosphorus.

(c) Carbon or silicon

Carbon is higher up on the Periodic Table and thus more electronegative than silicon.

Problem 1.4 Classify these bonds as nonpolar covalent, polar covalent, or ionic.
(a) S-H   (b) P-H   (c) C-F   (d) C-Cl

Using the rule that bonds formed from atoms with an electronegativity difference of 0.4 or less are nonpolar covalent, the following table can be constructed:

Bond	Differences in electronegativity	Type of bond
S-H	$2.5 - 2.1 = 0.4$	Nonpolar covalent
P-H	$2.1 - 2.1 = 0$	Nonpolar covalent
C-F	$4.0 - 2.5 = 1.5$	Polar covalent
C-Cl	$3.0 - 2.5 = 0.5$	Polar covalent

Problem 1.5 Indicate the direction of polarity in these polar covalent bonds using the symbols $\delta-$ and $\delta+$.
(a) C-N
$$\overset{\delta+}{C}-\overset{\delta-}{N}$$

Nitrogen is more electronegative than carbon

(b) N-O
$$\overset{\delta+}{N}-\overset{\delta-}{O}$$

Oxygen is more electronegative than nitrogen

(c) C-Cl
$$\overset{\delta+}{C}-\overset{\delta-}{Cl}$$

Chlorine is more electronegative than carbon

Problem 1.6 Draw Lewis structures, showing all valence electrons, for the following covalent molecules.
(a) C_2H_6 (b) CS_2 (c) HCN

Problem 1.7 Draw Lewis structures for these ions, and show which atom in each bears the formal charge.
(a) $CH_3NH_3^+$ (b) CO_3^{2-} (c) OH^-
   Methylammonium ion   Carbonate ion   Hydroxide ion

Problem 1.8 Draw a condensed structural formula for the one ether of molecular formula C_3H_8O.

$$CH_3-CH_2-O-CH_3$$

Problem 1.9 Draw condensed structural formulas for the three ketones of molecular formula $C_5H_{10}O$.

Problem 1.10 Draw condensed structural formulas for the two carboxylic acids of molecular formula $C_4H_8O_2$.

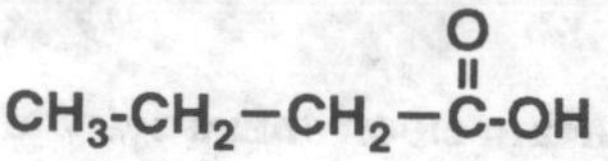

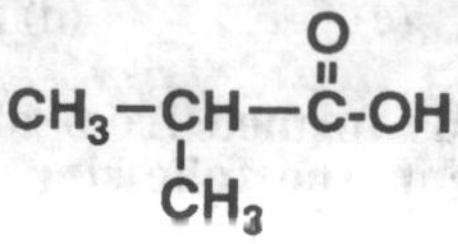

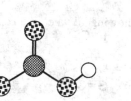

$$CH_3\text{-}CH_2\text{-}CH_2\text{-}\overset{\displaystyle O}{\overset{\displaystyle \|}{C}}\text{-OH} \qquad CH_3\text{-}\underset{\underset{\displaystyle CH_3}{|}}{CH}\text{-}\overset{\displaystyle O}{\overset{\displaystyle \|}{C}}\text{-OH}$$

Problem 1.11 Predict all bond angles for these molecules.
(a) CH₃OH

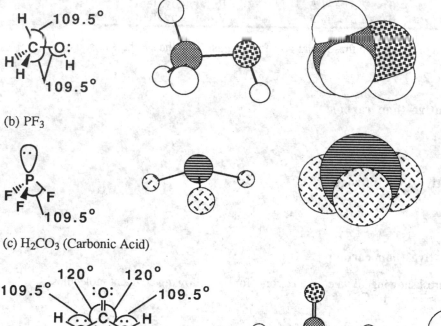

(b) PF₃

(c) H₂CO₃ (Carbonic Acid)

Problem 1.12 Which molecules are polar? For each molecule that is, specify the direction of its dipole moment.
(a) CH₂Cl₂

Recall that a molecular dipole moment is determined as the vector sum of the bond dipoles in three-dimensional space. Thus, by superimposing the bond dipoles on a three-dimensional drawing, the molecular dipole moment can be determined.

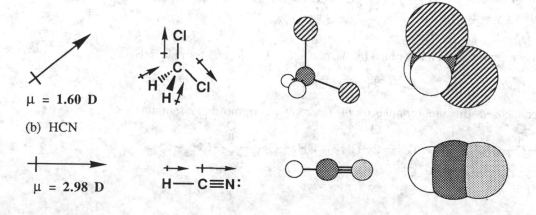

μ = 1.60 D

(b) HCN

μ = 2.98 D

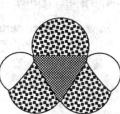

(c) H_2O_2

The H_2O_2 molecule can rotate around the O-O single bond, so we must consider the molecular dipole moments in the various possible conformations. Conformations such as the one depicted on the left below have a net molecular dipole moment, while conformations such as the one the right below do not. The presence of at least some conformations (such as that on the left) that have a molecular dipole moment means that the entire molecule must have an overall dipole moment, in this case $\mu = 2.2$ D.

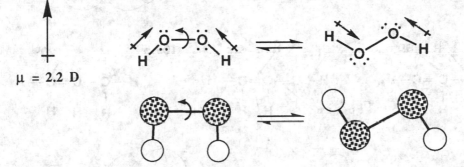

$\mu = 2.2$ D

Problem 1.13 Draw the contributing structure indicated by curved arrows. Be certain to show all formal charges.

(a) and (b) structures

(c) structures

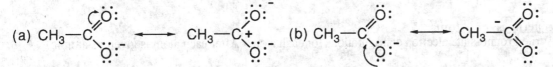

Problem 1.14 Which sets are pairs of contributing structures?

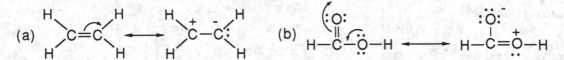

The set in (a) is a pair of contributing structures, while the set in (b) is not. The structure on the right in set (b) is not a viable contributing structure because there are five bonds to the carbon atom.

Problem 1.15 Estimate the relative contribution of the members in each set.

The first structure makes the greater contribution in (a) and (b). In both cases, the second contributing structure involves the disfavored creation and separation of unlike charges.

Problem 1.15 Describe the bonding in these molecules in terms atomic orbitals involved, and predict all bond angles.

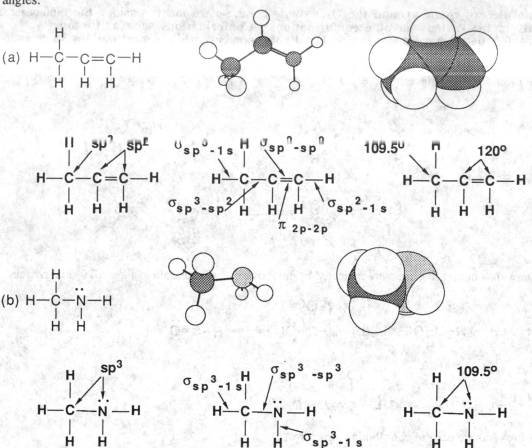

Electronic Structures of Atoms

Problem 1.17 Write ground-state electron configurations for each atom. After each atom is given its atomic number.

(a) Sodium (11)

Na (11 electrons) $1s^2 2s^2 2p^6 3s^1$

(b) Magnesium (12)

Mg (12 electrons) $1s^2 2s^2 2p^6 3s^2$

(c) Oxygen (8)

O (8 electrons) $1s^2 2s^2 2p^4$

(d) Nitrogen (7)

N (7 electrons) $1s^2 2s^2 2p^3$

Problem 1.18 Which atom has the ground-state electron configuration of

(a) $1s^2 2s^2 2p^6 3s^2 3p^4$

Sulfur (16) has this ground-state electron configuration

(b) $1s^2 2s^2 2p^4$

Oxygen (8) has this ground-state electron configuration

Problem 1.19 Define valence shell and valence electron.

The valence shell is the outermost occupied shell of an atom. A valence electron is an electron in the valence shell.

Problem 1.20 How many electrons are in the valence shell of each atom?
(a) Carbon

With a ground-state electron configuration of $1s^2 2s^2 2p^2$ there are four electrons in the valence shell of carbon.

(b) Nitrogen

With a ground-state electron configuration of $1s^2 2s^2 2p^3$ there are five electrons in the valence shell of nitrogen.

(c) Chlorine

With a ground-state electron configuration of $1s^2 2s^2 2p^6 3s^2 3p^5$ there are seven electrons in the valence shell of chlorine.

(d) Aluminum

With a ground-state electron configuration of $1s^2 2s^2 2p^6 3s^2 3p^1$ there are three electrons in the valence shell of aluminum.

Lewis Structures
Problem 1.21 Judging from their relative positions in the Periodic Table, which atom is more electronegative?
(a) Carbon or nitrogen

In general, electronegativity increases from left to right across a row and from bottom to top of a column in the Periodic Table. This is because electronegativity increases with increasing positive charge on the nucleus and with decreasing distance of the valence electrons from the nucleus. Nitrogen is farther to the right on the Periodic Table and thus more electronegative than carbon.

(b) Chlorine or bromine

Chlorine is higher up on the Periodic Table and thus more electronegative than bromine.

(c) Oxygen or sulfur

Oxygen is higher up on the Periodic Table and thus more electronegative than sulfur.

Problem 1.22 Which of these compounds have covalent bonds and which have ionic bonds?
(a) LiF   (b) CH_3F (c) $MgCl_2$ (d) HCl

Using the rule that an ionic bond is formed between atoms with an electronegativity difference of 1.9 or greater, the following table can be constructed:

Bond	Differences in electronegativity	Type of bond
Li-F	4.0 - 1.0 = 3.0	Ionic
C-H	2.5 - 2.1 = 0.4	Nonpolar covalent
C-F	4.0 - 2.5 = 1.5	Polar covalent
Mg-Cl	3.0 - 1.2 = 1.8	Polar covalent
H-Cl	3.0 - 2.1 = 0.9	Polar covalent

Based on these values, only LiF has an ionic bond, the other compounds have only covalent bonds.

Problem 1.23 Using the symbols δ- and δ+, indicate the direction of polarity, if any, in each covalent bond.
(a) C-Cl

δ+   δ-
C-Cl

Chlorine is more electronegative than carbon

(b) S-H

δ-   δ+
S-H

Sulfur is more electronegative than hydrogen

(c) C-S

Carbon and sulfur have the same electronegativities so there is no direction of polarity in a C-S bond

(d) P-H

Phosphorus and hydrogen have the same electronegativities, so there is no direction of polarity in a P-H bond

Problem 1.24 Write Lewis structures for these molecules. Be certain to show all valence electrons. None of these compounds contains a ring of atoms.

(a) H_2O_2
Hydrogen peroxide

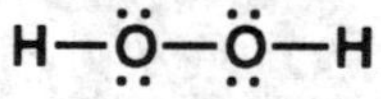

H—Ö—Ö—H

(b) N_2H_4
Hydrazine

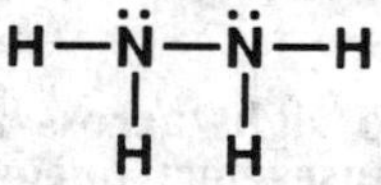

H—N̈—N̈—H
     | |
     H   H

(c) CH_3OH
Methanol

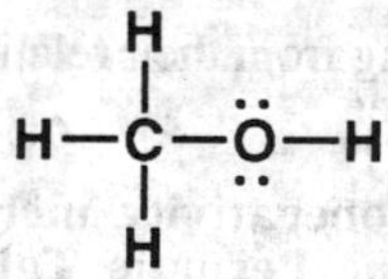

        H
        |
H—C—Ö—H
        |
        H

(d) CH_3SH
Methanethiol

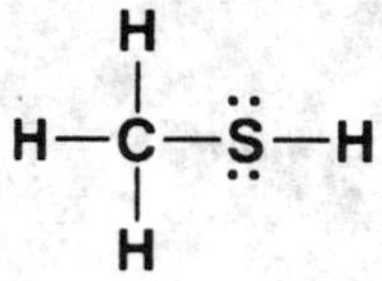

     H
     |
H—C—S̈—H
     |
     H

(e) CH_3NH_2
Methylamine

     H
     |
H—C—N̈—H
     | |
     H   H

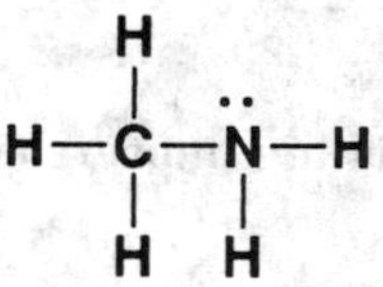

(f) CH_2Cl_2
Dichloromethane

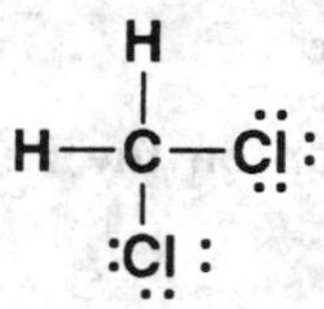

     H
     |
H—C—C̈l:
     |
    :C̈l:

(g) CH_3OCH_3
Dimethyl ether

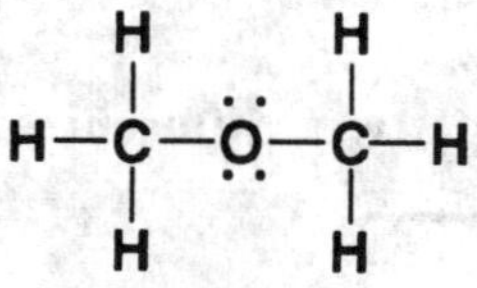

     H   H
     | |
H—C—Ö—C—H
     | |
     H   H

(h) H_2CO_3
Carbonic acid

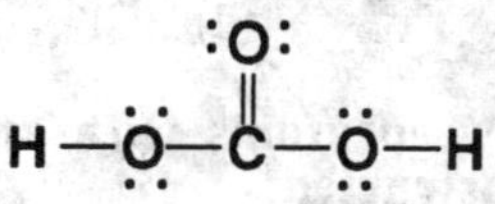

        :O:
         ‖
H—Ö—C—Ö—H

(i) CH_2O
Methanal

H
 \
  C=Ö
 /
H

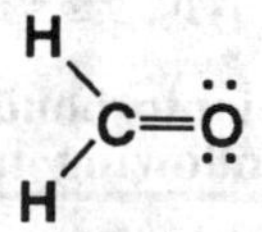

(j) CH_3CO_2H
Ethanoic acid

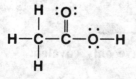

(k) CH_3COCH_3
Propanone

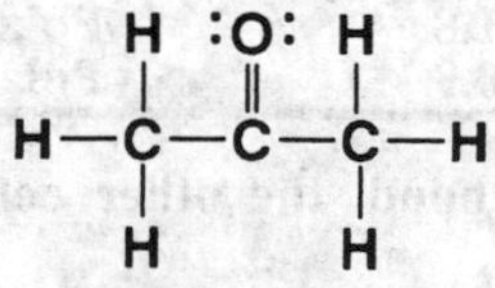

(l) HCN
Hydrogen cyanide

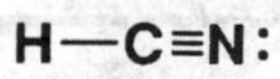

H—C≡N:

(m) HNO_3
   Nitric acid

(n) HNO_2
   Nitrous acid

(o) HCO_2H
   Methanoic acid

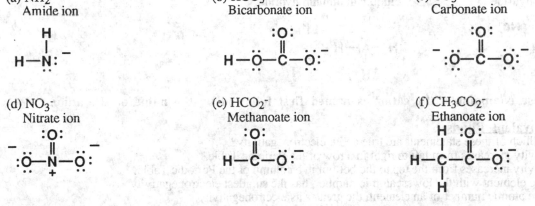

Problem 1.25 Why are the following molecular formulas impossible?
(a) CH_5

Carbon atoms can only accommodate four bonds, and each hydrogen atom can only accommodate one bond. Thus, there is no way for a stable bonding arrangement to be created that utilizes one carbon atom and all five hydrogen atoms.

(b) C_2H_7

Since hydrogen atoms can only accommodate one bond each, no single hydrogen atom can make stable bonds to both carbon atoms. Thus, the two carbon atoms must be bonded to each other. This means that each of the bonded carbon atoms can accommodate only three more bonds. Therefore, only six hydrogen atoms can be bonded to the carbon atoms, not seven hydrogen atoms.

Problem 1.26 Write Lewis structures for these ions. Show all valence electrons and all formal charges.
(a) NH_2^-
   Amide ion

(b) HCO_3^-
   Bicarbonate ion

(c) CO_3^{2-}
   Carbonate ion

(d) NO_3^-
   Nitrate ion

(e) HCO_2^-
   Methanoate ion

(f) $CH_3CO_2^-$
   Ethanoate ion

Problem 1.27 Following the rule that each atom of carbon, oxygen, and nitrogen reacts to achieve a complete outer shell of eight electrons, add unshared pairs of electrons as necessary to complete the valence shell of each atom in these ions. Then assign formal charges as appropriate.

The following structural formulas show all valence electrons and all formal charges.

Problem 1.28 Following are several Lewis structures showing all valence electrons. Assign formal charges to each structure as appropriate.

There is a formal positive charge in parts (a), (e), and (f). There is a formal negative charge in parts (b), (c), and (d).

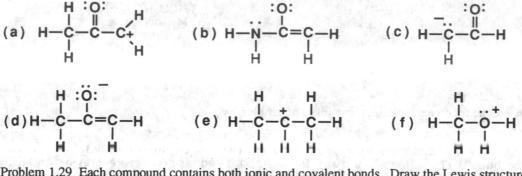

Problem 1.29 Each compound contains both ionic and covalent bonds. Draw the Lewis structure for each and show by dashes which are covalent bonds and, by indication of charges, which are ionic bonds.

(a) CH_3ONa
Sodium methoxide

(b) NH_4Cl
Ammonium chloride

(c) $NaHCO_3$
Sodium bicarbonate

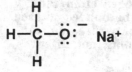

(d) $NaBH_4$
Sodium borohydride

(e) $LiAlH_4$
Lithium aluminum hydride

In naming these compounds, the cation is named first followed by the name of the anion.

Polarity of Covalent Bonds

Problem 1.30 Which of these statements are true about electronegativity?
(a) Electronegativity increases from left to right in a row of the Periodic Table.
(b) Electronegativity increases from the top to the bottom in a column of the Periodic Table.
(c) Hydrogen, the element with the lowest atomic number, has the smallest electronegativity.
(d) The higher the atomic number of an element, the greater its electronegativity.

Electronegativity increases from left to right across a row and from bottom to top of a column in the Periodic Table. Thus, statement (a) is true, but (b), (c), and (d) are false.

Problem 1.31 Why does fluorine, the element in the upper right corner of the Periodic Table, have the largest electronegativity of any element?

Electronegativity increases with increasing positive charge on the nucleus and with decreasing distance of the valence electrons from the nucleus. Fluorine is that element for which these two parameters lead to maximum electronegativity.

Problem 1.32 Arrange the single covalent bonds within each set in order of increasing polarity.
(a) C-H, O-H, N-H   (b) C-H, B-H, O-H   (c) C-H, C-Cl, C-I
C-H < N-H < O-H   B-H < C-H < O-H   C-I ≈ C-H < C-Cl
(0.4) (0.9) (1.4)   (0.1) (0.4) (1.4)   (0.4) (0.4) (0.9)

(d) C-S, C-O, C-N   (e) C-Li, C-B, C-Mg
C-S < C-N < C-O   C-B < C-Mg < C-Li
(0) (0.5) (1.0)   (0.5) (1.3) (1.5)

The difference in electronegativities is given in parentheses underneath each answer.

Problem 1.33 Using the values of electronegativity given in Table 1.5, predict which indicated bond in each set is more polar and using the symbols δ+ and δ-, show its direction of polarity
(a) CH_3-OH or CH_3O-H   (b) H-NH_2 or CH_3-NH_2

$$\overset{\delta-\quad\delta+}{CH_3O\text{-}H}$$ $$\overset{\delta+\quad\delta-}{H\text{-}NH_2}$$

(c) CH_3-SH or CH_3S-H   (d) CH_3-F or H-F

$$\overset{\delta-\quad\delta+}{CH_3S\text{-}H}$$ $$\overset{\delta+\quad\delta-}{H\text{-}F}$$

Problem 1.34 Identify the most polar bond in each molecule.
(a) $HSCH_2CH_2OH$ (b) $CHCl_2F$ (c) $HOCH_2CH_2NH_2$

The O-H bond   **The C-F bond**   **The O-H bond**
(1.4)   **(1.5)**   **(1.4)**

The difference in electronegativities is given in parentheses underneath each answer.

Problem 1.35 Predict whether the carbon-metal bond in these organometallic compounds is nonpolar covalent, polar covalent, or ionic. For polar covalent bonds, show the direction of polarity by the symbols δ+ and δ-.

(a) $\overset{\displaystyle (0.7)}{}$
$$\overset{\delta-\quad\delta+}{CH_3CH_2\text{---}Pb\text{-}CH_2CH_3}$$
with CH_2CH_3 above and CH_2CH_3 below the Pb

Tetraethyllead

(b)   **(1.3)**
$$\overset{\delta-\quad\delta+}{CH_3\text{---}Mg\text{---}Cl}$$
Methylmagnesium
chloride

(c)   **(0.6)**
$$\overset{\delta-\quad\delta+}{CH_3\text{---}Hg\text{---}CH_3}$$
Dimethylmercury

All of the above carbon-metal bonds are polar covalent because the difference in electronegativities is between 0.5 and 1.9. In each case, carbon is the more electronegative element. The difference in electronegativities is given above the carbon-metal bond in each answer.

Bond Angles and Shapes of Molecules
Problem 1.36 Use the VSEPR model to predict bond angles about each highlighted atom.

Approximate bond angles as predicted by the valence-shell electron-pair repulsion model are as shown.

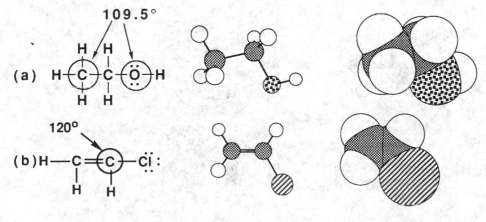

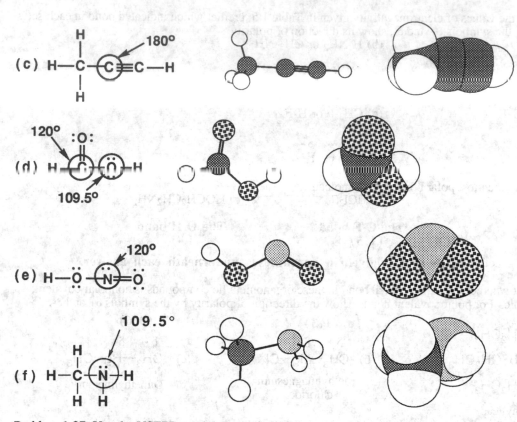

Problem 1.37 Use the VSEPR model to predict bond angles about each atom of carbon, nitrogen, and oxygen in these molecules. *Hint*; first add unshared pairs of electrons as necessary to complete the valence shell of each atom and then make your predictions.

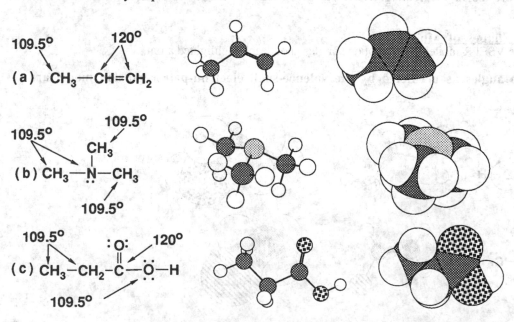

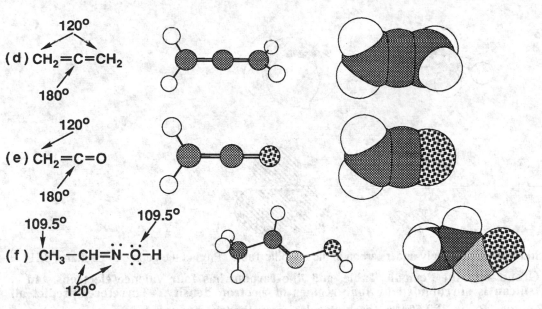

(d) CH₂=C=CH₂   120°   180°

(e) CH₂=C=O   120°   180°

(f) CH₃—CH=N—Ö—H   109.5°   109.5°   120°

Problem 1.38 Use VSEPR model to predict the geometry of the following ions:

(a) NH₂⁻

H—N: 109.5°
   |
   H

(b) NO₂⁻

⁻:Ö—N=Ö   120°

(c) NO₂⁺

Ö=N=Ö   180°
    +

(d) NO₃⁻

:O: 120°
⁻:Ö—N—Ö:⁻
    +

(e) CH₃CO₂⁻

109.5° H :O: 120°
        | ‖
    H—C—C—Ö:⁻
        |
        H

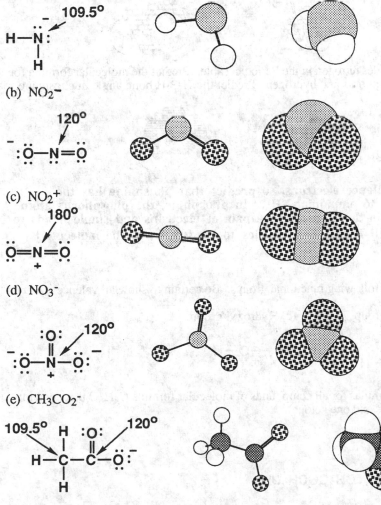

(f) CH_3^-

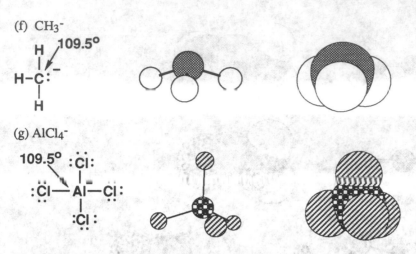

(g) $AlCl_4^-$

Problem 1.39 Silicon is immediately under carbon in the Periodic Table. Predict the geometry of silane, SiH_4.

Silicon is in Group 4 of the Periodic Table, and like carbon, has four valence electrons. In silane, SiH_4, silicon is surrounded by four regions of electron density. Therefore, predict all H-Si-H bond angles to be 109.5°, so the molecule is tetrahedral around Si.

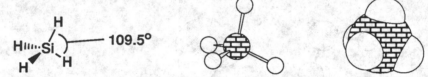

Problem 1.40 Phosphorus is immediately under nitrogen in the Periodic Table. Predict the molecular formula for phosphine, the compound formed from phosphorus and hydrogen. Predict the H-P-H bond angle in phosphine.

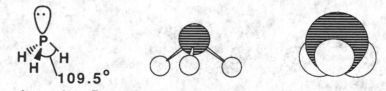

Like nitrogen, phosphorus has five valence electrons, so predict that phosphine has the molecular formula of PH_3 in analogy to ammonia, NH_3. In phosphine, the phosphorus atom is surrounded by four areas of electron density; one lone pair of electrons and single bonds to three hydrogen. Therefore, predict all H-P-H bond angles to be 109.5°, so the molecule is pyramidal.

Functional Groups
Problem 1.41 Draw Lewis structures for the following functional groups. Be certain to show all valence electrons on each.
(a) Carbonyl group   (b) Carboxyl group   (c) Hydroxyl group

Problem 1.42 Draw condensed structural formulas for all compounds of molecular formula C_4H_8O that contain:
(a) A carbonyl group (there are two aldehydes and one ketone).

Ketones

$CH_3-\overset{\overset{\text{O}}{\|}}{C}-CH_2-CH_3$ also written as $CH_3COCH_2CH_3$

Aldehydes

$$CH_3-CH_2-CH_2-\overset{\overset{\displaystyle O}{\|}}{C}-H \quad \text{also written as} \quad CH_3CH_2CH_2CHO$$

$$CH_3-\underset{\underset{\displaystyle CH_3}{|}}{CH}-\overset{\overset{\displaystyle O}{\|}}{C}-H \quad \text{also written as} \quad (CH_3)_2CHCHO$$

(b) A carbon-carbon double bond and an alcohol (there are eight)

There are three separate but related things to build into this answer; the carbon skeleton (the order of attachment of carbon atoms), the location of the double bond, and the location of the -OH group. Here, as in other problems of this type, it is important to have a system and to follow it. As one way to proceed, first decide the number of different carbon skeletons that are possible. A little doodling with paper and pencil should convince you that there are only two.

$$C-C-C-C \quad \text{and} \quad C-\overset{\overset{\displaystyle C}{|}}{C}-C$$

Next locate the double bond on these carbon skeletons. There are three possible locations for it.

$$C=C-C-C \quad \text{and} \quad C-C=C-C \quad \text{and} \quad C=\overset{\overset{\displaystyle C}{|}}{C}-C$$

Finally, locate the -OH group and then add the remaining seven hydrogens to complete each structural formula. For the first carbon skeleton, there are four possible locations of the -OH group; for the second carbon skeleton there are two possible locations; and for the third, there are also two possible locations. Four of these compounds (marked by an asterisk) are not stable and are in equilibrium with a more stable aldehyde or ketone. You need not be concerned, however, with this now. Just concentrate on drawing the required eight structural formulas.

$$H\overset{*}{O}-CH=CH-CH_2-CH_3 \qquad CH_2=\overset{\overset{\displaystyle *OH}{|}}{C}-CH_2-CH_3 \qquad CH_2=CH-\overset{\overset{\displaystyle OH}{|}}{CH}-CH_3$$

$$CH_2=CH-CH_2-CH_2-OH \qquad HO-CH_2-CH=CH-CH_3 \qquad CH_3-\overset{\overset{\displaystyle *OH}{|}}{C}=CH-CH_3$$

$$H\overset{*}{O}-CH=\overset{\overset{\displaystyle CH_3}{|}}{C}-CH_3 \qquad CH_2=\overset{\overset{\displaystyle CH_3}{|}}{C}-CH_2-OH$$

(c) A carbon-carbon double bond and an ether (there are four)

$$CH_3-O-CH_2-CH=CH_2 \quad CH_3-O-CH=CH-CH_3 \quad CH_3-CH_2-O-CH=CH_2$$

$$CH_3-O-\overset{\overset{\displaystyle CH_3}{|}}{C}=CH_2$$

Problem 1.43 Draw structural formulas for
(a) The eight alcohols of molecular formula $C_5H_{12}O$.

To make it easier for you to see the patterns of carbon skeletons and functional groups, only carbon atoms and hydroxyl groups are shown in the following solutions. To complete these structural formulas, you need only supply enough hydrogen atoms to complete the tetravalence of each carbon.

There are three different carbon skeletons on which the -OH group can be placed:

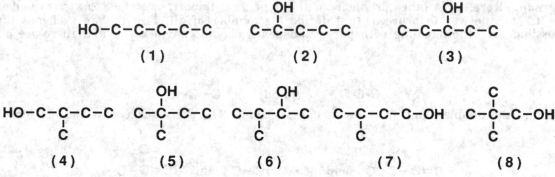

Three isomeric alcohols are possible from the first carbon skeleton, four from the second carbon skeleton, and one from the third carbon skeleton.

(b) The six ethers of molecular formula $C_5H_{12}O$.

Following are structural formulas for the six isomeric ethers of molecular formula $C_5H_{12}O$. They are drawn first with all possible combinations of $(C)_1$-O-$(C)_4$ and then all possible combinations of $(C)_2$-O-$(C)_3$.

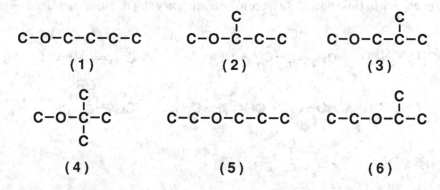

(c) The eight aldehydes of molecular formula $C_6H_{12}O$.

Following are structural formulas for the eight aldehydes of molecular formula $C_6H_{12}O$. They are drawn starting with the aldehyde group and then attaching the remaining five carbons in a chain (structure 1), then four carbons in a chain and one carbon as a branch on the chain (structures 2, 3, and 4) and finally three carbons in a chain and two carbons as branches (structures 5, 6, 7, and 8).

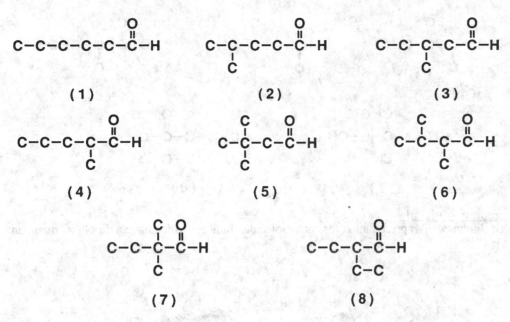

(d) The six ketones of molecular formula $C_6H_{12}O$.

Following are structural formulas for the six ketones of molecular formula $C_6H_{12}O$. They are drawn first with all combinations of one carbon to the left of the carbonyl group and four carbons to the right (structures 1, 2, 3, and 4) and then with two carbons to the left and three carbons to the right (structures 5 and 6).

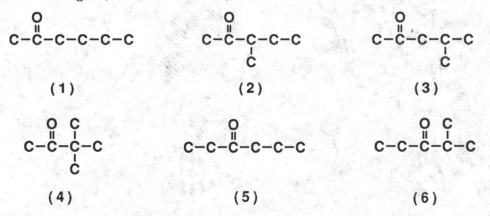

(e) The eight carboxylic acids of molecular formula $C_6H_{12}O_2$.

There are eight carboxylic acids of molecular formula $C_6H_{12}O_2$. They have the same carbon skeletons as the eight isomeric aldehydes of molecular formula $C_6H_{12}O$ shown in part (c) of this problem. In place of the aldehyde group, substitute a carboxyl group.

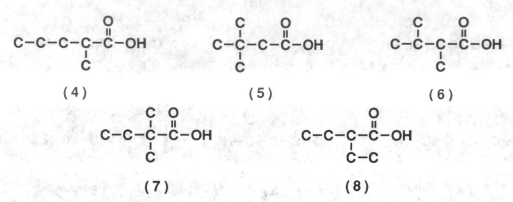

(4)   (5)   (6)

(7)   (8)

Polar and Nonpolar Molecules

1.44 Draw a three-dimensional representation for each molecule. Indicate which have a dipole moment, and in what direction it is pointing.

(a) CH_3Cl

(b) CH_2Cl_2

(c) CH_2ClBr

(d) $CHCl_3$

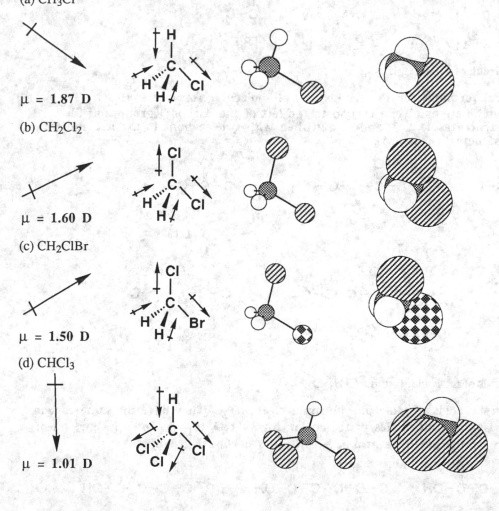

$\mu = 1.87$ D

$\mu = 1.60$ D

$\mu = 1.50$ D

$\mu = 1.01$ D

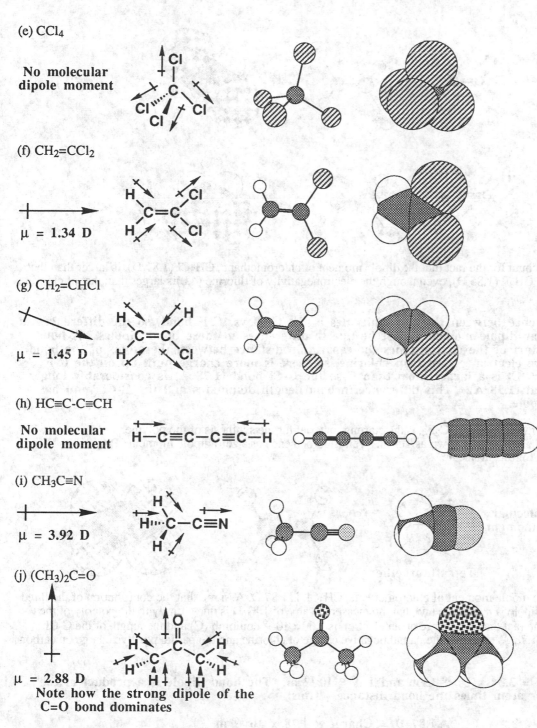

(e) CCl_4

No molecular dipole moment

(f) $CH_2=CCl_2$

$\mu = 1.34$ D

(g) $CH_2=CHCl$

$\mu = 1.45$ D

(h) $HC\equiv C\text{-}C\equiv CH$

No molecular dipole moment

(i) $CH_3C\equiv N$

$\mu = 3.92$ D

(j) $(CH_3)_2C=O$

$\mu = 2.88$ D
Note how the strong dipole of the C=O bond dominates

(k) BrCH=CHBr (two answers)

The two bromine atoms can either be on opposite sides or on the same side of the double bond. Recall that double bonds do not rotate.

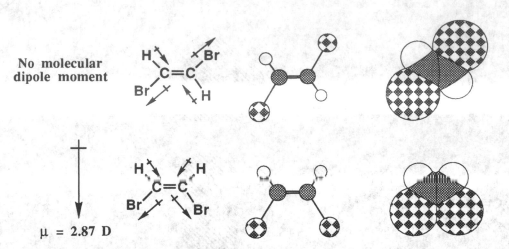

No molecular
dipole moment

μ = 2.87 D

Problem 1.45 Account for the fact that the dipole moment of chloromethane, CH_3Cl (1.87 D), is larger than that of fluoromethane, CH_3F (1.85 D), even though the electronegativity of fluorine (4.0) is larger than that of chlorine (3.5).

The only difference between the two molecules is the C-Cl vs. C-F bond, so the differences in overall molecular dipole moments must be due to differences in these bond dipoles. A bond dipole is a product of the charge times the separation distance between the atoms of the bond. Fluorine is more electronegative than chlorine, so there is more charge built up on the C-F bond. However, Cl is a larger atom than F so the C-Cl bond (1.78 Å) is considerably longer than a C-F bond (1.38 Å). This difference in bond length dominates and the C-Cl bond has a larger bond dipole.

Problem 1.46 Tetrafluoroethene, C_2F_4, is the starting material for the synthesis of the polymer poly(tetrafluoroethene), better known as Teflon. Tetrafluoroethene has a zero dipole moment. Propose a structural formula for this molecule.

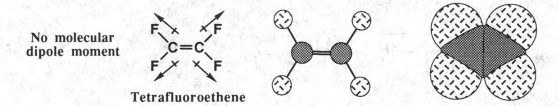

No molecular
dipole moment

Tetrafluoroethene

Problem 1.47 The dipole moment of chloromethane, CH_3Cl, is 1.87 D. Assume that the contribution of the three C-H bonds to its dipole is negligible and that the measured value of 1.87 D is due entirely to the polarity of the C-Cl bond. Given the fact that the charge on an electron is 1.60 x 10^{-19} coulomb (C) and the length of the C-Cl bond in CH_3Cl is 1.78 Å, calculate the partial negative charge of chlorine and the partial positive charge on carbon in this molecule.

Recall that 1 D = 3.34 x 10^{-30} C•m and 1 Å = 10^{-10} m. The bond dipole is a product of the charge on either atom times the bond distance. Thus:

$$1.87 \text{ D} = \text{Charge x } 1.78 \text{ x } 10^{-10} \text{ m}$$

rearranging and substituting the D unit conversion gives:

$$\text{Charge} = \frac{(1.87) \text{ x } (3.34 \text{ x } 10^{-30} \text{ C•m})}{1.78 \text{ x } 10^{-10} \text{ m}} = 3.51 \text{ x } 10^{-20} \text{ C}$$

Dividing by the charge on an electron gives the partial charge on each atom.

$$\text{Charge} = \frac{3.51 \times 10^{-20} \text{ C}}{1.6 \times 10^{-19} \text{ C}} = 0.22$$

Note that the charge on the carbon atom is + 0.22, and the charge on the chlorine atom is - 0.22.

Resonance and Contributing Structures

Problem 1.48 Which of these statements are true about resonance contributing structures?
(a) All contributing structures must have the same number of valence electrons.
(b) All contributing structures must have the same arrangement of atoms.
(c) All atoms in a contributing structure must have complete valence shells.
(d) All bond angles in sets of contributing structures must be the same.

For sets of contributing structures, electrons (usually π electrons or lone pair electrons) move, but the atomic nuclei maintain the same arrangement in space. Thus, statements (b) and (d) are true. In addition, the total number of electrons, valence and inner shell electrons, in each contributing structure must be the same, so statement (a) is also true. However, the movement of electrons often leaves one or more atoms without a filled valence shell in a given contributing structure, so statement (c) is false.

Problem 1.49 Draw the contributing structure indicated by the curved arrow(s). Assign formal charges as appropriate.

(a)

(b)

(c)

(d)

(e)

(f)

Problem 1.50 Using the VSEPR model, predict the bond angles about the carbon atom in each pair of contributing structures in problem 1.49. In what way do the bond angles change from one contributing structure to the other?

As stated in the answer to Problem 1.48, bond angles do not change from one contributing structure to the other.

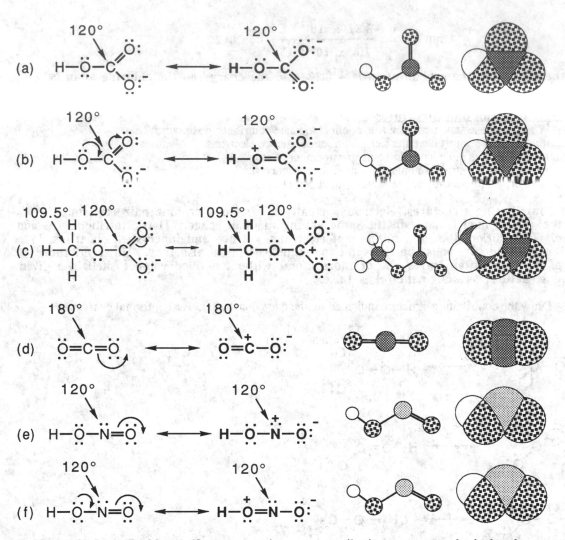

Problem 1.51 In the Problem 1.49, you were given one contributing structure and asked to draw another. Label pairs of contributing structures that are equivalent. For pairs of contributing structures that are not equivalent, label the more important contributing structure.

**(a) The two structures are equivalent because each involves a similar separation of charge.
(b, c, d, e, f) The first structure is more important, because the second involves creation and separation of unlike charges.**

Problem 1.52 Are the structures in each set valid contributing structures?

(a)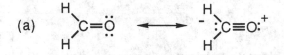

The structure on the right is not a valid contributing structure because there are 10 valence electrons around the carbon atom. Besides this, the two structures cannot be valid contributing structures of the same molecule because they have a different number of valence electrons. The molecule on the left has 12 valence electrons, and the molecule on the right has 14 valence electrons.

(b)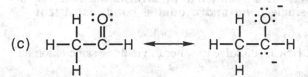

Both of these are valid contributing structures.

(c)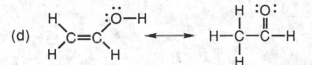

The structure on the right is not a valid contributing structure because there are two extra electrons and thus it is a completely different species.

(d)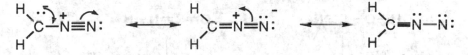

Although each is a valid Lewis structure, they are not valid contributing structures for the same resonance hybrid. An atomic nucleus, namely a hydrogen, has changed position. Later you will learn that these two molecules are related to each other, and are called tautomers.

Problem 1.53 Following are three contributing structures for diazomethane, CH_2N_2.

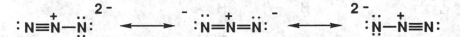

(a) Using curved arrows, show how each contributing structure is converted to the one on its right.

The arrows are indicated on the above structures.

(b) Which contributing structure makes the largest contribution to the hybrid.

The middle structure has filled valence shells, so this will make the largest contribution to the hybrid.

Problem 1.54 Draw a Lewis structure for the azide ion, N_3^-. (The order of attachment is N-N-N). How does the resonance model account for the fact that the lengths of the N-N bonds in this ion are identical.

It is not possible to draw a single Lewis structure that adequately describes the azide anion. Rather, it can be drawn as the three contributing structures shown below.

$$:N≡N-\overset{..}{\underset{..}{N}}:^{2-} \longleftrightarrow \ ^-:\overset{..}{N}=\overset{+}{N}=\overset{..}{N}:^- \longleftrightarrow \ ^{2-}:\overset{..}{\underset{..}{N}}-\overset{+}{N}≡N:$$

Taken together, the three contributing structures present a symmetric picture of the bonding, thus explaining why both N-N bonds are identical.

Problem 1.55 Draw a Lewis structure for the ozone molecule, O_3. (The order of attachment is O-O-O). How does the resonance model account for the fact that the length of each O-O bond in ozone (1.28Å) is shorter than the O-O single bond in hydrogen peroxide (1.47 Å), but longer than the O-O double bond in the oxygen molecule (1.23 Å).

It is not possible to draw a single Lewis structure that adequately describes the ozone molecule. Rather, it is better to draw ozone as two contributing structures.

Taken together, the two contributing structures present a symmetric picture of the bonding in which each O-O bond is intermediate between a single bond and a double bond, consistent with the measured bond lengths.

Problem 1.56 Cyanic acid, HOCN, and hydrocyanic acid, HNCO, dissolve in water to yield the same anion on loss of H⁺.
(a) Write a Lewis structure for cyanic acid.

H—Ö—C≡N:

(b) Write a Lewis structure for hydrocyanic acid.

Ö=C=N̈—H

(c) Account for the fact that each acid gives the same anion on loss of H⁺.

Loss of an H⁺ from the two different acids gives the same anion that can best be described by drawing the following two contributing structures.

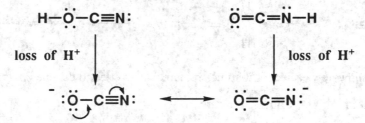

Molecular Orbital Theory
Problem 1.57 State the orbital hybridization of each circled atom.

Each circled atom is either sp, sp², or sp³ hybridized.

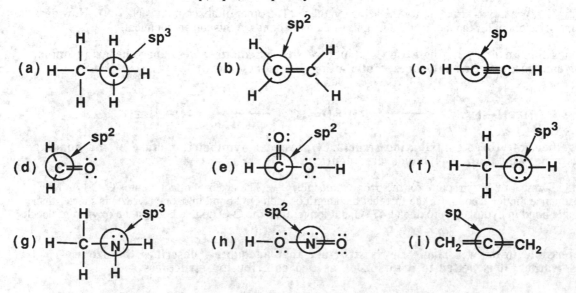

Problem 1.58 Describe each circled bond in terms of the overlap of atomic orbitals.

Shown is whether the bond is sigma or pi, as well as the orbitals used to form it.

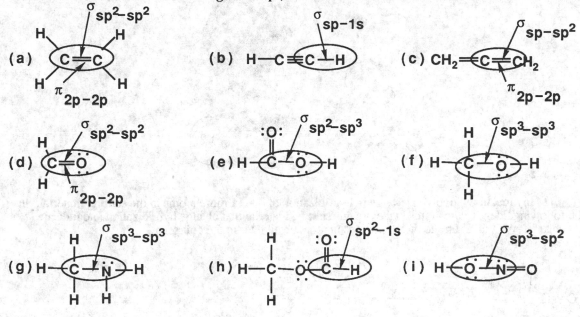

Problem 1.59 Following is the structural formula of benzene, C_6H_6.

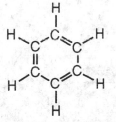

(a) Predict each H-C-C bond angle in benzene; predict each C-C-C bond angle.

Each carbon atom in benzene has three areas of electron density around it, so according to the VSEPR model, the carbon atoms are trigonal planar. Predict each H-C-C bond angle to be 120° and each C-C-C bond angle to be 120°.

(b) State the hybridization of each carbon atom in benzene.

Each carbon atom is sp^2 hybridized because each one makes three σ bonds and one π bond.

(c) Predict the shape of benzene.

Since all of the carbon atoms in the ring are sp^2 hybridized and thus trigonal planar, predict carbon atoms in benzene to form a flat hexagon in shape, with the hydrogen atoms in the same plane as the carbon atoms.

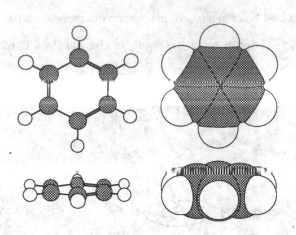

Problem 1.60 Many reactions involve a change in hybridization of one or more atoms in the starting material. In each of the following, identify the atoms in the organic starting material that change hybridization and indicate what the change is. We will examine these reactions in more detail later in the course.

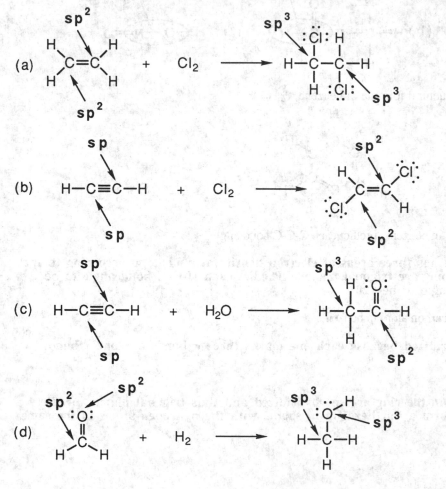

(e) 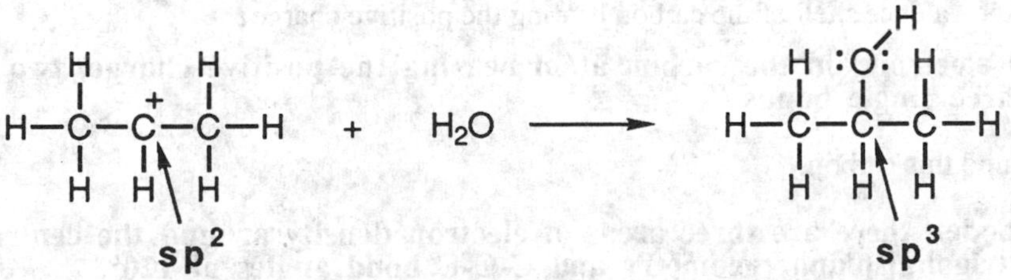

Problem 1.61 Following is the structural formula of famotidine, manufactured by Merck Sharpe & Dohme under the name Pepcid. The primary clinical use of Pepcid is for the treatment of active duodenal ulcers and benign gastric ulcers. Pepcid is a competitive inhibitor of histamine H_2 receptors and reduces both the gastric acid concentration and volume of gastric secretions.

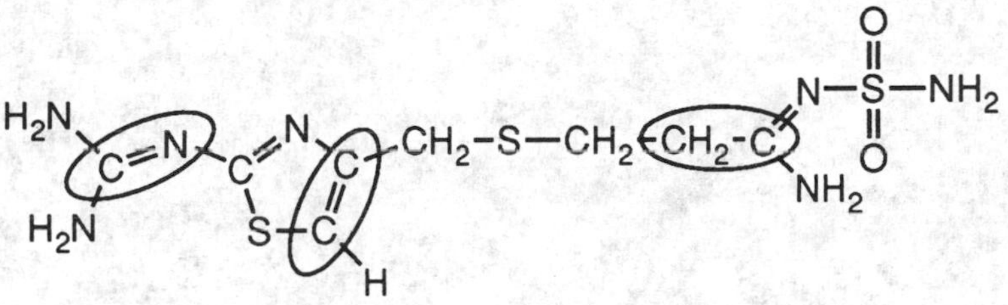

(a) Complete the Lewis structure of famotidine showing all valence electrons and any formal positive or negative charges.

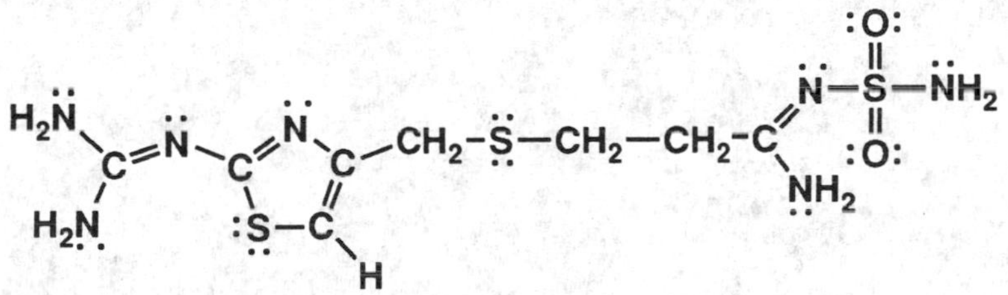

(b) Describe each circled bond in terms of the overlap of atomic orbitals.

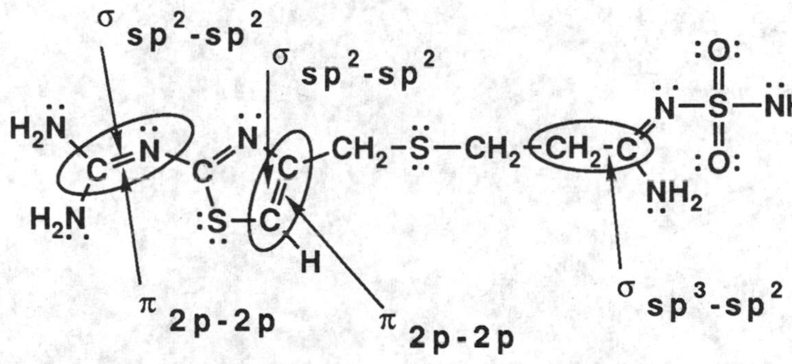

Problem 1.62 In Chapter 6, we study a group of organic cations called carbocations. Following is the structure of one such carbocation, the *tert*-butyl cation.

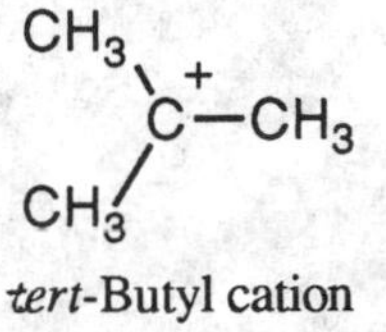

tert-Butyl cation

(a) How many electrons are in the valence shell of the carbon bearing the positive charge?

There are six valence shell electrons on the carbon atom bearing the positive charge, two contained in each of the three single bonds.

(b) Predict the bond angles around this carbon.

According to the VSEPR model, there are three areas of electron density around the central carbon atom, so predict a trigonal planar geometry and C-C-C bond angles of 120°.

(c) Given the bond angle you predicted in (b), what hybridization do you predict for this carbon?

Given the trigonal planar geometry predicted in (b), predict sp^2 hybridization of this carbon atom.

CHAPTER 2: ALKANES AND CYCLOALKANES

SUMMARY OF REACTIONS

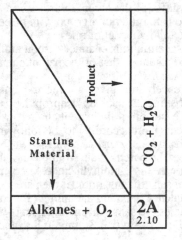

REACTION 2A: OXIDATION (Section 2.10)
- Alkanes react with O_2 to give CO_2, H_2O, and heat.

$$CH_4 + 2O_2 \longrightarrow CO_2 + 2H_2O$$

- This reaction is the basis for using alkanes as sources of heat and energy.

SUMMARY OF IMPORTANT CONCEPTS

2.0 OVERVIEW
• A **hydrocarbon** is a molecule that contains only carbon and hydrogen, and an **alkane** is a hydrocarbon that contains only single bonds. ✳

2.1 STRUCTURE OF ALKANES
• Alkanes have the general formula C_nH_{2n+2}.
• The carbon atoms of alkanes are sp^3 hybridized and thus tetrahedral, with bond angles of approximately 109.5°.

2.2 CONSTITUTIONAL ISOMERISM IN ALKANES
• **Constitutional isomers** are two or more molecules that have the same molecular formula but the atoms are attached to each other in different ways.
   - Constitutional isomers have different chemical properties.
   - For methane (CH_4), ethane (C_2H_6), and propane (C_3H_8) there is only one way to attach the carbon atoms to each other, hence there are no constitutional isomers of these alkanes. For alkanes with four or more carbon atoms, the number of constitutional isomers is counted using a type of mathematics called graph theory.
   - There is no foolproof way to find all constitutional isomers for a given molecular formula, so you must use a combination of a systematic method and creativity. *[This is harder than it looks and requires a great deal of practice.]*
   - A reasonable system to answer the constitutional isomer questions is to first write all possible carbon skeletons by starting with the straight chain alkane, then systematically adding appropriate branches.

2.3 NOMENCLATURE OF ALKANES
• To name organic compounds, chemists use systematic nomenclature rules established by the **International Union of Pure and Applied Chemistry (IUPAC)**. ✳
• For simple, unbranched alkanes the name consists of a **prefix** and a **suffix**. ✳
   - The prefix indicates the number of carbon atoms. For example, "**prop**" means three carbon atoms. For a more complete list of prefixes see Table 2.2 in the text. *[It is important to learn these now because the rest of the book assumes you are familiar with these names.]*

- Following the prefix, the suffix **ane** is used to designate that a compound is an alkane. For example, **propane** is an alkane with three carbon atoms ($CH_3CH_2CH_3$). It turns out that the **ane** suffix is actually composed of the so-called infix "**an**" and the true suffix "**e**."
- For substituted or **branched alkanes**, the nomenclature is based on viewing the molecule as a chain with substituents derived from an alkane, called **alkyl groups**. ✳
    - To name an **alkyl group**, the **ane** suffix of the parent hydrocarbon is dropped and is replaced by the suffix **yl**. For example, **propyl** is used to name the alkyl group with three carbon atoms ($CH_3CH_2CH_2$-).
- Branched alkanes are named using the following set of rules.
    - The alkane derived from the longest continuous chain of carbon atoms is taken as the **parent chain**. The **root** or **stem name** of the branched alkane is that of the parent chain. *[This can be tricky, especially when the parent chain is drawn in a crooked fashion.]*
    - Each substituent attached to the parent chain is given a name and a number. Certain common names (see below) can be used for naming substituents, such as "isopropyl."
    - The **substituent number** shows the carbon of the parent chain to which the substituent is attached. The numbers are designated on the parent chain according to the following rules. *[This numbering scheme is as complicated as it seems, and requires a lot of practice to master.]*
        If there is one substituent, number the parent chain from the end that gives the substituent the lower number. For example, a correct name is 2-methylhexane, *not* 5-methylhexane.
        If there are two or more *identical* substituents, number the parent chain from the end that gives the lower number to the substituent encountered first, and the number for each substituent is given in the final name. Indicate the number of times the same substituent occurs by a special set of prefixes. The prefixes **di**, **tri**, **tetra**, **penta**, or **hexa**. For example, 2,3-dimethylhexane has methyl groups at positions 2 and 3 on the parent hexane chain.
        If there are two or more *different* substituents, list them in **alphabetical order**, and number the parent chain from the end that gives the lower number to the substituent encountered first. For example, 4-ethyl-3-methyloctane is an acceptable name because <u>e</u>thyl comes before <u>m</u>ethyl in alphabetical order.
        If there are *different* substituents in equivalent positions on the parent chain, give the lower number to the substituent of lower alphabetical order.
        Hyphenated prefixes, for example, *sec-* and *tert-* are not considered when alphabetizing. The prefix **iso** is not a hyphenated prefix, and therefore is included when alphabetizing. The prefixes **di, tri, tetra, penta, etc.,** are also not included in the alphabetizing. Thus, ethyl comes before dimethyl, because it is ethyl and methyl that are actually being compared, not ethyl and dimethyl. ✳
- In spite of the precision of the IUPAC system, an unsystematic set of **common names** is still used for certain compounds. *[These names are deeply rooted in organic chemistry and are still widely used . Remember that it is always correct to use an IUPAC name. However, it is also important to learn how to use the common names, because you will run across them often.]*
    - In the **common nomenclature**, the total number of carbon atoms in an alkane, regardless of their arrangement, determines the name. The following terms are used in common nomenclature to indicate a few selected branching patterns.
        Iso is used to indicate that one end of an otherwise continuous chain terminates in a $(CH_3)_2CH$- group. For example isobutane, $(CH_3)_2CHCH_3$.
        Neo is used to indicate that one end of an otherwise continuous chain terminates in a $(CH_3)_3C$- group. For example, neopentane, $(CH_3)_4C$.
        More complicated patterns of branching cannot be accommodated by common nomenclature, so the IUPAC system must be used.
        You can use some common names such as *tert*-butyl and isopropyl for substituents, even though the rest of the molecule is named according to IUPAC rules.
- **Classify atoms** according to their **environment**. ✳
    - Classify a carbon atom in an alkane according to the number of alkyl groups bonded to it. *[This is very important when it comes to understanding relative reactivity.]* A carbon atom bonded to a single alkyl group is a **primary carbon** atom, a carbon atom bonded to two alkyl groups is a **secondary carbon** atom, a carbon atom bonded to three alkyl groups is a **tertiary carbon** atom, and a carbon atom bonded to four alkyl groups is a **quaternary carbon** atom.
    - **Hydrogen atoms** are also classified as primary, secondary, or tertiary when they are bonded to a primary, secondary, or tertiary carbon atom, respectively.
    - **Equivalent** hydrogen atoms have the same chemical environment. ✳ *[This concept is very important when it comes to spectroscopy, especially nuclear magnetic resonance (NMR) spectroscopy].*
        To determine which hydrogens in a molecule are equivalent, use the following procedure: In your mind, replace each hydrogen with a "**test atom**." If replacement of two different hydrogens by the "test atom"

gives the same compound, then the hydrogens are equivalent. If replacement of two different hydrogens by the "test atom" gives different compounds, then the hydrogens are not equivalent.

2.4 CYCLOALKANES
• Organic chemists use **line-angle drawings** as a simple way to represent complex molecules. In line-angle drawings, each line represents a C-C bond, each double line represents a C=C bond, and each triple line represents a C≡C bond. The vertex of each angle represents a carbon atom. In this way, only the **carbon framework** of the molecule is shown. It is understood that hydrogen atoms complete the tetravalence of the carbon atoms. For example, ∧∧ represents pentane, $CH_3CH_2CH_2CH_2CH_3$.
• A **cycloalkane** is an alkane in which there is a ring of carbon atoms. ✻
  - **IUPAC cycloalkane nomenclature** rules are as follows:
    Use the prefix **cyclo** in front of the name of the alkane with the same number of carbon atoms as the

    number of carbons in the ring. For example, cyclohexane is a six-membered ring, ⬡.
    List substituents on the ring by name and number as you would on an open-chain hydrocarbon.
    If there is only a single substituent on the ring, there is no need to give a number. If there are two or more substituents, give each substituent a number to indicate its location on the ring. Number the atoms of the ring beginning with the substituent of lowest alphabetical order.
• A **bicycloalkane** is a cycloalkane with two rings that share two or more atoms in common.
  - **IUPAC bicycloalkane nomenclature** rules are as follows:
    The parent name of a bicycloalkane is that of the alkane with the same number of carbon atoms as are in the bicyclic ring system.
    Numbering begins at one bridgehead carbon and proceeds along the longest bridge to the second bridgehead carbon, and then along the next longest bridge back to the original bridgehead carbon, and so forth until all atoms of the bicycloalkane rings are numbered.
    Ring sizes are shown by counting the number of carbon atom linked to the bridgeheads and placing them in decreasing order in brackets between the prefix **bicyclo** and the parent name. For example,

    bicyclo[2.2.1]heptane ⬡.
    Name and locate substituents by the rules already described in Section 2.3A.
• A **spiroalkane** is a cycloalkane in which the two rings share only one atom. Name spiroalkanes with the

    prefix **spiro**. For example, spiro[1.1]nonane ⬡. Numbering begins at the carbon atom on the shorter bridge nearest the spirocarbon atom, around the shorter bridge, through the spirocarbon atom, and around the longer bridge. Name the smaller bridge first as in spiro[4.5]decane.

2.5 THE IUPAC SYSTEMS-A GENERAL SYSTEM OF NOMENCLATURE
• The name assigned to any compound consists of at least three parts; **the prefix, the infix** and **the suffix**.
  - The prefix tells the number of carbon atoms in the parent chain. See Table 2.3 in the text for examples.
  - The infix (part of the name directly in front of the suffix) tells the nature of the carbon-carbon bonds in the parent chain.
    an means the compound has all C-C single bonds, **en** means one or more C=C double bond, and **yn** means there is one or more C≡C triple bond.
  - The suffix tells the class of the compound to which the substance belongs.
    The class of a compound is determined by the functional groups present.
    Important suffixes include **e** for hydrocarbons, **ol** for alcohols, **al** for aldehydes, **one** for ketones, and **oic acid** for carboxylic acids.

2.6 CONFORMATIONS OF ALKANES AND CYCLOALKANES
• The **conformation** of an alkane refers to the *three-dimensional* arrangement of atoms that results from rotation about carbon-carbon bonds. ✻
  - It is convenient to analyze alkane conformations using a **Newman projection**. Although there may be a number of C-C bonds in a molecule to analyze, you can look at only one C-C bond with each Newman projection.
    In a Newman projection, view the molecule along the axis of one C-C bond. ✻ *[Understanding this statement is the key to using Newman projections.]*
    Thus, a Newman projection examines how the different groups are distributed around only the two adjacent carbon atoms involved with the selected C-C bond.
    When drawing a Newman projection, orient the molecule so that the selected bond is parallel to your line of vision. Use a large circle to represent the rear carbon, and a dot to represent the front carbon. Show

the three groups bound to the carbon atom that are *nearer* your eye on lines extending from the dot at the center of the circle at angles of 120°. Show the three groups bound to the carbon atom that is *farther* from your eye on lines extending from the *circumference* of the circle at angles of 120°. *[Make sure you know how to go between a three-dimensional molecular model and a Newman projection. Figures 2.5 and 2.6 are especially helpful.]*

- At room temperature, the C-C bonds can rotate rapidly. Thus, an infinite number of conformations are possible around a C-C bond as it rotates. The two extreme conformations are named **eclipsed** and **staggered**.
  - In an **eclipsed conformation**, the groups on the near carbon atom are directly in front of the groups on the far carbon atom.

  - In a **staggered conformation**, the groups on the near carbon atom are as far apart as possible from the groups on the far carbon.

  - **A dihedral angle** is the angle between a given substituent on the near carbon atom and a given substituent on the far carbon atom of a Newman projection.
    For eclipsed conformations, the dihedral angles are thus 0°, 120°, 240°, for nearest groups, and for staggered conformations, the dihedral angles are thus 60°, 180°, 300° for nearest groups.
- The staggered conformations have lower **potential energy** than eclipsed conformations, probably due to the **repulsion** of **electron pairs** in the bonds resulting in their preferring to be as far apart as possible. Hydrogen atoms are probably not large enough to "crash" into each other even in an eclipsed conformation. ✳
  - This lower potential energy means that alkanes spend the majority of their time in a staggered conformation.
- Taking the bond between the carbons 2 and 3 as reference, there are two types of **staggered conformations for butane**, namely **gauche** and **anti**. ✳
  - The two **gauche** conformations have the two methyl groups adjacent, that is with dihedral angles of 60°.

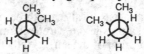

  - The **anti** conformation has the two methyl groups as far apart as possible, that is with a dihedral angle of 180°. *[If you do not fully understand this, review the preceding sections before going any further.]*

The methyl groups take up a large amount of space and thus the anti conformation is the most stable (lowest potential energy), because the methyl groups are farthest apart. The gauche conformations are the next most stable, and all of the eclipsed conformations are the least stable.
Other large groups are also more stable in the anti conformation. The larger the groups, the larger the preference for being anti. ✳
- The most stable three-dimensional arrangement of atoms in cycloalkanes minimizes **angle strain** and **nonbonded interaction strain**. ✳ *[Notice the discussion has turned back to cycloalkanes.]*
  - **Angle strain** arises because the geometry of certain cycloalkanes creates bond angles other than the ideal 109.5°.
  - **Nonbonded interaction strain** arises because the geometry of cycloalkanes forces nonbonded atoms or groups into close proximity. As expected, this type of strain is proportionately more important for larger atoms or groups.

- The three carbon atoms of **cyclopropanes** must necessarily lie in a plane.
  - There is a large amount of angle strain in cyclopropane because the bond angles are 60°, a long way from the preferred 109.5°. There is a large amount of nonbonded interaction strain because all of the groups bonded to the central carbon atoms are eclipsed. *[Use a model to prove this to yourself if necessary.]*
- To minimize steric strain, the **larger cycloalkanes** exist in a variety of **puckered,** nonplanar conformations.✻
- The most stable conformation of **cyclobutane** is slightly puckered ⟍⟋.
  - The puckered conformation of cyclobutane relieves nonbonded interaction strain , because the hydrogens are no longer fully eclipsed. Note that the puckering does cause a slight increase in angle strain because the angles are decreased to about 88°.

- **An envelope conformation** is the most stable conformation of **cyclopentane**. ⟍⟋. There are five possible envelope conformations, each with a different carbon atom that is out of the plane formed by the other four.
  - This puckering relieves nonbonded interaction strain by reducing the number of eclipsed hydrogen atoms in the molecule. In this case the puckering only causes a slight increase in angle strain, since the bond angles are 105°.
- There are a number of different puckered conformations of **cyclohexane,** by far the most important of which is a remarkably stable **chair conformation** ⟍⟋. ✻ *[Understanding the following details of cyclohexane chair conformations is very important, because you will need to use these ideas in the future when issues like the relative stabilities of reaction intermediates or carbohydrates are discussed.]*
  - The chair conformation is dramatically more stable than the planar form, because *all* the groups are perfectly staggered in the chair conformation. Furthermore, the chair conformation has *all* bond angles near the preferred 109.5°. Cyclohexane molecules therefore spend the great majority of their time in the chair conformation.
- In the chair conformation of cyclohexane, the 12 different hydrogen atoms attached to the six carbon atoms of the ring can be classified as one of two types, **axial** or **equatorial.** ✻
  - The **six axial positions** are **perpendicular** to the mean plane of the cyclohexane ring; three axial hydrogens point straight up, and three point straight down.
  - The **six equatorial positions** point roughly outward from the cyclohexane ring. *[Models will help you understand the difference between axial and equatorial positions.]*

    There are two different chair conformations of cyclohexane that are in equilibrium with each other. *[Using models, you should verify that interconverting the two possible chair cyclohexane conformations changes all of the axial hydrogens to equatorial hydrogens, and vice versa.]*
- There are several less stable puckered conformations of cyclohexane such as the **boat** and **twist boat.** These conformations are less stable than the chair conformation, because they have either some eclipsed hydrogen atoms or bond angles other than the optimum 109.5°. These less stable conformations of cyclohexane are intermediates in the interconversion of the two chair forms of cyclohexane.
- If one or more of the hydrogens of a cyclohexane are replaced by any larger atom or group, the more stable chair conformation is the one that places the larger atom or group in an equatorial position. ✻ *[This is perhaps the most important concept involved with cyclohexane chair conformations, and you will use it over and over again.]*
  - The larger the atom or group, the greater the preference for it to be in an equatorial position. This preference for large atoms or groups to be equatorial derives from a nonbonded interaction strain called **diaxial interactions.**

    Atoms or groups that are axial are relatively close to the two other atoms or groups that are also axial on the same side of the ring, thus groups will crash into these other axial substituents. This contact occurs between axial atoms or groups on the same side of the cyclohexane ring, hence the name **diaxial interactions.** *[Confirm this nonbonded interaction strain for yourself by making a model of methylcyclohexane, and make the chair conformation that places the methyl group in an axial position.]*
- Atoms or groups in equatorial positions are out away from other groups, thus minimizing nonbonded interaction strain. ✻ *[Confirm this by converting your model to the other chair conformation, and notice the methyl group, which is now equatorial, is relatively free from nonbonded interaction strain.]*
- Chemists have synthesized a variety of highly strained small-ring compounds including **propellanes, cubanes,** and **prismanes.**

2.7 *CIS-TRANS* ISOMERISM IN CYCLOALKANES AND BICYCLOALKANES
- Cycloalkanes with substituents on two or more carbons of the ring show a type of isomerism called *cis-trans* **isomerism.** ✻

- *Cis-trans* **isomerism** is a type of isomerism that depends on the placement of substituent groups on the atoms of a ring or on a double bond.

    Cis-trans isomerism can be understood by thinking of the cycloalkane as a planar structure. *[This is just a helpful trick. Of course the true cycloalkane structures for everything larger than cyclopropane are puckered.]*

    For a cycloalkane with two substituents, the *cis* isomer is the one in which the two substituents are on the same side of the ring plane.

    The *trans* **isomer** is the one in which the two substituents are on opposite sides of the ring plane. In other words, for a given constitutional isomer such as 1,2 dimethylcyclopentane, the two methyl groups can be either *cis* or *trans* with respect to each other. *[Note that in order for the cis and trans comparison to be valid, the same constitutional isomer must be considered in both cases.]*

    No matter how much the cycloalkane ring puckers or interchanges between conformations, the two methyl groups of the *cis* isomer will always be on the same side of the ring plane, and the two methyl groups of the *trans* isomer will always remain on opposite sides of the ring plane. Put another way, no amount of conformational change can convert the *cis* isomer into the *trans* isomer, and vice versa. *[Again, making models will save you a lot of time, and it will also make things more clear .]*

- The analysis of disubstituted cyclohexanes becomes more complicated in the context of chair conformations. For example, *trans*-1,4-dimethylcyclohexane can exist in two chair conformations. *[You should make a model to prove this to yourself.]*

    In one chair conformation, both methyl groups are axial. In the other chair conformation, both methyl groups are equatorial.

    The chair conformation with both methyl groups equatorial is more stable. *[It has fewer diaxial interactions]*

- For *cis*-1,4-dimethylcyclohexane, each chair conformation has one methyl group axial and one methyl group equatorial, so these conformations are equally stable. *[Again, a model will be very helpful here.]*

• **Bicycloalkanes** also exhibit *cis-trans* isomerism. *[You should practice analyzing compounds like cis- and trans-decalin with models using the same ideas just discussed for the cyclopentane and cyclohexane derivatives.]*

2.8 PHYSICAL PROPERTIES OF ALKANES AND CYCLOALKANES

• At room temperature, the simple alkanes the size of butane or smaller are **gases**, while pentane through decane are **liquids**.

• At lower temperatures, the alkanes can be frozen into **solids**.

• The fact that alkanes can exist as liquids and solids depends on the existence of **intermolecular coulombic forces of attraction** that can hold the alkane molecules together.

    - All intermolecular forces that hold ions and molecules together are **electrostatic** in nature, that is, they are based on attraction between groups with opposite full charges in the case of ions, or opposite partial charges associated with dipole moments in the case of overall neutral molecules. Important types of **intermolecular coulombic forces of attraction** include **ion-ion interactions, ion-dipole interactions, dipole-dipole interactions** and **hydrogen bonding**. These interactions will be discussed at a later time. ✳

    - **Dispersion forces** are the type of intermolecular attractive force most relevant to alkanes so they will be discussed now. Dispersion forces are the weakest of all the intermolecular forces and they are the result of electrostatic attraction between *temporary* dipole moments.

        Even molecules such as alkanes without large permanent dipole moments have **temporary dipole moments** caused by instantaneous fluctuations of electron density.

        Averaged over time, the electron distribution is symmetrical, but, at any instant, there are small dipole moments caused by small non-symmetrical shifts in electron density. Such an instantaneous dipole moment induces an equally instantaneous but opposite dipole moment in adjacent molecules or atoms. These weak but opposite temporary dipole moments form the basis of weak electrostatic attractions referred to as **dispersion forces**.

        The strength of dispersion forces depends on how easily an electron cloud is **polarized**. Small atoms with tightly held electrons show weaker dispersion forces than larger atoms or molecules with less tightly held electrons.

• Alkanes have very low **boiling points** because they are held together primarily by nothing but weak dispersion forces.

    -Larger alkanes (higher molecular weight) have higher boiling points than smaller alkanes (lower molecular weight), and unbranched alkanes have higher boiling points than branched constitutional isomers. The strength of dispersion forces is proportional to the surface area of contact between molecules. Branched molecules are more compact resulting in smaller surface areas than unbranched constitutional isomers. *[The above ideas explain a large amount of experimental data.]*

2.9 SOURCES OF ALKANES

- The three natural sources of alkanes are **natural gas**, **petroleum,** and **coal**.
  - **Natural gas** is mostly methane, with some ethane and a small amount of other small alkanes that are gases at room temperature.
  - **Petroleum** is an incredibly complex mixture of compounds, and the commercially important fractions are purified by large scale distillation towers. Presently, petroleum is by far the most important source of organic raw materials for products like fuels, lubricants, nylon, dacron, textile fibers, asphalt, and synthetic rubber.
  - **Coal** has an extremely complex structure, and a great deal of chemistry is required to produce useful alkanes such as fuels.

2.10 REACTIONS OF ALKANES

- Alkanes are relatively unreactive, having small permanent dipole moments and only the relatively strong sigma type of bonds.
- On the other hand, alkanes can react with oxygen and halogens under certain conditions. ✳

CHAPTER 2
Solutions to the Problems

Problem 2.1 Do the structural formulas in each set represent identical compounds or constitutional isomers?

(a) $CH_3-\overset{\overset{\displaystyle CH_2-CH_3}{|}}{CH}-\overset{\overset{\displaystyle }{|}}{\underset{\underset{\displaystyle CH_2-CH_3}{|}}{CH}}-CH_3$ and $CH_3-CH_2-\overset{\overset{\displaystyle CH_3}{|}}{CH}-CH_2-\overset{\overset{\displaystyle CH_3}{|}}{CH}-CH_3$

These molecules are constitutional isomers. Each has six carbons in the longest chain. The first has one-carbon branches on carbons 3 and 4 of the chain; the second has one-carbon branches on carbons 2 and 4 of the chain.

(b) $CH_3-\overset{\overset{\displaystyle CH_3}{|}}{CH}-\underset{\underset{\displaystyle CH_2-CH_3}{|}}{CH}-CH_3$ and $CH_3-\overset{\overset{\displaystyle CH_3}{|}}{CH}-\underset{\underset{\displaystyle CH_3}{|}}{CH}-CH_2-CH_3$

These molecules are identical. Each has five carbons in the longest chain, and one-carbon branches on carbons 2 and 3 of the chain.

Problem 2.2 Draw structural formulas for the three constitutional isomers of molecular formula C_5H_{12}.

$CH_3-CH_2-CH_2-CH_2-CH_3$ $CH_3-\overset{\overset{\displaystyle CH_3}{|}}{CH}-CH_2-CH_3$ $CH_3-\overset{\overset{\displaystyle CH_3}{|}}{\underset{\underset{\displaystyle CH_3}{|}}{C}}-CH_3$

Problem 2.3 Write IUPAC names for the following alkanes. Show that each name indicates the total number of carbons in the molecule.

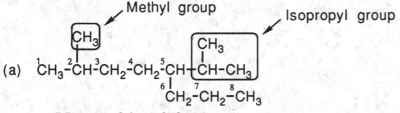

(a)

5-Isopropyl-2-methyloctane

There are 3 (isopropyl) + 1 (methyl) + 8 (octane) = 12 carbon atoms.

(b)

4-Isopropyl-4-propylheptane

There are 3(isopropyl) + 3(propyl) + 7(heptane) = 13 carbon atoms.

<u>Problem 2.4</u> State the number of sets of equivalent hydrogens in each compound and the number of hydrogens in each set.

(a) $CH_3-CH_2-\overset{\overset{\displaystyle CH_3}{|}}{CH}-CH_2-CH_3$

(b) $CH_3-\overset{\overset{\displaystyle CH_3}{|}}{CH}-CH_2-\overset{\overset{\displaystyle CH_3}{|}}{\underset{\underset{\displaystyle CH_3}{|}}{C}}-CH_3$

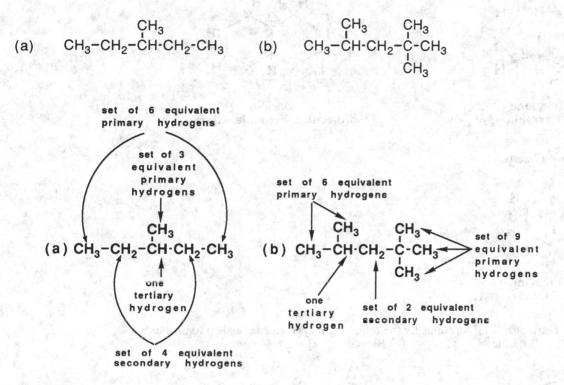

<u>Problem 2.5</u> Following are line-angle drawings for three cycloalkanes. Write a structural formula and molecular formula for each.

(a)

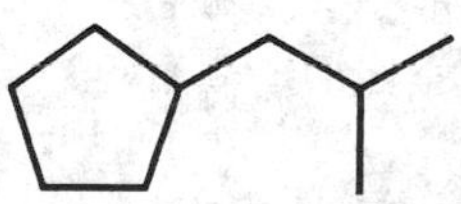

$(C_5H_9)CH_2CH(CH_3)_2$
Isobutylcyclopentane
Molecular Formula C_9H_{18}

(b)

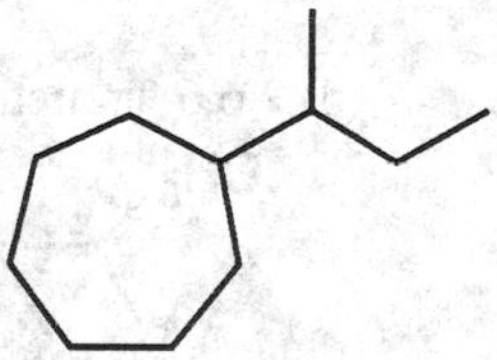

$(C_7H_{13})CH(CH_3)CH_2CH_3$
sec-Butylcycloheptane
Molecular Formula $C_{11}H_{22}$

(c)

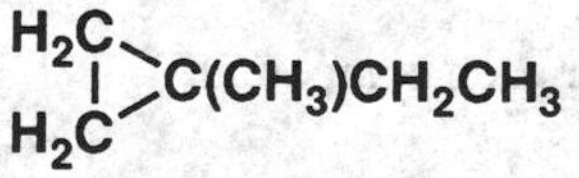

H_2C
$|\ \ \ \ \ C(CH_3)CH_2CH_3$
H_2C

1-Ethyl-1-methylcyclopropane
Molecular Formula C_6H_{13}

Problem 2.6 Write molecular formulas for each bicycloalkane, given its number of carbon atoms.
(a) Hydrindane (9 carbons)   (b) Decalin (10 carbons)

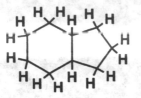

Hydrindane
Molecular Formula C_9H_{16}

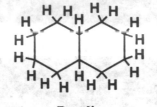

Decalin
Molecular Formula $C_{10}H_{18}$

(c) Norbornane (7 carbons)

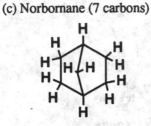

Norbornane
Molecular Formula C_7H_{12}

Problem 2.7 Draw structural formulas for the following bicycloalkanes and spiroalkanes.
(a) Bicyclo[3.1.0]hexane   (b) Bicyclo[2.2.2]octane

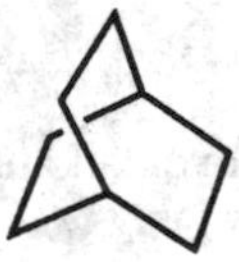

(c) Bicyclo[4.2.0]octane 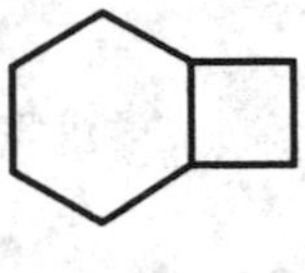   (d) 2,6,6-Trimethylbicyclo[3.1.1]heptane

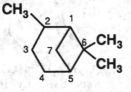

(e) Spiro[2.4]heptane 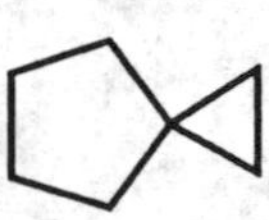   (f) Spiro[2.5]octane

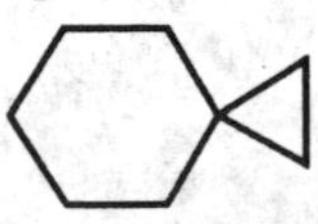

Problem 2.8 Combine the proper prefix, infix, and suffix and write the IUPAC name for each compound.

(a) CH₃-C-CH₃

2-Propanone

(b) CH₃-CH₂-CH₂-CH₂-C-H

Pentanal

(c)

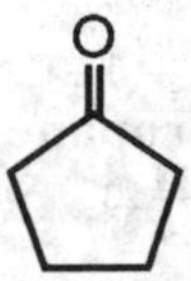

Cyclopentanone

(d)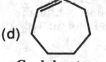

Cycloheptene

<u>Problem 2.9</u> From studies of dipole moments, it has been estimated that in the gas phase at room temperature, the ratio of molecules in the anti to gauche conformation is approximately 7.6 to 1. Calculate the difference in energy between these two conformations.

$$\Delta G^\circ = -RT\ln K_{eq}$$

$$K_{eq} = \frac{7.6}{1} = 7.6 \quad so \quad \ln K_{eq} = 2.03$$

Plugging in the gas constant ($R = 1.98$ cal·K^{-1}·mol^{-1}) and temperature ($T = 298$ K)

$$\Delta G^\circ = -(1.98 \text{ cal·K}^{-1}\text{·mol}^{-1})(298 \text{ K})(2.03) = \boxed{1200 \text{ cal/mol} = 1.2 \text{ kcal/mol}}$$

<u>Problem 2.10</u> Following is a chair conformation of cyclohexane with carbon atoms numbered 1 through 6.

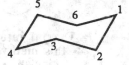

(a) Draw hydrogen atoms that are above the plane of the ring on carbons 1 and 2 and below the plane of the ring on carbon 4.
(b) Which of these hydrogens are equatorial; which are axial?
(c) Draw the other chair conformation. Now, which hydrogens are equatorial; which are axial?

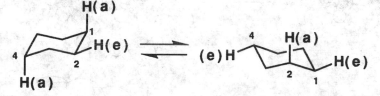

In the above figure (a) = axial and (e) = equatorial.

<u>Problem 2.11</u> Estimate the difference in energy between the alternative chair conformations of the trisubstituted cyclohexane given in Example 2.11.

Diaxial interactions between this methyl group and the circled axial hydrogens

Diaxial interactions between this methyl group and the circled axial hydrogens

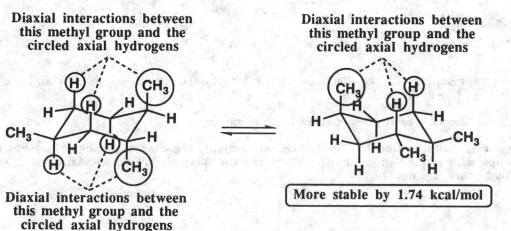

Diaxial interactions between this methyl group and the circled axial hydrogens

More stable by 1.74 kcal/mol

Shown above are the two alternative chair conformations of the trisubstituted cyclohexane given in Example 2.11. The conformation on the left has two axial methyl groups, while the conformation on the right only has one. Each axial methyl group adds 1.74 kcal/mol in energy, so the conformation on the right is more stable by 1.74 kcal/mol.

Problem 2.12 For 1,4-dimethylcyclohexane:
(a) Draw the chair conformation in which both methyl groups are equatorial.
(b) Draw the chair conformation in which both methyl groups are axial.

Diaxial interactions between this methyl group and the circled axial hydrogens

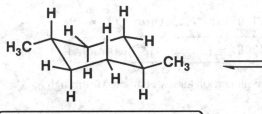

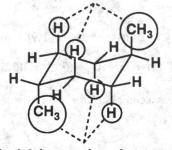

More stable by 3.48 kcal/mol

Diaxial interactions between this methyl group and the circled axial hydrogens

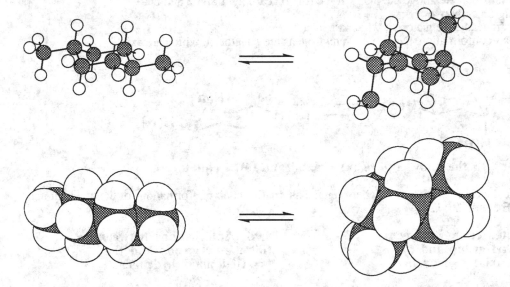

In the above representations, the diequatorial conformation is shown on the left and the diaxial conformation is shown on the right.

(c) Estimate the difference in energy between these two conformations.

The diequatorial conformation has no diaxial interactions, the diaxial conformation has two methyl groups with diaxial interactions. Therefore the diequatorial conformation is 2 x 1.74 = 3.48 kcal/mol more stable.

(d) Calculate the ratio of the diequatorial to the diaxial conformation for 1,4-dimethylcyclohexane at 25°C.

$$\text{Rearranging} \quad \Delta G° = -RT\ln K_{eq} \quad \text{gives} \quad \ln K_{eq} = \frac{-\Delta G°}{RT}$$

Plugging in the gas constant ($R = 1.98 \text{ cal} \cdot K^{-1} \cdot mol^{-1}$) and temperature ($T = 298$ K) as well as the value for $\Delta G°$ converted to cal/mol gives

$$\ln K_{eq} = \frac{3480 \text{ cal/mol}}{(1.98 \text{ cal} \cdot K^{-1} \cdot mol^{-1})(298 \text{ K})} = 6.0$$

$$K_{eq} = e^{6.0} = 400$$

So the ratio of diequatorial to diaxial is 400:1

Problem 2.13 Which cycloalkanes show *cis-trans* isomerism? For each that does, draw the *cis* and *trans* isomers.

(a)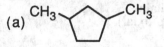

1,3-Dimethylcyclopentane shows *cis-trans* isomerism. In the following drawings, the ring is drawn as a planar pentagon with substituents above and below the plane of the pentagon.

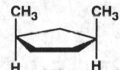   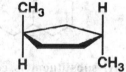

cis-1,3-Dimethyl-   *trans*-1,3-Dimethyl-
    cyclopentane   cyclopentane

(b)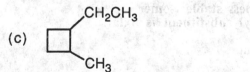

Ethylcyclopentane does not show *cis-trans* isomerism.

(c)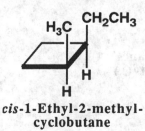

1-Ethyl-2-methylcyclobutane shows *cis-trans* isomerism.

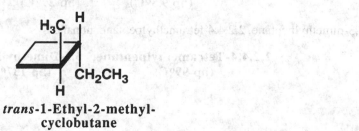

cis-1-Ethyl-2-methyl-   *trans*-1-Ethyl-2-methyl-
    cyclobutane   cyclobutane

Problem 2.14 Following is a planar hexagon representation for one isomer of 1,2,4-trimethylcyclohexane. Draw alternative chair conformations of this compound and state which is the more stable.

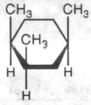

Following are alternative chair conformations for the all *cis* isomer of 1,2,4-trimethylcyclohexane. The alternative chair conformation on the right is the more stable because it has only one axial methyl group.

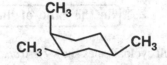
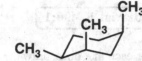

less stable chair   more stable chair
(two methyl groups axial)   (one methyl group axial)

Problem 2.15 Which stereoisomer is more stable?

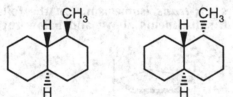

In the first isomer of *trans*-decalin, the methyl substituent is equatorial and in the second isomer, it is axial. The equatorial methyl isomer is more stable.

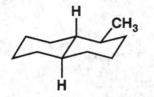
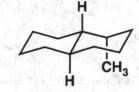

more stable isomer   less stable isomer
(methyl substituent is equatorial)   (methyl substituent is axial)

Problem 2.16 Arrange the following in order of increasing boiling point.
(a) 2-Methylbutane, 2,2-Dimethylpropane, Pentane

2,2-Dimethylpropane,   2-Methylbutane,   Pentane
(bp 9.5°C)   (bp 29°C)   (bp 36°C)

(b) 3,3-dimethylheptane, 2,2,4,4-tetramethylpentane, nonane

2,2,4,4-Tetramethylpentane,   3,3-Dimethylheptane,   Nonane
(bp 99°C)   (bp 137°C)   (bp 151°C)

Constitutional Isomerism

Problem 2.17 Which statements are true about constitutional isomers?
(a) They have the same molecular formula.
(b) They have the same molecular weight.
(c) They have the same order of attachment of atoms.
(d) They have the same physical properties.

Statements (a) and (b) are true, statements (c) and (d) are false.

Problem 2.18 Which of the following are identical compounds and which are constitutional isomers?

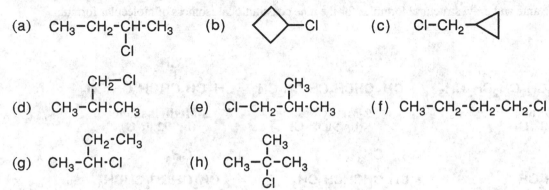

Following are names and molecular formulas of each
(a) 2-Chlorobutane; C_4H_9Cl
(b) Chlorocyclobutane; C_4H_7Cl
(c) Chloromethylcyclopropane; C_4H_7Cl
(d) 1-Chloro-2-methylpropane; C_4H_9Cl
(e) 1-Chloro-2-methylpropane; C_4H_9Cl
(f) 1-Chlorobutane; C_4H_9Cl
(g) 2-Chlorobutane; C_4H_9Cl
(h) 2-Chloro-2-methylpropane; C_4H_9Cl

The following are identical: (d),(e)   (a),(g)

The following compounds are constitutional isomers of molecular formula C_4H_9Cl: 2-chlorobutane (a)(g), 1-chloro-2-methylpropane (d)(e), and 2-chloro-2-methylpropane (h).

The following compounds are constitutional isomers of molecular formula C_4H_7Cl: chlorocyclobutane (b) and chloromethylcyclopropane (c).

Problem 2.19 Tell whether the compounds in each set are constitutional isomers.

(a) CH_3-CH_2-OH and CH_3-O-CH_3 C_2H_6O Yes

(b) $CH_3-\overset{O}{\overset{\|}{C}}-CH_3$ and $CH_3-CH_2-\overset{O}{\overset{\|}{C}}-H$ C_3H_6O Yes

(c) $CH_3-\overset{O}{\overset{\|}{C}}-O-CH_3$ and $CH_3-CH_2-\overset{O}{\overset{\|}{C}}-OH$ $C_3H_6O_2$ Yes

(d) $CH_3-\overset{OH}{\overset{|}{C}H}-CH_2-CH_3$ and $CH_3-\overset{O}{\overset{\|}{C}}-CH_2-CH_3$ No

$C_4H_{10}O$ C_4H_8O

C_5H_{10}

C_5H_{12} N_o

(e) [cyclopentane structure]   and $CH_3-CH_2-CH_2-CH_2-CH_3$

(f) [cyclopentane structure]   and $CH_2=CH-CH_2-CH_2-CH_3$ Yes

C_5H_{10} C_5H_{10}

Sets (a), (b), (c), and (f) contain constitutional isomers; sets (d) and (e) do not.

Problem 2.20   Name and draw structural formulas for the nine constitutional isomers of molecular formula C_7H_{16}.

$CH_3CH_2CH_2CH_2CH_2CH_2CH_3$

Heptane
(bp 94.8° C)

$CH_3CHCH_2CH_2CH_2CH_3$ with CH_3 substituent

2-Methylhexane
(bp 90.0° C)

$CH_3CH_2CHCH_2CH_2CH_3$ with CH_3 substituent

3-Methylhexane
(bp 92.0° C)

$CH_3CCH_2CH_2CH_3$ with two CH_3 substituents

2,2-Dimethylpentane
(bp 79.2° C)

$CH_3CHCHCH_2CH_3$ with two CH_3 substituents

2,3-Dimethylpentane
(bp 89.8° C)

$CH_3CHCH_2CHCH_3$ with two CH_3 substituents

2,4-Dimethylpentane
(bp 80.5° C)

$CH_3CH_2CCH_2CH_3$ with two CH_3 substituents

3,3-Dimethylpentane
(bp 86.1° C)

$CH_3CH_2CHCH_2CH_3$ with CH_2CH_3 substituent

3-Ethylpentane
(bp 93.5° C)

CH_3CHCCH_3 with H_3C, CH_3 and CH_3 substituents

2,2,3-Trimethyl-
butane
(bp 80.9° C)

Problem 2.21   Draw structural formulas for all of the following:
(a) Alcohols of molecular formula $C_4H_{10}O$.

$CH_3-CH_2-CH_2-CH_2-OH$ $CH_3-CH_2-CH-CH_3$ with OH substituent CH_3-C-OH with two CH_3 substituents

$CH_3-CH-CH_2-OH$ with CH_3 substituent

(b) Aldehydes of molecular formula C_4H_8O.

$CH_3-CH_2-CH_2-\overset{O}{\overset{\|}{C}}-H$ $CH_3-CH-\overset{O}{\overset{\|}{C}}-H$ with CH_3 substituent

(c) Ketones of molecular formula $C_5H_{10}O$.

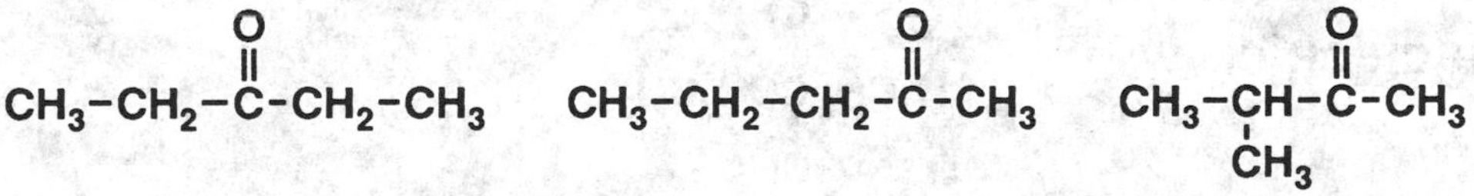

(d) Carboxylic acids of molecular formula $C_5H_{10}O_2$.

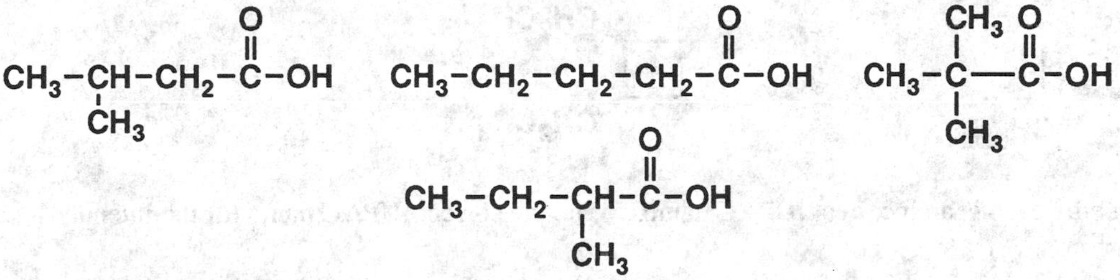

Nomenclature of Alkanes and Cycloalkanes

Problem 2.22 Write IUPAC names for these alkanes and cycloalkanes.

(a) 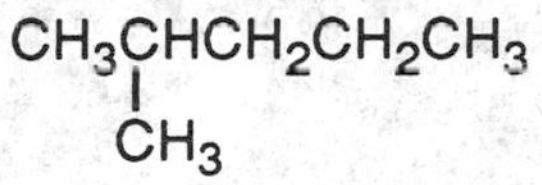

2-Methylpentane (isohexane)

(b) 

2,5-Dimethylhexane

(c) 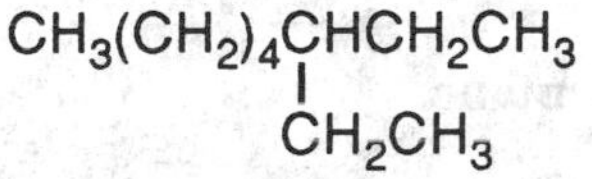

3-Ethyloctane

(d) $(CH_3)_2CHCH_2CH_2C(CH_3)_3$

2,2,5-Trimethylhexane

(e) 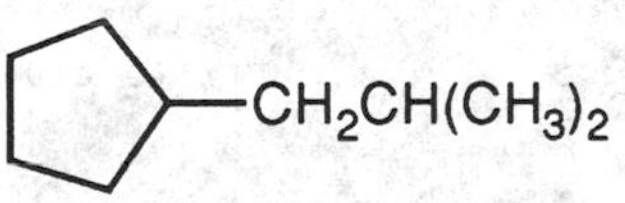

Isobutylcyclopentane

(f) 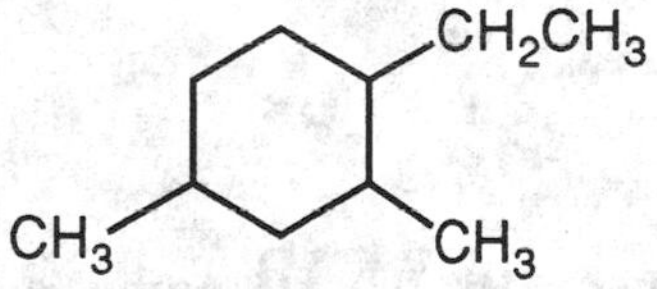

1-Ethyl-2,4-dimethylcyclohexane

Problem 2.23 Write structural formulas for these alkanes.

(a) 2,2,4-Trimethylhexane

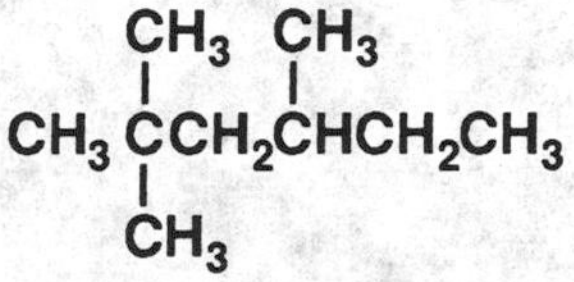

(b) 2,2-Dimethylpropane

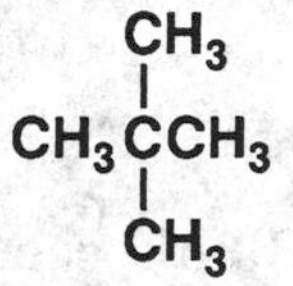

(c) 3-Ethyl-2,4,5-trimethyloctane

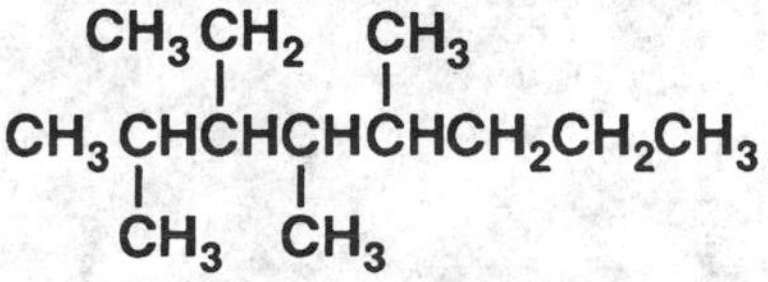

(d) 5-Butyl-2,2-dimethylnonane

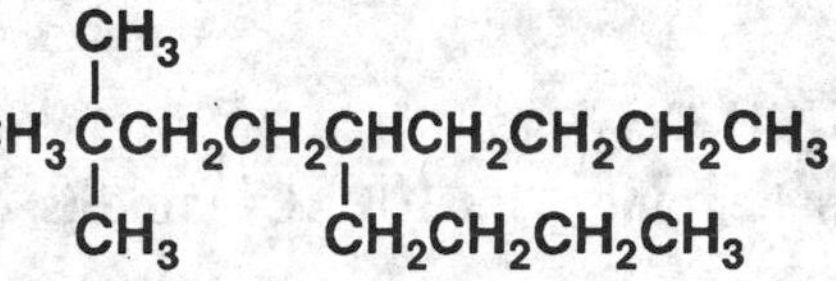

(e) 4-Isopropyloctane

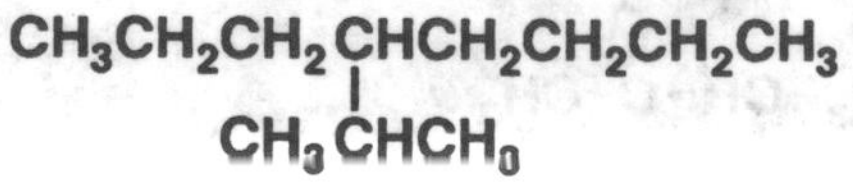

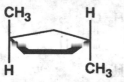

(f) 3,3-Dimethylpentane

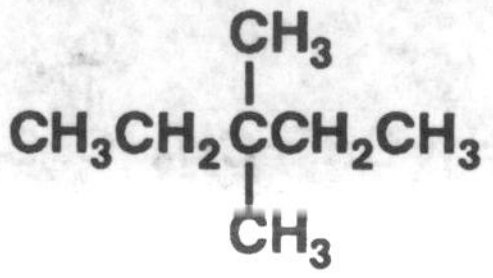

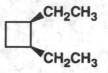

(g) *trans*-1,3-Dimethylcyclopentane

CH_3 H

H CH_3

(h) *cis*-1,2-Diethylcyclobutane

CH_2CH_3

CH_2CH_3

<u>Problem 2.24</u> Explain why each is an incorrect IUPAC name. Write the correct IUPAC name for the intended compound.

(a) 1,3-Dimethylbutane

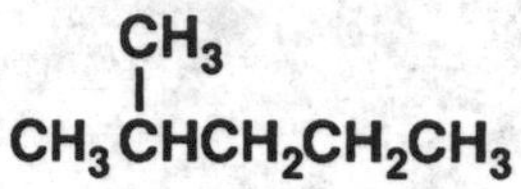

CH_3
|
$CH_3 CHCH_2CH_2CH_3$

The longest chain is pentane. Its IUPAC name is 2-methylpentane.

(b) 4-Methylpentane

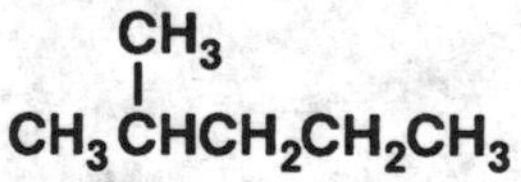

CH_3
|
$CH_3 CHCH_2CH_2CH_3$

The pentane is numbered incorrectly. Its IUPAC name is 2-methylpentane.

(c) 2,2-Diethylbutane

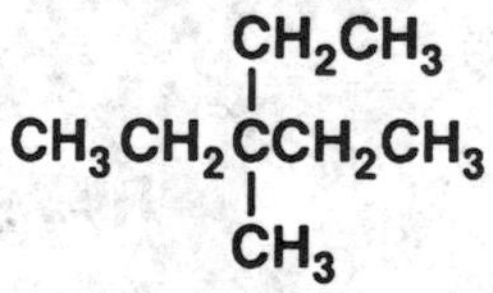

CH_2CH_3
|
$CH_3 CH_2 CCH_2CH_3$
|
CH_3

The longest chain is pentane. Its IUPAC name is 3-ethyl-3-methylpentane.

(d) 2-Ethyl-3-methylpentane

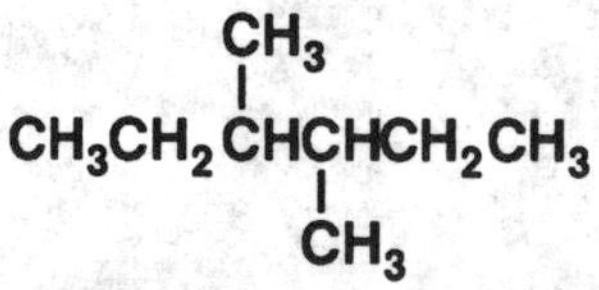

CH_3
|
$CH_3CH_2 CHCHCH_2CH_3$
|
CH_3

The longest chain is hexane. Its IUPAC name is 3,4-dimethylhexane.

(e) 2-Propylpentane

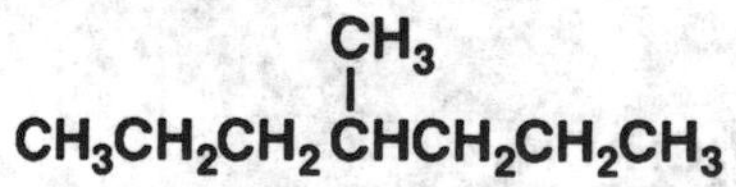

CH_3
|
$CH_3CH_2CH_2 CHCH_2CH_2CH_3$

The longest chain is heptane. Its IUPAC name is 4-methylheptane.

(f) 2,2-Diethylheptane

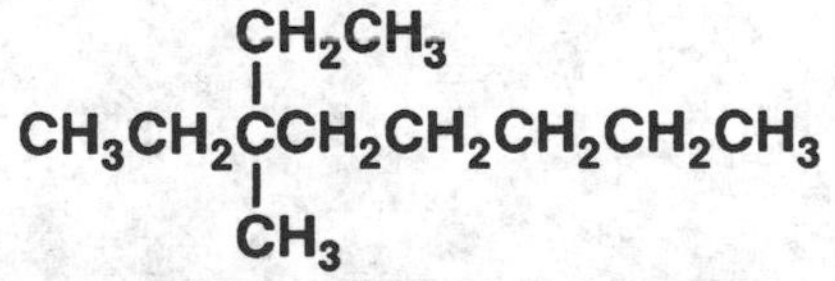

$$CH_3CH_2 \overset{\overset{\displaystyle CH_2CH_3}{|}}{\underset{\underset{\displaystyle CH_3}{|}}{C}} CH_2CH_2CH_2CH_2CH_3$$

The longest chain is octane.   Its IUPAC name is 3-ethyl-3-methyloctane.

(g) 2,2-Dimethylcyclopropane

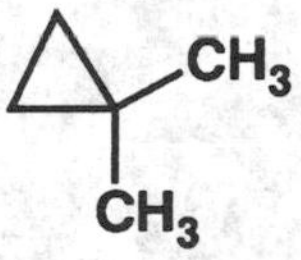

The ring is numbered incorrectly.   Its IUPAC name is 1,1-dimethylcyclopropane.

(h) 1-Ethyl-5-methylcyclohexane

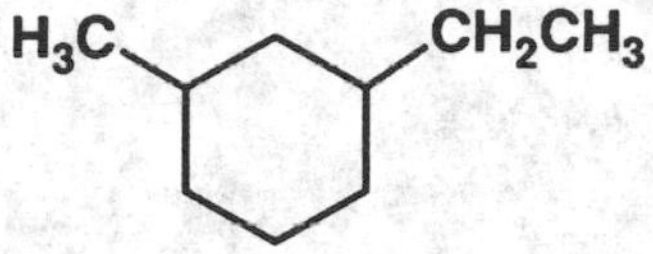

The ring is numbered incorrectly.   Its IUPAC name is 1-ethyl-3-methylcyclohexane.

The IUPAC System of Nomenclature

Problem 2.25 For each of the following IUPAC names, draw the corresponding structural formula for each compound.

(a) Butanone

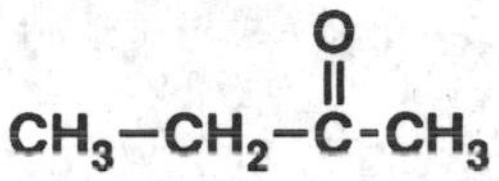

$$CH_3-CH_2-\overset{\overset{\displaystyle O}{||}}{C}-CH_3$$

(b) Butanal

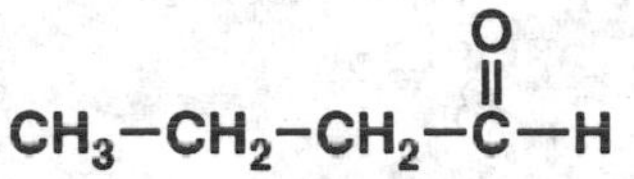

$$CH_3-CH_2-CH_2-\overset{\overset{\displaystyle O}{||}}{C}-H$$

(c) Butanoic acid

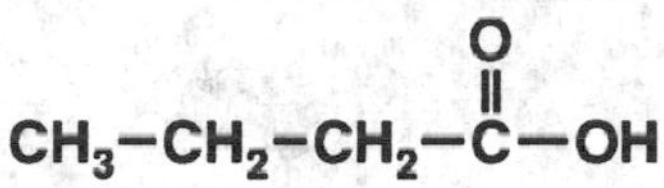

$$CH_3-CH_2-CH_2-\overset{\overset{\displaystyle O}{||}}{C}-OH$$

(d) Ethanoic acid

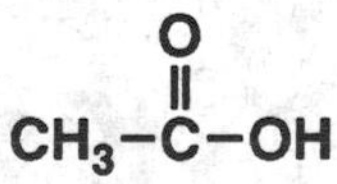

$$CH_3-\overset{\overset{\displaystyle O}{||}}{C}-OH$$

(e) Hexanoic acid

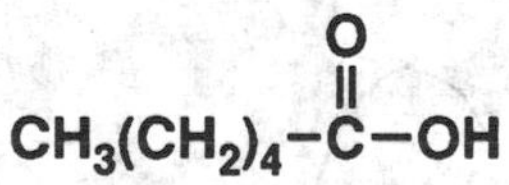

$$CH_3(CH_2)_4-\overset{\overset{\displaystyle O}{||}}{C}-OH$$

(f) Propanoic acid

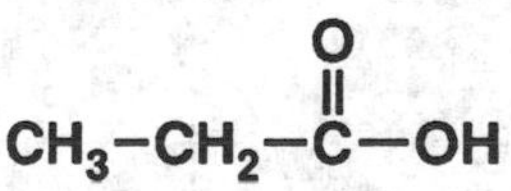

$$CH_3-CH_2-\overset{\overset{\displaystyle O}{||}}{C}-OH$$

(g) Propanal

$$CH_3-CH_2-\overset{\overset{\displaystyle O}{||}}{C}-H$$

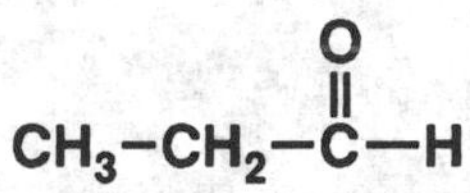

(h) Cyclopentene

(i) Cyclopentanol

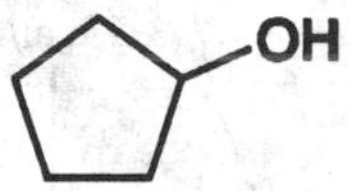

(j) Cyclopentanone

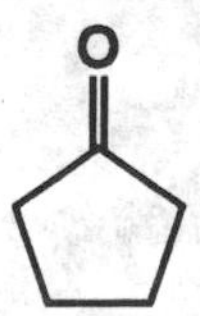

(k) Cyclopropanol

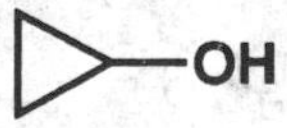

(l) Propanone

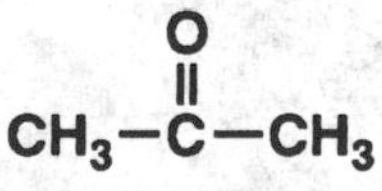

$$CH_3-\overset{\overset{\displaystyle O}{||}}{C}-CH_3$$

Problem 2.26 Write IUPAC names for these compounds.

(a) 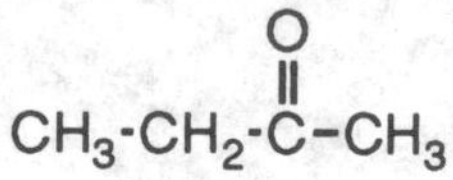 $CH_3\text{-}CH_2\text{-}\overset{\displaystyle O}{\overset{\|}{C}}\text{-}CH_3$

Butanone

(b) 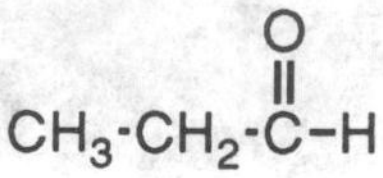 $CH_3\text{-}CH_2\text{-}\overset{\displaystyle O}{\overset{\|}{C}}\text{-}H$

Propanal

(c) 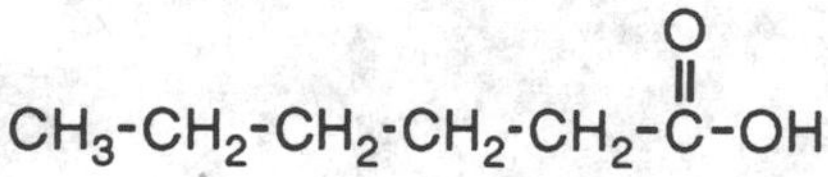 $CH_3\text{-}CH_2\text{-}CH_2\text{-}CH_2\text{-}CH_2\text{-}\overset{\displaystyle O}{\overset{\|}{C}}\text{-}OH$

Hexanoic acid

(d) 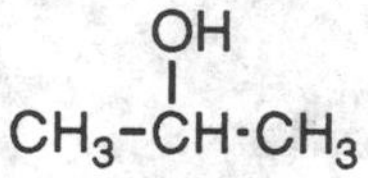 $CH_3\text{-}\overset{\displaystyle OH}{\overset{|}{C}H}\text{-}CH_3$

2-Propanol

(e)

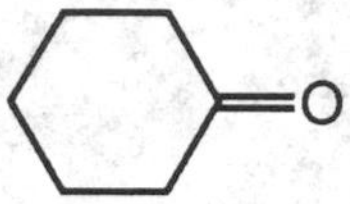

Cyclohexanone

(f)

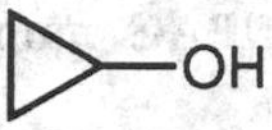

Cyclopropanol

(g) $CH_3\text{-}CH\text{=}CH_2$

Propene

(h)

Cyclohexene

Problem 2.27 Torsional strain resulting from eclipsed C-H bonds is approximately 1.0 kcal/mol (4.2 kJ/mol) and that for eclipsed C-H and C-CH$_3$ bonds is approximately 1.5 kcal/mol (6.2 kJ/mol). Given this information, sketch a graph of potential energy versus dihedral angle for propane.

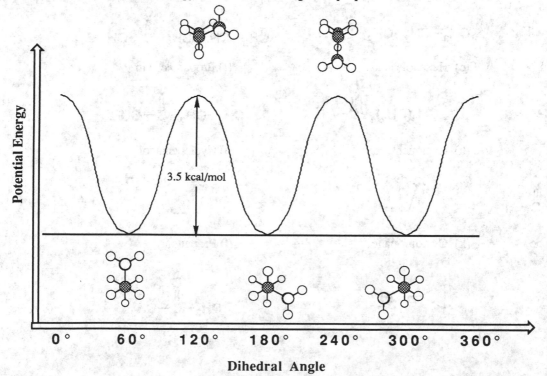

Notice that the energy of the eclipsed conformations is 3.5 kcal/mol higher in energy than the staggered conformations. This is because each eclipsed conformation has two C-H bonds

eclipsed with other C-H bonds (worth 1.0 kcal/mol each) and one C-H bond eclipsed to a C-CH_3 bond (worth 1.5 kcal).

Problem 2.28 How many different staggered conformations are there for 2-methylpropane? How many different eclipsed conformations are there?

Looking down any of the carbon-carbon bonds, there is one staggered and one eclipsed conformation of 2-methylpropane.

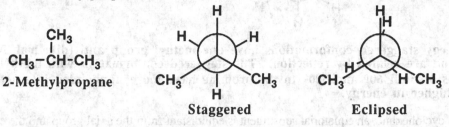

$$CH_3$$
$$CH_3-CH-CH_3$$
2-Methylpropane

Staggered   Eclipsed

Problem 2.29 Consider 1-bromopropane , $CH_3CH_2CH_2Br$.
(a) Draw a Newman projection in which -CH_3 and -Br are anti (dihedral angle 180°).

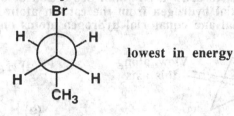

lowest in energy

(b) Draw Newman projections in which -CH_3 and -Br are gauche (dihedral angles 60° and 300°).

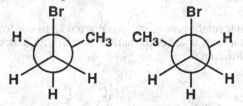

related by reflection

(c) Which of these is the lowest energy conformation.

The anti (dihedral angle 180°) is the lowest energy conformation.

(d) Which of these conformations, if any, are related by reflection?

The two gauche conformations are of equal energy, and are related by reflection.

Problem 2.30 Consider 1-bromo-2-methylpropane and draw the following:
(a) The staggered conformation(s) of lowest energy.

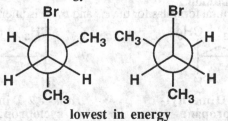

**lowest in energy
(related by reflection)**

(b) The staggered conformation(s) of highest energy.

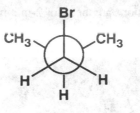

highest in energy

The lower energy staggered conformations have one methyl group anti (dihedral angle 180°) to the bromine and are related by reflection. The staggered conformation with methyl groups at dihedral angles of both 60° and 300° to the bromine have more nonbonded interaction strain and are thus higher in energy.

<u>Problem 2.31</u> In cyclohexane, an equatorial substituent is equidistant from the axial group and the equatorial group on an adjacent carbon. What is the simplest way to demonstrate this fact?

The best way to see this fact is to draw a Newman projection of one of the carbon-carbon bonds. As can be seen, the axial hydrogen from the carbon atom in the front is in between and thus equidistant to the axial and equatorial hydrogen atoms on the rear carbon atom.

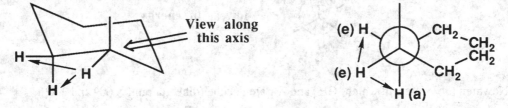

<u>Problem 2.32</u> *trans*-1,4-Di-*tert*-butylcyclohexane exists in a normal chair conformation. *cis*-1,4-Di-*tert*-butylcyclohexane, however, adopts a twist boat conformation. Draw both isomers and explain why the *cis* isomer is more stable in the twist boat-conformation.

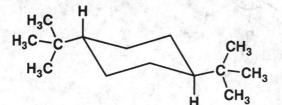
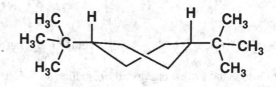

The *trans* isomer in the chair form   The *cis* isomer in the twist-boat form

The *cis* isomer adopts a twist boat conformation because each of the bulky *tert*-butyl groups can be in an pseudo-equatorial position as shown above. If the *cis* isomer existed in a normal chair conformation, then one *tert*-butyl group would be equatorial, while the other would be forced axial resulting in a large amount of nonbonded interaction strain.

Cis-Trans Isomerism in Cycloalkanes
<u>Problem 2.33</u> Name and draw structural formulas for the *cis* and *trans* isomers of 1,2-dimethylcyclopropane.

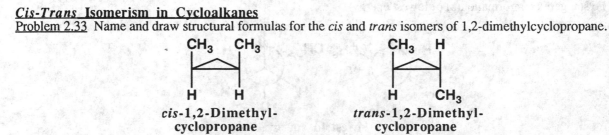

cis-1,2-Dimethyl-   trans-1,2-Dimethyl-
    cyclopropane   cyclopropane

Problem 2.34 Name and draw structural formulas for all cycloalkanes of molecular formula C_5H_{10}. Be certain to include *cis-trans* isomers as well as constitutional isomers.

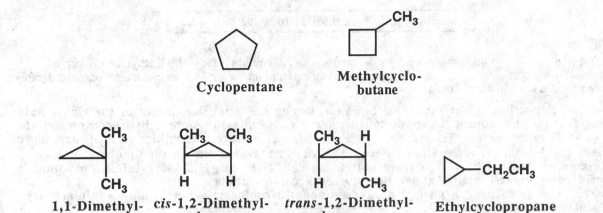

Cyclopentane

Methylcyclo-
butane

1,1-Dimethyl-
cyclopropane

cis-1,2-Dimethyl-
cyclopropane

trans-1,2-Dimethyl-
cyclopropane

Ethylcyclopropane

Problem 2.35 Using a planar pentagon representation for the cyclopentane ring, draw structural formulas for the *cis* and *trans* isomers of:
(a) 1,2-Dimethylcyclopentane

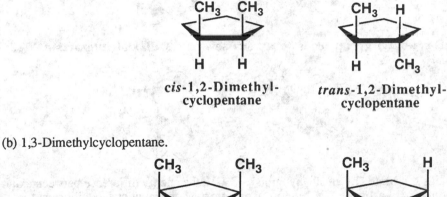

cis-1,2-Dimethyl-
cyclopentane

trans-1,2-Dimethyl-
cyclopentane

(b) 1,3-Dimethylcyclopentane.

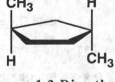

cis-1,3-Dimethyl-
cyclopentane

trans-1,3-Dimethyl-
cyclopentane

Problem 2.36 Energy differences between axial substituted and equatorial substituted cyclohexane chair conformations of cyclohexane were given in Table 2.4
(a) Calculate the ratio of equatorial *tert*-butylcyclohexane to axial *tert*-butylcyclohexane at 25°C.

According to the value given in Table 2.4, the equatorial *tert*-butylcyclohexane is 4.9 kcal/mol more stable than the axial conformation.

$$\text{Rearranging} \quad \Delta G° = -RT\ln K_{eq} \quad \text{gives} \quad \ln K_{eq} = \frac{-\Delta G°}{RT}$$

Plugging in the gas constant ($R = 1.98$ cal·K^{-1}·mol^{-1}) and temperature ($T = 298$ K) as well as the value for $\Delta G°$ converted to cal/mol gives

$$\ln K_{eq} = \frac{4900 \text{ cal/mol}}{(1.98 \text{ cal·}K^{-1}\text{·}mol^{-1})(298 \text{ K})} = 8.3$$

$$K_{eq} = e^{8.3} = 4000$$

So the ratio of equatorial to axial is 4000:1

(b) Explain by using molecular models why the conformational equilibria for methyl, ethyl, and isopropyl substituents are comparable, but why the conformational equilibrium for *tert*-butylcyclohexane lies considerably farther toward the equatorial conformation.

Rotation is possible about the single bond connecting the axial substituent to the ring. Axial methyl, ethyl and isopropyl groups can assume a conformation where a hydrogen creates the 1,3-diaxial interactions. With a *tert*-butyl substituent, however, a bulkier -CH₃ group must create the 1,3-diaxial interaction. Because of the increased steric strain (nonbonded interactions) created by the axial *tert*-butyl, the potential energy of the axial conformation is considerably greater than that for the equatorial conformation.

As seen below, an axial isopropyl group can adopt a conformation with only a minimal 1,3 diaxial interaction:

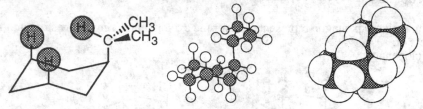

On the other hand, an axial *tert*-butyl group leads to a very severe 1,3 diaxial interaction:

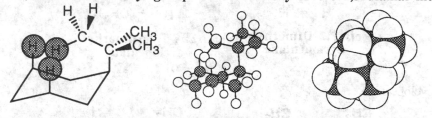

<u>Problem 2.37</u> When cyclohexane is substituted by an ethynyl group, -C≡CH, the energy difference between axial and equatorial conformations is only 0.41 kcal/mol (1.75 kJ/mol). Compare the conformational equilibrium for methylcyclohexane with that for ethynylcyclohexane and account for the difference between the two.

For the ethynyl case, using the same equations as in part (a) of 2.36 gives:

$$\ln K_{eq} = \frac{410 \text{ cal/mol}}{(1.98 \text{ cal} \cdot K^{-1} \cdot \text{mol}^{-1})(298 \text{ K})} = 0.70$$

$$K_{eq} = e^{0.7} = 2.0$$

So the ratio of equatorial to axial ethynyl is 2:1

Using the value of 1.74 kcal/mol given in Table 2.4 for the methyl case gives:

$$\ln K_{eq} = \frac{1740 \text{ cal/mol}}{(1.98 \text{ cal} \cdot K^{-1} \cdot \text{mol}^{-1})(298 \text{ K})} = 2.95$$

$$K_{eq} = e^{2.95} = 19$$

So the ratio of equatorial to axial methyl is 19:1

The above ratios make sense since as can be seen with the following structures, the bulkier methyl group is expected to have more severe 1,3 diaxial interactions than the linear -C≡CH group.

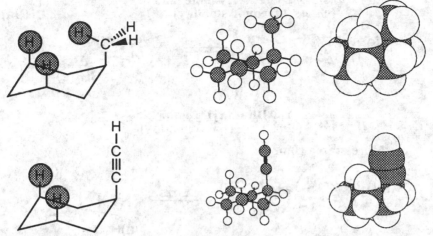

Problem 2.38 Draw the alternative chair conformations for the *cis* and *trans* isomers of 1,2-dimethylcyclohexane; 1,3-dimethylcyclohexane; and 1,4-dimethylcyclohexane.
(a) Indicate by a label whether each methyl group is axial or equatorial.
(b) For which isomer(s) are the alternative chair conformations of equal stability?
(c) For which isomer(s) is one chair conformation more stable than the other?

Cis and *trans* isomers are drawn as pairs. The more stable chair is labeled in cases where there is a difference.

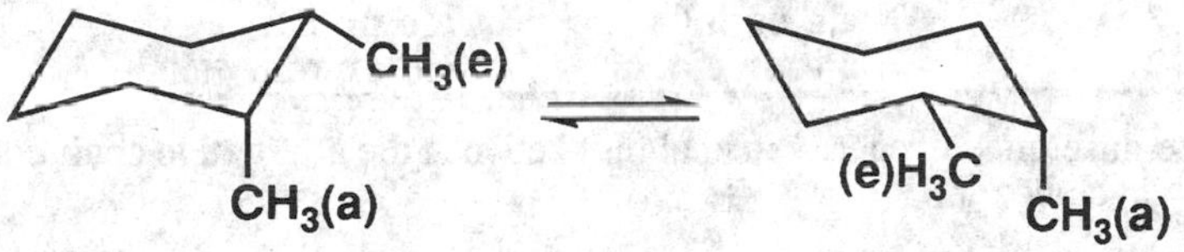

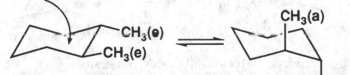

cis-1,2-Dimethylcyclohexane
(chairs of equal stability)

More stable chair

trans-1,2-Dimethylcyclohexane

More stable chair

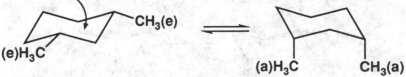

cis-1,3-Dimethylcyclohexane

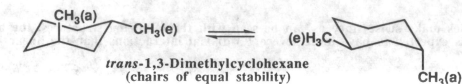

trans-1,3-Dimethylcyclohexane
(chairs of equal stability)

cis-1,4-Dimethylcyclohexane
(chairs of equal stability)

More stable chair

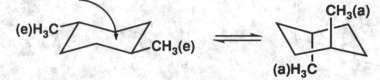

trans-1,4-Dimethylcyclohexane

<u>Problem 2.39</u> Use your answers from problem 2.38 to complete the table showing correlations between *cis*, *trans* and axial, equatorial for the disubstituted derivatives of cyclohexane.

These relationships are summarized in the following table.

Position of Substitution	cis	trans
1,4	a,e or e,a	e,e or a,a
1,3	e,e or a,a	a,e or e,a
1,2	a,e or e,a	e,e or a,a

<u>Problem 2.40</u> Calculate the difference in energy in kcal/mol between the alternative chair conformations of:
(a) *trans*-1-Chloro-4-methylcyclohexane

More stable chair

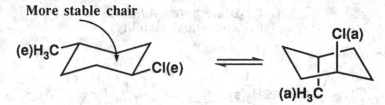

trans-1-Chloro-4-methylcyclohexane

The diequatorial conformation (on the left) will be 0.52 + 1.74 = 2.26 kcal/mol more stable.

(b) *cis*-4-Methylcyclohexanol

Slightly more stable chair

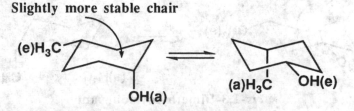

cis-4-Methylcyclohexananol

The conformation on the left is the slightly more stable because the bulkier methyl group is equatorial. The molecule on the right has the smaller -OH group equatorial. The difference in energy will be 1.74 - 0.95 = 0.79 kcal/mol

Problem 2.41 There are four *cis-trans* isomers of 2-isopropyl-5-methylcyclohexanol.

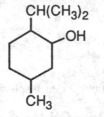

2-Isopropyl-5-methylcyclohexanol

(a) Using a planar hexagon representation for the cyclohexane ring, draw structural formulas for the four *cis-trans* isomers.
(b) Draw the more stable chair conformation for each of your answers in part (a).
(c) Of the four *cis-trans* isomers, which is the most stable? (If you answered this part correctly, you picked the isomer found in nature and given the name menthol)

Following are planar hexagon representations for the four *cis-trans* isomers. In each, the isopropyl group is shown by the symbol R-. One way to arrive at these structural formulas is to take one group as a reference and then arrange the other two groups in relation to it. In these drawings, -OH is taken as the reference and placed above the plane of the ring. Once -OH is fixed, there are only two possible arrangements for the isopropyl group on carbon 2; either *cis* or *trans* to -OH. Similarly, there are only two possible arrangements for the methyl group on carbon-5; either *cis* or *trans* to -OH. Note that even if you take another substituent as a reference, and even if you put the reference below the plane of the ring, there are still only four *cis-trans* isomers for this compound.

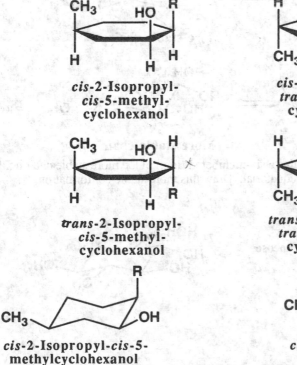

cis-2-Isopropyl-
cis-5-methyl-
cyclohexanol

cis-2-Isopropyl-
trans-5-methyl-
cyclohexanol

trans-2-Isopropyl-
cis-5-methyl-
cyclohexanol

trans-2-Isopropyl-
trans-5-methyl-
cyclohexanol

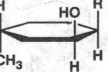

cis-2-Isopropyl-*cis*-5-
methylcyclohexanol

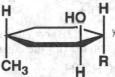

cis-2-Isopropyl-*trans*-5-
methylcyclohexanol

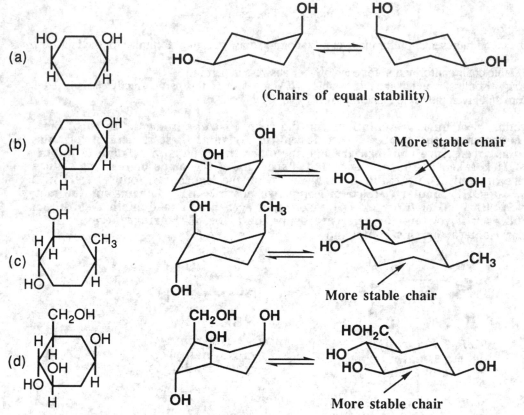

Most stable chair (all equatorial)

trans-2-Isopropyl-*cis*-5-
methylcyclohexanol

trans-2-Isopropyl-*trans*-5-
methylcyclohexanol

Problem 2.42 Draw alternative chair conformations for each substituted cyclohexane and state which chair is more stable.

(a)

(Chairs of equal stability)

(b)

More stable chair

(c)

More stable chair

(d)

More stable chair

Problem 2.43 Glucose (Section 24.2B) contains a six-membered ring. In the more stable chair conformation of this molecule, all substituents on the ring are equatorial. Draw this more stable conformation.

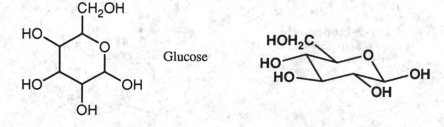

Glucose

Problem 2.44 1,2,3,4,5,6-Hexachlorocyclohexane shows *cis-trans* isomerism. At one time a crude mixture of these isomers was sold as the insecticide benzene hexachloride (BHC) under the trade names Kwell and Gammexane. The insecticidal properties of the mixture arise from one isomer known as the γ-isomer (gamma-isomer) which is *cis*-1,2,4,5-*trans*-3,6-hexachloro-cyclohexane.
(a) Draw a structural formula for 1,2,3,4,5,6-hexachlorocyclohexane disregarding for the moment the existence of *cis-trans* isomerism. What is the molecular formula of this compound?

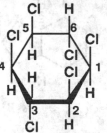

$C_6H_6Cl_6$

(b) Using a planar hexagon representation for the cyclohexane ring, draw a structural formula for the γ-isomer.

(c) Draw a chair conformation for the γ-isomer and show by labels which chlorine atoms are axial and which are equatorial.
(d) Draw the alternative chair conformation of the γ-isomer and again label which chlorine atoms are axial and which are equatorial.
(e) Which of the alternative chair conformations of the γ-isomer is more stable? Explain.

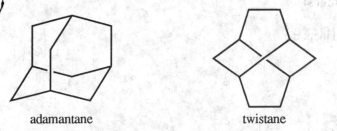

The two chairs are of equal stability; in each three -Cl atoms are axial and three are equatorial

Problem 2.45 What kinds of conformations do the 6 membered rings exhibit in adamantane and twistane? You will find it helpful to build molecular models, particularly of twistane.

adamantane   twistane

In adamantane, the cyclohexane rings all have chair conformations, and in twistane the cyclohexane rings all have twist-boat conformations.

Problem 2.46 Which of the following bicycloalkanes would you expect to show *cis-trans* isomerism? Explain. For each that does, draw suitable stereorepresentations of both *cis* and *trans* isomers.
(a) Bicyclo[2.2.2]octane

No *cis-trans* isomers. It is only possible to fuse a two-carbon bridge to carbons 1 and 4 of a cyclohexane ring if the two-carbon bridge is fused in a *cis* fashion. This molecule is drawn below in two different stereorepresentations.

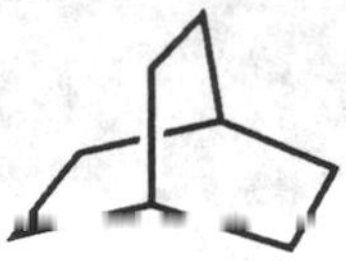

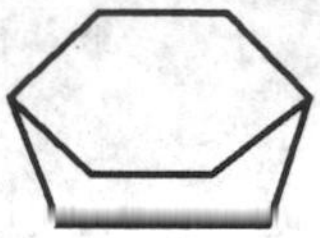

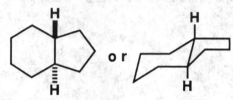

Bicyclo[2.2.2]octane
(no *cis-trans* isomers)

(b) Bicyclo[4.3.0]nonane

trans-**Bicyclo[4.3.0]nonane**   *cis*-**Bicyclo[4.3.0]nonane**

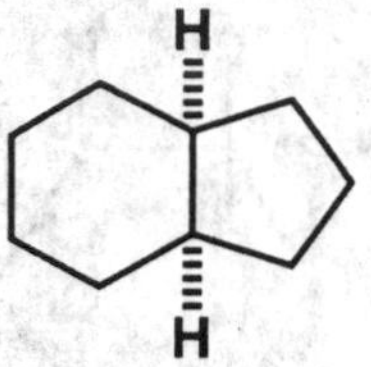

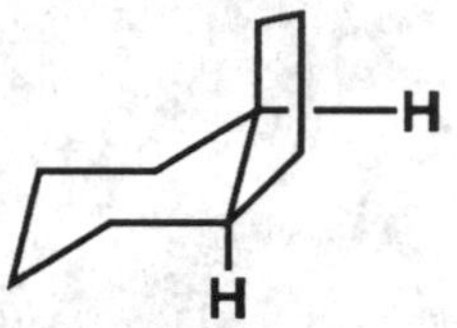

(c) 2-Methylbicyclo[2.2.1]heptane

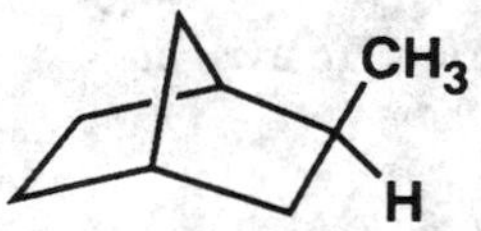

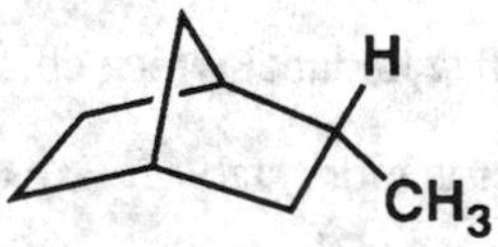

2-Methylbicyclo[2.2.1]heptane
(a *cis* and a *trans* isomer)

(d) 1-chlorobicyclo[2.2.1]heptane

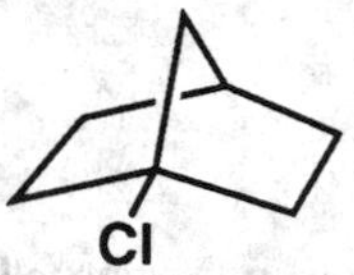

1-Chlorobicyclo[2.2.1]heptane
(no *cis-trans* isomers)

(e) 7-Chlorobicyclo[2.2.1]heptane

7-Chlorobicyclo[2.2.1]heptane
(no *cis-rans* isomers)

Physical Properties
Problem 2.47 In Problem 2.20, you drew structural formulas for all isomeric alkanes of molecular formula C_7H_{16}. Predict which isomer has the lowest boiling point and which has the highest boiling point.

Names and boiling points of these isomers are given in the solution to Problem 2.20 The isomer with the lowest boiling point is 2,2-dimethylpentane, bp 79.2°C. The isomer with the highest boiling point is heptane, bp 94.8°C.

Problem 2.48 What generalization can you make about the densities of alkanes relative to that of water?

All alkanes that are liquid at room temperature are less dense than water. This is why alkanes such as those in gasoline and petroleum float on water.

Problem 2.49 What unbranched alkane has about the same boiling point as water? (Refer to Table 2.6 on the physical properties of alkanes.) Calculate the molecular weight of this alkane and compare it with that of water.

Heptane, C_7H_{16}, has a boiling point of 98.4°C and a molecular weight of 100. Its molecular weight is approximately 5.5 times that of water. Although considerably smaller, the water molecules are held together by hydrogen bonding while the much larger heptane molecules are held together only by relatively weak dispersion forces.

Reactions of Alkanes
Problem 2.50 Complete and balance the following combustion reactions. Assume that each hydrocarbon is converted completely to carbon dioxide and water.

(a) Propane + O_2 $\longrightarrow$ $CH_3CH_2CH_3$ + $5 O_2$ $\longrightarrow$ $3 CO_2$ + $4 H_2O$

(b) Octane + O_2 $\longrightarrow$ $2 CH_3(CH_2)_6CH_3$ + $25 O_2$ $\longrightarrow$ $16 CO_2$ + $18 H_2O$

(c) Cyclohexane + O_2 $\longrightarrow$ ⬡ + $9 O_2$ $\longrightarrow$ $6 CO_2$ + $6 H_2O$

(d) 2-Methylpentane + O_2 $\longrightarrow$

$$2\ CH_3\overset{\overset{\displaystyle CH_3}{|}}{C}HCH_2CH_2CH_3\ +\ 19 O_2\ \longrightarrow\ 12 CO_2\ +\ 14 H_2O$$

Problem 2.51 Following are heats of combustion per mole for methane, propane, and 2,2,4-trimethylpentane. Each is a major source of energy. On a gram-for-gram basis, which of these hydrocarbons is the best source of heat energy.

Hydrocarbon	Component of	$\Delta H°$ [kcal/mol (kJ/mol)]
CH_4	natural gas	-212(886 kJ/mol)
$CH_3CH_2CH_3$	LPG	-531(2220 kJ/mol)
$CH_3CCH_2CHCH_3$ (with CH_3 groups)	gasoline	-1304(5451 kJ/mol)

On a gram-per-gram basis, methane is the best source of heat energy.

Hydrocarbon	Molecular Weight	Heat of Combustion (kcal/mol)	Heat of Combustion (kcal/gram)
Methane	16.04	-212	-13.3
Propane	44.09	-531	-12.0
2,2,4-Trimethylpentane	114.2	-1304	-11.4

Problem 2.52 Following are heats of combustion for ethane, propane, and pentane at 25°C.

Hydrocarbon	$\Delta H°$(kcal/mol)
ethane	-372.8
propane	-530.6
pentane	-845.2

(a) From these data, calculate the average heat of combustion of a methylene ($-CH_2-$) group in a gaseous hydrocarbon.

The average heat of combustion per $-CH_2-$ group is -157.5 kcal/mol. From more extensive data (not given in this problem), the average heat of combustion per methylene group in long chain, unstrained alkanes is -157.4 kcal/mol.

(b) Using the value of the heat of combustion of a methylene group calculated in part (a), estimate the heat of combustion of gaseous cyclopropane.

Using the value from part (a), calculate the heat of combustion of gaseous cyclopropane to be 3 x (-157.4 kcal/mol) = -472.2 kcal/mol.

(c) Compare your estimated value and the experimentally determined value of -499.9 kcal/mol. How might you account for the difference between the two values?

The heat of combustion of cyclopropane is -499.9 - (-472.2) = -27.7 kcal/mol greater. The increased heat of combustion is a measure of strain (angle strain and eclipsed hydrogen interactions) present in the cyclopropane molecule.

Problem 2.53 Using the value of -157.4 kcal/mol as the average heat of combustion of a methylene group:
(a) Calculate the heat of combustion for the following cycloalkanes.
(b) Calculate the total "strain energy" for each cycloalkane.
(c) Calculate the strain energy per methylene group.

Cycloalkane	Calculated Heat of Combustion (kcal/mol)	Observed Heat of Combustion (kcal/mol)	Calculated Total Strain Energy (kcal/mol)	Strain Energy Per -CH_2- Group (kcal/mol)
cyclopropane	-472.2	-499.9	27.7	9.23
cyclobutane	-629.6	-655.9	26.3	6.57
cyclopentane	-787.0	-793.5	6.5	1.30
cyclohexane	-944.4	-944.5	0.1	0.02
cycloheptane	-1,101.8	-1,108.2	6.4	0.91
cyclooctane	-1,259.2	-1,269.0	9.8	1.22
cyclononane	-1,416.6	-1,429.5	12.9	1.43
cyclodecane	-1,574.0	-1,586.0	12.0	1.20
cycloundecane	-1,731.4	-1,742.4	11.0	1.00
cyclododecane	-1,888.0	-1,891.2	2.4	0.20
cyclotetradecane	-2,203.6	-2,203.6	0.0	0.00

(d) Rank these cycloalkanes in order of most stable to least stable, based on strain energy per methylene group.

The rank order of alkane stabilities listed from most to least stable is as follows: cyclotetradecane, cyclohexane, cyclododecane, cycloheptane, cycloundecane, cyclodecane, cyclooctane, cyclopentane, cyclononane, cyclobutane, cyclopropane.

Molecular Modeling
Problem 2.54 Build a structural model of ethane in ChemDraw, import it into Chem3D, and minimize its energy. Then click on one carbon atom and rotate the carbon-carbon single bond to create both the staggered and eclipsed conformations. Visually measure the distance between adjacent hydrogens in the staggered conformation and in the eclipsed conformation and estimate the ratio of eclipsed/staggered distance.

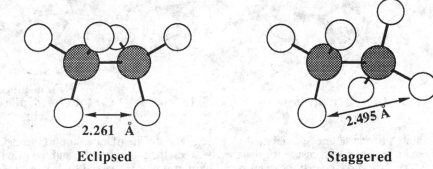

2.261 Å	2.495 Å
Eclipsed	**Staggered**

As shown in the above structures, the ratio of eclipsed/staggered distance is 2.261 Å/2.495 Å = 0.906.

Problem 2.55 Draw a line-angle structure for cyclopentane in ChemDraw, import it into Chem3D, and minimize its energy. Then measure all C-C-C bond angles and compare them with the value of 105° given in the text for the "envelope" conformation of cyclopentane. How do you account for the difference in these values? (*Hint:* Consider the torsional strain of eclipsed hydrogen interactions. You recognize them, but does ChemDraw?)

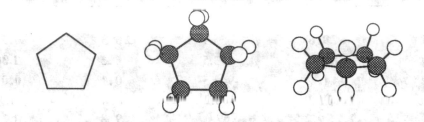

The version of the structure optimization routine used by Chem3D is not sophisticated enough to recognize the torsional strain of the eclipsed hydrogen interactions. Thus, the structure you see is a flat pentagon, as opposed to the correct molecular structure, the so-called "envelope" conformation.

Problem 2.56 Build a line-angle structure of a chair cyclohexane in ChemDraw, import it into Chem3D, and minimize its energy. Now rotate the model so you view the chair from above, from the side, from the foot piece to the head piece, and so on. As you do these rotations, convince yourself that the six axial C-H bonds are parallel and that they alternate up, down, and so on. Also convince yourself that opposite (in positions 1 and 4) equatorial C-H bonds are parallel, and that one of them is above the other below the plane of the ring.

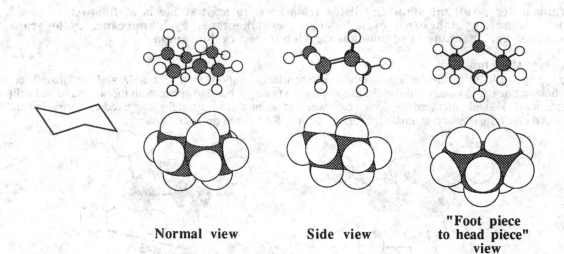

Normal view   Side view   "Foot piece to head piece" view

Problem 2.57 Build a line-angle structure of axial methylcyclohexane in ChemDraw, import it into Chem3D, minimize the energy, and use a ruler to measure the distance between the methyl group and the ring hydrogens on carbons 2,3,4,5, and 6. You should find that the axial methyl group is closer to axial hydrogens on carbons 3 and 5 than to any other ring hydrogens. Note that when you do this minimization, you find a "local minimum" because the energy required for ring flipping to the equatorial methyl conformation is too high.

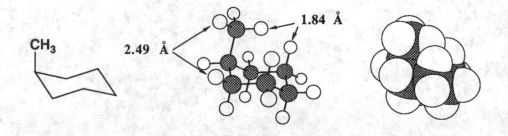

According to the Chem3D model, the closest contacts between a hydrogen atom on the axial methyl group and an axial hydrogen is 1.84 Å. This is much closer than the 2.49 Å distance between the other methyl group hydrogen atoms and the nearest equatorial ring hydrogen atom.

<u>Problem 2.58</u> Build a line-angle structure of spiro[5.5]undecane in ChemDraw, import it into Chem3D, minimize the energy. Show that both six-member rings can assume chair conformations. It may take a bit of rotation of the stereoview, but if you get it right, you will see both rings as strain-free chair conformations.

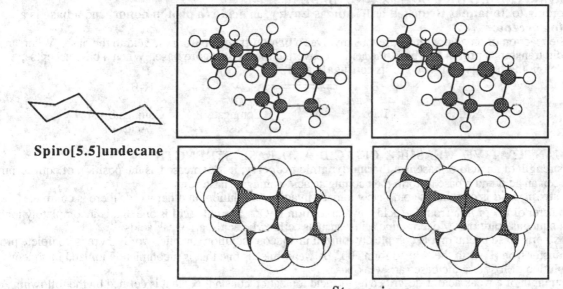

Spiro[5.5]undecane

Stereoviews

<u>Problem 2.59</u> A large number of ChemDraw figures are given in this chapter, all of which are good exercises for you to practice using ChemDraw and Chem3D and for the creation of stereoviews. You might try reproducing and studying them as three-dimensional objects.

CHAPTER 3: ACIDS AND BASES

3.0 OVERVIEW
- A large number of organic reactions involve acid-base reactions. These include acid-base reactions of important functional groups such as alcohols and carbonyl compounds. In addition, many reactions involve catalysis by protons or Lewis acids, such as $AlCl_3$.

3.1 BRØNSTED-LOWRY ACIDS AND BASES
- According to **Johannes Brønsted** and **Thomas Lowry**, an **acid** is a **proton donor**, and a **base** is a proton acceptor. ✶
  - The reaction of an acid with a base thus involves a **proton transfer** from the acid to the base. When an acid transfers a proton to a base, the acid is converted to its **conjugate base**. When a base accepts a proton, the base is converted to its **conjugate acid**.

$$H\text{-}A \quad + \quad B^- \rightleftharpoons A^- \quad + \quad H\text{-}B$$

| Acid | Base | Conjugate Base | Conjugate Acid |

3.2 QUANTITATIVE MEASURE OF ACID AND BASE STRENGTH
- The strength of an acid or base is a **thermodynamic property**; it is expressed as the position of equilibrium between an acid and a base to form their conjugate acid-conjugate base pairs. ✶
  - A **strong acid** is one that is completely ionized in aqueous solution, in other words there is complete transfer of the proton from the acid to water to form H_3O^+. A **weak acid** is one that is incompletely ionized in aqueous solution. Most organic acids, such as carboxylic acids, are weak acids.
  - A **strong base** is one that is completely ionized in aqueous solution, in other words there is complete proton transfer from H_2O to the base to form HO^-. A **weak base** is one that is incompletely ionized in aqueous solution. Most organic bases are weak bases.
- The strength of a weak acid is described by the acid ionization constant K_a that is defined by the following equation. ✶

$$K_a = K_{eq}[H_2O] = \frac{[H_3O^+][A^-]}{[HA]}$$

Here $[A^-]$ is the concentration of conjugate base, and $[HA]$ is the concentration of the acid HA, all at equilibrium.
  - The ionization constants for weak acids have negative exponents, so it is convenient to refer to the **pK_a** where $pK_a = -\log_{10}K_a$. In other words, a weak acid with a pK_a of 5.0 has a K_a of 1.0×10^{-5}. ✶
  - Because pK_a is defined as the *negative* log of the K_a, the larger the pK_a, the weaker the acid and the smaller the pK_a, the stronger the acid. Note, the strongest acids actually have negative pK_a values. ✶

3.3 MOLECULAR STRUCTURE AND ACIDITY
- The **relative acidity** of organic acids can be understood qualitatively in terms of some key molecular parameters. In an acid-base equilibrium reaction, it is usually helpful to focus attention on the charged species such as the conjugate base, A^-, produced by deprotonation of the acid, HA. The more stable the charged species (conjugate base), i.e. the more delocalized the negative charge, the more acidic the parent acid. ✶
  - **Acid strength within a row** of the Periodic Table **increases from left to right** due to increasing electronegativity. ✶
    More electronegative elements are more able to accommodate a negative charge, so the negatively charged conjugate bases of more electronegative elements are more stable. A more stable conjugate base within a period of the Periodic Table means the parent acid will be more acidic.
  - **Acid strength within a column** of the Periodic Table **increases from top to bottom** due to **decreasing bond strength**. ✶
    Elements with weaker bond strengths to hydrogen atoms (elements farther down in a column) are more acidic.
  - **For atoms of the same element, acid strength increases** with **increasing s character** of the hybridization. Thus, hydrogens bonded to sp atoms are more acidic than those bonded to sp^2 atoms, which are more acidic than those bonded to sp^3 atoms. This is a huge effect. ✶

The 2s orbital electrons are of lower energy than the 2p orbital electrons, so the more s character there is to an orbital, the lower the energy of electrons in that orbital. Electrons in an sp orbital (50% s) are of lower energy than electrons in an sp^2 orbital (only 33% s), etc. For this reason, the unshared pair of electrons in an alkyne anion (sp hybridized) are of lower energy (more stable) than the unshared pair of electrons in an alkene anion (sp^2 hybridized), which are lower in energy than the unshared pair of electrons in an alkane anion (sp^3 hybridized). Again, greater stability (lower energy) of the anionic conjugate base leads to greater acidity of the parent acid.

- **Acid strength increases** with **increased resonance delocalization** of the negative charge in the conjugate base. ✳

The stability of an anion is increased by increased delocalization of the negative charge. Resonance delocalization, such as that seen in the carboxylate ion, can increase charge delocalization, and thus acid strength, compared to functional groups such as alcohols for which there is no resonance delocalization.

- **Acid strength** is influenced by **inductive effects**. ✳

Nearby atoms of higher electronegativity can withdraw some of the electron density of a bond to hydrogen, making that hydrogen more acidic. This effect is transmitted through sigma bonds and is referred to as the **inductive effect**. The inductive effect also influences the stability of conjugate bases. Electronegative atoms adjacent to an anionic atom (made anionic by loss of the acidic proton) are stabilizing since the negative charge is partially removed by, and delocalized onto, the electronegative atoms. Greater stability of a conjugate base means that the parent acid will be more acidic.

3.4 THE POSITION OF EQUILIBRIUM IN ACID-BASE REACTIONS

• The conjugate base of a strong acid is a weak base, and the conjugate base of a weak acid is a strong base. ✳ *[This relationship will make the following rule easier to use]*

• For any proton transfer reaction, the **position of equilibrium** favors the side of the reaction equation that has the weaker acid and weaker base. ✳ *[With this extremely helpful rule, you can predict the outcome of virtually any proton transfer reaction as long as you know the relevant pK_a's.]*

$$\text{H-A} \quad + \quad \text{B}^- \quad \rightleftharpoons \quad \text{A}^- \quad + \quad \text{H-B}$$

| Stronger Acid | Stronger Base | Weaker Base | Weaker Acid |

Equilibrium Favors
This Side

- The **equilibrium constant (K_{eq})** is the ratio of the K_a for the acid species (H-A) on the left side of the equation as written divided by the K_a for the conjugate acid (H-B) species on the right side of the equation:

$$K_{eq} = \frac{K_a \text{ of acid on left side of equation}}{K_a \text{ of conjugate acid on right side of equation}}$$

- This equation is just a quantitative way to state that the position of equilibrium favors the weaker acid and weaker base. A K_{eq} greater than one favors the species on the right as written, and a K_{eq} less than one favors the species on the left as written.

3.5 LEWIS ACIDS AND BASES

• According to the definition first proposed by G. N. Lewis, a **Lewis Acid** is a species that forms a new covalent bond by accepting a pair of electrons. A **Lewis base** is a species that forms a new covalent bond by donating a pair of electrons. ✳ *[This concept is important, because it describes much more than just proton transfer reactions.]*

- The reaction of a Lewis acid with a Lewis base can be described by the following equation:

$$\text{A} + :\text{B} \longrightarrow \text{A}^- - \text{B}^+$$

- Here the "A" represents a Lewis acid and ":B" represents a Lewis base.

• The Lewis definitions of acids and bases are more general than the Brønsted-Lowry definitions: all Brønsted-Lowry acids (proton donors) are also Lewis acids and all Brønsted-Lowry bases (proton acceptors) are also Lewis bases. The Lewis definitions cover reactions other than proton transfer reactions, such as diethyl ether ($CH_3CH_2OCH_2CH_3$) reacting with boron trifluoride (BF_3).

CHAPTER 3
Solutions to the Problems

<u>Problem 3.1</u> Write these reactions as proton-transfer reactions. Label which reactant is the acid and which the base; which product is the conjugate base of the original acid and which the conjugate acid of the original base. Use curved arrows to show the flow of electrons in each reaction.

(a) CH_3-S-H + OH^- ⟶ CH_3-S^- + H_2O

CH_3-S-H + $:O-H$ ⟶ $CH_3-S:^-$ + $H-O-H$

  acid   base   conjugate   conjugate
                     base   acid

(b) CH_3-O-H + NH_2^- ⟶ CH_3-O^- + NH_3

CH_3-O-H + $:N-H$ (with H below) ⟶ $CH_3-O:^-$ + $H-N-H$ (with H below)

  acid   base   conjugate   conjugate
                     base   acid

<u>Problem 3.2</u> For each value of K_a, calculate the corresponding value of pK_a. Which compound is the stronger acid?
(a) Acetic acid, $K_a = 1.74 \times 10^{-5}$ (b) Water, $K_a = 2.00 \times 10^{-16}$

The pK_a is equal to $-\log_{10}K_a$. The pK_a of acetic acid is 4.76 and the pK_a of water is 15.7. Acetic acid, with the smaller pK_a value, is the stronger acid.

<u>Problem 3.3</u> Predict the position of equilibrium and calculate the equilibrium constant, K_{eq}, for these acid-base reactions. The pK_a of methylammonium is 10.64.

(a) CH_3NH_2 + CH_3CO_2H ⇌ $CH_3NH_3^+$ + $CH_3CO_2^-$
  methylamine   acetic acid   methylammonium   acetate
                            ion   ion

Acetic acid is the stronger acid; equilibrium lies to the right

$$CH_3NH_2 + CH_3CO_2H \rightleftharpoons CH_3NH_3^+ + CH_3CO_2^-$$

              pK_a 4.76 pK_a 10.64

    **(stronger   (stronger   (weaker   (weaker
     base)   acid)   acid)   base)**

$$K_{eq} = \frac{K_a \text{ of acid on left side of equation}}{K_a \text{ of conjugate acid on right side of equation}} = \frac{10^{-4.76}}{10^{-10.64}} = \frac{1.74 \times 10^{-5}}{2.29 \times 10^{-11}}$$

$$\boxed{K_{eq} = 7.60 \times 10^5}$$

(b) $CH_3CH_2O^-$ + NH_3 $\rightleftharpoons$ CH_3CH_2OH + NH_2^-

　　　ethoxide　　ammonia　　　　　　ethanol　　　amide
　　　　ion　　　　　　　　　　　　　　　　　　　ion

Ethanol is the stronger acid; equilibrium lies to the left.

$$CH_3CH_2O^- + NH_3 \rightleftharpoons CH_3CH_2OH + NH_2^-$$

　　　　　　　　　pK_a 33　　　　　pK_a 15.9

　　　　　　(weaker　　（weaker　　　（stronger　　（stronger
　　　　　　base)　　　acid)　　　　　acid)　　　　base)

$$K_{eq} = \frac{K_a \text{ of acid on left side of equation}}{K_a \text{ of conjugate acid on right side of equation}} = \frac{10^{-33}}{10^{-15.9}} = \frac{1.0 \times 10^{-33}}{1.3 \times 10^{-16}}$$

$$\boxed{K_{eq} = 7.9 \times 10^{-18}}$$

<u>Problem 3.4</u> Write the reaction between each Lewis acid/base pair, showing electron flow by means of curved arrows.

(a) $B(CH_3CH_2)_3$ + OH^- $\longrightarrow$

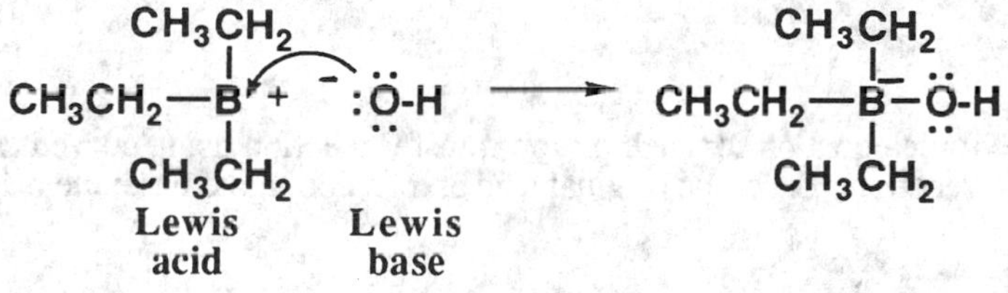

(b) CH_3Cl + $AlCl_3$ $\longrightarrow$

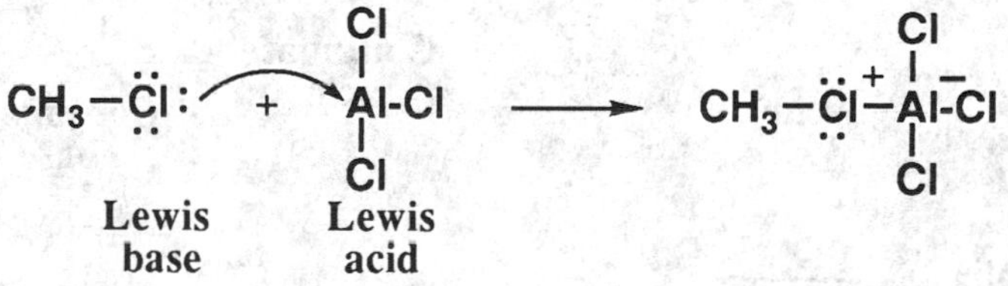

<u>Problem 3.5</u> Complete a net ionic equation for each proton-transfer reaction using curved arrows to show the flow of electron pairs in each reaction. In addition, write Lewis structures for all starting materials and products. Label the original acid and its conjugate base; the original base and its conjugate acid. If you are uncertain about which substance in each equation is the proton donor, refer to Table 3.1 for the relative strengths of proton acids.

(a) NH_3 + HCl $\longrightarrow$

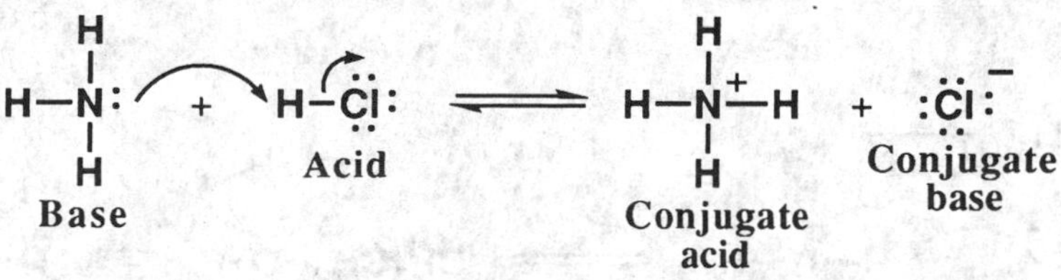

(b) $CH_3CH_2O^- + HCl \longrightarrow$

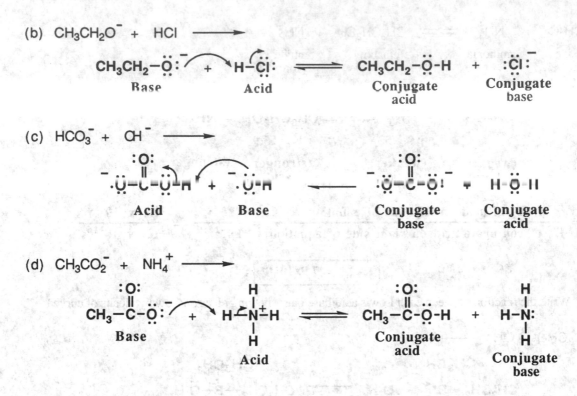

(c) $HCO_3^- + OH^- \longrightarrow$

(d) $CH_3CO_2^- + NH_4^+ \longrightarrow$

Problem 3.6 Complete a net ionic equation for each proton-transfer reaction using curved arrows to show the flow of electron pairs in each reaction. Label the original acid and its conjugate base; the original base and its conjugate acid.

(a) $NH_4^+ + OH^- \longrightarrow$

(b) $CH_3CO_2^- + CH_3NH_3^+ \longrightarrow$

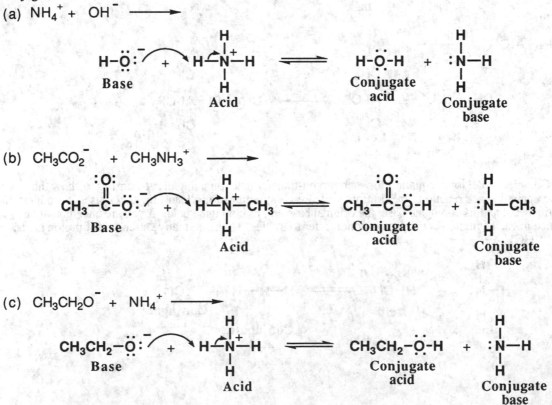

(c) $CH_3CH_2O^- + NH_4^+ \longrightarrow$

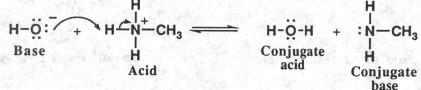

(d) $CH_3NH_3^+$ + OH^- ⟶

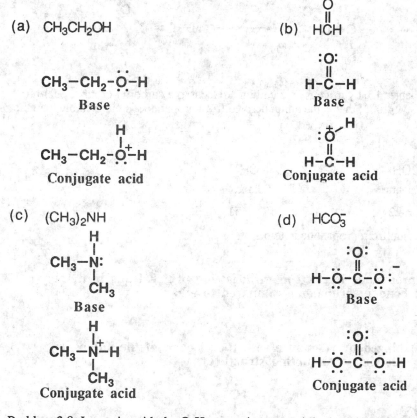

Problem 3.7 Each molecule and ion can function as a base. Write the structural formula of the conjugate acid formed by reaction of each with H^+.

(a) CH_3CH_2OH

$CH_3{-}CH_2{-}\overset{..}{\underset{..}{O}}{-}H$
Base

$CH_3{-}CH_2{-}\overset{\overset{H}{|}}{\underset{..}{O}}{}^{+}{-}H$
Conjugate acid

(b)
$$\overset{\overset{O}{\|}}{H\overset{..}{C}H}$$

$H{-}\overset{\overset{:O:}{\|}}{C}{-}H$
Base

$H{-}\overset{\overset{:\overset{+}{O}\diagup^{H}}{\|}}{C}{-}H$
Conjugate acid

(c) $(CH_3)_2NH$

$CH_3{-}\overset{\overset{H}{|}}{\underset{\underset{CH_3}{|}}{N}}{:}$
Base

$CH_3{-}\overset{\overset{H}{|}}{\underset{\underset{CH_3}{|}}{N}}{}^{+}{-}H$
Conjugate acid

(d) HCO_3^-

$H{-}\overset{..}{\underset{..}{O}}{-}\overset{\overset{:O:}{\|}}{C}{-}\overset{..}{\underset{..}{O}}{:}^{-}$
Base

$H{-}\overset{..}{\underset{..}{O}}{-}\overset{\overset{:O:}{\|}}{C}{-}\overset{..}{\underset{..}{O}}{-}H$
Conjugate acid

Problem 3.8 In acetic acid, the O-H proton is more acidic than the H_3C protons. Show how the concept of electronegativity can be used to account for this difference in acidity.

Within a row, acidity increases with increasing electronegativity of the atom attached to hydrogen. Oxygen is more electronegative than carbon, so the hydrogen on the oxygen atom is more acidic. Note that the same argument applies when the appropriate contributing structures are considered. When the proton is removed from the oxygen atom of a carboxylic acid and a negatively charged oxygen atom results, the negative charge can be shared with the adjacent oxygen atom of the carboxylic acid via the contributing resonance structures shown below:

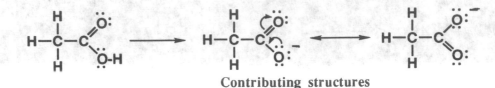

Contributing structures

In BOTH contributing structures, the negative charge is on a relatively electronegative oxygen atom.

When the same type of analysis is carried out on the anion produced by deprotonation of the -CH₃ group, it can be seen that this anion is less stable because one of the important contributing structures places the negative charge on the less electronegative carbon atom.

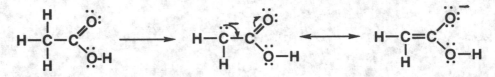

<u>Problem 3.9</u> As we shall see later in Chapter 15, hydrogens on a carbon adjacent to a carbonyl group are far more acidic than those not adjacent to a carbonyl group; that is the anion derived from propanone is more stable than the anion derived from ethane.

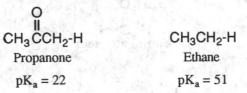

Account for the greater stability of the anion from propanone in terms of:
(a) The inductive effect

The adjacent carbonyl group of propanone is electron withdrawing, which leads to more inductive polarization of the C-H bond in propanone compared to ethane.

(b) The resonance effect.

The anion of propanone can be stabilized by resonance with the pi bond of the adjacent carbonyl group as shown in the following contributing structures.

In this way, the negative charge is delocalized significantly, leading to a more stable anion and thus propanone is more acidic. Note how the negative charge is placed on the more electronegative oxygen atom in the structure on the right. The anion of deprotonated ethane does not have any opportunities for resonance stabilization.

<u>Problem 3.10</u> Offer an explanation for the following observations.
(a) H_3O^+ is a stronger acid than NH_4^+.

Oxygen is more electronegative than nitrogen, so the proton is more acidic on H_3O^+.

(b) Nitric acid, HNO_3, is a stronger acid than nitrous acid, HNO_2.

There is one more electronegative oxygen atom on nitric acid to help stabilize the deprotonated anion through a resonance effect. In addition, the third oxygen atom of HNO_3 can provide additional inductive polarization to help weaken the O-H bond.

(c) Ethanol and water have approximately the same acidity.

Both have the proton attached to sp^3 hybridized oxygen atoms, and neither deprotonated species can be resonance stabilized.

(d) Trifluoroacetic acid is a stronger acid than trichloroacetic acid.

$$X-\overset{\overset{\displaystyle X}{|}}{\underset{\underset{\displaystyle X}{|}}{C}}-\overset{\overset{\displaystyle :O:}{\|}}{C}-\ddot{\overset{..}{O}}:^{-}$$

Fluorine is more electronegative than chlorine, so the above anion is more stable when X = F compared to when X = Cl due to increased inductive effects. In addition, the fluorine atoms can provide additional inductive polarization that weakens the O-H bond.

Quantitative Measure of Acid and Base Strength
Problem 3.11 Which has the larger numerical value:
(a) The pK_a of a strong acid or the pK_a of a weak acid?

The weaker acid will have the pK_a with a larger numerical value.

(b) The K_a of a strong acid or the K_a of a weak acid?

The stronger acid will have the K_a with a larger numerical value.

Problem 3.12 In each pair, select the stronger acid:
(a) Pyruvic acid (pK_a 2.49) or lactic acid (pK_a 3.85)

The stronger acid is the one with the smaller pK_a and, therefore, the larger value of K_a. Pyruvic acid is the stronger acid.

(b) Citric acid (pK_a 3.08) or phosphoric acid (pK_{a1} 2.10)

Phosphoric acid is the stronger acid.

(c) Nicotinic acid (niacin, K_a 1.4×10^{-5}) or acetylsalicylic acid (aspirin, K_a 3.3×10^{-4})

Acetylsalicylic acid is the stronger acid.

(d) Phenol (K_a 1.12×10^{-10}) or acetic acid (K_a 1.74×10^{-5})

Acetic acid is the stronger acid.

Problem 3.13 Arrange the compounds in each set in order of increasing acid strength. Consult Table 3.1 for pK_a values of each acid.

(a) CH_3CH_2OH $HOCO^-$ (with =O) C_6H_5COH (with =O)
        Ethanol   Bicarbonate ion   Benzoic acid
pK_a: 15.9   10.33   4.19

The compounds are already in order of increasing acid strength. Ethanol is the weakest acid, benzoic acid is the strongest acid, and bicarbonate ion is in between.

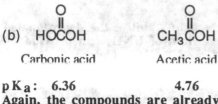

(b) HOCOH   CH$_3$COH   HCl

    Carbonic acid   Acetic acid   Hydrogen chloride

pK$_a$: 6.36   4.76   -7

Again, the compounds are already in order of increasing acid strength. Carbonic acid is the weakest acid, hydrogen chloride is the strongest acid, and acetic acid is in between.

Problem 3.14 Arrange the compounds in each set in order of increasing base strength. Consult Table 3.1 for pK$_a$ values for the conjugate acid of each base.

The weaker the conjugate acid (higher pK$_a$), the stronger the base.

(a)   NH$_3$ HOCO$^-$ CH$_3$CH$_2$O$^-$

      9.24   6.34   15.9   pK$_a$ of conjugate acid

Base strength increases in the order:

HOCO$^-$ <   NH$_3$ <   CH$_3$CH$_2$O$^-$

(b)   OH$^-$ HOCO$^-$ CH$_3$CO$^-$

     15.7   6.34   4.76   pK$_a$ of conjugate acid

Base strength increases in the order:

CH$_3$CO$^-$ <   HOCO$^-$ <   OH$^-$

(c)   H$_2$O   NH$_3$ CH$_3$CO$^-$

    -1.74   9.24   4.76   pK$_a$ of conjugate acid

Base strength increases in the order:

H$_2$O   <   CH$_3$CO$^-$ <   NH$_3$

(d) NH$_2$$^-$ CH$_3$CO$^-$ OH$^-$

    33   4.76   15.7   pK$_a$ of conjugate acid

Base strength increases in the order:

CH$_3$CO$^-$ <   OH$^-$ <   NH$_2$$^-$

Quantitative Position of Equilibrium in Acid-Base Reactions

Problem 3.15 Unless under pressure, carbonic acid in aqueous solution breaks down into carbon dioxide and water, and carbon dioxide is evolved as bubbles of gas. Write an equation for the conversion of carbonic acid to carbon dioxide and water.

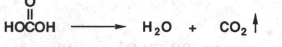

$$HOCOH \longrightarrow H_2O + CO_2 \uparrow$$

This relationship explains why carbonated drinks evolve CO_2 gas when they are opened and the pressure is released.

Problem 3.16 Will carbon dioxide be evolved when sodium bicarbonate is added to an aqueous solution of these compounds?
(a) Sulfuric acid   (b) Ethanol   (c) Ammonium chloride

In order for carbon dioxide to be evolved, the sodium bicarbonate must be protonated to give carbonic acid (Problem 3.15). The pK_a of carbonic acid is 6.36. The pK_a's for sulfuric acid, ethanol and ammonium chloride are -5.2, 15.9, and 9.24, respectively. Thus, sulfuric acid is the only acid strong enough to protonate sodium bicarbonate and evolve carbon dioxide.

Problem 3.17 Acetic acid, CH_3CO_2H, is a weak organic acid, pK_a 4.76. Write equations for the equilibrium reactions of acetic acid with each base. Which equilibria involving acetic acid lie considerably toward the left? Which lie considerably toward the right?
(a) $NaHCO_3$

$$CH_3CO_2H + HCO_3^- \, Na^+ \rightleftharpoons CH_3CO_2^- \, Na^+ + H_2CO_3$$
$$pK_a \; 4.76 \qquad\qquad\qquad\qquad\qquad pK_a \; 6.36$$

(b) NH_3

$$CH_3CO_2H + NH_3 \rightleftharpoons CH_3CO_2^- + NH_4^+$$
$$pK_a \; 4.76 \qquad\qquad\qquad\qquad pK_a \; 9.24$$

(c) H_2O

$$CH_3CO_2H + H_2O \rightleftharpoons CH_3CO_2^- + H_3O^+$$
$$pK_a \; 4.76 \qquad\qquad\qquad\qquad pK_a \; -1.74$$

(d) $NaOH$

$$CH_3CO_2H + HO^- \, Na^+ \rightleftharpoons CH_3CO_2^- \, Na^+ + H_2O$$
$$pK_a \; 4.76 \qquad\qquad\qquad\qquad pK_a \; 15.7$$

Equilibrium favors the direction that gives the weaker acid and weaker base. Therefore, equilibrium will favor formation of an acid with a pK_a value higher than 4.76, or formation of acetic acid if that is the weaker acid. Based on the pK_a values shown, reactions (a), (b), and (d) have equilibria that lie considerably to the right, while reaction (c) has an equilibrium that lies considerably to the left.

Problem 3.18 Alcohols are very weak organic acids, pK_a 16-18. The pK_a of ethanol, CH_3CH_2OH, is 15.9. Which equilibria involving ethanol lie considerably toward the left? Which lie considerably toward the right?
(a) HCO_3^-

$$CH_3CH_2OH + HCO_3^- \rightleftharpoons CH_3CH_2O^- + H_2CO_3$$
$$pK_a \; 6.36$$

(b) OH$^-$

$$CH_3CH_2OH \;+\; OH^- \;\rightleftharpoons\; CH_3CH_2O^- \;+\; H_2O$$

$$pK_a \; 15.7$$

(c) NH$_2^-$

$$CH_3CH_2OH \;+\; NH_2^- \;\rightleftharpoons\; CH_3CH_2O^- \;+\; NH_3$$

$$pK_a \; 33$$

(d) NH$_3$

$$CH_3CH_2OH \;+\; NH_3 \;\rightleftharpoons\; CH_3CH_2O^- \;+\; NH_4^+$$

$$pK_a \; 9.24$$

Equilibrium favors the direction that gives the weaker acid and weaker base. Therefore, equilibrium will favor formation of an acid with a pK_a value higher than 15.9, or formation of ethanol if that is the weaker acid. Based on the pK_a values shown, only reaction (c) has an equilibria that lies considerably to the right, while reactions (a) and (d) have equilibria that lie considerably to the left. Reaction (b) has an equilibrium that lies only slightly to the left because the pK_a for ethanol is only slightly higher than that for water.

Problem 3.19 Benzoic acid, $C_6H_5CO_2H$, is insoluble in water, but its sodium salt, $C_6H_5CO_2^-Na^+$, is quite soluble in water. Will benzoic acid dissolve in
(a) Aqueous sodium hydroxide?   (b) Aqueous sodium bicarbonate?
(c) Aqueous sodium carbonate?

The pK_a of benzoic acid is 4.19. The pK_a values for the conjugate acids of sodium hydroxide, sodium bicarbonate ($NaHCO_3$), and sodium carbonate (Na_2CO_3) are 15.7, 6.36, and 10.33, respectively. Thus, equilibrium will favor reaction of benzoic acid with all three of these bases to give the soluble $C_6H_5CO_2^-Na^+$. Therefore, benzoic acid will dissolve in aqueous solutions of all three bases.

Problem 3.20 Phenol, C_6H_5OH, is only slightly soluble in water, but its sodium salt, $C_6H_5O^-Na^+$, is quite soluble in water. Will phenol dissolve in:
(a) Aqueous NaOH?   (b) Aqueous $NaHCO_3$?
(c) Aqueous Na_2CO_3?

The pK_a of phenol is 9.95. The pK_a values for the conjugate acids of sodium hydroxide, sodium bicarbonate($NaHCO_3$), and sodium carbonate (Na_2CO_3) are 15.7, 6.36, and 10.33, respectively. Thus, equilibrium will favor reaction of phenol with only sodium hydroxide and sodium carbonate to give the soluble $C_6H_5O^-Na^+$. Phenol will dissolve in aqueous solutions of these two bases. Sodium bicarbonate is not a strong enough base to deprotonate phenol, so phenol will not dissolve in an aqueous solution of sodium bicarbonate.

Problem 3.21 For an acid-base reaction, one way to determine the predominant species at equilibrium is to say that the reaction arrow points to the acid with the higher value of pK_a. For example:

$$NH_4^+ \;+\; H_2O \;\longleftarrow\; NH_3 \;+\; H_3O^+$$

$$pK_a \; 9.24 \qquad\qquad\qquad\qquad pK_a \; \text{-}1.74$$

$$NH_4^+ \;+\; OH^- \;\longrightarrow\; NH_3 \;+\; H_2O$$

$$pK_a \; 9.24 \qquad\qquad\qquad\qquad pK_a \; 15.7$$

Explain why this rule works.

In acid-base reactions, the position of equilibrium favors reaction of the stronger acid and stronger base to give the weaker acid and weaker base. The acid with the higher pK_a is the weaker acid, so the arrow will point toward it.

<u>Problem 3.22</u> Will ethyne react with sodium amide according to the following equation to form a salt and ammonia. Calculate K_{eq} for this equilibrium.

$$HC \equiv CH \ + \ NH_2^- \ Na^+ \ \rightleftharpoons \ HC \equiv C^- \ Na^+ \ + \ NH_3$$

| Ethyne | Sodium amide | Sodium ethynide | Ammonia |

pK_a 25 pK_a 33

Since the pK_a for ammonia is larger than the pK_a for acetylene, equilibrium will be to the right, favoring formation of the salt and ammonia.

$$K_{eq} = \frac{K_a(Ethyne)}{K_a(Ammonia)} = \frac{1 \times 10^{-25}}{1 \times 10^{-33}} = \boxed{1 \times 10^8}$$

<u>Problem 3.23</u> Will ethene react with sodium amide according to the following equation to form a salt and ammonia. Calculate K_{eq} for this equilibrium.

$$H_2C = CH_2 \ + \ NH_2^- \ Na^+ \ \rightleftharpoons \ H_2C = CH^- \ Na^+ \ + \ NH_3$$

| Ethylene | Sodium amide | Sodium ethenide | Ammonia |

pK_a 44 pK_a 33

Since the pK_a for ammonia is smaller than the pK_a for ethylene, equilibrium will be to the left, favoring formation of the ethene and sodium amide.

$$K_{eq} = \frac{K_a(Ethylene)}{K_a(Ammonia)} = \frac{1 \times 10^{-44}}{1 \times 10^{-33}} = \boxed{1 \times 10^{-11}}$$

<u>Problem 3.24</u> Using pK_a values given in Table 3.1, predict the position of equilibrium in this acid-base reaction and calculate its K_{eq}.

$$H_3PO_4 \ + \ CH_3CH_2OH \ \rightleftharpoons \ H_2PO_4^- \ + \ CH_3CH_2OH_2^+$$

pK_a 2.1 pK_a -3.6

Since the pK_a for the ethyloxonium ion ($CH_3CH_2OH_2^+$) is smaller than the pK_a for phosphoric acid (H_3PO_4), equilibrium will be to the left, favoring formation of the phosphoric acid and ethanol.

$$K_{eq} = \frac{K_a(H_3PO_4)}{K_a(CH_3CH_2OH)} = \frac{7.9 \times 10^{-3}}{4.0 \times 10^3} = \boxed{2.0 \times 10^{-6}}$$

Problem 3.25 2,4-Pentanedione is a considerably stronger acid than is propanone (acetone). Write a structural formula for the conjugate base of each acid and account for the greater stability of the conjugate base from 2,4-pentanedione.

Propanone
$pK_a = 22$

2,4-Pentanedione
$pK_a = 9$

A carbonyl group adjacent to a negatively charged carbon atom, an arrangement that is referred to as an enolate ion, can lead to resonance stabilization by virtue of a resonance contributor that places the negative charge on oxygen along with formation of a carbon-carbon double bond. Thus, there are two important contributing structures that describe the resonance stabilization of the conjugate base of propanone.

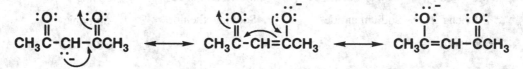

However, there are two adjacent carbonyl groups and thus three important contributing structures that describe the resonance stabilization of the conjugate base of 2,4-pentanedione.

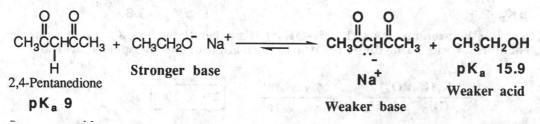

The added contributing structure provides additional resonance stabilization for the conjugate base of 2,4-pentanedione compared to propanone, making the corresponding 2,4-pentantedione acid the stronger acid.

In addition, carbonyl groups are electron-withdrawing, so the additional carbonyl group of 2,4-pentanedione weakens its C-H bond more than what is seen in propanone.

Problem 3.26 Write an equation for the acid-base reaction between 2,4-pentanedione and sodium ethoxide and calculate its equilibrium constant, K_{eq}. Label the stronger acid, stronger base and so on. The pK_a of 2,4-pentanedione is 9; that of ethanol is 15.9.

$CH_3CCHCCH_3$ + $CH_3CH_2O^-$ Na^+ $\longrightarrow$ $CH_3CCHCCH_3$ + CH_3CH_2OH

2,4-Pentanedione	Stronger base	pK_a 15.9
pK_a 9	Na^+	Weaker acid
	Weaker base	

Since the pK_a for 2,4-pentanedione is smaller than the pK_a for ethanol, equilibrium will be to the right, favoring formation of the 2,4-pentanedione salt and ethanol.

$$K_{eq} = \frac{K_a(2,4\text{-Pentandione})}{K_a(\text{Ethanol})} = \frac{1.0 \times 10^{-9}}{1.3 \times 10^{-16}} = \boxed{7.9 \times 10^6}$$

Lewis Acids and Bases

<u>Problem 3.27</u> For each equation, label the Lewis acid and Lewis base. In addition, use curved arrows to show the flow of electrons in each reaction.

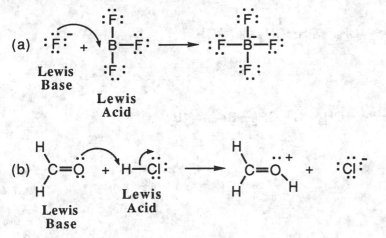

Problem 3.28 Complete the following reactions between Lewis acid-Lewis base pairs. Label which starting material is the Lewis acid, which the Lewis base, and use curved arrows to show the flow of electrons in each reaction. Note that in doing these problems, it is essential that you show all valence electrons for at least the atoms participating directly in each reaction.

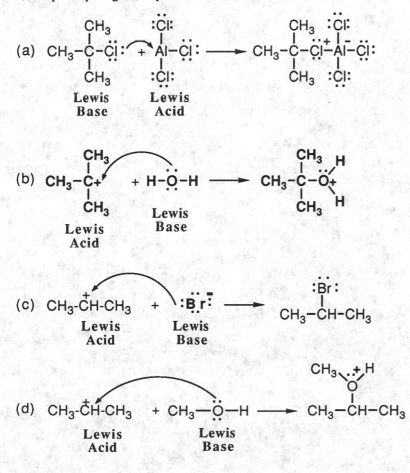

Problem 3.29 Each reaction can be written as a Lewis acid-Lewis base reaction. Label the Lewis acid, the Lewis base and use curved arrows to show the flow of electrons in each reaction. In doing this problem, it is essential that you show all valence electrons for all atoms participating in each reaction.

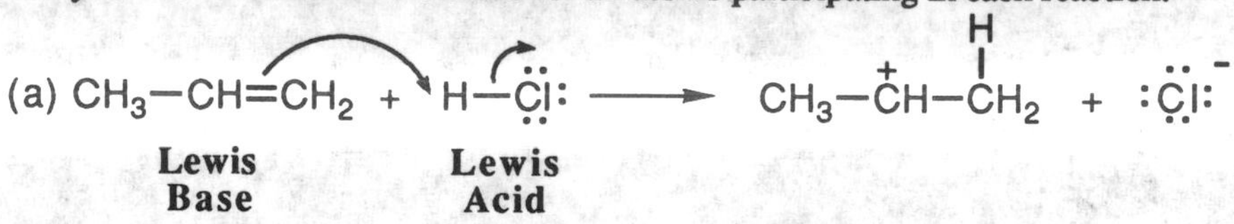

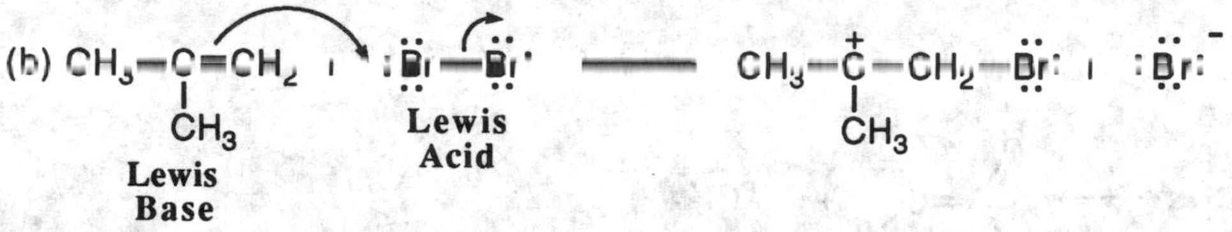

(a) $CH_3-CH=CH_2$ + H—$\overset{..}{\underset{..}{Cl}}$: $\longrightarrow$ $CH_3-\overset{+}{CH}-\overset{H}{\underset{}{CH_2}}$ + :$\overset{..}{\underset{..}{Cl}}$:$^-$

      Lewis   Lewis
      Base   Acid

(b) $CH_3-\underset{\underset{CH_3}{|}}{C}=CH_2$ + :$\overset{..}{\underset{..}{Br}}$—$\overset{..}{\underset{..}{Br}}$: $\longrightarrow$ $CH_3-\underset{\underset{CH_3}{|}}{\overset{+}{C}}-CH_2-\overset{..}{\underset{..}{Br}}$: + :$\overset{..}{\underset{..}{Br}}$:$^-$

                Lewis
                Acid
     Lewis
     Base

CHAPTER 4: STEREOCHEMISTRY

4.0 OVERVIEW
• Molecules are three-dimensional, and this chapter describes very important consequences of that three-dimensionality, namely stereochemistry and chirality.

4.1 ISOMERISM
• **Isomers** are *different* molecules that have the same molecular formula. ✳ *[This is the first of several very important definitions that should be learned now to avoid confusion later. They may sound relatively simple, but learning to apply them correctly in cases of complex molecules can be quite challenging.]*
• **Constitutional isomers** are different molecules with the same molecular formula, but with a different order of attachment of atoms. In other words, the atoms are connected to each other differently. For example, pentane ($\diagdown\diagup\diagdown$) and 2-methylbutane ($\curlyvee$) are constitutional isomers of molecular formula C_5H_{12}.
• **Stereoisomers** are different molecules with the same molecular formula, the same order of attachment of atoms, but a different **orientation** of those atoms or groups in space. ✳ *[This concept is subtle and may be confusing at first, but it is actually a very powerful concept that should be understood before moving on.]*
  - **Diastereomers** are stereoisomers that are *not* mirror images of each other. This definition is best understood in the context of the definition of enantiomers.
  - **Enantiomers** are stereoisomers that are mirror images of each other. ✳
  - A **mirror image** is what you see as the reflection of an object in a mirror.

4.2 CHIRALITY
• **Chirality** is a property of three-dimensional objects that is very important in chemistry. An object is **chiral** if it is not superposable on its mirror image. That is, a chiral object and its mirror image cannot be oriented in space so that all of their features (corners, edges, points, bonds, atoms, etc.) correspond exactly to each other.✳ *[This is a very difficult concept, but one that is absolutely central to the rest of this chapter and the study of stereochemistry. Chirality should be understood before proceeding.]*
  - Great examples of chiral objects are your hands. They are mirror images of each other (if you hold your left hand up to a mirror you see an image that looks like your right hand in the mirror), yet you cannot orient your two hands in space so that they are superposable. Try this for yourself if you are having trouble understanding chirality. Because your hands are chiral, they have "handedness", that is, the left hand is different in three-dimensional space than your right hand even though they are mirror images of each other and both have four fingers and a thumb. For this reason, a glove that fits your left hand well will not fit your right hand and you must buy one left-handed and one right-handed glove.
  - Other examples of chiral objects include your feet, an airplane propeller or ceiling fan, a wood screw or a drill bit. In fact, the vast majority of objects around you are chiral.
• An object is **achiral** if it is superposable on its mirror image. Examples of objects that are achiral include a perfect cube or a perfect sphere. To be achiral, the molecule must have symmetry.
  - Achiral objects can be identified when they possess a **plane of symmetry**.
    A **plane of symmetry** is an imaginary plane passing through an object dividing it such that one half is the mirror reflection of the other half. Only highly symmetric objects such as a perfect sphere have a plane of symmetry. ✳ *[This is an important concept. It may be helpful to carefully examine Figure 4.2 to make sure you understand when an object has a plane of symmetry.]*
• **Molecules can be chiral.** To be chiral, a molecule cannot be superposable upon its mirror image. In addition, a chiral molecule does not have a plane of symmetry. ✳
• Often, molecules are chiral because they contain a tetrahedral carbon atom with four different substituents. This is because a tetrahedral carbon atom that has four *different* substituents is **not** superposable on its mirror image.✳ *[This important concept is best understood using models. Try making a model of a tetrahedral carbon atom with four different substituents attached. Next, make a model of its mirror image, and try to superpose the two models. The fact that the two models cannot be superposed confirms that the molecules display chirality. Since the molecules are non-superposable mirror images of each other, they are **enantiomers**.]*
  - A **stereocenter** is a point in a molecule at which interchange of two atoms or groups of atoms bonded to that point produces a different stereoisomer. A carbon atom with four different substituents is a **tetrahedral stereocenter**. ✳

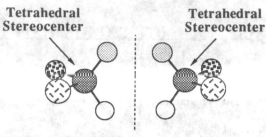

Tetrahedral Stereocenter   **Tetrahedral Stereocenter**

A pair of enantiomers

4.3 NAMING ENANTIOMERS : THE R-S SYSTEM

• The configuration of a tetrahedral stereocenter is assigned using the **R-S convention** first introduced by Cahn, Ingold, and Prelog. This convention is based on assigning **priorities** to the four different substituents around the stereocenter as follows:

- 1. Each atom attached directly to the stereocenter is assigned a priority based on atomic number. The higher the atomic number, the higher the priority. For example, -Br is assigned a higher priority than -Cl, which is assigned a higher priority than -OH, etc.
- 2. For isotopes, the higher the atomic mass of the isotope, the higher its priority.
- 3. If different priorities based on atomic number cannot be assigned to the atoms directly bonded to the stereocenter, then look at the next set of atoms and continue until different priorities can be assigned. **It is the first point of difference** that matters here.
- 4. Atoms of double or triple bonds are considered as if they are bonded to an equivalent number of similar atoms by single bonds. *[This is the hardest priority rule, and practice is usually needed to fully understand it.]*

• In order to assign the configuration of a tetrahedral stereocenter according to the R-S convention use the following procedure:

- 1. Locate the tetrahedral stereocenter and identify its four substituents.
- 2. Assign a different priority, 1,2,3, or 4, to each substituent using the rules listed above.
- 3. Orient the molecule in space such that the group with the lowest priority (4) is directed away from you. In other words, orient the molecule so that when you look at it, the lowest priority group is directly behind the carbon stereocenter. The three remaining groups (1-3) will be arranged like the legs of a tripod directed toward you.
- 4. Read the remaining three groups in order from highest (1) to lowest priority (3).
- 5. If reading the groups proceeds in a **clockwise** direction, the configuration is designated as **R**, if reading proceeds in a **counterclockwise** direction, the configuration is **S**. ✶ *[It is essential that you become very good at assigning stereochemistry, and practice is the best way to become very good.]*

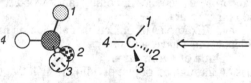

View from this direction

A carbon stereocenter with four different substituents (*1* highest priority, *4* lowest priority)

The view down the C-4 bond. Groups increase in priority in clockwise direction so the configuration is "R"

4.4 FISCHER PROJECTIONS

• **A Fisher projection** is used to indicate the configuration of chiral molecules. To write a Fisher projection, orient the molecule so that the **vertical bonds** of a stereocenter are **directed away from you** and the **horizontal bonds** are **directed toward you**. The Fisher projection is now drawn as a two dimensional figure, that can be related to the three-dimensional structure by the convention that the two horizontal lines represents bonds directed toward you, and the two vertical lines represent bonds directed away from you.

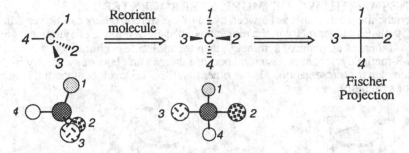

- You can only manipulate a Fischer projection in precise ways. For example, you can only rotate a Fischer projection in the plane of the paper by 180°, NOT 90°. If this rule is not followed, then the new Fischer projection may inadvertently depict a different stereoisomer than the original Fischer projection.

4.5 ACYCLIC MOLECULES WITH TWO OR MORE STEREOCENTERS

• For a molecule with **n** stereocenters, the maximum number of stereoisomers possible is 2^n. This is because each stereocenter can be either R or S.

• A good way to learn about the stereochemical consequences of having more than one stereocenter in the same molecule is to examine all four possible stereoisomers that arise when there are two stereocenters in the same molecule. For example, consider the four stereoisomers of 2,3,4-trihydroxybutanal drawn below:

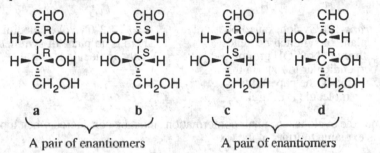

- From right to left, the four isomers are the R,R (**a**); S,S (**b**); R,S (**c**) and S,R (**d**) stereoisomers. When examined in pairs, the following relationships become apparent:

     a,b and **c,d** are non-superposable mirror images of each other; thus they are pairs of **enantiomers**. *[This would be a good time to review the definition of enantiomers if necessary.]*
     a,c and **a,d** and **b,c** and **b,d** are all pairs of stereoisomers that are not mirror images of each other, thus these are all pairs of **diastereomers**.

• *[Hint: when asked to identify the stereochemical relationships between pairs of stereoisomers, it is helpful to first assign the configuration (R or S) to each stereoisomer then compare these designations, instead of trying to compare the molecules directly.]* ✳

• Certain molecules have special symmetry properties that reduce the number of stereoisomers to fewer than the predicted 2^n. In these cases, some of the stereoisomers contain a plane of symmetry so the molecules are achiral. Molecules or ions that contain two or more tetrahedral stereocenters but are achiral because of a plane of symmetry are called **meso compounds**. ✳

- **Meso compounds** will always be the R,S/S,R isomer of a molecule that has two stereocenters. For example, the R,S/S,R isomer of tartaric acid is a meso compound. Of course, not all R,S isomers of molecules with two stereocenters are meso compounds. To be a meso compound there must be a plane of symmetry in the molecule.

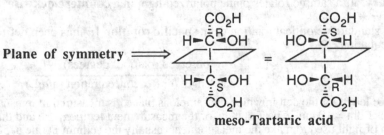

meso-Tartaric acid

4.6 CYCLOALKANES WITH TWO OR MORE STEREOCENTERS

• Understanding stereochemical relationships between cycloalkane stereoisomers can be very difficult. The key is being able to identify planes of symmetry in the molecules. If there is a plane of symmetry present, then the molecule is achiral. If there is no plane of symmetry, then the molecule is chiral. ✶
  - For example, *cis*-3-methylcyclopentanol is chiral because there is no plane of symmetry in the molecule as can be seen by the planar representation. On the other hand, *cis*-1,3-cyclopentanediol is achiral because there is a plane of symmetry.

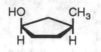   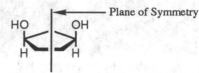

Plane of Symmetry

cis-3-Methylcyclopentanol   *cis*-1,3-Cyclopentanediol
No plane of symmetry; chiral   Plane of symmetry; achiral

  - With disubstituted cyclohexane derivatives, it is especially important to keep track of the substitution pattern. For example, *trans*-1,4-cyclohexanediol has a plane of symmetry and is thus achiral, while *trans*-1,3-cyclohexanediol does not have a plane of symmetry and is chiral.

trans-1,4-Cyclohexanediol   *trans*-1,3-Cyclohexanediol
A plane of symmetry extends   There is no plane of symmetry;
through both OH groups   the molecule is chiral
perpendicular to ring (in the
plane of the paper); the
molecule is achiral.

  - Note that sometimes the cyclohexane chair conformations must be considered when looking for planes of symmetry. This is explained further in the text.

4.7 PROPERTIES OF STEREOISOMERS

• In general, **enantiomers have identical physical properties** when those properties are **measured in an achiral environment**. For example, two enantiomers have the same boiling points, melting points, solubilities in achiral solvents, the same pK_a values, etc. On the other hand, two **diastereomers have different physical properties** like melting points or boiling points. ✶

4.8 OPTICAL ACTIVITY - HOW CHIRALITY IS MEASURED IN THE LABORATORY

• Although enantiomers have identical physical properties when those properties are measured in an achiral way, they are different compounds. Therefore, they have different physical properties when the measurements are made in a chiral way such as interactions with plane polarized light. Each member of a pair of enantiomers rotates the plane of plane polarized light, so they are called **optically active**.
  - Normal light consists of waves oscillating in all planes perpendicular to its path. Certain materials such as Polaroid sheets allow only waves oscillating in a single plane to pass through. The resulting light is thus called **plane polarized light**, because all of the resulting light is oscillating in the same plane.
  - Samples of enantiomers rotate the plane of plane polarized light. A **polarimeter** is the instrument used in the laboratory to measure the direction and magnitude of rotation of plane polarized light.
    If a sample of an enantiomer rotates plane polarized light in a **clockwise** direction, it is called **dextrorotary**. If a sample rotates plane polarized light in a **counterclockwise** direction, it is called **levorotary**. ✶
    In order to standardize optical rotation data, **specific rotation [α]** has been defined according to the following equation:

$$[\alpha]_{\lambda}^{T} = \frac{\text{observed rotation (degrees)}}{\text{cell length (dm) x concentration (g/mL)}}$$

Note that the length of the cell in which the sample is placed is measured in unusual units, namely decimeters. 1 dm = 10 cm. The T stands for the measurement temperature and the λ stands for the wavelength of light used to make the measurement (usually the sodium D line at 589 nm).

By convention, a dextrorotary compound is designated with a plus sign (+) and a levorotary compound is designated with a minus sign (-).

- Two enantiomers rotate plane polarized light by the same number of degrees, but with opposite sign. Meso compounds and all other achiral molecules do not rotate plane polarized light. ✳

- An equimolar mixture of two enantiomers is called a **racemic mixture**. For racemic mixtures, the rotations of plane polarized light exactly cancel and there is no overall rotation. If two enantiomers in a mixture are in unequal amounts, the entire mixture will have a net rotation, and the magnitude and direction of that rotation can be used to determine the exact ratio of the two enantiomers in the mixture according to the following equation:

$$\% \text{ optical purity} = \frac{[\alpha]_{obs}}{[\alpha]_{pure\ enantiomer}} \times 100$$

  - **Enantiomeric excess**, abbreviated **ee**, is equal to the percent optical purity, and is often referred to when enantiomers are made in different amounts in the same reaction. **Enantiomeric excess** is defined according to the following equation:

$$\% \text{ enantiomeric excess (ee)} = \left(\frac{\text{mol one enantiomer - mol other enantiomer}}{\text{mol both enantiomers}} \right) \times 100$$

- There is **no** necessary **relationship between** a **configuration** (R and S) and the **sign of rotation** (+ and -). For some pairs of stereoisomers, the "R" enantiomer has a (+) rotation, while the "S" enantiomer has a (-) rotation, yet for other pairs of stereoisomers, the "S" enantiomer has a (+) rotation and the "R" enantiomer has a (-) rotation. This makes sense since the R and S designations are based on artificial nomenclature rules, while the (+) and (-) rotations are the result of actual physical measurements. ✳

4.9 SEPARATION OF ENANTIOMERS-RESOLUTION

- **Resolution** is the process whereby a racemic mixture is separated into its enantiomers. A pair of enantiomers is difficult to separate, because they both have identical physical properties such as melting points, boiling points, etc. A common method of resolution involves combining the racemic mixture with another compound that is a single enantiomer. The combination results in the production of two diastereomers that can usually be separated because diastereomers have different physical properties. Following resolution, the enantiomers are recovered. ✳

  - A reaction that lends itself to resolution is salt formation. For example, one enantiomer of a chiral compound, such as an amine like (+)-cinchonine, can be used to form diastereomeric salts with a racemic mixture of a chiral carboxylic acid. Following resolution, the pure enantiomers of the carboxylic acid are recovered by acid precipitation.

4.10 THE SIGNIFICANCE OF CHIRALITY IN THE BIOLOGICAL WORLD

- Almost all of the molecules of living systems are chiral. Even though all of these chiral molecules could in theory exist as a mixture of stereoisomers, almost invariably only one stereoisomer is found in nature.

  - Enzymes, Mother Nature's molecular machines, are composed of chiral amino acids. As a result, the enzymes are themselves chiral. For that reason, enzymes can be thought of as a chiral environment. Enzymes are therefore able to distinguish substrate enantiomers, so enzymes can be used for chiral resolutions.

CHAPTER 4
Solutions to the Problems

<u>Problem 4.1</u> Each molecule has one stereocenter. Draw stereorepresentations for enantiomers of each.

Each part has a tetrahedral stereocenter. The stereocenters are labeled with an asterisk.

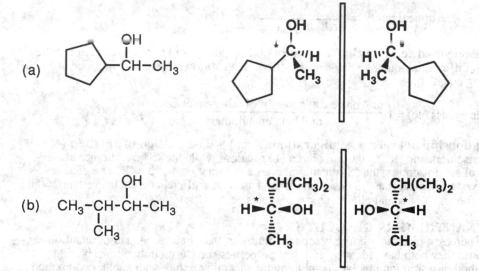

(a)

(b)

<u>Problem 4.2</u> Assign priorities to the groups in each set.
(a) $-CH_2OH$ and $-CH_2CH_2OH$

The $-CH_2OH$ group has higher priority because the FIRST point of difference is the underlined O atom of $-CH_2\underline{O}H$ that takes priority over the underlined C atom of $-CH_2\underline{C}H_2OH$.

(b) $-CH_2OH$ and $-CH=CH_2$

$$\overset{1}{-CH}=\overset{2}{CH_2} \quad \underrightarrow{\text{is treated as}} \quad \begin{matrix} C & C \\ \| 1 & \| 2 \\ -C-C-H \\ | & | \\ H & H \end{matrix}$$

Nevertheless, the FIRST point of difference is the underlined O atom of $-CH_2\underline{O}H$ that takes priority over any of the atoms attached to carbon 1 of $-CH=CH_2$. Thus, the $-CH_2OH$ group takes priority over the $-CH=CH_2$ group.

<u>Problem 4.3</u> Assign an R or S configuration to each stereocenter.

The drawings underneath each molecule show the order of priority, the perspective from which to view the molecule, and the R,S designation for the configuration.

(a)

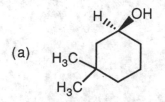

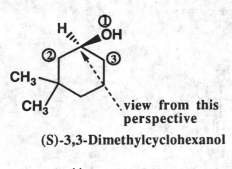

view from this perspective

(S)-3,3-Dimethylcyclohexanol

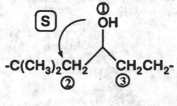

If you view from the perspective shown, this is what you see

(b)

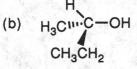

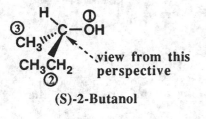

view from this perspective

(S)-2-Butanol

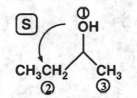

If you view from the perspective shown, this is what you see

(c)

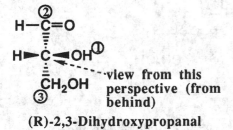

view from this perspective (from behind)

(R)-2,3-Dihydroxypropanal

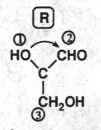

If you view from the perspective shown, this is what you see

Problem 4.4 We said that rotation of a Fischer projection by 90° in the plane of the paper gives a different molecule. Show that this manipulation of (S)-2-butanol gives (R)-2-butanol?

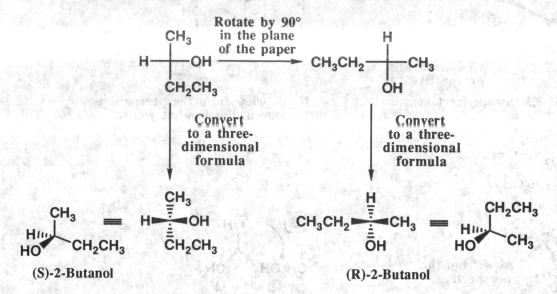

Rotation of the Fischer projection of (S)-2-butanol by 90° in the plane of the paper gives the enantiomer, (R)-2-butanol.

Problem 4.5 Convert each three-dimensional formula to a Fischer projection. In so doing, orient the carbon chain vertically. Assign R,S configurations to each stereoisomer.

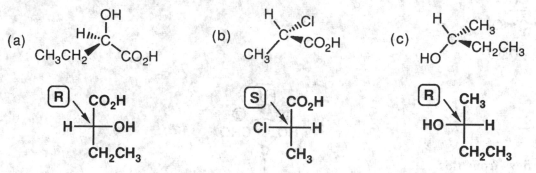

Note that other Fischer projections are also possible.

Problem 4.6 Following are Fisher projection formulas for the four stereoisomers of 3-chloro-2-butanol.

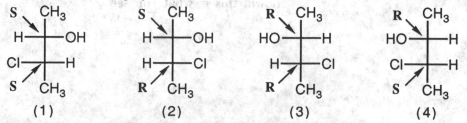

(a) Show the R,S configuration of each stereoisomer.

The configuration of each stereocenter has been labeled on the above structures. This labeling often helps when trying to establish stereochemical relationships between molecules.

(b) Which compounds are enantiomers?

Enantiomers are stereoisomers that are mirror images of each other. The pairs of enantiomers are structures (1) and (3) (S,S and R,R) as well as structures (2) and (4) (S,R and R,S).

(c) Which compounds are diastereomers?

Diastereomers are stereoisomers that are not mirror images of each other. Therefore, there are several sets of diastereomers. The following chart describes the relationship between any pair of molecules

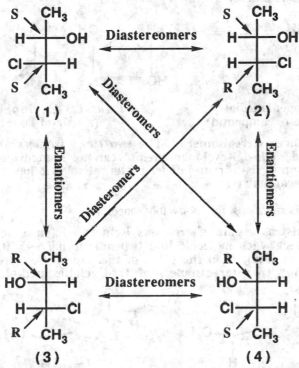

Problem 4.7 Following are four Newman projection formulas for tartaric acid.

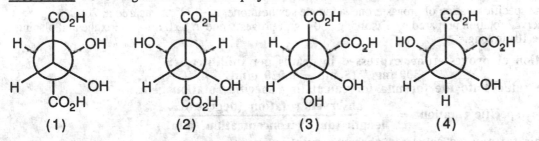

(a) Which represent the same compound?

Compounds (1) and (4) are the same compound having the configuration (2R,3R). Compounds (2) and (3) are also the same compound, having the configuration (2R,3S).

(b) Which are enantiomers?

There are no sets of enantiomers in this set.

(c) Which represent a meso compound?

The meso compound of tartaric acid has the (2R,3S) configuration so compounds (2) and (3) are the meso compound.

(d) Which are diastereomers?

Compounds (1) and (4) are diastereomers of compounds (2) and (3).

Problem 4.8 How many stereoisomers exist for 1,3-cyclopentanediol?

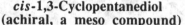

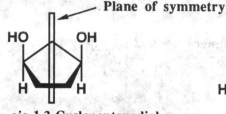 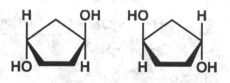

cis-1,3-Cyclopentanediol   *trans*-1,3-Cyclopentanediol
(achiral, a meso compound)   (a pair of enantiomers)

1,3-Cyclopentanediol has three stereoisomers. The two *trans* isomers are enantiomers, the *cis* isomer is a meso compound. *Cis*-1,3-cyclopentanediol can be recognized as a meso compound because it is superposable upon its mirror image. Alternatively, it has a plane of symmetry that bisects it into two mirror halves.

Problem 4.9 How many stereoisomers exist for 1,4-cyclohexanediol?

1,4-Cyclohexanediol can exist as a pair of *cis-trans* isomers. Each is achiral because of a plane of symmetry that bisects each molecule into two mirror halves. In the figure below, the plane of symmetry in each molecule is in the plane of the paper. As a result of each isomer being achiral, there are only two stereoisomers of 1,4-Cyclohexanediol.

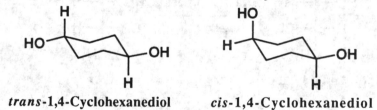

trans-1,4-Cyclohexanediol   *cis*-1,4-Cyclohexanediol

Problem 4.10 The specific rotation of progesterone, a female sex hormone, is +172°, measured at 20°C. Calculate the observed rotation prepared by dissolving 300 mg of progesterone in 15.0 mL of dioxane and placing it in a sample tube 10 cm long.

The concentration of progesterone, expressed in grams per milliliter is:
$$300 \text{ mg } / 15 \text{ mL} = 0.020 \text{ g/mL}$$
Inserting these values into the formula for calculating specific rotation gives:

$$\text{specific rotation } = \frac{\text{observed rotation (degrees)}}{\text{length (dm) x concentration (g/mL)}}$$

Rearranging this formula to solve for observed rotation gives:
   observed rotation (degrees) = specific rotation x length (dm) x concentration (g/mL)
Plugging in the experimental values gives the final answer.

   observed rotation (degrees) =   +172° x 1.00 dm x 0.020 g/mL = $\boxed{+3.4°}$

<u>Problem 4.11</u> One commercial synthesis of naproxen (the active ingredient in Aleve and a score of other over-the-counter and prescription nonsteroidal antiinflammatory drug preparations) gives the enantiomer shown in 97% enantiomeric excess.

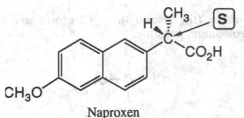

Naproxen
(A nonsteroidal antiinflammatory drug)

(a) Assign an R,S configuration to this enantiomer of naproxen.

The stereocenter has the S configuration.

(b) What are the percentages of R and S enantiomers in the mixture?

$$\text{Enantiomeric excess (ee)} = \frac{S - R}{S + R} \times 100 = \%S - \%R = 97\%$$

solving the above equation gives

$$\boxed{98.5\% \ S \ \text{and} \ 1.5\% \ R}$$

Chirality

<u>Problem 4.12</u> Think about the helical coil of a telephone cord or a spiral binding and suppose that you view the spiral from one end and find that it is a left-handed twist. If you view the same spiral from the other end, is it a right-handed twist, or a left-handed twist from that end as well?

A helical coil has the same handedness viewed from either end.

<u>Problem 4.13</u> Next time you have the opportunity to view a collection of whelks, augers, or other sea shells that have a helical twist, study the chirality of their twists. Do you find an equal number of left-handed and right-handed whelks, for example, or are they all or mostly all of one chirality? What about the chirality of whelks compared with augers and other spiral shells?

This question was just meant to make you think about chirality in nature, but if you do know the answer please share it with your class.

<u>Problem 4.14</u> One reason we can be sure that sp^3-hybridized carbon atoms are tetrahedral is the number of stereoisomers that can exist for different organic compounds.
(a) How many stereoisomers are possible for $CHCl_3$, CH_2Cl_2, and $CHClBrF$ if the four bonds to carbon have a tetrahedral arrangement?

Both tetrahedral $CHCl_3$ and tetrahedral CH_2Cl_2 are achiral, so there is only one stereoisomer of either.

On the other hand, tetrahedral CHBrClF is chiral so there are two stereoisomers possible.

(b) How many stereoisomers are possible for each of these compounds if the four bonds to the carbon have a square planar geometry?

Even with a square planar geometry (the H and three Cl atoms are in the same plane as the C atom), there is only one stereoisomer possible.

$$
\begin{array}{c}
\text{H} \\
| \\
\text{Cl}-\text{C}-\text{Cl} \\
| \\
\text{Cl}
\end{array}
$$

There are two possible stereoisomers of CH_2Cl_2, one with the Cl atoms adjacent to each other, and another with the Cl atoms that are opposite each other.

$$
\begin{array}{c}
\text{H} \\
| \\
\text{H}-\text{C}-\text{Cl} \\
| \\
\text{Cl}
\end{array}
\qquad\qquad
\begin{array}{c}
\text{H} \\
| \\
\text{Cl}-\text{C}-\text{Cl} \\
| \\
\text{H}
\end{array}
$$

There are three possible stereoisomers of a square planar CHBrClF as shown.

$$
\begin{array}{c}
\text{Br} \\
| \\
\text{H}-\text{C}-\text{Cl} \\
| \\
\text{F}
\end{array}
\qquad
\begin{array}{c}
\text{Cl} \\
| \\
\text{H}-\text{C}-\text{Br} \\
| \\
\text{F}
\end{array}
\qquad
\begin{array}{c}
\text{Br} \\
| \\
\text{H}-\text{C}-\text{F} \\
| \\
\text{Cl}
\end{array}
$$

Enantiomers

Problem 4.15 Which compounds contain stereocenters?

(a) 2-Chloropentane

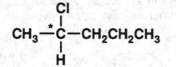

$$
\begin{array}{c}
\text{Cl} \\
| \\
\text{CH}_3-\overset{*}{\text{C}}-\text{CH}_2\text{CH}_2\text{CH}_3 \\
| \\
\text{H}
\end{array}
$$

(b) 3-Chloropentane

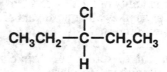

$$
\begin{array}{c}
\text{Cl} \\
| \\
\text{CH}_3\text{CH}_2-\text{C}-\text{CH}_2\text{CH}_3 \\
| \\
\text{H}
\end{array}
$$

(c) 3-Chloro-1-pentene

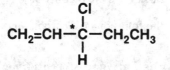

$$
\begin{array}{c}
\text{Cl} \\
| \\
\text{CH}_2=\text{CH}-\overset{*}{\text{C}}-\text{CH}_2\text{CH}_3 \\
| \\
\text{H}
\end{array}
$$

(d) 1,2-Dichloropropane

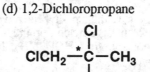

$$
\begin{array}{c}
\text{Cl} \\
| \\
\text{ClCH}_2-\overset{*}{\text{C}}-\text{CH}_3 \\
| \\
\text{H}
\end{array}
$$

Problem 4.16 Using only C, H, and O, write structural formulas for the lowest molecular weight chiral

(a) Alkane

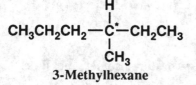

$$
\begin{array}{c}
\text{H} \\
| \\
\text{CH}_3\text{CH}_2\text{CH}_2-\overset{*}{\text{C}}-\text{CH}_2\text{CH}_3 \\
| \\
\text{CH}_3
\end{array}
$$

3-Methylhexane

(b) Alcohol

$$
\begin{array}{c}
\text{H} \\
| \\
\text{HO}-\overset{*}{\text{C}}-\text{CH}_2\text{CH}_3 \\
| \\
\text{CH}_3
\end{array}
$$

2-Butanol

(c) Aldehyde

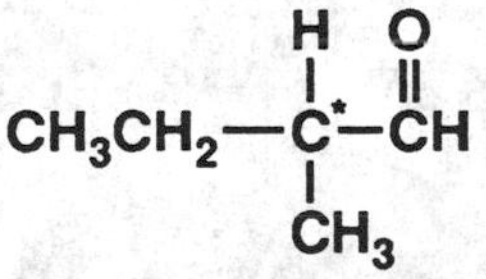

2-Methylbutanal

(d) Ketone

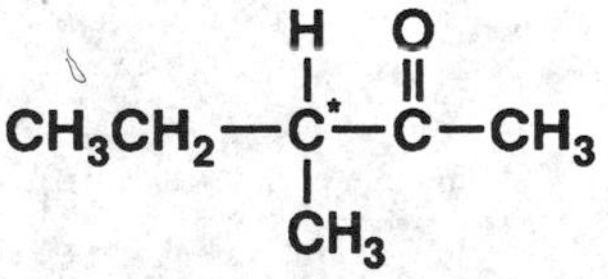

3-Methyl-2-pentanone

(e) Carboxylic acid

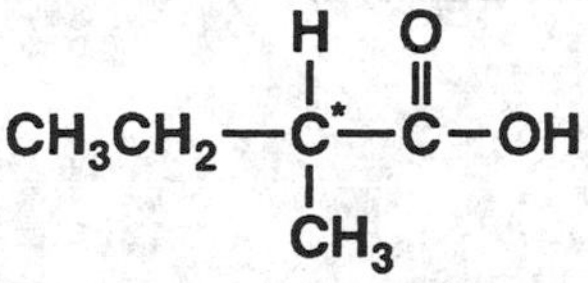

2-Methylbutanoic acid

Problem 4.17 Draw mirror images for these molecules.

The mirror images are shown in bold.

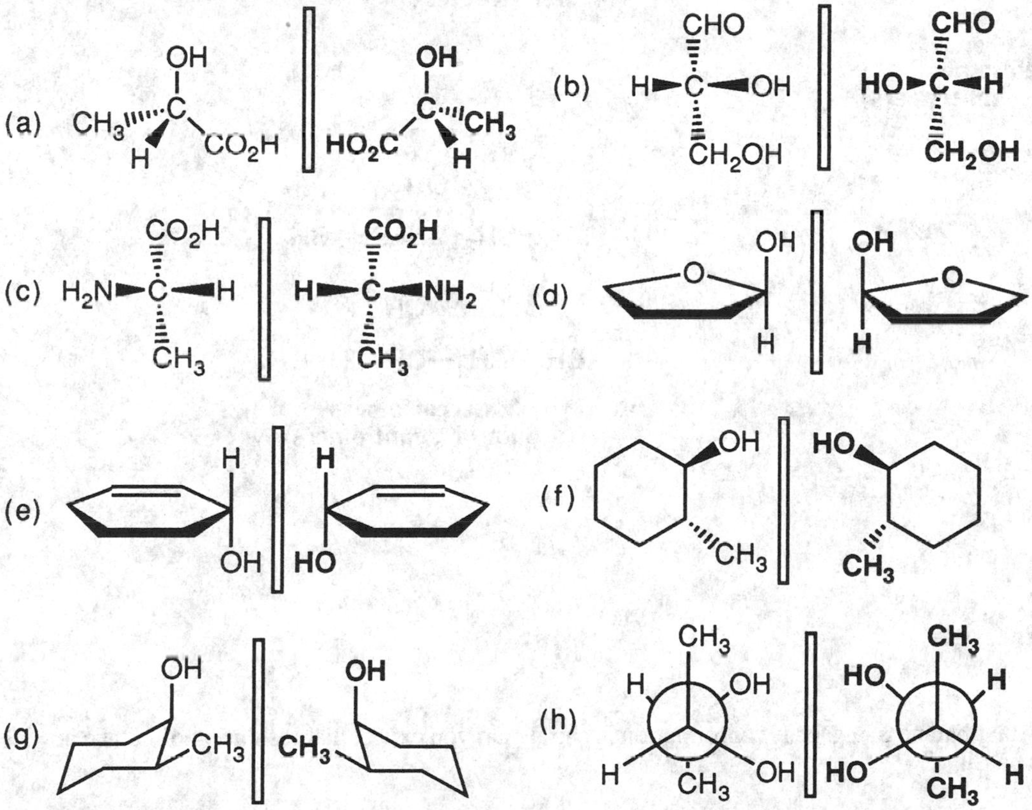

Problem 4.18 Following are several stereorepresentations for lactic acid. Take (a) as a reference structure. Which of the stereorepresentations are identical with (a) and which are mirror images of (a)?

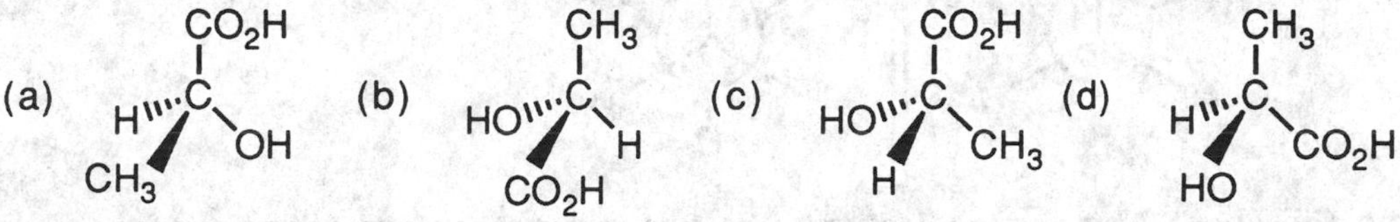

All of the above stereorepresentations have the (S)-configuration so they are identical.

<u>Problem 4.19</u> Mark each stereocenter in the following molecules with an asterisk. How many stereoisomers are possible for each molecule?

(a)
$$CH_3-\underset{\underset{OH}{|}}{\overset{\overset{CH_3}{|}}{C}}-CH=CH_2$$

No stereocenters

(b)
$$H-\underset{\underset{CH_3}{|}}{\overset{\overset{CO_2H}{|}}{C}}-OH$$

$$H-\underset{\underset{CH_3}{|}}{\overset{\overset{CO_2H}{|}}{\overset{*}{C}}}-OH$$

2 Stereoisomers
(a pair of enantiomers)

(c)
$$CH_3-\underset{\underset{NH_2}{|}}{\overset{\overset{CH_3}{|}}{CH}}-CH-CO_2H$$

$$CH_3-\underset{\underset{NH_2}{|}}{\overset{\overset{CH_3}{|}}{CH}}-\overset{*}{CH}-CO_2H$$

2 stereoisomers
(a pair of enantiomers)

(d)
$$CH_3-\overset{\overset{O}{\|}}{C}-CH_2-CH_3$$

No stereocenters

(e)
$$H-\underset{\underset{CH_2OH}{|}}{\overset{\overset{CH_2OH}{|}}{C}}-OH$$

No stereocenters

(f)
$$CH_3-CH_2-\underset{\underset{}{\overset{\overset{OH}{|}}{CH}}}-CH=CH_2$$

$$CH_3-CH_2-\overset{*}{\underset{}{\overset{\overset{OH}{|}}{CH}}}-CH=CH_2$$

2 stereoisomers
(a pair of enantiomers)

(g)
$$H-\underset{\underset{CH_2\cdot CO_2H}{|}}{\overset{\overset{CH_2\cdot CO_2H}{|}}{C}}-CO_2H$$

No Stereocenters

<u>Problem 4.20</u> Show that butane in a gauche conformation is chiral. Do you expect that resolution of butane at room temperature is possible?

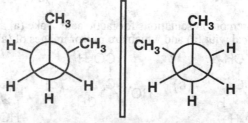

As can be seen from the above Newman projections, the two gauche conformations are non-superposable mirror images of each other. However, these conformations rapidly interconvert at room temperature through rotation around the central C-C bond, so they cannot be resolved.

Designation of Configuration: The R-S Convention

Problem 4.21 Assign the priorities to the groups in each set.

The groups are ranked from highest to lowest under each problem. Remember that priority is assigned at the first point of difference.

(a) -H -CH$_3$ -OH -CH$_2$OH   (b) -CH$_2$CH=CH$_2$ -CH=CH$_2$ -CH$_3$ -CH$_2$CO$_2$H
-OH > -CH$_2$OH > -CH$_3$ > -H   **-CH$_2$CO$_2$H > -CH=CH$_2$ > -CH$_2$CH=CH$_2$ > -CH$_3$**

(c) -CH$_3$ -H -CO$_2^-$ -NH$_3^+$ (d) -CH$_3$ -CH$_2$SH -NH$_3^+$ -CO$_2^-$
-NH$_3^+$ > -CO$_2^-$ > -CH$_3$ > -H   **-NH$_3^+$ > -CH$_2$SH > -CO$_2^-$ > -CH$_3$**

Problem 4.22 Following are structural formulas for the enantiomers of carvone. Each has a distinctive odor characteristic of the source from which it is isolated. Assign R,S configurations to each enantiomer.

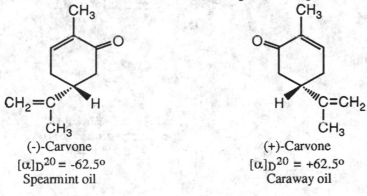

(-)-Carvone
$[\alpha]_D^{20} = -62.5°$
Spearmint oil

(+)-Carvone
$[\alpha]_D^{20} = +62.5°$
Caraway oil

Following are R-S designations for each enantiomer.

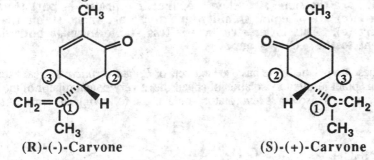

(R)-(-)-Carvone   (S)-(+)-Carvone

Problem 4.23 Following is a staggered conformation for one of the enantiomers of 2-butanol.

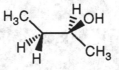

(a) Is this (R)-2-butanol or (S)-2-butanol?

The structure drawn is (S)-2-butanol.

(b) Draw a Newman projection for this enantiomer, viewed along the bond between carbons 2 and 3.

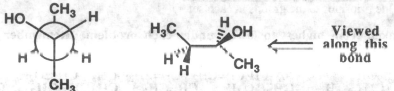

(c) Draw a Newman projection for two more staggered conformations of this molecule. Which of your conformations is the more stable? Assume that -OH and -CH₃ are comparable in size.

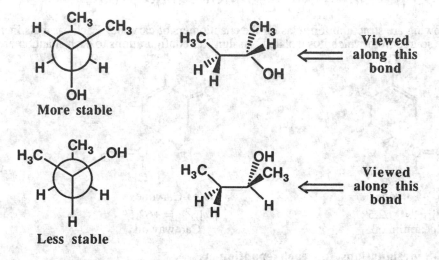

Assuming that -OH and -CH₃ are the same size, then the structure drawn in part (b) and the upper structure shown in part (c) are of equal stability, and these are more stable than the lower structure shown in part (c). This lower structure is less stable because both the -OH and -CH₃ groups are adjacent to the -CH₃ group.

<u>Problem 4.24</u> For centuries, Chinese herbal medicine has used extracts of *Ephedra sinica* to treat asthma. Phytochemical investigation of this plant resulted in isolation of ephedrine, a very potent dilator of the air passages of the lungs. The naturally occurring stereoisomer is levorotatory and has the following structure. Assign R or S configuration to each stereocenter.

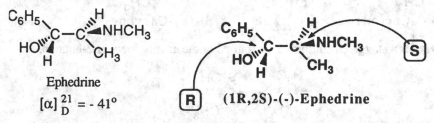

<u>Problem 4.25</u> When oxaloacetic acid and acetyl-coenzyme A (acetyl-CoA) labeled with radioactive carbon-14 in position 2 are incubated with citrate synthase, an enzyme of the TCA cycle, only the following enantiomer of [2-^{14}C]-citric acid is formed. Note that citric acid containing only ^{12}C is achiral. Assign an R or S configuration to this enantiomer of [2-^{14}C] citric acid. *Note:* Carbon-14 has a higher priority than carbon-12.

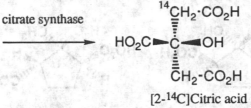

Oxaloacetic acid   Acetyl-CoA   [2-^{14}C]Citric acid

This enantiomer is S-(2-^{14}C)citric acid.

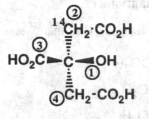   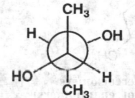

S-[2-^{14}C]Citric acid   **If you view from the proper perspective, this is what you see**

Molecules With Two Or More Stereocenters

Problem 4.26 Draw Newman projections for the three stereoisomers of 2,3-butanediol, showing the methyl groups anti (dihedral angle 180°).

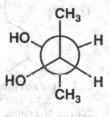

(2S,3S)-2,3-Butanediol   **(2R,3R)-2,3-Butanediol**   **(2R,3S)-2,3-Butanediol (meso)**

Problem 4.27 Draw Fischer projections for the four stereoisomers of 3-chloro-2-butanol, showing the carbon chain vertical and -H, -OH, and -Cl horizontal.

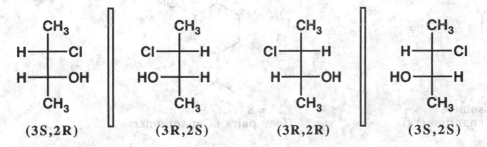

      (3S,2R)   **(3R,2S)**   **(3R,2R)**   **(3S,2S)**

Problem 4.28 Draw stereorepresentations for all stereoisomers of this compound. Label those that are meso compounds; those which are pairs of enantiomers.

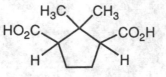

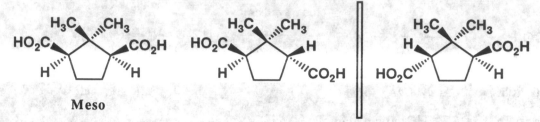

Meso

A pair of enantiomers

Problem 4.29 Mark each stereocenter in the following molecules with an asterisk. How many stereoisomers are possible for each molecule?

(a) 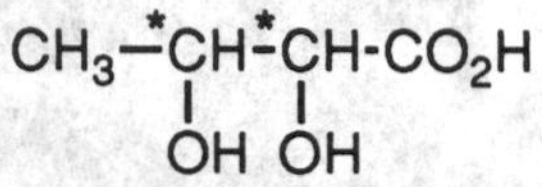 $CH_3-\overset{*}{C}H-\overset{*}{C}H-CO_2H$
              $\underset{OH}{|}\;\;\underset{OH}{|}$

2^2 = 4 Stereoisomers
(two pairs of enantiomers)

(b)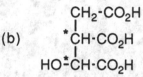
      $CH_2\text{-}CO_2H$
      $\overset{*}{C}H\text{-}CO_2H$
      $HO\overset{*}{-}CH\text{-}CO_2H$

2^2 = 4 Stereoisomers
(two pairs of enantiomers)

(c)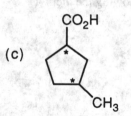

2^2 = 4 Stereoisomers
(two pairs of enantiomers)

(d)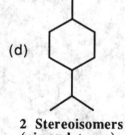

2 Stereoisomers
(cis and trans)
Note: there is a plane
of symmetry down the
center of this one

(e)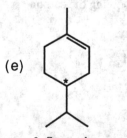

2 Stereoisomers
(one pair of enantiomers)

(f)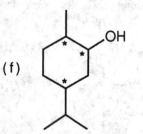

2^3 = 8 Stereoisomers
(four pairs of enantiomers)

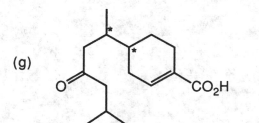

(g)

2^2 = 4 Stereoisomers
(two pairs of enantiomers)

(h)

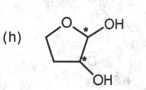

2^2 = 4 Stereoisomers
(two pairs of enantiomers)

(i)

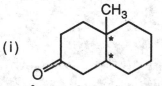

2^2 = 4 Stereoisomers
(two pairs of enantiomers)

<u>Problem 4.30</u> How many stereoisomers are possible for this compound, which is an aggregating pheromone for the Norway spruce beetle?

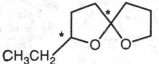

CH_3CH_2

There are two stereocenters in the molecule. As a result, there are 2^2 = 4 possible stereoisomers.

<u>Problem 4.31</u> Mark all stereocenters and state the maximum number of stereoisomers possible for each molecule.

Each stereocenter is marked with an asterisk.

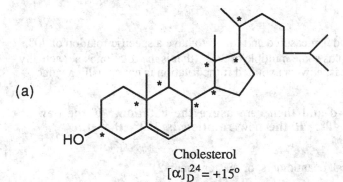

(a)

HO

Cholesterol
$[\alpha]_D^{24}$ = +15°

2^8 = 256 Stereoisomers (128 pairs of enantiomers)

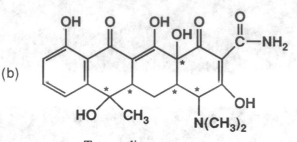

(b)

Tetracycline

$[\alpha]_D^{25} = +225°$

32 stereoisomers (16 pairs of enantiomers)

(c)

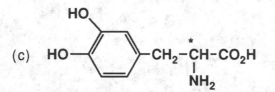

L-Dopa

3-(3,4-dihydroxyphenyl)alanine

$[\alpha]_D^{27} = -11.5°$

2 stereoisomers (1 pair of enantiomers)

(d)

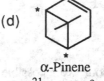

α-Pinene

$[\alpha]_D^{21} = +50.7°$

**4 stereoisomers
(2 pairs of enantiomers)**

(e)   CH₃—CH—CH—CO₂H

$$CH_3-\overset{*}{C}H-\overset{*}{C}H-CO_2H$$

Threonine

$[\alpha]_D^{20} = -27.4°$

**4 stereoisomers
(2 pairs of enantiomers)**

Problem 4.32 If the optical rotation of a new compound is measured and found to have a specific rotation of 40°, how can you tell if the actual rotation is not really 40° plus some multiple of 360° (that is the rotation is not actually 40 + (n x 360)°, where n has only integer values. That is, how can you tell if the rotation is not actually a value such as 400° or 760°?

You could dilute the solution by a factor of two and then remeasure the rotation. If the new rotation is 20°, then the original rotation was 40°. If the new rotation is 200°, then the original rotation was 400°.

Problem 4.33 Are the formulas within each set identical, enantiomers, or diastereomers?

(a)

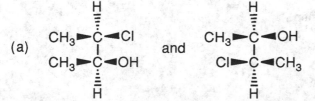

and

Diastereomers.   Configurations   are   (2S,3R)-3-chloro-2-butanol   and   (2R,3R)-3-chloro-2-butanol.

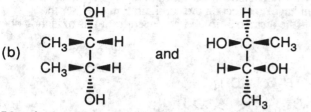

(b)   and

Identical. They are both (2R,3S)-2,3-butanediol, a meso compound.

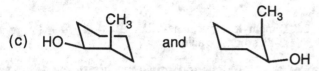

(c)   HO   and

Identical. They are both (1S,2R)-*cis*-2-methylcyclohexanol.

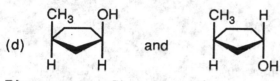

(d)   and

Diastereomers. *Cis* and *trans* isomers.

Problem 4.34 Which are meso compounds?

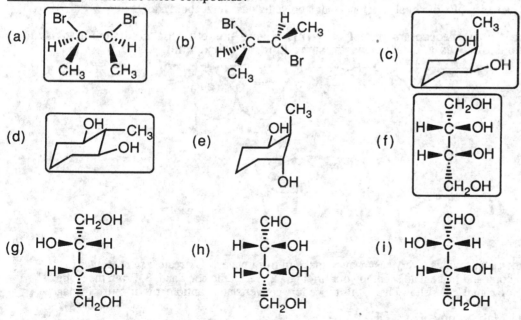

The meso compounds have a plane of symmetry: (a), (c), (d), and (f).

Problem 4.35 Vigorous oxidation of the following achiral bicycloalkene gives 2,2-dimethylcyclopentane-1,3-dicarboxylic acid. Assume that the conditions of oxidation have no effect on the configuration of either the starting bicycloalkene or the resulting dicarboxylic acid. Is the dicarboxylic acid produced from this oxidation one enantiomer, a racemic mixture, or a meso compound?

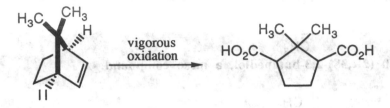

(1R,4S)-7,7-Dimethyl-bicyclo[2.2.1]hept-2-ene

2,2-Dimethylcyclopentane-1,3-dicarboxylic acid

The two carboxyl groups, derived from oxidation of the double bond, must be *cis* to each other. Therefore, the compound is meso with the following configuration.

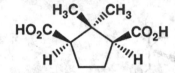

meso-2,2-Dimethylcyclopentane-1,3-dicarboxylic acid

Problem 4.36 A long polymer chain, such as polyethylene ($-CH_2CH_2-)_n$, can potentially exist in solution as a chiral object. Give two examples of chiral structures that a polyethylene chain could adopt.

Although there are no stereocenters in polyethylene, a long polyethylene chain could exist in chiral conformations such as a chiral helix or some kind of chiral knot.

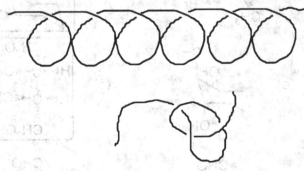

Molecular Modeling
Problem 4.37 ChemDraw provides a very easy way to make mirror images. 1. Create a stereocenter in ChemDraw, make a copy and place it adjacent to your original. 2. Select the copy and, 3. From the "Object" menu, select "Flip Horizontal". As shown here, this procedure converts an enantiomer to its mirror image.

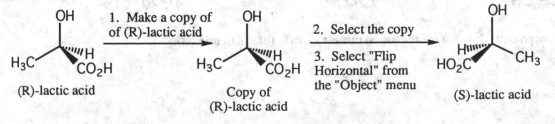

(R)-lactic acid

1. Make a copy of of (R)-lactic acid

Copy of (R)-lactic acid

2. Select the copy

3. Select "Flip Horizontal" from the "Object" menu

(S)-lactic acid

Now try these procedures with these molecules chosen from the text.

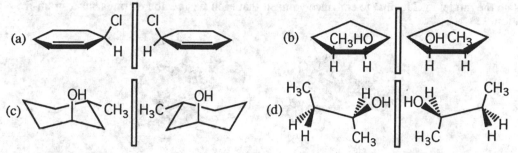

<u>Problem 4.38</u> Chem3D provides a particularly effective way to create and view stereoisomers. (1) Create a stereoisomer, for example, (R)-lactic acid, in ChemDraw and then import it into Chem3D. (2) Under "Analyze", select "Minimize" to minimize the energy of the stereoisomer you have drawn. (3) Under "Build", select "Reflect in the Y-Z plane". You will now have the enantiomer, in this case (S)-lactic acid, of your original stereoisomer.

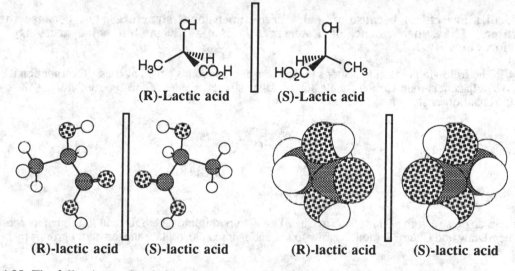

<u>Problem 4.39</u> The following molecule is an attractant pheromone for the olive fly.

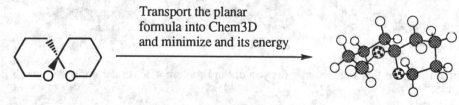

Transport the planar
formula into Chem3D
and minimize and its energy

(a) Build a line-angle structure of this molecule in ChemDraw (as shown on the left). Using the directions given in Problem 4.37, create its mirror image. Are they superposable?

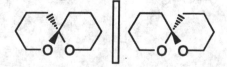

No, careful inspection verifies that they are not superposable mirror images of each other.

(b) Select your ChemDraw structure, paste it into Chem3D, and minimize its energy to give the model on the right. Rotate the model in Chem3D to convince yourself that each six-membered rings has a strain-free chair conformation.

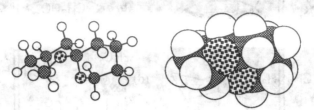

(c) This molecule has no stereocenter and yet it is chiral. From examination of the three-dimensional model in Chem3D, convince yourself that it has no plane or center of symmetry and that it is, in fact, chiral.

(d) "The presence of a stereocenter in an organic molecule is a sufficient condition for chirality, but it is not a necessary condition." Explain.

Some molecules are chiral because of their three-dimensional structures, not because they have a stereocenter. Thus, the presence of a stereocenter is a sufficient but not a necessary condition for chirality.

<u>Problem 4.40</u> The following molecule belongs to the class of compounds called allenes. The functional group of an allene is two adjacent carbon-carbon double bonds. Disubstituted allenes of this type are chiral. The specific rotation of the enantiomer shown is -314.

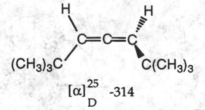

(a) Build a line-angle structure of this allene in ChemDraw, import it into Chem3D, and minimize its energy. To make the ChemDraw and Chem3D models easier to see and manipulate, replace the tert-butyl groups by methyl groups.

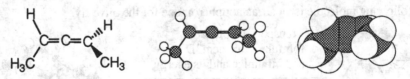

(b) Make the mirror image of this stereoisomer (as you did in Problem 4.38) and convince yourself that the two are not superposable.

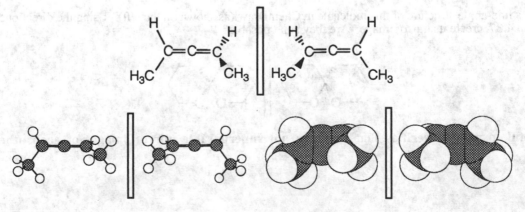

CHAPTER 5: ALKENES I

5.0 OVERVIEW
- **Unsaturated hydrocarbons** are hydrocarbons that contain one or more carbon-carbon double or triple bonds.
    - **Alkenes** are unsaturated hydrocarbons with one or more carbon-carbon double bonds.
    - **Alkynes** are unsaturated hydrocarbons with one or more carbon-carbon triple bonds.
    - **Aromatic hydrocarbons** are hydrocarbons that have cyclic structures with special patterns of alternating double and single bonds.

5.1 STRUCTURE OF ALKENES
- A carbon-carbon double bond consists of **one sigma bond** formed by the overlap of sp^2 hybridized orbitals of adjacent carbon atoms and **one pi bond** formed by the overlap of unhybridized 2p orbitals. ✳ *[This picture of the orbitals involved with the carbon-carbon double bond is crucial to your understanding of the reactions and properties of alkenes, so make sure you understand Figure 5.1 in the book.]*
    - Each carbon atom in a carbon-carbon double bond is sp^2 hybridized so its geometry is trigonal planar with bond angles near 120°. Notice that this means the carbon atoms and the atoms attached directly to them are in the same plane.
    - Carbon-carbon triple bonds are shorter than carbon-carbon double bonds, which are shorter than carbon-carbon single bonds. This is because electrons in sp orbitals that overlap to form the sigma portion of a triple bond are held closest to the nuclei, because they have a higher percentage of s character. By the same logic, electrons in the sp^2 orbitals that overlap to form the sigma portion of a double bond are held closer to the nuclei, because they have a higher percentage s character than sp^3 orbitals.
    - Not surprisingly, carbon-carbon triple bonds are stronger than carbon-carbon double bonds which are stronger than carbon-carbon single bonds.
    - The most important implication of this model of carbon-carbon double bonds is that they **cannot rotate** because rotation would decrease the extent of 2p-2p overlap. ✳
- Because carbon-carbon double bonds cannot rotate, alkenes display a kind of **stereoisomerism** called *cis-trans* **isomerization**.
    - A *cis* alkene is one in which the main carbon chain stays on the **same side** of the double bond. For example, *cis*-2-butene. ✳

$$H_3C\diagdown \diagup CH_3$$
$$C=C$$
$$H \diagup \diagdown H$$

cis-2-butene

    - A *trans* alkene is one in which the main carbon chain crosses over to the **opposite side** of the double bond. For example, *trans*-2-butene. ✳

$$H_3C\diagdown \diagup H$$
$$C=C$$
$$H \diagup \diagdown CH_3$$

trans-2-butene

- *cis* Alkenes are less stable than an analogous *trans* alkene because of non-bonded interaction strain that occurs in the *cis* alkene that is absent in the *trans* isomer.

5.2 NOMENCLATURE OF ALKENES
- Form IUPAC names by changing the **an** infix of the parent alkane to the infix **en**. For example ethene and propene are alkenes with two and three carbon atoms respectively. ✳
    - Form names for more complicated alkenes by choosing the longest carbon chain with the carbon-carbon double bond as the parent alkane.
        Number the chain to give the double bond the smallest number, indicate the position of the double bond by using the number of the **first atom of the double bond**.
        Name branched or **substituted alkenes** according to the same rules discussed for alkanes.
        For **cycloalkenes**, number the carbon atoms of the double bond 1 and 2 in the direction that gives the substituent encountered first the smallest number.
    - Alkenes that contain more than one carbon-carbon double bond are called **dienes**, **trienes**, and so forth.
    - **Unconjugated double bonds** are separated by at least one sp^3 hybridized atom. (⌇⌇, for example).

- **Conjugated double bonds** are on adjacent pairs of atoms, that is at least two adjacent double bonds with no sp³ hybridized atoms between them. (∿∿∿ , for example).
- A molecule has **cumulated double bonds** if two double bonds are on the same carbon atom ($CH_2=C=CH_2$, for example).
    - Several alkenes are known by their common names including **ethylene**, **propylene**, **isobutylene**, and **butadiene**. Substituents are also given common names such as **methylene**, **vinyl**, and **allyl**. *[These must be learned because their use is so widespread.]*
- The **E,Z system** provides an alternative way to name alkenes. The **E,Z system** developed by Cahn, Ingold, and Prelog is a comprehensive system of nomenclature. The **E,Z system** uses priority rules to rank the two substituents on each carbon atom of the double bond:
    - Each atom is assigned a priority based on atomic number; the higher the atomic number the higher the priority.
    - If you cannot assign priority differences to the two substituents by comparing the first atoms, then continue down the chains until the first point of difference is reached. *[This can be confusing. Notice the total size of the substituents attached to the double bond is not important, it is the __first point of difference in priority__ that matters. For example, a -CH₂Cl group has a higher priority than CH₂CH₂CH₂CH₂CH₃ because the first point of difference , Cl , has a higher priority than any atom on the first carbon of the larger alkyl group.]*
    - In the case of double and triple bonds, count the atoms participating in the double or triple bond as if they are bonded to an equivalent number of similar atoms.

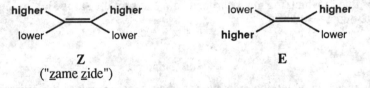

- For example, a -CH=CH₂ group is counted as and a -C≡CH group is counted as . *[The only way to get good at this is to practice.]*
- To assign an alkene as **E** or **Z**, use the following rules:
    If the atoms or groups of atoms of higher priority are on the **same side** of the double bond, it is a **Z** alkene. It is easy to remember this as Z for "zame zide."
    If the higher priority substituents are on **opposite** sides, it is an **E** alkene. *[Caution! Because the letter Z has a zig-zag shape, some students find it tempting to assume that the Z stands for higher priority groups on opposite sides of the alkene making a zig-zag shape around the double bond. This is not the way to assign structures, because E is used for higher priority groups on opposite sides of alkenes.]*

higher ⟍ ╱ higher   lower ⟍ ╱ higher
lower ╱ ⟍ lower   higher ╱ ⟍ lower

Z   **E**
("zame zide")

For alkenes with more than one double bond, each double bond is named as E or Z as applicable.
- Only cycloalkenes with 8 or more carbon atoms in the ring can have a *trans* geometry, otherwise the angle strain will only allow for *cis* double bonds.
- For alkenes with **n** double bonds, there are up to 2^n possible *cis-trans* isomers. There will be fewer if the molecule contains any symmetry.

5.3 PHYSICAL PROPERTIES OF ALKENES:
- Because alkenes are nonpolar, the only interactions between alkene molecules are dispersion forces, thus their properties are very similar to those of alkanes.

5.4 NATURALLY OCCURRING ALKENES-TERPENE HYDROCARBONS
- Alkenes are common in nature, and comprise a very important set of biological molecules. **Terpenes** are one group of biological alkene molecules that have some interesting features. **Terpenes** are based on carbon skeletons that can be divided into two or more units that have the same carbon skeleton of **isoprene**.

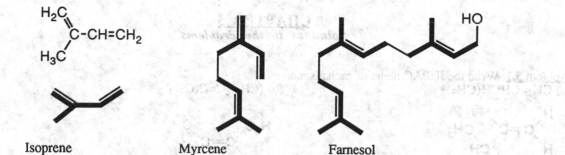

Isoprene   Myrcene   Farnesol
(The isoprene units of these terpenes are shown in bold)

• In nature, terpenes are not synthesized from isoprene, but from the pyrophosphate ester of 3-methyl-3-butene-1-ol.

CHAPTER 5
Solutions to the Problems

<u>Problem 5.1</u> Write the IUPAC name of each alkane.
(a) $CH_2=CHCH(CH_3)_2$ (b) $(CH_3)_2C=C(CH_3)_2$

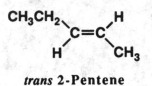

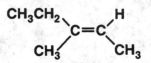

3-Methyl-1-butene   **2,3-Dimethyl-2-butene**

<u>Problem 5.2</u> Which alkenes show *cis-trans* isomerism? For each alkene that does, draw the *trans* isomer.
(a) 2-Pentene   (b) 2-Methyl-2-pentene   (c) 3-Methyl-2-pentene

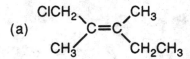

No *cis-trans* isomers since there are two methyl groups on one end of the double bond.

trans **2-Pentene**   ***trans*-3-Methyl-2-pentene**

<u>Problem 5.3</u> Name each alkene and specify its configuration by the E-Z system.

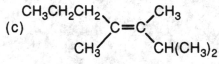

(a)   (b)

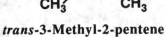

(E)-1-Chloro-2,3-dimethyl-2-pentene   **(Z)-1-Bromo-1-chloropropene**

(c)

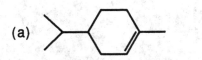

(E)-2,3,4-Trimethyl-3-heptene

<u>Problem 5.4</u> Write the IUPAC name of each cycloalkene

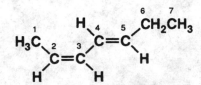
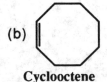
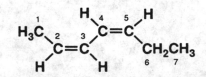

4-Isopropyl-1-methyl-cyclohexene   **Cyclooctene**   **4-*tert*-Butylcyclohexene**

<u>Problem 5.5</u> Draw structural formulas for the other two stereoisomers for 2,4-heptadiene.

***cis,trans*-2,4-Heptadiene**   ***cis,cis*-2,4-Heptadiene**

<u>Problem 5.6</u> The sex pheromone from the silkworm is (10E, 12Z)-10,12-hexadecadiene-1-ol. Draw a structural formula for this compound.

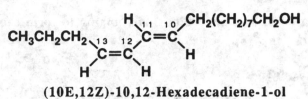

(10E,12Z)-10,12-Hexadecadiene-1-ol

Structure of Alkenes
<u>Problem 5.7</u> Predict all bond angles about each highlighted carbon atom. To make these predictions, use the valence shell electron-pair repulsion model (Section 1.4).

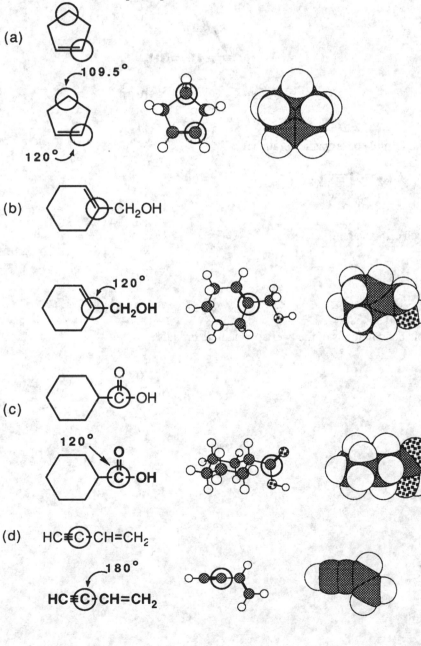

Problem 5.8 For each highlighted carbon atom in Problem 5.7, identify which atomic orbitals are used to form each sigma bond and which are used to form each pi bond.

Each bond is labeled sigma or pi and the orbitals overlapping to form each bond are shown.

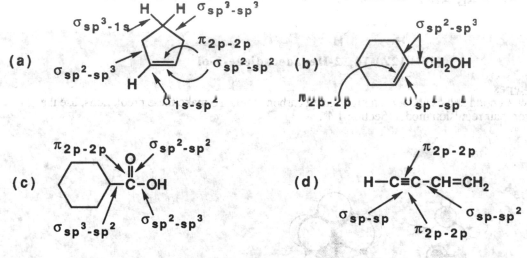

(a)

(b)

(c)

(d)

Problem 5.9 Following is the structural formula of propadiene (allene)

$$CH_2{=\!\!=}C{=\!\!=}CH_2$$

Propadiene
(Allene)

(a) State the orbital hybridization of each carbon atom.

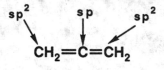

(b) Describe each carbon-carbon double bond in terms of the overlap of atomic orbitals.

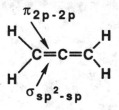

(c) Predict all bond angles in allene.

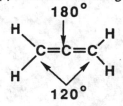

(d) Draw a stereorepresentation showing the shape of this molecule.

The central carbon atom of allene is sp hybridized with bond angles of 180° about it. The terminal carbons are sp^2 hybridized with bond angles of 120° about each. The planes created by H-C-H bonds at the ends of the molecule are perpendicular to each other.

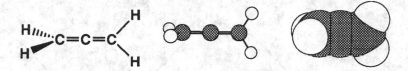

Problem 5.10 Following are lengths for a series of C-C single bonds. Propose an explanation for the differences in bond lengths.

Structure	Length of C-C single bond (nm)
CH_3-CH_3	0.1537
$CH_2=CH-CH_3$	0.1510
$CH_2=CH-CH=CH_2$	0.1465
$HC\equiv C-CH_3$	0.1459

The s electrons are on average held closer to atomic nuclei than p electrons. Thus, hybrid orbitals with higher s character have the electrons held closer to the nucleus and thus make bonds that are shorter. As shown in the table, a σ bond formed from overlap of an sp^3 orbital with an sp^2 orbital is shorter than a σ bond formed from overlap of an sp^3 orbital with another sp^3 orbital. Similarly, sp^3-sp overlap produces a bond that is shorter than that produced by sp^3-sp^2 overlap.

Problem 5.11 The best overlap between two adjacent p orbitals takes place when their axes are parallel. The overlap, and thus the strength of the pi component of a double bond, decreases approximately as $\cos^2 \theta$, where θ is the angle the axes of two p orbitals make with each other. How does the overlap decrease for p orbitals twisted 10°, 20°, 30°, 45°, 60°, and 90°?

Lets give the pi component of a double bond a relative bond strength of 1.00 when the angle of axes of the two p orbitals is the ideal 0°. The following table lists the relative bond strength values observed with the various angles listed.

p Orbital angle	Relative bond strength value
0°	1.00
10°	0.97
20°	0.88
30°	0.75
45°	0.50
60°	0.25
90°	0

<u>Problem 5.12</u> Prepare a plot of potential energy versus angle of rotation about the carbon-carbon double bond in ethylene? How does the energy scale for this plot compare with the energy scale for a plot of potential energy versus angle of rotation about the carbon-carbon single bond in butane?

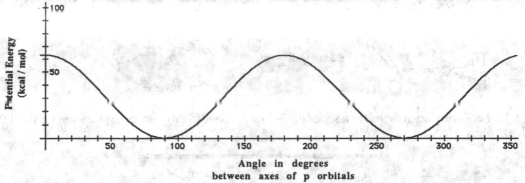

For the carbon 2-carbon 3 bond in butane, the only barrier to rotation is the relatively weak nonbonded interaction strain, so the total potential energy barrier is only ~5 kcal/mol. As shown on the above graph, a carbon-carbon pi bond is worth much more, namely ~63 kcal/mol, because it involves an actual bonding interaction between overlapping 2p orbitals.

Nomenclature of Alkenes
<u>Problem 5.13</u> Draw structural formulas for these alkenes.

(a) *trans*-2-Methyl-3-hexene

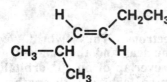

(b) 2-Methyl-2-hexene

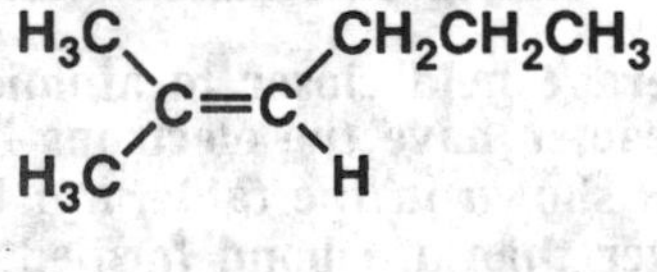

(c) 2-Methyl-1-butene

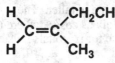

(d) 3-Ethyl-3-methyl-1-pentene

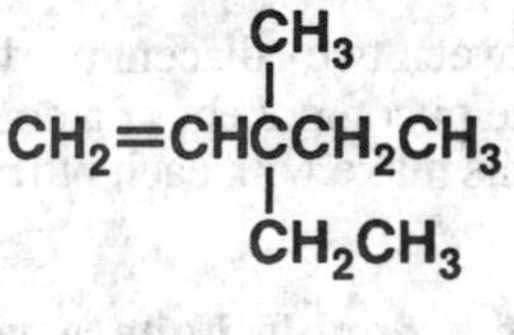

$CH_2=CHCCH_2CH_3$

(e) 2,3-Dimethyl-2-butene

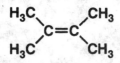

(f) *cis*-2-Pentene

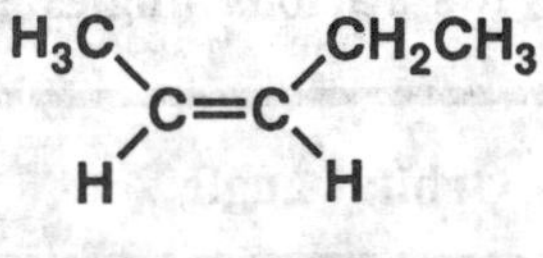

(g) (Z)-1-Chloropropene

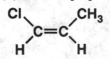

(h) 3-Methylcyclohexene

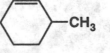

(i) 1-Isopropyl-4-methylcyclohexene

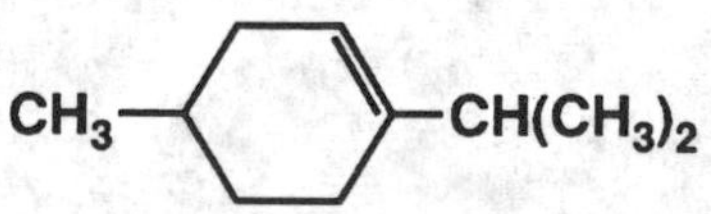

(j) (6E)-2,6-Dimethyl-2,6-octadiene

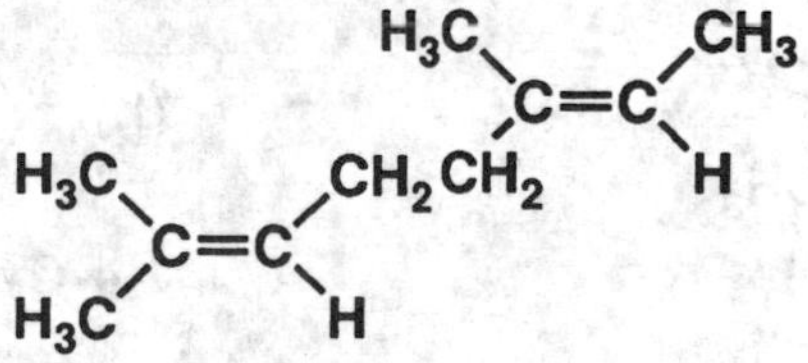

(k) Allylcyclopropane

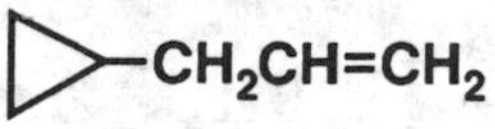

(l) Vinylcyclopropane

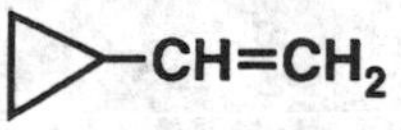

(m) 2-Chloropropene

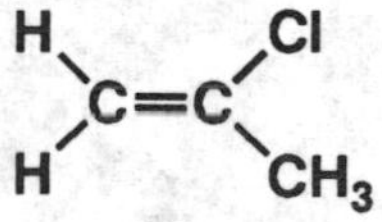

(n) Tetrachloroethylene

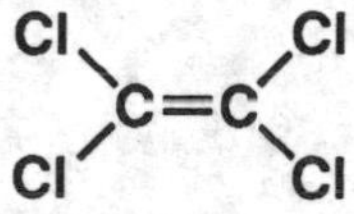

(o) 1-Chlorocyclohexene

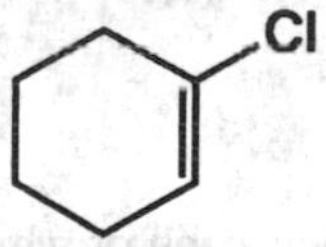

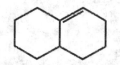

(p) Bicyclo[2.2.1]-2-heptene

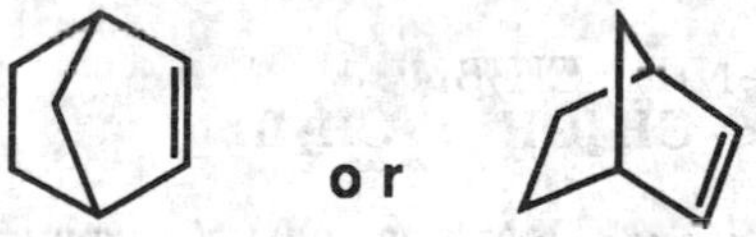

o r

(q) Bicyclo[4.4.0]-1-decene

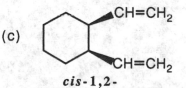

Problem 5.14 Name these alkenes and cycloalkenes.

(a) $CH_2=C$ with $(CH_2)_4CH_3$ and $CH_2CH(CH_3)_2$

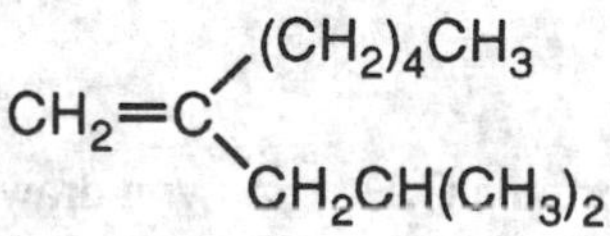

2-Isobutyl-1-heptene

(b)

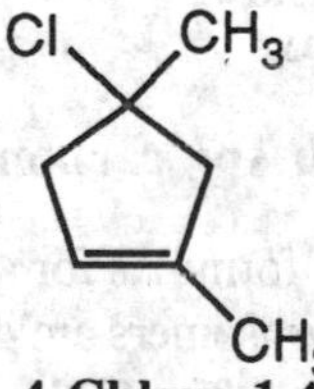

4-Chloro-1,4-dimethylcyclopentene

(c)

cis-1,2-Divinylcyclohexane

(d) $(CH_3)_2CHCH=C(CH_3)_2$

2,4-Dimethyl-2-pentene

(e)

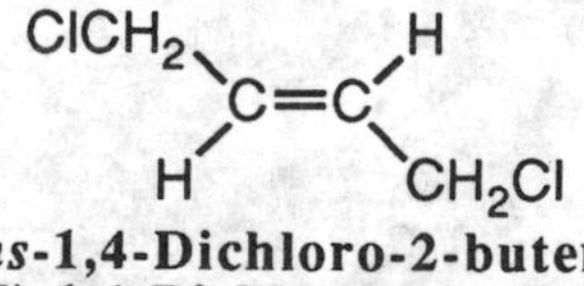

trans-1,4-Dichloro-2-butene
(E)-1,4-Dichloro-2-butene

(f)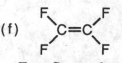

Tetrafluoroethene
Tetrafluoroethylene

(g)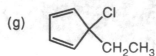

5-Chloro-5-ethyl-
1,3-cyclopentadiene

(h)

1,4-Cyclohexadiene

(i)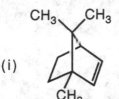

1,7,7-Trimethyl-
bicyclo[2.2.1]-2-heptene

Problem 5.15   Arrange the following groups in order of increasing priority.

(a) $-CH_3$ $-H$ $-Br$ $-CH_2CH_3$ (b) $-OCH_3$ $-CH(CH_3)_2$ $-B(CH_2CH_3)_2$ $-H$

   $-H$ < $-CH_3$ < $-CH_2CH_3$ < $-Br$ $-H$ < $B(CH_2CH_3)_2$ < $-CH(CH_3)_2$ < $-OCH_3$

(c) $-CH_3$ $-CH_2OH$ $-CH_2NH_2$ $-CH_2Br$

   $-CH_3$ < $-CH_2NH_2$ < $-CH_2OH$ < $-CH_2Br$

Problem 5.16   Assign an E-Z configuration and a *cis-trans* configuration to these carboxylic acids, each of which is an intermediate in the tricarboxylic acid cycle. Following each is given its common name.

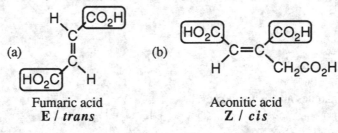

(a)   (b)

Fumaric acid   Aconitic acid
E / *trans*   **Z / *cis***

The highest priority group on each sp² carbon atom is circled.

Problem 5.17   Name and draw structural formulas for all alkenes of molecular formula C_5H_{10}. As you draw these alkenes, remember that *cis* and *trans* isomers are different compounds and must be counted separately in drawing all alkenes possible for this molecular formula.

Four alkenes of molecular formula C_5H_{10} do not show *cis-trans* isomerism.

                                               CH_3
                                               |
$CH_2=CHCH_2CH_2CH_3$ $CH_2=CCH_2CH_3$

           1-Pentene   **2-Methyl-1-butene**

                CH_3 CH_3
                | |
$CH_2=CHCHCH_3$ $CH_3C=CHCH_3$

      3-Methyl-1-butene   **2-Methyl-2-butene**

One alkene of molecular formula C_5H_{10} shows *cis-trans* isomerism.

trans-2-Pentene 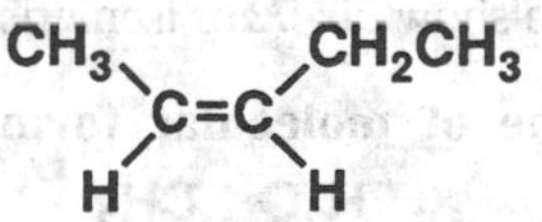   *cis*-2-Pentene

<u>Problem 5.18</u> For each molecule that shows *cis-trans* isomerism, draw the *cis* isomer.

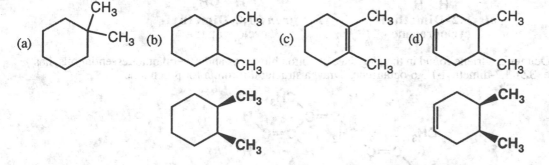

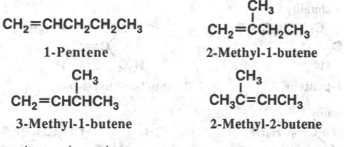

<u>Problem 5.19</u> Draw structural formulas for all compounds of molecular formula C_5H_{10} that are:
(a) Alkenes that do not show *cis-trans* isomerism.

Four alkenes of molecular formula C_5H_{10} do not show *cis-trans* isomerism.

$CH_2{=}CHCH_2CH_2CH_3$

1-Pentene

CH_3
$|$
$CH_2{=}CCH_2CH_3$

2-Methyl-1-butene

CH_3
$|$
$CH_2{=}CHCHCH_3$

3-Methyl-1-butene

CH_3
$|$
$CH_3C{=}CHCH_3$

2-Methyl-2-butene

(b) Alkenes that do show *cis-trans* isomerism.

One alkene of molecular formula C_5H_{10} shows *cis-trans* isomerism.

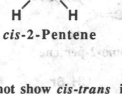

trans-2-Pentene 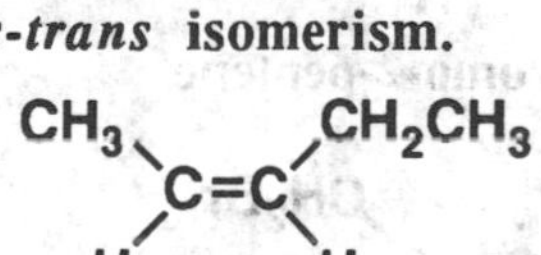   *cis*-2-Pentene

(c) Cycloalkanes that do not show *cis-trans* isomerism.

Four cycloalkanes of molecular formula C_5H_{10} do not show *cis-trans* isomerism.

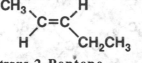

CH_2CH_3 CH_3 CH_3

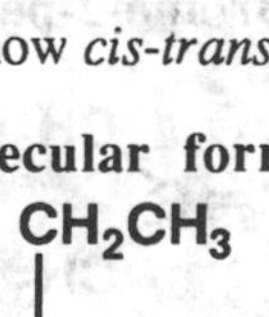

Ethylcyclopropane 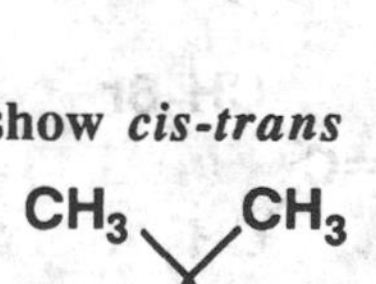   1,1-Dimethylcyclopropane

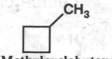

CH_3

Methylcyclobutane   Cyclopentane

(d) Cycloalkanes that do show *cis-trans* isomerism.

Only one cycloalkane of molecular formula C₅H₁₀ shows *cis-trans* isomerism.

cis-**1,2-Dimethyl-**
cyclopropane

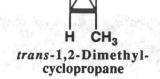

trans-**1,2-Dimethyl-**
cyclopropane

<u>Problem 5.20</u> β-Ocimene, a triene found in the fragrance of cotton blossoms and several other essential oils, has the IUPAC name (3Z)-3,7-dimethyl-1,3,6-octatriene. Draw a structural formula for β-ocimene.

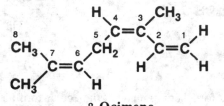

β-Ocimene
(3Z)-3,7-Dimethyl-1,3,6-octatriene

<u>Problem 5.21</u> Draw the structural formula for at least one bromoalkene of molecular formula C₅H₉Br that shows:
(a) Neither E,Z isomerism nor chirality.

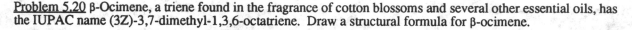

or

5-Bromo-1-pentene   **1-Bromo-3-methyl-2-butene**

(b) E,Z isomerism but not chirality.

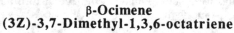

or

(E)-5-Bromo-2-pentene   **(Z)-5-Bromo-2-pentene**

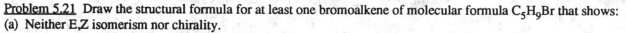

or

(E)-1-Bromo-2-pentene   **(Z)-1-Bromo-2-pentene**

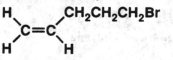

or

(E)-1-Bromo-2-methyl-2-butene   **(Z)-1-Bromo-2-methyl-3-buttene**

(c) Chirality but not E,Z isomerism.

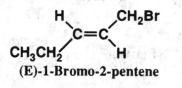

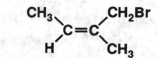

or

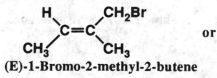

4-Bromo-1-pentene   **3-Bromo-1-pentene**

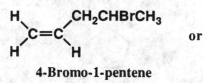

(d) Both chirality and E,Z isomerism.

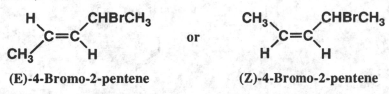

(E)-4-Bromo-2-pentene   or   (Z)-4-Bromo-2-pentene

Problem 5.22 Following are structural formulas and common names for four molecules that contain both a carbon-carbon double bond and another functional group. Give each an IUPAC name.

(a) 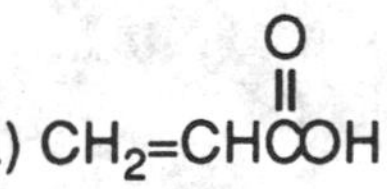 $CH_2=CH\overset{\displaystyle O}{\overset{\displaystyle \|}{C}}OH$

Acrylic acid

2-Propenoic acid

(b) 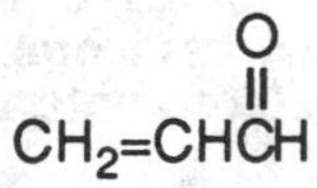 $CH_2=CH\overset{\displaystyle O}{\overset{\displaystyle \|}{C}}H$

Acrolein

2-Propenal

(c)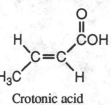

Crotonic acid

(E)-2-Butenoic acid

(d) 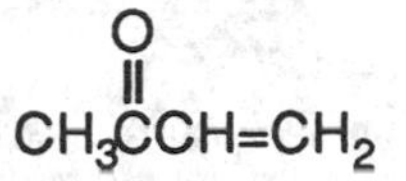 $CH_3\overset{\displaystyle O}{\overset{\displaystyle \|}{C}}CH=CH_2$

Methyl vinyl ketone

3-Buten-2-one

Problem 5.23 *Trans*-cyclooctene has been resolved, and its enantiomers are stable at room temperature. *Trans*-cyclononene has also been resolved, but it racemizes with a half-life of 4 min at 0°C. How can racemization of this cycloalkene take place without breaking any bonds? Why does *trans*-cyclononene racemize under these conditions but not *trans*-cyclooctene? You will find it especially helpful to build molecular models of these cycloalkenes.

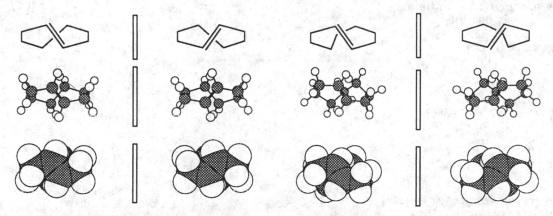

Enantiomers of *trans*-cyclooctene   Enantiomers of *trans*-cyclononene

The enantiomers of *trans*-cyclooctene and *trans*-cyclononene are shown above. It may be helpful to construct molecular models and prove to yourself that the two different configurations of the ring are in fact non-superposable mirror images of each other. For both *trans*-cyclooctene and *trans*-cyclononene, the enantiomers are configurational isomers. The enantiomers are interconverted by a change in configuration that is analogous to the chair flipping of chair cyclohexane. The *trans* double bond adds a considerable degree of rigidity to the ring, since the other atoms must exist in a slightly "stretched" configuration to accommodate the *trans* geometry. Nevertheless, the *trans*-cyclononene ring has more carbon atoms, so it is more flexible and can undergo the configurational interconversion more readily.

Problem 5.24 How many stereoisomers are possible for the following natural products?
(a) Geraniol (Figure 5.5)   (b) Limonene (Figure 5.5)

CH₂OH

← about this
double bond

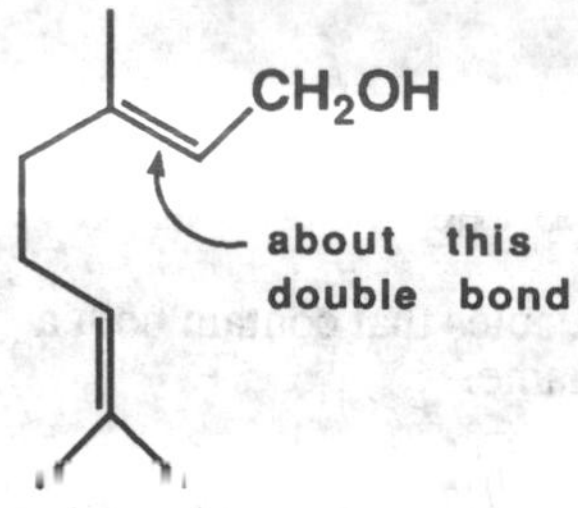

Geraniol: 2 stereoisomers
One pair of *cis-trans* isomers;
(the *trans* isomer is shown)
no stereocenters

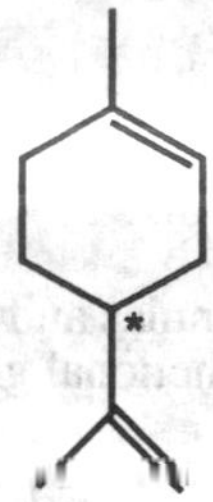

Limonene: 2 stereoisomers
No *cis-trans* isomers,
one stereocenter.

(c) α-Pinene (Figure 5.5)   (d) Farnesol (Figure 5.6)

CH₂OH

← about these
double bonds

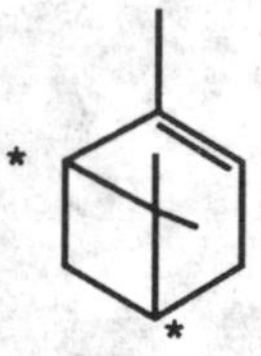

α-Pinene: 2 stereoisomers
No *cis-trans* isomers
two stereocenters, but
because of ring constraints
only one pair of enantiomers
is possible.

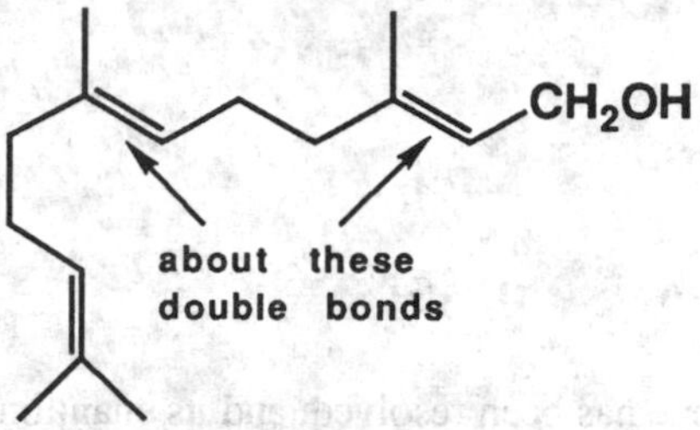

Farnesol: 4 stereoisomers
Four *cis-trans* isomers; (the
***trans-trans* isomer is shown)**
and no stereocenters

(e) Zingiberene (Figure 5.6)

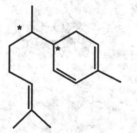

Zingiberene: 4 stereoisomers
No *cis-trans* isomers, but
two stereocenters.

Problem 5.25 Which alkenes exist as pairs of *cis-trans* isomers? For each alkene that does, draw the *trans* isomer.

For an alkene to exist as a pair of *cis-trans* isomers, both carbon atoms of the double bond must have two different substituents. Thus, (b), (c), and (e) exist as a pair of *cis-trans* isomers. The *trans* isomer for each alkene is drawn under its respective condensed molecular formula.

(a) CH_2=CHBr   (b) CH_3CH=CHBr   (c) BrCH=CHBr

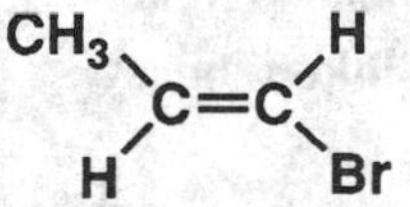

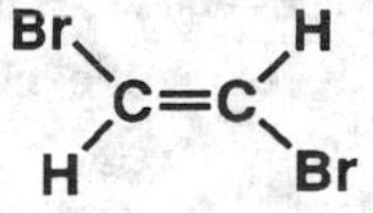

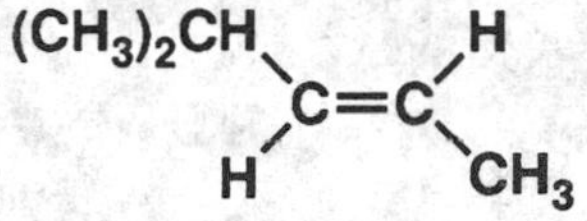

Problem 5.26 Four stereoisomers exist for 3-penten-2-ol.

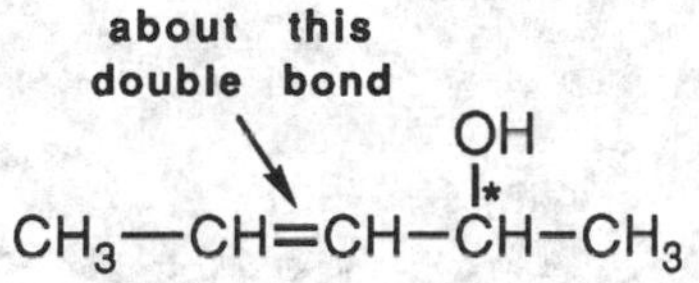

about this
double bond

CH_3—CH=CH—CH—CH_3

3-Penten-2-ol

(a) Explain how these four stereoisomers arise.

There is one double bond that provides for *cis-trans* isomers, and one stereocenter in 3-penten-2-ol.

(b) Draw the stereoisomer having the E configuration about the carbon-carbon double bond and the R configuration at the stereocenter.

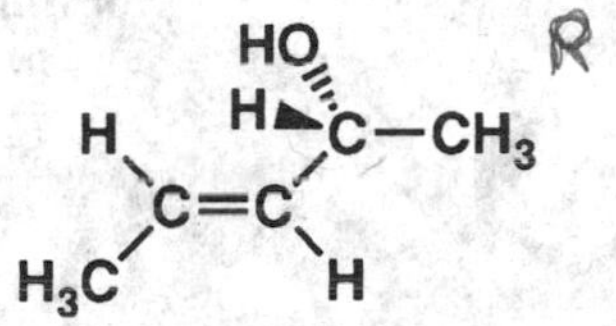

Terpenes

Problem 5.27 Show how the carbon skeleton of farnesol can be coiled and then cross-linked to give the carbon skeleton of zingiberene (Figure 5.6).

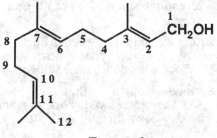

Farnesol

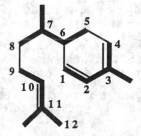

Zingiberene

Problem 5.28 Show that the structural formula of vitamin A (Section 5.2G) can be divided into four isoprene units joined by head-to-tail linkages and cross-linked at one point to form the six-membered ring.

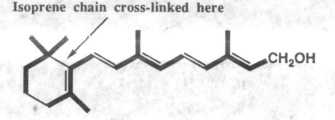

Problem 5.29 Following is the structural formula of lycopene, a deep-red compound that is partially responsible for the red color of ripe fruits, especially tomatoes. Approximately 20 mg of lycopene can be isolated from 1 kg of ripe tomatoes. See The Merck Index, 12th edition, #5650.

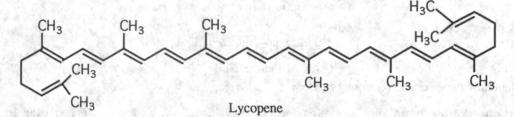

(a) Show that lycopene is a terpene, that is, its carbon skeleton can be divided into two sets of four isoprene units with the units in each set joined head-to-tail.

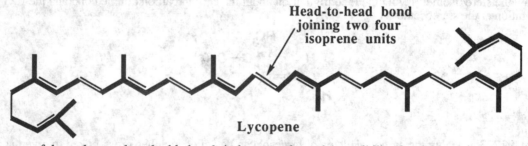

(b) How many of the carbon-carbon double bonds in lycopene have the possibility for *cis,trans* isomerism? Lycopene is the all-*trans* isomer.

The double bonds on the two ends of the molecule cannot show *cis-trans* isomerism. The other 11 double bonds can show *cis-trans* isomerism.

Problem 5.30 The structural formula of β-carotene, precursor to vitamin A, is given in Section 25.6A. As you might suspect, it was first isolated from carrots. Dilute solutions of β-carotene are yellow, hence its use as a food coloring. Compare the carbon skeletons of β-carotene and lycopene. What are the similarities? What are the differences?

The main structural difference between β-carotene and lycopene is that β-carotene has six-membered rings on the ends, not an open chain. On the other hand, both β-carotene and lycopene can be divided into two sets of four isoprene units as shown below, and all of the double bonds are *trans* in both molecules.

Isoprene chain cross-linked at these two points

Head-to-head bond joining two four isoprene units

Problem 5.31 Following is the structural formula of warburganal, a crystalline solid isolated from the plant *warburgia ugandensis, Canellaceae*. An important use of warburganal is its antifeeding activity against the African army worm. In addition, it acts as a plant growth regulator and has cytotoxic, antimicrobial and molluscicidal properties. See The Merck Index, 12 edition, # 10173.

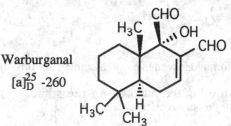

Warburganal
$[a]_D^{25}$ -260

(a) Show that warburganal is a terpene.

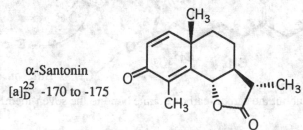

Warburganal

(b) Label each stereocenter and specify the number of stereoisomers possible for a molecule of this structure.

There are three stereocenters, so there are 2 x 2 x 2 = 8 stereoisomers possible for warburgnal.

Problem 5.32 α-Santonin, $C_{15}H_{18}O_3$, isolated from the flower heads of certain species of Artemisia, is an anthelmintic, that is, a drug used to rid the body of worms (helminths). It has been estimated that over one third of the world's population is infested with these parasites. α-Santonin in oral doses of 60 mg is used as an anthelmintic for roundworms (*Ascaris lumbricoides*). See The Merck Index, 12 edition, #8509.

α-Santonin
$[a]_D^{25}$ -170 to -175

(a) Locate the three isoprene units in santonin and show how the carbon skeleton of farnesol might be coiled and then cross-linked to give α-santonin. Two different coiling patterns of the carbon skeleton of farnesol can lead to α-santonin. Try to find them both.

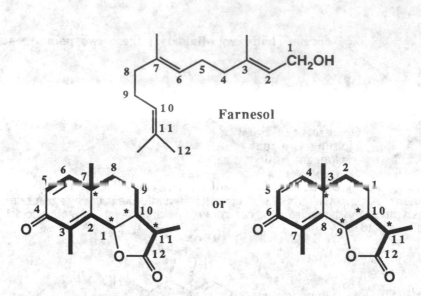

Farnesol

or

(b) Label all stereocenters in α-santonin. How many stereoisomers are possible for this molecule?

The four stereocenters of α-santonin are marked on the structures above. There are 2 x 2 x 2 x 2 = 16 stereoisomers possible for α-santonin.

Problem 5.33 In many parts of South America, extracts of the leaves and twigs of *Montanoa tomentosa* are brewed with water to make a "tea" used to stimulate menstruation, to facilitate labor, and as an abortifacient. Phytochemical investigations of this plant have resulted in isolation of a very potent fertility-regulating compound called zoapatanol. See The Merck Index, 12 edition, #10318.

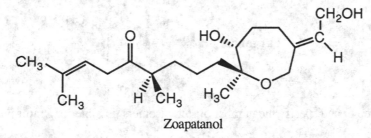

Zoapatanol

(a) Show that the carbon skeleton of zoapatanol can be divided into four isoprene units bonded head-to-tail and then cross-linked in one point along the chain.

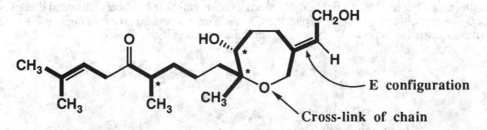

E configuration

Cross-link of chain

(b) Specify the configuration about the carbon-carbon double bond to the seven-membered ring according to the E,Z system.

The double bond in question has the E configuration, because the hydroxymethyl group is on the side of the double bond opposite the higher priority carbon atom that is linked to the ether oxygen.

(c) How many stereoisomers are possible for this molecule? In answering this problem, you must consider both E,Z isomerism and R,S isomerism.

There is just the one double bond capable of E,Z isomerism and four stereocenters as shown on the above structure. Thus, there are a total of 2 x 2 x 2 x 2 x 2 = 32 stereoisomers possible.

Problem 5.34 Pyrethrin II and pyrethrosin are two natural products isolated from plants of the chrysanthemum family. Pyrethrin II is a natural insecticide and is marketed as such. (a) Label all stereocenters in each molecule and all carbon-carbon double bonds about which there is the possibility for *cis,trans* isomerism. See The Merck Index, 12 edition, #s 8148 and 8149.

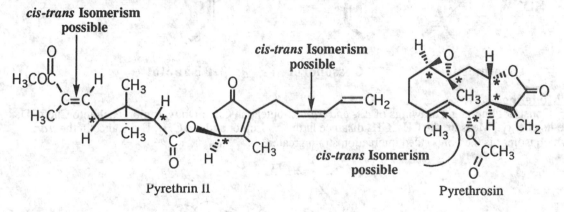

Pyrethrin II   Pyrethrosin

(a) State the number of stereoisomers possible for each molecule.

For pyrethin II there are two double bonds capable of *cis-trans* isomerism and three stereocenters for a total of 2 x 2 x 2 x 2 x 2 = 32 possible stereoisomers. For pyrethrosin there is one double bond capable of *cis-trans* isomerism and five stereocenters for a total of 2 x 2 x 2 x 2 x 2 x 2 = 64 possible stereoisomers.

(b) Show that the bicyclic ring system of pyrethrosin is composed of three isoprene units.

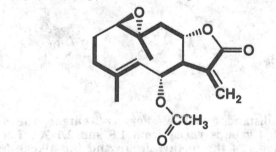

<u>Problem 5.35</u> Show that the carbon skeletons of the three terpenes drawn in the Chemistry in Action box <u>Terpenoids of the Cotton Plant</u> can be divided into three isoprene units bonded head-to-tail and then cross-linked at appropriate carbons.

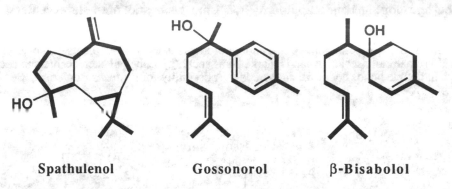

Spathulenol   **Gossonorol**   **β-Bisabolol**

Molecular Modeling

<u>Problem 5.36</u> Construct line-angle drawings of *cis*- and *trans*-2-butene in ChemDraw, import each into Chem3D, and minimize its energy. Measure the CH_3, CH_3 distance in the *cis* isomer and the CH_3, H distance in the *trans* isomer. In which isomer is the nonbonded interaction strain greater?

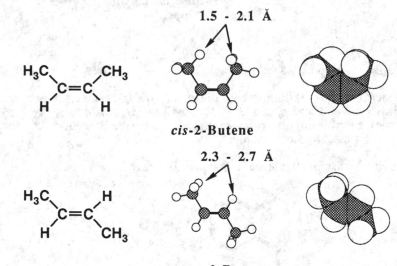

cis-**2-Butene**

trans-**2-Butene**

As the methyl groups rotate, the distances change. For *cis*-2-butene, the distance between the hydrogen atoms on adjacent methyl groups varies from 1.5 and 2.1 Å. For *trans*-2-butene, the distance between the hydrogen atoms of the methyl group and the alkene hydrogen atom varies from 2.3 and 2.7 Å. Clearly, there is greater nonbonded interaction strain in *cis*-2-butene.

<u>Problem 5.37</u> Construct line-angle drawings of the *cis*- and *trans* isomers of 2,2,5,5-tetramethyl-3-hexene in ChemDraw, import each into Chem3D, and minimize its energy. Compare the degree of nonbonded interaction strain in these two isomers. Also compare the degree of nonbonded interaction strain in *cis*-2-butene and *cis*-2,2,5,5-tetramethyl-3-hexene.

cis-**2,2-5,5-Tetramethyl-3-hexene**

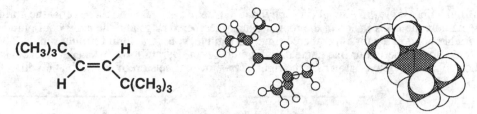

trans-2,2-5,5-Tetramethyl-3-hexene

Again the *trans* isomer has significantly less nonbonded interaction strain because the methyl groups are farther apart. Also, the additional methyl groups of *cis*-2,2,5,5-tetramethyl-3-hexene add significant nonbonded interaction strain compared to *cis*-2-butene.

<u>Problem 5.38</u> Build a line-angle drawing of cyclohexene in ChemDraw, import it into Chem3D, and minimize its energy. Compare the calculated Chem3D bond angles with those predicted by the VSEPR model. Explain any differences.

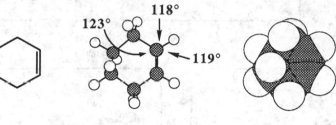

Cyclohexene

As can be seen on the above structures, the angles predicted by Chem3D are not exactly 120° as predicted by the VSEPR model for an sp^2 hybridized carbon atom. This difference is the result of strain introduced by the ring of cyclohexene.

<u>Problem 5.39</u> Build line-angle drawings of *cis*- and *trans*-cyclooctene in ChemDraw, import each into Chem3D, and minimize its energy. Compare the calculated Chem3D bond angles with those predicted by the VSEPR model. In which isomer are deviations from the VSEPR model predictions greater?

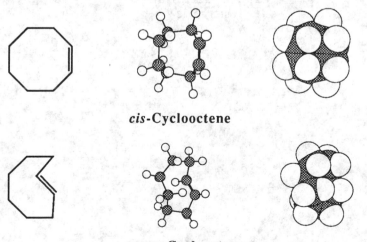

cis-Cyclooctene

trans-Cyclooctene

Although they are similar, the angles deviate slightly more in the *trans* isomer compared to the *cis* isomer.

Problem 5.40 Build a line-angle drawing of caryophyllene in ChemDraw. Be certain to show the correct stereochemistry, namely the *trans* fusion of the four and nine-membered rings and the *trans* configuration of the carbon-carbon double bond in the nine-membered ring. As a guide, your structural formula should look like the formula on the left. Now import your line-angle drawing into Chem3D, minimize its energy, and construct a stereoview. One such stereoview is shown on the right. Try showing the stereoview with and without hydrogen atoms.

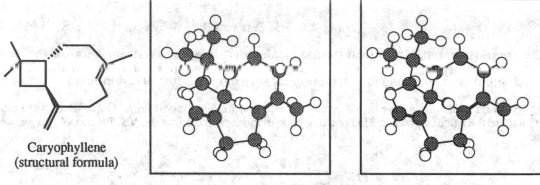

Caryophyllene
(structural formula)

Caryophyllene
(Stereoview)

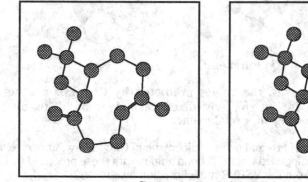

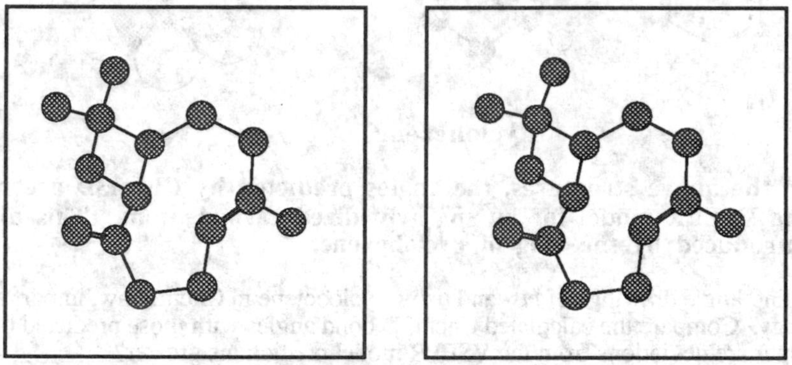

Caryophyllene (No H's)
(Stereoview)

CHAPTER 6: ALKENES II

SUMMARY OF REACTIONS

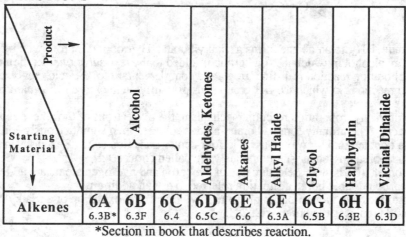

	Alcohol			Aldehydes, Ketones	Alkanes	Alkyl Halide	Glycol	Halohydrin	Vicinal Dihalide
Alkenes	**6A** 6.3B*	**6B** 6.3F	**6C** 6.4	**6D** 6.5C	**6E** 6.6	**6F** 6.3A	**6G** 6.5B	**6H** 6.3E	**6I** 6.3D

*Section in book that describes reaction.

REACTION 6A: ACID-CATALYZED HYDRATION (Section 6.3B)

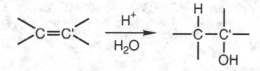

- In the presence of an acid catalyst like sulfuric acid, water adds to alkenes to give alcohols.
- The reaction mechanism involves formation of a carbocation intermediate from protonation of the pi bond, followed by nucleophilic attack on the resulting electrophilic carbocation by water.
- The water can attack from either side of the trigonal planar carbocation so the reaction is not stereoselective.
- This reaction is **regioselective**, that is one constitutional isomer is produced in preference to other possible constitutional isomers.
- **Markovnikov's rule** is followed so the hydrogen ends up on the carbon atom that already has more hydrogen atoms attached. This is because the water attacks the more stable carbocation, namely the one that is more highly substituted.

REACTION 6B: OXYMERCURATION / REDUCTION (Section 6.3F)

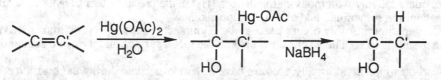

- Hydration of alkenes can be accomplished by treatment of an alkene with mercury(II) acetate in water followed by reduction with $NaBH_4$.
- The reaction mechanism involves formation of a **mercurinium ion intermediate** that is then attacked by water. The reaction is completed by adding $NaBH_4$ that results in displacement of the mercury atom by hydride ion.
- Oxymercuration is both regioselective and stereoselective because **Markovnikov's rule is followed** and **anti addition predominates**. This means the mercurinium ion intermediate has a partially bridged structure that prevents attack of the nucleophilic water from the side with the mercury ion. There is some carbocation character to the intermediate because the carbon atom that forms the more stable carbocation is preferentially attacked by the water. Note that the intermediate is not a full blown carbocation since it does not rearrange.

REACTION 6C: HYDROBORATION (Section 6.4)

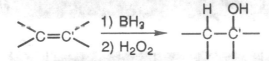

- Addition of borane, BH_3, to an alkene forms a trialkylborane. Hydroboration followed by treatment with peroxide gives an alcohol in which the -OH group is added to the *less-substituted* carbon of the alkene. Thus, the hydroboration reaction is distinct from, and complementary to, the acid-catalyzed hydration and oxymercuration reactions in which the OH group adds primarily on the more substituted carbon in agreement with Markovnikov's rule.
- In the first step of the hydroboration reaction mechanism, the pi electrons of the alkene react with the boron atom, a Lewis acid. The initially formed borane-alkene complex then simultaneously adds boron and hydrogen to the double bond by way of a cyclic four-center transition state. Reactions in which bond-forming and bond-breaking occur simultaneously are called **concerted**.
- Note that addition of BH_3 is stereoselective in that the boron and hydrogen atoms are added to the same side of the carbon-carbon double bond, a situation referred to as **syn** addition.
- The reaction continues two more times to give a trialkylborane, and oxidation of a trialkylborane with hydrogen peroxide replaces the boron with OH.

REACTION 6D: OZONOLYSIS (Section 6.5C)

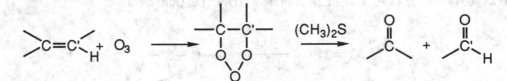

- Ozone reacts with an alkene to form an ozonide intermediate that can be cleaved by the addition of $(CH_3)_2S$ into the two carbonyl species, namely ketones and aldehydes.
- Ozonolysis can produce an aldehyde product if there was a hydrogen on one of the original alkene sp^2 carbon atoms.

REACTION 6E: CATALYTIC HYDROGENATION (Section 6.6)

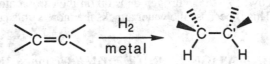

- Alkenes react quantitatively with molecular hydrogen (H_2) in the presence of a transition metal catalyst (platinum, palladium, ruthenium, and nickel) to give alkanes.
- Catalytic hydrogenation probably occurs because the H_2 molecule splits apart and makes two metal-hydrogen bonds on the metal surface. The alkene adsorbs onto the metal surface and then two sequential C-H bonds are made.
- The carbon-carbon sigma bond usually does not have a chance to rotate during the reaction so both hydrogen atoms are added to the same face of the alkene. This is referred to as **syn addition**, which leads to products with *cis* stereochemistry.

REACTION 6F: HYDROHALOGENATION (Section 6.3A)

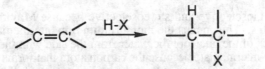

-HF, HCl, HBr, and HI can add to alkenes to give alkyl halides.

- Like reaction 6A, protonation of the pi bond results in formation of a carbocation intermediate, then the halide anion is the nucleophile that reacts with the carbocation.
- The reaction follows Markovnikov's rule, so the hydrogen ends up on the carbon atom that has the greater number of hydrogens already attached to it.
- Because there is no bridged intermediate, the halide anion can attack from either side of the trigonal planar (sp^2 hybridized) carbocation. Therefore, addition of hydrogen halides to alkenes is not a stereoselective reaction.

REACTION 6G: OXIDATION TO GLYCOLS (Section 6.5B)

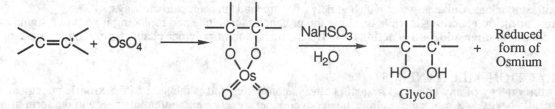

- In these reactions, a cyclic intermediate is formed from the alkene and then treated with a reducing agent, such as NaHSO$_3$ to yield a **glycol** (two OH groups on adjacent carbon atoms).
- Note how the cyclic intermediate ensures that both oxygen atoms are added to the same face of the alkene (syn addition).
- The OsO$_4$ reagent can also be used catalytically with the addition of other oxidizing reagents such as hydroperoxides (ROOH).

REACTION 6H: HALOHYDRIN FORMATION (Section 6.3E)

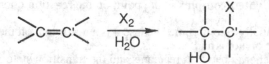

- Treatment of an alkene with Br$_2$ or Cl$_2$ in the presence of water results in addition of HO- and Br-, or HO- and Cl- to the alkene. The resulting compounds are called **halohydrins**, either **bromohydrins** or **chlorohydrins**.
- The reaction involves initial formation of a bridged bromonium or chloronium ion intermediate, followed by nucleophilic attack of water.
- The bridged halonium intermediate is analogous to the mercurinium ion intermediate, displaying both bridged and partial carbocation characteristics. As a result, the nucleophilic attack of water occurs on the side of the intermediate opposite the halogen, and at the site of the more stable carbocation. Consistent with this, **anti stereochemistry** is observed, and the HO- ends up on the carbon atom that is more highly substituted (makes the more stable carbocation).

REACTION 6I: FORMATION OF VICINAL DIHALIDES (Section 6.3D)

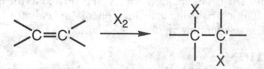

- **Bromination** and **chlorination** involve the addition of Br$_2$ and Cl$_2$ to an alkene, respectively.
- In these reactions, one of the halogen atoms acts as an electrophile, breaking the halogen-halogen bond. This creates a positively-charged intermediate and a halide anion. The positively-charged intermediate has a unique bridged structure and is referred to as a bridged halonium ion. The halide anion then completes the reaction by creating a bond to the positively-charged species from the side of the molecule opposite the halogen bridge.
- The halogen bridge blocks the top of the structure, so the halide anion *must* attack from the side opposite the bridging group. The net result is that the two halogens end up on opposite faces of the molecule. This **stereochemical orientation** is referred to as **anti addition**.

SUMMARY OF IMPORTANT CONCEPTS

6.0 OVERVIEW
- **Reaction mechanisms** describe how chemical bonds are formed and broken during the course of a reaction, the order in which the bonds are broken and formed, the rates at which these processes occurs, and the role of solvent or a catalyst if any. Mechanisms provide a theoretical framework within which to organize a great deal of descriptive chemistry.

6.1 REACTIONS OF ALKENES: AN OVERVIEW
- In contrast to alkanes, alkenes react with a variety of compounds in characteristic ways:
  - First, **addition reactions** involve breaking the pi bond of an alkene and replacing it with two sigma bonds.
  - Second, **polymer addition reactions** involve the formation of polymer chains from monomer alkene molecules.

6.2 REACTION MECHANISMS
- The **total energy** of any chemical system is always conserved, and is the sum of the **kinetic energy** and **potential energy**. As molecules collide they convert kinetic energy into potential energy in the form of bond vibrational energy.
- **Exothermic reactions** are ones in which the products are of lower potential energy than the reactants. Exothermic reactions release heat corresponding to this difference in energy, which is referred to as the **heat of reaction**. **Endothermic reactions** are ones in which the products are of **higher potential energy** than the reactants.
- A **reaction coordinate** is a plot of the position of atoms associated with changing energy as reactants proceed to products during a reaction.
- For simple **one step reactions**, reaction occurs if sufficient potential energy becomes concentrated in the proper bonds.
  - The **transition state** or **activated complex** is the point on the reaction coordinate where the potential energy is a maximum.
  - An transition state has essentially zero lifetime because it is a maximum on the energy diagram, yet it does have a definite arrangement of atoms and electrons.
  - The difference in potential energy between reactants and the transition state is called the **activation energy, E_a**. A molecule must have more potential energy than the activation energy to proceed from starting materials to products.
- In multi-step reactions, each step has its own transition state and activation energy.
  - An **intermediate** is a potential energy minimum between two transition states on a reaction coordinate for a multi-step reaction. Reactive intermediates are rarely present in appreciable concentrations because the activation energy for their conversion back to reactants or on to products is so small.
  - The slowest step is the one that crosses the highest potential energy barrier and is called the **rate-limiting step**. *[This is a very important concept in the study of reaction mechanisms. Notice that the overall rate of a multi-step reaction cannot be faster than the rate-limiting step.]*
  - The relationship between the **rate constant, k,** and the activation energy for most chemical reactions is given by the **Arrhenius equation**.

$$k = \text{reaction rate constant} = Ae^{-E_a/RT}$$

In the Arrhenius equation, A = the frequency factor in sec^{-1} and is related to the probability that there will be a collision with successful orientations, E_a is the activation energy in $kcal \cdot mol^{-1}$, R is the gas constant ($1.987 \times 10^{-3}\ kcal \cdot mol^{-1} \cdot deg^{-1}$) and T is temperature in degrees Kelvin.

6.3 ELECTROPHILIC ADDITIONS
- The details of **alkene addition reactions** can best be understood by considering the mechanism of the reaction as well as the structure of the alkene.
  - The electrons of the alkene pi bond are located relatively far from the atomic nuclei, so they can act as a type of nucleophile with extremely electron deficient chemical species, referred to as **electrophiles**. *[This is the key idea of the chapter, and all of the following reactions should be thought of as starting with the weakly nucleophilic pi electrons attacking an electrophilic species.]*
  - When the pi electrons react with an electrophile, the pi bond is broken and a new sigma bond is formed with the electrophile. This creates a positively charged intermediate that is itself attacked by a nucleophile to form another new sigma bond, thereby completing the reaction.

The key to understanding the details of these reactions is to keep track of the electrophile, the nucleophile, and the structure of the positively charged intermediate.

- The bottom line is that the pi bond is replaced by two new sigma bonds, one to an electrophile and one to a nucleophile.
• **Markovnikov's rule** can be used to understand the regiochemistry of addition reactions such as the addition of H-X to alkenes. **Markovnikov's rule** states that in the addition of H-X the **H atom goes onto the atom that already has the greater number of hydrogen atoms.** The mechanistic basis of Markovnikov's rule can be understood by considering the structure of the carbocation intermediate formed during the addition reaction.

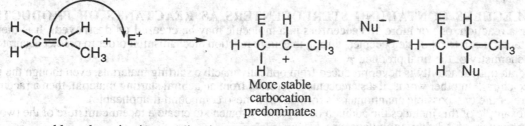

More stable
carbocation
predominates

- The more stable carbocation intermediate is produced predominantly, and this more stable cation intermediate is the predominant (Markovnikov) product. The more stable carbocation has more alkyl groups attached to the positively charged carbon atom.
  The alkyl groups stabilize an adjacent carbocation through two effects. **The inductive effect** involves polarization of the electron density of adjacent sigma bonds by the electron-withdrawing cationic carbon. **Hyperconjugation** involves donation of electron density from adjacent sigma bonds due to overlap with the empty $2p$ orbital of the cationic carbon.
- The bottom line is that a tertiary (3°) carbocation is more stable than a secondary (2°) carbocation, which is more stable than a primary (1°) carbocation, which is more stable than a methyl cation.
• A characteristic of carbocations is that they can **rearrange** if transfer of a **hydride ion** (a hydrogen atom plus the two bonding electrons) or alkyl group can create a carbocation or equal or greater stability.

A 2° carbocation can rearrange to a 3° carbocation

6.5 OXIDATION OF ALKENES
• **Oxidation/reduction reactions** are a very important class of reactions in organic chemistry in which electrons are lost or gained by a reactant during the course of a reaction. Oxidation/reduction reactions can be recognized by writing **balanced half-reactions**.
  - To write a balanced half-reaction, first write a half-reaction showing the organic reactants and products. Complete a material balance using H^+ and H_2O for reactions carried out in acid, or OH^- and H_2O for reactions carried out in base. Finally complete the charge balance by adding electrons on one side or the other.
• An **oxidation** is defined as a reaction in which electrons are lost from a reactant being transformed into products.
• A **reduction** is defined as a reaction in which electrons are gained by a reactant being transformed into products.

6.6 ADDITION OF HYDROGEN-CATALYTIC REDUCTION

• **Heat of hydrogenation** for an alkene is defined as the change in enthalpy, $\Delta H°$, for the reaction between an alkene and hydrogen to form an alkane.

  - Heats of hydrogenation are negative. In other words reduction of an alkene to an alkane is an exothermic process, because the reaction involves the breaking of a C-C pi bond (pi bonds are relatively weak) and a sigma bond (the H-H bond) to from two stronger sigma (C-H) bonds.

  - An alkene with a lower heat of hydrogenation (less exothermic) is the more stable alkene. This makes sense because you expect to get less energy out of a molecule that is already more stable.

  - From the comparison of heats of hydrogenation of a variety of molecules, the following general conclusions can be reached:

    More highly substituted alkenes are more stable than less highly substituted alkenes. ✳

    Trans alkenes are more stable than *cis* alkenes. This is a **steric affect** in that the *cis* substituents are so close that there is a net repulsion between their electron clouds. ✳

    Conjugated dienes, those in which pi bonds are adjacent to each other, are more stable than dienes that are not conjugated. The stability of conjugated dienes results from overlap of the pi orbitals of the adjacent double bonds. ✳

6.7 MOLECULES CONTAINING STEREOCENTERS AS REACTANTS OR PRODUCTS

• During a reaction, one or more **stereocenters** in a molecule may be **created or destroyed**. It is therefore important to keep track of stereocenters during the entire reaction mechanism in order to predict accurately the stereochemistry of the final product. ✳

• In general, optical activity is never produced from optically inactive starting materials, even though the products may be chiral. In other words, if a stereocenter is created from an achiral starting material, then a racemic mixture of the two possible enantiomers is formed (or a meso compound if applicable).

  - An example of this includes the addition of Br_2 to *cis*-2-butene to create a racemic mixture of the two enantiomers of 2,3-dibromobutane.

• Alternatively, optical activity is generated in a reaction only if at least one of the reactants itself is chiral, or if the reaction is carried out in the presence of a catalyst that is itself chiral.

CHAPTER 6
Solutions to the Problems

<u>Problem 6.1</u> Suppose that the activation energy for a particular chemical reaction is 25.2 kcal/mol. By what factor is the rate of reaction increased when the reaction takes place at 35°C compared with the rate at 25°C?

$$\frac{k_2}{k_1} = \frac{A\,e^{-Ea/RT_2}}{A\,e^{-Ea/RT_1}}$$

Taking the logarithm of each side, converting to base 10, and rearranging gives:

$$\log\frac{k_2}{k_1} = \frac{E_a}{2.303\,R}\left(\frac{1}{T_1} - \frac{1}{T_2}\right)$$

Plugging in the actual values gives:

$$\log\frac{k_2}{k_1} = \frac{25.2\ \text{(kcal/mol)}}{2.303\,(1.987 \times 10^{-3}\ \text{kcal/mol K})}\left(\frac{1}{298\ \text{K}} - \frac{1}{308\ \text{K}}\right)$$

Solving the equation gives:

$$\log\frac{k_2}{k_1} = (5.507 \times 10^3\ \text{K})(1.09 \times 10^{-4}\ \text{K}^{-1}) = 0.600$$

$$\boxed{\frac{k_2}{k_1} = 10^{(0.600)} = 3.98}$$

<u>Problem 6.2</u> Complete the first three entries in this table for reactions taking place at 25°C. Given the pattern of these first three entries, estimate the approximate values for the remaining two entries. How many kilocalories per mol in activation energy corresponds to a power of 10 in relative rates?

Plugging in the appropriate values gives:

$$\Delta E_a = -2.303\,RT\log\frac{k_2}{k_1} = (-1.36\ \text{kcal/mol})\,(\log\frac{k_2}{k_1})$$

ΔE_a (kcal/mol)	$\dfrac{k_2}{k_1}$
0	1
-1.36	10
-2.72	100
-4.08	1000
-5.44	10000

<u>Problem 6.3</u> Name and draw the structural formula for the product of each alkene addition reaction.

(a) $CH_3-CH=CH_2$ + HI $\longrightarrow$ CH_3CHCH_3

2-Iodopropane

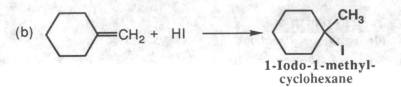

(b)

1-Iodo-1-methyl-
cyclohexane

Problem 6.4 Arrange these carbocations in order of increasing stability.

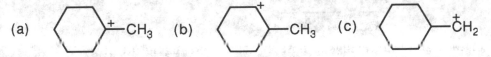

(a) $+$—CH_3 (b)   —CH_3 (c)   —$\overset{+}{C}H_2$

The order of increasing stability of carbocations is methyl < primary < secondary < tertiary.
Thus the three carbocations can be ranked as follows:

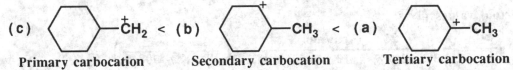

(c)   —$\overset{+}{C}H_2$ < (b)   —CH_3 < (a) $+$—CH_3

Primary carbocation   Secondary carbocation   Tertiary carbocation

Problem 6.5 Propose a mechanism for addition of HI to 1-methylcyclohexene to give 1-iodo-1-
methylcyclohexane. Which step in your mechanism is rate limiting?

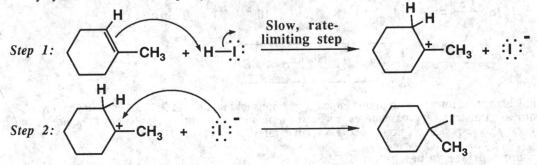

Step 1:

Slow, rate-
limiting step

Step 2:

Problem 6.6 Draw the structural formula for the product of each alkene hydration reaction.

(a) 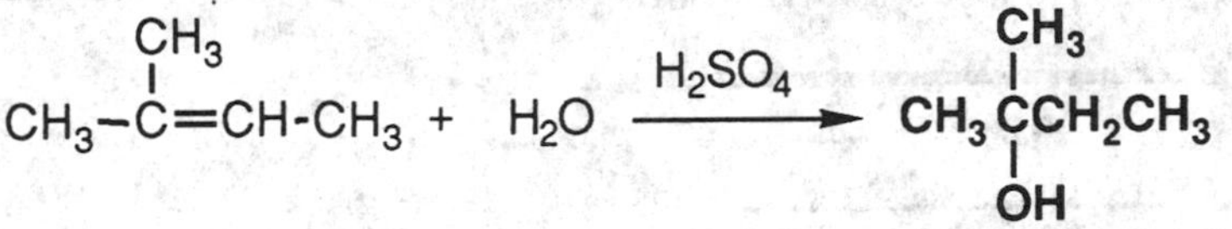 $CH_3-\overset{\overset{\displaystyle CH_3}{|}}{C}=CH-CH_3$ + H_2O $\xrightarrow{H_2SO_4}$ $CH_3\underset{\underset{\displaystyle OH}{|}}{\overset{\overset{\displaystyle CH_3}{|}}{C}}CH_2CH_3$

2-Methyl-2-butanol

(b) 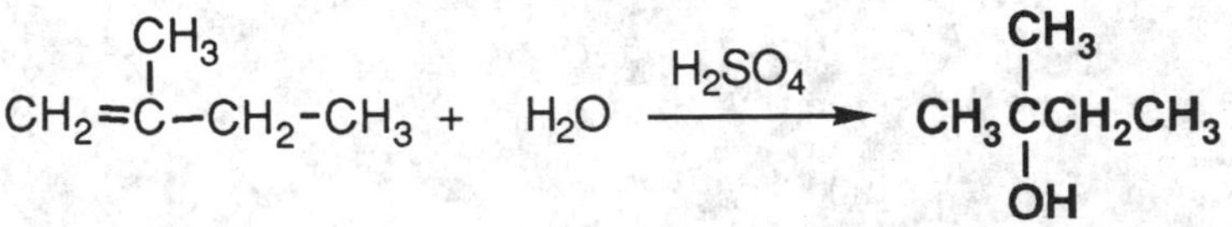 $CH_2=\overset{\overset{\displaystyle CH_3}{|}}{C}-CH_2-CH_3$ + H_2O $\xrightarrow{H_2SO_4}$ $CH_3\underset{\underset{\displaystyle OH}{|}}{\overset{\overset{\displaystyle CH_3}{|}}{C}}CH_2CH_3$

2-Methyl-2-butanol

Problem 6.7 Propose a mechanism for the acid-catalyzed hydration of 1-methylcyclohexene to give 1-methylcyclohexanol. Which step in your mechanism is the rate-limiting step?

Step 1:

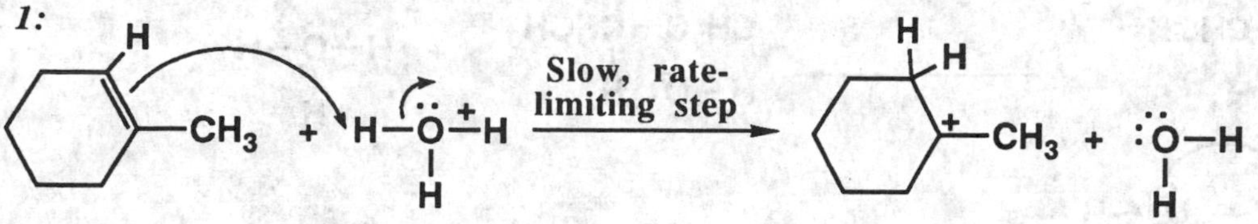

Step 2:

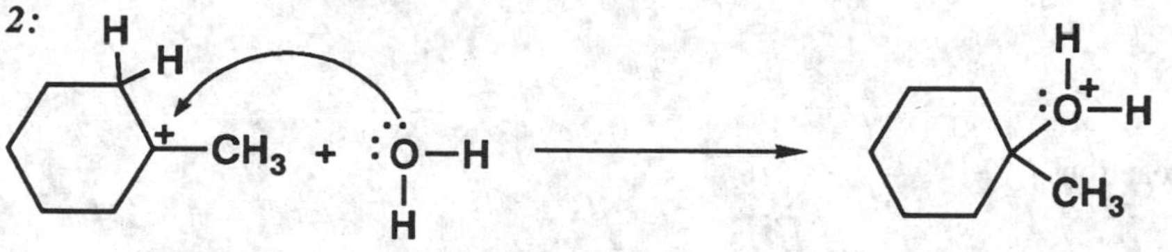

Step 3:

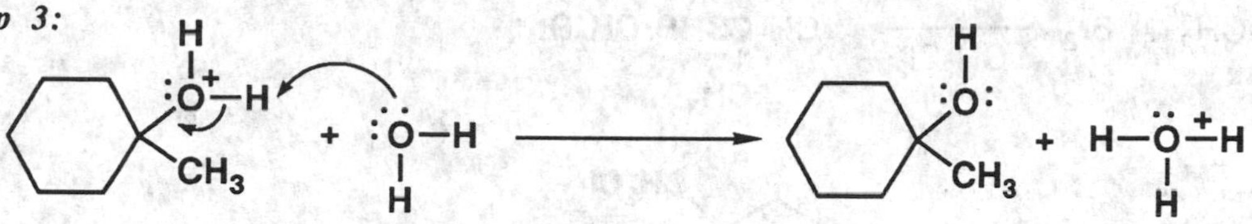

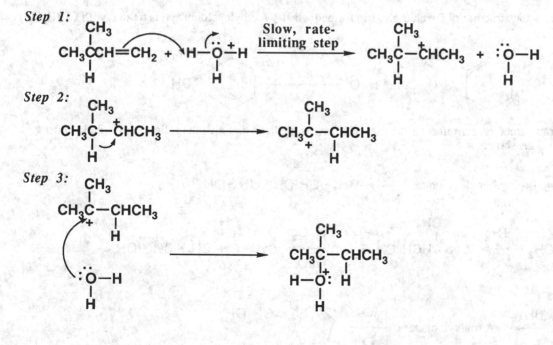

Problem 6.8 Acid-catalyzed hydration of 3-methyl-1-butene gives 2-methyl-2-butanol as the major product. Propose a mechanism for formation this rearranged alcohol.

Step 1:

Step 2:

Step 3:

Step 4:

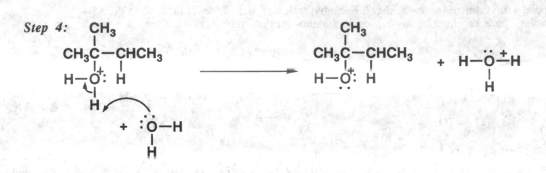

Problem 6.9 Complete these reactions.

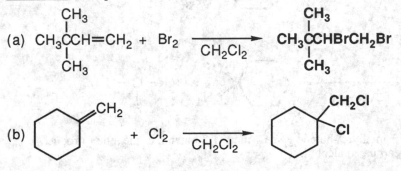

Problem 6.10 Draw the structure of the chlorohydrin formed when 1-methylcyclohexene is treated with Cl_2/H_2O.

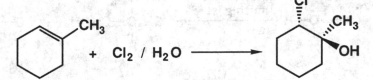

Problem 6.11 Draw structural formulas for the trialkylborane and alkene that give the following alcohols under the reaction conditions shown.

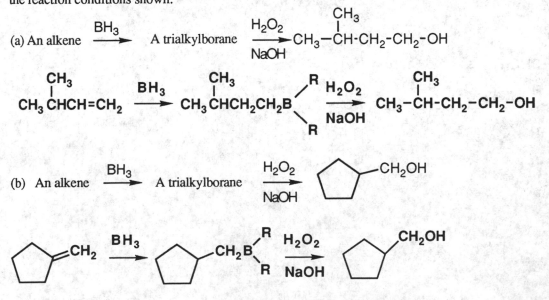

Problem 6.12 Use a balanced half-reaction to show that each transformation involves a reduction.

(a)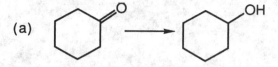

Two hydrogens are required to produce the product alcohol from the ketone. Therefore, the balanced half-reaction needs two protons and two electrons (for charge balance) on the left-hand side. Since the electrons are on the left-hand side of the equation, the reaction is a two-electron reduction.

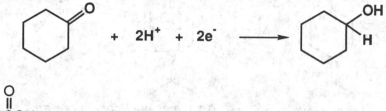

(b) $CH_3\text{-}CH_2\text{-}\overset{\displaystyle O}{\overset{\|}{C}}OH \longrightarrow CH_3CH_2CH_2OH$

Two hydrogens are required to produce the product alcohol from the carboxylic acid. Therefore, the balanced half-reaction needs two protons and two electrons (for charge balance) on the left-hand side. Additionally, the product alcohol has one less oxygen atom than the carboxylic acid starting material, so there must be an H_2O molecule added to the right side of the equation to balance the oxygen atoms. This H_2O molecule has two more hydrogens that must be balanced by adding two more protons and electrons to the left-hand side of the equation, giving a total of four protons and four electrons on the left-hand side. Since the electrons are on the left-hand side of the equation, the reaction is a four-electron reduction.

$$CH_3\text{-}CH_2\text{-}\overset{\displaystyle O}{\overset{\|}{C}}OH + 4H^+ + 4e^- \longrightarrow CH_3CH_2CH_2OH + H_2O$$

Problem 6.13 What alkene of molecular formula C_6H_{12}, when treated with ozone and then dimethyl sulfide, gives the following product(s)?

(a) $C_6H_{12} \xrightarrow[\text{2. } (CH_3)_2S]{\text{1. } O_3} CH_3CH_2\overset{\displaystyle O}{\overset{\|}{C}}H$ (only product)

$$CH_3CH_2CH=CHCH_2CH_3$$
$$(\textit{cis or trans})$$

(b) $C_6H_{12} \xrightarrow[\text{2. } (CH_3)_2S]{\text{1. } O_3} CH_3\overset{\displaystyle O}{\overset{\|}{C}}H + CH_3\overset{\displaystyle O}{\overset{\|}{C}}CH_2CH_3$ (equal moles of each)

$$CH_3CH=\overset{\displaystyle CH_3}{\overset{|}{C}}CH_2CH_3$$
$$(\textit{cis or trans})$$

(c)   C₆H₁₂ 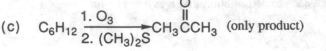 CH₃CCH₃   (only product)

Problem 6.14 Which of these terpenes (Figure 5.5) contains conjugated double bonds?

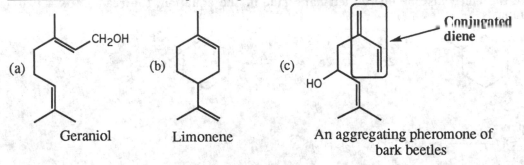

(a)   (b)   (c)

Geraniol   Limonene   An aggregating pheromone of
                                                              bark beetles

Problem 6.15 Estimate the stabilization gained due to conjugation when 1,4-pentadiene is converted to *trans*-1,3-pentadiene. Note that the answer is not as simple as comparing the heats of hydrogenation of 1,4-pentadiene and *trans*-1,3-pentadiene because, although the double bonds are moved from unconjugated to conjugated, the degree of substitution of one of the double bonds is also changed, in this case from a monosubstituted double bond to a *trans* disubstituted double bond. To answer this question, you must separate the effect due to conjugation from that due to change in degree of substitution.

1,4-Pentadiene   ***trans*-1,3-Pentadiene**

Using the values from Table 6.3, the heats of hydrogenation for 1,4-Pentadiene and *trans*-1,3-pentadiene are -60.8 kcal/mol and -54.1 kcal/mol, respectively, giving a difference of 6.7 kcal/mol. This must be corrected for the fact that 1,4-pentadiene has two terminal (monosubstituted) double bonds, while *trans*-1,3-pentadiene has one terminal (monosubstituted) and one internal (*trans* disubstituted) double bond. This correction can be approximated by the difference in heats of hydrogenation observed for 1-butene (-30.3 kcal/mol) and *trans*-2-butene (-27.6 kcal/mol) equal to 2.7 kcal/mol. Thus, the stabilization gained in the above transformation is 6.7 kcal/mol + 2.7 kcal/mol = 9.4 kcal/mol.

Energetics of Chemical Reactions
Problem 6.16 Most chemical reactions occur as written if they are exothermic, that is, if the bonds that are formed in the products are stronger than the ones broken in the starting materials. To determine if a reaction is exothermic as written, add the bond dissociation energies of all bonds broken in the starting materials (it costs energy to break bonds). Subtract from this the total of bond dissociation energies of all bonds formed in the products (formation of bonds liberates energy). If the sum of these numbers is negative, the reaction is exothermic (energy is liberated) and the reaction proceeds to the right as written. If the sum of these numbers is positive, the reaction is endothermic (it requires energy) and it does not proceed to the right as written. Using the table of bond dissociation energies at 25°C, determine which of the following reactions are energetically favorable at room temperature, that is, if a suitable catalyst could be found, which would proceed to the right as written?

Bond	Bond dissociation energy (kcal/mol)	Bond	Bond dissociation energy (kcal/mol)
H-H	104 (535)	C-Si	72 (201)
O-H	110.6 (462.8)	C=C	146 (611)
C-H	98.7 (413)	C=O (aldehyde)	174 (728)
N-H	93.4 (391)	C=O (CO$_2$)	192 (803)
Si-H	76 (318)	C≡O	257 (1075)
C-C	82.6 (346)	N≡N	227 (950)
C-N	73 (305)	C≡C	200 (837)
C-O	85.5 (358)	O=O	119 (498)
C-I	51 (213)		

The following reactions can only occur to a significant extent as written if they are exothermic, that is, if the bonds that are formed are stronger than the ones that are broken in the reaction. Recall that a catalyst increases the rate, but does not change the overall thermodynamics of a reaction.

To find out if a reaction is exothermic, the dissociation energy of all the bonds in the molecules on each side of the equation are added together. If the bond dissociation energy total from the right side of the equation is higher than the total from the left side of the equation, then the reaction is exothermic (ΔH for the reaction is negative).

(a) $CH_2=CH_2 + H_2 + N_2 \longrightarrow H_2N\text{-}CH_2\text{-}CH_2\text{-}NH_2$

The bond dissociation energies from the left side of the equation:
146 + (4 x 98.7) + 104 + 227 = 871.8 kcal/mol
(C=C)   (4 C-II)   (H-H) (N≡N)

The bond dissociation energies from the right side of the equation:
(4 x 93.4) + (2 x 73) + 82.6 + (4 x 98.7) = 997 kcal/mol
(4 N-H)   (2 C-N)   (C-C)   (4 C-H)

This reaction is exothermic because 997 is larger than 871.8.

(b) $CH_2=CH_2 + CH_4 \longrightarrow H\text{-}CH_2\text{-}CH_2\text{-}CH_3$

The bond dissociation energies from the left side of the equation:
146 + (4 x 98.7) + (4 x 98.7) = 935.6 kcal/mol
(C=C)   (4 C-H)   (4 C-H)

The bond dissociation energies from the right side of the equation:
(3 x 82.6) + (8 x 98.7) = 1037.4 kcal/mol
(3 C-C)   (8 C-H)

This reaction is exothermic because 1037.4 is larger than 934.6.

(c) $CH_2=CH_2 + (CH_3)_3SiH \longrightarrow H\text{-}CH_2\text{-}CH_2\text{-}Si(CH_3)_3$

The bond dissociation energies from the left side of the equation:
146 + (4 x 98.7) + (9 x 98.7) + (3 x 72) + 76 = 1721.1 kcal/mol
(C=C)   (4 C-H)   (9 C-H)   (3 C-Si) (Si-H)

The bond dissociation energies from the right side of the equation:
(82.6) + (5 x 98.7) + (9 x 98.7) + (4 x 72) = 1752.4 kcal/mol
(C-C)   (5 C-H)   (9 C-H)   (4 C-Si)

This reaction is exothermic because 1752.4 is larger than 1721.1.

(d) $CH_2=CH_2 + CHI_3 \longrightarrow H\text{-}CH_2\text{-}CH_2\text{-}C(I)_3$

The bond dissociation energies from the left side of the equation:
$$146 + (5 \times 98.7) + (3 \times 51) = 792.5 \text{ kcal/mol}$$
(C–C)   (5 C-H)   (3 C-I)

The bond dissociation energies from the right side of the equation:
$$(3 \times 82.6) + (5 \times 98.7) + (3 \times 51) = 894.3 \text{ kcal/mol}$$
(3 C-C)   (5 C-H)   (3 C-I)

> **This reaction is exothermic because 894.3 is larger than 792.5.**

(e) $CH_2=CH_2 + CO + H_2 \longrightarrow H\text{-}CH_2\text{-}CH_2\overset{\displaystyle O}{\overset{\|}{-}}CH$

The bond dissociation energies from the left side of the equation:
$$146 + (4 \times 98.7) + 257 + 104 = 901.8 \text{ kcal/mol}$$
(C=C)   (4 C-H)   (C≡O) (H-H)

The bond dissociation energies from the right side of the equation:
$$(3 \times 82.6) + (6 \times 98.7) + 174 = 1014 \text{ kcal/mol}$$
(3 C-C)   (6 C-H)   (C=O)

> **This reaction is exothermic because 1014 is larger than 901.8.**

(f)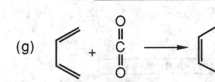

The bond dissociation energies from the left side of the equation:
$$(3 \times 146) + 82.6 + (10 \times 98.7) = 1507.6 \text{ kcal/mol}$$
(3 C=C)   (C-C)   (10 C-H)

The bond dissociation energies from the right side of the equation:
$$146 + (5 \times 82.6) + (10 \times 98.7) = 1546 \text{ kcal/mol}$$
C=C   (5 C-C)   (10 C-H)

> **This reaction is exothermic because 1546 is larger than 1507.6.**

(g)

The bond dissociation energies from the left side of the equation:
$$(2 \times 146) + 82.6 + (2 \times 192) + (6 \times 98.7) = 1350.8 \text{ kcal/mol}$$
(2 C=C)   (C-C) (2 C=O)   (6 C-H)

The bond dissociation energies from the right side of the equation:
$$146 + (3 \times 82.6) + (6 \times 98.7) + (2 \times 85.5) + 174 = 1331 \text{ kcal/mol}$$
C=C   (3 C-C)   (6 C-H)   (2 C-O) (C=O)

> **This reaction is endothermic because 1331 is smaller than 1350.8.**

(h)   HC≡CH + O$_2$ ⟶ H-C-C-H

The structure shown has two C=O groups (O on top of each C).

The bond dissociation energies from the left side of the equation:
   200 + (2 x 98.7) + 119 = 516.4 kcal/mol
   (C≡C)   (2 C-H)   (O=O)

The bond dissociation energies from the right side of the equation:
   82.6 + (2 x 174) + (2 x 98.7) = 628 kcal/mol
   (C-C)   (2 C=O)   (2 C-H)

$$\boxed{\text{This reaction is exothermic because 628 is larger than 516.4.}}$$

(i) 2CH$_4$ + O$_2$ ⟶ 2CH$_3$OH

The bond dissociation energies from the left side of the equation:
   2(4 x 98.7) + 119 = 908.6 kcal/mol
   2(4 C-H)   (O=O)

The bond dissociation energies from the right side of the equation:
   2(3 x 98.7) + 2(85.5) + 2(110.6) = 984.4 kcal/mol
   2(3C-H)   2(C-O)   2(O-H)

$$\boxed{\text{This reaction is endothermic because 908.6 is smaller than 984.4.}}$$

Electrophilic Additions

Problem 6.17 Draw structural formulas for the isomeric carbocations formed by the addition of H$^+$ to each alkene. Label each carbocation primary, secondary, or tertiary and state which of the isomeric carbocations is formed more readily.

(a)
$$CH_3-CH_2-\underset{\underset{CH_3}{|}}{C}=CH-CH_3$$

$$CH_3-CH_2-\overset{\overset{CH_3}{|}}{\underset{+}{C}}-CH_2-CH_3 \quad + \quad CH_3-CH_2-\overset{\overset{CH_3}{|}}{CH}-\underset{+}{CH}-CH_3$$

Tertiary   Secondary
(Formed more readily)   (Less stable)

(b)   CH$_3$-CH$_2$-CH=CH-CH$_3$

$$CH_3-CH_2-\overset{+}{C}H-CH_2-CH_3 \quad + \quad CH_3-CH_2-CH_2-\overset{+}{C}H-CH_3$$

Both secondary carbocations
(Formed at equal rates)

(c)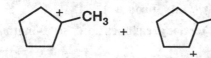

Tertiary   Secondary
(Formed more readily)   (Less stable)

(d)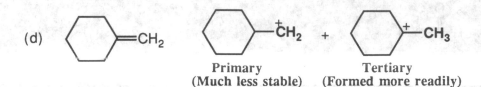

Primary   Tertiary
(Much less stable)   (Formed more readily)

Problem 6.18 Arrange the alkenes in each set in order of increasing rate of reaction with HI. Draw the structural formula of the major product formed in each case, and explain the basis for your ranking.

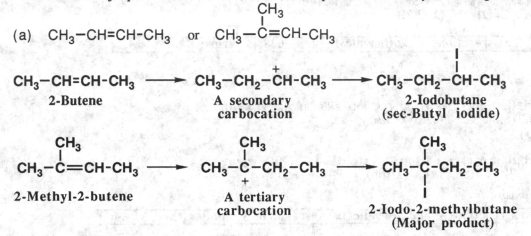

(a) $CH_3-CH=CH-CH_3$ or $CH_3-\overset{\underset{\displaystyle CH_3}{|}}{C}=CH-CH_3$

$CH_3-CH=CH-CH_3 \longrightarrow CH_3-CH_2-\overset{+}{C}H-CH_3 \longrightarrow CH_3-CH_2-\overset{\underset{\displaystyle CH_3}{|}}{C}H-CH_3$

2-Butene   A secondary   2-Iodobutane
                              carbocation   (sec-Butyl iodide)

$CH_3-\overset{\underset{\displaystyle CH_3}{|}}{C}=CH-CH_3 \longrightarrow CH_3-\overset{\underset{\displaystyle +}{\overset{\displaystyle CH_3}{|}}}{C}-CH_2-CH_3 \longrightarrow CH_3-\overset{\underset{\displaystyle CH_3}{|}}{C}-CH_2-CH_3$

2-Methyl-2-butene   A tertiary
                              carbocation   2-Iodo-2-methylbutane
                                                    (Major product)

The reaction of 2-methyl-2-butene is the only one that can form a tertiary carbocation, so 2-methyl-2-butene is the compound that reacts faster with HI.

(b) 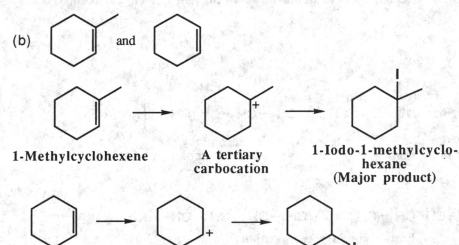   and

1-Methylcyclohexene   A tertiary   1-Iodo-1-methylcyclo-
                        carbocation   hexane
                                              (Major product)

Cyclohexene   A secondary   Iodocyclohexane
                      carbocation   (Only product)

Only 1-methylcyclohexene can form a tertiary carbocation, so 1-methylcyclohexene reacts faster with HI.

Problem 6.19 Predict the organic product(s) of the reaction of 2-butene with each reagent.
(a) H$_2$O (H$_2$SO$_4$)   (b) Br$_2$ (c) Cl$_2$

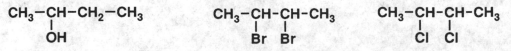

$CH_3-\overset{\underset{\displaystyle OH}{|}}{C}H-CH_2-CH_3$ $CH_3-\overset{\underset{\displaystyle Br}{|}}{C}H-\overset{\underset{\displaystyle Br}{|}}{C}H-CH_3$ $CH_3-\overset{\underset{\displaystyle Cl}{|}}{C}H-\overset{\underset{\displaystyle Cl}{|}}{C}H-CH_3$

(d) Br$_2$ in H$_2$O

CH$_3$-CH-CH-CH$_3$
      | |
      Br OH

(e) HI

CH$_3$-CH-CH$_2$-CH$_3$
      |
      I

(f) Cl$_2$ in H$_2$O

CH$_3$-CH-CH-CH$_3$
      | |
      Cl OH

(g) Hg(OAc)$_2$, H$_2$O

CH$_3$-CH-CH-CH$_3$
      | |
      OH HgOAc

(h) the product in (g) + NaBH$_4$

CH$_3$-CH-CH$_2$-CH$_3$
      |
      OH

<u>Problem 6.20</u> Draw a structural formula of an alkene that undergoes acid-catalyzed hydration to give these alcohols as the major product. More than one alkene may give each alcohol as the major product.

(a) 3-Hexanol

CH$_3$CH$_2$CH=CHCH$_2$CH$_3$

(b) 1-Methylcyclobutanol

[cyclobutane with =CH$_2$]   or   [cyclobutene with CH$_3$]

(c) 2-Methyl-2-butanol

H$_2$C=CHCH$_2$CH$_3$ (with CH$_3$)   or   CH$_3$C=CHCH$_3$ (with CH$_3$)

(d) 2-Propanol

CH$_3$CH=CH$_2$

<u>Problem 6.21</u> Reaction of 2-methyl-2-pentene with each reagent shows a high regioselectivity. Draw a structural formula for the product of each reaction, and account for the observed regioselectivity.

In each case, the reaction mechanism involves formation of a tertiary carbocation, that then reacts with a nucleophile to give the product shown.

(a) HI

CH$_3$-C(CH$_3$)(I)-CH$_2$-CH$_2$-CH$_3$

(b) HBr

CH$_3$-C(CH$_3$)(Br)-CH$_2$-CH$_2$-CH$_3$

(c) H$_2$O in the presence of H$_2$SO$_4$

CH$_3$-C(CH$_3$)(OH)-CH$_2$-CH$_2$-CH$_3$

(d) Br$_2$ in H$_2$O

CH$_3$-C(CH$_3$)(HO)-CH(Br)-CH$_2$-CH$_3$

(e) Hg(OAc)$_2$ in H$_2$O

CH$_3$-C(CH$_3$)(HO)-CH(HgOAc)-CH$_2$-CH$_3$

<u>Problem 6.22</u> Reaction of 1-methylcyclopentene with each reagent shows a high degree of regioselectivity and stereoselectivity. Account for the observed regioselectivity and stereoselectivity.
(a) BH$_3$

Attack of the borane occurs in a concerted fashion, simultaneously forming both the new C-H and C-B bonds on the same face of the double bond, that is syn. Largely for steric reasons, the H atom ends up on the less-hindered carbon atom.

**Bond forming and
bond breaking is concerted**

(b) Br_2 in H_2O

**Attack by H_2O on the carbon bearing the methyl group followed by loss of a proton gives the
trans bromohydrin.**

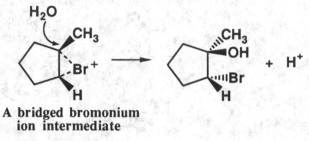

**A bridged bromonium
ion intermediate**

(c) $Hg(OAc)_2$ in H_2O

**Attack by water on the bridged mercurinium ion intermediate followed by loss of a proton
results in -OH *trans* to -HgOAc.**

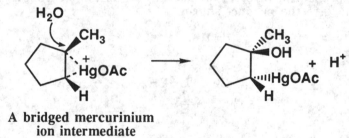

**A bridged mercurinium
ion intermediate**

Problem 6.23 Draw a structural formula for an alkene with the indicated molecular formula that gives the
compound shown as the major product. Note that more than one alkene may give the same compound as the
major product.

(a) C_5H_{10} + H_2O $\xrightarrow{H_2SO_4}$
$$CH_3-\overset{\overset{\displaystyle CH_3}{|}}{\underset{\underset{\displaystyle OH}{|}}{C}}-CH_2-CH_3$$

$$\underset{\underset{\displaystyle CH_2=\overset{\overset{\displaystyle CH_3}{|}}{C}-CH_2-CH_3}{}}{} \text{ o r } CH_3-\overset{\overset{\displaystyle CH_3}{|}}{C}=CH-CH_3$$

(b) C_5H_{10} + Br_2 $\longrightarrow$ $CH_3-\overset{\overset{\displaystyle CH_3}{|}}{CH}-\underset{\underset{\displaystyle Br}{|}}{CH}-\underset{\underset{\displaystyle Br}{|}}{CH_2}$

$$CH_3-\overset{\overset{\displaystyle CH_3}{|}}{CH}-CH=CH_2$$

(c) C_7H_{12} +   HCl   ⟶

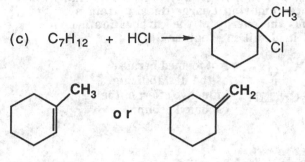

Problem 6.24 Account for the fact that addition of HCl to 1-bromopropene gives exclusively 1-bromo-1-chloropropane.

$$CH_3CH=CHBr \quad + \quad HCl \quad \longrightarrow \quad CH_3CH_2CHBrCl$$

1-Bromopropene   1-Bromo-1-chloropropane

The exclusive product must be derived from the significantly more stable carbocation. In this case, the significantly more stable carbocation is the one with the positive charge on the carbon atom attached to the bromine atom, despite the fact that this carbocation is primary versus the alternative secondary carbocation. Thus, the bromine atom must be able to stabilize an adjacent cationic carbon atom. It turns out that the stabilization is primarily a resonance effect, involving the lone pairs of the bromine atom as shown. Note how the resonance structure on the right illustrates how the positive charge is partially delocalized onto the bromine atom. This type of interaction will be discussed in detail in Chapter 20.

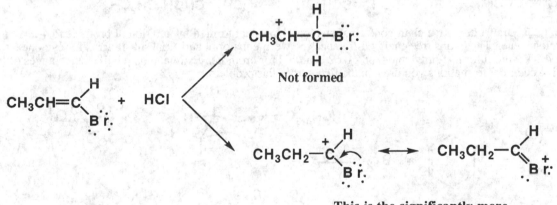

This is the significantly more
stable cation due to resonance
stabilization as shown

Problem 6.25 Propenoic acid (acrylic acid) reacts with HCl to give 3-chloropropanoic acid. It does not produce 2-chloropropanoic acid. Account for this result.

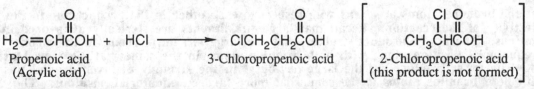

Propenoic acid   3-Chloropropenoic acid   2-Chloropropenoic acid
(Acrylic acid)   (this product is not formed)

The exclusive product must be derived from the significantly more stable carbocation. In this case, the significantly more stable carbocation is the one with the positive charge on the terminal carbon atom, despite the fact that this carbocation is primary versus the alternative secondary carbocation. Thus, the carbonyl group attached to the internal carbon atom must be destabilizing to an adjacent cationic carbon atom. It turns out that the destabilization is primarily an inductive effect, based on the fact that a carbonyl group is electron–withdrawing. An electron–withdrawing group is destabilizing, since removing charge density from a carbocation increases the charged character and thus the energy of the carbocation even further. This type of interaction will be discussed in detail in Chapter 20.

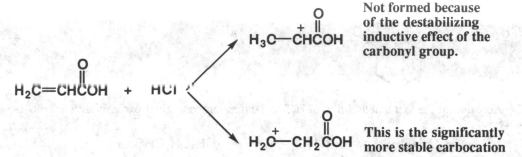

Not formed because of the destabilizing inductive effect of the carbonyl group.

This is the significantly more stable carbocation

Problem 6.26 Draw a structural formula for the alkene of molecular formula C_5H_{10} that reacts with Br_2 to give each product.

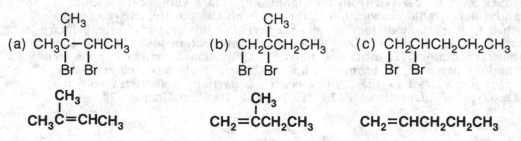

Problem 6.27 Draw alternative chair conformations for the product formed by addition of bromine to 4-*tert*-butylcyclohexene. The Gibbs free energy differences between equatorial and axial substituents for cyclohexane ring are 4.9 kcal/mol for *tert*-butyl and 0.48 - 0.62 kcal/mol for bromine. Estimate the relative percentages of the alternative chair conformations you drew in the first part of this problem.

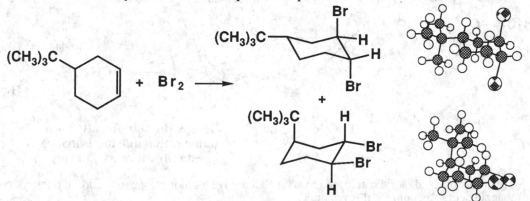

Note that the bromine atoms are *trans* with respect to each other in the product due to the stereoselectivity of the reaction. Recall that large substituents are sterically disfavored in axial positions. The upper product structure has the large *tert*-butyl group in the strongly favored equatorial position along with both bromine atoms in the somewhat disfavored axial positions. The lower product has the large *tert*-butyl in the strongly disfavored axial position along with both bromine atoms in the somewhat more favored equatorial positions. The conformational energy difference based on the axial vs. equatorial *tert*-butyl group is 4.9 kcal/mol, and we will use an intermediate value of 0.55 kcal/mol for the conformational energy difference for the axial vs. equatorial bromine atoms. Thus, the relative conformation energy can be estimated as being <u>favorable for the upper structure</u> by an amount equal to the value that is favorable for the *tert*-butyl group (4.9 kcal/mol) minus the disfavorable contributions of the two bromine atoms (2 x +0.55 kcal/mol) for a total of 3.8 kcal/mol.

At equilibrium the relative amounts of each form are given by the equation:

$$\Delta G^\circ = -RT\ln K_{eq}$$

Here K_{eq} refers to the ratio of the alternative chair conformations. Converting to base 10 and rearranging gives

$$\log K_{eq} = \frac{-\Delta G^\circ}{(2.303)RT}$$

Taking the antilog of both sides gives:

$$K_{eq} = 10^{\left(\frac{-\Delta G^\circ}{(2.303)RT}\right)}$$

Plugging in the values for ΔG°, R, and 298 K gives the final answer:

$$\boxed{K_{eq} = 10^{\left(\frac{-(-3.8\ \text{kcal/mol})}{(2.303)(1.987\ \times\ 10^{-3}\ \text{kcal/mol K})(298\ \text{K})}\right)} = 10^{2.79} = 6.1\ \times\ 10^{2}}$$

Thus, the structure with the *tert*-butyl group equatorial will be favored by about 610 to 1 at equilibrium at room temperature.

<u>Problem 6.28</u> Draw a structural formula for the cycloalkene of molecular formula C_6H_{10} that reacts with Cl_2 to give each compound.

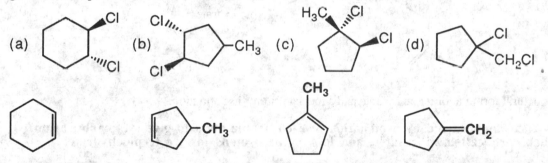

<u>Problem 6.29</u> Reaction of this bicycloalkene with bromine in carbon tetrachloride gives a *trans*-dibromide. In both (a) and (b), the bromine atoms are *trans* to each other. However, only one of these products is formed. Which *trans* dibromide is formed and how do you account for the fact that it is formed to the exclusion of the other *trans* dibromide?

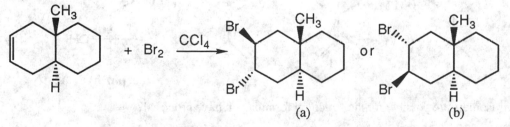

Product (a) is formed. Electrophilic addition of bromine to an alkene occurs via a bridged bromonium ion intermediate and anti addition of the two bromine atoms. In a cyclohexane ring, anti addition corresponds to *trans* and diaxial addition. Only in formula (a) are the two added bromines *trans* and diaxial. In (b) they are *trans*, but diequatorial, so this isomer cannot be formed.

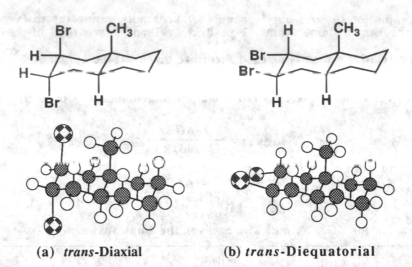

(a) *trans*-Diaxial   (b) *trans*-Diequatorial

Problem 6.30 Terpin hydrate is prepared commercially by the addition of two mol of water to limonene (Figure 5.5) in the presence of dilute sulfuric acid. Terpin hydrate is used medicinally as an expectorant for coughs. It may be given as terpin hydrate and codeine.

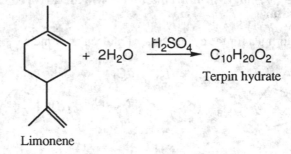

Limonene

(a) Propose a structural formula for terpin hydrate and a mechanism for its formation.

Add water to each double bond by protonation of each double bond to give a 3° carbocation, reaction of each carbocation with water, and loss of the protons to give terpin hydrate.

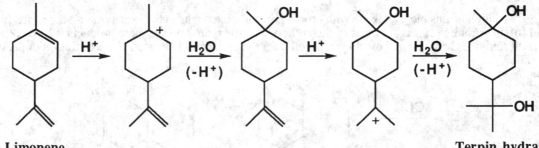

Limonene   Terpin hydrate

(b) How many *cis-trans* isomers are possible for the structural formula you have proposed?

There are two *cis-trans* isomers, shown here as chair conformations with the $(CH_3)_2COH$-side chain equatorial.

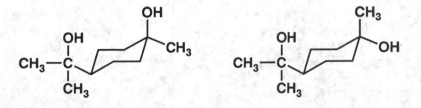

Problem 6.31 Propose a mechanism for this reaction. In so doing, account for its regioselectivity.

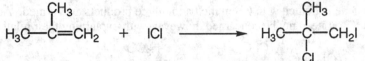

Because the Cl atom ends up on the tertiary carbon atom, the I atom must be transferred first as shown. This makes sense because Cl is more electronegative than I.
Step 1:

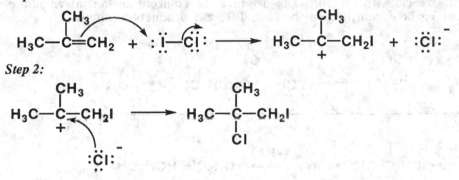

Step 2:

Problem 6.32 Treatment of 2-methylpropene with methanol in the presence of sulfuric acid gives *tert*-butyl methyl ether. Propose a mechanism for the formation of this product.

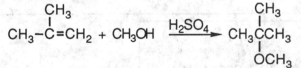

Reaction of the alkene with a proton gives a tertiary carbocation intermediate. Reaction of this intermediate with the oxygen atom of methanol followed by loss of a proton gives *tert*-butyl methyl ether.
Step 1:

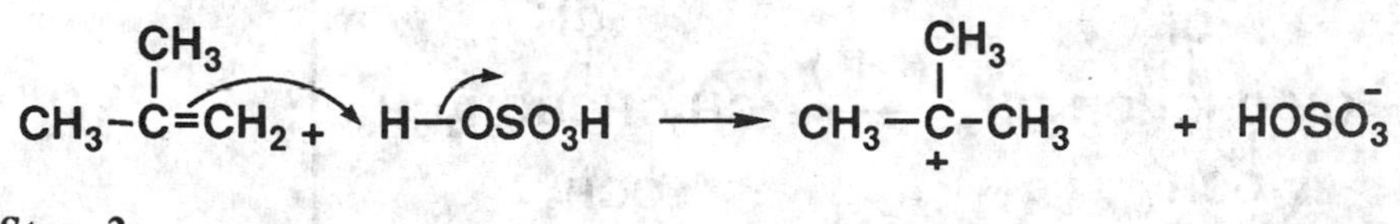

$$CH_3-\overset{\overset{\displaystyle CH_3}{|}}{C}=CH_2 + H\text{---}OSO_3H \longrightarrow CH_3-\overset{\overset{\displaystyle CH_3}{|}}{\underset{+}{C}}-CH_3 + HOSO_3^-$$

Step 2:

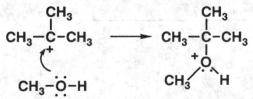

Step 3:

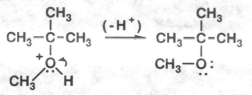

<u>Problem 6.33</u> When 2-pentene is treated with Cl_2 in methanol, three products are formed. Account for the formation of each product. (You need not be concerned, however, with explaining their relative percentages.)

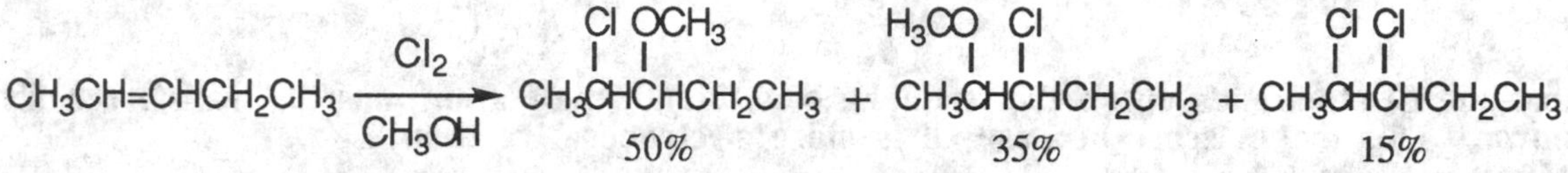

$$CH_3CH=CHCH_2CH_3 \xrightarrow[CH_3OH]{Cl_2}$$

                    Cl OCH₃   H₃CO Cl   Cl Cl

$$CH_3CHCHCH_2CH_3 + CH_3CHCHCH_2CH_3 + CH_3CHCHCH_2CH_3$$

                        50%   35%   15%

In this reaction, the chlorine reacts with the alkene to produce the chloronium ion intermediate that can then react at either carbon atom to give the four different products as shown.

Step 1:

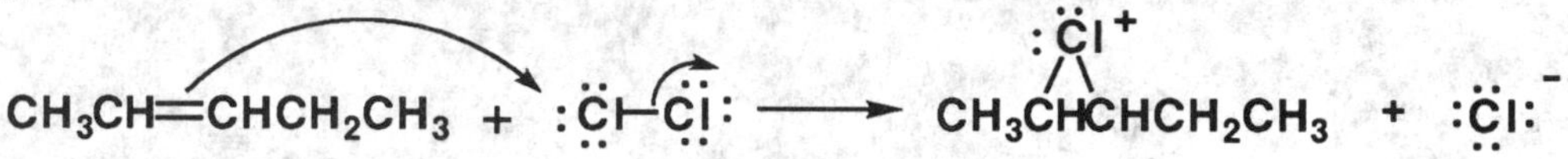

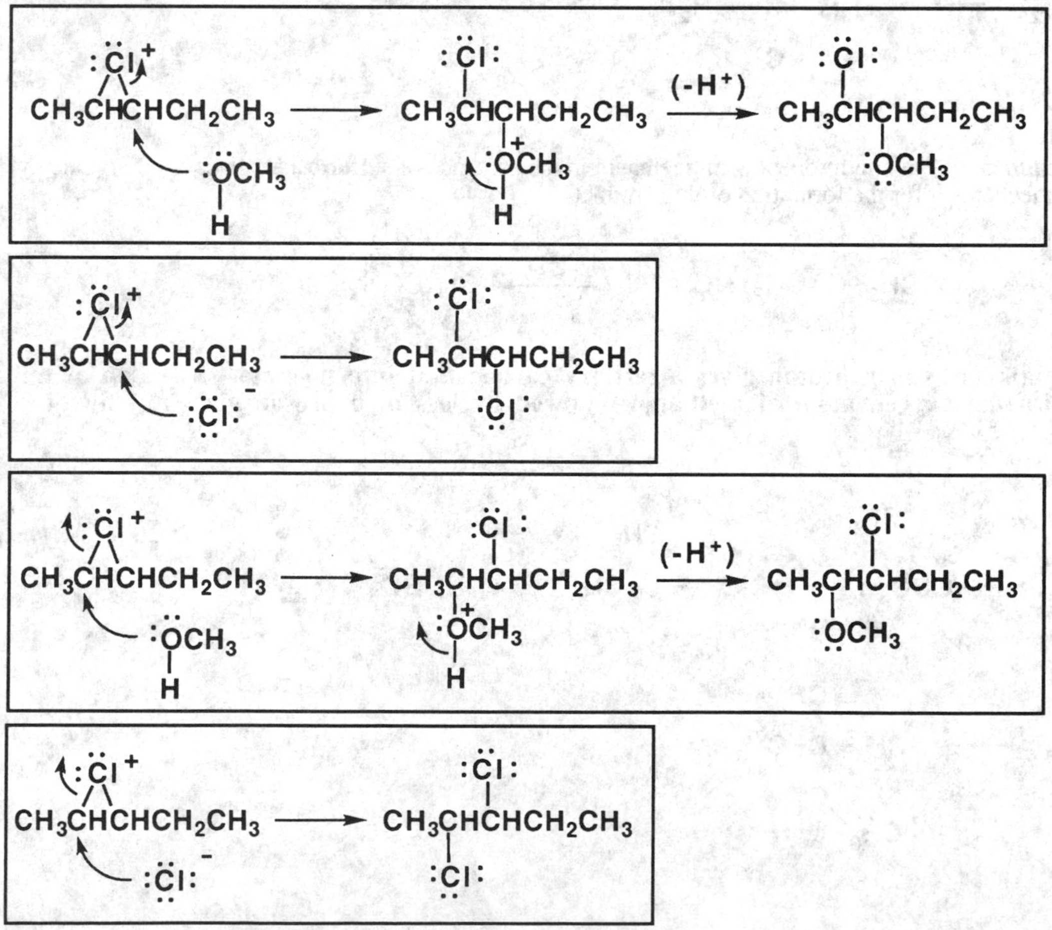

$$CH_3CH=CHCH_2CH_3 + :\ddot{C}l\!-\!\ddot{C}l: \longrightarrow CH_3CHCHCH_2CH_3 + :\ddot{C}l:^-$$

$$CH_3CHCHCH_2CH_3 \longrightarrow CH_3CHCHCH_2CH_3 \xrightarrow{(-H^+)} CH_3CHCHCH_2CH_3$$

$$CH_3CHCHCH_2CH_3 \longrightarrow CH_3CHCHCH_2CH_3$$

$$CH_3CHCHCH_2CH_3 \longrightarrow CH_3CHCHCH_2CH_3 \xrightarrow{(-H^+)} CH_3CHCHCH_2CH_3$$

$$CH_3CHCHCH_2CH_3 \longrightarrow CH_3CHCHCH_2CH_3$$

Problem 6.34 Treatment of cyclohexene with HBr in the presence of acetic acid gives bromocyclohexane (85%) and cyclohexyl acetate (15%). Propose a mechanism for the formation of the latter product.

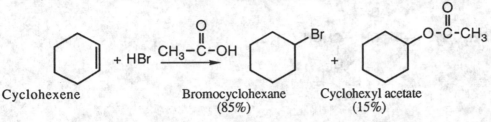

Cyclohexene   Bromocyclohexane   Cyclohexyl acetate
                               (85%)   (15%)

Reaction of cyclohexene with a proton gives a secondary carbocation intermediate. Reaction of this intermediate with an oxygen atom of acetic acid, followed by loss of a proton gives cyclohexyl acetate.

Step 1:

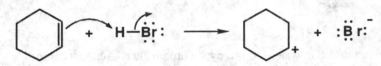

Step 2:

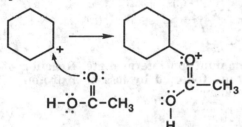

Step 3:

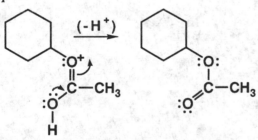

Problem 6.35 Propose a mechanism for this reaction:

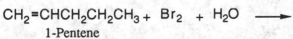

$$CH_2=CHCH_2CH_2CH_3 + Br_2 + H_2O \longrightarrow \underset{\text{1-Bromo-2-pentanol}}{\overset{\overset{\displaystyle Br \quad\, OH}{|\quad\;\; |}}{CH_2-CHCH_2CH_2CH_3}} + HBr$$

1-Pentene

Reaction of 1-pentene with bromine gives a bridged bromonium ion intermediate. Anti attack of water on this intermediate at the more substituted secondary carbon, followed by loss of a proton, gives 1-bromo-2-pentanol.

Step 1:

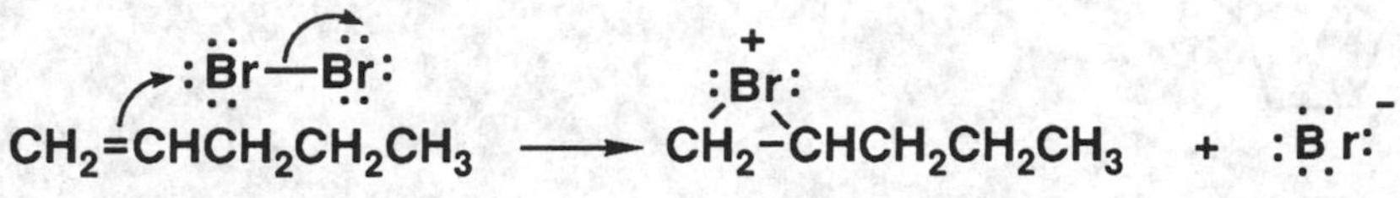

$$CH_2=CHCH_2CH_2CH_3 \quad \overset{\displaystyle :\!\ddot{B}r\!-\!\ddot{B}r\!:}{\longrightarrow} \quad \overset{+}{\underset{CH_2-CHCH_2CH_2CH_3}{:\!\ddot{B}r\!:}} \quad + \quad :\!\ddot{B}r\!:^-$$

Step 2:

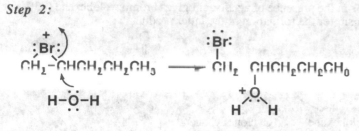

Step 3:

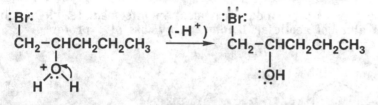

<u>Problem 6.36</u> Treatment of 4-penten-1-ol with bromine in water forms a cyclic bromoether. Account for the formation of this product rather than a bromohydrin as was formed in Problem 6.35.

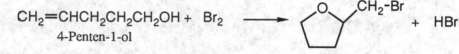

$$CH_2=CHCH_2CH_2CH_2OH + Br_2 \longrightarrow$$
4-Penten-1-ol

Reaction of the alkene with bromine gives a bridged bromonium ion intermediate. Reaction of this intermediate with the oxygen atom of the hydroxyl group followed by loss of a proton gives the observed cyclic ether, a derivative of tetrahydrofuran.

Step 1:

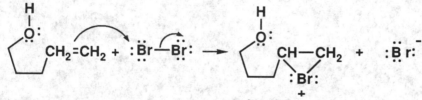

A bridged bromonium
ion intermediate

Step 2:

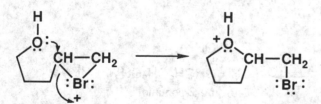

Step 3:

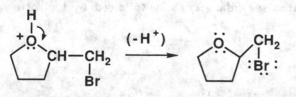

Problem 6.37 Provide a mechanism for each reaction:

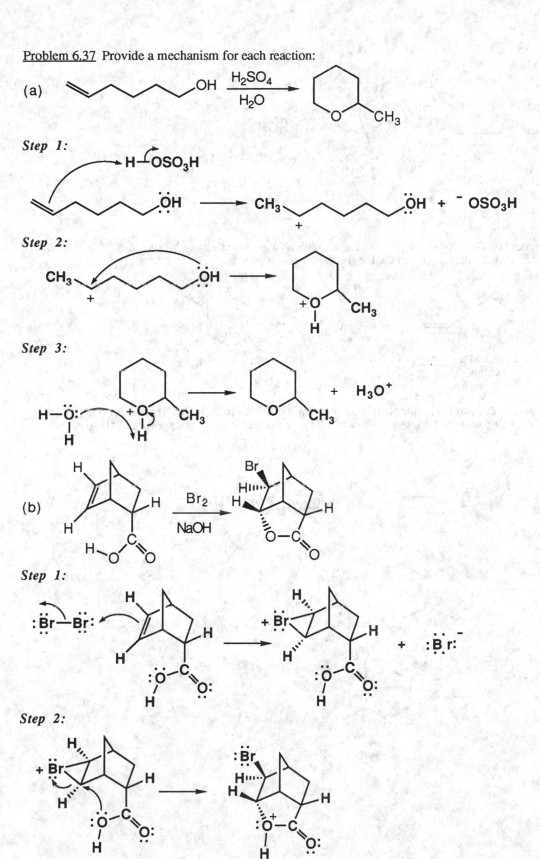

(a)

Step 1:

Step 2:

Step 3:

(b)

Step 1:

Step 2:

Step 3:

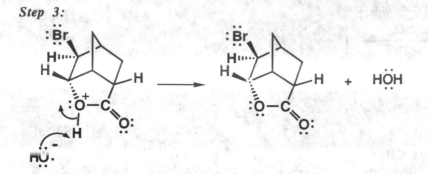

Problem 6.38 Treatment of 1-methyl-1-vinylcyclopentane with HCl gives mainly 1-chloro-1,2-dimethylcyclohexane. Propose a mechanism for the formation of this product.

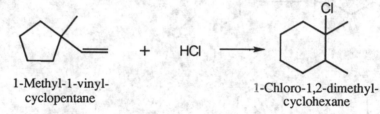

1-Methyl-1-vinyl-
cyclopentane

1-Chloro-1,2-dimethyl-
cyclohexane

The initially formed secondary carbocation can rearrange to form a more stable tertiary carbocation and a six-membered ring as shown. This new cation reacts with Cl- to give the final product.

Step 1:

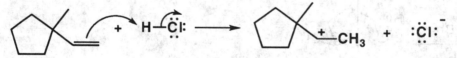

Step 2:

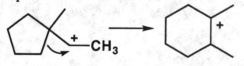

Step 3:

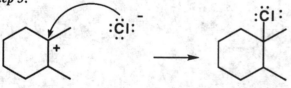

Hydroboration

Problem 6.39 Each alkene is treated with diborane in tetrahydrofuran (THF) to form a trialkylborane which is then oxidized with hydrogen peroxide in aqueous sodium hydroxide. Draw a structural formula of the alcohol formed in each case. Specify stereochemistry where appropriate.

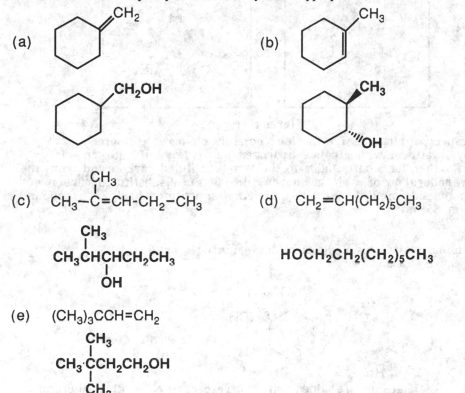

(a)

(b)

(c) $CH_3-\overset{\underset{\displaystyle CH_3}{|}}{C}=CH-CH_2-CH_3$

(d) $CH_2=CH(CH_2)_5CH_3$

$CH_3\overset{\underset{\displaystyle OH}{|}}{CH}\overset{\underset{\displaystyle }{|}}{\underset{CH_3}{}}CHCH_2CH_3$

$HOCH_2CH_2(CH_2)_5CH_3$

(e) $(CH_3)_3CCH=CH_2$

$CH_3\overset{\underset{\displaystyle CH_3}{|}}{\underset{\underset{\displaystyle CH_3}{|}}{C}}CH_2CH_2OH$

Problem 6.40 Reaction of α-pinene with diborane followed by treatment of the resulting trialkylborane with alkaline hydrogen peroxide gives an alcohol with the following structural formula.

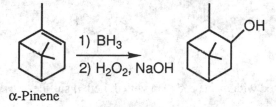

α-Pinene

1) BH₃
2) H₂O₂, NaOH

(a) Four *cis,trans* isomers are possible for this bicyclic alcohol. Draw formulas for all four.

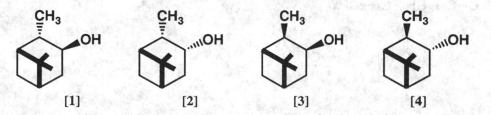

[1]   [2]   [3]   [4]

(b) Of the four possible *cis,trans* isomers, one is formed in over 85% yield. Which isomer is formed, and how do you account for this stereoselectivity? *Hint:* Make a line-angle drawing of α-pinene in ChemDraw, import it into Chem3D, and determine from the three-dimensional model from which face of the double bond borane is more likely to approach.

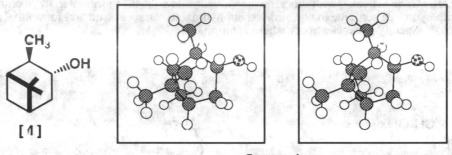

Stereoview

Shown in part (a) are perspective formulas for the four possible *cis-trans* isomers. Hydroboration followed by treatment with alkaline hydrogen peroxide results in syn (*cis*) addition of -H and -OH. Furthermore, boron adds to the less-substituted carbon and from the least hindered side. In hydroboration of α-pinene, boron adds to the disubstituted carbon of the double bond and from the side opposite the bulky dimethyl substituted bridge. Compound [4] is the product formed in 85% yield.

Oxidation

Problem 6.41 Write structural formulas for the major organic product(s) formed by reaction of 1-methylcyclohexene with each oxidizing agent.

(a) H_2O_2/OsO_4

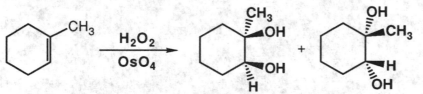

Note that even though the two OH groups are added syn with respect to each other, there are still two enantiomers produced in the reaction.

(b) O_3 followed by $(CH_3)_2S$

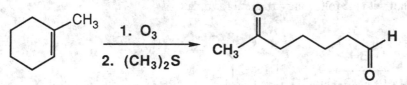

Problem 6.42 Each alkene is treated with ozone and then with dimethyl sulfide. Draw the structural formula of the organic product(s) formed from each.

(a)

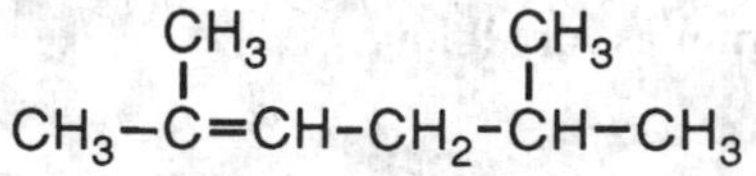

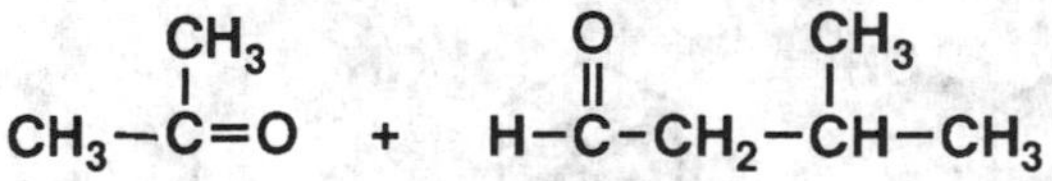

$$CH_3-\overset{\overset{\displaystyle CH_3}{|}}{C}=CH-CH_2-\overset{\overset{\displaystyle CH_3}{|}}{CH}-CH_3$$

$$CH_3-\overset{\overset{\displaystyle CH_3}{|}}{C}=O \quad + \quad H-\overset{\overset{\displaystyle O}{\|}}{C}-CH_2-\overset{\overset{\displaystyle CH_3}{|}}{CH}-CH_3$$

(b)   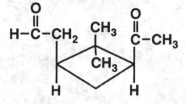 $CH_3-\overset{\overset{\displaystyle CH_3}{|}}{C}=CH-CH_2-CH=\overset{\overset{\displaystyle CH_3}{|}}{C}CH_2CH_3$

$\overset{\overset{\displaystyle CH_3}{|}}{CH_3-C}=O$ + $H-\overset{\overset{\displaystyle O}{||}}{C}-CH_2-\overset{\overset{\displaystyle O}{||}}{C}-H$ + $H_3C-\overset{\overset{\displaystyle O}{||}}{C}-CH_2CH_3$

(c) α-Pinene (Fig. 5.5)

$H-\overset{\overset{\displaystyle O}{||}}{C}-CH_2$ $\overset{\overset{\displaystyle CH_3}{|}}{\underset{\underset{\displaystyle CH_3}{|}}{}}$ $\overset{\overset{\displaystyle O}{||}}{C}-CH_3$

(d) Limonene (Fig. 5.5)

$CH_3-\overset{\overset{\displaystyle O}{||}}{C}-CH_2-CH_2-\overset{\overset{\displaystyle CH_2-\overset{\displaystyle O}{\overset{||}{C}}-H}{|}}{\underset{\underset{\displaystyle O}{||}}{C}}H-C-CH_3$ + $H-\overset{\overset{\displaystyle O}{||}}{C}-H$

(e) Zingiberene (Fig. 5.6)

$CH_3-\overset{\overset{\displaystyle O}{||}}{C}-CH_3$ + $H-\overset{\overset{\displaystyle O}{||}}{C}-CH_2-CH_2-\overset{\overset{\displaystyle CH_3}{|}}{C}H-\underset{\underset{\displaystyle \overset{|}{\underset{\overset{\displaystyle C-H}{||}}{O}}}{}}{C}H-CH_2-\overset{\overset{\displaystyle O}{||}}{C}-H$ + $H-\overset{\overset{\displaystyle O}{||}}{C}-\overset{\overset{\displaystyle O}{||}}{C}-CH_3$

(f) Caryophyllene (Fig. 5.6)

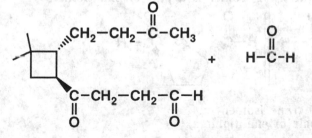

Problem 6.43 Draw the structural formula of the alkene that reacts with ozone followed by dimethyl sulfide to give each product or set of products.

(a) C_7H_{12} $\xrightarrow[\text{2. }(CH_3)_2S]{\text{1. }O_3}$ $CH_3\overset{\overset{\displaystyle O}{||}}{C}CH_2CH_2CH_2\overset{\overset{\displaystyle O}{||}}{C}CH_3$

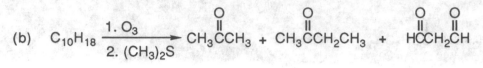

(b) $C_{10}H_{18}$ $\xrightarrow[\text{2. (CH}_3)_2\text{S}]{\text{1. O}_3}$ $CH_3\overset{O}{\overset{\|}{C}}CH_3$ + $CH_3\overset{O}{\overset{\|}{C}}CH_2CH_3$ + $H\overset{O}{\overset{\|}{C}}CH_2C\overset{O}{\overset{\|}{}}H$

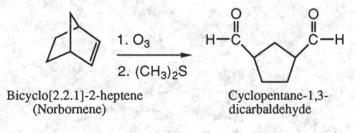

(c) $C_{10}H_{18}$ $\xrightarrow[\text{2. (CH}_3)_2\text{S}]{\text{1. O}_3}$ $CH_3\overset{CH_3}{\underset{|}{C}}HCH_2\overset{O}{\overset{\|}{C}}CH_2CH_2CH_2CH_2C\overset{O}{\overset{\|}{}}H$

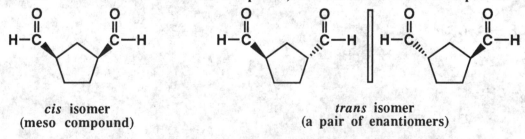

Problem 6.44 Bicyclo[2.2.1]-2-heptene (norbornene) is oxidized by ozone/dimethyl sulfide to cyclopentane-1,3-dicarbaldehyde.

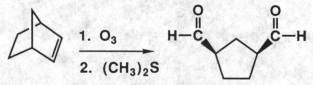

Bicyclo[2.2.1]-2-heptene   Cyclopentane-1,3-
(Norbornene)   dicarbaldehyde

(a) How many stereoisomers are possible for this dicarbaldehyde?

Three. The *cis* isomer is a meso compound, and the *trans* isomer is a pair of enantiomers.

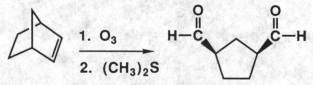

        cis isomer   *trans* isomer
   (meso compound)   (a pair of enantiomers)

(b) Which of the possible stereoisomers is/are formed by ozonolysis of norbornene?

Only the *cis* isomer is formed. Because of the geometry of the bicycloalkene, the two carbon atoms of the alkene double bond must be fused *cis* to each other.

<u>Problem 6.45</u> (a) Draw a structural formula for the bicycloalkene of molecular formula C_8H_{12} that, on treatment with ozone followed by dimethyl sulfide, gives cyclohexane-1,4-dicarbaldehyde.

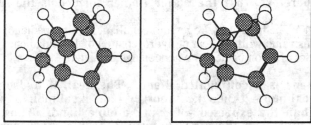

Cyclohexane-1,4-dicarbaldehyde

Following are two stereorepresentations for the bicycloalkene.

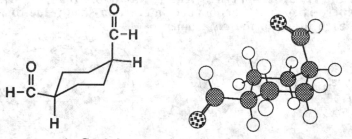

Bicyclo[2.2.2]-2-heptene

Stereoview

(b) Do you predict the product to be the *cis* isomer, the *trans* isomer, or a mixture of *cis* and *trans* isomers? Explain.

The product is the *cis* isomer. In either of the alternative chair conformations of the product, one carbaldehyde group is axial and the other is equatorial.

(c) Draw a suitable stereorepresentation for the more stable chair conformation of the dicarbaldehyde formed in this oxidation.

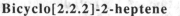

cis-Cyclohexane-1,4-dicarbaldehyde

Reduction
<u>Problem 6.46</u> Predict the major organic product(s) of the following reactions. Show stereochemistry where appropriate.

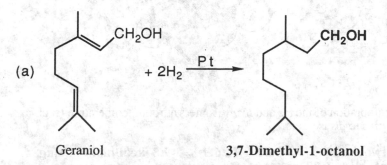

Geraniol   **3,7-Dimethyl-1-octanol**

Reduction of geraniol adds hydrogen atoms to each carbon-carbon double bond. There is no possibility for *cis-trans* isomerism in the product.

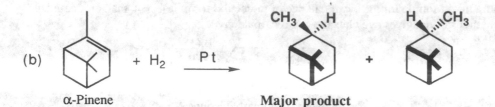

(b) α-Pinene   + H₂ $\xrightarrow{\text{Pt}}$ Major product   +

Reduction of α-pinene adds hydrogen atoms preferentially from the less hindered side of the double bond, namely the side opposite the one-carbon bridge bearing the two methyl groups. Predict, therefore, that the major isomer formed is the first one shown.

Problem 6.47 The heat of hydrogenation of allene (1,2-propadiene) to propene is -35.3 kcal/mol (177 kJ/mol). Compare this value with the heat of hydrogenation of 1,3-butadiene to 1-butene. Does allene have the characteristics of a conjugated or a nonconjugated diene?

1,3-Butadiene is a conjugated diene. The heat of hydrogenation of 1,3-butadiene to 1-butene is -56.6 kcal/mol - (-30.3 kcal/mol) = -26.3 kcal/mol; a value that is 4.0 kcal/mol less negative than the expected -30.3 per double bond. Thus, the conjugation adds stability of ~ 4.0 kcal/mol. On the other hand, the heat of hydrogenation of allene (1,2-propadiene) to propene is -35.3 kcal/mol, a value that is substantially more negative than the expected -30.3 kcal/mol. Thus, allene does not have the characteristics of a conjugated diene, and is even less stable than a standard non-conjugated diene.

Problem 6.48 The heat of hydrogenation of *cis*-di-*tert*-butylethylene is -36.7 kcal/mol (154 kJ/mol) while that of the *trans* isomer is only -26.9 kcal/mol (113 kJ/mol).
(a) Why is the heat of hydrogenation of the *cis* isomer so much larger than that of the *trans* isomer?

A larger value means the *cis*-di-*tert*-butylethylene is less stable than the *trans* isomer. This makes sense because there is so much nonbonded interaction strain due to the *tert*-butyl groups smashing into each other in the *cis* isomer.

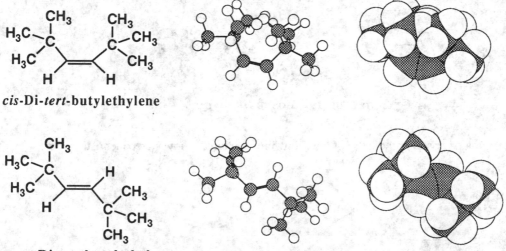

cis-Di-*tert*-butylethylene

trans-Di-*tert*-butylethylene

(b) If a catalyst could be found that allowed equilibration of the *cis* and *trans* isomers at room temperature (such catalysts do exist), what would be the ratio of *trans* to *cis* isomers?

The difference in energy between the two isomers is -36.7 - (-26.9) = -9.8 kcal/mol. Using the equation derived in the answer to problem 6.27 gives:

$$K_{eq} = 1 \, 0^{\left(\frac{-\Delta G^\circ}{(2.303)R\,T}\right)} = 1 \, 0^{\left(\frac{-(-9.8 \text{ kcal/mol})}{(2.303)(1.987 \times 10^{-3} \text{ kcal/mol K})(298 \text{ K})}\right)} = 1 \, 0^{7.2} = 1.5 \times 10^7$$

Thus, at room temperature, the ratio would be 1.5×10^7 to 1 in favor of the *trans* isomer.

Synthesis

Problem 6.49 Show how to convert ethylene to these compounds.

(a) Ethane

$$\text{H}_2\text{C}=\text{CH}_2 + \text{H}_2 \xrightarrow[\text{catalyst}]{\text{Transition metal}} \text{CH}_3\text{CH}_3$$
Ethane

(b) Ethanol

$$\text{H}_2\text{C}=\text{CH}_2 + \text{H}_2\text{O} \xrightarrow{\text{H}_2\text{SO}_4} \text{CH}_3\text{CH}_2\text{OH}$$
Ethanol

or

$$\text{H}_2\text{C}=\text{CH}_2 \xrightarrow[\text{2. NaBH}_4]{\text{1. Hg(OAc)}_2, \text{H}_2\text{O}} \text{CH}_3\text{CH}_2\text{OH}$$
Ethanol

or

$$\text{H}_2\text{C}=\text{CH}_2 \xrightarrow[\text{2. H}_2\text{O}_2, \text{NaOH}]{\text{1. BH}_3} \text{CH}_3\text{CH}_2\text{OH}$$
Ethanol

(c) Bromoethane

$$\text{H}_2\text{C}=\text{CH}_2 + \text{HBr} \longrightarrow \text{CH}_3\text{CH}_2\text{Br}$$
Bromoethane

(d) 2-Chloroethanol

$$\text{H}_2\text{C}=\text{CH}_2 \xrightarrow{\text{Cl}_2 / \text{H}_2\text{O}} \text{CH}_2\text{ClCH}_2\text{OH}$$
2-Chloroethanol

(e) 1,2-Dibromoethane

$$\text{H}_2\text{C}=\text{CH}_2 + \text{Br}_2 \xrightarrow{\text{CCl}_4} \text{CH}_2\text{BrCH}_2\text{Br}$$
1,2-Dibromoethane

(f) 1,2-Ethanediol

$$\text{H}_2\text{C}=\text{CH}_2 \xrightarrow[\text{H}_2\text{O}_2]{\text{OsO}_4} \text{CH}_2\text{OHCH}_2\text{OH}$$
1,2-Ethanediol

(g) Chloroethane

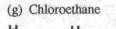

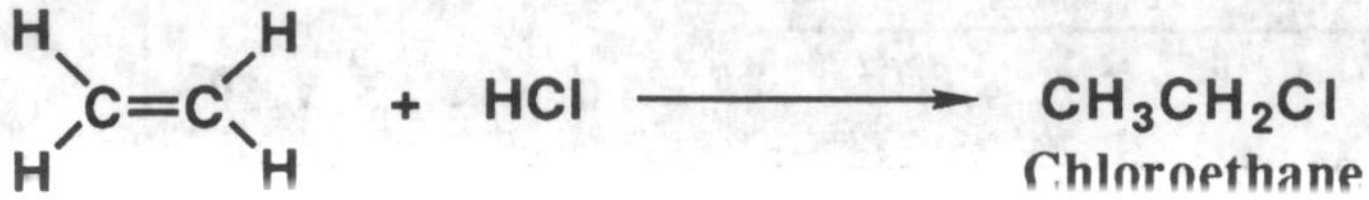

$$H_2C=CH_2 \quad + \quad HCl \quad \longrightarrow \quad CH_3CH_2Cl$$
Chloroethane

Problem 6.50 Show how to convert cyclopentene into these compounds.
(a) *trans*-1,2-Dibromocyclopentane

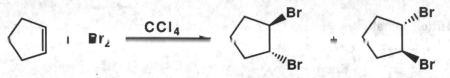

trans-**1,2-Dibromocyclopentane**

Note that the product of the reaction is actually a pair of enantiomers.

(b) *cis*-1,2-Cyclopentanediol

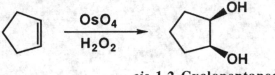

cis-**1,2-Cyclopentanediol**

(c) Cyclopentanol

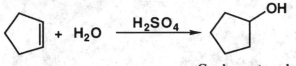

Cyclopentanol

or

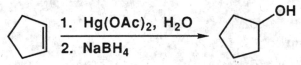

Cyclopentanol

or

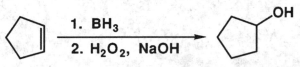

Cyclopentanol

(d) Iodocyclopentane

Iodocyclopentane

(e) Cyclopentane

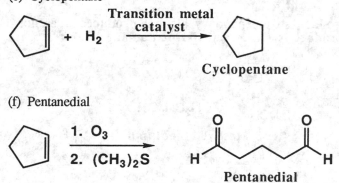

Cyclopentane

(f) Pentanedial

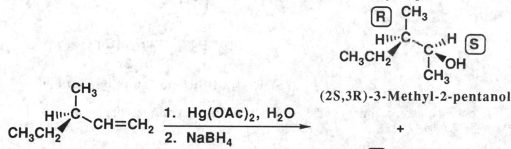

Pentanedial

Reactions that Produce Chiral Compounds

Problem 6.51 State the number and kind of stereoisomers formed when (R)-3-methyl-1-pentene is treated with these reagents.

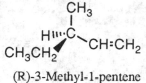

(R)-3-Methyl-1-pentene

(a) Hg(OAc)$_2$, H$_2$O followed by NaBH$_4$

The alcohol produced in this reaction has a new stereocenter, so the net result is a pair of diastereomers; (2S,3R)-3-methyl-2-pentanol and (2R,3R)-3-methyl-2-pentanol.

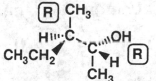

(2S,3R)-3-Methyl-2-pentanol

+

(2R,3R)-3-Methyl-2-pentanol

(b) H$_2$/Pt

The alkane produced in this reaction does not have any new stereocenters, and the only product, 3-methylpentane, is not chiral.

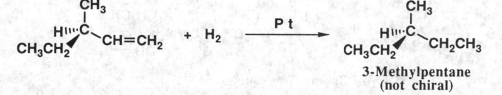

3-Methylpentane
(not chiral)

(c) BH$_3$ followed by H$_2$O$_2$ in NaOH

These reagents give the non-Markovnikov product, so the primary alcohol produced does not have any new stereocenters, and the only product is (R)-3-methyl-1-pentanol.

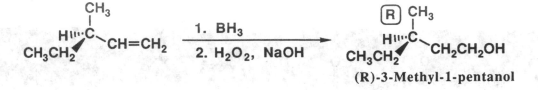

(R)-3-Methyl-1-pentanol

(d) Br$_2$ in CCl$_4$

The dibromide produced in this reaction has a new stereocenter, so the net result is a pair of diastereomers; (2R,3R)-1,2-dibromo-3-methylpentane and (2S,3R)-1,2-dibromo-3-methylpentane.

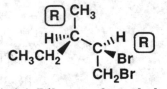

(2R,3R)-1,2-Dibromo-3-methylpentane

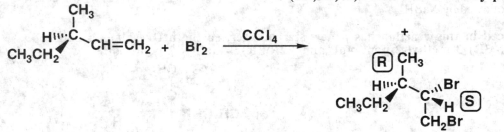

(2S,3R)-1,2-Dibromo-3-methylpentane

Problem 6.52 Describe the stereochemistry of the bromohydrin formed in each reaction.
(a) *cis*-3-Hexene + Br$_2$/H$_2$O

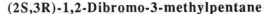

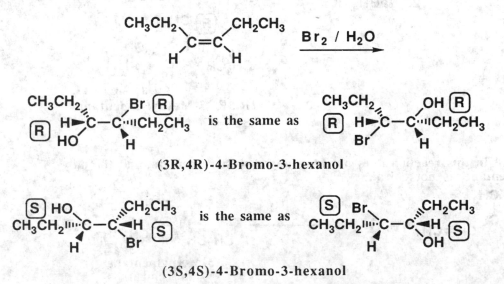

(3R,4R)-4-Bromo-3-hexanol

is the same as

(3S,4S)-4-Bromo-3-hexanol

Remember that the anti stereochemistry of addition means that the Br and OH groups add to opposite faces of the double bond. There are only two products because of symmetry, the (3R,4R)-4-bromo-3-hexanol and (3S,4S)-4-bromo-3-hexanol.

(b) *trans*-3-Hexene + Br$_2$/H$_2$O

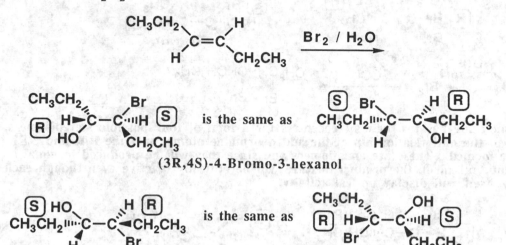

(3R,4S)-4-Bromo-3-hexanol

(3S,4R)-4-Bromo-3-hexanol

Remember that the anti stereochemistry of addition means that the Br and OH groups add to opposite faces of the double bond. There are only two products because of symmetry, the (3R,4S)-4-bromo-3-hexanol and (3S,4R)-4-bromo-3-hexanol.

Problem 6.53 In each of these reactions, the organic starting material is achiral. The structural formula of the product is given. For each reaction state:
(1) How many stereoisomers are possible for the product.
(2) Which of the possible stereoisomers is/are formed in the reaction shown.
(3) Whether the product is optically active or optically inactive.

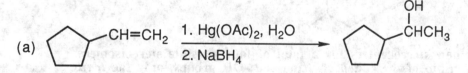

The product molecule has one new stereocenter, and both enantiomers will be formed in equal amounts (racemic mixture). Thus, the product mixture will be optically inactive even though each enantiomer by itself will display optical activity.

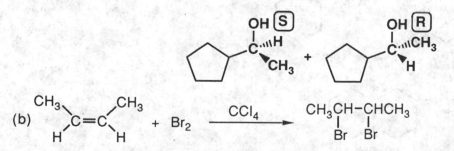

The product molecule has two new stereocenters for a total of three possible stereoisomers, a meso compound (*cis* addition) and a pair of enantiomers (*trans* addition). However, due to the anti relationship of the added bromine atoms, only the R,R and S,S isomers are formed.

These are enantiomers and since they will be produced in equal amounts (racemic mixture), the product mixture will be optically inactive even though each enantiomer by itself will display optical activity.

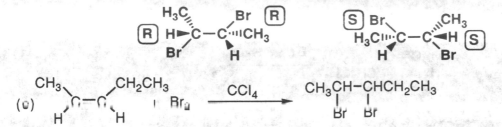

The product molecule has two new stereocenters for a total of four possible stereoisomers. However, due to the anti relationship of the added bromine atoms, only the R,R and S,S isomers can be formed. These are enantiomers and since they will be produced in equal amounts (racemic mixture), the product mixture will be optically inactive even though each enantiomer by itself will display optical activity.

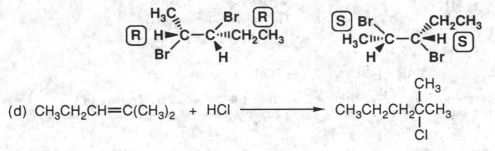

(d) $CH_3CH_2CH=C(CH_3)_2$ + HCl ⟶

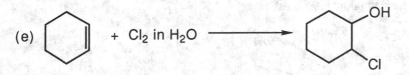

The product molecule has no stereocenters so no stereoisomerism is possible. The product mixture will therefore be optically inactive.

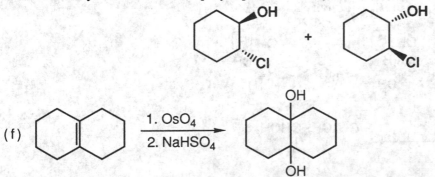

The product molecule has two new stereocenters for a total of four possible stereoisomers. However, due to the anti relationship of the added chlorine and -OH groups, only the *trans* isomers can be formed. These are enantiomers and since they will be produced in equal amounts (racemic mixture), the product mixture will be optically inactive even though each enantiomer by itself would display optical activity.

Because of symmetry in the molecule, the product has no stereocenters. However, there is still the possibility of *cis-trans* isomers, so two products are possible. Because of the mechanism of the OsO_4 reaction, only the *cis* isomer will be formed. The product is achiral so there is no optical activity.

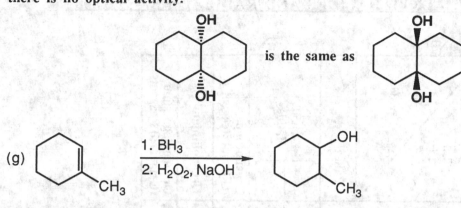

is the same as

(g)

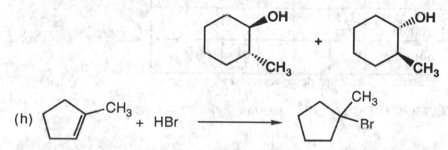

1. BH_3
2. H_2O_2, NaOH

The product molecule has two new stereocenters for a total of four possible stereoisomers. However, due to the syn addition involved with the reaction of borane, the -OH and methyl groups must be *trans* leading to only the two products shown. These are enantiomers and since they will be produced in equal amounts (racemic mixture), the product mixture will be optically inactive even though each enantiomer by itself will display optical activity.

(h)

The product has no stereocenters so no stereoisomerism is possible. The product mixture will therefore be optically inactive.

CHAPTER 7: ALKYL HALIDES AND RADICAL REACTIONS

SUMMARY OF REACTIONS

Starting Material \ Product →	Alkanes	Alkyl Halides	Organocopper (Gilman) Reagent	Organolithium	Organomagnesium (Grignard) Reagents
Alkanes		7A 7.4*			
Alkenes		7B 7.6			
Alkyl Halides				7C 7.7A	7C 7.7A
Gilman Reagent Alkyl Halides	7D 7.7C				
Organolithium				7E 7.7C	

*Section of book that describes reaction.

REACTION 7A: HALOGENATION OF ALKANES (Section 7.4)

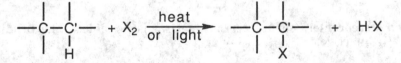

- **Halogenation of alkanes** occurs by a radical mechanism. The reaction begins with an **initiation** step involving the homolytic cleavage of a halogen such as chlorine with heat or light to produce two halogen radicals. ✳
  - **Propagation** of the chain reaction occurs when the halogen radical abstracts an H• from the alkane, leaving an alkyl radical while making H-X. The alkyl radical reacts with another molecule of the halogen, X_2, to create the alkyl halide product and another halogen radical that reacts with another alkane to continue the chain reaction. **Chain termination** occurs when two radical species react with each other to produce a new covalent bond thereby quenching both radicals.
  - Halogenation of alkanes is **regioselective**. That is, reactions using a variety of different alkanes demonstrate that tertiary hydrogen atoms are replaced in preference to secondary hydrogen atoms which are replaced in preference to primary hydrogen atoms. The regioselectivity is derived from the fact that the tertiary radicals are more stable than secondary radicals, secondary radicals are more stable than primary radicals, and methyl radicals are the least stable of all. The carbon atom having the unpaired electron is almost always sp^2 hybridized in alkyl radicals. *[Thus, alkyl radical structure and order of stabilities are largely analogous to alkyl cation structure and order of stabilities.]*
  - Halogenation using bromine is significantly more regioselective than chlorination. In order to understand the detailed regioselectivity observed with halogenation of alkanes, it is helpful to consider **Hammond's postulate**. According to Hammond's postulate, the *structure* of a transition state more closely resembles the stable species (reactants or products) that is closest in energy to the transition state. In other words, the transition state of an endothermic reaction resembles products while the transition state for an exothermic reaction resembles reactants. ✳ *[This concept is useful in a number of situations, so it is important to make sure it is understood before going on.]*

- Abstraction of hydrogen by chlorine or bromine radicals is the first propagation step of the radical chain reaction with alkanes. This step has a higher activation energy for bromine relative to chlorine, therefore, according to Hammond's postulate, the transition state for the bromine reaction resembles the free radical product of this step. For the exothermic chlorination reaction, the transition state is more reactant like, and has little free radical character. In other words, the transition state for the bromine reaction has more radical character and, therefore, a higher degree of selectivity is observed with bromine in favor of the more stable radicals such as allyl or tertiary radicals. *[This is a complex argument ; make sure you understand it before moving on.]*

REACTION 7B: ALLYLIC HALOGENATION (Section 7.6)

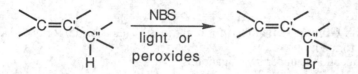

- When different conditions are used, alkenes can react with radical species by a chain reaction mechanism called **allylic substitution**. ✷
- In allylic substitution, radicals such as •Br at low concentration react with an alkene to form an **allylic radical**.
- The •Br could be derived by heat or light causing the homolytic bond dissociation of Br_2, or by using a reagent such as *N*-bromosuccinimide in the presence of either heat or peroxides.
- The allylic radical then reacts with Br_2 to generate the bromoalkene and •Br. The •Br can then continue the chain propagation step by reacting with another molecule of alkene.
- The allylic radical, although still highly reactive, is relatively stable for a radical because the unpaired electron is delocalized. ✷

$$CH_2=CH-CH_2 \longleftrightarrow CH_2-CH=CH_2$$

- The radical delocalization is facilitated by the sp^2 hybridization of the radical carbon, since the 2p orbital containing the unpaired electron can overlap with the 2p orbitals of the pi bond of the alkene.
- For unsymmetrical molecules, the reaction is regioselective. For example, secondary allylic hydrogens are substituted in preference to primary allylic hydrogens, etc.
- Allylic substitution can also occur with Cl_2 and light or heat.

REACTION 7C: FORMATION OF ORGANOLITHIUM AND GRIGNARD REAGENTS
(Section 7.7A)

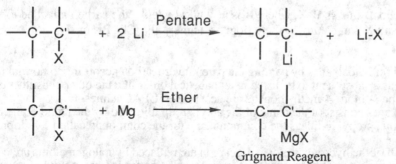

Grignard Reagent

- Alkyl halides react with two mol of lithium metal to form **organolithium reagents** and Li-X. ✷
- Alkyl hallides react with magnesium metal in ether to form **Grignard reagents**.
- Both organolithium reagents and Grignard reagents have **polar C-metal bonds**. Since the metals are less electronegative than carbon, the **carbon atom** has the negative end of the bond dipole moment and thus a **partial negative charge**. The bottom line is that these carbon atoms with a partial negative charge can act as **carbon nucleophiles** with electrophilic species such as carbonyl groups. *[This ability of carbon to act as a nucleophile is important because it provides a convenient way to make carbon-carbon bonds.]* ✷
- Organolithium and Grignard reagents are **very strong bases** and react with any functional group that is a stronger acid (proton donor) than the alkane from which the organolithium or Grignard reagent was derived.

Thus, organolithium and Grignard reagents react with functional groups such as carboxylic acids, thiols, alcohols (and water!), terminal alkynes, amines, etc. In these proton transfer reactions, the parent alkane is generated from the organolithium or Grignard reagent.

REACTION 7D: REACTION OF GILMAN REAGENTS AND ALKYL HALIDES TO PRODUCE ALKANES (Section 7.7C)

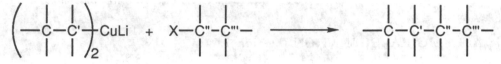

Gilman Reagent

- **Alkyl halides** react with **Gilman reagents** to form **alkanes**. This is an important reaction because a carbon-carbon bond is formed. ✳
- The mechanism of this reaction is under investigation.

REACTION 7E: FORMATION OF GILMAN REAGENTS (Section 7.7C)

Gilman Reagent

- Two mol of **organolithium reagents** react with **CuI** to produce lithium dialkylcopper reagents that are referred to as **Gilman reagents**. ✳
- Gilman reagents are important because they can react with alkyl halides to create carbon-carbon bonds (Reaction 7D, Section 7.7C).

SUMMARY OF IMPORTANT CONCEPTS

7.0 OVERVIEW
- **Haloalkanes**, also known as **alkyl halides** in the common nomenclature, are compounds that contain a **halogen atom** attached to an **sp^3 carbon atom**. The terms haloalkane and alkyl halide are used interchangeably for the rest of the chapter. They can be prepared from and converted into a variety of different compounds. Thus, alkyl halides are an important and versatile class of molecules for organic synthesis. ✳

7.1 STRUCTURE
- The general symbol for haloalkanes is **R-X**, where **R** is **any alkyl group** (the carbon attached to the halogen must be sp^3 hybridized) and **X** can be any of the **halogens**, namely -F, -Cl, -Br, or -I.

7.2 NOMENCLATURE
- IUPAC names are derived for haloalkanes by naming the parent hydrocarbon according to normal rules, and treating the halogen atom as a substituent to be listed in alphabetical order like the other substituents. For example, 2-bromobutane or 3-fluoro-4-methylnonane are acceptable IUPAC names.
- Common names of haloalkanes consist of the name of the alkyl group followed by the name of the halide as a separate word. For example, propyl iodide is the common name for the compound called 1-iodopropane in the IUPAC nomenclature.
  - Haloalkanes in which all of the hydrogens of a hydrocarbon are replaced by halogen atoms are called perhaloalkanes. For example, perchloropropane is the common name for the compound of molecular formula C_3Cl_8.

7.3 PHYSICAL PROPERTIES
- Electronegativity increases in the order I<Br<Cl<F, but bond length increases in the opposite order, namely F<Cl<Br<I. The maximal combination of electronegativity and bond length occurs with C-Cl bonds, so chloroalkanes are the most polar, having the largest bond dipole moments. ✳
  - As liquids, haloalkane molecules are attracted to each other by a combination of dipole-dipole interactions and dispersion forces.

- The **boiling points** of haloalkanes are generally higher than comparable alkanes of similar size and shape. This is because of the dipole-dipole interactions that are possible with the haloalkanes, as well as the increased **polarizability** of halogen atoms compared with hydrocarbons. Large atoms, such as bromine or iodine with lone pairs of electrons, are highly polarizable, that is their electron density can be temporarily "moved around," which increases the strength of induced dipole interactions (dispersion forces) between molecules.
- The **densities** of haloalkanes are higher than hydrocarbons because of the relatively high mass to volume ratios of the halogens, especially bromine and iodine.
- As shown in Table 7.6 in the text, the larger the halogen, the longer and weaker the C-X bond. Only C-F bonds are stronger than C-H bonds.

CHAPTER 7
Solutions to the Problems

<u>Problem 7.1</u> Write the IUPAC name, and where possible, the common name of each compound.

(a)

$$CH_3-\overset{\overset{\displaystyle CH_3}{|}}{CH}\cdot CH_2\text{-}Cl$$

1-Chloro-2-methylpropane
(Isobutyl chloride)

(b)

$$\underset{H}{\overset{H_3C}{>}}C=C\underset{CH_3}{\overset{Cl}{<}}$$

(Z)-2-Chloro-2-propene

(c)

$$\text{(cyclohexane with Cl and } C(CH_3)_3)$$

cis-1-tert-Butyl-4-
chlorocyclohexane

(d)

$$CH_2=\overset{\overset{\displaystyle Cl}{|}}{C}CH=CH_2$$

2-Chloro-1,3-Butadiene

<u>Problem 7.2</u> Name and draw structural formulas for all monochlorination products formed by treatment of butane with Cl_2. Predict the major product based on the regioselectivity of the reaction of Cl_2 with alkanes.

$$CH_3CH_2CH_2CH_3 \ + \ Cl_2 \ \xrightarrow[\text{or light}]{\text{heat}} \ \text{monochlorobutanes} \ + \ HCl$$

$$\underset{\textbf{2-Chlorobutane}}{CH_3\overset{\overset{\displaystyle Cl}{|}}{CH}CH_2CH_3} \qquad\qquad \underset{\textbf{1-Chlorobutane}}{CH_2ClCH_2CH_2CH_3}$$

There are 4 secondary hydrogen atoms and 6 primary hydrogen atoms on the molecule. The ratio of reactivity for 2°:1° chlorination is 4:1. Therefore, the predominant product will be the 2-chlorobutane, formed in approximately:

$$\frac{4 \times 4}{(4 \times 4) + (6 \times 1)} \times 100 = \boxed{73\%}$$

<u>Problem 7.3</u> Using tables of bond dissociation energies (Appendix 3), calculate $\Delta H°$ for bromination of propane to give 1-bromopropane and hydrogen bromide.

$$CH_3CH_2CH_3 \ + \ Br_2 \ \longrightarrow \ CH_3CH_2CH_2Br \ + \ HBr$$

$\Delta H°$ equals the difference between the bond dissociation energies of bonds made vs. bonds broken in the reaction. One C-H bond (+ 100 kcal/mol) and one Br-Br bond (+ 46 kcal./mol) was broken while one C-Br (-68 kcal/mol) bond and one H-Br bond (-88 kcal/mol) was made in this reaction. Thus, for the whole reaction $\Delta H°$ = 100 + 46 - 68 - 88 = - 10 kcal/mol.

<u>Problem 7.4</u> Write a pair of chain propagation steps for the radical bromination of propane to give 1-bromopropane, and calculate $\Delta H°$ for each propagation step, and for the overall reaction.

Following is a pair of chain propagation steps for this reaction. Of these steps, the first involving hydrogen abstraction, has the higher activation energy.

$$CH_3-CH_2-CH_3 \;+\; \cdot Br \longrightarrow CH_3-CH_2-\overset{\cdot}{C}H_2 \;+\; H-Br$$

$$\begin{array}{ccccc} +100 & & & -88 & +12 \end{array}$$

$$\overset{\Delta H^\circ}{(kcal/mol)}$$

$$CH_3-CH_2-\overset{\cdot}{C}H_2 \;+\; Br_2 \longrightarrow CH_3-CH_2-\overset{\overset{\displaystyle Br}{|}}{C}H_2 \;+\; \cdot Br$$

$$\begin{array}{ccccc} +46 & & & -68 & -22 \end{array}$$

$$CH_3-CH_2-CH_3 \;+\; Br_2 \longrightarrow CH_3-CH_2-\overset{\overset{\displaystyle Br}{|}}{C}H_2 \;+\; H-Br \quad \boxed{-10}$$

Problem 7.5 Given the solution to Example 7.5, predict the structure of the product(s) formed when 3-hexene is treated with NBS?

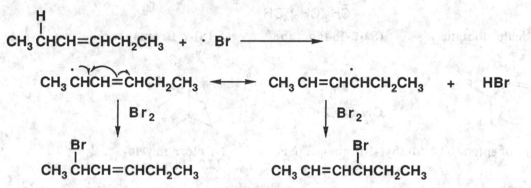

Problem 7.6 Explain how these Grignard reagents will react with molecules of their own kind to "self-destruct."

Grignard reagents are strong bases. In both cases the Grignard reagent will self-destruct due to an acid-base reaction.

(a) $HOCH_2CH_2CH_2MgBr$

$$HOCH_2CH_2CH_2MgBr \;\rightleftharpoons\; (MgBr)^+ \; ^-OCH_2CH_2CH_3$$

(b) $HC\equiv C(CH_2)_4CH_2MgBr$

$$HC\equiv C(CH_2)_4CH_2MgBr \;\rightleftharpoons\; (MgBr)^+ \; ^-C\equiv C(CH_2)_4CH_3$$

Problem 7.7 Show how to bring about each conversion using a lithium diorganocopper reagent.

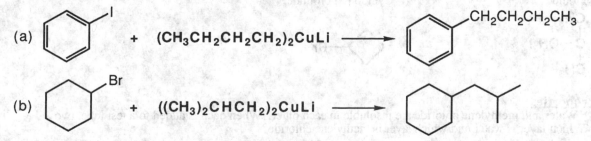

Nomenclature
Problem 7.8 Give IUPAC names for the following compounds. Where stereochemistry is shown, include a designation of configuration in your answer.

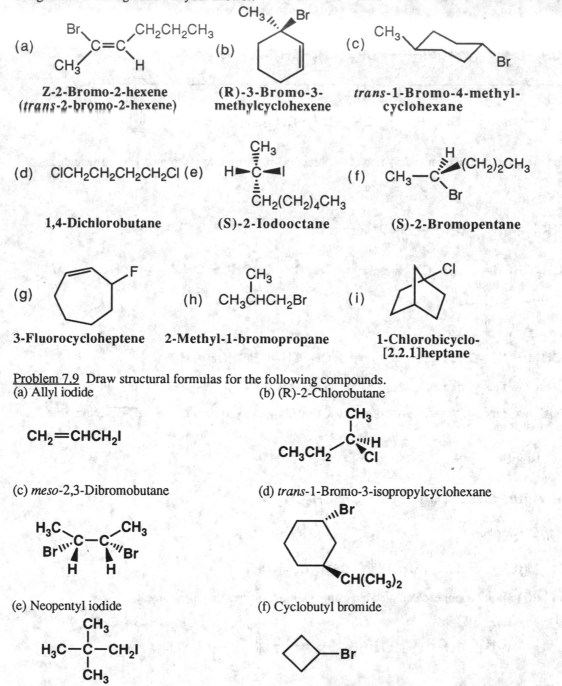

(a)

Z-2-Bromo-2-hexene
(*trans*-2-bromo-2-hexene)

(b)

(R)-3-Bromo-3-
methylcyclohexene

(c)

***trans*-1-Bromo-4-methyl-**
cyclohexane

(d)   ClCH₂CH₂CH₂CH₂Cl

1,4-Dichlorobutane

(e)

(S)-2-Iodooctane

(f)

(S)-2-Bromopentane

(g)

3-Fluorocycloheptene

(h)   CH₃CHCH₂Br

2-Methyl-1-bromopropane

(i)

1-Chlorobicyclo-
[2.2.1]heptane

Problem 7.9 Draw structural formulas for the following compounds.
(a) Allyl iodide   (b) (R)-2-Chlorobutane

CH₂=CHCH₂I

(c) *meso*-2,3-Dibromobutane   (d) *trans*-1-Bromo-3-isopropylcyclohexane

(e) Neopentyl iodide   (f) Cyclobutyl bromide

Physical Properties
Problem 7.10 Water and methylene chloride are insoluble in each other. When each is added to a test tube, two layers form. Which layer is water and which layer is methylene chloride?

The densities of water and methylene chloride are 1.00 and 1.327 g/mL, respectively. The increased density of the methylene chloride is a consequence of the relatively high atomic

weight of the chlorine atoms compared to oxygen, hydrogen, and carbon. Thus, methylene chloride will be the bottom layer.

Problem 7.11 The boiling point of methylcyclohexane (C_7H_{14}, MW 98.2) is 101°C. The boiling point of perfluoromethylcyclohexane (C_7F_{14}, MW 350) is 76°C. Account for the fact that although the molecular weight of perfluoromethylcyclohexane is over 3 times that of methylcyclohexane, its boiling point is lower than that of methylcyclohexane.

This difference is due to the low polarizability of fluorine that is attributed to its small size and the tightness with which its electrons are held.

Problem 7.12 Account for the fact that among the chlorinated derivatives of methane, chloromethane has the largest dipole moment and tetrachloromethane has the smallest dipole moment.

Name	Molecular Formula	Dipole Moment (debyes: D)
chloromethane	CH_3Cl	1.87
dichloromethane	CH_2Cl_2	1.60
trichloromethane	$CHCl_3$	1.01
tetrachloromethane	CCl_4	0

Each C-Cl bond is polar covalent with carbon bearing a partial positive charge and chlorine bearing a partial negative charge. Recall that molecular dipole moments are the vector sum of all the individual bond dipole moments. As shown on the structures below, adjacent C-Cl bond dipoles actually cancel each other to some extent in dichloromethane and trichloromethane, and completely in tetrachloromethane.

Vector sum

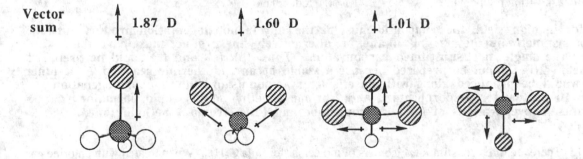

1.87 D   1.60 D   1.01 D

Chloromethane   Dichloromethane   Trichloromethane   Tetrachloromethane

Halogenation of Alkanes
Problem 7.13 Name and draw structural formulas for all possible monohalogenation products that might be formed in the following reactions.

(a)

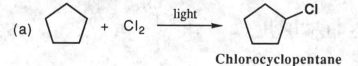

Chlorocyclopentane

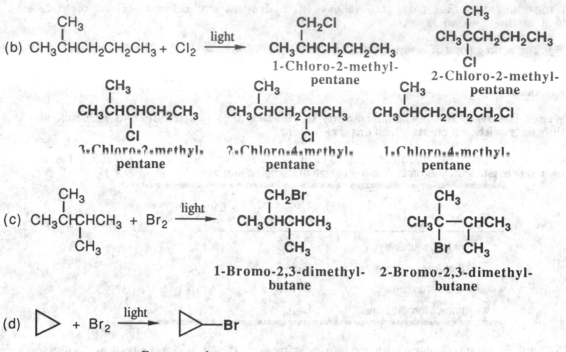

1-Bromo-2,3-dimethyl-
butane

2-Bromo-2,3-dimethyl-
butane

Bromocyclopropane

Problem 7.14 Which compounds can be prepared in high yield by regioselective halogenation of an alkane?
(a) 2-Chloropentane   (b) Chlorocyclopentane
(c) 2-Bromo-2-methylheptane   (d) 2-Bromo-3-methylbutane
(e) 2-Bromo-2,4,4-trimethylpentane   (f) Iodoethane

To be made in high yield, the compound must be the only monohalogentation product possible because of symmetry in the starting alkane, or, alternatively, the product must have the halogen on the single most-substituted carbon atom. Thus, (b), (c), and (e) could be prepared in high yield.   (f) Cannot be prepared because it would be an endothermic reaction. The other products would be produced along with unsatisfactory amounts of other monohalogenation products. In particular, 3-chloropentane and 2-bromo-2-methylbutane would be major contaminants in preparations of (a) 2-chloropentane and (d) 2-bromo-3-methylbutane, respectively.

Problem 7.15 There are three constitutional isomers of molecular formula C_5H_{12}. When treated with chlorine gas at 300°C, isomer A gives a mixture of four monochlorination products. Under the same conditions, isomer B gives a mixture of three monochlorination products and isomer C gives only one monochlorination product. From this information, assign structural formulas to isomers A, B, and C.

Structural formulas for the three alkanes of are:

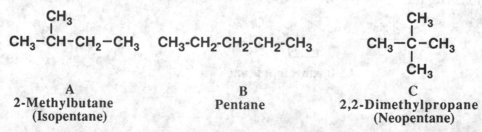

A
2-Methylbutane
(Isopentane)

B
Pentane

C
2,2-Dimethylpropane
(Neopentane)

To arrive at the correct assignments of structural formulas, first write formulas for all monochloroalkanes possible from each structural formula. Then compare these numbers with those observed for A, B, and C. Because isomer B gives three monochlorination products, it must be pentane. By the same reasoning, A must be 2-methylbutane, and C must be 2,2-dimethylpropane.

Problem 7.16 Following is a balanced equation for bromination of propane.

$$CH_3CH_2CH_3 + Br_2 \longrightarrow CH_3\overset{Br}{\underset{|}{C}}HCH_3 + HBr$$

(a) Using the values for bond dissociation energies given in Appendix A, calculate $\Delta H°$ for this reaction.

Formation of 2-bromopropane (isopropyl bromide) by radical bromination of propane is exothermic by 14 kcal/mol.

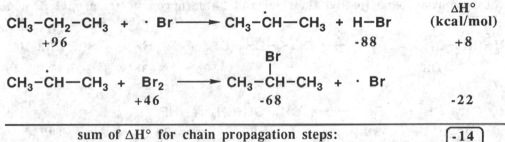

| | | | | | $\Delta H°$ |
| | | | | | (kcal/mol) |

$$CH_3-CH_2-CH_3 + Br_2 \longrightarrow CH_3-\overset{Br}{\underset{|}{C}}H-CH_3 + H-Br$$

+96   +46   -68   -88   -14

(b) Propose a pair of chain propagation steps and show that they add up to the observed reaction.

Following is a pair of chain propagation steps for this reaction. Of these steps, the first involving hydrogen abstraction has the higher activation energy.

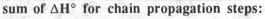

$$CH_3-CH_2-CH_3 + \cdot Br \longrightarrow CH_3-\overset{\cdot}{C}H-CH_3 + H-Br \qquad \begin{array}{c}\Delta H°\\(kcal/mol)\end{array}$$

+96   -88   +8

$$CH_3-\overset{\cdot}{C}H-CH_3 + Br_2 \longrightarrow CH_3-\overset{Br}{\underset{|}{C}}H-CH_3 + \cdot Br$$

+46   -68   -22

sum of $\Delta H°$ for chain propagation steps: $\boxed{-14}$

Following is an alternative pair of chain propagation steps. Because of the considerably higher activation energy of the first of these steps, the rate of chain propagation by this mechanism is so low that it is not competitive with the chain mechanism first proposed.

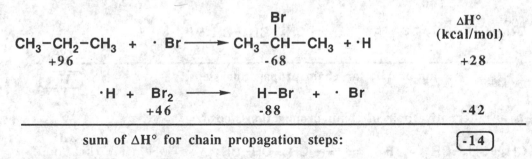

$$CH_3-CH_2-CH_3 + \cdot Br \longrightarrow CH_3-\overset{Br}{\underset{|}{C}}H-CH_3 + \cdot H \qquad \begin{array}{c}\Delta H°\\(kcal/mol)\end{array}$$

+96   -68   +28

$$\cdot H + Br_2 \longrightarrow H-Br + \cdot Br$$

+46   -88   -42

sum of $\Delta H°$ for chain propagation steps: $\boxed{-14}$

(c) Calculate $\Delta H°$ for each chain propagation step.

See answer to part (b)

(d) Which propagation step is rate-limiting, that is, which crosses the higher potential energy barrier?

See answer to part (b)

Problem 7.17 Write a balanced equation and calculate $\Delta H°$ for reaction of CH_4 and I_2 to give CH_3I and HI. Explain why this reaction cannot be used as a method of preparation of iodomethane.

The reaction is not a useful preparation method, because formation of iodomethane (methyl iodide) by radical iodination is *endothermic* by 13 kcal/mol. It will not occur spontaneously.

$$CH_4 \;+\; I_2 \;\longrightarrow\; CH_3I \;+\; HI \qquad \begin{matrix}\Delta H° \\ (kcal/mol)\end{matrix}$$

| +105 | +36 | | -57 | -71 | +13 |

Problem 7.18 Following are balanced equations for fluorination of propane to produce a mixture of 1-fluoropropane and 2-fluoropropane.

$$CH_3CH_2CH_3 \;+\; F_2 \;\longrightarrow\; CH_3CH_2CH_2F \;+\; HF$$
$$\text{Propane} \qquad\qquad\qquad \text{1-Fluoropropane}$$

$$CH_3CH_2CH_3 \;+\; F_2 \;\longrightarrow\; CH_3\overset{\displaystyle F}{\overset{|}{C}}HCH_3 \;+\; HF$$
$$\text{Propane} \qquad\qquad\qquad \text{2-Fluoropropane}$$

Assume that each product is formed by a radical chain mechanism.
(a) Calculate $\Delta H°$ for each reaction.

Formation of 1-fluoropropane (propyl fluoride) and 2-fluoropropane (isopropyl fluoride) are both exothermic.

$$CH_3-CH_2-CH_3 \;+\; F_2 \;\longrightarrow\; CH_3-CH_2-CH_2-F \;+\; HF \qquad \begin{matrix}\Delta H° \\ (kcal/mol)\end{matrix}$$

| | +100 | +38 | | -106 | -136 | -104 |

$$CH_3-CH_2-CH_3 \;+\; F_2 \;\longrightarrow\; CH_3-\overset{\displaystyle F}{\overset{|}{C}}H-CH_3 \;+\; HF$$

| | +96 | +38 | | -107 | -136 | -109 |

(b) Propose a pair of chain propagation steps for each reaction, and calculate ΔH for each step.

$$CH_3-CH_2-CH_3 \;+\; \cdot F \;\longrightarrow\; CH_3-CH_2-\overset{\displaystyle \cdot}{C}H_2 \;+\; HF \qquad \begin{matrix}\Delta H° \\ (kcal/mol)\end{matrix}$$

| | +100 | | | | -136 | -36 |

$$CH_3-CH_2-\overset{\displaystyle \cdot}{C}H_2 \;+\; F_2 \;\longrightarrow\; CH_3-CH_2-CH_2-F \;+\; \cdot F$$

| | | +38 | | -106 | | -68 |

sum of $\Delta H°$ for chain propagation steps: $\boxed{-104}$

Following is an alternative pair of chain propagation steps.

$$CH_3-CH_2-CH_3 \;+\; \cdot F \;\longrightarrow\; CH_3-\overset{\displaystyle \cdot}{C}H-CH_3 \;+\; HF \qquad \begin{matrix}\Delta H° \\ (kcal/mol)\end{matrix}$$

| | +96 | | | | -136 | -40 |

$$CH_3-\overset{\displaystyle \cdot}{C}H-CH_3 \;+\; F_2 \;\longrightarrow\; CH_3-\overset{\displaystyle F}{\overset{|}{C}}H-CH_3 \;+\; \cdot F$$

| | | +38 | | -107 | | -69 |

sum of $\Delta H°$ for chain propagation steps: $\boxed{-109}$

(c) Reasoning from the Hammond postulate, predict the regioselectivity of radical fluorination relative to that of radical chlorination and bromination.

Because the hydrogen abstraction step in each radical fluorination sequence is highly exothermic, the transition state is reached very early in hydrogen abstraction and the intermediate in this step has very little radical character. Therefore, the relative stabilities of primary versus secondary radicals is of little importance in determination of product. Accordingly predict very low regioselectivity for fluorination of hydrocarbons.

Problem 7.19 As you demonstrated in Problem 7.18, radical fluorination of alkanes is highly exothermic. As per Hammond's postulate, assume that the transition state for radical fluorination is almost identical to starting material. With this assumption, estimate the fraction of each monofluoro product formed in the fluorination of 2-methylbutane.

If it is assumed that the transition state for radical flourination is almost identical to starting materials, then relative radical stabilities are not important. Thus, all types of carbon atoms (3°, 2°, and 1°) will react with equal rate. The fraction of each monofluoro product will then be determined by the number of each kind of hydrogen present as shown.

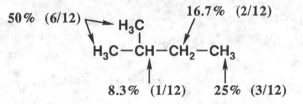

50% (6/12)   16.7% (2/12)

8.3% (1/12)   25% (3/12)

Problem 7.20 Cyclobutane reacts with bromine to give bromocyclobutane, but bicyclo[1.1.0]butane reacts with bromine to give 1,3-dibromocyclobutane. Account for the differences between the reactions of these two compounds.

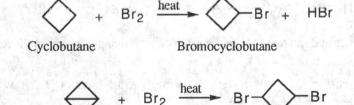

Cyclobutane   Bromocyclobutane

Bicyclo[1.1.0]butane   1,3-Dibromocyclobutane

The upper reaction follows the normal radical chain mechanism. The propagation steps consist of hydrogen atom abstraction followed by reaction with Br_2 to generate the product.

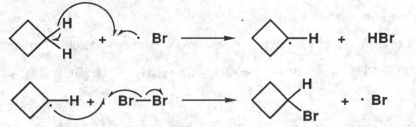

The lower reaction can be explained by a mechanism in which the highly strained bridging bond reacts with the Br radical during the first propagation step as shown below. Note that the extreme ring strain of the bicyclic molecule provides the driving force for the first propagation step.

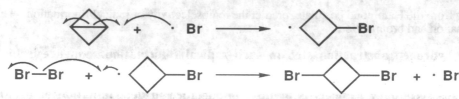

Problem 7.21 The first chain propagation step of all radical halogenation reactions we considered in Section 7.5B is abstraction of hydrogen by the halogen atom to give an alkyl radical and HX, as for example

$$CH_3CH_3 + \cdot Br \longrightarrow CH_3CH_2\cdot + HBr$$

Suppose, instead, that radical halogenation occurs by an alternative pair of chain propagation steps, beginning with this step:

$$CH_3CH_3 + \cdot Br \longrightarrow CH_3CH_2Br + H\cdot$$

(a) Propose a second chain propagation step. Remember that a characteristic of chain propagation steps is that they add to the observed reaction.

$$Br\!-\!Br + \,\,\overset{\frown}{\,\,}H \longrightarrow HBr + \cdot Br$$

(b) Calculate the heat of reaction, $\Delta H°$, for the two steps.

		$\Delta H°$ (kcal/mol)

$$CH_3CH_3 + \cdot Br \longrightarrow CH_3CH_2Br + \cdot H$$
$$+100 \qquad\qquad\qquad -68 \qquad\qquad\qquad +32$$

$$Br\!-\!Br + \cdot H \longrightarrow HBr + \cdot Br$$
$$+46 \qquad\qquad -88 \qquad\qquad -42$$

$$\overline{CH_3CH_3 + Br_2 \longrightarrow CH_3CH_2Br + HBr \qquad \boxed{-10}}$$

(c) Compare the energetics and relative rates of the set of chain propagation steps in Section 7.5B with the set proposed here.

Here, the first propagation step has a very high activation barrier, so it would not compete with the first propagation step proposed in Section 7.5B.

Allylic Halogenation
Problem 7.22 Following is a balanced equation for the allylic bromination of propene.

$$CH_2\!=\!CHCH_3 + Br_2 \longrightarrow CH_2\!=\!CHCH_2Br + HBr$$

(a) Calculate the heat of reaction, ΔH, for this conversion.

See the answer to part (b)

(b) Propose a pair of chain propagation steps and show that they add up to the observed stoichiometry.

$$CH_2\!=\!CH\!-\!CH_3 + \cdot Br \longrightarrow CH_2\!=\!CH\!-\!\overset{\cdot}{C}H_2 + H\!-\!Br \qquad \text{(kcal/mol)}$$
$$+86 \qquad\qquad\qquad\qquad -88 \qquad\qquad -2$$

$$CH_2\!=\!CH\!-\!\overset{\cdot}{C}H_2 + Br_2 \longrightarrow CH_2\!=\!CH\!-\!CH_2Br + \cdot Br$$
$$+46 \qquad\qquad\qquad -55 \qquad\qquad -9$$

$$\overline{\qquad\qquad\qquad\qquad\qquad\qquad\qquad\qquad\qquad\qquad\qquad}$$

sum of $\Delta H°$ for chain propagation steps: $\boxed{-11}$

(c) Calculate the ΔH° for each chain propagation step and show that they add up to the observed ΔH° for the overall reaction.

See the answer to part (b).

<u>Problem 7.23</u> Using the table of bond dissociation energy (Appendix 3), estimate the bond dissociation energy of each indicated bond in cyclohexene.

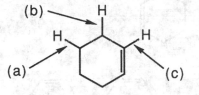

Estimate (a) the bond dissociation energy (BDE) of this secondary site to be 96 kcal/mol, (b) BDE of this allylic site to be 86 kcal/mol, and (c) BDE of this vinylic site to be 106 kcal/mol.

<u>Problem 7.24</u> Propose a series of chain initiation, propagation, and termination steps for this reaction and estimate its heat of reaction.

Initiation:

$$Br_2 \xrightarrow{\text{light}} 2\ Br\cdot$$

Propagation:

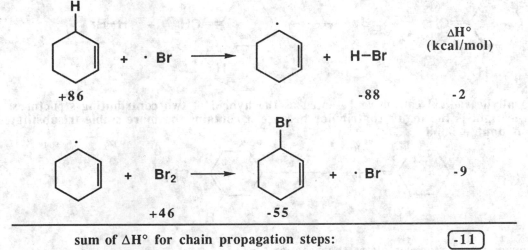

sum of ΔH° for chain propagation steps: -11

Termination:

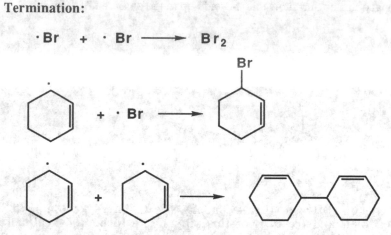

Problem 7.25 The major product formed when methylenecyclohexane is treated with NBS in carbon tetrachloride is 1-bromomethylcyclohexene. Account for the formation of this product.

Recall that NBS can be considered a source of Br radicals and Br$_2$.

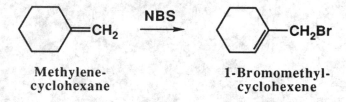

Methylene-
cyclohexane

1-Bromomethyl-
cyclohexene

The above reaction can be explained by a first propagation step involving a hydrogen atom abstraction.

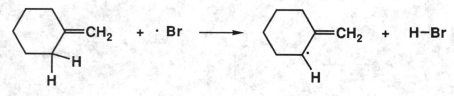

The resulting allylic radical can be represented as the hybrid of two contributing structures. The one on the right is the major contributor because it contains the more stable trisubstituted carbon-carbon double bond.

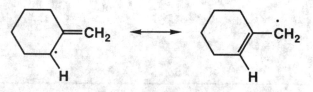

The second propagation step completes the reaction.

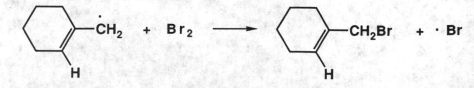

Problem 7.26 Draw the structural formula of the products formed when each alkene is treated with one equivalent of NBS in CCl$_4$ in the presence of benzoyl peroxide. (There are two possible products from each alkene.)
(a) CH$_3$CH=CHCH$_2$CH$_3$

There are two possible allylic radicals that could be produced, giving a total of three possible products:

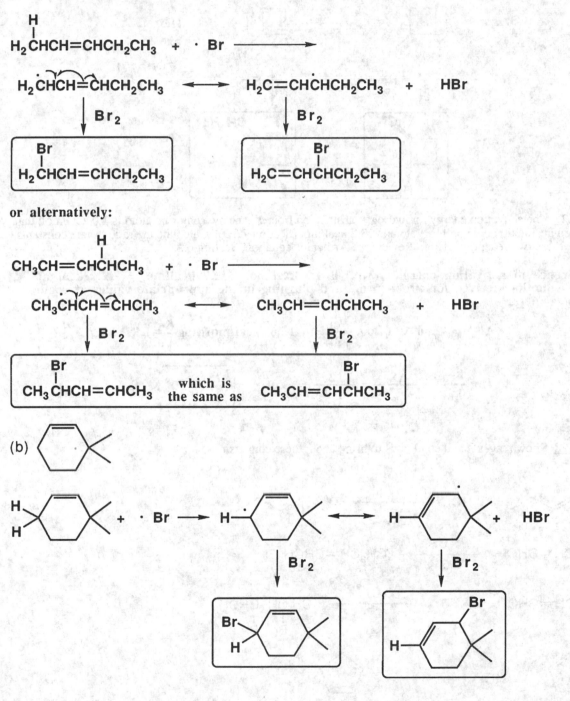

(c)

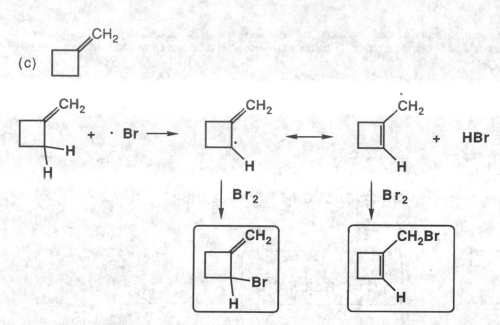

<u>Problem 7.27</u> The activation energy for hydrogen abstraction from ethane by a chlorine atom is 1.0 kcal/mol; that for hydrogen abstraction by a bromine atom is 13.2 kcal/mol (Section 7.5D). Calculate the ratio of rate constants, k_{Cl}/k_{Br}, for these two reactions. Hint: Review the Arrhenius equation, Section 6.2.

The difference in activation energies (ΔE_a) is 1.0 kcal/mol - 13.2 kcal/mol = -12.2 kcal/mol. Using the equation derived for Problem 6.2 and plugging in the appropriate values at room temperature gives:

$$\Delta E_a = -2.303 \text{ RT } \log \frac{k_{Cl}}{k_{Br}} = (-1.36 \text{ kcal/mol})(\log \frac{k_{Cl}}{k_{Br}})$$

$$\log \frac{k_{Cl}}{k_{Br}} = \frac{\Delta E_a}{-1.36} = \frac{-12.2}{-1.36} = 9.0 \qquad \boxed{\frac{k_{Cl}}{k_{Br}} = 10^9}$$

Synthesis
<u>Problem 7.28</u> Show reagents and conditions to bring about these conversions.

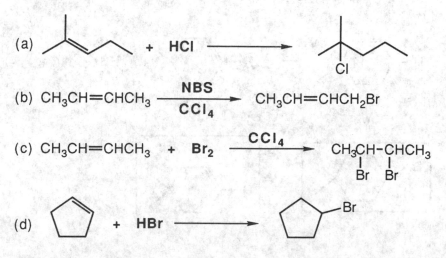

(a)   + HCl ⟶

(b) $CH_3CH=CHCH_3$ $\xrightarrow[\text{CCl}_4]{\text{NBS}}$ $CH_3CH=CHCH_2Br$

(c) $CH_3CH=CHCH_3$ + Br_2 $\xrightarrow{\text{CCl}_4}$ $CH_3CH-CHCH_3$
                                                          | |
                                                         Br   Br

(d)   + HBr ⟶

(e)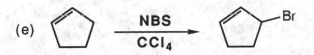

Problem 7.29 Complete these reactions involving lithium diorganocopper (Gilman) reagents.

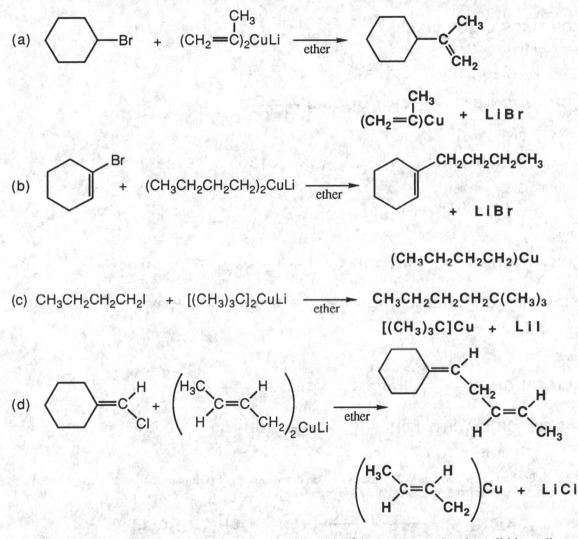

Problem 7.30 Show how to convert 1-bromopentane to each of these compounds using a lithium diorganocopper (Gilman) reagent. Write an equation, showing structural formulas, for each synthesis.
(a) Nonane

$$CH_3CH_2CH_2CH_2CH_2Br + (CH_3CH_2CH_2CH_2)_2CuLi \xrightarrow{ether}$$

$$CH_3(CH_2)_7CH_3 + (CH_3CH_2CH_2CH_2)Cu + LiBr$$
Nonane

(b) 3-Methyloctane

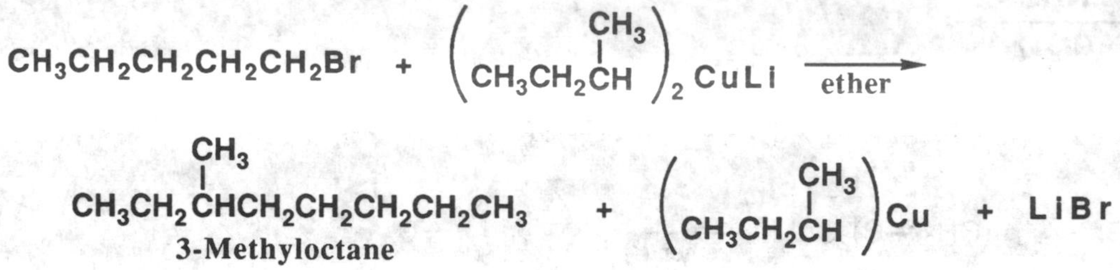

$$CH_3CH_2CH_2CH_2CH_2Br \ + \ \left(\underset{\underset{CH_3CH_2CH}{|}}{\overset{CH_3}{}} \right)_2 CuLi \ \xrightarrow{\text{ether}}$$

$$\underset{\text{3-Methyloctane}}{\overset{CH_3}{\underset{|}{CH_3CH_2CHCH_2CH_2CH_2CH_2CH_3}}} \ + \ \left(\underset{\underset{CH_3CH_2CH}{|}}{\overset{CH_3}{}} \right)Cu \ + \ LiBr$$

(c) 2,2-Dimethylheptane

$$CH_3CH_2CH_2CH_2CH_2Br \ + \ [(CH_3)_3C]_2CuLi \xrightarrow{\text{ether}}$$

$$\underset{\text{2,2-Dimethylheptane}}{\overset{CH_3}{\underset{\underset{CH_3}{|}}{\overset{|}{CH_3CCH_2CH_2CH_2CH_2CH_3}}}} \ + \ [(CH_3)_3C]Cu \ + \ LiBr$$

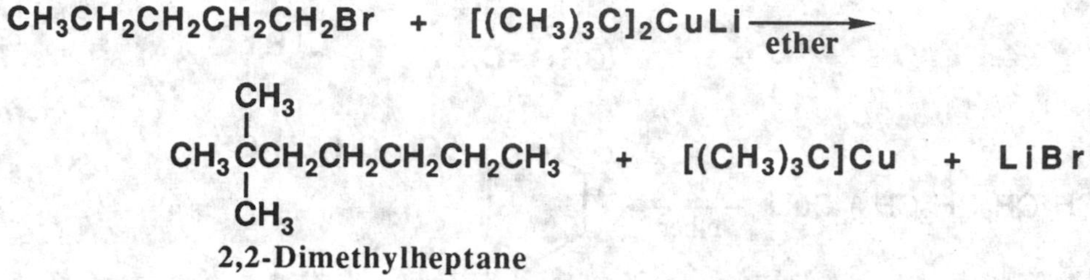

(d) 1-Heptene

$$CH_3CH_2CH_2CH_2CH_2Br \ + \ (CH_2=CH)_2CuLi \xrightarrow{\text{ether}}$$

$$\underset{\text{1-Heptene}}{CH_2=CHCH_2CH_2CH_2CH_2CH_3} \ + \ (CH_2=CH)Cu \ + \ LiBr$$

(e) 1-Octene

$$CH_3CH_2CH_2CH_2CH_2Br \ + \ (CH_2=CHCH_2)_2CuLi \xrightarrow{\text{ether}}$$

$$\underset{\text{1-Octene}}{CH_2=CHCH_2CH_2CH_2CH_2CH_2CH_3} \ + \ (CH_2=CHCH_2)Cu \ + \ LiBr$$

Problem 7.31 In Problem 7.30, you used a series of lithium diorganocopper (Gilman) reagents. Show how to prepare each Gilman reagent from an appropriate alkyl or vinylic halide.

(a) $2 \ CH_3CH_2CH_2CH_2X \xrightarrow{Li} \xrightarrow[\text{ether}]{CuI} (CH_3CH_2CH_2CH_2)_2CuLi$

(b) 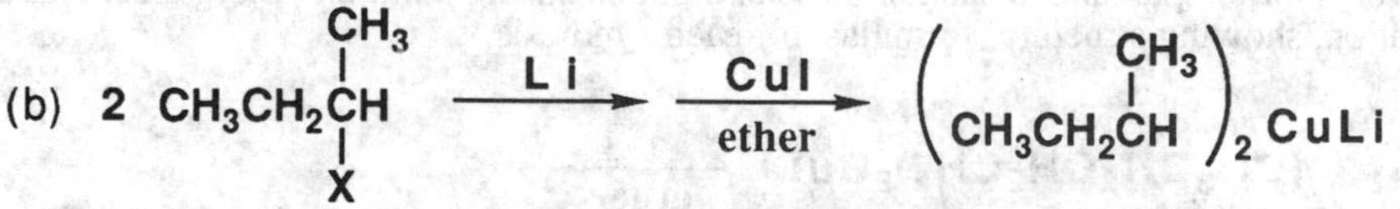 $2 \ \underset{\underset{X}{|}}{\overset{\overset{CH_3}{|}}{CH_3CH_2CH}} \xrightarrow{Li} \xrightarrow[\text{ether}]{CuI} \left(\underset{\underset{CH_3CH_2CH}{|}}{\overset{CH_3}{}} \right)_2 CuLi$

(c) $2 \ (CH_3)_3CX \xrightarrow{Li} \xrightarrow[\text{ether}]{CuI} [(CH_3)_3C]_2CuLi$

(d) 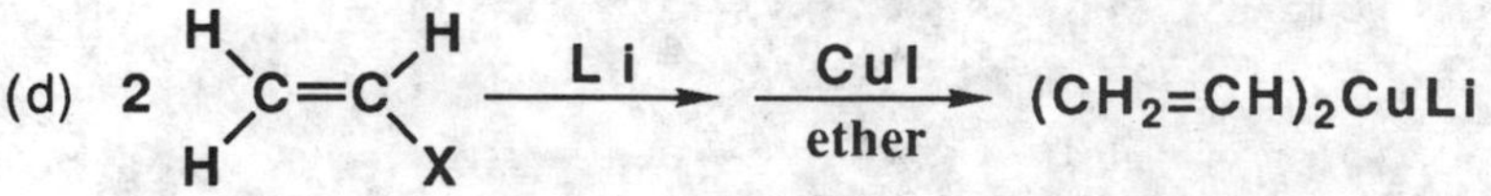 $2 \ \underset{H}{\overset{H}{}}C=C\underset{X}{\overset{H}{}} \xrightarrow{Li} \xrightarrow[\text{ether}]{CuI} (CH_2=CH)_2CuLi$

(e)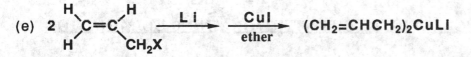

Problem 7.32 Show how to prepare each compound from the given starting compound through the use of a lithium diorganocopper (Gilman) reagent.

(a) 4-Methylcyclopentene from 4-bromocyclopentene

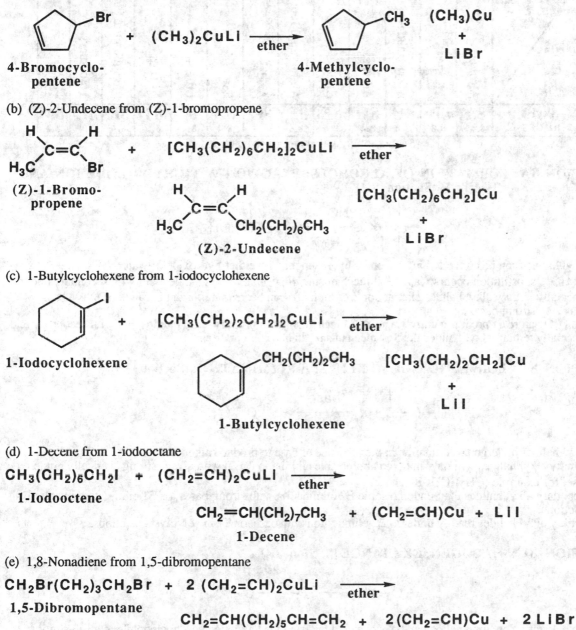

(b) (Z)-2-Undecene from (Z)-1-bromopropene

(c) 1-Butylcyclohexene from 1-iodocyclohexene

(d) 1-Decene from 1-iodooctane

$$CH_3(CH_2)_6CH_2I \ + \ (CH_2{=}CH)_2CuLi \xrightarrow{\text{ether}}$$

1-Iodooctene

$$CH_2{=}CH(CH_2)_7CH_3 \ + \ (CH_2{=}CH)Cu \ + \ LiI$$

1-Decene

(e) 1,8-Nonadiene from 1,5-dibromopentane

$$CH_2Br(CH_2)_3CH_2Br \ + \ 2\,(CH_2{=}CH)_2CuLi \xrightarrow{\text{ether}}$$

1,5-Dibromopentane

$$CH_2{=}CH(CH_2)_5CH{=}CH_2 \ + \ 2\,(CH_2{=}CH)Cu \ + \ 2\,LiBr$$

1,8-Nonadiene

CHAPTER 8: NUCLEOPHILIC SUBSTITUTION AND β-ELIMINATION

SUMMARY OF REACTIONS

Starting Material → Product	Alcohols	Alkenes	Alkyl Halides	Alkynes	Amines	Ammonium Ions	Azides	Esters	Ethers	Nitriles	Phosphonium Salts	Thioethers	Thiols
Alkyl Halides	**8A** 8.4*	**8B** 8.8	**8C** 8.4	**8D** 8.4	**8E** 8.4	**8F** 8.4	**8G** 8.4	**8H** 8.4	**8I** 8.4	**8J** 8.6	**8K** 8.4	**8L** 8.4	**8M** 8.4

*Section of book that describes reaction.

REACTION 8A: FORMATION OF ALCOHOLS: REACTION WITH HYDROXIDE ION AND WATER (Section 8.4)

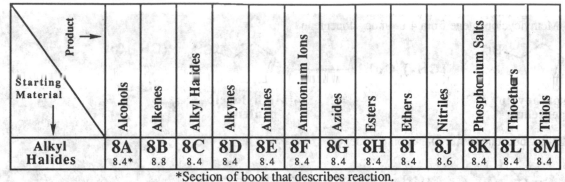

- Alkyl halides can be converted into alcohols by treatment with either hydroxide or water.
- For HO⁻, substitution occurs via an S_N2 mechanism with primary alkyl halides, but with secondary and especially tertiary alkyl halides elimination (reaction 8B) can become important because HO⁻ is also a relatively strong base.
- The H_2O can react predominantly via an S_N2 mechanism with primary alkyl halides, but with secondary and especially tertiary alkyl halides the S_N1 mechanism can become important.

REACTION 8B: FORMATION OF ALKENES: β-ELIMINATION (Section 8.8)

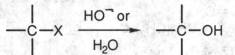

- Alkyl halides undergo β-elimination in the presence of base to produce alkenes.
- Primary alkyl halides will only undergo appreciable elimination (E2) with a very strong, sterically hindered base, for example $(CH_3)_3CO^- K^+$.
- Secondary alkyl halides may undergo some E2 elimination with strong bases, or E1 elimination in solvolysis reactions.
- Tertiary alkyl halides readily undergo E2 elimination with base, or E1 in solvolysis reactions.

REACTION 8C: HALOGEN EXCHANGE (Section 8.4)

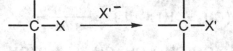

- The halogen of an alkyl halide can be exchanged by using the halide ion as a nucleophile in a substitution reaction.
- Like most of the non-basic nucleophiles, the reaction will take place via an S_N2 mechanism for methyl, primary, and secondary alkyl halides, but the S_N1 mechanism is important for tertiary alkyl halides.

REACTION 8D: ALKYLATION OF ALKYNES: REACTION WITH TERMINAL ALKYNE ANIONS (Section 8.4)

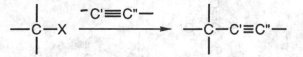

- Alkyne anions react with certain alkyl halides to generate other alkynes. The alkyne anions are produced by deprotonation of terminal alkynes.
- Alkyne anions are such strong bases, that substitution is important only for methyl or primary alkyl halides. Elimination is an important side reaction for secondary alkyl halides and the only reaction for tertiary alkyl halides.

REACTION 8E: ALKYLATION OF AMINES (Section 8.4)

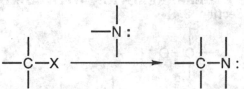

- Amines react with alkyl halides to produce alkylated amines.

REACTION 8F: ALKYLATION OF TERTIARY AMINES (Section 8.4)

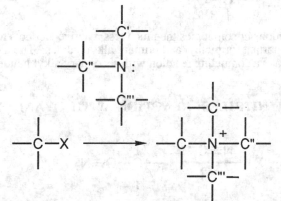

- Tertiary amines also react with alkyl halides under forcing conditions to produce tetraalkylammonium ions.

REACTION 8G: FORMATION OF ALKYL AZIDES: REACTION WITH THE AZIDE ANION (Section 8.4)

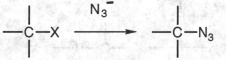

- Alkyl halides react with the azide anion to produce alkyl azides.

REACTION 8H: FORMATION OF ESTERS: REACTION WITH CARBOXYLATE ANIONS AND CARBOXYLIC ACIDS (Section 8.4)

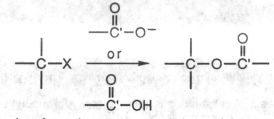

- Alkyl halides react with carboxylate anions or sometimes the much less reactive carboxylic acids to produce esters.

REACTION 8I: FORMATION OF ETHERS: REACTION WITH ALKOXIDE ANIONS AND ALCOHOLS (Section 8.4)

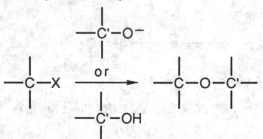

- Alkyl halides react with alkoxide anions or sometimes the much less reactive alcohols to produce ethers.
- Reaction with alkoxides is most important for methyl and primary alkyl halides. The alkoxide anions are such strong bases that β-elimination is a competing reaction with secondary alkyl halides, and the only reaction with tertiary alkyl halides.

REACTION 8J: FORMATION OF NITRILES: REACTION WITH CYANIDE ANION (Section 8.4)

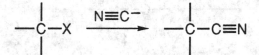

- Alkyl halides react with the cyanide anion to produce nitriles.

REACTION 8K: FORMATION OF PHOSPHONIUM SALTS: REACTION WITH PHOSPHINES (Section 8.4)

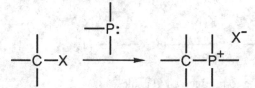

- Alkyl halides react with phosphines to produce phosphonium ions.

REACTION 8L: FORMATION OF THIOETHERS: REACTION WITH THIOLATE ANIONS AND THIOLS Section 8.4)

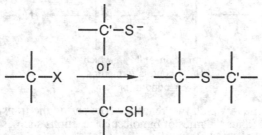

- Alkyl halides react with the thiolate anions or the less nucleophilic thiols to produce thioethers.

REACTION 8M: FORMATION OF THIOLS: REACTION WITH HS⁻ AND H₂S (Section 8.4)

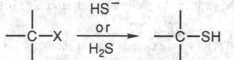

- Alkyl halides react with HS⁻ or the less nucleophilic H_2S to produce thiols.

SUMMARY OF IMPORTANT CONCEPTS

8.1 NUCLEOPHILIC ALIPHATIC SUBSTITUTION
• **Nucleophilic substitution reactions** are reactions in which one nucleophile is substituted for another. They are very important reactions for alkyl halides, because a wide variety of different functional groups can be prepared in this way. In these reactions, the **halogen atom** is **replaced by** the **nucleophile**. ✳
  - For example, Table 8.4 in the text lists a number of negatively-charged and neutral nucleophiles that can react with alkyl halides such as methyl bromide to produce numerous types of molecules.

8.2 SOLVENTS FOR NUCLEOPHILIC SUBSTITUTION REACTIONS
• The solvent can have a strong influence on nucleophilic aliphatic substitution reactions. Two different types of solvents are used in substitution reactions, **protic solvents** and **aprotic solvents**. **Protic solvents** are solvents that contain a functional group such as -OH that can act a hydrogen bond donor. **Aprotic solvents** do not have any functional group that can act as a hydrogen bond donor.
• Solvents are further classified as **polar** and **nonpolar**. **Polar** solvents interact strongly with ions and polar molecules, while **nonpolar** solvents do not interact with ions and polar molecules. **Dielectric constant** is a common measure of solvent polarity and is defined as the amount of electrostatic insulation provided by molecules placed between two charges.
• Water, formic acid, methanol, and ethanol are considered to be **polar protic solvents**. Important **polar aprotic** solvents include dimethyl sulfoxide, acetonitrile, dimethylformamide, and acetone while **nonpolar aprotic** solvents include dichloromethane, diethyl ether, and hexane. It is helpful to categorize solvents this way, because solvents in a given category influence reactions between nucleophiles and alkyl halides in similar ways.

8.3 MECHANISMS OF NUCLEOPHILIC ALIPHATIC SUBSTITUTION
• There are two different limiting mechanisms for nucleophilic aliphatic substitution reactions. ✳
• The term S_N2 stands for Substitution reaction, Nucleophilic, 2nd order (also called bimolecular). According to the S_N2 mechanism, bond-breaking and bond-forming occur at the same time. Thus, both the nucleophile and alkyl halide are involved in the rate-limiting step, hence this is a **bimolecular reaction**. The reaction takes place in a single step, so there is only a single transition state, not any intermediates. In this case, the departing halogen atom is called the **leaving group**. ✳

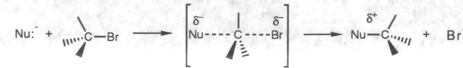

Transition state in which
Nu-C bond is formed as
C-Br bond is broken.

- A **bimolecular reaction** is one in which two reactants take part in the transition state of the slow or rate-limiting step of the reaction. Thus the rates of bimolecular reactions such as S_N2 reactions are proportional to the concentration of both the alkyl halide and nucleophile.
- Since the nucleophile is involved in the rate-limiting step of the S_N2 reaction, stronger nucleophiles react with a faster rate. Stronger nucleophiles are said to have increased **nucleophilicity**.
    In general, within a period of the periodic table, nucleophilicity increases from right to left. Furthermore, for different reagents with the same nucleophilic atom, an anion is a better nucleophile than a neutral species.
• Solvents have a dramatic effect on nucleophilicity.
    -**Polar aprotic** solvents are good at solvating cations but not anions, so anionic nucleophiles participate readily in nucleophilic substitution reactions in polar aprotic solvents.
- **Polar protic** solvents greatly inhibit S_N2 reactions with negatively-charged nucleophiles, because the nucleophile is so highly solvated and thus unreactive. As a result, S_N2 reactions are dramatically faster in polar aprotic solvents, such as acetonitrile (CH_3CN), compared with polar protic solvents like water.
- If the halide leaving group is attached to a stereocenter, then the **configuration** of the stereocenter is **inverted** during an S_N2 **reaction**. This is because the nucleophile enters from the **opposite side** of the molecule **as the departing leaving group**, thus the molecule inverts analogous to the way an umbrella is inverted in the wind.
- S_N2 reactions are particularly **sensitive** to **steric factors**, since they are greatly retarded by steric hindrance (crowding) at the site of reaction.
- Since there is no carbocation produced in S_N2 reactions, there are no skeletal rearrangement observed.
• The term S_N1 stands for $\underline{S}$ubstitution reaction, $\underline{N}$ucleophilic, $\underline{1}$st order (also called unimolecular). According to the S_N1 mechanism, there are two steps. The carbon-halogen bond breaks in the rate-limiting first step, creating a carbocation intermediate that then forms a new bond to the nucleophile in the second step. Only the alkyl halide is involved with the rate-limiting step, thus the reaction is **unimolecular**. Since the reaction involves two steps, there are two transition states and one intermediate. ✳

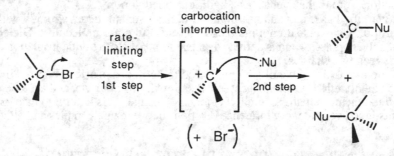

- A **unimolecular reaction** is one in which only one reactant takes part in the transition state of the rate-limiting step. Thus the rates of unimolecular reactions such as S_N1 reactions are proportional to the concentration of the alkyl halide only.
- Since nucleophiles are not involved in the rate-limiting step, stronger nucleophiles do not react faster in S_N1 reactions.
- Because the S_N1 mechanism involves creation and separation of unlike charges in order to form the carbocation intermediate, polar solvents that can stabilize these charges by solvation greatly accelerate S_N1 reactions. For example, S_N1 reactions are much faster in water than in ethanol.
- If the leaving group is attached to a stereocenter, then the **configuration** of the stereocenter is **racemized** during an S_N1 **reaction**. This is because the carbocation intermediate is achiral and the nucleophile can approach from either side, leading to both possible enantiomers as products. In theory, an S_N1 reaction will result in complete racemization, but in fact only partial racemization is observed. This is accounted for by proposing that while bond-breaking between carbon and the leaving group is complete, the leaving group

remains associated for some period of time with the carbocation as an ion pair. To the extent that the leaving group remains associated as an ion pair, it hinders approach of the nucleophile from that face, favoring attack from the opposite face resulting in an excess of inversion.
  - S_N1 reactions are greatly accelerated by electronic factors that stabilize carbocations.
  - Since there is a carbocation intermediate in S_N1 reactions, **skeletal rearrangements** are observed if they produce another carbocation of equal or greater stability.
• The **structure** of the **alkyl halide** greatly **influences** which **mechanism** will be followed. The order of reactivity for the S_N2 mechanism increases in the order: 3° < 2° < 1° allylic = 1° < methyl, since steric hindrance is highest for 3° alkyl halides and least for methyl halides. On the other hand, the order of reactivity for the S_N1 mechanism increases in the order: methyl < 1° < 2° = 1° allylic < 3°, since 3° carbocations are most stable and methyl carbocations are least stable. ✳
  - The net result is that when nucleophilic substitution reactions occur, **methyl** and **1° alkyl** halides react **exclusively** by the S_N2 mechanism. **1° allylic** and **2° alkyl** halides can react by **either** the S_N2 or S_N1 mechanism. **3° alkyl and allylic halides** react **exclusively** by the S_N1 mechanism. ✳
  - Other reaction mechanisms such as β-elimination reactions can take place when a nucleophile reacts with an alkyl halide, and the structure of the alkyl halide greatly influences which of these reactions occurs.
• The leaving group develops a partial negative charge as it is departing by either the S_N1 or S_N2 mechanism. Thus, the lower the basicity, the better a halide is able to function as a leaving group. I^- is the best leaving group, and leaving group ability increases in the order: $F^- < Cl^- < Br^- < I^-$.

8.5 NEIGHBORING GROUP PARTICIPATION
• **Alkyl halides** with a **nucleophilic atom**, usually an N, O, or S atom, β **to an alkyl halide** are **highly reactive.** This is because the neighboring nucleophilic atom aids in the departure of the halide leaving group, producing a reactive, highly strained, three membered ring intermediate that then reacts with an external nucleophile to complete the reaction. Examples include the poisonous mustard gases used in World War 1.

8.7 PHASE TRANSFER CATALYSIS
• Numerous nucleophiles are anions so they are usually soluble in water, but not the organic solvents in which alkyl halides are soluble. To make matters worse, most organic solvents are not miscible with water. In order to get these two reactants together in the same solvent, a **phase transfer catalyst** can be used. A phase transfer catalyst is a **cation** that is surrounded by **hydrophobic** (dissolves in organic solvent) **groups.** An example is the the tetrabutylammonium cation. This cation makes an ionic bond with the anionic nucleophile and brings it into the organic solvent where it can react with the alkyl halide in a nucleophilic substitution reaction.

8.8 ß-ELIMINATION
• Most nucleophiles are also bases and alkyl halides are predisposed to ß-elimination, so this must always be considered as a possible competing reaction for substitution. **ß-Elimination** involves **loss of a leaving group such as the halide ion** and a **proton from a ß-carbon atom** (the carbon adjacent to the one with the halide). The stronger the base, the higher the percentage of ß-elimination product formed in a reaction. Note that when there is more than one ß carbon atom with a hydrogen atom attached, multiple alkene products are possible.

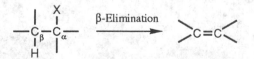

• There are two mechanisms for the ß-elimination reaction of alkyl halides, called **E1** and **E2**. These are analogous in some ways to S_N1 and S_N2 mechanisms.
  - **E1 reactions** involve departure of a leaving group, such as a halide ion, to create a carbocation (analogous to the first step of the S_N1 reaction), followed by departure of a hydrogen atom on a ß-carbon to yield the final product. Like the S_N1 reaction, the carbocation is an intermediate.

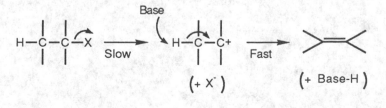

The rate-limiting step in an E1 reaction is loss of the halide to generate the carbocation. Thus, the **E1 reaction is unimolecular** (first order) since the rate only depends on the concentration of alkyl halide. **E1 reactions** give predominantly the **Zaitsev elimination product**, namely the **more highly substituted alkene**. This is because the product determining step has partial double bond character, so the transition state with lower energy is the one with the more stable partial double bond.

- **E2 reactions** are concerted in that the base removes the ß-hydrogen at the same time the C-X bond is broken.

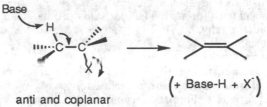

anti and coplanar
geometry of H and X

The only step in the E2 reaction involves both the base and the alkyl halide. Thus, the **E2 reaction is bimolecular** (second order), since the rate depends on both the concentration of base and the concentration alkyl halide.

E2 reactions also give predominantly the **Zaitsev elimination product**, since there is significant partial double bond character in the transition state.

E2 reactions proceed preferentially when the **ß-hydrogen atom** removed by the base and the **departing X atom** are oriented **anti** and **coplanar** to one another. This is particularly important for cyclohexane derivatives, where the ß-hydrogen and X atom must both be axial (one pointing up, one pointing down) to satisfy the anti and coplanar arrangement.

8.11 SUBSTITUTION VERSUS ELIMINATION

• In the absence of any base, tertiary alkyl halides in polar solvents undergo unimolecular reactions to give a combination of substitution (S_N1) and elimination (E1). Although the exact ratios are hard to predict, the amount of substitution can be increased by increasing the concentration of non-basic nucleophile.

• In general, for bimolecular reactions, increased steric hindrance increases the ratio of elimination to substitution products. This is because steric hindrance interferes with the approach of the nucleophile to the backside of the C-X bond, thus impeding the substitution reaction.

- **Tertiary halides** react with all basic reagents to give **elimination products**. There is too much steric hindrance for substitution to compete effectively with elimination.

- **Secondary alkyl halides** have an intermediate amount of steric hindrance and are **borderline**. Substitution or elimination may predominate depending on the particular nucleophile/base, solvent, and temperature of the reaction. Strongly basic nucleophiles such as alkoxides favor E2 reactions, but weakly basic strong nucleophiles favor substitution.

- **Primary alkyl halides** and **methyl halides** have very little steric hindrance, so they react with all nucleophiles, even strongly basic nucleophiles like hydroxide ions and alkoxides ions, to give predominantly **substitution products**.

CHAPTER 8
Solutions to the Problems

<u>Problem 8.1</u> Draw structural formulas for the products of the following nucleophilic aliphatic substitution reactions.

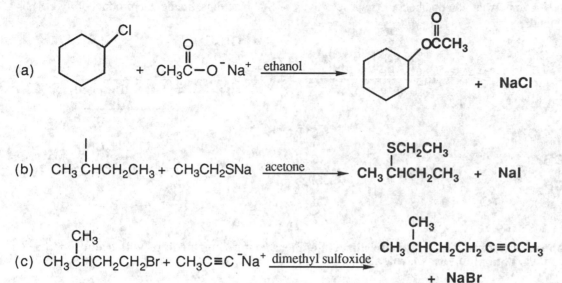

(a)   + $CH_3C-O^-Na^+$ <u>ethanol</u>   + **NaCl**

(b) $CH_3 CHCH_2CH_3$ + CH_3CH_2SNa <u>acetone</u> → $CH_3 CHCH_2CH_3$ + **NaI**

(c) $CH_3 CHCH_2CH_2Br$ + $CH_3C≡C^-Na^+$ <u>dimethyl sulfoxide</u> → $CH_3 CHCH_2CH_2 C≡CCH_3$ + **NaBr**

<u>Problem 8.2</u> Complete these S_N2 reactions, showing the configuration of the product.

In both cases, the stereochemistry at the site of reaction is due to the nucleophile's backside attack that occurs during an S_N2 reaction.

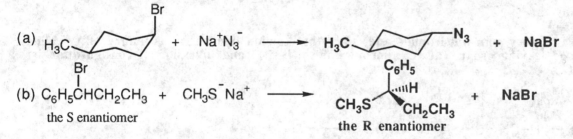

(a) H_3C + $Na^+N_3^-$ → H_3C N_3 + **NaBr**

(b) $C_6H_5CHCH_2CH_3$ + $CH_3S^-Na^+$ →   + **NaBr**

the S enantiomer   the R enantiomer

<u>Problem 8.3</u> Write an additional contributing structure for each carbocation and state which of the two makes the greater contribution to the resonance hybrid.

A more highly substituted carbocation is more stable, so the contributing structure that has the more highly substituted carbocation will make the greater contribution to the resonance hybrid.

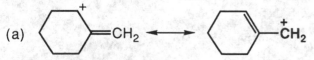

(a) $=CH_2$ ⟷ $-\overset{+}{C}H_2$

Greater contribution
(2° carbocation)

(b)

These are both 2° carbocations.
They make equivalent contributions.

<u>Problem 8.4</u> Knowing what you do about the regioselectivity of S_N2 reactions, predict the product of hydrolysis of this compound.

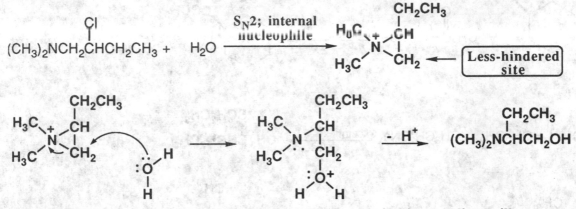

The nucleophilic attack by water on the three-membered ring intermediate will occur on the less-hindered carbon atom as shown above.

<u>Problem 8.5</u> Write the expected substitution product(s) for each reaction and predict the mechanism by which each product is formed.

(a)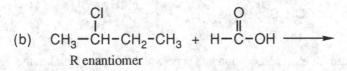

The SH⁻ is a very good nucleophile and, since the reaction involves a secondary alkyl halide with a good leaving group, the reaction mechanism is S_N2 and inversion of configuration is observed.

(b)
$$CH_3-\overset{\overset{Cl}{|}}{CH}-CH_2-CH_3 + H-\overset{\overset{O}{||}}{C}-OH \longrightarrow$$
R enantiomer

The alkyl halide is secondary and chloride is a good leaving group. Formic acid is an excellent ionizing solvent and a poor nucleophile. Therefore, substitution takes place by an S_N1 mechanism and leads to racemization.

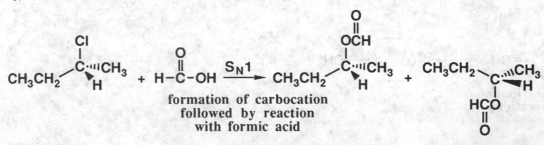

formation of carbocation
followed by reaction
with formic acid

<u>Problem 8.6</u> Predict the β-elimination product(s) formed when each chloroalkane is treated with sodium ethoxide in ethanol. If two or more products might be formed, predict which is the major product.

When there is a choice, the more highly substituted alkene will be the major product, as predicted by Zaitsev's rule.

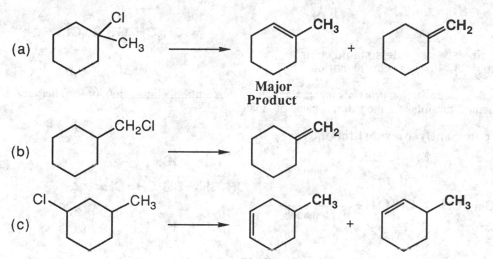

<u>Problem 8.7</u> 1-Chloro-4-isopropylcyclohexane exists as two stereoisomers: one *cis* and one *trans*. Treatment of either isomer with sodium ethoxide in ethanol gives 4-isopropylcyclohexene by an E2 reaction.

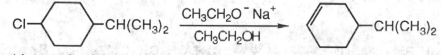

1-Chloro-4-isopropylcyclohexane   4-isopropylcyclohexene

The *cis* isomer undergoes E2 reaction several orders of magnitude faster than the *trans* isomer. How do you account for this experimental observation?

The isopropyl group is the larger substituent on the cyclohexane ring. In the more stable chair conformation of both the *cis* and *trans* isomers, it will be in an equatorial position. In the more stable chair conformation of the *cis* isomer, -Cl is axial and coplanar to -H on adjacent carbons. This chair conformation undergoes β-elimination by an E2 mechanism.

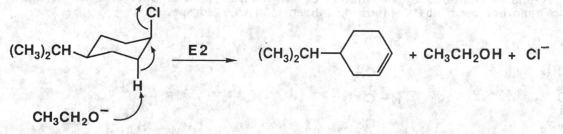

In the more stable chair conformation of the *trans* isomer, chlorine is equatorial and not coplanar to either -H on an adjacent carbon. Interconversion from this chair to the less stable chair results in the -Cl becoming axial and coplanar to an -H atom. It is this conformation that undergoes E2 elimination to give the cycloalkene. The bottom line is that the *cis* isomer undergoes E2 reaction more slowly because of the energy required to convert the more stable chair, but E2-unreactive chair, to the less stable, but E2-reactive, chair.

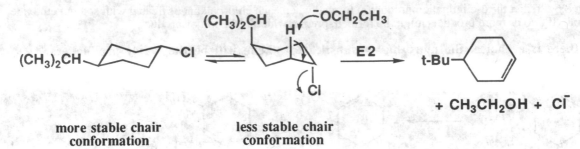

more stable chair   less stable chair
conformation   conformation

Problem 8.8 Predict whether each reaction proceeds predominantly by substitution, elimination, or whether the two compete. Write structural formulas for the major organic product(s).

All will proceed predominantly by substitution.

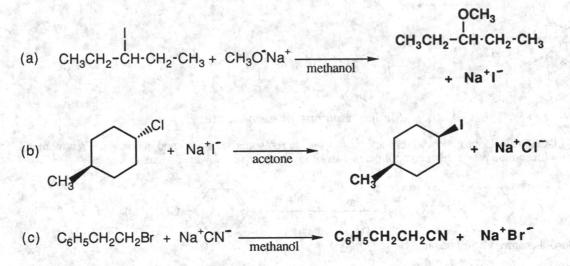

(a) $CH_3CH_2\text{-}CH\text{-}CH_2\text{-}CH_3$ + $CH_3O^-Na^+$ $\xrightarrow{\text{methanol}}$ $CH_3CH_2\text{-}CH\text{-}CH_2\text{-}CH_3$ (OCH$_3$)

     + Na^+I^-

(b) + Na^+I^- $\xrightarrow{\text{acetone}}$ + Na^+Cl^-

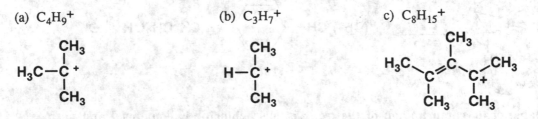

(c) $C_6H_5CH_2CH_2Br$ + Na^+CN^- $\xrightarrow{\text{methanol}}$ $C_6H_5CH_2CH_2CN$ + Na^+Br^-

Nucleophilic Aliphatic Substitution
Problem 8.9 Draw a structural formula for the most stable carbocation of each molecular formula and indicate how each might be formed.

For (a) and (b), the most stable cations are the most highly substituted alkyl cations, while for (c) the most stable cation is the most highly substituted allylic cation. For (d), the most stable cation is resonance stabilized by an adjacent oxygen atom as shown. Spreading the charge over more than one atom has a stabilizing influence on charged species.

(a) $C_4H_9^+$ (b) $C_3H_7^+$ c) $C_8H_{15}^+$

$H_3C-\overset{CH_3}{\underset{CH_3}{\overset{|}{\underset{|}{C}}}}{}^+$ $H-\overset{CH_3}{\underset{CH_3}{\overset{|}{\underset{|}{C}}}}{}^+$ $H_3C-\overset{CH_3}{\underset{CH_3}{\overset{|}{C}}}=\overset{}{\underset{CH_3}{\overset{|}{C}}}{}^+CH_3$

(d) $C_3H_7O^+$

$$H_3C-\overset{+}{\underset{\underset{H}{|}}{C}}-\overset{..}{\underset{..}{O}}CH_3 \quad \longleftrightarrow \quad H_3C-\underset{\underset{H}{|}}{C}=\overset{+}{\underset{..}{O}}CH_3$$

Problem 8.10 Reaction of 1-bromopropane and sodium hydroxide in ethanol follows an S_N2 mechanism. What happens to the rate of this reaction if
(a) the concentration of NaOH is doubled?

The rate of a bimolecular reaction such as the S_N2 reaction is proportional to the concentrations of both the hydroxide and alkyl halide. Thus, if the concentration of hydroxide is doubled, the rate doubles.

(b) the concentrations of both NaOH and 1-bromopropane are doubled?

The rate of a bimolecular reaction, such as the S_N2 reaction, is proportional to the concentrations of both the hydroxide and alkyl halide. Thus, if the concentration of both hydroxide and alkyl halide are doubled, then the rate quadruples.

(c) the volume of the solution in which the reaction is carried out is doubled?

Doubling the volume lowers the concentration of each reactant by a factor of two, so the rate is slower by a total of a factor of four.

Problem 8.11 From each pair, select the stronger nucleophile.

(a) H_2O or OH^- (b) $CH_3\overset{\overset{O}{\|}}{C}O^-$ or OH^- (c) CH_3SH or CH_3S^-

    $OH^- > H_2O$ $OH^- > CH_3\overset{\overset{O}{\|}}{C}O^-$ $CH_3S^- > CH_3SH$

(d) Cl^- or I^- (e) Cl^- or I^- (f) CH_3OCH_3 or CH_3SCH_3
    in DMSO   in methanol

    $Cl^- > I^-$ $I^- > Cl^-$ $CH_3SCH_3 > CH_3OCH_3$

Problem 8.12 Draw the structural formula for the product of each S_N2 reaction. Where configuration of the starting material is given, show the configuration of the product.

(a) $CH_3CH_2CH_2Cl + CH_3CH_2ONa \xrightarrow[\text{ethanol}]{} CH_3CH_2CH_2OCH_2CH_3 + NaCl$

(b) $(CH_3)_3N: + CH_3I \xrightarrow[\text{acetone}]{} (CH_3)_4N^+ \; I^-$

(c)   ⌬—CH_2Br + NaCN $\xrightarrow[\text{acetone}]{}$ ⌬—CH_2CN + NaBr

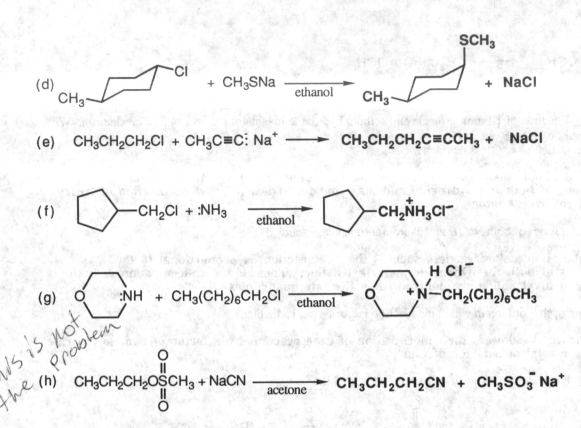

(e) $CH_3CH_2CH_2Cl + CH_3C\equiv C:^- Na^+ \longrightarrow CH_3CH_2CH_2C\equiv CCH_3 + NaCl$

This is not the problem

Problem 8.13 You were told that each reaction in the previous problem proceeds by an S_N2 mechanism. Suppose you were not told the mechanism. Describe how you could conclude from the structure of the alkyl halide, the nucleophile, and the solvent that each reaction is in fact an S_N2 reaction.

(a)   A primary halide, strong nucleophile/strong base in ethanol, a moderately ionizing solvent all favor S_N2.
(b)   Trimethylamine is a moderate nucleophile. A methyl halide in acetone, a weakly ionizing solvent, all work together to favor S_N2.
(c)   Cyanide is a strong nucleophile. A primary halide in acetone, a weakly ionizing solvent, all work together to favor S_N2.
(d)   The alkyl chloride is secondary, so either an S_N1 or S_N2 mechanism is possible. Ethylsulfide ion is a strong nucleophile, but weak base. It therefore reacts by an S_N2 pathway.
(e)   The sodium salt of the terminal alkyne is a moderate nucleophile, but also a strong base. Because the halide is primary, an S_N2 pathway is favored.
(f)   Ammonia is a weak base and moderate nucleophile, and the halide is primary. Therefore S_N2 is favored.
(g)   The major factor here favoring an S_N2 pathway is that the leaving group is halide and on a primary carbon.
(h)   The cyanide anion is a strong nucleophile and mesylate is a good leaving group on a primary carbon. Therefore S_N2 is favored.

Problem 8.14 Treatment of 1,3-dichloropropane with potassium cyanide results in formation of pentanedinitrile (1,3-dicyanopropane). The rate of this reaction is about 1000 times greater in DMSO than it is in ethanol. Account for this difference in rate.

$$Cl\text{-}CH_2CH_2CH_2\text{-}Cl + 2\ K^+CN^- \longrightarrow NC\text{-}CH_2CH_2CH_2\text{-}CN + 2\ K^+Cl^-$$

1,3-Dichloropropane   1,3-Dicyanopropane

The hydroxyl H atom of ethanol is a hydrogen bond donor, so in ethanol the CN⁻ is strongly solvated via hydrogen bonding. This strong solvation slows down the CN⁻ reaction with the alkyl halide. DMSO cannot act as a hydrogen bond donor, so the CN⁻ is not strongly solvated thereby allowing faster reaction with the alkyl halide.

Problem 8.15 Treatment of 1-aminoadamantane, $C_{10}H_{17}N$, with methyl 2,4-dibromobutanoate involves two successive S_N2 reactions and gives compound A, an intermediate in the synthesis of carmantidine. Propose a structural formula for this intermediate. Carmantidine has been used in treating the spasms associated with Parkinson's disease.

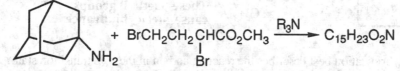

There are two successive S_N2 displacement reactions to give the four-membered ring.

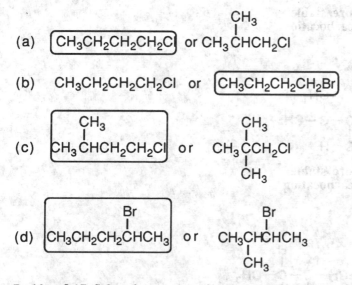

Problem 8.16 Select the member of each pair that shows the faster rate of S_N2 reaction with KI in acetone.

The relative rates of S_N2 reactions for pairs of molecules in this problem depend on two factors: (1) bromine is a better leaving group than chlorine, and (2) a primary carbon without β-branching is less hindered and more reactive toward S_N2 substitution than a primary carbon with one, two, or three branches on the β-carbon atom. The molecule that reacts faster is circled.

(a) $\boxed{CH_3CH_2CH_2CH_2Cl}$ or $CH_3\overset{\underset{\displaystyle CH_3}{|}}{C}HCH_2Cl$

(b) $CH_3CH_2CH_2CH_2Cl$ or $\boxed{CH_3CH_2CH_2CH_2Br}$

(c) $\boxed{CH_3\overset{\underset{\displaystyle CH_3}{|}}{C}HCH_2CH_2Cl}$ or $CH_3\overset{\underset{\displaystyle CH_3}{|}}{\underset{\underset{\displaystyle CH_3}{|}}{C}}CH_2Cl$

(d) $\boxed{CH_3CH_2CH_2\overset{\underset{\displaystyle Br}{|}}{C}HCH_3}$ or $CH_3\overset{\underset{\displaystyle Br}{|}}{C}H\overset{\underset{\displaystyle CH_3}{|}}{C}HCH_3$

Problem 8.17 Select the member of each pair that shows the faster rate of S_N2 reaction with KN_3 in acetone.

The compound with the least steric hindrance will reacts the faster. In both pairs, it is the molecule on the left.

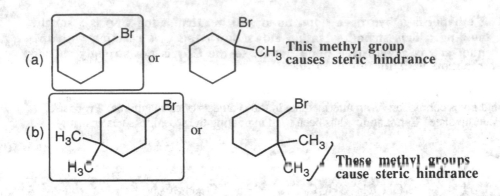

Problem 8.18 What hybridization best describes the reacting carbon in the S_N2 transition state? Would electron-withdrawing groups or electron-donating groups stabilize the transition state better?

The hybridization state of the reacting carbon is best described as sp^2 in the S_N2 transition state. Since the reacting carbon has only minimal net charge, neither electron-withdrawing groups or electron-donating groups have a significant influence on stability of the S_N2 transition state.

Problem 8.19 Each carbocation is capable of rearranging to a more stable carbocation. Limiting yourself to a single 1,2-shift, suggest a structure for the rearranged carbocation.

(a) $(CH_3)_2CH\overset{+}{C}HCH_3$

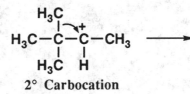

(b) $(CH_3)_3\overset{+}{C}CHCH_3$

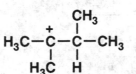

(c) $CH_2=CHCH_2\overset{+}{C}HCH_2CH_3$

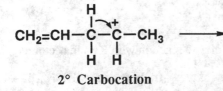

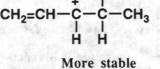

(d) CH₃OCH₂ĊHC(CH₃)₃

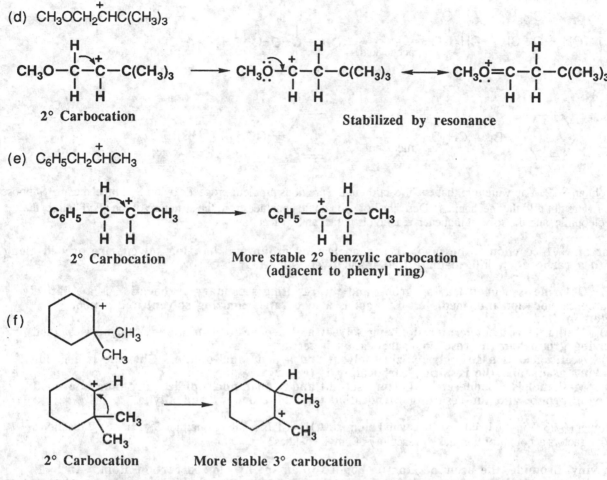

2° Carbocation   Stabilized by resonance

(e) C₆H₅CH₂ĊHCH₃

2° Carbocation   More stable 2° benzylic carbocation
                        (adjacent to phenyl ring)

(f)

2° Carbocation   More stable 3° carbocation

Problem 8.20 Attempts to prepare optically active iodides by nucleophilic displacement on optically active compounds with I⁻ normally produce racemic alkyl iodides. Why are the product alkyl iodides racemic?

Iodide is a good nucleophile as well as a good leaving group. The alkyl iodide that is formed will therefore react with other iodide nucleophiles according to an S_N2 mechanism. The resulting repeated inversion of stereochemistry leads to full racemization of the product.

Problem 8.21 Draw a structural formula for the product of each S_N1 reaction. Where configuration of the starting material is given, show the configuration of the product.

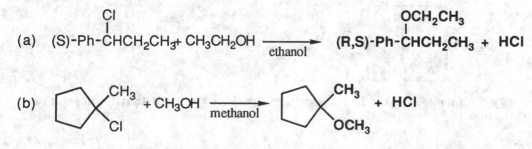

(a) (S)-Ph-ĊHCH₂CH₃ + CH₃CH₂OH $\xrightarrow{\text{ethanol}}$ (R,S)-Ph-ĊHCH₂CH₃ + HCl

(b) [cyclopentane with CH₃ and Cl] + CH₃OH $\xrightarrow{\text{methanol}}$ [cyclopentane with CH₃ and OCH₃] + HCl

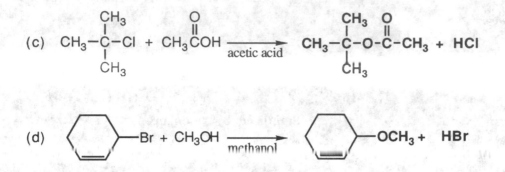

(c)

(d)

Problem 8.22 You were told that each reaction in the previous problem proceeds by an S_N1 mechanism. Suppose you were not told the mechanism. Describe how you could conclude from the structure of the alkyl halide, the nucleophile, and the solvent that each reaction is in fact an S_N1 reaction.

For an S_N1 reaction to be favored, a good leaving group and ionizing solvent are needed along with a carbocation intermediate that is relatively stable.

(a) Chlorine is a good leaving group and the resulting secondary carbocation is a relatively stable carbocation intermediate. Ethanol is a moderately ionizing solvent and a poor nucleophile.
(b) Methanol is a moderately ionizing solvent and a poor nucleophile. Chlorine is a good leaving group and the resulting carbocation is tertiary.
(c) Acetic acid is a strongly ionizing solvent and a poor nucleophile. Chlorine is a good leaving group and the resulting carbocation is tertiary.
(d) Methanol is a moderately ionizing solvent and a poor nucleophile. Bromine is a good leaving group, and the resulting carbocation is both secondary and allylic.

Problem 8.23 Vinylic halides such as vinyl bromide, $CH_2=CHBr$, undergo neither S_N1 nor S_N2 reactions. What factors account for this lack of reactivity of vinylic halides?

In vinyl bromide, the bromine atom is bonded to an sp^2 hybridized carbon atom. An S_N1 reaction would give a vinyl carbocation, but such a carbocation is high in energy and very difficult to generate. In order to undergo an S_N2 reaction, the nucleophile must approach in a direction opposite the C-X bond. This trajectory is not possible for a vinyl halide.

Problem 8.24 Select the member of each pair that undergoes S_N1 solvolysis in aqueous ethanol more rapidly.

Relative rates for each pair of compounds listed in this problem depend on a combination of two factors: (1) bromine is a better leaving group than chlorine and (2) the stability of the resulting carbocation. The molecule that reacts faster is circled.

(a) $CH_3CH_2CH_2CH_2Cl$ or

The activation energy for formation of a tertiary carbocation is lower than that for formation of a primary carbocation.

(b) 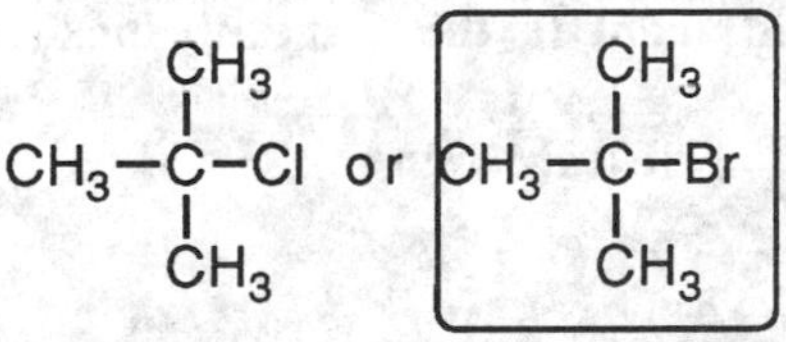 $CH_3-\underset{\underset{CH_3}{|}}{\overset{\overset{CH_3}{|}}{C}}-Cl$ or $\boxed{CH_3-\underset{\underset{CH_3}{|}}{\overset{\overset{CH_3}{|}}{C}}-Br}$

Bromine is a better leaving group than chlorine.

(c) $\boxed{CH_2=CHCH_2Cl}$ or $CH_3CH_2CH_2Cl$

The activation energy for formation of a resonance-stabilized 1° allylic carbocation is lower than that for formation of a primary carbocation.

(d) 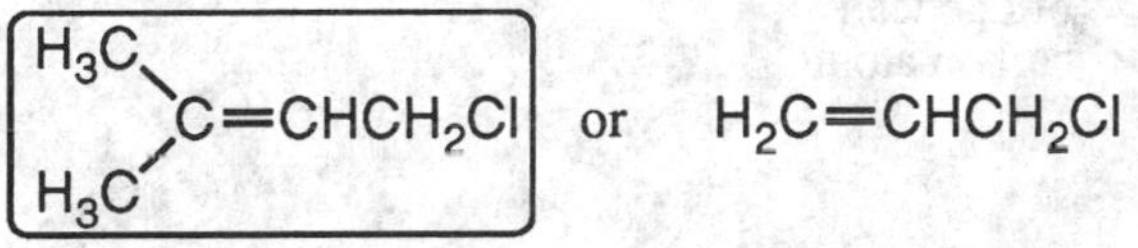 $\boxed{\underset{H_3C}{\overset{H_3C}{>}}C=CHCH_2Cl}$ or $H_2C=CHCH_2Cl$

The activation energy for formation of the dialkyl allylic carbocation is lower than that for formation of an unsubstituted allylic carbocation.

(e) 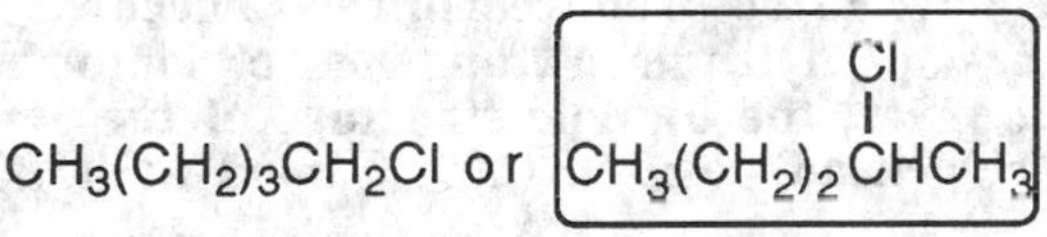 $CH_3(CH_2)_3CH_2Cl$ or $\boxed{CH_3(CH_2)_2\overset{\overset{Cl}{|}}{C}HCH_3}$

The activation energy for formation of a secondary carbocation is lower than that for formation of a primary carbocation.

(f)   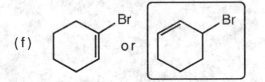   or

The activation energy for formation of an allylic carbocation is lower than that for formation of a vinylic carbocation.

<u>Problem 8.25</u> Account for the following relative rates of solvolysis under experimental conditions favoring S_N1 reaction.

	$CH_3\overset{..}{\underset{..}{O}}CH_2CH_2Cl$	$CH_3CH_2CH_2CH_2Cl$	$CH_3CH_2\overset{..}{\underset{..}{O}}CH_2Cl$
relative rate of solvolysis (S_N1)	0.2	1	10^9

1-Chloro-1-ethoxymethane (chloromethyl ethyl ether) reacts the fastest by far in a solvolysis reaction because the carbocation produced by loss of the chlorine atom is stabilized by the adjacent ether oxygen atom. Thus, the activation energy for this reaction is significantly lower than for the other two molecules. The most important contributing structures are shown below for the stabilized carbocation.

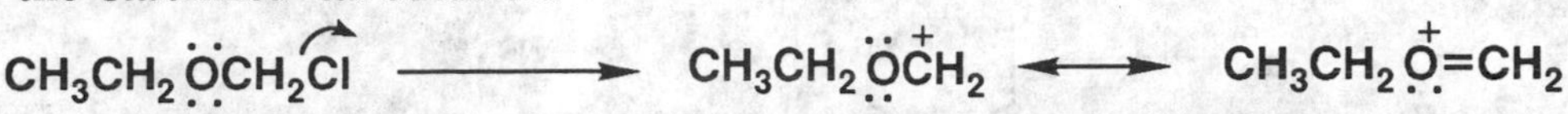

$$CH_3CH_2\overset{..}{\underset{..}{O}}CH_2\overset{\frown}{Cl} \longrightarrow CH_3CH_2\overset{..}{\underset{..}{O}}\overset{+}{C}H_2 \longleftrightarrow CH_3CH_2\overset{+}{\underset{..}{O}}=CH_2$$

1-Chloro-2-methoxyethane (2-chloroethyl methyl ether) reacts the slowest because the carbocation produced during the reaction is somewhat destabilized by the ether oxygen atom that is two carbon atoms away. This is because oxygen is more electronegative than carbon, so there is a partial positive charge on the carbon atoms bonded to the oxygen. Thus, the

carbocation produced by departure of the chlorine atom is adjacent to this partially positive carbon atom; a destabilizing arrangement.

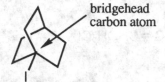

$$CH_3 - \overset{\delta-}{\underset{..}{O}} - CH_2CH_2Cl \longrightarrow CH_3 - \overset{\delta-}{\underset{..}{O}} - CH_2CH_2^+$$

Please note that in the case of 1-chloro-2-methoxyethane, there is no way to produce contributing structures with any positive charge on the oxygen atom like that shown for 1-chloro-1-ethoxymethane above.

Problem 8.26 Not all tertiary halides undergo S_N1 reactions readily. For example, 1-iodobicyclo-[2.2.2]octane is very unreactive under S_N1 conditions. What feature of this molecule is responsible for such lack of reactivity?

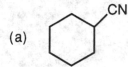

bridgehead carbon atom

1-Iodobicyclo[2.2.2]octane

In order to form a cation, great angle strain would have to be produced in the molecule. This is because carbocations prefer to be trigonal planar (sp^2 hybridized), and loss of iodine would place a carbocation at the bridgehead position. However, the bicyclic structure of the molecule enforces a tetrahedral geometry at the bridgehead position (109.5° bond angles), thus preventing formation of the carbocation.

Problem 8.27 Show how you might synthesize the following compounds from an alkyl halide and a nucleophile:

(a)

CN

Treatment of a halocyclohexane with cyanide.

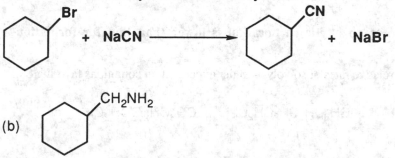

$$\text{(Br-cyclohexane)} + NaCN \longrightarrow \text{(CN-cyclohexane)} + NaBr$$

(b)

CH$_2$NH$_2$

Treatment of chloromethylcyclohexane with two mol ammonia. The first mol ammonia is for displacement of chlorine. The second mol ammonia is to neutralize the HCl formed in the substitution reaction.

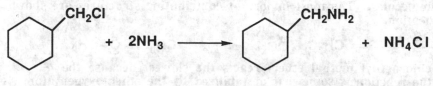

$$\text{(CH}_2\text{Cl-cyclohexane)} + 2NH_3 \longrightarrow \text{(CH}_2\text{NH}_2\text{-cyclohexane)} + NH_4Cl$$

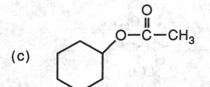

(c)

Treatment of a halocyclohexane with the sodium salt of acetic acid.

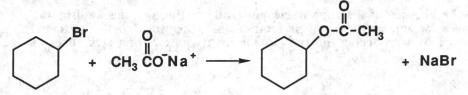

(d) $CH_3(CH_2)_3CH_2SH$

Treatment of a 1-halopentane with sodium hydrosulfide.

$CH_3(CH_2)_3CH_2Br + HS^-Na^+ \longrightarrow CH_3(CH_2)_3CH_2SH + NaBr$

(e) $CH_3(CH_2)_5C\equiv CH$

Treatment of a 1-halohexane with the sodium salt of acetylene.

$CH_3(CH_2)_4CH_2Br + HC\equiv C^-Na^+ \longrightarrow CH_3(CH_2)_4C\equiv CH + NaCl$

(f) $CH_3CH_2OCH_2CH_3$

Treatment of a haloethane with sodium or potassium ethoxide in ethanol.

$CH_3CH_2O^-Na^+ + CH_3CH_2I \xrightarrow{CH_3CH_2OH} CH_3CH_2OCH_2CH_3 + NaI$

(g)

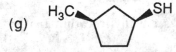

Treatment of the appropriate *trans*-halocyclopentane with the thiol anion.

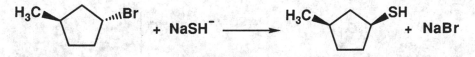

Problem 8.28 3-Chloro-1-butene reacts with sodium ethoxide in ethanol to produce 3-ethoxy-1-butene. The rate of this reaction is second order; first order in 3-chloro-1-butene and first order in sodium ethoxide. In the absence of sodium ethoxide, 3-chloro-1-butene reacts with ethanol to produce both 3-ethoxy-1-butene and 1-ethoxy-2-butene. Explain these results.

For the second order reaction with sodium ethoxide, the mechanism is S_N2 so the $-OCH_2CH_3$ group ends up only where the $-Cl$ leaving group was attached.

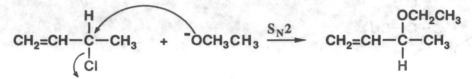

3-Chloro-1-butene   3-Ethoxy-1-butene

For the second reaction, the absence of a strong nucleophile allows the S_N1 mechanism to operate. The allylic carbocation intermediate can be attacked at either the 1 or 3 positions, a fact that is readily explained by considering the two predominant contributing structures of this intermediate.

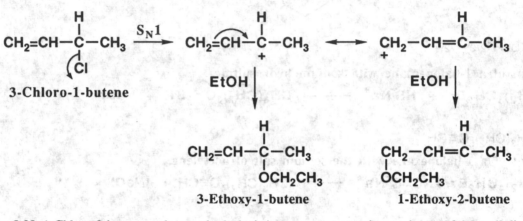

3-Chloro-1-butene

3-Ethoxy-1-butene   1-Ethoxy-2-butene

Problem 8.29 1-Chloro-2-butene undergoes hydrolysis in warm water to give a mixture of these allylic alcohols. Propose a mechanism for their formation.

$$CH_3CH=CHCH_2Cl \xrightarrow{H_2O} CH_3CH=CHCH_2OH + CH_3\overset{\underset{|}{OH}}{C}HCH=CH_2$$

1-Chloro-2-butene   2-Buten-1-ol   3-Buten-2-ol

The two products can be explained by an S_N1 mechanism that produces an allylic cation that can react with water at either the 1 or 3 position.

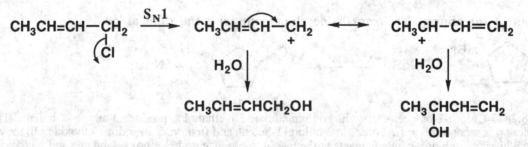

Problem 8.30 In the following reaction, nucleophilic substitution occurs with rearrangement. Suggest a mechanism for formation of the observed product. If the starting material is optically active, the product is also optically active. (*Hint:* an intermediate, $C_8H_{18}NCl$, can, with care, be isolated from the reaction mixture. This intermediate is water soluble.)

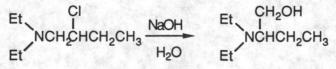

An internal nucleophilic reaction leads to the three-membered ring intermediate shown, that is then attacked by hydroxide at the less-hindered site to yield the product.

Step 1:

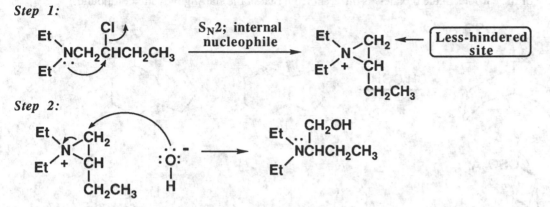

Step 2:

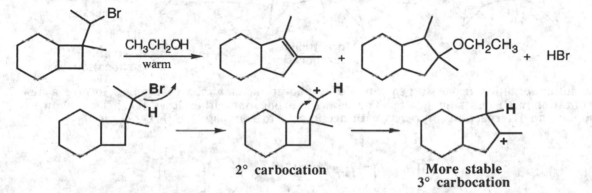

<u>Problem 8.31</u> Propose a mechanisms for the formation of these products in the solvolysis of this alkyl bromide.

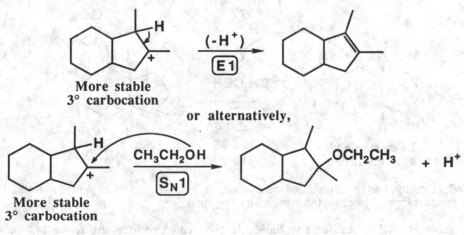

The bromide leaves to generate a 2° carbocation, that then rearranges to give the more stable 3° carbocation. Note how this new 3° carbocation with a five-membered ring also has much less ring strain than the strained four-membered ring 2° carbocation. The new 3° carbocation either loses the proton shown to complete an E1 reaction, or reacts with ethanol to complete the S_N1 reaction.

Problem 8.32 Solvolysis of the following bicyclic compound in acetic acid gives a mixture of products, two of which are shown. The leaving group is the anion of a sulfonic acid, Ar-SO₃H. A sulfonic acid is a strong acid and its anion, $ArSO_3^-$, is a weak base and a good leaving group. Propose a mechanism for this reaction. (*Hint:* The connectivity of the carbons in the products is different from that in the starting bicyclic compound.)

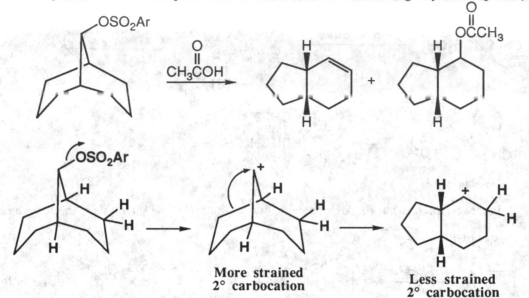

The sulfonate anion leaves to generate a 2° carbocation, that then rearranges to give a new 2° carbocation with less ring strain. The new 2° carbocation either loses the proton shown to complete an E1 reaction, or reacts with acetic acid to complete the S_N1 reaction.

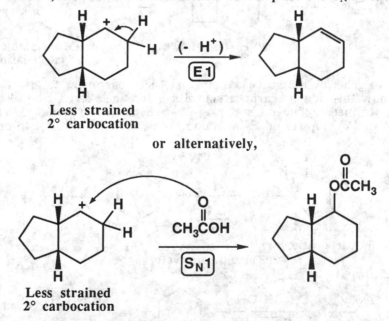

Problem 8.33 Which compound in each set undergoes more rapid solvolysis when heated at reflux in ethanol. Show the major product formed from the more reactive compound.

More rapid solvolysis will occur for the molecule that can produce the more stable carbocation or all things being equal, the one with the better leaving group.

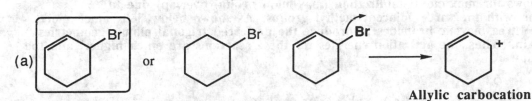

The molecule on the left can make a more stable allylic carbocation.

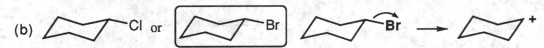

Bromide is a better leaving group than chloride.

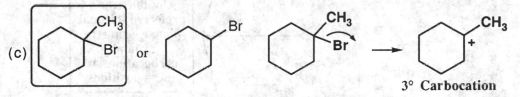

The molecule on the left can make a more stable 3° carbocation.

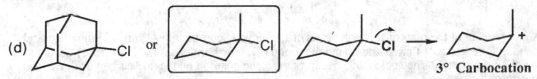

The molecule on the right can make a stable 3° carbocation. In fact, the molecule on the left cannot adopt the favored trigonal planar geometry so it reacts much slower than expected.

<u>Problem 8.34</u> Account for the relative rates of solvolysis of these compounds in aqueous acetic acid.

$(CH_3)_3CBr$

1	10^{-2}	10^{-7}	10^{-12}

More rapid solvolysis will occur for the molecule that can produce the more stable carbocation.

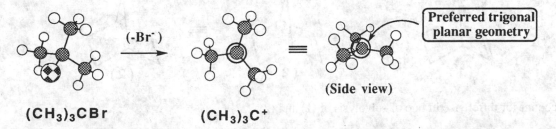

$(CH_3)_3CBr$ $(CH_3)_3C^+$ (Side view)

The *tert*-butyl carbocation produced upon solvolysis of *tert*-butyl bromide can easily adopt the preferred trigonal planar geometry as shown above. A trigonal planar geometry is preferred

because this allows for maximum stabilization (maximum orbital overlap) due to hyperconjugation with the three adjacent methyl groups. As shown below, it is much more difficult for the three bicyclic bromides to produce the preferred trigonal planar geometries due to ring strain. Thus, the activation energies for these reactions are much higher, and the reactions are slower.

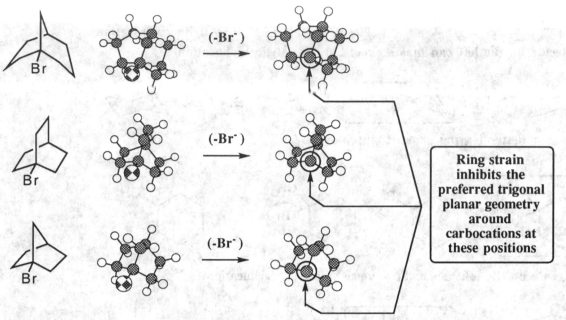

Ring strain inhibits the preferred trigonal planar geometry around carbocations at these positions

Problem 8.35 A comparison of the rates of S_N1 solvolysis of the bicyclic compounds (1) and (2) in acetic acid shows that compound (1) reacts 10^{11} times faster than compound (2). Furthermore, solvolysis of (1) occurs with complete retention of configuration: The nucleophile occupies the same position on the one-carbon bridge as did the leaving OSO$_2$Ar group.

(a) Draw structural formulas for the products of solvolysis of each compound.

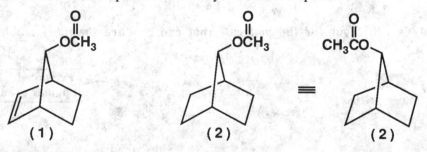

(b) Account for difference in rate of solvolysis of (1) and (2).

The pi bond of (1) is in position to not only assist in the departure of the -OSO$_2$Ar leaving group, but also stabilize the resulting carbocation intermediate. This stabilization can be visualized by considering the filled pi bonding orbital of the double bond overlapping with the empty 2p orbital of the cation. This interaction has the effect of placing some pi electron density into the positively charged carbon 2p orbital, thereby stabilizing the positive charge.

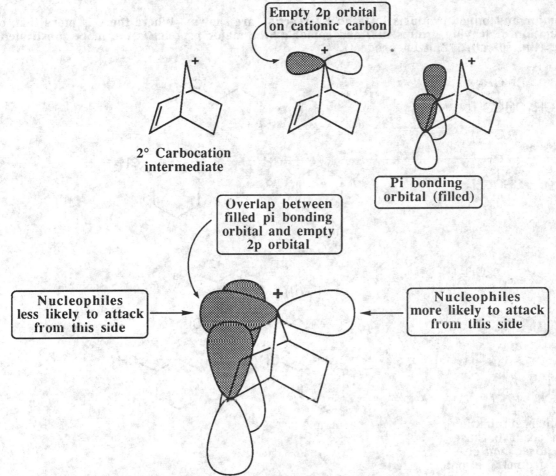

(c) Account for complete retention of configuration in the solvolysis of (1).

As shown on the figure above, the pi bond will block a nucleophile from attacking the cation on one face. Thus, the nucleophile will approach from the other side, the side that the -OSO$_2$Ar departed from, leading to predominantly retention of configuration.

β-Eliminations

<u>Problem 8.36</u> Draw structural formulas for the alkene(s) formed by treatment of each alkyl halide with sodium ethoxide in ethanol. Assume that elimination occurs by an E2 mechanism.

The major and minor products for each E2 reaction are shown. Where there is more than one combination of leaving groups *anti* and coplanar, the major product is the more substituted alkene (the so-called Zaitzev product).

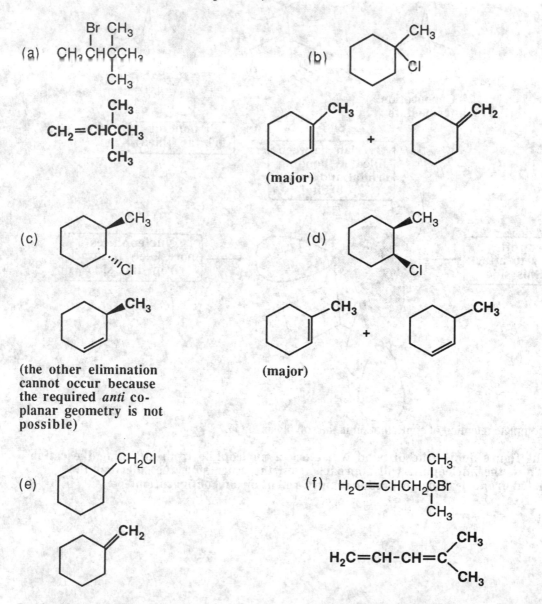

(a)

(b)

(major)

(c)

(the other elimination cannot occur because the required *anti* coplanar geometry is not possible)

(d)

(major)

(e)

(f)

<u>Problem 8.37</u> Draw the structural formula of all chloroalkanes that undergo dehydrohalogenation when treated with KOH to give each alkene as the major product. For some parts, only one chloroalkane gives the desired alkene as the major product. For other parts, two chloroalkanes may give the desired alkene as the major product.

Recall that the alkene shown must be the most highly substituted of the possible elimination products for the starting chloroalkane(s).

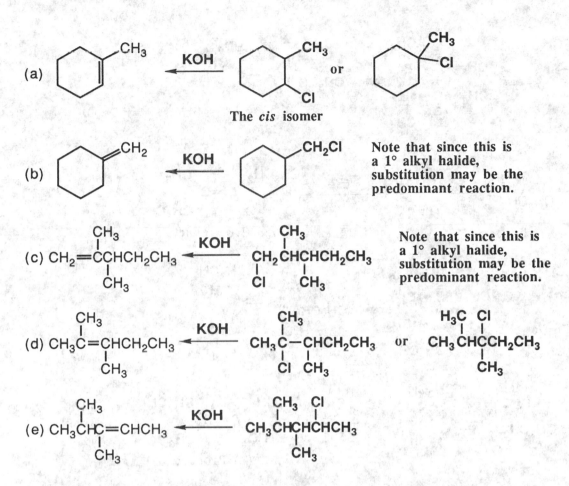

(a)   KOH   ←   or

The *cis* isomer

(b)   KOH   ←   Note that since this is
                                                         a 1° alkyl halide,
                                                         substitution may be the
                                                         predominant reaction.

(c) $CH_2=CCHCH_2CH_3$ KOH   ← $CH_2CHCHCH_2CH_3$ Note that since this is
                                                        a 1° alkyl halide,
                                                        substitution may be the
                                                        predominant reaction.

(d) $CH_3C=CHCH_2CH_3$ KOH   ← $CH_3C-CHCH_2CH_3$ or $CH_3CHCCH_2CH_3$

(e) $CH_3CHC=CHCH_3$ KOH   ← $CH_3CHCHCHCH_3$

Problem 8.38 Following are diastereomers (A) and (B) of 3-bromo-3,4-dimethylhexane. On treatment with sodium ethoxide in ethanol, each gives 3,4-dimethyl-3-hexene as the major product. One of these diastereomers gives the (E)-alkene, and the other gives the (Z)-alkene. Which diastereomer gives which alkene? Account for the stereoselectivity of each β-elimination.

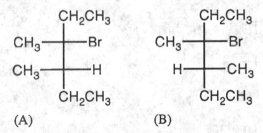

(A)   (B)

Rotate the given conformation of each stereoisomer into a conformation in which Br and H are anti and coplanar and then made to undergo E2 elimination. You will find that diastereomer (A) gives the (E)-isomer and diastereomer (B) gives the (Z)-isomer.

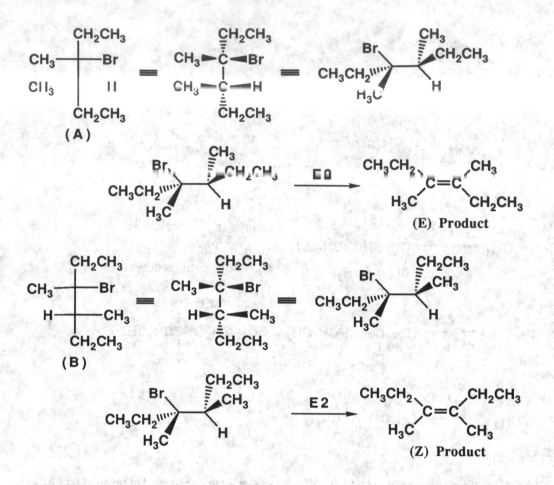

(A)

(E) Product

(B)

(Z) Product

Problem 8.39 Treatment of the following stereoisomer (Fischer projection) of 1-bromo-1,2-diphenylpropane with sodium ethoxide in ethanol gives a single stereoisomer of 1,2-diphenylpropene. Knowing what you do about the stereoselectivity of E2 reactions, predict whether the product has the E configuration or the Z configuration.

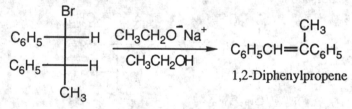

1-Bromo-1,2-
diphenylpropane

1,2-Diphenylpropene

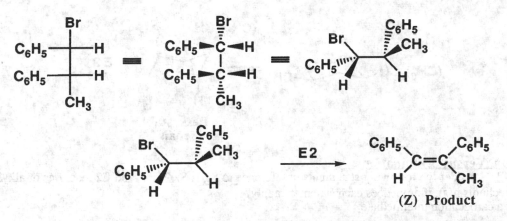

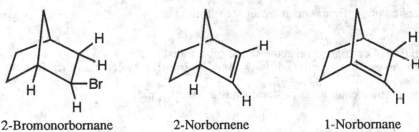

(Z) Product

Problem 8.40 Elimination of HBr from 2-bromonorbornane gives only 2-norbornene and no 1-norbornene. How do you account for the regioselectivity of this dehydrohalogenation? In answering this question, you may find it helpful to make molecular models of both 1-norbornene and 2-norbornene and analyze the angle strain in each.

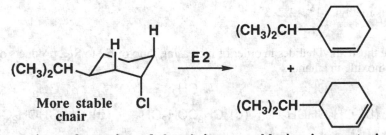

2-Bromonorbornane   2-Norbornene   1-Norbornane

1-Norbornene is not formed for at least two reasons. First, it is not possible to achieve the preferred *anti* and coplanar geometry of the hydrogen and Br atom that would lead to 1-norbornene. Second, the alkene in 1-norbornene has considerably more angle strain compared with 2-norbornene.

Problem 8.41 Which compound reacts faster when heated at reflux with potassium *tert*-butoxide in *tert*-butyl alcohol, *cis*-1-bromo-3-isopropylcyclohexane or *trans*-1-bromo-3-isopropylcyclohexane? Draw the structure of the expected product from the faster reacting compound.

The isopropyl group is the larger substituent on the cyclohexane ring. In the more stable chair conformation of both the *cis* and *trans* isomers, it will be in an equatorial position. In the more stable chair conformation of the *trans* isomer, -Cl is axial and coplanar to -H on adjacent carbons. This chair conformation readily undergoes β-elimination by an E2 mechanism so this is the fastest reaction.

More stable chair   E2

In the more stable chair conformation of the *cis* isomer, chlorine is equatorial and not coplanar to either -H on an adjacent carbon. Interconversion from this chair to the less stable chair results in the -Cl becoming axial and coplanar to an -H on an adjacent carbon atom. It is this conformation that undergoes E2 elimination to give the cycloalkene. The bottom line is that the *cis* isomer undergoes E2 reaction more slowly because of the energy required to convert the more stable chair, but E2-unreactive chair, to the less stable, but E2-reactive, chair.

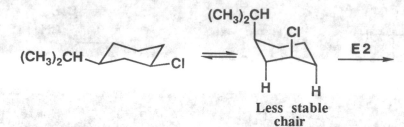

Substitution Versus Elimination

Problem 8.42 Consider the following statements in reference to S_N1, S_N2, E1, and E2 reactions of alkyl halides. To which mechanism(s), if any, does each statement apply?

(a) Involves a carbocation intermediate
S_N1, E1

(b) Is first-order in alkyl halide and first-order in nucleophile
S_N2

(c) Involves inversion of configuration at the site of substitution
S_N2

(d) Involves retention of configuration at the site of substitution
None. S_N1, however, may proceed with predominantly racemization, but some retention.

(e) Substitution at a stereocenter gives predominantly a racemic product
S_N1

(f) Is first-order in alkyl halide and zero-order in base
E1

(g) Is first-order in alkyl halide and first-order in base
E2

(h) Is greatly accelerated in protic solvents of increasing polarity
S_N1, E1

(i) Rearrangements are common
S_N1, E1

(j) Order of reactivity is 3° > 2° > 1° > methyl
S_N1, E2, E1

(k) Order of reactivity is methyl > 1° > 2° > 3°
S_N2

Problem 8.43 Arrange these alkyl halides in order of increasing ratio of E2 to S_N2 products observed on reaction of each with sodium ethoxide in ethanol.

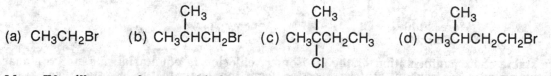

(a) CH_3CH_2Br (b) $CH_3\overset{\underset{|}{CH_3}}{CH}CH_2Br$ (c) $CH_3\overset{\underset{|}{CH_3}}{\underset{\underset{Cl}{|}}{C}}CH_2CH_3$ (d) $CH_3\overset{\underset{|}{CH_3}}{CH}CH_2CH_2Br$

More E2 will occur the more hindered the site of reaction and/or the more substituted (stable) the product alkene. Thus, listed in order from least to most E2 product:

$$CH_3CH_2Br \quad < \quad CH_3\underset{\underset{CH_3}{|}}{C}HCH_2CH_2Br \quad < \quad CH_3\underset{\underset{CH_3}{|}}{C}HCH_2Br \quad < \quad CH_3\underset{\underset{Cl}{|}}{\overset{\overset{CH_3}{|}}{C}}CH_2CH_3$$

<u>Problem 8.44</u> Draw a structural formula for the major organic product of each reaction and specify the most likely mechanism for formation of the product you have drawn.

The substitution and elimination products for the reactions are given in bold. In each case, the different parameters discussed in the chapter are considered including the type of alkyl halide (primary, secondary, tertiary, etc.) and the relative strength of the nucleophile/base.

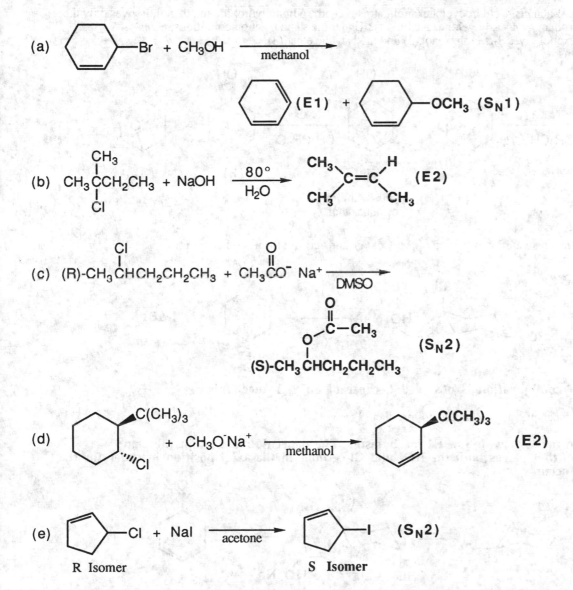

(a) [cyclohexenyl]—Br + CH₃OH $\xrightarrow{\text{methanol}}$ [cyclohexadiene] **(E1)** + [cyclohexenyl]—OCH₃ **(S_N1)**

(b) $CH_3\underset{\underset{Cl}{|}}{\overset{\overset{CH_3}{|}}{C}}CH_2CH_3$ + NaOH $\xrightarrow[\text{H}_2\text{O}]{80^\circ}$ [alkene] **(E2)**

(c) (R)-$CH_3\underset{\underset{Cl}{|}}{C}HCH_2CH_2CH_3$ + $CH_3\overset{\overset{O}{||}}{C}O^-$ Na⁺ $\xrightarrow{\text{DMSO}}$ (S)-$CH_3CHCH_2CH_2CH_3$ with $O-\overset{\overset{O}{||}}{C}-CH_3$ group **(S_N2)**

(d) [trans-1-tert-butyl-2-chlorocyclohexane] + $CH_3O^-Na^+$ $\xrightarrow{\text{methanol}}$ [4-tert-butylcyclohexene] **(E2)**

(e) [cyclopentenyl]—Cl + NaI $\xrightarrow{\text{acetone}}$ [cyclopentenyl]—I **(S_N2)**

R Isomer   S Isomer

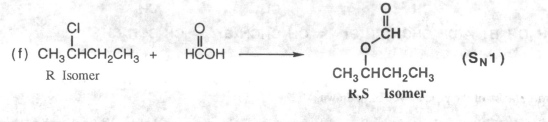

(f) $CH_3CHCH_2CH_3$ + $HCOH$ ⟶ ... **(S$_N$1)**

R Isomer

R,S   Isomer

(g) $CH_3CH_2O^- Na$ + $CH_2=CHCH_2Cl$ ⟶$_{ethanol}$ $CH_3CH_2OCH_2CH=CH_2$ **(S$_N$2)**

<u>Problem 8.45</u> When *cis*-4-chlorocyclohexanol is treated with sodium hydroxide in ethanol, it gives only the substitution product *trans*-1,4-cyclohexanediol (1). Under the same reaction conditions, *trans*-4-chlorocyclohexanol gives 3-cyclohexenol (2) and the bicyclic ether (3).

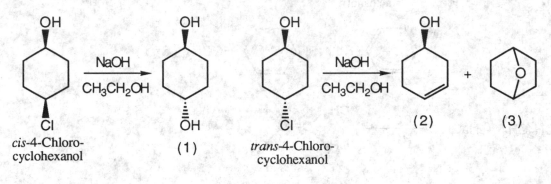

cis-4-Chloro-   (1)   *trans*-4-Chloro-   (2)   (3)
cyclohexanol   cyclohexanol

(a) Propose a mechanism for formation of product (1), and account for its configuration.

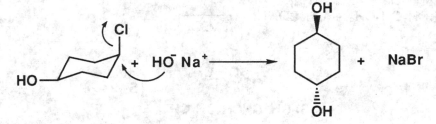

Inversion of configuration is observed because of an S$_N$2 mechanism.

(b) Propose a mechanism for formation of product (2).

The reaction takes place by an E2 mechanism. The molecule must adopt the chair conformation that places both the HO- and Cl- groups in the axial position in order for the reaction to occur.

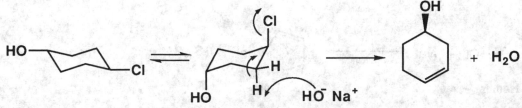

(c) Account for the fact that the bicyclic ether (3) is formed from the *trans* isomer but not from the *cis* isomer.

The bicyclic ether product (3) is formed from an intramolecular backside attack of the deprotonated axial hydroxyl group upon an axial chlorine atom. Only the *trans* isomer can adopt the diaxial orientation necessary for this process.

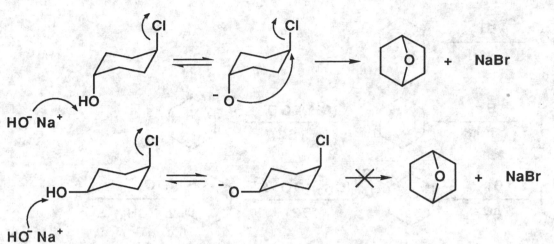

Synthesis

Problem 8.46 Show how to convert the given starting material into the desired product. Note that some syntheses require only one step whereas others require two or more steps.

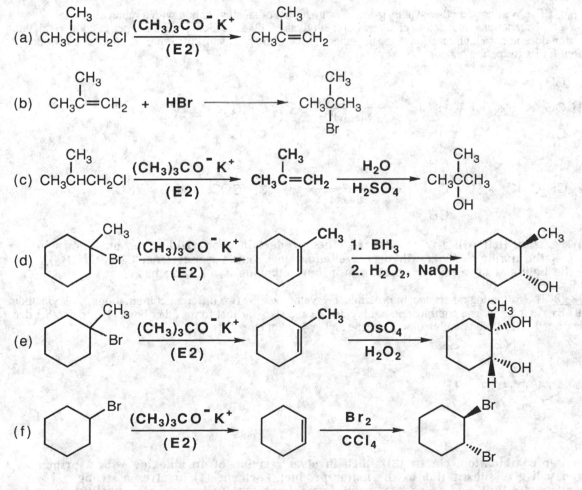

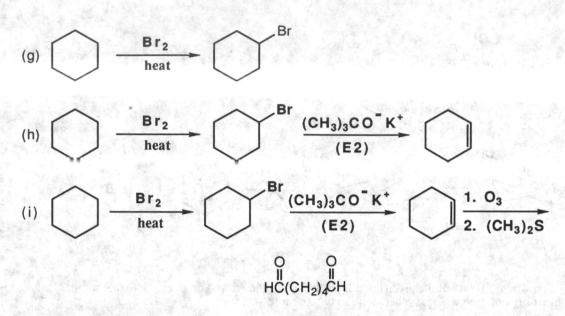

(g)

(h)

(i)

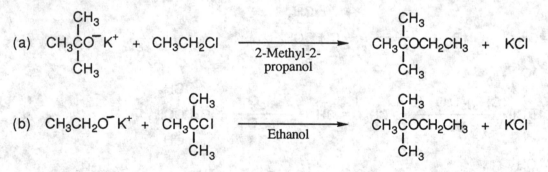

Problem 8.47 The Williamson ether synthesis involves treatment of an alkyl halide with a metal alkoxide. Following are two reactions intended to give *tert*-butyl ethyl ether. One reaction gives the ether in good yield, the other reaction does not. Which reaction gives the ether? What is the product of the other reaction, and how do you account for its formation?

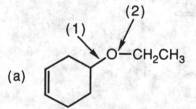

(a) $CH_3\underset{\underset{CH_3}{|}}{\overset{\overset{CH_3}{|}}{C}}O^- K^+$ + CH_3CH_2Cl $\xrightarrow{\text{2-Methyl-2-propanol}}$ $CH_3\underset{\underset{CH_3}{|}}{\overset{\overset{CH_3}{|}}{C}}OCH_2CH_3$ + KCl

(b) $CH_3CH_2O^- K^+$ + $CH_3\underset{\underset{CH_3}{|}}{\overset{\overset{CH_3}{|}}{C}}Cl$ $\xrightarrow{\text{Ethanol}}$ $CH_3\underset{\underset{CH_3}{|}}{\overset{\overset{CH_3}{|}}{C}}OCH_2CH_3$ + KCl

The only reaction that will give the desired ether product in good yield is the one shown in (a). In (b), the major product will be the elimination product isobutylene, $CH_2=C(CH_3)_2$, because the halide is on a tertiary carbon atom and ethoxide is a strong base.

Problem 8.48 The following ethers can, in principle, be synthesized by two different combinations of alkyl halide and metal alkoxide. Show one combination of alkyl halide and alkoxide that forms ether bond (1) and another that forms ether bond (2). Which combination gives the higher yield of ether?

(a)

As the better combination, choose (2) which involves reaction of an alkoxide with a primary halide and will give substitution as the major product. Scheme (1) involves a strong base/strong nucleophile and secondary halide, conditions that will give both substitution and elimination products.

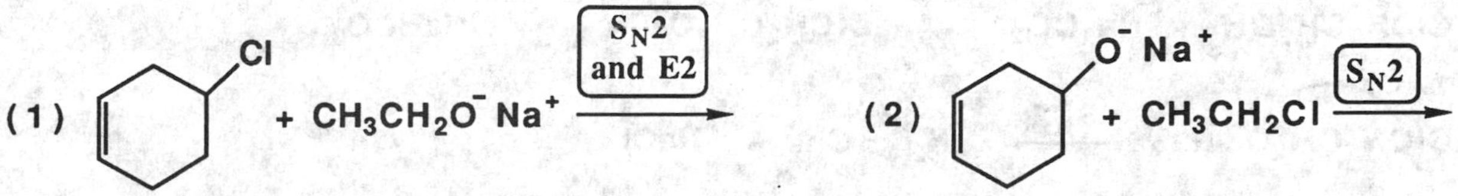

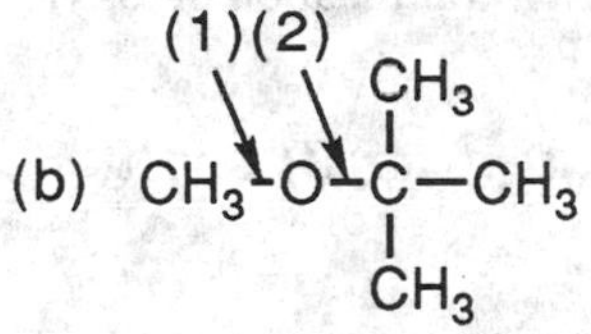

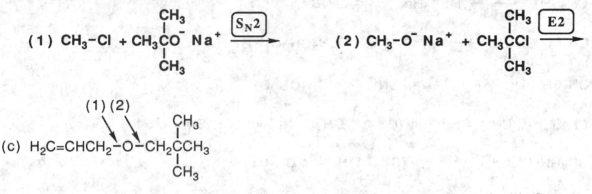

(b)

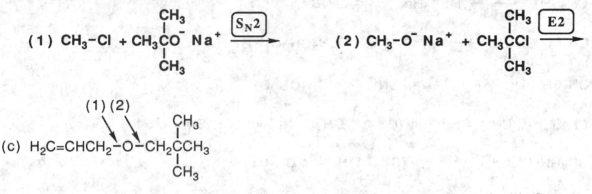

Because of the high degree of branching in the haloalkane in (2), S_N2 substitution by this pathway is impossible. Therefore, choose (1) as the only reasonable alternative.

(1) CH_3-Cl + $CH_3\underset{\underset{CH_3}{|}}{\overset{\overset{CH_3}{|}}{C}}O^-$ Na^+ $\xrightarrow{\boxed{S_N2}}$ **(2)** CH_3-O^- Na^+ + $CH_3\underset{\underset{CH_3}{|}}{\overset{\overset{CH_3}{|}}{C}}Cl$ $\xrightarrow{\boxed{E2}}$

(c)

Because of the high degree of branching on the β-carbon in the haloalkane in (2), S_N2 substitution by this pathway is prevented. Therefore, choose (1) as the only reasonable alternative.

(1) $H_2C=CHCH_2Cl$ + $CH_3\underset{\underset{CH_3}{|}}{\overset{\overset{CH_3}{|}}{C}}CH_2O^-$ Na^+ $\xrightarrow{\boxed{S_N2}}$

(2) $H_2C=CHCH_2O^-Na^+$ + $CH_3\underset{\underset{CH_3}{|}}{\overset{\overset{CH_3}{|}}{C}}CH_2Cl$ $\xrightarrow{\boxed{\text{No reaction}}}$

<u>Problem 8.49</u> Propose a mechanism for this reaction.

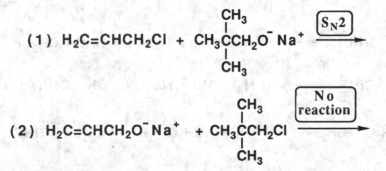

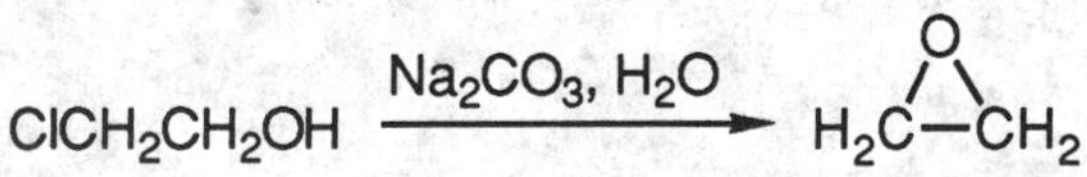

$$ClCH_2CH_2OH \xrightarrow{Na_2CO_3,\ H_2O} H_2C\overset{O}{-}CH_2$$

The mechanism of this reaction involves an initial deprotonation of the hydroxyl group, followed by an intramolecular S_N2 reaction to give the epoxide.

$$CICH_2CH_2OH + Na_2CO_3 \rightleftharpoons CICH_2CH_2O^- Na^+ + NaHCO_3$$

$$CICH_2CH_2O^- Na^+ \xrightarrow{S_N2} H_2C\overset{O}{\overbrace{}}CH_2 + NaCl$$

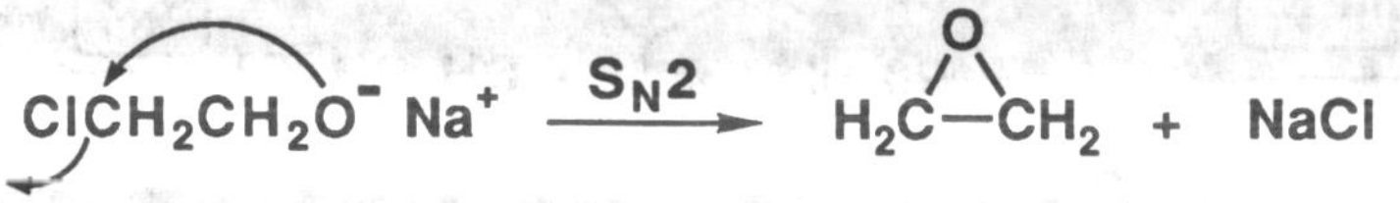

Problem 8.50 Each of these compounds can be synthesized by an S_N2 reaction. Suggest a combination of alkyl halide and nucleophile that will give each product.

In the following reactions, the Br atom could be replaced with Cl or I, except for (f) and (k) in which Cl is required.

(a) $CH_3OCH_3 \longleftarrow CH_3O^- Na^+ + CH_3Br$

(b) $CH_3SH \longleftarrow HS^- Na^+ + CH_3Br$

(c) $CH_3CH_2CH_2PH_2 \longleftarrow PH_3 + CH_3CH_2CH_2Br$

(d) $CH_3CH_2CN \longleftarrow Na^+ CN^- + CH_3CH_2Br$

(e) $CH_3SCH_2C(CH_3)_3 \longleftarrow (CH_3)_3CCH_2S^- Na^+ + CH_3Br$

(f) $(CH_3)_3NH^+ Cl^- \longleftarrow (CH_3)_2NH + CH_3Cl$

(g) $C_6H_5\overset{O}{\overset{\|}{C}}OCH_2C_6H_5 \longleftarrow C_6H_5\overset{O}{\overset{\|}{C}}O^- Na^+ + C_6H_5CH_2Br$

(h) $(R)\text{-}CH_3\overset{N_3}{\underset{|}{C}}HCH_2CH_2CH_3 \longleftarrow Na^+ N_3^- + (S)\text{-}CH_3\overset{Br}{\underset{|}{C}}HCH_2CH_2CH_3$

(i) $CH_2=CHCH_2OCH(CH_3)_2 \longleftarrow (CH_3)_2CHO^- Na^+ + CH_2=CHCH_2Br$

(j) $CH_2=CHCH_2OCH_2CH=CH_2 \longleftarrow CH_2=CHCH_2O^- Na^+ + CH_2=CHCH_2Br$

(k) 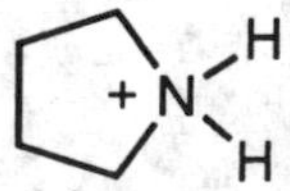 $\longleftarrow NH_3 + CICH_2CH_2CH_2Cl$

(l) 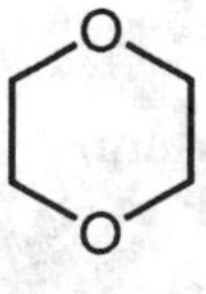 $\longleftarrow HOCH_2CH_2OH + BrCH_2CH_2Br$

CHAPTER 9: ALCOHOLS AND THIOLS

SUMMARY OF REACTIONS

Starting Material \ Product →	Aldehydes	Alkenes	Alkyl Halides		Carboxylic Acids	Disulfides	Ketones	Ketones/Aldehydes	Metal Alkoxides	Sulfonate Esters
Alcohols		9A 9.7	9B 9.6A	9C 9.6B,C					9D 9.5	9E 9.6C
Alcohols (Primary)	9F 9.9				9G 9.9					
Alcohols (Secondary)							9H 9.9			
Thiols						9I 9.11C				
Vicinal Diols							9J 9.8	9K 9.10		

*Section of book that describes reaction.

REACTION 9A: ACID-CATALYZED DEHYDRATION (Section 9.7)

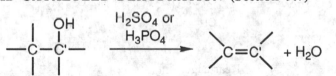

- Alcohols can be heated with H_3PO_4 or H_2SO_4 to generate an alkene. The net result of this process is the removal of H_2O from the alcohol, thus the process is called **dehydration**. ✳
- When more than one alkene can be formed from a reaction such as dehydration, the more stable alkene is formed in larger amounts. This generalization is known as Zaitsev's rule. In general, the more stable alkene is the one that is more highly substituted, that is, the one with more alkyl groups on the sp^2 carbon atoms.
- The mechanism dehydration of a secondary (2°) or tertiary (3°) alcohol involves protonation of the oxygen atom of the -OH group, followed by loss of water to form a carbocation. Since there is now no good nucleophile to react with the carbocation, a different reaction takes place, namely loss of H^+ to give the alkene. In the case of primary (1°) alcohols, loss of water and H^+ occur simultaneously.
- Note how this dehydration reaction is the reverse of acid-catalyzed hydration of an alkene.

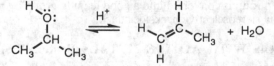

- The reaction conditions determine the position of this equilibrium. Large amounts of water favor formation of the alcohol, removing all traces of water favors formation of the alkene.
- The mechanism of forming the alkene from the alcohol with acid catalysis is <u>exactly</u> the reverse of the mechanism of forming an alcohol from the alkene under acid catalysis. This is a good illustration of the important concept of **microscopic reversibility**; which says that for any equilibrium reaction, the transition states and intermediates for the forward reaction are exactly the same as the transition states and intermediates of the backward reaction. In other words, the reactions proceed via the same mechanism in both directions.

REACTION 9B: REACTION WITH H-X (Section 9.6A)

- **Alcohols** react with **H-X** to form **alkyl halides**. In these reactions, the **-OH group** is turned into a much better leaving group (H_2O) via **protonation** to give an **oxonium ion** intermediate. The H_2O either departs on its own to create a carbocation (S_N1) that reacts with halide or is displaced by the nucleophilic halide (S_N2). The pathway followed depends on the nature of the alcohol (methyl, $1°$, $2°$, or $3°$). ✳
- Tertiary alcohols are converted to alkyl halides by an S_N1 **mechanism**. Substitution of X for -OH occurs via a Nucleophilic attack by X^- onto the carbocation and there is only **1** reactant in the rate-limiting step (the protonated oxonium species that simply loses water without reacting with any other molecules). In general, tertiary alcohols react faster by the S_N1 mechanism because carbocation formation is involved in the rate-limiting step and tertiary carbocations are the most stable. Primary alcohols form the least stable carbocations, so they rarely if ever undergo an S_N1 reaction. Secondary alcohols fall somewhere in-between in both carbocation stability and S_N1 reactivity.
- On the other hand, **primary alcohols** are most likely to react by the S_N2 **mechanism**, because primary alcohols cannot form stable carbocations and they have less steric hindrance to interfere with nucleophilic attack. This S_N2 reaction involves Substitution of the X for -OH via a Nucleophilic attack by X^- onto the backside of the C-O bond and the rate-limiting step involves **2** reagents, the X^- and the protonated alcohol.
- **Secondary alcohols** fall somewhere in between tertiary and primary alcohols in terms of reactivity and mechanism.
- **Secondary and primary alcohols with branching at the β carbon** can rearrange because a carbocation intermediate is involved. Carbocations are notorious for rearrangements, which involve the migration of an alkyl group to create a new carbocation of equal or greater stability.

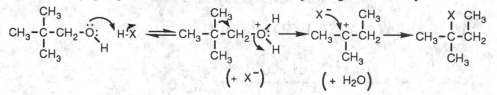

Rearrangement

REACTION 9C: REACTION WITH PBr₃ OR SOCl₂ (Section 9.6B,C)

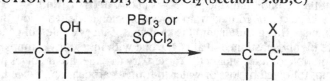

- The **-OH group** of alcohols can also be **replaced** with a **halide** using **PBr₃ or SOCl₂** to produce alkyl bromides or chlorides, respectively. The other products of the reaction are H_3PO_3 or SO_2 and HCl, respectively. ✳
- In these reactions, the -OH group is converted into a good leaving group by reaction with PBr₃ or SOCl₂, so the halide can displace it via stereoselective backside attack.

REACTION 9D: REACTION WITH ACTIVE METALS (Section 9.5)

$$-\overset{|}{\underset{|}{C}}-OH \xrightarrow{\ M\ } -\overset{|}{\underset{|}{C}}-O^- M^+ \ + H_2$$

- **Alcohols** react with active metals such as **Li, Na, and K** to produce hydrogen gas (H_2) and a **metal alkoxide**. ✳
- Metal alkoxides are slightly more basic than hydroxide, HO^-. This means that alcohols are slightly weaker acids than water [Section 3.3].

- The sodium alkoxides can also be produced using sodium hydride (NaH).

REACTION 9E: CONVERSION TO SULFONATE ESTERS (Section 9.6C)

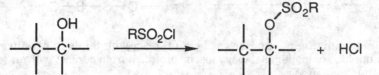

- **Alcohols** react with **sulfonyl chloride reagents** to produce **sulfonate esters**. Common derivatives include R = methyl and R = *p*-tolyl. ✶
- **Sulfonate esters**, readily prepared from alcohols, are **good leaving groups**. Thus, production of sulfonate esters provides a convenient way of turning -OH groups into good leaving groups in preparation for reaction with nucleophiles.

REACTION 9F: OXIDATION OF PRIMARY ALCOHOLS TO ALDEHYDES (Section 9.9)

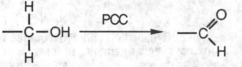

- A special reagent called **pyridinium chlorochromate** reacts with **primary alcohols** and **stops at the aldehyde**, without reacting further to the carboxylic acid. ✶
- The mechanism of this and all other chromium promoted oxidations involves formation of a chromate ester, followed by deprotonation of the carbon atom that was once attached to the OH group of the alcohol. The chromium ester decomposes with the chromium species taking two electrons to give the oxidized product. PCC stops at the aldehyde stage because no water is present with this reagent to hydrate the aldehyde carbonyl group.

REACTION 9G: OXIDATION OF PRIMARY ALCOHOLS TO CARBOXYLIC ACIDS (Section 9.9)

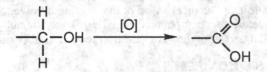

- **Primary alcohols are oxidized** all the way **to carboxylic acids** when aqueous solutions of various forms of chromium(VI) such as CrO_3, $K_2Cr_2O_7$, and especially H_2CrO_4. The H_2CrO_4 is prepared from CrO_3 or $K_2Cr_2O_7$ and H_2SO_4. In this case, an aldehyde is initially formed, but the presence of water causes the aldehyde carbonyl to be hydrated to give a geminal diol that is oxidized further to a carboxylic acid before it can be isolated. ✶

REACTION 9H: OXIDATION OF SECONDARY ALCOHOLS TO KETONES (Section 9.9)

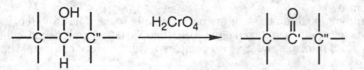

- **Secondary alcohols** can be **oxidized to ketones**. Oxidizing agents such as PCC can be used, but the use of H_2CrO_4 is also common. ✶

REACTION 9I: OXIDATION OF THIOLS TO DISULFIDES (Section 9.11C)

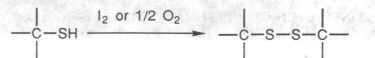

- Even relatively mild **oxidizing agents** such as I_2 or O_2 react with thiols to **produce disulfides**. In fact, thiols are so susceptible to oxidation that they must be protected from contact with air (the O_2) to avoid spontaneous disulfide formation. ✳
- Disulfide bonds are especially important in proteins, where they help stabilize three-dimensional structure.

REACTION 9J: THE PINACOL REARRANGEMENT OF GLYCOLS (Section 9.8)

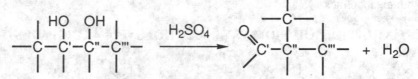

- Glycols such as pinacol (2,3-dimethyl-2,3-butanediol) undergo a unique reaction under acid catalyzed dehydration conditions called a **pinacol rearrangement**. In this reaction, there is loss of water like a normal dehydration, yet the intermediate carbocation rearranges via migration of an alkyl group, leading to formation of the ketone or aldehyde product. ✳

REACTION 9K: CLEAVAGE OF GLYCOLS WITH PERIODIC ACID (Section 9.10)

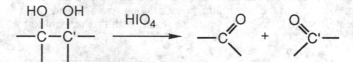

- **Periodic acid (H_5IO_6 or $HIO_4 \cdot 2H_2O$)** can cleave **glycols** to give carbonyl compounds, either aldehydes or ketones depending on the starting material. Recall that **glycols** are compounds with -OH groups on adjacent carbon atoms. The mechanism of the reaction involves a cyclic periodate ester that decomposes to give the two carbonyl compounds and HIO_3. Thus, the net reaction is a two-electron oxidation of the glycol, and a corresponding two-electron reduction of periodic acid. ✳

SUMMARY OF IMPORTANT CONCEPTS

9.0 OVERVIEW
• Alcohols and thiols are important functional groups that are involved in a number of characteristic reactions, and are very common in nature.

9.1 STRUCTURE OF ALCOHOLS AND THIOLS
• An **alcohol** is a molecule that contains an -OH (**hydroxyl**) group attached to an sp^3 hybridized carbon atom.✳
• A **thiol** is analogous to an alcohol, except that a thiol has an -SH (**sulfhydryl**) group attached to an sp^3 hybridized carbon atom. ✳
• Both oxygen and sulfur are in Column 6A of the periodic table so each has **6 valence electrons**. The **sulfur atom** is significantly **larger** than the oxygen atom (0.104 nm vs. 0.066 nm), because the valence electrons in the sulfur reside in the third principle energy level compared with the second principle energy level for the valence electrons of oxygen.

9.2 NOMENCLATURE
• All alcohols and thiols can be named according to IUPAC rules, but numerous common names are still used for the simpler ones so these must be learned as well.
• In the **IUPAC system**, an alcohol is named by selecting the parent chain as the longest chain that contains the -OH group. The suffix **e** is changed to **ol** and a number is added to designate the position of the -OH group. The location of the **-OH group takes precedence** over **alkyl groups and halogens**.

- When there is a choice, the chain is numbered to give the -OH group the lowest number. This includes cyclic and bicyclic alcohols. Examples of IUPAC names for alcohols include 2-pentanol (not 4-pentanol!) and cyclohexanol.
- A molecule that has more than one -OH group is called a **diol** if it has two, or a **triol** if it has three, etc. These molecules are named by adding the suffix **diol** or **triol**, etc., after the suffix **e** and then providing a number for each carbon atom having an -OH group attached. Again the chain is numbered to give the lowest possible numbers to -OH groups.

  Examples of IUPAC names for diols or triols include 1,2-pentanediol or 2,3,5-octanetriol.
- Diols that contain -OH groups on adjacent carbon atoms are still referred to with the common name of **glycols**. For example ethylene glycol is really 1,2-ethanediol, the major component of antifreeze.
- Compounds that contain an -OH group and a C=C bond are named as an alcohol, with the parent chain numbered so that the -OH group is assigned the lowest number. The infix **en** is used in place of **an** and the suffix **ol** is used in place of the suffix **e**. Numbers are assigned as normal.

  For example, 3-penten-2-ol is the correct IUPAC name for CH_3-CH=CH-CHOH-CH_3.
- According to the IUPAC rules, a thiol is named the same as an alcohol, except the suffix **thiol** is placed *after* the suffix **e**. For example 2-butanethiol is a correct IUPAC name. In the common nomenclature rules, compounds containing an -SH group are called **mercaptans**. For example, butyl mercaptan is the common name for 1-butanethiol.
- The alcohol group has higher nomenclature priority than the thiol group, so molecules that contain both an -OH group and an -SH group are named as alcohols. In this case, the -SH group is referred to as **mercapto**. For example, 4-mercapto-2-butanol is the correct IUPAC name for $HSCH_2$-CH_2-CHOH-CH_3. *[This concept of functional group priority for nomenclature will become more important as we learn about more functional groups.]*
- Alcohols are characterized according to how many alkyl groups are attached to the carbon atom bearing the -OH group. A **primary (1°) alcohol** is one in which the -OH group is attached to a primary carbon atom, for example 1-propanol. A **secondary (2°) alcohol** is one in which the -OH group is attached to a secondary carbon atom, for example 2-propanol. A **tertiary (3°) alcohol** is one in which the -OH group is attached to a tertiary carbon atom, for example 2-methyl-2-propanol. ✷ *[Characterizing an alcohol as primary, secondary, or tertiary will be the key to understanding some of the reactions of alcohols.]*

1-propanol (1°)   2-propanol (2°)   2-methyl-2-propanol (3°)

9.3 PHYSICAL PROPERTIES

- Alcohols and thiols have very different physical properties, because the O-H bond is much more polar than the S-H bond.
- Because oxygen is more electronegative than either carbon or hydrogen, there is increased electron density and thus a partial negative charge on the oxygen atom of an alcohol. Similarly, since hydrogen is so much less electronegative than oxygen, the hydrogen atom has relatively little electron density and thus a partial positive charge. ✷ *[This picture of the hydroxyl group having partial positive charge on hydrogen and partial negative charge on oxygen is very important and explains most of the reactions of alcohols. Better yet, thinking of the hydroxyl group in this way will allow you to <u>predict</u> successfully these reactions.]*

- As the result of the polarization of the O-H bond, there is a permanent dipole moment in the molecule. Two molecules with permanent dipole moments are attracted to each other since the negative end of one dipole is attracted to the positive end of the other dipole, and *vice versa*. This attraction is called **dipole-dipole interaction**, an example of a weak intermolecular force that holds different molecules together. ✷
- A **hydrogen bond** is a special type of dipole-dipole interaction that occurs when the positive end of one of the dipoles is a hydrogen that is bonded to a very electronegative element (O, N, F). For example, hydrogen bonds that occur between different alcohol molecules involve the hydrogen of one -OH group interacting with the lone pair of electrons on the oxygen atom of another -OH group.

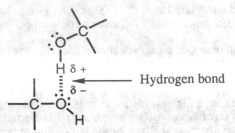

Hydrogen bond

- A hydrogen bond has a bond strength of about 5 kcal/mol in water. This is small compared to the strength of covalent bonds that are about 100 kcal/mol, but when several hydrogen bonds are working together, they are strong enough to hold large linear chains in precise, three dimensional structures such as those found in proteins and nucleic acids.
- The hydroxyl group can take part in hydrogen bonding, so molecules with -OH groups can stick to each other. As a result, alcohols have boiling points and melting points that are significantly higher than the corresponding alkanes or alkenes of similar molecular weight.
- Because water (H_2O) molecules are hydrogen bonded to each other in solution, only molecules that can themselves take part in dipole-dipole interactions like hydrogen bonds can separate the water molecules and thus dissolve. For that reason, the -OH group on small alcohols allows them to dissolve in water.
  The longer the non-polar alkyl chain on an alcohol, the lower the solubility of the alcohol in water and the higher the solubility in non-polar solvents like hexane and benzene.
• The **physical properties of thiols** are completely **different** than those of alcohols. The most noticeable physical property of thiols is their stench. The **S-H bond** is **not** as **polar** as the O-H bond. This is because sulfur and hydrogen atoms are similar in electronegativity, so these atoms do not have significant partial charges in the -SH group. As a result, thiols show little association by hydrogen bonding so they have lower boiling/melting points and are less soluble in water compared with analogous alcohols. ✻ *[This correlation between hydrogen bonding ability, water solubility and boiling/melting points is a good illustration of how understanding molecular structure can lead to accurate predictions of physical properties.]*

9.4 ACIDITY AND BASICITY OF ALCOHOLS
• **Alcohols** are **weakly acidic (pK$_a$ 16-18)** in aqueous solution, the relative acidities of which can be determined by the degree of solvation of the respective alkoxides. ✻
  - The bulkier the alkyl group, the lesser the ability of water to solvate the alkoxide, so the less acidic the alcohol. Thus 2-methyl-2-propanol is less acidic than methanol.
• The oxygen atom of an alcohol is a weak Lewis base and can be protonated by extremely strong acids to generate oxonium ions.

9.11 THIOLS
• **Thiols** are **considerably more acidic (pK$_a$ 8-9) than** the corresponding **alcohols (pK$_a$ 16-18)**. In other words, they are easier to deprotonate in base.
• **Thiols** readily **form insoluble salts with** most **heavy metal ions**, especially mercury and lead. ✻
• **Thiols** are **easily oxidized** to a number of higher oxidation states. As a result, thiols can be considered as mild reducing agents. ✻ *[Recall that when a molecule is oxidized, it gives up electrons. Thus an easily oxidized substance can be considered a reducing agent since it likes to give away electrons.]*
  - Thiols (R-SH) can be oxidized to disulfides (R-S-S-R), sulfinic acids (R-SO$_2$H), and sulfonic acids (R-SO$_3$H).

CHAPTER 9
Solutions to the Problems

Problem 9.1 Write IUPAC names for these alcohols:

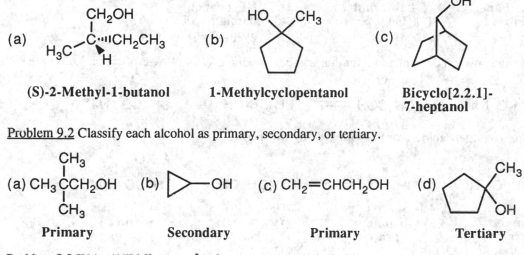

(a)

(S)-2-Methyl-1-butanol

(b)

1-Methylcyclopentanol

(c)

**Bicyclo[2.2.1]-
7-heptanol**

Problem 9.2 Classify each alcohol as primary, secondary, or tertiary.

(a) $CH_3\overset{\overset{\displaystyle CH_3}{|}}{\underset{\underset{\displaystyle CH_3}{|}}{C}}CH_2OH$ (b) ▷—OH   (c) $CH_2=CHCH_2OH$ (d)

Primary   **Secondary**   **Primary**   **Tertiary**

Problem 9.3 Write IUPAC names for these unsaturated alcohols.

(a) $CH_2=CHCH_2CH_2OH$ (b) $CH_3\overset{\overset{\displaystyle OH}{|}}{C}HCH=CH_2$

     3-Buten-1-ol   **3-Buten-2-ol**

Problem 9.4 Write IUPAC names for these thiols.

(a) $(CH_3)_2CHCH_2CH_2SH$ (b)

   3-Methyl-1-butanethiol   **(Z)-4-Hexene-2-thiol**

Problem 9.5 Arrange these compounds in order of increasing boiling point.
$CH_3CH_2CH_2CH_2CH_2CH_3$ $HOCH_2CH_2CH_2CH_2OH$ $CH_3CH_2CH_2CH_2CH_2OH$

In order of increasing boiling point they are:

$CH_3CH_2CH_2CH_2CH_2CH_3$ $CH_3CH_2CH_2CH_2CH_2OH$ $HOCH_2CH_2CH_2CH_2OH$

        Hexane   **1-Pentanol**   **1,4-Butanediol**
        bp 69°C   **bp 137°C**   **bp 235°C**

Hydrogen bonding, or lack of it, is the key. Both 1-pentanol and 1,4-butanediol can associate by hydrogen bonding, so their boiling points are substantially higher than hexane. Because 1,4-butanediol has more sites for hydrogen bonding, it has a higher boiling point than 1-pentanol.

Problem 9.6 Arrange these compounds in order of increasing solubility in water.

$$CICH_2CH_2Cl \qquad CH_3CH_2CH_2OH \qquad CH_3CH_2CH_2CH_2OH$$

In order of increasing solubility in water, they are:

$CICH_2CH_2Cl$	$CH_3CH_2CH_2CH_2OH$	$CH_3CH_2CH_2OH$
1,2-Dichloroethane	1-Butanol	1-Propanol
slightly soluble	8 g/100 g H_2O	soluble in all proportions

Problem 9.7 Predict the position of equilibrium for this acid-base reaction. (*Hint:* Review Section 3.4.)

$$CH_3CH_2O^- Na^+ \;+\; CH_3\overset{\displaystyle O}{\overset{\|}{C}}OH \;\rightleftharpoons\; CH_3CH_2OH \;+\; CH_3\overset{\displaystyle O}{\overset{\|}{C}}O^- Na^+$$

Acetic acid is the stronger acid; equilibrium lies to the right.

$$CH_3CH_2O^- Na^+ \;+\; CH_3\overset{\displaystyle O}{\overset{\|}{C}}OH \;\rightleftharpoons\; CH_3CH_2OH \;+\; CH_3\overset{\displaystyle O}{\overset{\|}{C}}O^- Na^+$$

	pK_a 4.76	pK_a 15.9	
(stronger base)	(stronger acid)	(weaker acid)	(weaker base)

Problem 9.8 Show how you might convert (R)-2-butanol to (S)-2-butanethiol via a tosylate.

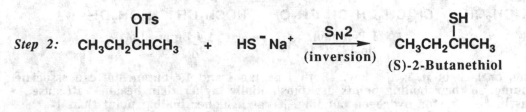

In the first step, the -OH group is converted to the good leaving group -OTs by treatment with tosyl chloride (Ts-Cl) in pyridine.

Step 1:
$$\underset{\text{(R)-2-Butanol}}{CH_3CH_2\overset{\displaystyle OH}{\overset{|}{C}}HCH_3} \;+\; \text{Ts-Cl} \;\xrightarrow{\text{Pyridine}}\; CH_3CH_2\overset{\displaystyle OTs}{\overset{|}{C}}HCH_3$$

Next the strong nucleophile HS- is used to carry out an S_N2 reaction and displace the -OTs group. Note that the inversion of stereochemistry seen with S_N2 reactions insures that the desired (S) isomer of 2-butanethiol is produced.

Step 2:
$$CH_3CH_2\overset{\displaystyle OTs}{\overset{|}{C}}HCH_3 \;+\; HS^- Na^+ \;\xrightarrow[\text{(inversion)}]{S_N2}\; \underset{\text{(S)-2-Butanethiol}}{CH_3CH_2\overset{\displaystyle SH}{\overset{|}{C}}HCH_3}$$

<u>Problem 9.9</u> Draw structural formulas for the alkenes formed by acid-catalyzed dehydration of these alcohols. Where isomeric alkenes are possible, predict which is the major product.

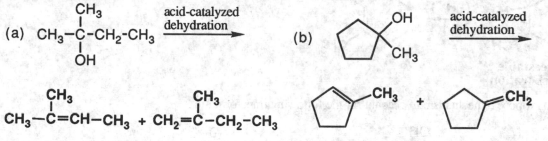

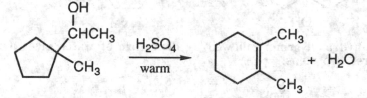

Major product   Major product

<u>Problem 9.10</u> Propose a mechanism to account for this reaction.

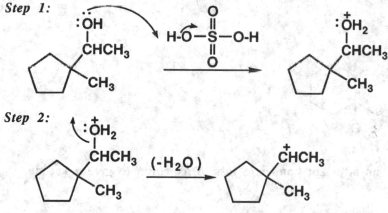

This reaction can best be explained as an acid catalyzed E1 reaction with a carbocation rearrangement. The first steps involve protonation of the -OH group, followed by loss of H_2O to give a 2° carbocation.

Step 1:

Step 2:

2° Carbocation

This 2° carbocation then rearranges to a more stable 3° carbocation by expanding to a six-membered ring.
The 3° carbocation completes the E1 reaction by losing a proton to generate the product alkene.

Step 3:

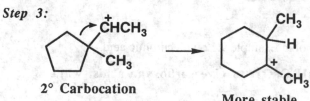

2° Carbocation

More stable
3° carbocation

Step 4:

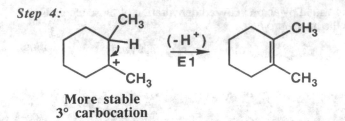

**More stable
3° carbocation**

<u>Problem 9.11</u> Propose a mechanism to account for the following transformation:

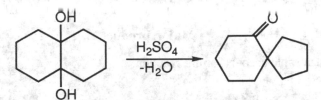

Protonation of either hydroxyl group followed by departure of water gives a tertiary carbocation in the first two steps.

Step 1:

Step 2:

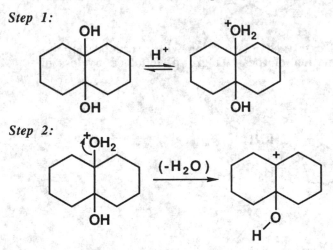

Migration of a pair of electrons from an adjacent bond and loss of a proton to give a ketone gives the observed product.

Step 3:

<u>Problem 9.12</u> Draw the product of treatment of each alcohol in Example 9.12 with chromic acid.

Primary alcohols are oxidized by chromic acid (H_2CrO_4) to give carboxylic acids, while secondary alcohols give ketones.

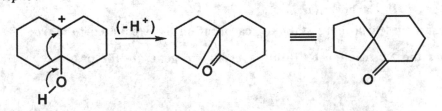

(a) $CH_3(CH_2)_4CH_2OH$ + H_2CrO_4 $\longrightarrow$ $CH_3(CH_2)_4\overset{\displaystyle O}{\overset{\displaystyle \|}{C}}OH$

   1-Hexanol   **Hexanoic acid**

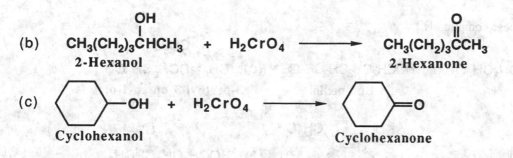

(b)   2-Hexanol + H_2CrO_4 ⟶ 2-Hexanone

(c)   Cyclohexanol + H_2CrO_4 ⟶ Cyclohexanone

<u>Problem 9.13</u> α-Hydroxyketones and α-hydroxyaldehydes are also cleaved by treatment with periodic acid. It is the not the α-hydroxyketone or aldehyde, however, that undergoes reaction with periodic acid, but rather the compound formed by addition of a molecule water to the carbonyl group of the α-hydroxyketone or aldehyde. Draw the structural formula for the product formed by addition of a molecule of H_2O to the carbonyl group of the following compound and write a mechanism for the reaction of this product with HIO_4. (Hint: Notice the similarity of this oxidation to the oxidation of an aldehyde by chromic acid (Section 9.9).

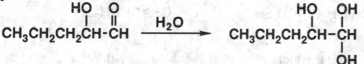

$$CH_3CH_2CH_2CH\text{-}CH \xrightarrow[\text{2. } HIO_4]{\text{1. } H_2O}$$

The aldehyde group reacts with water to give the following molecule, a geminal diol.
Step 1:

$$CH_3CH_2CH_2CH\text{-}CH \xrightarrow{H_2O} CH_3CH_2CH_2CH\text{-}CH(OH)$$

This geminal diol then reacts with HIO_4 according to the usual mechanism to give butanal plus formic acid.

Step 2:

$$CH_3CH_2CH_2CH\text{-}CH(OH) + O{=}I{-}OH \longrightarrow CH_3CH_2CH_2CH\text{-}CH$$

Step 3:

$$CH_3CH_2CH_2CH\text{-}CH \longrightarrow CH_3CH_2CH_2CH + HCOH + HIO_3$$

Butanal   Formic acid

<u>Structure and Nomenclature</u>
<u>Problem 9.14</u> Which are secondary alcohols?

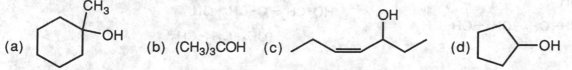

(a)   (b) $(CH_3)_3COH$ (c)   (d)

The secondary alcohols are (c) and (d). Molecules (a) and (b) are tertiary alcohols.

Problem 9.15 Name these compounds.

(a) CH₃CH₂CH₂CH₂CH₂OH (b) HOCH₂CH₂CH₂CH₂OH (c) CH₂=CCH₂CH₂OH
                                                          |
                                                          CH₃
1-Pentanol   **1,4-Butanediol**   **3-Methyl-3-buten-1-ol**

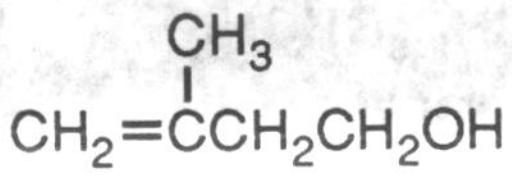

(d) 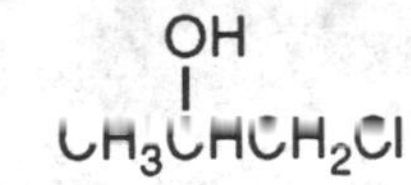   CH₃CHCH₂Cl
            |
           OH
1-Chloro-2-propanol

(e) 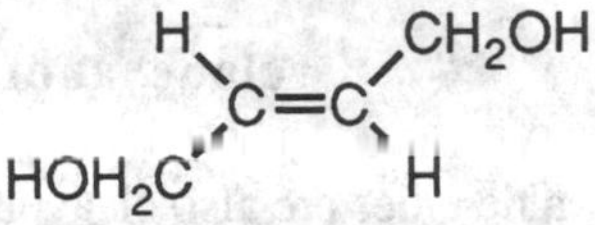 (E)-2-butene-1,4-diol

(f) 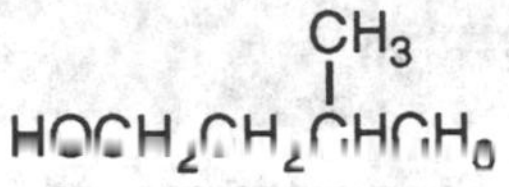 HOCH₂CH₂CHCH₃
                 |
                CH₃
3-Methyl-1-butanol
(Isopentyl alcohol)

(g) 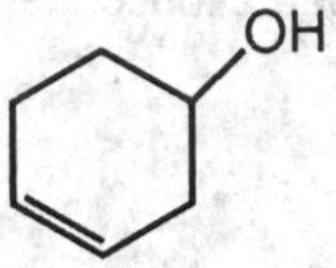 **3-Cyclohexenol**

(h) 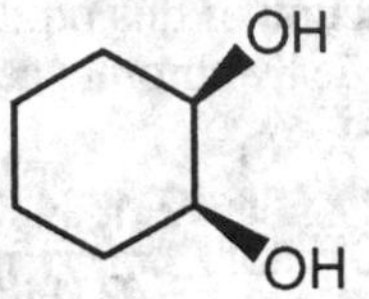 **cis-1,2-Cyclohexanediol**

(i) 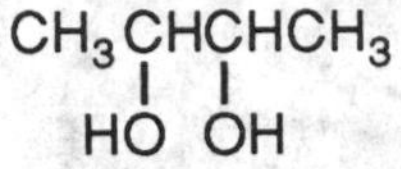 CH₃CHCHCH₃
         | |
        HO OH
2,3-Butanediol

(j) 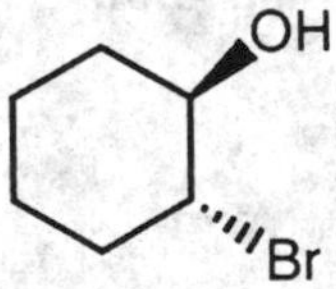 **trans-2-Bromocyclohexanol**

(k) CH₃CH₂CH₂CH₂SH
1-Butanethiol

(l) CH₃CH=CHCH₂SH
2-Buten-1-thiol

(m)   HSCH₂CH₂SH
1,2-Ethanedithiol

(n) 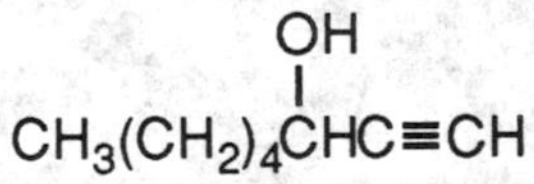 CH₃(CH₂)₄CHC≡CH
                  |
                 OH
1-Octyn-3-ol

(o) 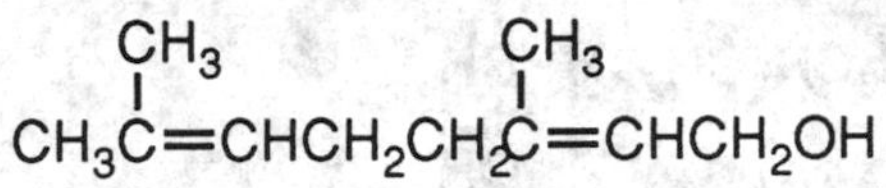 CH₃C=CHCH₂CH₂C=CHCH₂OH
          | |
         CH₃   CH₃
3,7-Dimethyl-2,6-octadien-1-ol

Problem 9.16 Write structural formulas for the following:
(a) Isopropyl alcohol

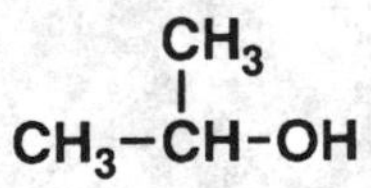

CH₃
|
CH₃−CH−OH

(b) Propylene glycol

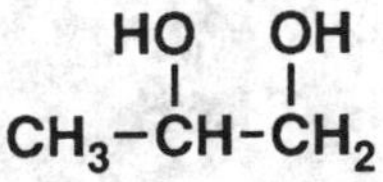

HO OH
 | |
CH₃−CH−CH₂

(c) 5-Methyl-2-hexanol

OH   CH₃
 | |
CH₃−CH−CH₂−CH₂−CH−CH₃

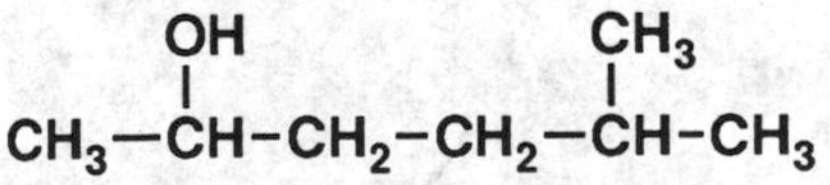

(d) 2-Methyl-2-propyl-1,3-propanediol

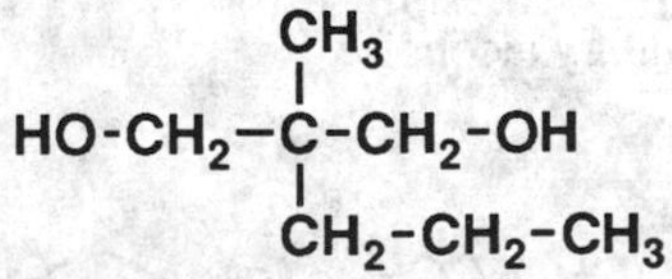

CH₃
|
HO-CH₂−C−CH₂−OH
|
CH₂−CH₂−CH₃

(e) 1-Chloro-2-hexanol

$$Cl-CH_2-\underset{\underset{OH}{|}}{CH}-CH_2-CH_2-CH_2-CH_3$$

(f) *cis*-3-Isobutylcyclohexanol

[cyclohexane ring with OH wedge at top and $CH_2CH(CH_3)_2$ at lower right]

(g) 2,2-Dimethyl-1-propanol

$$CH_3-\underset{\underset{CH_3}{|}}{\overset{\overset{CH_3}{|}}{C}}-CH_2-OH$$

(h) 2-Mercaptoethanol

$$HS-CH_2-CH_2-OH$$

(i) Allyl alcohol

$$CH_2=CH-CH_2-OH$$

(j) *trans*-2-Vinylcyclohexanol

[cyclohexane ring with OH and $CH=CH_2$ substituents]

(k) (Z)-5-Methyl-2-hexen-1-ol

$$HO-CH_2\underset{\overset{|}{H}}{\overset{\diagdown}{C}}=\underset{\overset{|}{H}}{\overset{\diagup}{C}}CH_2-\overset{\overset{CH_3}{|}}{CH}-CH_3$$

(l) 2-Propyn-1-ol

$$HC\equiv C-CH_2OH$$

(m) 3-Chloro-1,2-propanediol

$$HO-CH_2-\underset{\underset{OH}{|}}{CH}-CH_2-Cl$$

(n) *cis*-3-Pentene-1-ol

$$HO-CH_2-CH_2\underset{\overset{|}{H}}{\overset{\diagdown}{C}}=\underset{\overset{|}{H}}{\overset{\diagup}{C}}CH_3$$

(o) Bicyclo[2.2.1]-heptan-7-ol

Problem 9.17 Name and draw structural formulas for the eight isomeric alcohols of molecular formula $C_5H_{12}O$. Classify each as primary, secondary, or tertiary. Which are chiral?

The eight isomeric alcohols of molecular formulas $C_5H_{12}O$ are grouped by carbon skeleton. First, the three alcohols derived from pentane, then the four derived from 2-methylbutane (isopentane), and then the single alcohol derived from 2,2-dimethylpropane (neopentane). Each is given an IUPAC name. Common names, where appropriate, are given in parentheses.

$$\underset{\underset{\begin{array}{c}\text{1-Pentanol (pentyl alcohol)}\\\text{(primary)}\end{array}}{}}{CH_3CH_2CH_2CH_2\overset{\overset{OH}{|}}{CH_2}}$$

$$\underset{\underset{\begin{array}{c}\text{2-Pentanol}\\\text{(secondary)}\\\text{(chiral)}\end{array}}{}}{CH_3CH_2CH_2\overset{\overset{OH}{\overset{|}{*}}}{CH}CH_3}$$

$$\underset{\underset{\begin{array}{c}\text{3-Pentanol}\\\text{(secondary)}\end{array}}{}}{CH_3CH_2\overset{\overset{OH}{|}}{CH}CH_2CH_3}$$

OH
|
CH₃CHCH₂CH₂
|
CH₃

3-Methyl-1-butanol
(Isopentyl alcohol)
(primary)

OH
|*
CH₃CHCHCH₃
|
CH₃

3-Methyl-2-butanol
(secondary)
(chiral)

OH
|
CH₃CCH₂CH₃
|
CH₃

2-Methyl-2-butanol
(*tert*-Pentyl alcohol)
(tertiary)

OH
|*
CH₂CHCH₂CH₃
|
CH₃

2-Methyl-1-butanol
(primary)
(chiral)

CH₃
|
CH₃CCH₂OH
|
CH₃

2,2-Dimethyl-1-propanol
(Neopentyl alcohol)
(primary)

Physical Properties of Alcohols and Thiols

Problem 9.18 Arrange these compounds in order of increasing boiling point. (Values in °C are -42, 78, 117, and 198.)

(a) $CH_3CH_2CH_2CH_2OH$ (b) CH_3CH_2OH (c) $HOCH_2CH_2OH$ (d) $CH_3CH_2CH_3$

In order of increasing boiling point they are:

$CH_3CH_2CH_3$	CH_3CH_2OH	$CH_3CH_2CH_2CH_2OH$	$HOCH_2CH_2OH$
Propane	Ethanol	1-Butanol	Ethylene glycol
bp -42°C	bp 78°C	bp 117°C	bp 198°C

The keys for this problem are hydrogen bonding and size. Propane cannot make any hydrogen bonds, so it has the lowest boiling point by far. Ethanol and 1-butanol can each make hydrogen bonds through their single -OH group. However, 1-butanol is larger, so it will have a higher boiling point than ethanol. Ethylene glycol has two -OH groups per molecule with which to make hydrogen bonds, so it will have the highest boiling point of this set.

Problem 9.19 Arrange these compounds in order of increasing boiling point. (Values in °C are -42, -24, 78, and 118.)

(a) CH_3CH_2OH (b) CH_3OCH_3 (c) $CH_3CH_2CH_3$ (d) CH_3CO_2H

In order of increasing boiling point they are:

$CH_3CH_2CH_3$	CH_3OCH_3	CH_3CH_2OH	CH_3CO_2H
Propane	Dimethyl ether	Ethanol	Acetic acid
bp -42°C	bp -24°C	bp 78°C	bp 118°C

The keys for this problem are hydrogen bonding, polarity, and size. We know from the last problem that propane and ethanol have boiling points of -42°C and 78°C, respectively. Dimethyl ether is polar, but cannot make hydrogen bonds. Therefore, it makes sense that dimethyl ether has a boiling point (-24°C) that is higher than propane, but lower than ethanol. Acetic acid can make strong hydrogen bonds and has a higher molecular weight than ethanol, so it makes sense that acetic acid has the highest boiling point of this set.

Problem 9.20 Compounds that contain an N-H group associate by hydrogen bonding.
(a) Do you expect this association to be stronger or weaker than that of compounds containing an O-H group?

Weaker. The O-H bond is more polar, because the difference in electronegativity between N and H is less than the difference between O and H. Thus, the degree of intermolecular interaction between compounds containing an N-H group is less than that between compounds containing an -OH group.

Bond	Difference in electronegativity
N-H	3.0 - 2.1 = 0.9
O-H	3.5 - 2.1 = 1.4

(b) Based on your answer to part (a), which would you predict to have the higher boiling point, 1-butanol or 1-butanamine?

$$CH_3CH_2CH_2CH_2OH \qquad CH_3CH_2CH_2CH_2NH_2$$
1-Butanol   1-Butanamine
bp 117°C   bp -78°C

The stronger the hydrogen bonds, the higher the boiling point since hydrogen bonds in the liquid state must be broken upon boiling. Therefore, 1-butanol, with the stronger hydrogen bonds, will have the higher boiling point.

Problem 9.21 Which compounds can participate in hydrogen bonding with water? For each that can, indicate which site(s) can function as a hydrogen bond acceptor and which can function as a hydrogen bond donor.

The molecules in bold can participate in hydrogen bonding with water. The hydrogen bond donor and acceptor sites are labeled.

(a) $CH_3CH_2CH_2OCH_2CH_2OH$

acceptor   acceptor

$$CH_3CH_2CH_2-O-CH_2CH_2-O-H$$

donor

(b) $(CH_3CH_2)_2NH$

acceptor   donor

$$CH_3CH_2-N-H$$
$$|$$
$$CH_2CH_3$$

(c) $CH_3CH=CHCH_3$

Does not participate

(d) $CH_3\overset{O}{\underset{\|}{C}}CH_3$

acceptor

$$CH_3-\overset{O}{\underset{\|}{C}}-CH_3$$

(e) $CH_3\overset{O}{\underset{\|}{S}}CH_3$

acceptor

$$CH_3\overset{O}{\underset{\|}{S}}CH_3$$

(f) $CH_3CH_2\overset{O}{\underset{\|}{C}}OH$

acceptor   acceptor

$$CH_3CH_2\overset{O}{\underset{\|}{C}}-O-H$$

donor

Problem 9.22 From each pair of compounds, select the one that is more soluble in water.
(a) CH_2Cl_2 or CH_3OH

Methanol, CH_3OH, is soluble in all proportions in water. Dichloromethane, CH_2Cl_2, is insoluble. The highly polar -OH group of methanol is capable of participating both as a hydrogen bond donor and hydrogen bond acceptor with water and, therefore, interacts strongly with water by intermolecular association. No such interaction is possible with dichloromethane.

(b)   CH₃CCH₃   or   CH₃CCH₃

Propanone (acetone), CH_3COCH_3, is soluble in water in all proportions. 2-Methylpropene (isobutylene) is insoluble in water. Acetone has a large dipole moment and can function as a hydrogen bond acceptor from water.

(c) CH_3CH_2Cl or   NaCl

NaCl is the more soluble. Chloroethane is insoluble in water. Following is a review of some of the general water solubility rules developed in General Chemistry. For these rules, soluble is defined as dissolving greater than 0.10 mol/L. Slightly soluble is dissolving between 0.01 mol/L and 0.10 mol/L.
1. Sodium, potassium, and ammonium salts of halogens or nitrates are soluble.
2. Silver, lead, and mercury(I) salts of halogens are insoluble.
Thus, applying Rule 1, NaCl is soluble in water. Chloroethane (ethyl chloride) is a nonpolar organic compound and insoluble in water.

(d) $CH_3CH_2CH_2SH$ or $CH_3CH_2CH_2OH$

The alcohol is more soluble in water. Sulfur is less electronegative than oxygen, so an S-H bond is less polarized than an O-H bond. Hydrogen bonding is therefore weaker with thiols than alcohols, so the alcohol is more able to interact with water molecules through hydrogen bonding.

(e) CH₃CH₂CHCH₂CH₃   or   CH₃CH₂CCH₂CH₃

The alcohol is more able to interact with water through hydrogen bonding, and is more soluble in water. The hydroxyl group has both a hydrogen bond donor and acceptor (the oxygen and hydrogen atoms of the -OH group, respectively), while the carbonyl group has only a hydrogen bond acceptor (the oxygen atom).

Problem 9.23 Arrange the compounds in each set in order of decreasing solubility in water.
(a) Ethanol; butane; diethyl ether

CH_3CH_2OH $CH_3CH_2OCH_2CH_3$ $CH_3CH_2CH_2CH_3$

soluble in all   8 g/100 mL water   insoluble in water
  proportions

In general, the more strongly a molecule can take part in hydrogen bonding with water, the greater the molecule is able to interact with the water molecules and dissolve. Only ethanol can act both as a donor and acceptor of hydrogen bonds with water. Diethyl ether can act as an acceptor of hydrogen bonds. Butane can act as neither a donor nor an acceptor of hydrogen bonds.

(b) 1-Hexanol; 1,2-hexanediol; hexane

$$CH_2CHCH_2CH_2CH_2CH_3 \quad\quad CH_3CH_2CH_2CH_2CH_2CH_2OH \quad\quad CH_3CH_2CH_2CH_2CH_2CH_3$$
$$OH\ OH$$

1,2-Hexanediol molecules can take part in more hydrogen bonds with water than the 1-hexanol, since the diol has two -OH groups. Hexane has no polar bonds and thus cannot take part in any dipole-dipole interactions with water molecules.

Problem 9.24 Each compound given in this problem is a common organic solvent. From each pair of compounds, select the solvent with the greater solubility in water.

Solubility in water increases with increasing hydrogen bonding ability and decreases with increasing surface area of hydrophobic groups such as alkyl groups.

(a) CH_2Cl_2 or CH_3CH_2OH (b) $CH_3CH_2OCH_2CH_3$ or CH_3CH_2OH

  CH_3CH_2OH CH_3CH_2OH

(c) $CH_3\overset{\displaystyle O}{\overset{\|}{C}}CH_3$ or $CH_3CH_2OCH_2CH_3$ (d) $CH_3CH_2OCH_2CH_3$ or $CH_3(CH_2)_3CH_3$

   $CH_3\overset{\displaystyle O}{\overset{\|}{C}}CH_3$ $CH_3CH_2OCH_2CH_3$

Problem 9.25 The decalinols A and B can be equilibrated using aluminum isopropoxide in 2-propanol (isopropyl alcohol) containing a small amount of acetone. Assume a value of ΔG° (equatorial - axial) for cyclohexanol is 0.95 kcal/mol (4.0 kJ/mol). Calculate the percent of each isomer in the equilibrium mixture at 25°C.

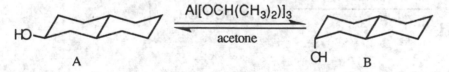

$$\text{HO} \xrightarrow[\text{acetone}]{Al[OCH(CH_3)_2)]_3}$$

A   OH   B

At equilibrium, the relative amounts of A and B are given by the equation:

$$\Delta G^\circ = -RT\ln K_{eq} \quad\quad \text{where } K_{eq} = \frac{[B]}{[A]}$$

$$\log K_{eq} = \frac{-\Delta G^\circ}{(2.303)RT}$$

Taking the antilog of both sides gives:

$$K_{eq} = 10^{\left(\frac{-\Delta G^\circ}{(2.303)RT}\right)}$$

Plugging in the values for ΔG°, R, and 298 K gives the final answer:

$$\boxed{K_{eq} = 10^{\left(\frac{-\,0.95\ \text{kcal/mol}}{(2.303)(1.987\ \times\ 10^{-3}\ \text{kcal/mol K})(298\ \text{K})}\right)} = 10^{-0.697} = 0.201}$$

Thus, the molecule with an equatorial hydroxyl group (A) will predominate in a 4.97 to 1 ratio.

Synthesis of Alcohols

We have encountered four reactions for the synthesis of alcohols, including glycols:
1. Acid-catalyzed hydration of alkenes (Section 6.3B)
2. Oxymercuration/reduction of alkenes (Section 6.3E)
3. Hydroboration/oxidation of alkenes (Section 6.4)
4. Oxidation of alkenes by osmium tetroxide (Section 6.5B)

Problem 9.26 Alkenes can be hydrated to form alcohols by (1) hydroboration followed by oxidation and (2) oxymercuration followed by reduction. Compare the products formed from these alkenes by sequence (1) compared with sequence (2).

Recall that hydroboration followed by oxidation gives predominantly non-Markovnikov regiochemistry and *syn* stereochemistry of addition. Oxymercuration followed by reduction gives predominantly Markovnikov regiochemistry and *anti* stereochemistry of addition.

(a) Propene

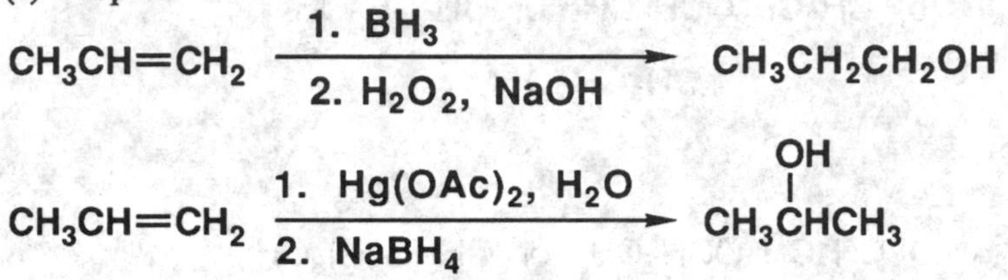

$$CH_3CH=CH_2 \xrightarrow[\text{2. } H_2O_2, \ NaOH]{\text{1. } BH_3} CH_3CH_2CH_2OH$$

$$CH_3CH=CH_2 \xrightarrow[\text{2. } NaBH_4]{\text{1. } Hg(OAc)_2, \ H_2O} CH_3\overset{\overset{\displaystyle OH}{|}}{C}HCH_3$$

(b) *cis*-2-Butene

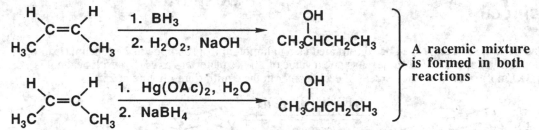

A racemic mixture is formed in both reactions

(c) *trans*-2-Butene

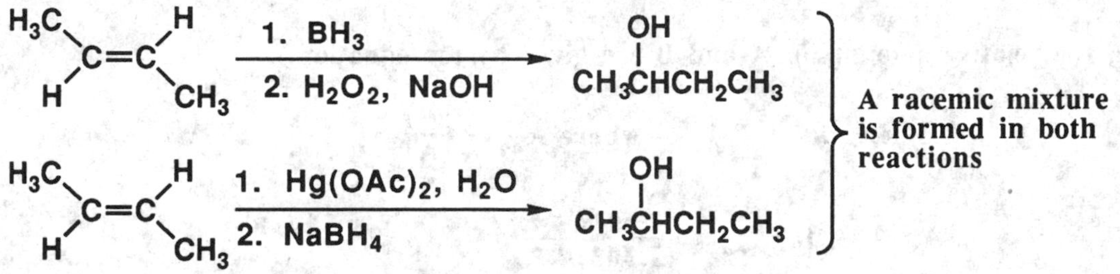

A racemic mixture is formed in both reactions

(d) Cyclopentene

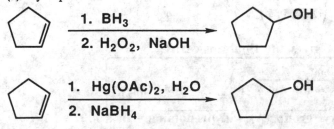

(e) 1-Methylcyclohexene

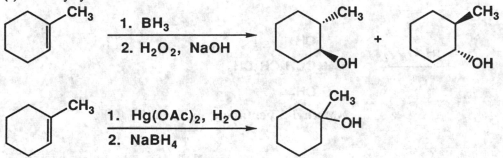

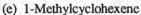

Problem 9.27 Give the structural formula of an alkene or alkenes from which each alcohol or glycol can be prepared in good yield.
(a) 2-Butanol

$CH_3CH_2CH=CH_2$

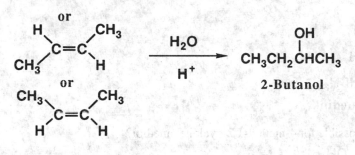

(b) 1-Methylcyclohexanol

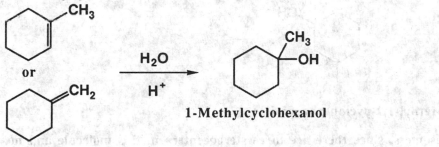

(c) 3-Hexanol

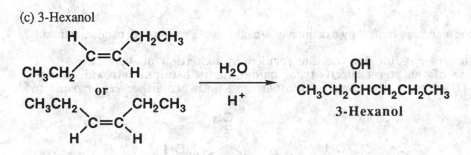

(d) 2-Methyl-2-pentanol

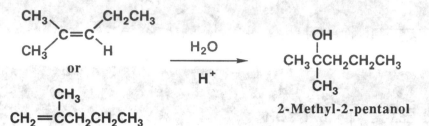

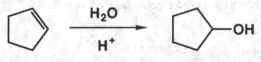

2-Methyl-2-pentanol

(e) Cyclopentanol

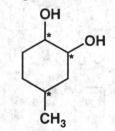

Cyclopentanol

(f) 1,2-Propanediol

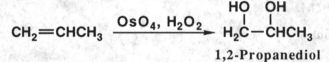

1,2-Propanediol

<u>Problem 9.28</u> (a) How many stereoisomers are possible for 4-methyl-1,2-cyclohexanediol?

4-Methyl-1,2-cyclohexanediol

There are **2³ or eight stereoisomers, since there are three stereocenters in this molecule and no planes of symmetry.**

(b) Which of the possible stereoisomers are formed by oxidation of 4-methylcyclohexene with osmium tetroxide?

There are two *cis-trans* isomers possible for the diol formed by oxidation of 4-methylcyclohexene. Because of the stereoselectivity of oxidation by osmium tetroxide under these conditions, the two -OH groups are *cis* to each other, and both are either *cis* or *trans* to the methyl group.

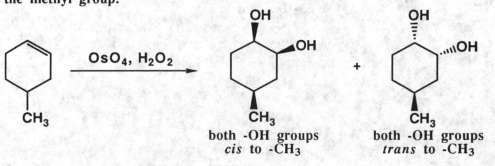

both -OH groups
cis to -CH₃

both -OH groups
trans to -CH₃

(c) Is the product formed in part (b) optically active or optically inactive?

The products will be optically active if the starting material is either pure 4R or 4S enantiomer, but the problem doesn't specify configuration so assume 4R,4S (racemic) and therefore inactive product.

Acidity of Alcohols
Problem 9.29 Complete the following acid-base reactions. In addition, show all valence electrons on the interacting atoms and show by the use of curved arrows the flow of electrons in each reaction.

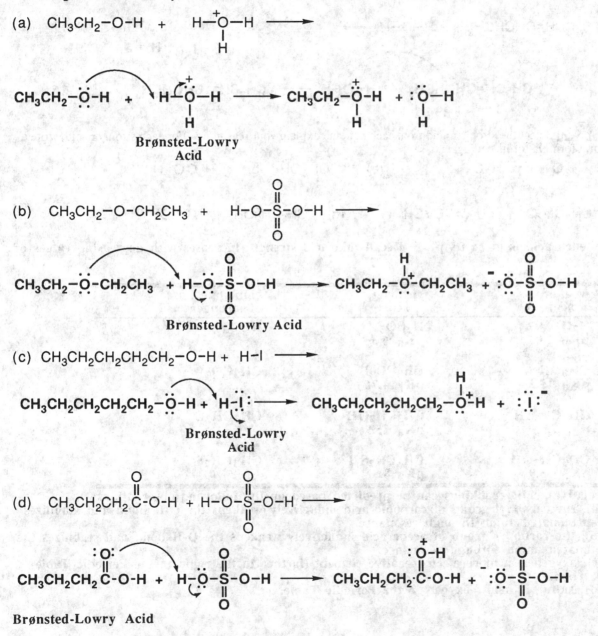

(a) CH_3CH_2-O-H + $H-\overset{+}{O}-H$ $\longrightarrow$
                                         |
                                         H

$CH_3CH_2-\overset{..}{\underset{..}{O}}-H$ + $H-\overset{+}{O}-H$ $\longrightarrow$ $CH_3CH_2-\overset{+}{\underset{..}{O}}-H$ + $:\overset{..}{O}-H$
                                | | |
                                H   H   H
                      Brønsted-Lowry
                            Acid

(b) $CH_3CH_2-O-CH_2CH_3$ + $H-O-\overset{\overset{O}{||}}{\underset{\underset{O}{||}}{S}}-O-H$ $\longrightarrow$

$CH_3CH_2-\overset{..}{\underset{..}{O}}-CH_2CH_3$ + $H-\overset{..}{\underset{..}{O}}-\overset{\overset{O}{||}}{\underset{\underset{O}{||}}{S}}-O-H$ $\longrightarrow$ $CH_3CH_2-\overset{\overset{H}{|}}{\underset{..}{\overset{+}{O}}}-CH_2CH_3$ + $\overset{-}{:}\overset{..}{\underset{..}{O}}-\overset{\overset{O}{||}}{\underset{\underset{O}{||}}{S}}-O-H$
                                               Brønsted-Lowry Acid

(c) $CH_3CH_2CH_2CH_2CH_2-O-H$ + $H-I$ $\longrightarrow$

$CH_3CH_2CH_2CH_2CH_2-\overset{..}{\underset{..}{O}}-H$ + $H-\overset{..}{\underset{..}{I}}:$ $\longrightarrow$ $CH_3CH_2CH_2CH_2CH_2-\overset{\overset{H}{|}}{\overset{+}{O}}-H$ + $:\overset{..}{\underset{..}{I}}:^{-}$
                                   Brønsted-Lowry
                                        Acid

(d) $CH_3CH_2CH_2\overset{\overset{O}{||}}{C}-O-H$ + $H-O-\overset{\overset{O}{||}}{\underset{\underset{O}{||}}{S}}-O-H$ $\longrightarrow$

$CH_3CH_2CH_2\overset{\overset{:\overset{..}{O}:}{||}}{C}-O-H$ + $H-\overset{..}{\underset{..}{O}}-\overset{\overset{O}{||}}{\underset{\underset{O}{||}}{S}}-O-H$ $\longrightarrow$ $CH_3CH_2CH_2\overset{\overset{:\overset{+}{O}-H}{||}}{C}-O-H$ + $\overset{-}{:}\overset{..}{\underset{..}{O}}-\overset{\overset{O}{||}}{\underset{\underset{O}{||}}{S}}-O-H$

Brønsted-Lowry Acid

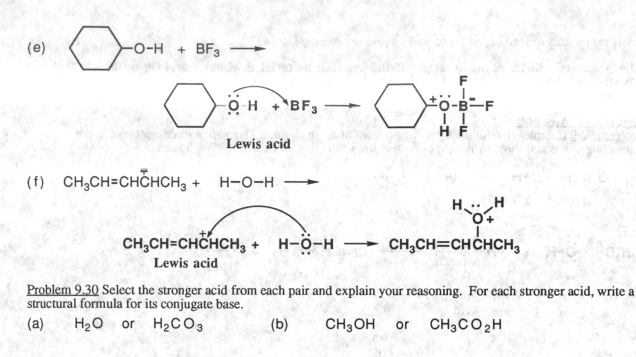

Problem 9.30 Select the stronger acid from each pair and explain your reasoning. For each stronger acid, write a structural formula for its conjugate base.

(a) H_2O or H_2CO_3 (b) CH_3OH or CH_3CO_2H

(c) CH_3CH_2OH or $CH_3C\equiv CH$ (d) CH_3CH_2OH or CH_3CH_2SH

Under each acid is given its pK_a. Recall that acid strength increases with decreasing values of pK_a.

	weaker acid	stronger acid	conjugate base of stronger acid
(a)	H_2O pK_a 15.7	H_2CO_3 pK_a 6.36	HCO_3^-
(b)	CH_3OH pK_a 15.5	CH_3CO_2H pK_a 4.76	$CH_3CO_2^-$
(c)	$CH_3C\equiv CH$ pK_a 25	CH_3CH_2OH pK_a 15.9	$CH_3CH_2O^-$
(d)	CH_3CH_2OH pK_a 15.9	CH_3CH_2SH pK_a 8.5	$CH_3CH_2S^-$

The relative acidity rankings can be predicted based on the following observations:
For (a), the carbonyl group of carbonic acid inductively weakens the O-H bond and stabilizes the deprotonated anion through resonance.
For (b), the carbonyl group of acetic acid inductively weakens the O-H bond and stabilizes the deprotonated anion through resonance.
For (c), oxygen is a more electronegative element (farther to the right on the Periodic Table) than carbon.
For (d), sulfur is below oxygen in the Periodic Table.

Problem 9.31 From each pair, select the stronger base. For each stronger base, write the structural formula of its conjugate acid.

Recall that the stronger base will have the weaker conjugate acid.

(a) OH^- or CH_3O^- (each in H_2O

$$HO^- \longrightarrow HOH$$

stronger   weaker
base   acid

This is a close one, but HOH is a weaker acid than CH_3OH because HOH has a less-electronegative H atom attached to the OH group.

b) $CH_3CH_2O^-$ or $CH_3C\equiv C^-$

$$CH_3C\equiv C^- \longrightarrow CH_3C\equiv CH$$

stronger   weaker
base   acid

Carbon is less electronegative (farther to the left on the Periodic Table) than oxygen, so the alkyne is the weaker acid.

(c) $CH_3CH_2S^-$ or $CH_3CH_2O^-$

$$CH_3CH_2O^- \longrightarrow CH_3CH_2OH$$

stronger   weaker
base   acid

Oxygen is above sulfur in the Periodic Table, so the alcohol is the weaker acid.

(d) $CH_3CH_2O^-$ or NH_2^-

$$NH_2^- \longrightarrow NH_3$$

stronger   weaker
base   acid

Nitrogen is less electronegative (farther to the left on the Periodic Table) than oxygen, so ammonia is the weaker acid.

Problem 9.32 In each equilibrium, label the stronger acid, the stronger base, the weaker acid, and the weaker base. Also estimate the position of each equilibrium.

(a) $CH_3CH_2O^-$ + $CH_3C\equiv CH$ $\rightleftharpoons$ CH_3CH_2OH + $CH_3C\equiv C^-$

$$CH_3CH_2O^- + CH_3C\equiv CH \longleftarrow CH_3CH_2OH + CH_3C\equiv C^- \quad K_{eq} = 10^{-9.1}$$

　　　　　　 pKa 25　　　　　　 pKa 15.9
weaker base　 weaker acid　　 stronger acid　 stronger base

Oxygen is a more electronegative element (farther to the right on the Periodic Table) than carbon, so the alcohol is the stronger acid.

(b) $CH_3CH_2O^- + HCl \rightleftharpoons CH_3CH_2OH + Cl^-$

$\underset{\substack{\text{stronger} \\ \text{base}}}{CH_3CH_2O^-} + \underset{\substack{pK_a -7 \\ \text{stronger} \\ \text{acid}}}{HCl} \longrightarrow \underset{\substack{pK_a 15.5 \\ \text{weaker} \\ \text{acid}}}{CH_3CH_2OH} + \underset{\substack{\text{weaker} \\ \text{base}}}{Cl^-} \quad K_{eq} = 10^{22.5}$

Chlorine is more electronegative (farther to the right on the Periodic Table) than oxygen, so HCl is the stronger acid.

(c) $CH_3\overset{\overset{\text{O}}{\|}}{C}\text{-OH} + CH_3CH_2O^- \rightleftharpoons CH_3\overset{\overset{\text{O}}{\|}}{C}\text{-O}^- + CH_3CH_2OH$

$\underset{\substack{pK_a 4.76 \\ \text{stronger} \\ \text{acid}}}{CH_3\overset{\overset{\text{O}}{\|}}{C}\text{-OH}} + \underset{\substack{\text{stronger} \\ \text{base}}}{CH_3CH_2O^-} \longrightarrow \underset{\substack{\text{weaker} \\ \text{base}}}{CH_3\overset{\overset{\text{O}}{\|}}{C}\text{-O}^-} + \underset{\substack{pK_a 15.9 \\ \text{weaker} \\ \text{acid}}}{CH_3CH_2OH} \quad K_{eq} = 10^{11.1}$

The carbonyl group of acetic acid inductively weakens the O-H bond and stabilizes the deprotonated anion through resonance.

Reactions of Alcohols

Problem 9.33 Write equations for the reaction of 1-butanol with each reagent. Where you predict no reaction, write NR.

(a) Na metal

$2\ CH_3CH_2CH_2CH_2OH + 2\ Na \longrightarrow CH_3CH_2CH_2CH_2O^-\ Na^+ + H_2$

(b) HBr, heat

$CH_3CH_2CH_2CH_2OH + HBr \xrightarrow{heat} CH_3CH_2CH_2CH_2Br + H_2O$

(c) HI heat

$CH_3CH_2CH_2CH_2OH + HI \xrightarrow{heat} CH_3CH_2CH_2CH_2I + H_2O$

(d) PBr$_3$

$3\ CH_3CH_2CH_2CH_2OH + PBr_3 \longrightarrow 3\ CH_3CH_2CH_2CH_2Br + P(OH)_3$

(e) SOCl$_2$, pyridine

$CH_3CH_2CH_2CH_2OH + SOCl_2 \xrightarrow{pyridine} CH_3CH_2CH_2CH_2Cl + SO_2 + HCl$

(f) K$_2$Cr$_2$O$_7$, H$_2$SO$_4$, heat

$CH_3CH_2CH_2CH_2OH + K_2Cr_2O_7 \xrightarrow[heat]{H_2SO_4} CH_3CH_2CH_2\overset{\overset{\text{O}}{\|}}{C}OH + Cr^{3+}$

(g) HIO$_4$

   NR

(h) PCC

$$CH_3CH_2CH_2CH_2OH + PCC \longrightarrow CH_3CH_2CH_2\overset{\overset{\displaystyle O}{\|}}{C}H + Cr^{3+}$$

(i) CH_3SO_2Cl, pyridine

$$CH_3CH_2CH_2CH_2OH + Cl\overset{\overset{\displaystyle O}{\|}}{\underset{\underset{\displaystyle O}{\|}}{S}}CH_3 \xrightarrow{\text{pyridine}} \begin{array}{c} CH_3CH_2CH_2CH_2O\overset{\overset{\displaystyle O}{\|}}{\underset{\underset{\displaystyle O}{\|}}{S}}CH_3 \\ + \\ HCl \end{array}$$

<u>Problem 9.34</u> Write equations for the reaction of 2-butanol with each reagent listed in Problem 9.33. Where you predict no reaction, write NR.

(a) Na metal

$$2 \ CH_3CH_2\overset{\overset{\displaystyle OH}{|}}{C}HCH_3 + 2 \ Na \longrightarrow 2 \ CH_3CH_2\overset{\overset{\displaystyle O^- \ Na^+}{|}}{C}HCH_3 + H_2$$

(b) HBr, heat

$$CH_3CH_2\overset{\overset{\displaystyle OH}{|}}{C}HCH_3 + HBr \xrightarrow{\text{heat}} CH_3CH_2\overset{\overset{\displaystyle Br}{|}}{C}HCH_3 + H_2O$$

(c) HI, heat

$$CH_3CH_2\overset{\overset{\displaystyle OH}{|}}{C}HCH_3 + HI \xrightarrow{\text{heat}} CH_3CH_2\overset{\overset{\displaystyle I}{|}}{C}HCH_3 + H_2O$$

(d) PBr$_3$

$$3 \ CH_3CH_2\overset{\overset{\displaystyle OH}{|}}{C}HCH_3 + PBr_3 \longrightarrow 3 \ CH_3CH_2\overset{\overset{\displaystyle Br}{|}}{C}HCH_3 + P(OH)_3$$

(e) SOCl$_2$, pyridine

$$CH_3CH_2\overset{\overset{\displaystyle OH}{|}}{C}HCH_3 + SOCl_2 \xrightarrow{\text{pyridine}} CH_3CH_2\overset{\overset{\displaystyle Cl}{|}}{C}HCH_3 + SO_2 + HCl$$

(f) $K_2Cr_2O_7$, H_2SO_4, heat

$$CH_3CH_2\overset{\overset{\displaystyle OH}{|}}{C}HCH_3 + K_2Cr_2O_7 \xrightarrow[\text{heat}]{H_2SO_4} CH_3CH_2\overset{\overset{\displaystyle O}{\|}}{C}CH_3 + Cr^{3+}$$

(g) HIO$_4$
    NR

(h) PCC

$$CH_3CH_2\overset{\overset{\displaystyle OH}{|}}{C}HCH_3 + PCC \longrightarrow CH_3CH_2\overset{\overset{\displaystyle O}{\|}}{C}CH_3 + Cr^{3+}$$

(i) CH_3SO_2Cl, pyridine

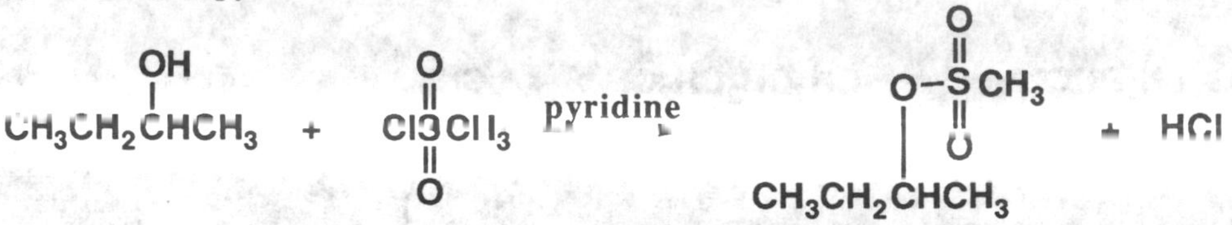

$$\underset{\substack{| \\ OH}}{CH_3CH_2CHCH_3} \quad + \quad \underset{\substack{\| \\ O}}{\overset{O}{Cl\overset{\|}{S}ClI_3}} \quad \xrightarrow{\text{pyridine}} \quad \underset{CH_3CH_2CHCH_3}{\overset{\overset{O}{\|}}{\underset{\underset{O}{\|}}{O-S-CH_3}}} \quad + \quad HCl$$

Problem 9.35 Complete these equations. Show structural formulas for the major products, but do not balance.

(a) $CH_3CH_2CH_2OH + H_2CrO_4 \longrightarrow CH_3CH_2\overset{\overset{O}{\|}}{C}OH + Cr^{3+}$

(b) $\underset{\substack{| \\ CH_3}}{CH_3CHCH_2CH_2OH} + SOCl_2 \longrightarrow \underset{\substack{| \\ CH_3}}{CH_3CHCH_2CH_2Cl} + SO_2 + HCl$

(c) $+ HCl \longrightarrow$ $+ H_2O$

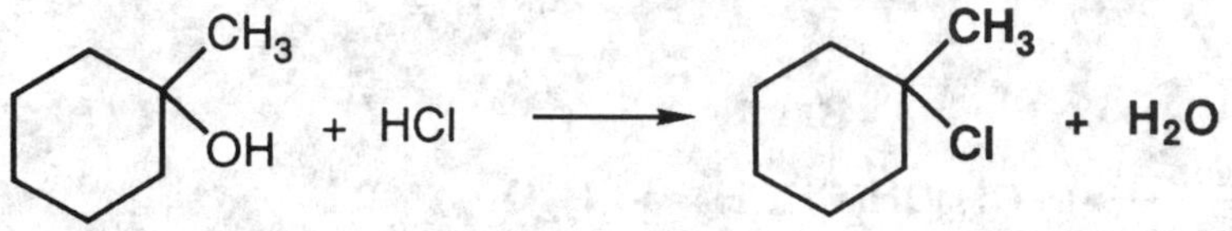

(d) $HOCH_2CH_2CH_2CH_2OH + HBr \longrightarrow BrCH_2CH_2CH_2CH_2Br + 2H_2O$

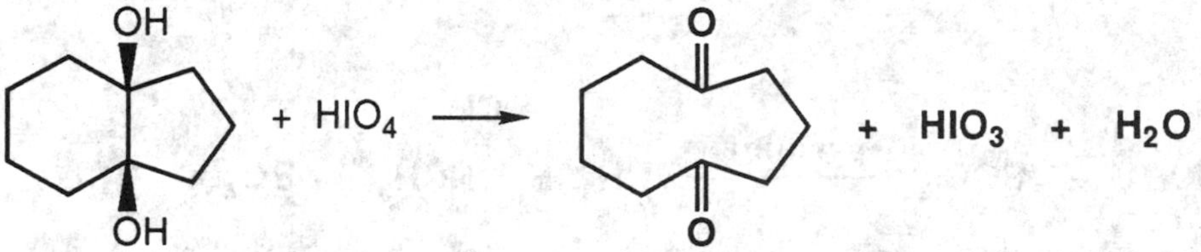

(e) $+ H_2CrO_4 \longrightarrow$ $+ Cr^{3+}$

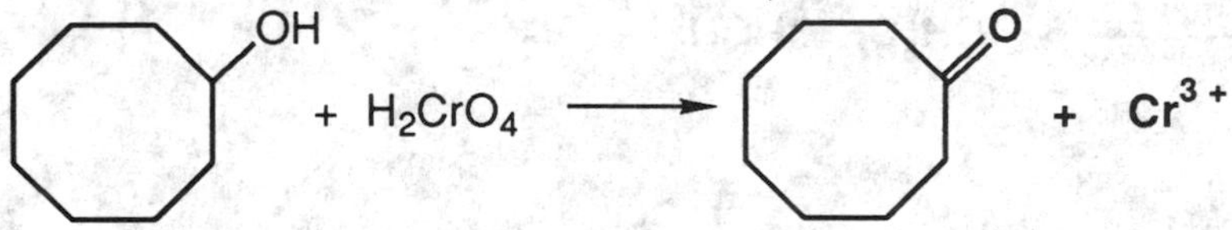

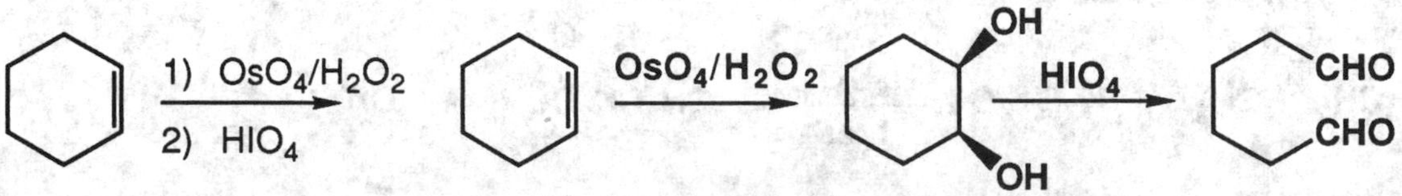

(f) $+ HIO_4 \longrightarrow$ $+ HIO_3 + H_2O$

(g) $\xrightarrow[\text{2) } HIO_4]{\text{1) } OsO_4/H_2O_2}$ $\xrightarrow{OsO_4/H_2O_2}$ $\xrightarrow{HIO_4}$

(h) $-OH + SOCl_2 \longrightarrow$ $-Cl + SO_2 + HCl$

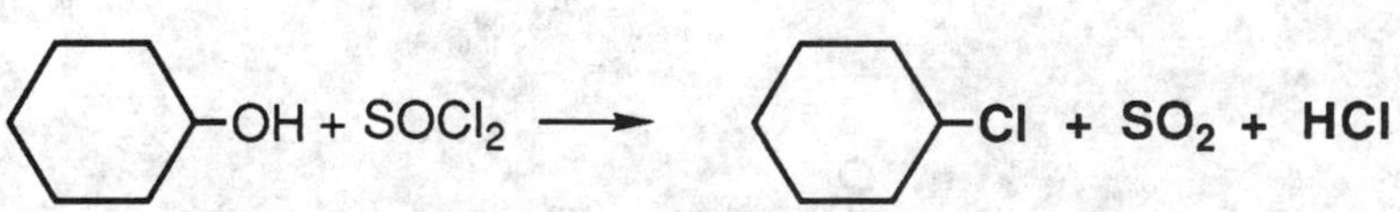

Problem 9.36 When (R)-2-butanol is left standing in aqueous acid, it slowly loses its optical activity. Account for this observation.

In aqueous acid, (R)-2-butanol is in equilibrium with a small amount of the protonated form, that subsequently loses H_2O to create an achiral cation. The achiral cation reacts with another

water molecule to give back 2-butanol. The key to this question is that the achiral cation can react with H_2O on either face of the ion, to give either (R)-2-butanol or (S)-2-butanol. Thus, a pure sample of (R)-2-butanol will gradually turn into a racemic mixture of (R)-2-butanol and (S)-2-butanol, thereby losing optical activity.

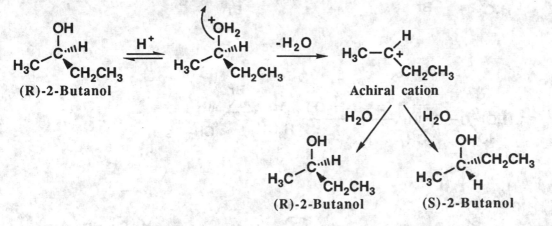

Problem 9.37 Two diastereomeric sets of enantiomers, A/B and C/D, exist for 3-bromo-2-butanol. When enantiomer A or B is treated with HBr, only racemic 2,3-dibromobutane is formed; no meso isomer is formed. When enantiomer C or D is treated with HBr, only meso 2,3-dibromobutane is formed; no racemic 2,3-dibromobutane formed. Account for these observations. (*Hint:* Consider neighboring group participation (Section 8.5) and the type of intermediate that could produce this stereoselectivity).

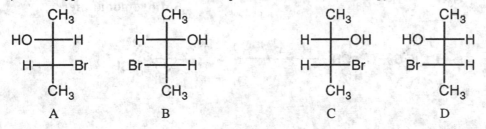

The -OH group of the starting materials will be protonated as usual, and H_2O will depart to produce a carbocation. The key to this problem is realizing that a bromine atom adjacent to a carbocation will form a three-membered ring bromonium ion intermediate that then dictates an *anti* geometry for the incoming bromide nucleophile as shown.

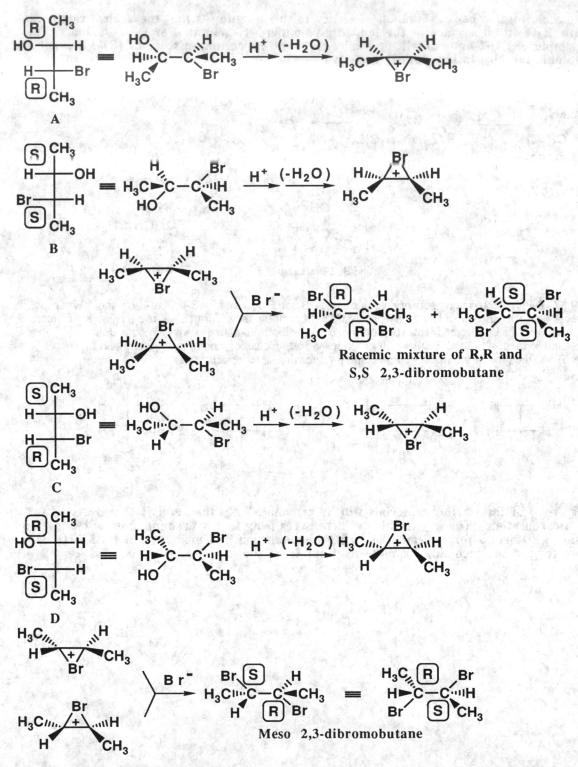

<u>Problem 9.38</u> Acid-catalyzed dehydration of 3-methyl-2-butanol gives three alkenes: 2-methyl-2-butene, 3-methyl-1-butene, and 2-methyl-1-butene. Propose a mechanism to account for the formation of each product.

The 3-methyl-2-butanol is protonated on the -OH group, followed by loss of water to produce a 2° carbocation.

Step 1:

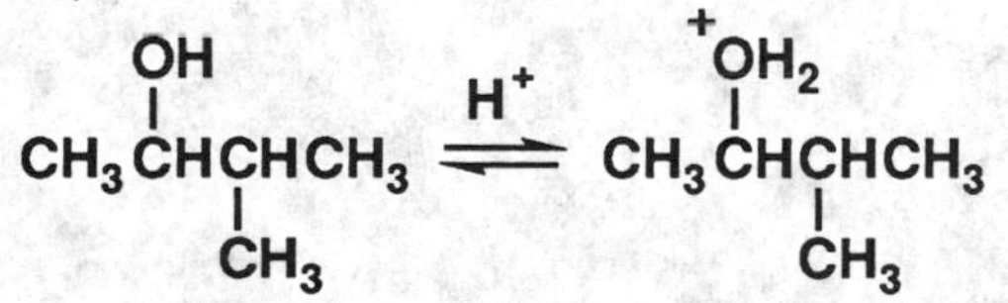

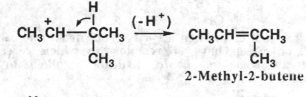

Step 2:

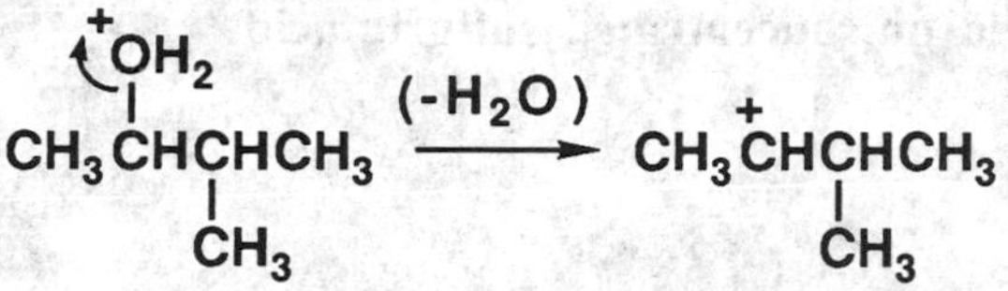

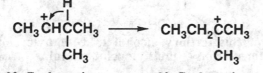

2° Carbocation

This 2° carbocation can lose either of two protons to give 2-methyl-2-butene or 3-methyl-1-butene.

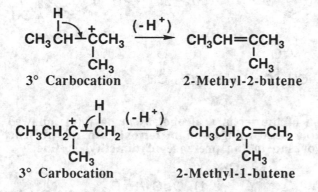

2-Methyl-2-butene

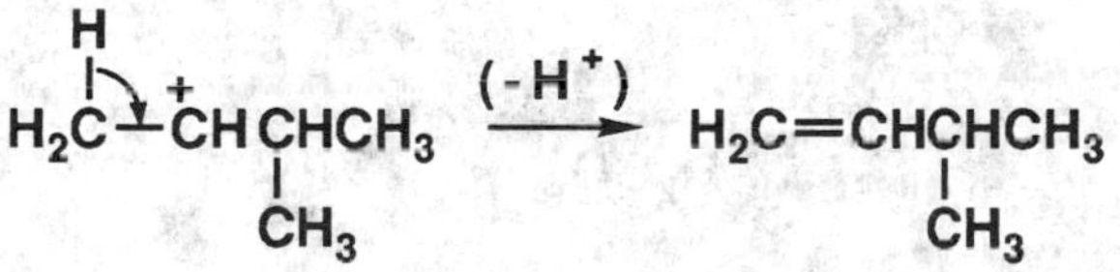

3-Methyl-1-butene

Alternatively, the 2° carbocation can rearrange to give a more stable 3° carbocation.

2° Carbocation   3° Carbocation

This 3° carbocation can lose either of 2 protons to give 2-methyl-2-butene or 2-methyl-1-butene.

3° Carbocation   2-Methyl-2-butene

3° Carbocation   2-Methyl-1-butene

Problem 9.39 Show how you might bring about the following conversions. For any conversion involving more than one step, show each intermediate compound formed.

(a)   CH₃CHCH₂OH ⟶ CH₃C=CH₂
with CH₃ groups

The most common laboratory methods for dehydration of an alcohol to an alkene involve heating the alcohol with either 85% phosphoric acid or concentrated sulfuric acid.

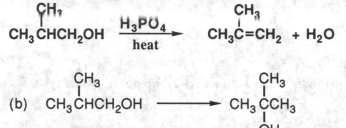

$$CH_3CHCH_2OH \xrightarrow[\text{heat}]{H_3PO_4} CH_3C=CH_2 + H_2O$$

(b)   CH₃CHCH₂OH ⟶ CH₃CCH₃ (OH)

We have not seen any reaction that can switch an -OH group from one carbon atom to an adjacent carbon atom. We have seen, however, reactions by which we can (1) bring about dehydration of an alcohol to an alkene and then (2) hydrate the alkene to an alcohol in the following way.

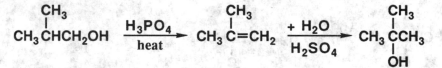

$$CH_3CHCH_2OH \xrightarrow[\text{heat}]{H_3PO_4} CH_3C=CH_2 \xrightarrow[\text{H}_2\text{SO}_4]{+ H_2O} CH_3CCH_3\ (OH)$$

(c)

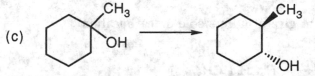

Acid-catalyzed dehydration of this tertiary alcohol to 1-methylcyclohexene followed by hydroboration/oxidation to form *trans*-2-methylcyclohexanol.

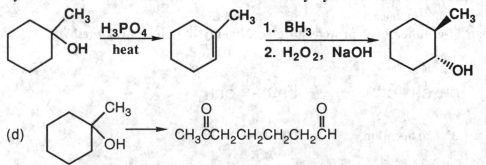

(d)

$$\rightarrow CH_3CCH_2CH_2CH_2CH_2CH$$ (dialdehyde/ketone)

Acid-catalyzed dehydration of the tertiary alcohol as in part (c) followed by oxidative cleavage of the carbon-carbon double bond using osmium tetroxide in the presence of periodic acid, or ozonolysis followed by work-up in the presence of dimethyl sulfide.

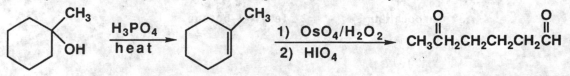

(e)

Hydroboration of the alkene followed by oxidation of the resulting organoborane with alkaline hydrogen peroxide gives a primary alcohol. Reaction of this alcohol with $SOCl_2$ gives the desired primary chloride.

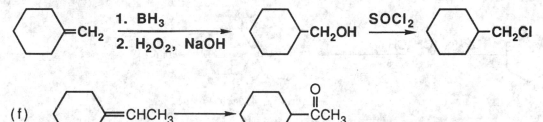

(f)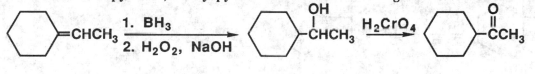

Hydroboration of the alkene followed by oxidation in alkaline hydrogen peroxide gives a secondary alcohol. Oxidation of this alcohol with chromic acid in aqueous sulfuric acid, by chromic acid in pyridine, or by pyridinium chlorochromate gives the desired ketone.

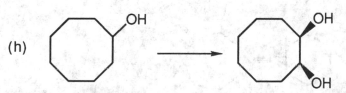

(g) $CH_3(CH_2)_6CH_2OH \longrightarrow CH_3(CH_2)_6\overset{\displaystyle O}{\overset{\displaystyle \|}{C}}H$

This conversion can be accomplished through oxidation with PCC.

$$CH_3(CH_2)_6CH_2OH \xrightarrow{\text{PCC}} CH_3(CH_2)_6\overset{\displaystyle O}{\overset{\displaystyle \|}{C}}H$$

(h)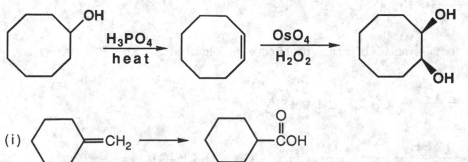

Acid-catalyzed dehydration of the secondary alcohol to an alkene followed by oxidation to the *cis*-glycol using osmium tetroxide in the presence of hydrogen peroxide.

(i)

Hydroboration of the alkene followed by oxidation of the resulting organoborane with alkaline hydrogen peroxide gives a primary alcohol. Oxidation of this alcohol with H_2CrO_4 gives the desired carboxylic acid.

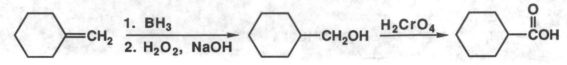

Pinacol Rearrangement

Problem 9.40 Propose a mechanism for the following pinacol rearrangement catalyzed by boron trifluoride etherate:

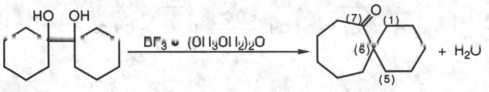

Spiro[5.6]dodecan-7-one

Step 1:

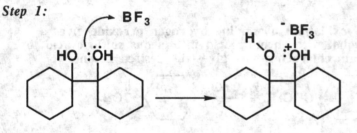

Step 2:

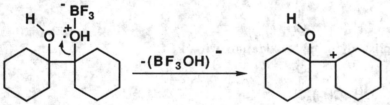

Step 3:

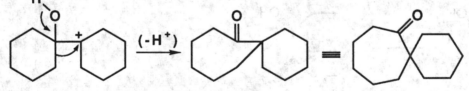

Step 4:

$$H^+ + H\ddot{O}-\bar{B}F_3 \longrightarrow HF + H\ddot{O}-BF_2$$

Synthesis

Problem 9.41 Give reactions for the synthesis of each alcohol from a suitable alkene.

(a) 2-Pentanol

$$CH_3CH_2CH_2CH{=}CH_2 \xrightarrow{H_3O^+} CH_3CH_2CH_2\overset{\overset{\displaystyle OH}{|}}{C}HCH_3$$

(b) 1-Pentanol

$$CH_3CH_2CH_2CH{=}CH_2 \xrightarrow[\text{2. } H_2O_2, \text{ NaOH}]{\text{1. } BH_3} CH_3CH_2CH_2CH_2CH_2OH$$

(c) 2-Methyl-2-pentanol

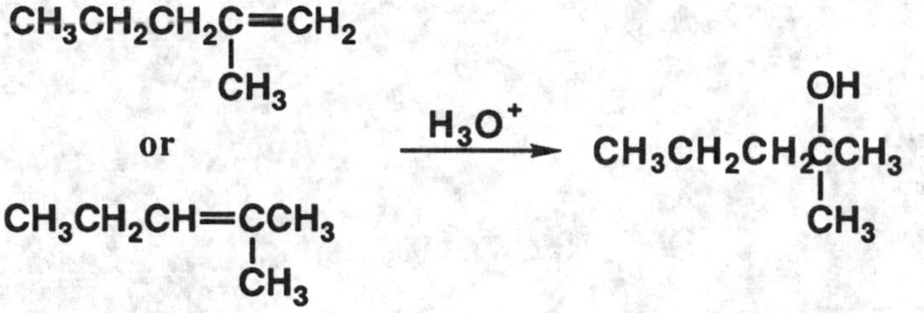

$CH_3CH_2CH_2\overset{\displaystyle |}{C}=CH_2$
$\overset{|}{CH_3}$

or

$CH_3CH_2CH=\overset{\displaystyle |}{C}CH_3$
$\overset{|}{CH_3}$

$\xrightarrow{\text{H}_3\text{O}^+}$

$CH_3CH_2CH_2\overset{\displaystyle OH}{\underset{\displaystyle CH_3}{\overset{|}{\underset{|}{C}}}}CH_3$

(d) 2-Methyl-2-butanol

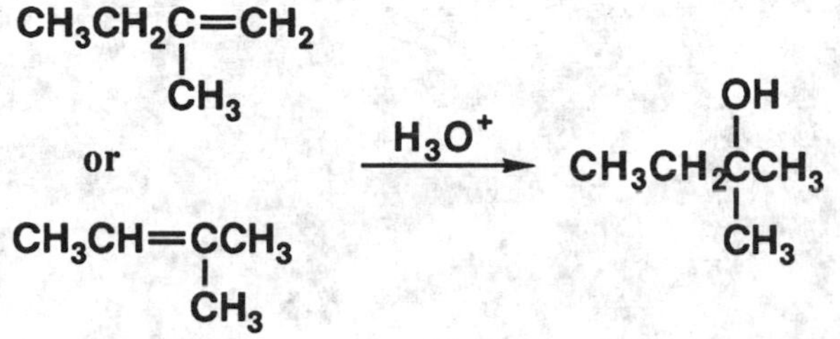

$CH_3CH_2\overset{\displaystyle |}{C}=CH_2$
$\overset{|}{CH_3}$

or

$CH_3CH=\overset{\displaystyle |}{C}CH_3$
$\overset{|}{CH_3}$

$\xrightarrow{\text{H}_3\text{O}^+}$

$CH_3CH_2\overset{\displaystyle OH}{\underset{\displaystyle CH_3}{\overset{|}{\underset{|}{C}}}}CH_3$

(e) 3-Pentanol

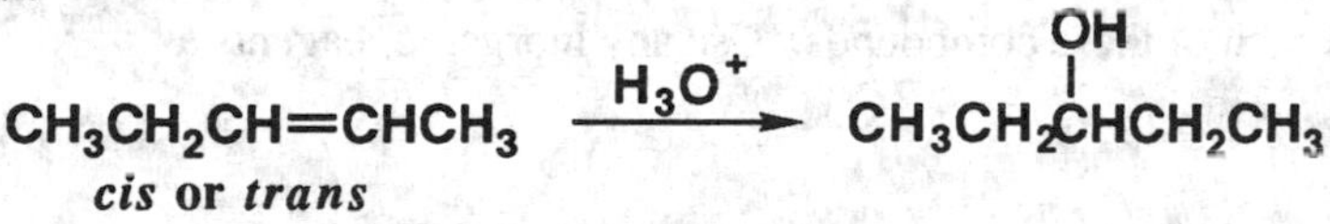

$CH_3CH_2CH=CHCH_3$ $\xrightarrow{\text{H}_3\text{O}^+}$ $CH_3CH_2\overset{\displaystyle OH}{\overset{|}{C}H}CH_2CH_3$
cis or *trans*

(f) 3-Ethyl-3-pentanol

$(CH_3CH_2)_2C=CHCH_3$ $\xrightarrow{\text{H}_3\text{O}^+}$ $CH_3CH_2\overset{\displaystyle OH}{\underset{\displaystyle CH_2CH_3}{\overset{|}{\underset{|}{C}}}}CH_2CH_3$

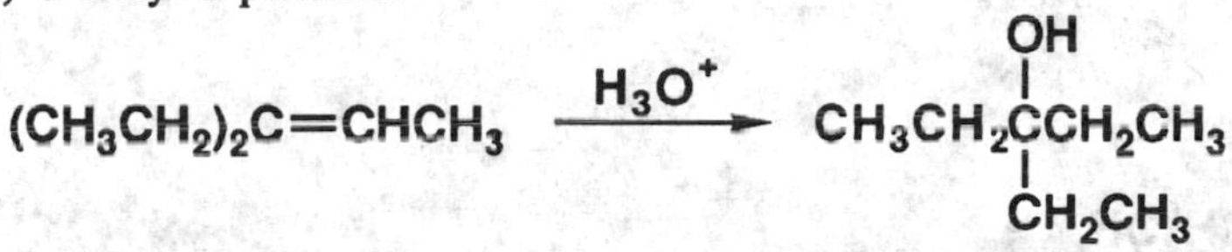

<u>Problem 9.42</u> Dihydropyran is synthesized by treating tetrahydrofurfuryl alcohol with acid. Propose a mechanism for this conversion.

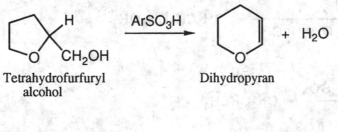

Tetrahydrofurfuryl   Dihydropyran
alcohol

Step 1:

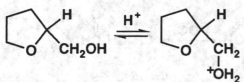

Step 2:

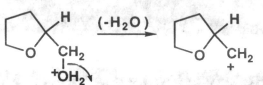

Step 3:

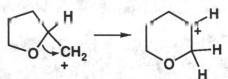

Step 4:

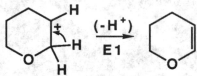

<u>Problem 9.43</u> Show how to convert propene to each of these compounds. Use any inorganic reagents as necessary.

(a) Propane

$$CH_3CH=CH_2 \ + \ H_2 \xrightarrow[\text{catalyst}]{\text{Transition metal}} CH_3CH_2CH_3$$

(b) 1,2-Propanediol

$$CH_3CH=CH_2 \xrightarrow[H_2O_2]{OsO_4} CH_3\overset{\overset{\displaystyle OH}{|}}{C}HCH_2OH$$

(c) 1-Propanol

$$CH_3CH=CH_2 \xrightarrow[\text{2. } H_2O_2, \ NaOH]{\text{1. } BH_3} CH_3CH_2CH_2OH$$

(d) 2-Propanol

$$CH_3CH=CH_2 \xrightarrow[]{H_2O, \ H_2SO_4} CH_3\overset{\overset{\displaystyle OH}{|}}{C}HCH_3$$

(e) Propanal

$$CH_3CH=CH_2 \xrightarrow[\text{2. } H_2O_2, \ NaOH]{\text{1. } BH_3} CH_3CH_2CH_2OH \xrightarrow{PCC} CH_3CH_2\overset{\overset{\displaystyle O}{\|}}{C}H$$

(f) Propanone

$$CH_3CH=CH_2 \xrightarrow{H_2O, \ H_2SO_4} CH_3\overset{\overset{\displaystyle OH}{|}}{C}HCH_3 \xrightarrow{H_2CrO_4} CH_3\overset{\overset{\displaystyle O}{\|}}{C}CH_3$$

(g) Propanoic acid

$$CH_3CH{=}CH_2 \xrightarrow[\text{2. } H_2O_2, \text{ NaOH}]{\text{1. } BH_3} CH_3CH_2CH_2OH \xrightarrow{H_2CrO_4} CH_3CH_2\overset{\overset{\textstyle O}{\|}}{C}OH$$

(h) 1-Bromo-2-propanol

$$CH_3CH{=}CH_2 \xrightarrow{Br_2 \,/\, H_2O} CH_3\overset{\overset{\textstyle OH}{|}}{C}HCH_2Br$$

(i) 3-Chloropropene

$$CH_3CH{=}CH_2 \xrightarrow[\substack{\text{heat} \\ \text{(Allylic} \\ \text{halogenation)}}]{Cl_2} ClCH_2CH{=}CH_2$$

(j) 1,2,3-trichloropropane

$$CH_3CH{=}CH_2 \xrightarrow[\substack{\text{heat} \\ \text{(Allylic} \\ \text{halogenation)}}]{Cl_2} ClCH_2CH{=}CH_2 \xrightarrow[\substack{\text{(Normal} \\ \text{halogenation)}}]{Cl_2} ClCH_2CHClCH_2Cl$$

(k) 1-Chloropropane

$$CH_3CH{=}CH_2 \xrightarrow[\text{2. } H_2O_2, \text{ NaOH}]{\text{1. } BH_3} CH_3CH_2CH_2OH \xrightarrow[\text{Pyridine}]{SOCl_2} CH_3CH_2CH_2Cl$$

(l) 2-Chloropropane

$$CH_3CH{=}CH_2 \xrightarrow{H_2O, \, H_2SO_4} CH_3\overset{\overset{\textstyle OH}{|}}{C}HCH_3 \xrightarrow[\text{Pyridine}]{SOCl_2} CH_3\overset{\overset{\textstyle Cl}{|}}{C}HCH_3$$

(m) 2-Propen-1-ol

$$CH_3CH{=}CH_2 \xrightarrow[\substack{\text{heat} \\ \text{(Allylic} \\ \text{halogenation)}}]{Cl_2} CH_2ClCH{=}CH_2 \xrightarrow[S_N2]{NaOH} CH_2OHCH{=}CH_2$$

(n) Propenal

$$CH_3CH{=}CH_2 \xrightarrow[\substack{\text{heat} \\ \text{(Allylic} \\ \text{halogenation)}}]{Cl_2} CH_2ClCH{=}CH_2 \xrightarrow[S_N2]{NaOH} \overset{\overset{\textstyle OH}{|}}{C}H_2CH{=}CH_2$$

$$\xrightarrow{PCC} \overset{\overset{\textstyle O}{\|}}{H}CCH{=}CH_2$$

Problem 9.44 Show how to bring about this conversion in good yield.

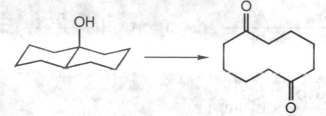

The alcohol is first converted to the alkene by treatment with acid, then the resulting alkene is treated with O_3. Alternatively, (not shown) the alkene could be treated with OsO_4 / H_2O_2 followed by HIO_4.

Problem 9.45 The tosylate of a primary alcohol normally undergoes an S_N2 reaction with hydroxide ion to give a primary alcohol. Reaction of this tosylate, however, gives a compound of molecular formula $C_7H_{12}O$. Propose a structural formula for this compound and a mechanism for its formation.

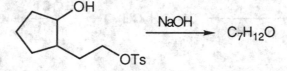

In this reaction, the HO⁻ is acting as a base, not a nucleophile. This makes sense since proton transfer reactions are generally so fast. The alkoxide produced by deprotonation of the alcohol carries out an intramolecular S_N2 attack to give the bicyclic product, $C_7H_{12}O$.

Step 1:

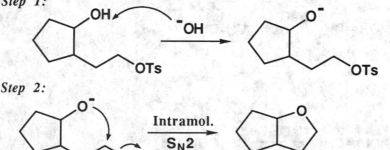

Step 2:

Problem 9.46 Show how to convert cyclohexene to each compound in good yield.

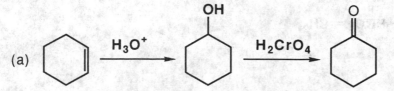

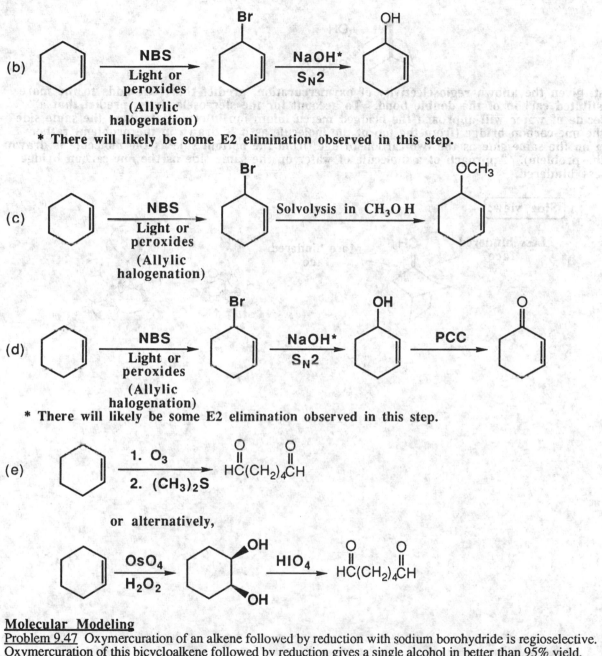

(b) $\xrightarrow[\text{peroxides}]{\text{NBS} \atop \text{Light or}}$ (Allylic halogenation) $\xrightarrow[S_N 2]{\text{NaOH*}}$

* There will likely be some E2 elimination observed in this step.

(c) $\xrightarrow[\text{peroxides}]{\text{NBS} \atop \text{Light or}}$ (Allylic halogenation) $\xrightarrow{\text{Solvolysis in } CH_3OH}$

(d) $\xrightarrow[\text{peroxides}]{\text{NBS} \atop \text{Light or}}$ (Allylic halogenation) $\xrightarrow[S_N 2]{\text{NaOH*}}$ $\xrightarrow{\text{PCC}}$

* There will likely be some E2 elimination observed in this step.

(e) $\xrightarrow[\text{2. } (CH_3)_2 S]{\text{1. } O_3}$ $HC(CH_2)_4 CH$

or alternatively,

$\xrightarrow[H_2O_2]{OsO_4}$ $\xrightarrow{HIO_4}$ $HC(CH_2)_4 CH$

Molecular Modeling

Problem 9.47 Oxymercuration of an alkene followed by reduction with sodium borohydride is regioselective. Oxymercuration of this bicycloalkene followed by reduction gives a single alcohol in better than 95% yield. Propose a structural formula for the alcohol formed. *Hint:* Build a line angle structure in ChemDraw, import it into Chem3D, minimize its energy, and then see if you can determine which face of the double bond is more accessible to the oxymercuration reagent.

Following is the structural formula of the bicyclic alcohol formed by oxymercuration/reduction.

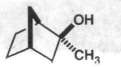

First, given the known regioselectivity of oxymercuration, predict that -OH adds to the more substituted carbon of the double bond. To account for the stereoselectivity, predict that a molecule of water will approach the bridged mercurinium ion intermediate from the same side as the one-carbon bridge (from the top of the molecule as it is drawn in the problem) rather than on the same side as the two-carbon bridge (from the bottom side as the molecule is drawn in the problem). Approach of a molecule of water on the same side as the one-carbon bridge is less hindered.

Side view:

Less hindered
face   CH₂   More hindered
                                face

CHAPTER 10: ALKYNES

SUMMARY OF REACTIONS

Starting Material \ Product →	Acetylide Anions	Alkanes	Alkenes (cis)		Alkenes (trans)	Alkynes	Dihaloalkanes	Dihaloalkenes	Haloalkenes	Ketones		Tetrahaloalkanes	Vinyl Esters
Acetylide Anions						10A 10.5*							
Alkynes		10B 10.7A	10C 10.7A	10D 10.8	10E 10.7B		10F 10.9B	10G 10.9A	10H 10.9B	10I 10.8	10J 10.8	10K 10.9A	10L 10.9D
Haloalkenes					10M 10.6								
Terminal Alkynes	10N 10.4												
Vicinal Dihalides					10O 10.6								

*Section in book that describes reaction.

REACTION 10A: ALKYLATION OF ACETYLIDE ANIONS (Section 10.5)

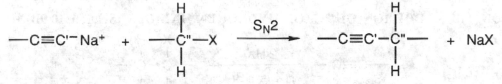

- **Acetylide anions (Reaction 10N, Section 10.4)** are strong bases and good nucleophiles. They react with primary alkyl halides via an S_N2 process to create **alkylated acetylenes**. This reaction is important because carbon-carbon bonds are formed. ✳
- Acetylide anions are such strong bases that secondary and especially tertiary alkyl halides give primarily E2 elimination.

REACTION 10B: CATALYTIC HYDROGENATION (Section 10.7A)

$$-C\equiv C'- \xrightarrow[\text{Pd, Pt, Ni}]{2H_2} -CH_2-C'H_2-$$

- **Catalytic hydrogenation** of an alkyne with hydrogen in the presence of a palladium, platinum, or nickel catalyst results in addition of two mol hydrogen to the alkyne to produce an alkane. ✳

REACTION 10C: CATALYTIC HYDROGENATION, LINDLAR CATALYST (Section 10.7A)

$$-C\equiv C'- \xrightarrow[\substack{\text{Pd/CaCO}_3 \\ \text{(Lindlar} \\ \text{catalyst)}}]{H_2} \underset{H}{\overset{}{\diagup}}C=C'\underset{H}{\overset{}{\diagdown}}$$

- If a special catalyst called the **Lindlar catalyst** is used, the reduction stops after the addition of one mol hydrogen to give a *cis* **alkene** in high yield. The *cis* stereoselectivity is observed because the reaction apparently involves the simultaneous or nearly simultaneous transfer of two hydrogen atoms from the surface of the metal catalyst to the alkyne. ✳

REACTION 10D: ADDITION OF DIBORANE: HYDROBORATION (Section 10.8)

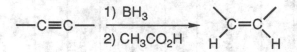

- Borane adds to internal alkynes to give a trialkenylborane. The hydrogen and boron end up on the same side of the double bond in the trialkenylborane (syn addition). ✳
- Treatment of the trialkenylborane with acetic acid replaces the boron with hydrogen to produce a *cis* **alkene**.
- For terminal alkynes, a special sterically hindered borane reagent, such as **(sia)$_2$BH,** is used instead of borane itself to prevent the unwanted addition of a second borane to make a dihydroborated alkane. The steric bulk of the (sia)$_2$BH prevents the second borane from reacting. Note that for terminal alkynes, the boron ends up on the less substituted carbon atom.

REACTION 10E: CHEMICAL REDUCTION (Section 10.7B)

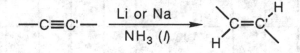

- **Chemical reduction** of an alkyne with sodium or lithium metal in liquid ammonia results in formation of a *trans* **alkene**. This is complementary to the use of the Lindlar catalyst that produces the *cis* **alkene** product. ✳

REACTION 10F: ADDITION OF 2 MOL HYDROGEN HALIDES (Section 10.9B)

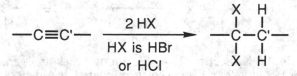

- Alkynes add two mol of **HBr** and **HCl.** ✳
- For terminal alkynes, both the first and second mol hydrogen halide follow Markovnikov's rule.
- Addition of **HCl** to acetylene yields **chloroethylene**, a compound of considerable industrial importance for the manufacturing of plastics such as **polyvinyl chloride** that is used for pipes and other fittings. Today, cheaper routes are used for the production of chloroethylene involving ethylene, chlorine, and heat.

REACTION 10G: ADDITION OF 1 MOL BROMINE OR CHLORINE (Section 10.9A)

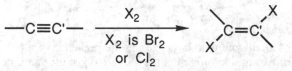

- Alkynes react with 1 mol of Cl$_2$ or Br$_2$ to give *anti* **addition** and thus the *trans* dihaloalkene. This can be isolated or reacted with another mol of the halogen to give the tetrahaloalkane (reaction 10K). ✳

REACTION 10H: ADDITION OF 1 MOL HYDROGEN HALIDES (Section 10.9B)

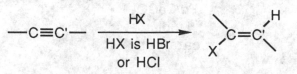

- Alkynes add one mol of **HBr** and **HCl**. The addition is stereoselective and results in *anti* **addition**. ✱
- For terminal alkynes, the addition of hydrogen halide follows Markovnikov's rule.

REACTION 10I: REACTION WITH DIBORANE AND PEROXIDE (Section 10.8)

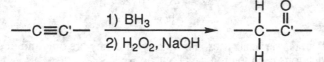

- Treatment with diborane results in formation of an alkenylborane. Note that for terminal alkynes, the boron ends up on the less substituted carbon atom. Reaction of this alkenyl borane with hydrogen peroxide in aqueous sodium hydroxide replaces the boron atom with -OH. The alcohol that is formed is called an **enol** , and it is not stable because it can rearrange to the more stable **ketone** or **aldehyde**. This process of rearrangement is called **tautomerization**. ✱
- The position of equilibrium for most keto-enol tautomer pairs lies very far to the side of the keto form.

REACTION 10J ADDITION OF WATER: HYDRATION (Section 10.9C)

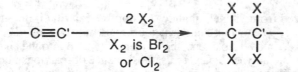

- Addition of **water** occurs in the presence of strong acids and Hg(II) salts, the resulting enol then undergoes keto-enol tautomerism to give a ketone. For terminal alkynes, the -OH group adds to the carbon atom with the alkyl group (according to Markovnikov's rule). ✱

REACTION 10K: ADDITION OF 2 MOL BROMINE OR CHLORINE (Section 10.9A)

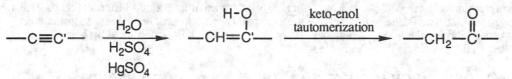

- Alkynes react with one mol of Cl₂ or Br₂ to give *anti* addition and thus the *trans* dihaloalkene (reaction 6F). This can be isolated or reacted with another mol of the halogen to give the tetrahaloalkane. ✱

REACTION 10L: ADDITION OF CARBOXYLIC ACIDS - FORMATION OF VINYL ESTERS (Section 10.9D)

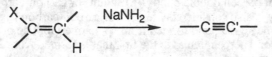

- Addition of **carboxylic acids** occurs in the presence of strong acids and Hg(II) salts to give **vinyl esters**. An important example involves the addition of acetic acid to acetylene to give vinyl acetate. Polymerization gives **poly(vinyl acetate)**, which is used in adhesives. ✱

REACTION 10M: DEHYDROHALOGENATION OF HALOALKENES (Section 10.6)

- Reaction of haloalkenes with a strong base such as $NaNH_2$ results in formation of an alkyne.✱
- Dehydrohalogenation of a haloalkene with at least one hydrogen on each adjacent carbon atom can also form a side product with cumulated double bonds called an **allene**.
- Allenes are compounds that have two carbon-carbon double bonds adjacent to each other.

$$H_2C = C = CH_2$$

Allene

- Allenes are less stable than isomeric alkynes. For this reason, they are only minor side products during the dehydrohalogenation reactions used to produce alkynes.

REACTION 10N: DEPROTONATION OF TERMINAL ALKYNES (Section 10.4)

$$-C \equiv C'-H \quad \xrightarrow{NaNH_2} \quad -C \equiv C'^- Na^+ \quad + \quad NH_3$$

- One of the major differences between the chemistry of alkynes and alkenes is that the **hydrogen attached to a carbon-carbon triple bond** is sufficiently **acidic** that it can be removed by a strong base, usually sodamide ($NaNH_2$), sodium hydride (NaH) or lithium diisopropylamide (LDA). ✱
- The pK_a for a terminal alkyne is 25, compared with nearly 44 for an alkene and 48 for an alkane. This low pK_a for the terminal alkyne is due to the fact that the lone pair on carbon produced upon deprotonation resides in an sp orbital for the alkyne which has 50% s character and is thus stabilized by being held relatively close to the positively charged nucleus.

REACTION 10O: DEHYDROHALOGENATION OF VICINAL DIHALIDES (Section 10.6)

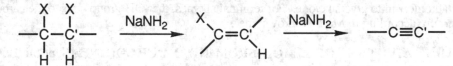

- Alkynes can be synthesized from vicinal dihalides by **double dehydrohalogenation** with a very strong base such as sodium amide. The dehydrohalogenation occurs in two steps. The first dehydrohalogenation converts the vicinal dihalide into a haloalkene, that is then converted into the alkyne. ✱
- Alkenes can be converted into alkynes by a very useful sequence of reactions involving initial reaction with a halogen to form a vicinal dihalide, that is then converted to the alkyne by addition of two mol of strong base like $NaNH_2$.

$$R-CH=C'H-R \quad \xrightarrow{Br_2} \quad R-\underset{\underset{Br}{|}}{\overset{\overset{H}{|}}{C}}-\underset{\underset{Br}{|}}{\overset{\overset{H}{|}}{C'}}-R \quad \xrightarrow{2\,NaNH_2} \quad R-C \equiv C'-R$$

Vicinal dibromide

SUMMARY OF IMPORTANT CONCEPTS

10.0 OVERVIEW
• **Alkynes** are molecules with a carbon-carbon triple bond composed of one sigma bond and two pi bonds. As with alkenes, the chemistry of alkynes is largely determined by the pi bonds, so in many ways the chemistry of alkynes is similar to that of alkenes. ✱

10.1 STRUCTURE
• Because carbon-carbon triple bonds are composed of one sigma bond and two pi bond, the bond length (0.121 nm) is shorter than for carbon-carbon double bonds (0.134 nm) or carbon-carbon single bonds (0.154 nm). The bond dissociation energy is also considerably higher for carbon-carbon triple bonds. ✱
  - A triple bond consists of one sigma bond formed from sp-sp overlap, one pi bond formed from $2p_y$-$2p_y$ overlap, and a second pi bond perpendicular to the first that is formed from $2p_z$-$2p_z$ overlap. See Figure 1.21 in the text for a diagram.

- The shorter carbon-carbon bond length in alkynes is due to the sp-sp overlap of the sigma bond. The sp orbitals are 50% s in character and the larger the percent s character, the shorter the bond.

10.2 NOMENCLATURE
• According to the **IUPAC system**, the infix **yn** is used to show the presence of a carbon-carbon triple bond. ✳
- The longest chain that contains the triple bond is numbered from the end that gives the triply bonded carbons the lower numbers. The location of the triple bond is indicated by the number of the first carbon of the triple bond.
- If more than one triple bond is present in the chain, then the infixes **adiyn**, **atriyn**, etc. are used.
- If a double and triple bond are found in the same molecule, then **en** and **yn** are both used. The triple bond takes precedence, so it is written last and the numbering scheme is chosen that gives the triple bond the lower number. For example, 4-hexen-1-yne is a correct name, not 2-hexen-5-yne, for ⌇≡.
- The IUPAC system retains the name acetylene for ethyne (HC≡CH).
• **Common names** for simple alkynes are derived from that of acetylene by prefixing the names of substituents on the carbon-carbon triple bonds to the name **acetylene**. For example, dimethylacetylene is the common name of 2-butyne.

10.3 PHYSICAL PROPERTIES
• The **physical properties** of alkynes are similar to those of alkanes and alkenes with similar carbon skeletons. For example, alkynes with four or less carbon atoms are gases at room temperature while those with five or more carbon atoms are liquids or solids at room temperature.
• When acetylene is passed through a solution of certain catalysts, it polymerizes to make a shiny polymer of **polyacetylene** that contains alternating double and single bonds. This polymer is very interesting because it can be doped (an electron can be removed or added to the pi bonding system) and then it becomes a **conductor** of electricity. When stretched, the polymer chains become more ordered and the conductivity is greater along the chains than perpendicular to them. This and other properties of polyacetylene may lead to some important applications in the future.

CHAPTER 10
Solutions to the Problems

Problem 10.1 Write the IUPAC name for each compound.

(a) $CH_3(CH_2)_5C\equiv CH$
       1-Octyne

(b) $CH_3\underset{\underset{CH_3}{|}}{\overset{\overset{OH}{|}}{C}}C\equiv CH$ (c) $CH_3\underset{\underset{CH_3}{|}}{\overset{\overset{CH_3}{|}}{C}}C\equiv CH$

2-Methyl-3-butyn-2-ol   **3,3-Dimethyl-1-butyne**

Problem 10.2 Write the common name for each compound.

(a) $CH_3\underset{\underset{}{|}}{\overset{\overset{CH_3}{|}}{C}}HC\equiv C\underset{\underset{}{|}}{\overset{\overset{CH_3}{|}}{C}}HCH_3$

(b)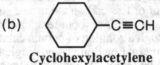

Diisopropylacetylene   **Cyclohexylacetylene**

(c) $HC\equiv CCH_2CH_2CH_2CH_3$
       Butylacetylene

Problem 10.3 Draw a structural formula for an alkene and dichloroalkane of the given molecular formula that yields the indicated alkyne by each reaction sequence.

The key to this problem is to realize that wherever the triple bond is in the carbon chain, that is where the double bond was.

(a) $C_6H_{12} \xrightarrow{Cl_2} C_6H_{12}Cl_2 \xrightarrow{2NaNH_2} CH_3CH_2C\equiv CCH_2CH_3$

$CH_3CH_2CH=CHCH_2CH_3 \xrightarrow{Cl_2} CH_3CH_2\underset{\underset{Cl}{|}}{C}H-\underset{\underset{Cl}{|}}{C}HCH_2CH_3 \xrightarrow{2NaNH_2}$
       (*cis* or *trans*)

(b) $C_7H_{14} \xrightarrow{Cl_2} C_7H_{14}Cl_2 \xrightarrow{2NaNH_2} CH_3C\equiv C\underset{\underset{CH_3}{|}}{\overset{\overset{CH_3}{|}}{C}}CH_3$

$CH_3-CH=CH-\underset{\underset{CH_3}{|}}{\overset{\overset{CH_3}{|}}{C}}-CH_3 \xrightarrow{Cl_2} CH_3-\underset{\underset{Cl}{|}}{C}H-\underset{\underset{Cl}{|}}{C}H-\underset{\underset{CH_3}{|}}{\overset{\overset{CH_3}{|}}{C}}-CH_3 \xrightarrow{2NaNH_2}$
       (*cis* or *trans*)

<u>Problem 10.4</u> Draw the structural formula for a hydrocarbon of the given molecular formula that undergoes hydroboration-oxidation to give the indicated product:

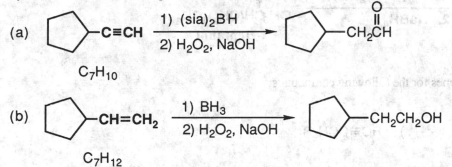

(a)

C_7H_{10}

(b)

C_7H_{12}

<u>Problem 10.5</u> Aicd-catalyzed hydration of 2-pentyne gives a mixture of two ketones, each of molecular formula $C_5H_{10}O$. Propose structural formulas for these two ketones and for the enol from which each is derived.

The two ketones are 2-pentanone and 3-pentanone.

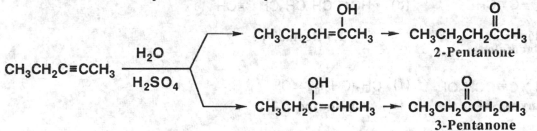

<u>Problem 10.6</u> Show how the synthetic scheme in Example 10.6 might be modified to give
(a) 1-Heptanol

A reduction step can be included using the Lindlar catalyst to give an alkene, that can then be converted to the primary alcohol on the less-substituted carbon of the double bond using hydroboration.

$$HC\equiv CH \xrightarrow[\text{2. } CH_3(CH_2)_3CH_2Br]{\text{1. } NaNH_2} HC\equiv C(CH_2)_4CH_3 \xrightarrow[\substack{\text{Pd/CaCO}_3 \\ \text{(Lindlar catalyst)}}]{H_2}$$

$$\underset{\substack{H \\ }}{\overset{\substack{H \\ }}{C}}=\underset{H}{\overset{CH(CH_2)_4CH_3}{C}} \xrightarrow[\text{2. } H_2O_2,\ NaOH]{\text{1. } BH_3} HOCH_2(CH_2)_5CH_3$$
                                                                    1-Heptanol

(b) 2-Heptanol

This is the same as part (a), but the last step uses oxymercuration/reduction to place the -OH group on the more substituted carbon of the double bond.

$$HC\equiv CH \xrightarrow[\text{2. } CH_3(CH_2)_3CH_2Br]{\text{1. } NaNH_2} HC\equiv C(CH_2)_4CH_3 \xrightarrow[\substack{\text{Pd/CaCO}_3 \\ \text{(Lindlar catalyst)}}]{H_2}$$

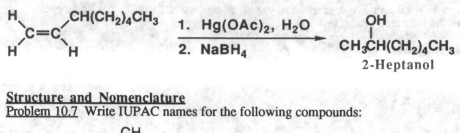

Structure and Nomenclature

Problem 10.7 Write IUPAC names for the following compounds:

(a) 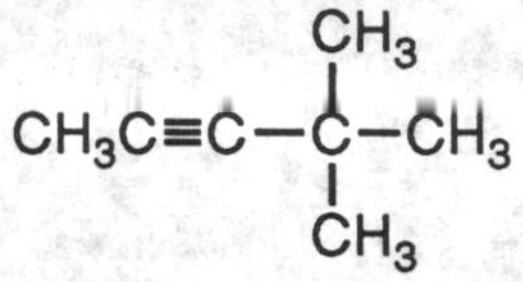 $CH_3C\equiv C-\underset{\underset{CH_3}{|}}{\overset{\overset{CH_3}{|}}{C}}-CH_3$

4,4-Dimethyl-2-pentyne
(*tert*-**Butylmethylacetylene**)

(b) $HC\equiv CCH_2Br$

3-Bromopropyne

(c) 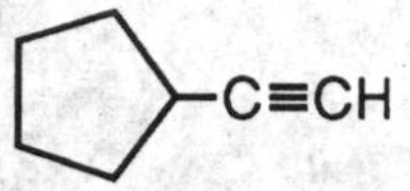 [cyclopentane]—$C\equiv CH$

Ethynylcyclopentane
(**Cyclopentylacetylene**)

(d) $HC\equiv CCH_2CH_2CH_2C\equiv CH$

1,6-Heptadiyne

(e) $CH_3(CH_2)_5C\equiv CCH_2OH$
2-Nonyn-1-ol

(f) $CH_3(CH_2)_6C\equiv CH$
1-Nonyne

(g) $CH_3C\equiv CCH_2OH$
2-Butyn-1-ol

(h) $CH_3(CH_2)_7C\equiv C(CH_2)_7CO_2H$
9-Octadecynoic acid

Problem 10.8 Draw structural formulas for the following compounds:

(a) 3-Hexyne

$CH_3CH_2C\equiv CCH_2CH_3$

(b) Vinylacetylene

$CH_2=CH-C\equiv CH$

(c) 3-Chloro-1-butyne

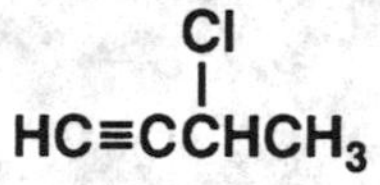

$HC\equiv C\underset{\underset{CH_3}{}}{\overset{\overset{Cl}{|}}{C}}HCH_3$

(d) 5-Isopropyl-3-octyne

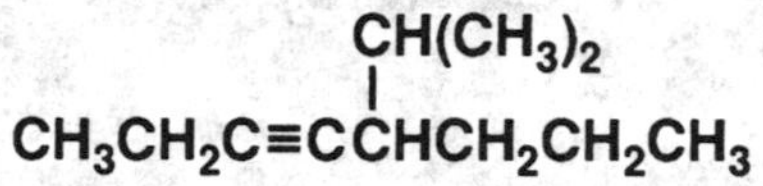

$CH_3CH_2C\equiv C\underset{}{\overset{\overset{CH(CH_3)_2}{|}}{C}}HCH_2CH_2CH_3$

(e) 3-Pentyn-2-ol

$CH_3C\equiv C\underset{\underset{OH}{|}}{C}HCH_3$

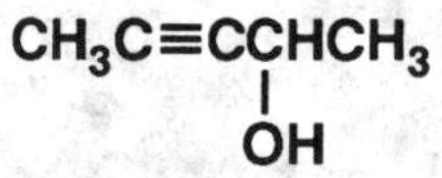

(f) 2-Butyne-1,4-diol

$HOCH_2C\equiv CCH_2OH$

(g) Diisopropylacetylene

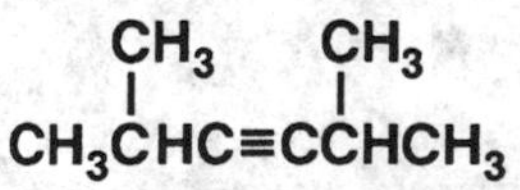

$CH_3\underset{\overset{|}{CH_3}}{C}HC\equiv C\underset{\overset{|}{CH_3}}{C}HCH_3$

(h) *tert*-Butylmethylacetylene

$CH_3C\equiv C\underset{\underset{CH_3}{|}}{\overset{\overset{CH_3}{|}}{C}}CH_3$

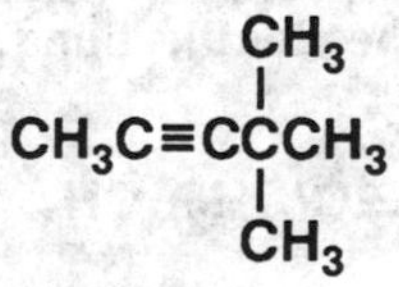

(i) Cyclodecyne

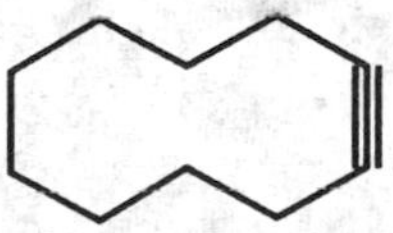

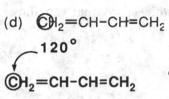

Problem 10.9 Predict all bond angles about each circled atom.

(a)   CH₃—C≡Ⓒ—CH₃

180°

CH₃—C≡Ⓒ—CH₃

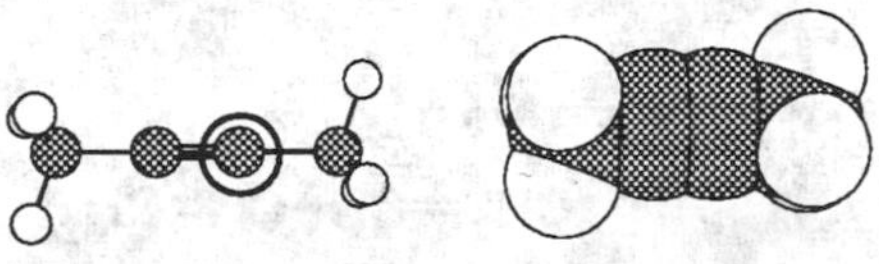

(b)   CH₂=Ⓒ H—C≡CH

120°

CH₂=Ⓒ H—C≡CH

(c)   CH₂=Ⓒ=CH-CH₃

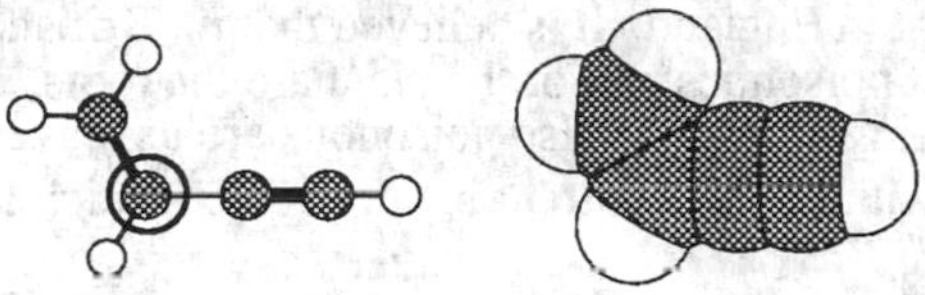

180°

CH₂=Ⓒ=CH-CH₃

(d)   Ⓒ H₂=CH—CH=CH₂

120°

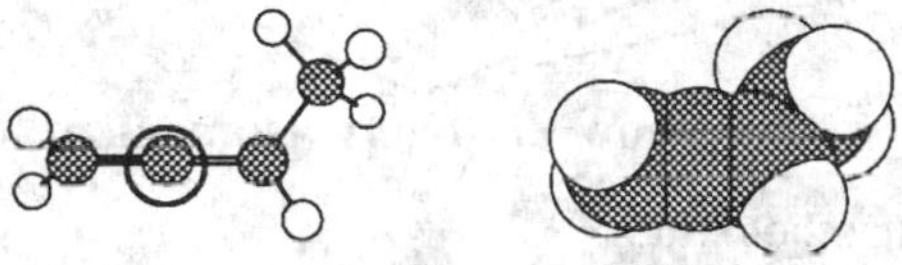

Ⓒ H₂=CH—CH=CH₂

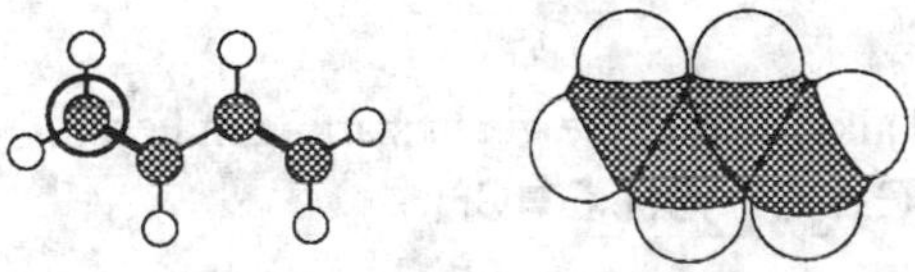

Problem 10.10 State the orbital hybridization of each circled atom.

(a)   CH₃—C≡Ⓒ—CH₃   (b)   CH₂=Ⓒ H—C≡CH

sp   **sp²**

CH₃—C≡Ⓒ—CH₃   CH₂=Ⓒ H—C≡CH

(c)   CH₂=Ⓒ=CH-CH₃   (d)   O=Ⓒ=O

sp   **sp**

CH₂=Ⓒ=CH-CH₃   O=Ⓒ=O

Problem 10.11 Describe each circled carbon-carbon bond in terms of the overlap of atomic orbitals.

(a) $CH_3-C\equiv C-CH_3$

$\sigma sp\text{-}sp^3$

$CH_3-C\equiv C-CH_3$

(b) $CH_2=CH-C\equiv CH$

$\sigma sp^2\text{-}sp$

$CH_2=CH-C\equiv CH$

(c) $CH_2=C=CH\text{-}CH_3$

$\sigma sp\text{-}sp^2$

$CH_2=C=CH\text{-}CH_3$

$\pi 2p\text{-}2p$

(d) $CH_2=CH-CH=CH_2$

$\sigma sp^2\text{-}sp^2$

$CH_2=CH\text{-}CH=CH_2$

Problem 10.12 Enanthotoxin is an extremely poisonous organic compound found in hemlock water dropwart, which is reputed to be the most poisonous plant in England. It is believed that no British plant has been responsible for more fatal accidents. The most poisonous part of the plant are the roots, which resemble small white carrots, giving the plant the name "five finger death." Also poisonous are its leaves, which look like parsley. Enanthotoxin is thought to interfere with the Na^+ current in nerve cells, which leads to convulsions and death. See the Merck Index, 12th Ed., #3608.

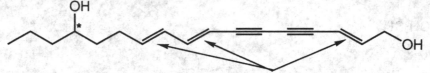

Can show *cis-trans* isomerism

How many stereoisomers are possible for enanthotoxin?

There is one tetrahedral stereocenter (marked with an *) and three double bonds that can show *cis-trans* isomerism, so there are 2^4 or 16 possible stereoisomers.

Preparation of Alkynes
Problem 10.13 Show how to prepare each alkyne from the given starting material.

(a) $CH_3CH_2CH_2CH=CH_2 \longrightarrow CH_3CH_2CH_2C\equiv CH$

First treat 1-pentene with either bromine (Br_2) or chlorine (Cl_2) to form a 1,2-dihalopentane. Then carry out a double dehydrohalogenation with two mol sodium amide ($NaNH_2$) to form 1-pentyne.

$$CH_3(CH_2)_2CH=CH_2 \xrightarrow{Br_2} CH_3(CH_2)_2\overset{Br}{\underset{}{C}}H\text{-}\overset{Br}{\underset{}{C}}H_2 \xrightarrow{2NaNH_2} CH_3CH_2CH_2C\equiv CH$$

(b) $CH_3(CH_2)_5\overset{}{\underset{Cl}{C}}HCH_3 \longrightarrow CH_3(CH_2)_4C\equiv CCH_3$

First convert 2-chlorooctane to 2-octene by dehydrohalogenation with sodium amide. This is an example of a β-elimination reaction. Then follow the procedure in (a) to convert an alkene to an alkyne, namely addition of Br_2 or Cl_2 followed by a double dehydrohalogenation.

$$CH_3(CH_2)_5\overset{}{\underset{Cl}{C}}HCH_3 \xrightarrow{NaNH_2} CH_3(CH_2)_4CH=CHCH_3 \xrightarrow[\text{2) } 2NaNH_2]{\text{1) } Br_2} CH_3(CH_2)_4C\equiv CCH_3$$

(c) $CH_3CH_2CH_2C\equiv CH \longrightarrow CH_3CH_2CH_2C\equiv CD$

First form the acetylide anion with sodium amide or sodium hydride (NaH) and then react the anion with a deuterium donor such as D_2O or CH_3CH_2OD.

$$CH_3CH_2CH_2C\equiv CH \xrightarrow{NaNH_2} CH_3CH_2CH_2C\equiv C^- \ Na^+ \xrightarrow{D_2O} CH_3CH_2CH_2C\equiv CD$$

Problem 10.14 If a catalyst could be found that would establish an equilibrium between 1,2-butadiene and 2-butyne, what would be the ratio of the more stable isomer to the less stable isomer at 25°C?

$$CH_2=C=CHCH_3 \rightleftharpoons CH_3C\equiv CCH_3 \qquad \Delta G° = -4.0 \text{ kcal/mol}$$

At equilibrium the relative amounts of each form are given by the equation:

$$\Delta G° = -RT\ln K_{eq} \qquad \text{where } K_{eq} = \frac{[\text{2-butyne}]}{[\text{1,2-butadiene}]}$$

Converting to base 10 and rearranging gives

$$\log K_{eq} = \frac{-\Delta G°}{(2.303)RT}$$

Taking the antilog of both sides gives:

$$K_{eq} = 10^{\left(\frac{-\Delta G°}{(2.303)RT}\right)}$$

Plugging in the values for $\Delta G°$, R, and 298 K gives the final answer:

$$K_{eq} = 10^{\left(\frac{-(-4.0 \text{ kcal/mol})}{(2.303)(1.987 \times 10^{-3} \text{ kcal/mol K})(298 \text{ K})}\right)} = 10^{2.93} = 8.6 \times 10^2$$

Since $\Delta G°$ is negative, it means the alkyne is favored, and if a catalyst could be found for the interconversion, the ratio of alkyne to allene would be 860 to 1.

Reactions of Alkynes
Problem 10.15 Complete these acid-base reactions and predict whether the position of equilibrium lies toward the left or toward the right.

Recall that equilibrium favors formation of the weaker acid, weaker base pair.

(a) $CH_3C\equiv CH + (CH_3)_3CO^- K^+ \underset{(CH_3)_3COH}{\rightleftharpoons} CH_3C\equiv C^- K^+ + (CH_3)_3COH$

    pKa 25   (weaker   (stronger   pKa 18
    (weaker   base)   base)   (stronger
    acid)   acid)

Oxygen is more electronegative (farther to the right on the Periodic Table) than carbon so the alcohol is the stronger acid and equilibrium lies toward the left.

(b) $CH_2=CH_2 + Na^+NH_2^- \underset{NH_3 (l)}{\rightleftharpoons} CH_2=CH^- Na^+ + NH_3$

    pKa 44   (weaker   (stronger   pKa 33
    (weaker   base)   base)   (stronger
    acid)   acid)

Nitrogen is more electronegative (farther to the right on the Periodic Table) than carbon so ammonia is the stronger acid and equilibrium lies toward the left.

(c) $CH_3C{\equiv}CCH_2OH$ + $Na^+NH_2^-$ $\xrightleftharpoons{\quad NH_3\ (l)\quad}$ $CH_3C{\equiv}CCH_2O^-\ Na^+$ + NH_3

| pK_a ~16 (stronger acid) | (stronger base) | (weaker base) | pK_a 33 (weaker acid) |

Oxygen is more electronegative (farther to the right on the Periodic Table) than nitrogen so the alcohol is the stronger acid and equilibrium lies toward the right.

__Problem 10.16__ Draw structural formulas for the major product(s) formed by reaction of 3-hexyne with each of these reagents. Where you predict no reaction, write NR.

(a) H_2(excess) / Pt

$CH_3CH_2CH_2CH_2CH_2\ CH_3$

(b) H_2 / Lindlar catalyst

CH_3CH_2 and CH_2CH_3 on C=C with H, H (cis alkene)

(c) Na in NH_3 (liquid)

CH_3CH_2 and H / C=C / H and CH_2CH_3 (trans alkene)

(d) BH_3 followed by H_2O_2 / NaOH

$CH_3CH_2\underset{\underset{\displaystyle O}{\|}}{C}CH_2CH_2CH_3$

(e) BH_3 followed by CH_3CO_2H

CH_3CH_2 and CH_2CH_3 / C=C / H and H

(f) BH_3 followed by CH_3CO_2D

CH_3CH_2 and CH_2CH_3 / C=C / H and D

(g) Cl_2 (1 mol)

CH_3CH_2 and Cl / C=C / Cl and CH_2CH_3

(h) $NaNH_2$ in NH_3 (liquid)

No Reaction

(i) HBr (1 mol)

CH_3CH_2 and Br / C=C / H and CH_2CH_3

(j) HBr (2 mol)

$CH_3CH_2\underset{\underset{\displaystyle Br}{|}}{\overset{\overset{\displaystyle Br}{|}}{C}}CH_2CH_2CH_3$

(k) H_2O in H_2SO_4 / $HgSO_4$

$CH_3CH_2\underset{\underset{\displaystyle O}{\|}}{C}CH_2CH_2CH_3$

(l) CH_3CO_2H in H_2SO_4 / $HgSO_4$

CH_3CH_2 and $O-\underset{\underset{\displaystyle O}{\|}}{C}CH_3$ / C=C / H and CH_2CH_3

Problem 10.17 Draw the structural formula of the enol formed in each alkyne hydration reaction and then draw the structural formula of the carbonyl compound with which each enol is in equilibrium.

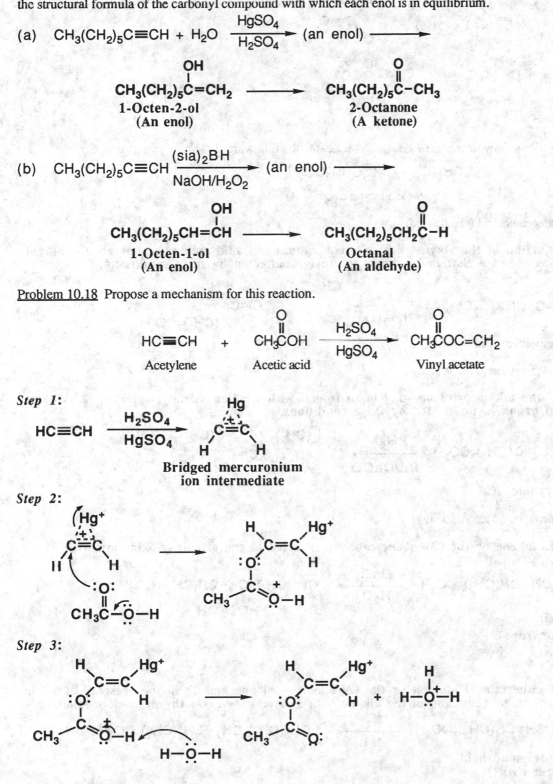

(a) $CH_3(CH_2)_5C{\equiv}CH$ + H_2O $\xrightarrow[H_2SO_4]{HgSO_4}$ (an enol) $\longrightarrow$

$$\underset{\substack{\textbf{1-Octen-2-ol}\\ \textbf{(An enol)}}}{CH_3(CH_2)_5\overset{OH}{\underset{|}{C}}{=}CH_2} \longrightarrow \underset{\substack{\textbf{2-Octanone}\\ \textbf{(A ketone)}}}{CH_3(CH_2)_5\overset{\overset{O}{\|}}{C}{-}CH_3}$$

(b) $CH_3(CH_2)_5C{\equiv}CH$ $\xrightarrow[NaOH/H_2O_2]{(sia)_2BH}$ (an enol) $\longrightarrow$

$$\underset{\substack{\textbf{1-Octen-1-ol}\\ \textbf{(An enol)}}}{CH_3(CH_2)_5\overset{OH}{\underset{|}{CH}}{=}CH} \longrightarrow \underset{\substack{\textbf{Octanal}\\ \textbf{(An aldehyde)}}}{CH_3(CH_2)_5CH_2\overset{\overset{O}{\|}}{C}{-}H}$$

Problem 10.18 Propose a mechanism for this reaction.

$$\underset{\textbf{Acetylene}}{HC{\equiv}CH} + \underset{\textbf{Acetic acid}}{CH_3\overset{\overset{O}{\|}}{C}OH} \xrightarrow[HgSO_4]{H_2SO_4} \underset{\textbf{Vinyl acetate}}{CH_3\overset{\overset{O}{\|}}{C}OC{=}CH_2}$$

Step 1:

Step 2:

Step 3:

Step 4: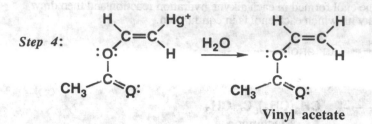

Vinyl acetate

Syntheses

Problem 10.19 Show how to convert 9-octadecynoic acid to the following:

$$CH_3(CH_2)_7C\equiv C(CH_2)_7CO_2H$$

9-Octadecynoic acid

(a) (E)-9-Octadecenoic acid (eliadic acid)

Chemical reduction of the alkyne with two mol sodium in liquid ammonia converts the alkyne to an (E)-alkene. Note that the carboxyl group is unaffected by these conditions.

$$CH_3(CH_2)_7C\equiv C(CH_2)_7CO_2H \xrightarrow[\text{NH}_3(\text{l})]{\text{2Na}}$$

9-Octadecynoic acid

$$\begin{array}{c} CH_3(CH_2)_7 \\ \diagdown \\ C=C \\ H \diagup \end{array} \begin{array}{c} H \\ \diagup \\ \diagdown \\ (CH_2)_7CO_2H \end{array}$$

(b) (Z)-9-Octadecenoic acid (oleic acid)

Reduction of the alkyne with one mol of hydrogen with Lindlar's catalyst gives a (Z)-alkene. The carboxyl group is unaffected by these conditions.

$$CH_3(CH_2)_7C\equiv C(CH_2)_7CO_2H \xrightarrow[\text{Pd/CaCO}_3]{\text{H}_2}$$

9-Octadecynoic acid

$$\begin{array}{c} CH_3(CH_2)_7 \\ \diagdown \\ C=C \\ H \diagup \end{array} \begin{array}{c} (CH_2)_7CO_2H \\ \diagup \\ \diagdown \\ H \end{array}$$

(c) 9,10-Dihydroxyoctadecanoic acid

Either the (E)-alkene or the (Z)-alkene can be converted to the glycol by oxidation with OsO$_4$/H$_2$O$_2$.

$$CH_3(CH_2)_7CH=CH(CH_2)_7CO_2H \xrightarrow[\text{H}_2\text{O}_2]{\text{OsO}_4} CH_3(CH_2)_7CH-CH(CH_2)_7CO_2H$$
$$\qquad\qquad\qquad\qquad\qquad\qquad\qquad\qquad\qquad\qquad\quad |\quad\ |$$
$$\qquad\qquad\qquad\qquad\qquad\qquad\qquad\qquad\qquad\qquad\ HO\ \ OH$$

9-Octadecenoic acid
(*cis* or *trans*)

(d) Octadecanoic acid

Reduction of either the (E)-alkene or the (Z)-alkene with one mol H$_2$ in the presence of a Ni, Pd, or Pt catalyst, or reduction of the alkyne with two mol H$_2$ gives the desired product.

$$CH_3(CH_2)_7CH=CH(CH_2)_7CO_2H \xrightarrow[\text{N i}]{\text{H}_2} CH_3(CH_2)_7CH_2-CH_2(CH_2)_7CO_2H$$

9-Octadecenoic acid
(*cis* or *trans*)

$$CH_3(CH_2)_7C{\equiv}C(CH_2)_7CO_2H \xrightarrow[\text{N i}]{2H_2} CH_3(CH_2)_7CH_2-CH_2(CH_2)_7CO_2H$$

9-Octadecynoic acid

Problem 10.20 For small-scale and consumer welding applications, many hardware stores sell cylinders of MAAP gas, which is a mixture of propyne (methylacetylene) and 1,2-propadiene (allene), with other hydrocarbons. How would you prepare the methylacetylene/allene mixture in the laboratory?

This gas mixture could be prepared from a double dehydrohalogenation of 1,2-dibromopropane. As described in section 10.6, the 1,2-propadiene (allene) is a side product of the reaction, being derived from β-elimination from the intermediate 2-bromopropene.

$$CH_3-CHBr-CH_2Br \xrightarrow{NaNH_2} CH_3-C{\equiv}CH + CH_2{=}C{=}CH_2$$

1,2-Dibromopropane   **Propyne**   **1,2-Propadiene**
                                                                       (Allene)

Problem 10.21 Show reagents and experimental conditions you might use to convert propyne into each product. Some of these syntheses can be done in one step. Others require two or more steps.

(a) $CH_3-C{\equiv}CH \longrightarrow$ CH$_3$-C-CH (with Br, Br on top and Br, Br on bottom)

Addition of two mol Br$_2$ to propyne gives 1,1,2,2-tetrabromopropane.

$$CH_3-C{\equiv}CH \xrightarrow{Br_2} CH_3C{=}CH \xrightarrow{Br_2} CH_3-C-CH$$

(with Br, Br above CH=CH and Br, Br on the final structure)

(b) $CH_3-C{\equiv}CH \longrightarrow$ CH$_3$-C-CH$_3$ (with Br above and Br below)

Addition of two mol HBr occurs by electrophilic addition gives first 2-bromopropene and then 2,2-dibromopropane.

$$CH_3-C{\equiv}CH \xrightarrow{HBr} CH_3C{=}CH_2 \xrightarrow{HBr} CH_3-C-CH_3$$

(with Br above CH$_3$C=CH$_2$ and Br above and below the final structure)

(c) $CH_3C{\equiv}CH \longrightarrow CH_3-\overset{O}{\overset{\|}{C}}-CH_3$

Acid-catalyzed hydration of the alkyne gives an enol which is in equilibrium, by keto-enol tautomerism, with the isomeric ketone, in this case propanone (acetone).

$$CH_3C{\equiv}CH + H_2O \xrightarrow[\text{HgSO}_4]{H_2SO_4} \left[CH_3\overset{OH}{\overset{|}{C}}{=}CH_3 \right] \longrightarrow CH_3\overset{O}{\overset{\|}{C}}-CH_3$$

(d) $CH_3C\equiv CH \longrightarrow CH_3CH_2-\overset{\overset{\displaystyle O}{\|}}{C}-H$

Hydroboration with $(sia)_2BH$ or other hindered derivative of borane followed by oxidation with alkaline hydrogen peroxide gives an enol which is in equilibrium, by keto-enol tautomerism, with the isomeric aldehyde, in this case propanal.

$$CH_3C\equiv CH \xrightarrow[\text{2) NaOH/H}_2\text{O}_2]{\text{1) (sia)}_2\text{B H}} \left[CH_3CH=\overset{\overset{\displaystyle OH}{|}}{CH} \right] \longrightarrow CH_3CH_2\overset{\overset{\displaystyle O}{\|}}{CH}$$

<u>Problem 10.22</u> Show reagents and experimental conditions you might use to convert each starting material into the desired product. Some of these syntheses can be done in one step. Others require two or more steps.

(a) $CH_3CH_2CH_2C\equiv CCH_3 \longrightarrow$

$$\underset{H}{\overset{CH_3CH_2CH_2}{>}}C=C\underset{CH_3}{\overset{H}{<}}$$

Chemical reduction of the alkyne with sodium in liquid ammonia gives (E)-2-hexene.

$$CH_3CH_2CH_2C\equiv CCH_3 \xrightarrow[\text{NH}_3\text{(l)}]{\text{2Na}} \underset{H}{\overset{CH_3CH_2CH_2}{>}}C=C\underset{CH_3}{\overset{H}{<}}$$

(b) $CH_3CH_2CH_2C\equiv CCH_3 \longrightarrow$

$$\underset{H}{\overset{CH_3CH_2CH_2}{>}}C=C\underset{H}{\overset{CH_3}{<}}$$

Hydroboration of the internal alkyne followed by reaction of the organoborane with acetic acid gives (Z)-2-hexene.

$$CH_3CH_2CH_2C\equiv CCH_3 \xrightarrow[\text{2) CH}_3\text{CO}_2\text{H}]{\text{1) BH}_3} \underset{H}{\overset{CH_3CH_2CH_2}{>}}C=C\underset{H}{\overset{CH_3}{<}}$$

Alternatively, catalytic reduction with hydrogen in the presence of Lindlar catalyst gives the (Z)-alkene.

$$CH_3CH_2CH_2C\equiv CCH_3 + H_2 \xrightarrow[\substack{\text{Lindlar}\\\text{catalyst}}]{\text{Pd/CaCO}_3} \underset{H}{\overset{CH_3CH_2CH_2}{>}}C=C\underset{H}{\overset{CH_3}{<}}$$

(c) $CH_3(CH_2)_4C\equiv CH \longrightarrow CH_3(CH_2)_4C\equiv C:^- \ Na^+$

The anion can be formed using sodium amide, $NaNH_2$, or sodium hydride, NaH.

$$CH_3(CH_2)_4C\equiv CH + NaNH_2 \longrightarrow CH_3(CH_2)_4C\equiv C:^- \ Na^+ + NH_3$$

(d)   CH₃CH₂C≡CH ⟶ CH₃CH₂C≡CD

Formation of the terminal acetylide anion followed by reaction with a deuterium donor such as D₂O gives 1-deutero-1-butyne.

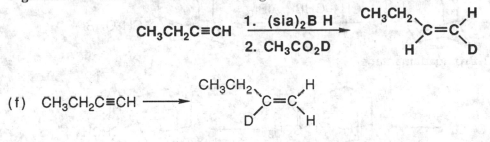

$$CH_3CH_2C{\equiv}CH \xrightarrow{\text{NaNH}_2} CH_3CH_2C{\equiv}C{:}^{-}\;Na^{+} \xrightarrow{\text{D}_2\text{O}} CH_3CH_2C{\equiv}CD$$

(e) 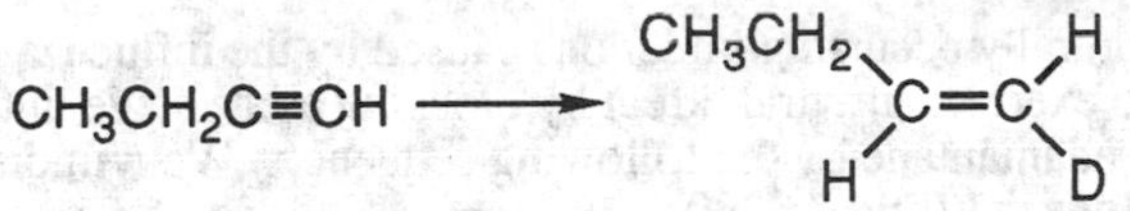   CH₃CH₂C≡CH ⟶

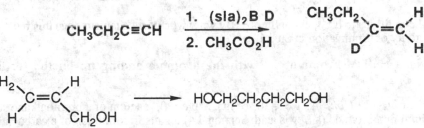

Hydroboration with this disubstituted derivative of diborane followed by reaction of the organoborane with deuteroacetic acid gives the desired 1-deutero-1-butene.

$$CH_3CH_2C{\equiv}CH \xrightarrow[\text{2. CH}_3\text{CO}_2\text{D}]{\text{1. (sia)}_2\text{B H}}$$

(f)   CH₃CH₂C≡CH ⟶

Hydroboration of the terminal alkyne with deuteroborane adds deuterium to the more substituted carbon of the alkyne. Reaction of the deuterated organoborane with acetic acid gives the 2-deutero-1-butene.

$$CH_3CH_2C{\equiv}CH \xrightarrow[\text{2. CH}_3\text{CO}_2\text{H}]{\text{1. (sia)}_2\text{B D}}$$

(g)

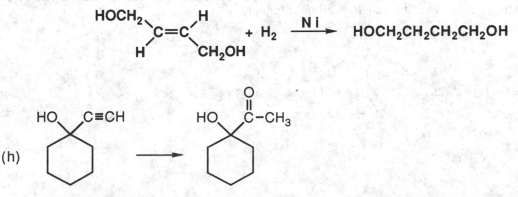

Catalytic reduction of the carbon-carbon double bond with one mol of hydrogen in the presence of a transition metal catalyst gives 1,4-butanediol.

(h)

Acid-catalyzed hydration of the carbon-carbon triple bond followed by keto-enol tautomerism of the resulting enol gives the desired ketone. Note that the tertiary alcohol in the starting material is unaffected by these conditions.

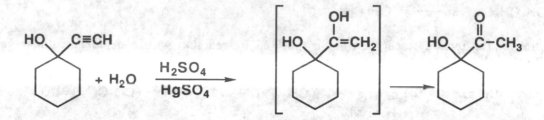

<u>Problem 10.23</u> Rimantadine is effective in preventing naturally occurring infections caused by the influenza A virus and in treating established illnesses. It is thought to exert its antiviral effect by blocking a late stage in the assembly of the virus. Rimantadine is synthesized from adamantane by the following sequence. We will discuss the chemistry of Step 5 in Chapter 15. See the Merck Index, 12Ed., # 8390.

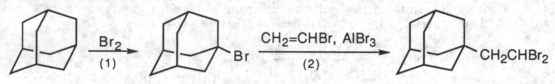

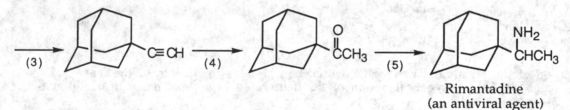

(a) Describe experimental conditions to bring about Step 1. By what type of mechanism does this reaction occur? Account for the regioselectivity of bromination in Step 1.

This reaction occurs via a radical bromination, with the bromine ending up on the tertiary carbon atom.

(b) Propose a mechanism for Step 2. *Hint:* As we shall see in Chapter 20, reaction of a bromoalkane such as 1-bromoadamantane with aluminum bromide (a Lewis acid, Section 3.4) results in formation of a carbocation and $AlBr_4^-$. Assume that adamantyl cation is formed in Step 2 and proceed from there to describe a mechanism.

The bromoethene pi electrons attack the admantyl cation, to create a new cation that captures a bromide from $AlBr_4^-$ to yield dibromoethyl-adamantane.

Step 1:

Step 2:

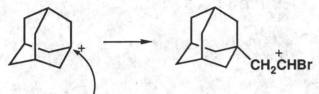

Step 3:

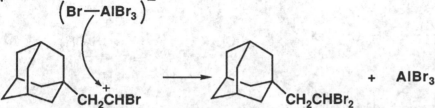

(c) Account for the regioselectivity of carbon-carbon bond formation in Step 2.

The carbon atom without the halogen ends up attached to the adamantane group, because this allows formation of the more stable intermediate with the cation adjacent to the halogen atom. The formation of this cation has a lower activation energy because of resonance stabilization of the cation provided by the halogen.

(d) Describe experimental conditions to bring about Step 3.

This reaction occurs via double dehydrohalogenation using a strong base such as NaNH$_2$.

(e) Describe experimental conditions to bring about Step 4.

This transformation occurs via an acid-catalyzed hydration of the alkyne using H$_2$O, H$_2$SO$_4$, and HgSO$_4$. The initially formed enol equilibrates to the more stable keto form.

<u>Problem 10.24</u> Show reagents and experimental conditions required to bring about the following transformations.

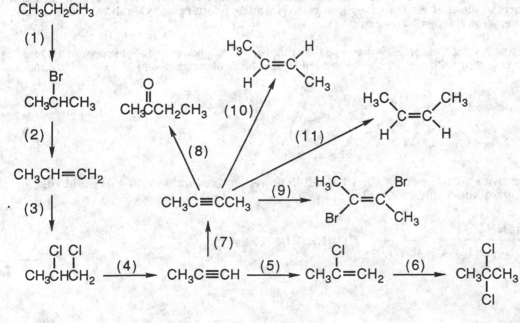

(1) **Br$_2$ with hν or heat**   (2) **(CH$_3$)$_3$CO$^-$K$^+$ (E2)**
(3) **Cl$_2$**   (4) **2 NaNH$_2$**
(5) **HCl**   (6) **HCl**
(7) **1. NaNH$_2$; 2. CH$_3$I**   (8) **HgSO$_4$, H$_2$SO$_4$, H$_2$O**
(9) **Br$_2$**   (10) **Li or Na in liquid NH$_3$**
(11) **H$_2$, Pd, CaCO$_3$ (Lindlar catalyst)**

<u>Problem 10.25</u> Show reagents to bring about each conversion.

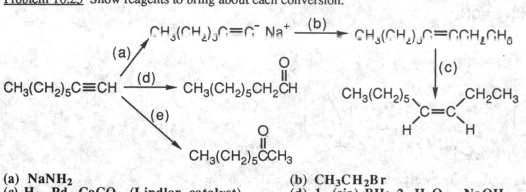

(a) **NaNH$_2$**   (b) **CH$_3$CH$_2$Br**
(c) **H$_2$, Pd, CaCO$_3$ (Lindlar catalyst)** (d) **1. (sia)$_2$BH; 2. H$_2$O$_2$, NaOH**
(e) **HgSO$_4$, H$_2$SO$_4$, H$_2$O**

<u>Problem 10.26</u> Which of these alkynes can be prepared in good yield by monoalkylation or dialkylation of acetylene? For each that cannot, explain why not.
(a) 3-Methyl-1-butyne   (b) 4,4-Dimethyl-1-pentyne   (c) 2-Octyne

The acetylide anion is a strong base as well as a good nucleophile, so good yields can only be obtained when acetylide anions react with methyl or primary alkyl halides. Secondary or tertiary alkyl halides give large amounts of elimination via the E2 mechanism. The preparation of (a) could therefore not be carried out in high yield, because preparation of 3-methyl-1-butyne would require reaction of an acetylide anion with a 2-halopropane, a secondary alkyl halide. On the other hand, (b) could be made in high yield from the acetylide anion reacting with 2,2-dimethyl-1-halopropane, a primary alkyl halide that cannot undergo E2 elimination, since there are no hydrogens beta to the halogen atom. The molecule in (c) could also be made in high yield, since sequential reactions with a methyl and a primary alkyl halide, halomethane and 1-halopentane, respectively, are required.

<u>Problem 10.27</u> Propose a synthesis for (Z)-9-tricosene (muscalure), the sex pheromone for the common house fly (*Musca domestica*) starting with acetylene and alkyl halides as sources of carbon atoms. See the Merck Index, 12th Ed., # 6388.

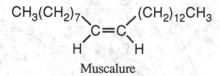

Muscalure

Alkylation of acetylene with the two alkyl halides followed by reduction using H$_2$ and the Lindlar catalyst gives muscalure.

$$HC \equiv CH \xrightarrow[\text{2. } CH_3(CH_2)_6CH_2Br]{\text{1. } NaNH_2} CH_3(CH_2)_7C \equiv CH$$

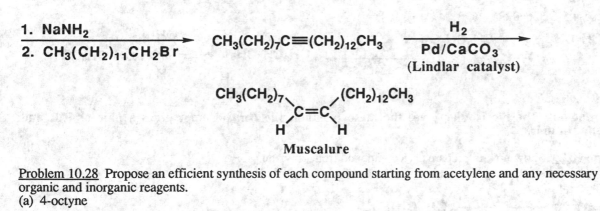

$$\xrightarrow[\text{2. } CH_3(CH_2)_{11}CH_2Br]{\text{1. } NaNH_2} \quad CH_3(CH_2)_7C\equiv(CH_2)_{12}CH_3 \quad \xrightarrow[\substack{Pd/CaCO_3 \\ \text{(Lindlar catalyst)}}]{H_2}$$

$$\underset{\substack{| \\ H}}{CH_3(CH_2)_7}C=C\underset{\substack{| \\ H}}{(CH_2)_{12}CH_3}$$

Muscalure

Problem 10.28 Propose an efficient synthesis of each compound starting from acetylene and any necessary organic and inorganic reagents.
(a) 4-octyne

$$HC\equiv CH \xrightarrow[\text{2. } CH_3CH_2CH_2Br]{\text{1. } NaNH_2} HC\equiv CCH_2CH_2CH_3 \xrightarrow[\text{2. } CH_3CH_2CH_2Br]{\text{1. } NaNH_2}$$

$$CH_3CH_2CH_2C\equiv CCH_2CH_2CH_3$$

(b) 4-Octanone

$$CH_3CH_2CH_2C\equiv CCH_2CH_2CH_3 \xrightarrow[HgSO_4]{H_2O, H_2SO_4} CH_3CH_2CH_2CH_2\overset{\overset{\displaystyle O}{\|}}{C}CH_2CH_2CH_3$$
(From (a))

(c) cis-4-Octene

$$CH_3CH_2CH_2C\equiv CCH_2CH_2CH_3 \xrightarrow[\substack{Pd/CaCO_3 \\ \text{(Lindlar catalyst)}}]{H_2} \underset{\substack{| \\ H}}{CH_3CH_2CH_2}C=C\underset{\substack{| \\ H}}{CH_2CH_2CH_3}$$
(From (a))

(d) trans-4-Octene

$$CH_3CH_2CH_2C\equiv CCH_2CH_2CH_3 \xrightarrow[NH_3 \ (l)]{\text{Li or Na}} \underset{\substack{| \\ H}}{CH_3CH_2CH_2}C=C\underset{\substack{| \\ CH_2CH_2CH_3}}{H}$$
(From (a))

(e) 4-Octanol

$$\underset{\substack{| \\ H}}{CH_3CH_2CH_2}C=C\underset{\substack{| \\ H}}{CH_2CH_2CH_3}$$
(From (c))

or

$$\xrightarrow[\text{2. } NaBH_4]{\text{1. } Hg(OAc)_2, H_2O} CH_3CH_2CH_2CH_2\overset{\overset{\displaystyle OH}{|}}{C}HCH_2CH_2CH_3$$

$$\underset{\substack{| \\ H}}{CH_3CH_2CH_2}C=C\underset{\substack{| \\ CH_2CH_2CH_3}}{H}$$
(From (d))

(f) meso-4,5-Octanediol

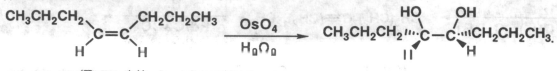

(From (c))

Note that only the *cis* isomer gives the meso product, the *trans* isomer gives a pair of R,R and S,S enantiomers.

Problem 10.29 Show how to prepare each compound from ethylene
(a) 1,2-Dichloroethane

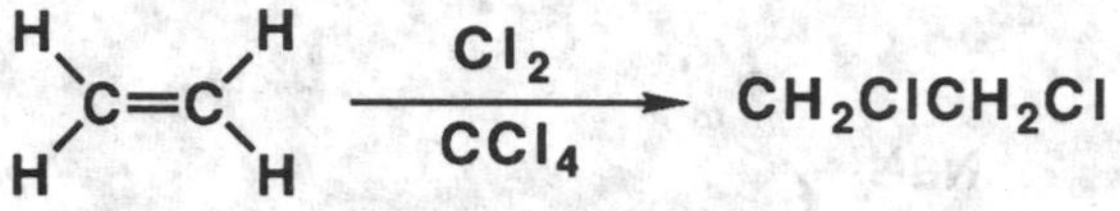

$$\underset{H}{\overset{H}{>}}C=C\underset{H}{\overset{H}{<}} \quad \xrightarrow[CCl_4]{Cl_2} \quad CH_2ClCH_2Cl$$

(b) Chloroethylene (vinyl chloride)

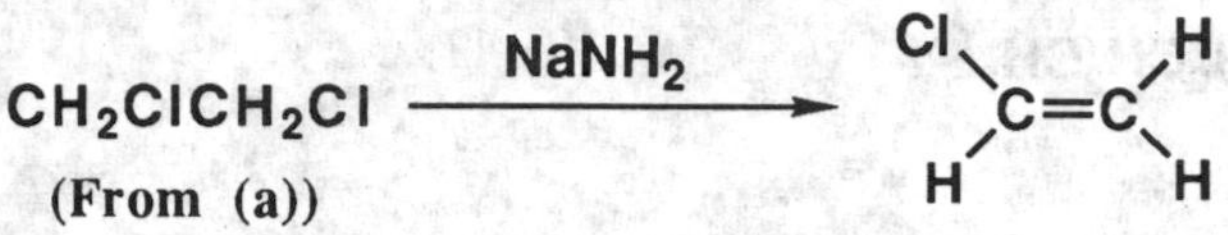

$$CH_2ClCH_2Cl \xrightarrow{NaNH_2} \underset{H}{\overset{Cl}{>}}C=C\underset{H}{\overset{H}{<}}$$

(From (a))

(c) 1,1-Dichloroethane

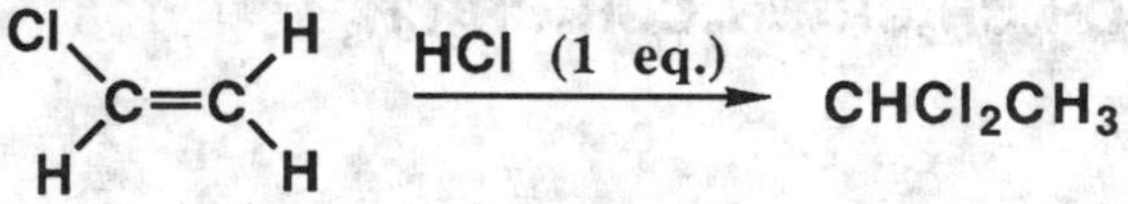

$$\underset{H}{\overset{Cl}{>}}C=C\underset{H}{\overset{H}{<}} \quad \xrightarrow{HCl\ (1\ eq.)} \quad CHCl_2CH_3$$

(From (b))

Problem 10.30 Show how to bring about this conversion.

The alkene is first converted to the dibromide, which is converted to the alkyne. The alkyne is reacted with base and the resulting acetylide anion treated with methyl iodide to give the alkyne with the proper number of carbon atoms. Metal reduction is used to give the desired *trans* final product.

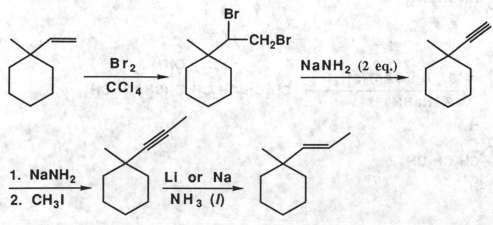

CHAPTER 11: ETHERS AND EPOXIDES

SUMMARY OF REACTIONS

Starting Material ↓ / Product →	Alcohols	Alkyl Halides	β–Alkynyl Alcohols	β–Amino Alcohols	β–Cyano Alcohols	Epoxides	Ethers	Glycols	Hydroperoxides	β–Hydroxy Ethers	β–Mercapto Alcohols	Silyl Ethers	Sulfones	Sulfoxides	Tetrahydropyranyl Ethers	Thioethers	Thiols
Alcohols							11A 11.4D*								11B 11.7		
Alcohols Alkenes							11C 11.4C										
Alcohols Chlorosilanes												11D 11.7					
Alkenes						11E 11.9B											
Alkyl Halides																11F 11.5	11G 11.5
Alkyl Halides Alkoxides							11H 11.4A										
Alkoxides Epoxides										11I 11.10B							
Alkyne Anions Epoxides			11J 11.10B														
Amines Epoxides				11K 11.10B													
Epoxides	11L 11.10B			11M 11.10B	11N 11.10A						11O 11.10B						
Epoxides Gilman Reag.	11P 11.10B																
Ethers		11Q 11.6A							11R 11.6B								
Halohydrins						11S 11.9B											
Sulfoxides													11T 11.6C				
Thioethers													11U 11.6C	11V 11.6C			

*Section in book that describes reaction.

REACTION 11A: ACID-CATALYZED DEHYDRATION OF ALCOHOLS (Section 11.4B)

$$2 \quad -\overset{|}{\underset{|}{C}}-OH \quad \xrightarrow{H_2SO_4} \quad -\overset{|}{\underset{|}{C}}-O-\overset{|}{\underset{|}{C}}- \quad + \quad H_2O$$

- Ethers can be produced via acid-catalyzed dehydration of primary alcohols. In this reaction, the -OH group is protonated to give a good leaving group (H_2O) that is displaced in an S_N2 process by another alcohol molecule. ∗

REACTION 11B: TETRAHYDROPYRANYL ETHER FORMATION AND CLEAVAGE
(Section 11.7)

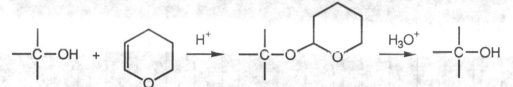

- **Dihydropyran reacts with primary alcohols** in acid to give tetrahydropyranyl ethers. These ethers are stable to neutral and basic solutions, but the original alcohol is regenerated by treatment with aqueous acid. Thus, tetrahydropyranyl ethers are useful protecting groups for hydroxyl groups. *

REACTION 11C: ACID-CATALYZED ADDITION OF ALCOHOLS TO ALKENES (Section 11.4C)

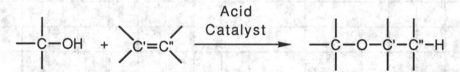

- Ethers can also be prepared through an **acid-catalyzed addition of alcohols to alkenes.** In these reactions, the acid reacts with the alkene to create a carbocation that is attacked by the nucleophilic oxygen atom of the alcohol. *
- Since a carbocation is involved, the reaction gives good yields only with alkenes such as isobutylene in which a tertiary carbocation is produced upon protonation, and when the alcohol is primary and thus resistant to dehydration.

REACTION 11D: PREPARATION AND HYDROLYSIS OF SILYL ETHER PROTECTING
GROUPS (Section 11.7)

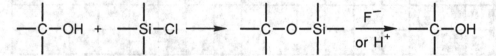

- **Alcohols** react with **chlorotrialkylsilanes** to give **silyl ethers. Silyl ethers react with F⁻ or aqueous acid** to give back the original **alcohol.** Thus, this system can be used as a **protecting group** for hydroxyl groups. *

REACTION 11E: OXIDATION OF ALKENES WITH PEROXYCARBOXYLIC ACIDS
(PERACIDS) (Section 11.9B)

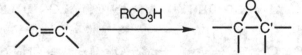

- The most common laboratory **synthesis of epoxides** is **from alkenes, using peroxyacids** as oxidizing agents. Commonly used reagents include 3-chloroperoxybenzoic acid and the magnesium salt of monoperoxyphthalic acid. The mechanism of this reaction is thought to be concerted. * *[This is a very interesting mechanism that is worth a close look because it can be trickier than it might first appear.]*

REACTION 11F: PREPARATION OF SULFIDES FROM ALKYL HALIDES (Section 11.5)

- Thioethers can be prepared from an S_N2 reaction between thiolate anions and alkyl halides. Symmetrical thioethers can be prepared from 2 equivalents of primary alkyl halides and one equivalent of sulfide ion (Na_2S).
- The same reaction can be used to prepare cyclic thioethers when an alkyl dihalide such as 1,5-dichloropentane or 1,4-dichlorobutane is used.

REACTION 11G: PREPARATION OF THIOLS FROM ALKYL HALIDES (Section 11.5)

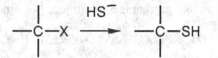

- Certain **thiols can be prepared** using an S_N2 reaction between an **alkyl halide and the hydrosulfide ion**. Elimination is a concern here since the hydrosulfide ion is basic, so primary alkyl halides give the highest yields. Yields are lower with secondary halides, and elimination predominates with tertiary alkyl halides. ✳

REACTION 11H: THE WILLIAMSON ETHER SYNTHESIS (Section 11.4A)

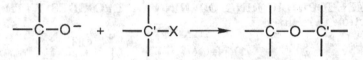

- The most common method used to synthesize ethers is the **Williamson ether synthesis**, which is nothing more than an S_N2 reaction between a metal alkoxide and alkyl halide. ✳
- Because the metal alkoxide is basic, an E2 reaction is always a concern with a Williamson ether synthesis. As a result, it is important to choose carefully which piece will be the alkyl halide. In other words, a primary alkyl halide is used if possible. Secondary alkyl halides give lower yields of ethers and tertiary alkyl halides give almost exclusively the elimination product.
- Alkyl tosylates may be used in place of alkyl halides.

REACTION 11I: NUCLEOPHILIC RING OPENING OF EPOXIDES WITH ALKOXIDES (Section 11.10B)

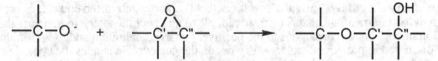

- **Strong nucleophiles** such as alkoxides can add directly to epoxides via an S_N2 **mechanism**. ✳
- Epoxide reactions are versatile because of the large number of different nucleophiles that can be used including alkyne anions, amines, hydride reagents, cyanide anion, and thiolates. These reactions are described as reactions 11I, 11J, 11K, 11M, and 11O, respectively. ✳
- As expected for an S_N2 reaction, in each case, the **nucleophile adds preferentially** to the **less hindered** epoxide **carbon atom**. ✳

REACTION 11J: NUCLEOPHILIC RING OPENING OF EPOXIDES WITH ALKYNE ANIONS (Section 11.10B)

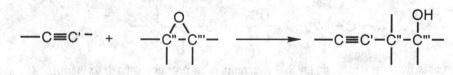

- **Alkyne anions** react with epoxides via an S_N2 mechanism to produce hydroxy alkynes. ✳

REACTION 11K: NUCLEOPHILIC RING OPENING OF EPOXIDES WITH AMINES (Section 11.10B)

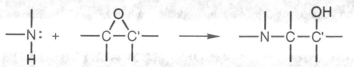

- **Amines** react with epoxides via an S_N2 mechanism to produce β-**aminoalcohols**. ✱

REACTION 11L: REDUCTION OF EPOXIDES TO CREATE ALCOHOLS (Section 11.10B)

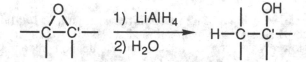

- **Hydride reagents** such as $LiAlH_4$ react with epoxides to produce **alcohols**. ✱
- The hydrogen ends up on the less hindered carbon atom, consistent with an S_N2 mechanism.

REACTION 11M: NUCLEOPHILIC RING OPENING OF EPOXIDES WITH CYANIDE (Section 11.10B)

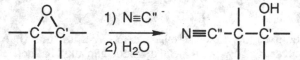

- The **cyanide anion** reacts with epoxides via an S_N2 mechanism to produce β-**hydroxy-nitriles**. ✱

REACTION 11N: ACID-CATALYZED RING OPENING OF EPOXIDES (Section 11.10A)

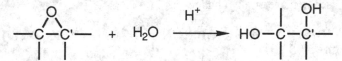

- In the presence of an **acid catalyst** (usually perchloric acid), the oxygen of the epoxide is protonated to form an oxonium ion intermediate, that is then susceptible to nucleophilic attack to generate the product. ✱
- The oxonium ion intermediate is analogous to the positively-charged halonium and mercurinium ion intermediates discussed in previous chapters. As a result, the nucleophile attacks the epoxide carbon atom that corresponds to the more stable carbocation, usually the more substituted epoxide carbon atom. Note that this is in distinct contrast to simple nucleophilic attack of epoxides by an S_N2 mechanism (reactions 11H, 11I, 11J, 11K, 11M, and 11O) in which the nucleophile preferentially attacks the less-substituted epoxide carbon atom.
- The nucleophile attacks anti and coplanar to the oxygen atom of the oxonium atom. For epoxides derived from cycloalkenes, this means the addition results in formation of a *trans* diol product.

REACTION 11O: NUCLEOPHILIC RING OPENING OF EPOXIDES WITH HYDROSULFIDE ION (Section 11.10B)

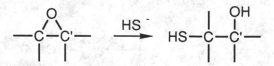

- The **HS⁻ ion** reacts with epoxides via an S_N2 mechanism to produce β-**hydroxythiols**. ✱

REACTION 11P: NUCLEOPHILIC RING OPENING OF EPOXIDES WITH GILMAN REAGENTS (Section 11.10B)

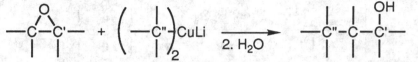

- **Gilman reagents, (R)$_2$CuLi,** react with epoxides to give **alcohols.** The reaction is important because a carbon-carbon bond is formed. ✳

REACTION 11Q: ACID-CATALYZED CLEAVAGE OF ETHERS WITH H-X (Section 11.6A)

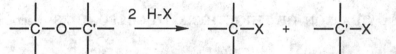

- **Ethers are very robust functional groups,** failing to react with reagents such as oxidizing agents or strong bases. On the other hand, ethers can be cleaved by concentrated aqueous HI or HBr. The mechanism of this reaction involves protonation of the ether oxygen atom to create a positively-charged oxonium ion intermediate. If both of the attached alkyl groups are primary, then cleavage occurs by an S$_N$2 mechanism with the halide ion acting as the nucleophile. If one of the alkyl groups is tertiary, then cleavage is by an S$_N$1 mechanism to create a carbocation that can react with halide, rearrange, or undergo elimination depending on the details of the reaction. ✳

REACTION 11R: FORMATION OF HYDROPEROXIDES FROM ETHERS (Section 11.6B)

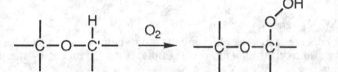

- Two **hazards** exist when working with low-molecular-weight ethers. First, they are extremely **flammable** and care should be taken to avoid flames and sparks when working with them. Second, ethers **react slowly** with molecular **oxygen to form explosive hydroperoxides** via a radical process, so they must be stored carefully and disposed of before a problem arises. ✳

REACTION 11S: INTERNAL NUCLEOPHILIC SUBSTITUTION IN HALOHYDRINS TO CREATE EPOXIDES (Section 11.9C)

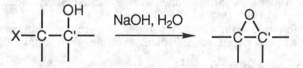

- **Epoxides** can also be **produced** by treatment of a **halohydrin with base.** The mechanism of the reaction involves an internal nucleophilic attack and loss of the halogen leaving group. This is a useful reaction because halohydrins can be produced from reaction of an alkene with aqueous Cl$_2$ or Br$_2$ (Reaction 5K, described in Section 5.3F). ✳

REACTION 11T: OXIDATION OF SULFOXIDES TO SULFONES (Section 11.6C)

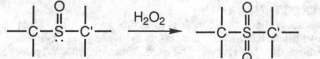

- **Sulfoxides** can be **easily oxidized to sulfones.** ✳

REACTION 11U: OXIDATION OF THIOETHERS TO SULFONES (Section 11.6C)

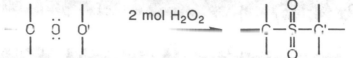

- **Thioethers** can be **easily oxidized** to two higher oxidation states, sulfoxides, and sulfones. Several oxidizing agents can be used to carry out these transformations including aqueous hydrogen peroxide (H_2O_2), sodium metaperiodate ($NaIO_4$), and air oxidation in the presence of oxides of nitrogen. ✳
- The extent of oxidation depends on the amount of oxidizing agent added. Two mol of an oxidizing agent such as H_2O_2 results in formation of a sulfone.

REACTION 11V: OXIDATION OF THIOETHERS TO SULFOXIDES (Section 11.6C)

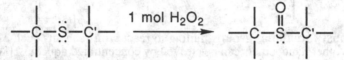

- Reaction of thioethers with one mol H_2O_2 results in formation of a sulfoxide. ✳

SUMMARY OF IMPORTANT CONCEPTS

11.1 STRUCTURE OF ETHERS
- The **ether functional group** is composed of an **sp³ hybridized oxygen atom** bonded to **two carbon atoms**. There are two lone pairs of electrons on the oxygen atom, and these provide many of the reactivity characteristics of ethers. ✳

11.2 NOMENCLATURE OF ETHERS
- In the IUPAC system, ethers are named by choosing the longest alkyl chain as the parent chain. The remaining -OR group is named as an alkoxy substituent. For example, ethoxyethane and 3-methoxyheptane are acceptable IUPAC names.

Ethoxyethane 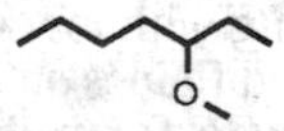   3-Methoxyheptane

- Low-molecular-weight ethers have common names that are often used. The common names are constructed by listing the two groups attached to oxygen in alphabetical order followed by the word ether. For example, diethyl ether or methyl propyl ether are acceptable common names.
- Some important ethers are cyclic, and they are given special names such as tetrahydrofuran or 1,4-dioxane.
- Several low molecular weight ethers are useful solvents, such as diethyl ether, 2-methoxyethanol, 2-ethoxyethanol, and diethylene glycol dimethyl ether
- Sulfur analogs of ethers are referred to as thioethers or sulfides, and are named in common nomenclature by using the word sulfide in place of ether. For example diethyl sulfide is the sulfur analog of diethyl ether.

11.3 PHYSICAL PROPERTIES OF ETHERS
- Ethers are polar molecules since the oxygen atom possesses a partial negative charge and each attached carbon atom possesses a partial positive charge. However, there is only limited dipole-dipole interaction between molecules because there is too much steric hindrance around the carbon atoms with the partial positive charges. As a result, the boiling points of ethers are not that much higher than those of similar hydrocarbons. On the other hand, the oxygen atom of ethers can accept hydrogen bonds, so ethers have higher solubilities in water than the corresponding hydrocarbons. ✳

11.7 ETHERS AS PROTECTING GROUPS
- Sometimes the acidity or nucleophilicity of alcohols is undesirable in a given reaction. In this case a so-called **protecting group** is used to block temporarily the reactivity of the hydroxyl group. The protecting group must be stable to the conditions of the reactions to be used, but easily cleaved after the reaction. Useful ether protecting groups for alcohols include the *tert*-butyl ether, silyl ethers, and tetrahydropyranyl ethers. ✳

11.8 EPOXIDES
• An epoxide is a cyclic ether in which oxygen is one atom in a three-membered ring. Epoxides can be named in a number of ways.
  - In the IUPAC system, the epoxide is named by listing the substituents on the ring as prefixes to the parent name **oxirane**. For example, ethyloxirane is an acceptable IUPAC name.
  - Two different systems are used to assign common names. In the first, the two atoms of the parent chain attached to the oxygen atoms of the epoxide are listed along with the prefix **epoxy**. For example, 1,2-epoxybutane is an acceptable common name. In another system, the alkene from which the epoxide could have been derived is named followed by the word **oxide**. For example, 1-butene oxide is an acceptable common name.

11.10 REACTIONS OF EPOXIDES
• **Epoxides** are **more reactive than normal ethers** because of the strain present in the three-membered ring. In particular, an epoxide is susceptible to nucleophilic attack with the oxygen atom acting as a leaving group (S_N2 mechanism; reactions 11H, 11I, 11J, 11K, 11M, and 11O), or via an oxonium intermediate with acid catalysis (reaction 11N). ✳

11.11 CROWN ETHERS
• **Crown ethers** are cyclic polyethers derived from ethylene glycol, or a related glycol, that have four or more ether linkages and twelve or more total atoms in the ring. ✳
  - Crown ethers are usually named by a short-hand nomenclature utilizing **crown** as the parent name. This is preceded by the total number of atoms in the ring and followed by the number of oxygen atoms in the ring. For example, 18-crown-6 is a common crown ether.
  - The cavity of certain crown ethers is the correct size to place an alkali metal inside. Since there are lone pairs of electrons on the oxygen atoms, there is a strong electrostatic attraction between the positively-charged ion and the crown ether. In general, the better the fit between crown ether interior size and ionic radius of the ion, the tighter the binding. For example, 18-crown-6 binds K^+ very tightly, but larger or smaller cations bind less tightly.

  Because the exteriors of crown ethers, such as 18-crown-6, are relatively hydrophobic, they can solubilize ions in organic solvents. The bound ion remains in the interior cavity of the crown ether molecule, away from the hydrophobic solvent. This is useful because crown ethers can be used to increase dramatically the solubility of ionic compounds in organic solvents.

CHAPTER 11
Solutions to the Problems

Problem 11.1 Write IUPAC and common names for these ethers.

(a)
$$\begin{array}{c} CH_3 \\ | \\ CH_3CHCH_2OCH_2CH_3 \end{array}$$

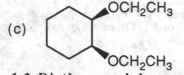

1-Ethoxy-2-methylpropane
(Ethyl isobutyl ether)

(b) $CH_3OCH_2CH_2OCH_3$

1,2-Dimethoxyethane
(Ethylene glycol dimethyl ether)

(c)

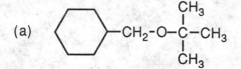

***cis*-1,2-Diethoxycyclohexane**

Problem 11.2 Arrange these compounds in order of increasing boiling point.

$CH_3OCH_2CH_2OCH_3$ $HOCH_2CH_2OH$ $CH_3OCH_2CH_2OH$

Boiling point increases with an increasing number of hydrogen bonding groups. In order of increasing boiling point, they are:

$CH_3OCH_2CH_2OCH_3$	$CH_3OCH_2CH_2OH$	$HOCH_2CH_2OH$
84°C	**125°C**	**198°C**

Problem 11.3 Show how you might use the Williamson ether synthesis to prepare these ethers:

(a)

There is only one combination of alkyl halide and metal alkoxide that gives the desired ether in good yield.

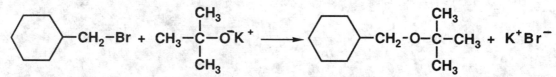

If this synthesis were attempted with the alkoxide derived from methylcyclohexyl alcohol and a *tert*-butyl halide, the reaction would produce an alkene by an E2 pathway.

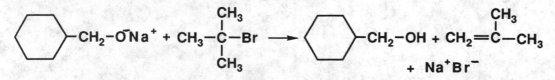

(b) $(CH_3CH_2CH_2CH_2)_2O$

Treatment of 1-bromobutane with sodium butoxide gives dibutyl ether.

$$CH_3CH_2CH_2CH_2O^-Na^+ \; + \; CH_3CH_2CH_2CH_2Br \longrightarrow \begin{array}{c} (CH_3CH_2CH_2CH_2)_2O \\ + \; Na^+Br^- \end{array}$$

<u>Problem 11.4</u> Show how ethyl hexyl ether might be prepared by a Williamson ether synthesis.

Using the Williamson ether synthesis, ethyl hexyl ether could be synthesized by either of the two following routes:

$$CH_3CH_2Br \; + \; CH_3(CH_2)_4CH_2O^-Na^+$$

or $\longrightarrow CH_3CH_2OCH_2(CH_2)_4CH_3$

$$CH_3(CH_2)_4CH_2Br \; + \; CH_3CH_2O^-Na^+$$

<u>Problem 11.5</u> Account for the fact that treatment of *tert*-butyl methyl ether with a limited amount of concentrated HI gives methanol and *tert*-butyl iodide rather than methyl iodide and *tert*-butyl alcohol.

The first step in the reaction involves protonation of the ether oxygen to give an oxonium ion intermediate. Cleavage of the oxonium ion intermediate on one side gives methanol and the *tert*-butyl cation. Cleavage on the other side gives a methyl cation and *tert*-butyl alcohol. Because of the greater stability of the tertiary carbocation, cleavage to give methanol and *tert*-butyl cation is favored. Reaction of the tertiary cation with the iodide completes the reaction.

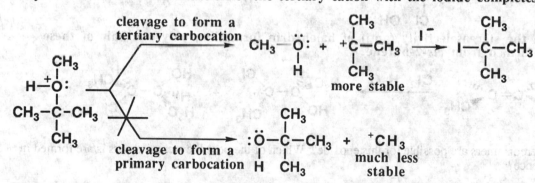

There is an alternative pathway for formation of product, namely reaction of the oxonium ion with iodide ion by an S_N2 pathway on the less-hindered methyl carbon to give iodomethane and the 2-methyl-2-propanol (*tert*-butyl alcohol). The fact is that the S_N1 pathway by way of a tertiary carbocation has a lower activation energy (a faster rate) than reaction of the oxonium ion with iodide ion by an S_N2 pathway so the S_N1 reaction predominates.

<u>Problem 11.6</u> Draw structural formulas for the major products of these reactions:

(a) $H_3C-\overset{\overset{\displaystyle CH_3}{|}}{\underset{\underset{\displaystyle CH_3}{|}}{C}}-O-CH_3 \; + \; \underset{\text{(excess)}}{HBr} \longrightarrow H_3C-\overset{\overset{\displaystyle CH_3}{|}}{\underset{\underset{\displaystyle CH_3}{|}}{C}}-Br \; + \; CH_3Br \; + \; H_2O$

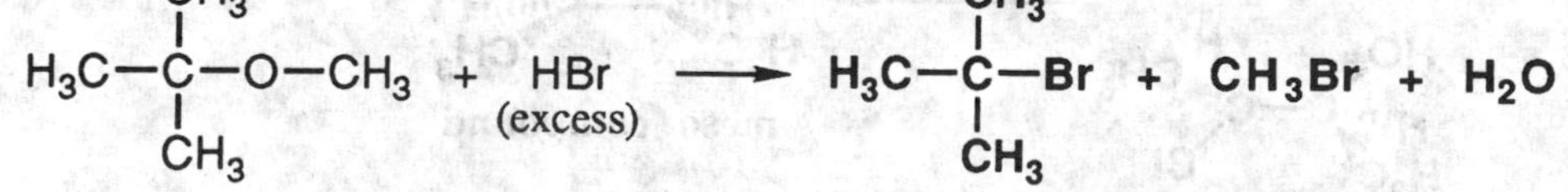

(b) 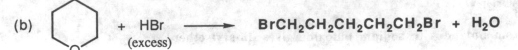 + HBr → $BrCH_2CH_2CH_2CH_2CH_2Br$ + H_2O
(excess)

Problem 11.7 Why is the use of the *tert*-butyl protecting group limited to protection of primary alcohols?

The *tert*-butyl ether is prepared from the primary alcohol and isobutylene in the presence of acid catalysis. If a secondary or tertiary alcohol were used in this reaction, dehydration of the alcohol to give an alkene would be a major side reaction, preventing a high yield.

Problem 11.8 Consider the possibilities for stereoisomerism in the halohydrin and epoxide formed in Example 11.8.

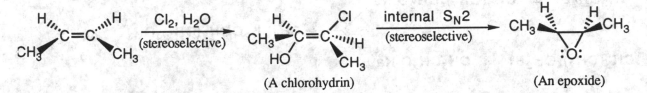

(A chlorohydrin)   (An epoxide)

(a) How many stereoisomers are possible for the chlorohydrin? Which of the possible chlorohydrins are formed by reaction of *cis*-2-butene with HOCl?

This chlorohydrin has two stereocenters and, by the 2^n rule, there are four possible stereoisomers - two pairs of enantiomers.

$$CH_3-\overset{*}{C}H-\overset{*}{C}H-CH_3$$
$$\overset{|}{Cl}\quad\overset{|}{OH}$$

However, given the stereoselectivity (anti) of halohydrin formation, only one pair of these enantiomers is formed from *cis*-2-butene.

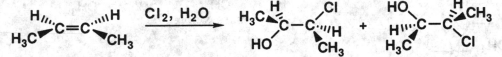

(b) How many stereoisomers are possible for this epoxide? Which of the possible stereoisomers is/are formed in this reaction sequence?

There are three stereoisomers possible for this epoxide; a pair of enantiomers and a meso compound. However, because of the stereoselectivity (requirement for anti, planar geometry) of the β-elimination reaction, there is only one stereoisomer produced; the meso compound.

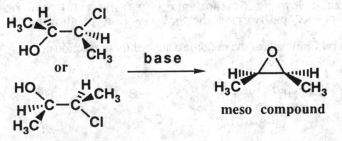

meso compound

<u>Problem 11.9</u> Show how to prepare each Gilman reagent in Example 11.7 from an appropriate alkyl halide.

$$CH_3CH_2Br \xrightarrow{\text{2 Li}} CH_3CH_2Li + LiBr$$

$$2\ CH_3CH_2Li \xrightarrow{\text{CuI}} [(CH_3CH_2)_2Cu]Li + LiI$$

or alternatively,

$$C_6H_5CH_2Br \xrightarrow{\text{2 Li}} C_6H_5CH_2Li + LiBr$$

$$2\ C_6H_5CH_2Li \xrightarrow{\text{CuI}} [(C_6H_5CH_2)_2Cu]Li + LiI$$

Structure and Nomenclature
<u>Problem 11.10</u> Write names for these compounds. Where possible, write both IUPAC and common names.

(a)

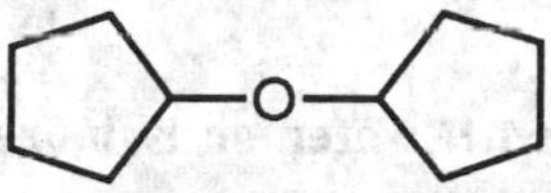

Cyclopentoxycyclopentane
Dicyclopentyl ether

(b)

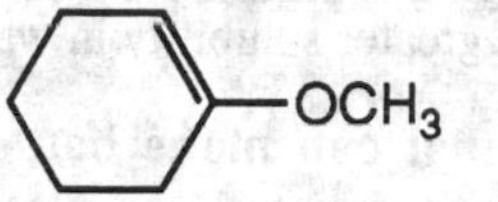

1-Methoxycyclohexene
1-Cyclohexenyl methyl ether

(c) $CH_3CH_2OCH_2CH_2OH$

2-Ethoxyethanol

(d) $CH_3CH_2OCH_2CH_2OCH_2CH_3$

1,2-Dithoxyethoxycthane
Ethylene glycol diethyl ether

(e)

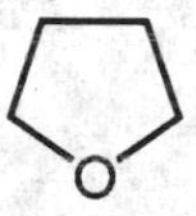

Tetrahydrofuran

(f)

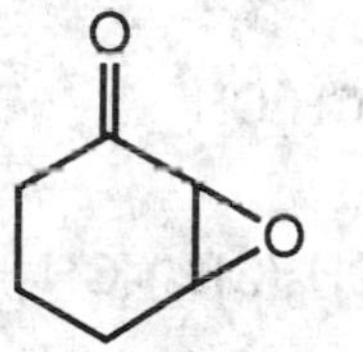

2,3-Epoxycyclohexanone

(g) 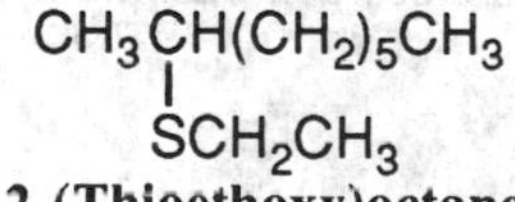 $CH_3CH(CH_2)_5CH_3$
            $|$
         SCH_2CH_3

2-(Thioethoxy)octane

(h) $[CH_3(CH_2)_4]_2O$

Pentoxypentane
Dipentyl ether

<u>Problem 11.11</u> Draw structural formulas for these compounds.
(a) Diisopropyl ether

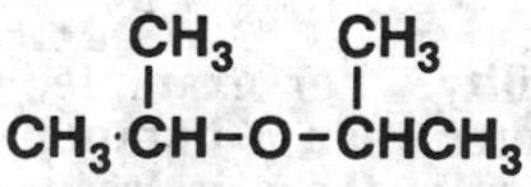

$$CH_3 \quad CH_3$$
$$|\qquad\quad |$$
$$CH_3\text{-}CH\text{-}O\text{-}CHCH_3$$

(b) *trans*-2,3-Diethyloxirane

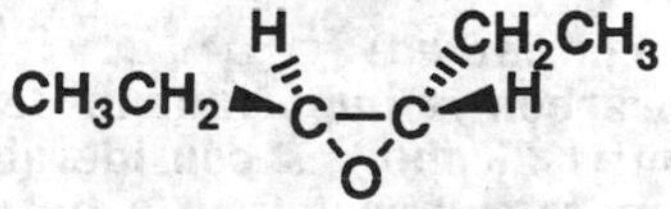

$$CH_3CH_2\text{---}C\text{---}C\text{---}H$$

(c) *trans*-2-Ethoxycyclopentanol

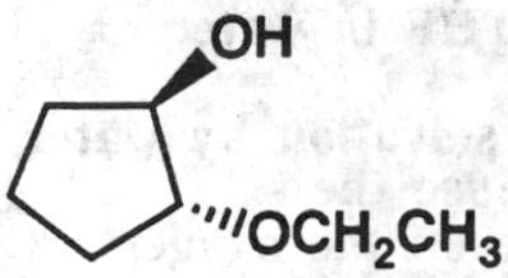

OH

''''OCH_2CH_3

(d) Divinyl ether

$$CH_2=CH\text{-}O\text{-}CH=CH_2$$

(e) Cyclohexene oxide

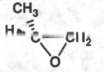

(f) Allyl cyclopropyl ether

$$H_2C \underset{H_2C}{\overset{}{\diagdown}} CH-O-CH_2-CH=CH_2$$

(g) (R)-2-Methyloxirane

$$\underset{\quad H}{\overset{CH_3}{\diagup}} \quad C - CH_2$$

(h) 1,1-Dimethoxycyclohexane

Physical Properties

Problem 11.12 Each compound given in this problem is a common organic solvent. From each pair of compounds, select the solvent with the greater solubility in water.

In each case select the molecule that can make better hydrogen bonds with water or is more polar.

(a) CH_2Cl_2 and CH_3CH_2OH CH_3CH_2OH

(b) $CH_3CH_2OCH_2CH_3$ and CH_3CH_2OH CH_3CH_2OH

(c) $CH_3\overset{O}{\overset{\|}{C}}CH_3$ and $CH_3CH_2OCH_2CH_3$ $CH_3\overset{O}{\overset{\|}{C}}CH_3$

(d) $CH_3CH_2OCH_2CH_3$ and $CH_3(CH_2)_3CH_3$ $CH_3CH_2OCH_2CH_3$

Problem 11.13 Following are structural formulas, boiling points, and solubilities in water for diethyl ether and tetrahydrofuran (THF). Account for the fact that tetrahydrofuran is much more soluble in water than diethyl ether.

$$CH_3CH_2-O-CH_2CH_3$$

Diethyl ether
bp 35°C
8 g/100 mL water

$$H_2C \overset{\displaystyle -CH_2}{\underset{O}{\diagup}} CH_2$$

Tetrahydrofuran
bp 67°C
very soluble in water

There are two factors to be considered:
(1) The shape of the hydrocarbon is important in determining water solubility. For example, 2-methyl-2-propanol (*tert*-butyl alcohol) is considerably more soluble in water than 1-butanol. A *tert*-butyl group is much more compact than a butyl group and, consequently, there is less disruption of water hydrogen bonding when *tert*-butyl alcohol is dissolved in water than when 1-butanol is dissolved in water. Similarly, the hydrocarbon portion of tetrahydrofuran (THF) is more compact than that of diethyl ether which increases the solubility of THF in water compared to diethyl ether.
(2) The oxygen atom of THF is more accessible for hydrogen bonding and solvation by water than the oxygen atom of diethyl ether. This greater accessibility arises because the hydrocarbon chains bonded to oxygen in THF are "tied back" whereas those on the oxygen atom of diethyl ether have more degrees of freedom and consequently present more steric hindrance to solvation.

Problem 11.14 Because of the Lewis base properties of ether oxygen atoms, crown ethers are excellent complexing agents for Na⁺, K⁺, and NH₄⁺. What kind of molecule might serve as a complexing agent for Cl⁻ or Br⁻?

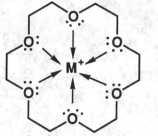

18-crown-6 with a generic metal M⁺

To build an analogous system for complexation of Cl⁻ it is necessary to first realize that the chloride ion is a Lewis base. Instead of a cavity with electron pair donors (as in crown ethers) we need a system with electron pair acceptors (or Lewis acid sites), such as the hydrogen atoms of protonated amines.

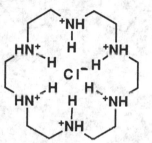

The size of the cavity will determine which anions will be complexed preferentially.

Preparation of Ethers

Problem 11.15 Write equations to show a combination of reactants to prepare each ether. Which ethers can be prepared in good yield by a Williamson ether synthesis? If there are any that cannot be prepared by the Williamson method, explain why not.

In each case, choose reagents that utilize a methyl halide (best), primary alkyl halide (2nd best), or a secondary alkyl halide (3rd best).

(a) CH₃CH₂OCHCH₃ (with CH₃ branch)

Na⁺ ⁻OCHCH₃ (with CH₃ branch) + CH₃CH₂Br ⟶ CH₃CH₂OCHCH₃ (with CH₃ branch) + Na⁺Br⁻

(b) CH₃COCH₂CH₂CH₃ (with two CH₃ branches on central C)

CH₃CO⁻K⁺ (with two CH₃ branches) + CH₃CH₂CH₂Br ⟶ CH₃COCH₂CH₂CH₃ (with two CH₃ branches) + K⁺Br⁻

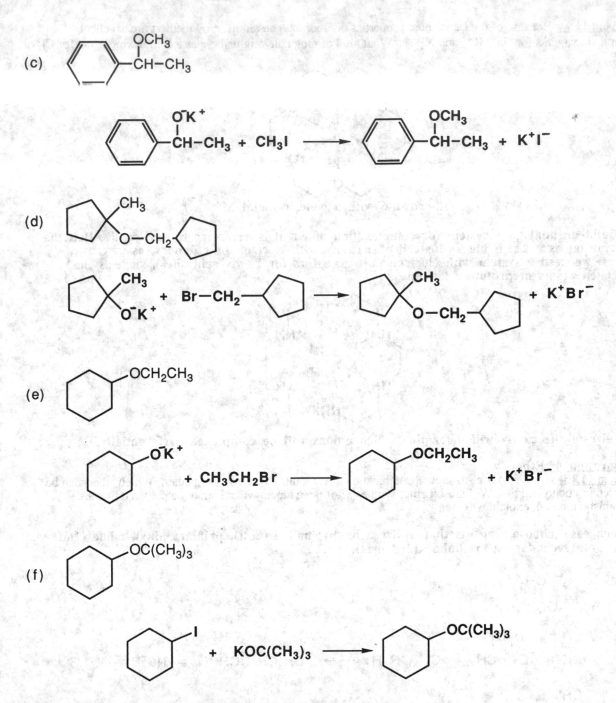

(c)

(d)

(e)

(f)

Some elimination will be observed, since a seconday alkyl iodide is being used.

<u>Problem 11.16</u> Propose a mechanism for this reaction.

The following mechanism is in three steps: (1) protonation of the alkene to form a resonance-stabilized carbocation, (2) reaction of the carbocation with methanol to give an oxonium ion, and (3) loss of a proton to give the ether.

Step 1:

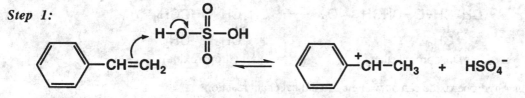

Step 2:

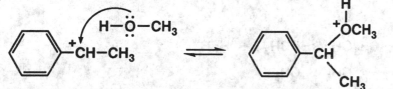

Step 3:

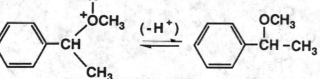

Reactions of Ethers

<u>Problem 11.17</u> Draw structural formulas for the products formed when each compound is refluxed in concentrated HI.

Since an excess of HI is used, you can assume that the alcohol products initially produced will be converted to alkyl iodides under these conditions.

(a) $CH_3CH_2OCH_2CH_2CH_3 \xrightarrow{HI} CH_3CH_2CH_2I + CH_3CH_2I$

(b) 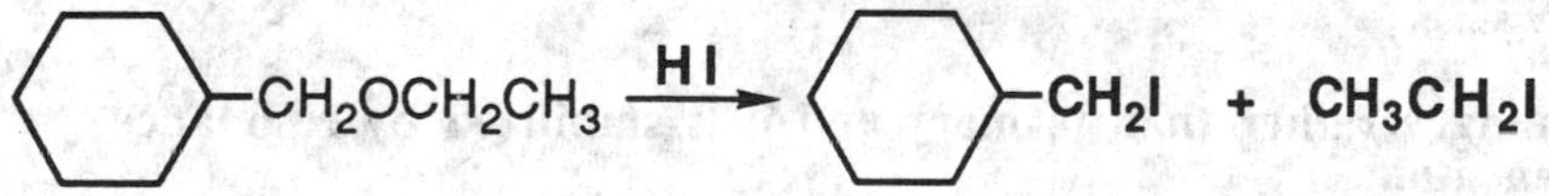 [cyclohexyl]$-CH_2OCH_2CH_3 \xrightarrow{HI}$ [cyclohexyl]$-CH_2I + CH_3CH_2I$

(c) 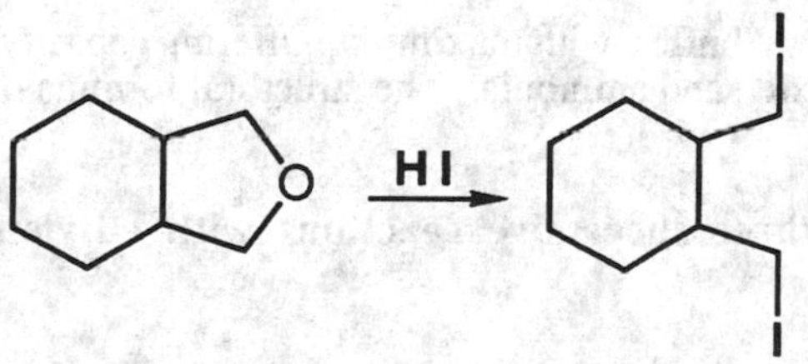 [bicyclic tetrahydrofuran] $\xrightarrow{HI}$ [cyclohexane with two CH₂I groups]

(d) 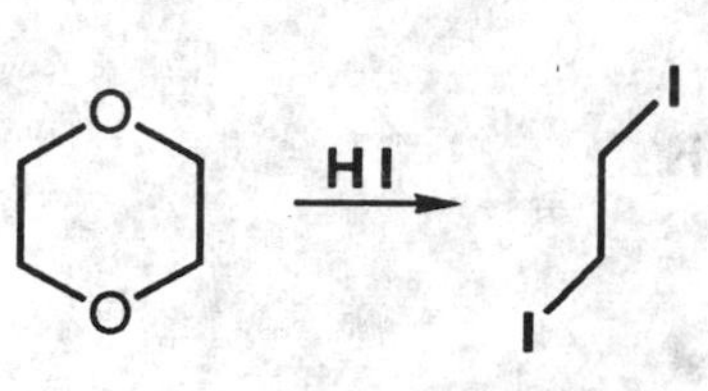 [1,4-dioxane] $\xrightarrow{HI}$ [diiodo compound]

<u>Problem 11.18</u> Following is an equation for the reaction of diisopropyl ether and oxygen to form a hydroperoxide.

$$CH_3CH-O-CHCH_3 + O_2 \longrightarrow CH_3CH-O-\overset{OOH}{\underset{|}{C}}CH_3$$
$$\underset{CH_3}{|} \quad \underset{CH_3}{|} \qquad\qquad \underset{CH_3}{|} \quad \underset{CH_3}{|}$$

Diisopropyl ether   (A hydroperoxide)

Formation of an ether hydroperoxide can be written as a radical chain reaction.
(a) Write a pair of chain propagation steps that accounts for the formation of this ether hydroperoxide. Assume that initiation is by an unknown radical, R•.

Initiation:

$$CH_3CH-O-CHCH_3 + R• \longrightarrow CH_3CH-O-\overset{•}{C}CH_3 + H-R$$
$$\underset{CH_3}{|} \quad \underset{CH_3}{|} \qquad\qquad \underset{CH_3}{|} \quad \underset{CH_3}{|}$$

Propagation:

$$CH_3CH-O-\overset{•}{C}CH_3 + O_2 \longrightarrow CH_3CH-O-\overset{O-O•}{\underset{|}{C}}CH_3$$
$$\underset{CH_3}{|} \quad \underset{CH_3}{|} \qquad\qquad\qquad \underset{CH_3}{|} \quad \underset{CH_3}{|}$$

$$CH_3CH-O-\overset{O-O•}{\underset{|}{C}}CH_3 + CH_3CH-O-CHCH_3 \longrightarrow$$
$$\underset{CH_3}{|} \quad \underset{CH_3}{|} \qquad \underset{CH_3}{|} \quad \underset{CH_3}{|}$$

$$CH_3CH-O-\overset{OOH}{\underset{|}{C}}CH_3$$
$$\underset{CH_3}{|} \quad \underset{CH_3}{|}$$
$$+$$
$$CH_3CH-O-\overset{•}{C}CH_3$$
$$\underset{CH_3}{|} \quad \underset{CH_3}{|}$$

(b) Account for the fact that hydroperoxidation of ethers is regioselective, that it is occurs preferentially at a carbon adjacent to the ether oxygen.

The radical formed is secondary rather than primary and it is stabilized by resonance interaction with the oxygen atom.

Synthesis and Reactions of Epoxides
<u>Problem 11.19</u> Triethanolamine (TEA) is a widely used biological buffer, with maximum buffering capacity at pH 7.8. Propose a synthesis of this compound from ethylene oxide and ammonia. The structural formula of triethanolamine is $(HOCH_2CH_2)_3N$.

Triethanolamine can be prepared from ammonia by three successive reactions with ethylene oxide.

$$NH_3 + \underset{\underset{O}{\diagdown\diagup}}{H_2C-CH_2} \longrightarrow HOCH_2CH_2NH_2$$

$$HOCH_2CH_2NH_2 + \underset{\underset{O}{\diagdown\diagup}}{H_2C-CH_2} \longrightarrow (HOCH_2CH_2)_2NH$$

$$(HOCH_2CH_2)_2NH + \underset{\underset{O}{\diagdown\diagup}}{H_2C-CH_2} \longrightarrow (HOCH_2CH_2)_3N$$
                                                                    Triethanolamine

<u>Problem 11.20</u> Ethylene oxide is the starting material for the synthesis of both methyl cellosolve and cellosolve, two important industrial solvents. Propose a mechanism for these reactions.

· H$_2$C—CH$_2$ + CH$_3$OH $\xrightarrow{\text{H}_2\text{SO}_4}$ CH$_3$OCH$_2$CH$_2$OH

       Oxirane   2-Methoxyethanol
 (Ethylene oxide)   (Methyl Cellosolve)

· H$_2$C—CH$_2$+ CH$_3$CH$_2$OH $\xrightarrow{\text{H}_2\text{SO}_4}$ CH$_3$CH$_2$OCH$_2$CH$_2$OH

                                 2-Ethoxyethanol
                                 (Ethyl Cellosolve)

For each reaction, protonation of the epoxide gives a cyclic oxonium ion. Reaction of this oxonium ion with the alcohol opens the epoxide ring. Loss of a proton completes the reaction.

Step 1:

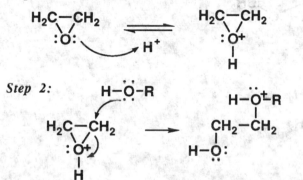

Step 2:

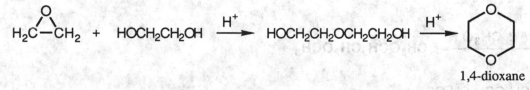

Step 3:

H—O$^+$—R $\xrightarrow{(-\text{H}^+)}$ ROCH$_2$CH$_2$OH
CH$_2$—CH$_2$
H—O:

<u>Problem 11.21</u> Ethylene oxide is the starting material for the synthesis of 1,4-dioxane. Propose a mechanism for each step in this synthesis.

H$_2$C—CH$_2$ + HOCH$_2$CH$_2$OH $\xrightarrow{\text{H}^+}$ HOCH$_2$CH$_2$OCH$_2$CH$_2$OH $\xrightarrow{\text{H}^+}$ 1,4-dioxane

The mechanism for this reaction involves an acid-catalyzed attack of the diol on the epoxide followed by protonation of one of the terminal hydroxyl groups, and displacement of water by the other alcohol.

Step 1:

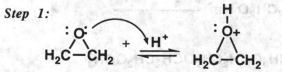

Step 2:

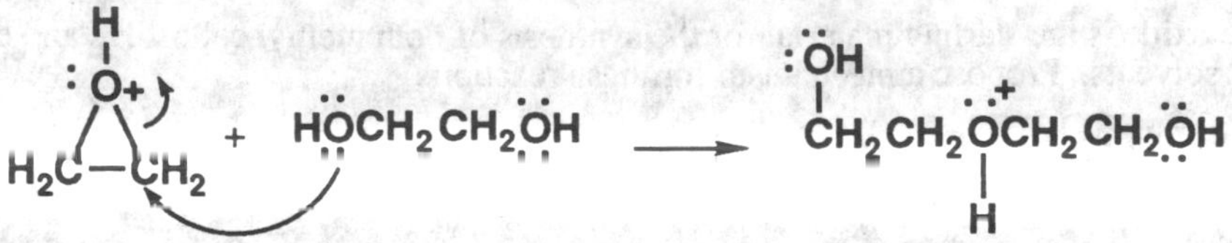

Step 3:

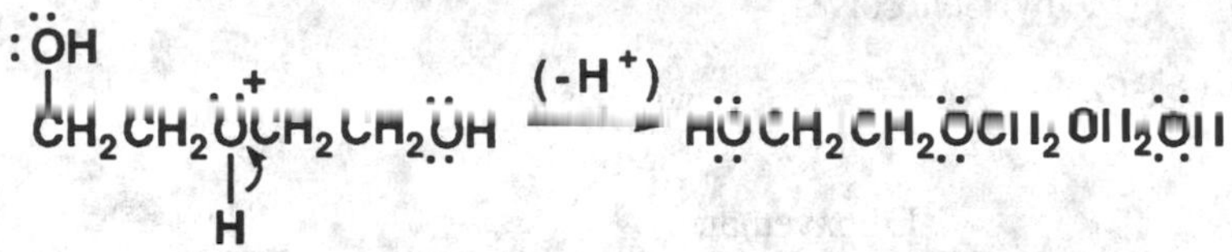

Step 4:

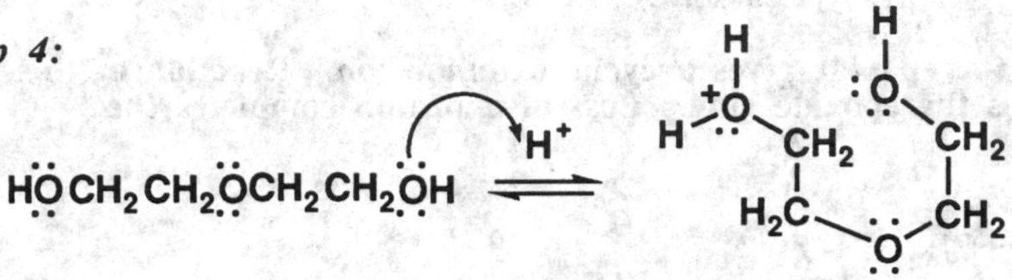

Step 5:

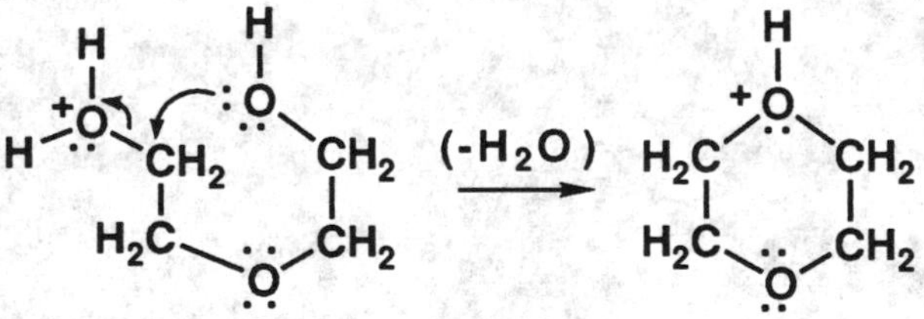

Step 6:

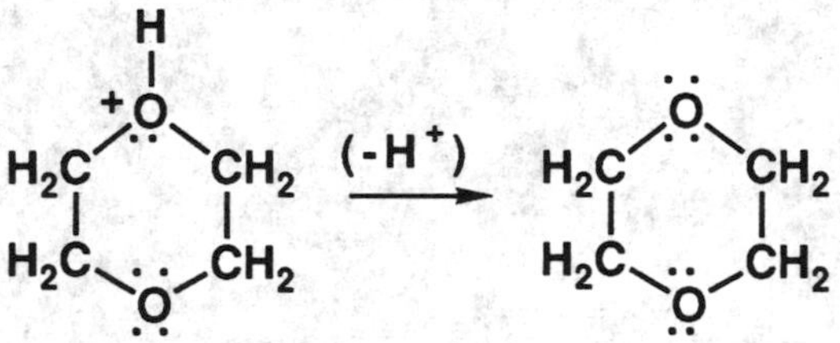

<u>Problem 11.22</u> Propose a synthesis for each ether starting with ethylene oxide and any readily available alcohols.

(a)   CH₃OCH₂CH₂OCH₃

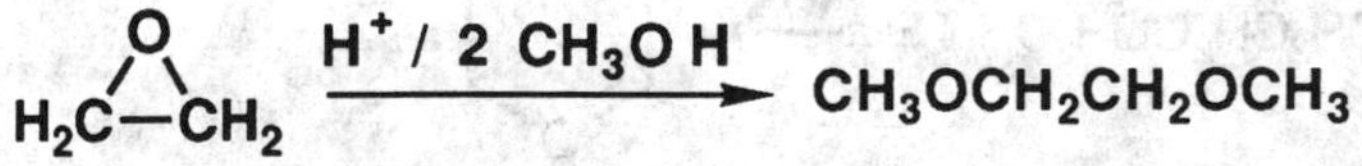

$$H_2C\text{---}CH_2 \quad \xrightarrow{H^+ / 2\ CH_3OH} \quad CH_3OCH_2CH_2OCH_3$$

(b)   CH₃OCH₂CH₂OCH₂CH₂OCH₃

$$H_2C\text{---}CH_2 + CH_3O^-Na^+ \xrightarrow{CH_3OH} CH_3OCH_2CH_2O^-Na^+ \xrightarrow{H_2C\text{---}CH_2}$$

$$CH_3OCH_2CH_2OCH_2CH_2O^-Na^+ \xrightarrow{H^+ / CH_3OH} CH_3OCH_2CH_2OCH_2CH_2OCH_3$$

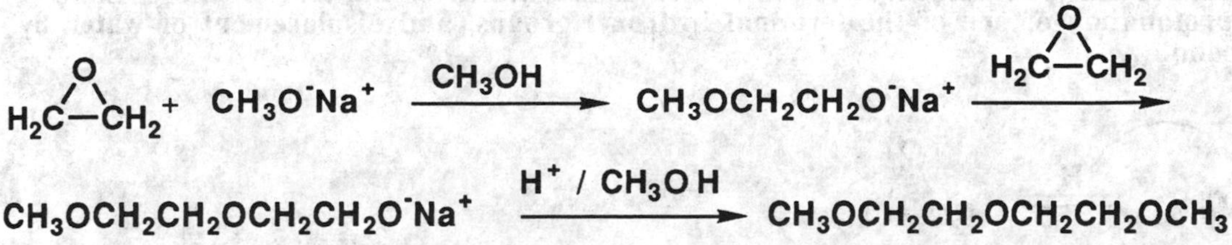

Problem 11.23 Propose a synthesis for 18-crown-6. If a base is used in your synthesis, does it make a difference if it is lithium hydroxide or potassium hydroxide ?

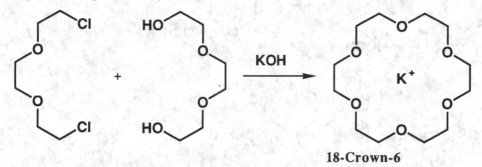

Since 18-crown-6 binds K+ the best, KOH should be used so that the crown ether can form around the K+ ion. This type of strategy is called the template approach and is used in a variety of similar situations.

The above two pieces can be synthesized as follows:

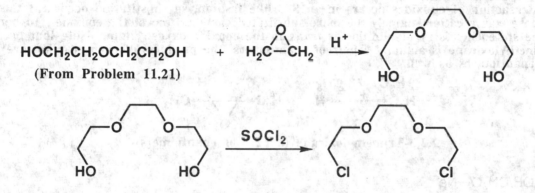

Problem 11.24 Propose a mechanism for the reaction of a primary alcohol, RCH_2OH, with dihydropyran to give a tetrahydropyranyl ether.

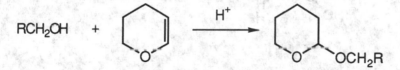

The first step in the mechanism involves protonation of the dihydropyran to give a carbocation. Note how the positive charge is located on the carbon next to the O atom, because the lone pairs on O can stabilize an adjacent positive charge through resonance. The nucleophilic alcohol attacks the carbocation, and the product tetrahydropyranyl ether is obtained following the loss of a proton.

Step 1:

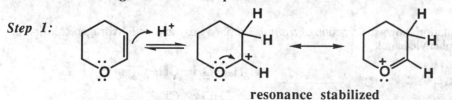

resonance stabilized

Step 2:

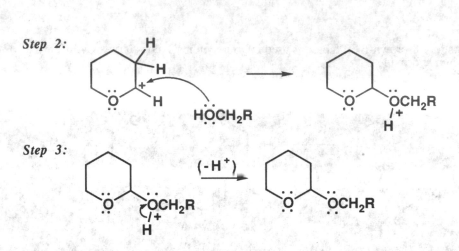

Step 3:

Problem 11.25 Predict the structural formula of the major product of the reaction of 2,2,3-trimethyloxirane with each set of reagents.

Nucleophiles in the absence of acid catalysis react via an S_N2 mechanism, so the nucleophile ends up on the less hindered, that is less substituted, carbon atom of the epoxide. The acid-catalyzed reaction of epoxides occurs in an S_N1-like fashion, so substitution occurs at the site of the more stable cation, namely the more substituted carbon atom of the epoxide. In both cases, the stereochemistry of addition is *trans* to the epoxide oxygen atom. Note that because 2,2,3-trimethyloxirane is actually a pair of enantiomers, the products of each reaction are a pair of enantiomers as well.

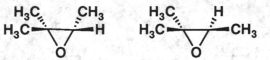

2,2,3-Trimethyloxirane (a pair of enantiomers)

(a) $CH_3OH/CH_3O^- Na^+$

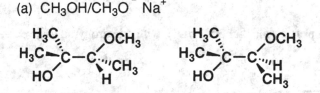

(b) CH_3OH/H^+

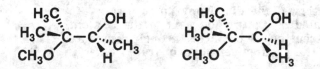

Problem 11.26 The following equation shows the reaction of *trans*-2,3-diphenyloxirane with hydrogen chloride in benzene to form 2-chloro-1,2-diphenylethanol.

trans-2,3-Diphenyloxirane   2-Chloro-1,2-diphenylethanol

(a) How many stereoisomers are possible for 2-chloro-1,2-diphenylethanol?

There are two stereocenters in the molecule so there are 2^2 or 4 possible stereoisomers.

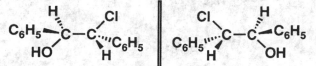

A pair of enantiomers
(from addition of HCl to *trans*-2,3-diphenyloxirane)

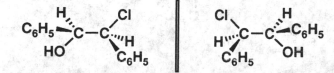

A pair of enantiomers
(from addition of HCl to *cis*-2,3-diphenyloxirane)

(b) Given that opening of the epoxide ring in this reaction is stereoselective, predict which of the possible stereoisomers of 2-chloro-1,2-diphenylethanol is/are formed in the reaction.

Only the upper pair of enantiomers is formed.

<u>Problem 11.27</u> Propose a mechanism to account for this rearrangement.

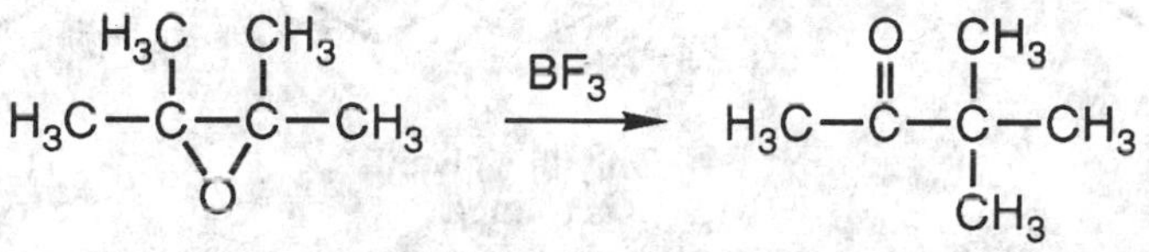

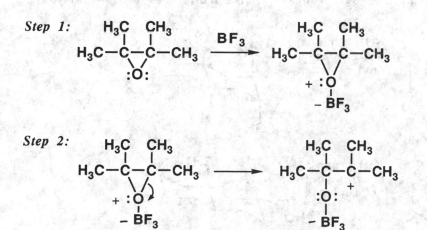

Reaction between boron trifluoride, a Lewis acid, and the epoxide oxygen, a Lewis base, forms an oxonium ion. The ring then opens followed by rearrangement of the resulting secondary carbocation and subsequent loss of boron trifluoride to give the observed ketone.

Step 1:

Step 2:

Step 3:

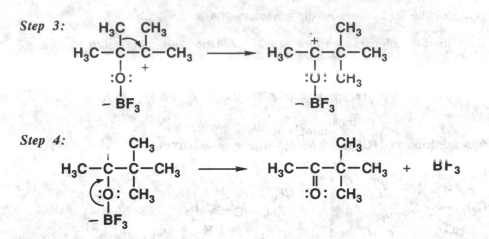

Step 4:

<u>Problem 11.28</u> Following is the structural formula for an epoxide derived from 9-methyldecalin. Acid-catalyzed hydrolysis of this epoxide gives a *trans*-diol. Of the two possible *trans*-diols, only one is formed. How do you account for this stereoselectivity? *Hint:* Begin by drawing *trans*-decalin with each six-membered ring in the more stable chair conformation (Section 2.7B), and then determine whether each substituent in the isomeric glycols is axial or equatorial.

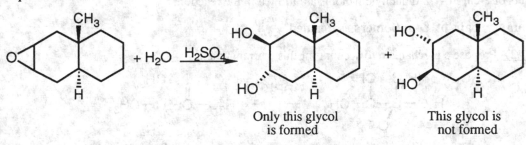

The key to this problem is that opening of the epoxide is stereospecific; the incoming nucleophile and the leaving protonated epoxide oxygen must be anti coplanar. In a cyclohexane ring, anti coplanar corresponds to *trans* and diaxial. An accurate model of the glycol formed will show that in it, the two -OH groups are diaxial and, therefore, *trans* and coplanar. In the alternative glycol, the -OH groups are also *trans*, but because they are diequatorial, they are not coplanar.

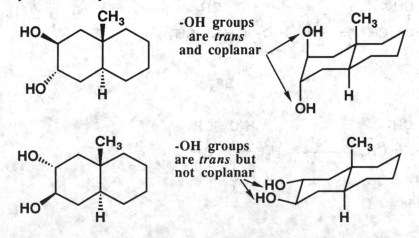

Problem 11.29 Following are two reaction sequences for converting 1,2-diphenylethylene into 2,3-diphenyloxirane.

$$Ph-CH=CH-Ph \xrightarrow{\text{ArCO}_3\text{H}} Ph-CH-CH-Ph$$

1,2-Diphenylethylene　　　　　　　　　　　　　　2,3-Diphenyloxirane

$$Ph-CH=CH-Ph \xrightarrow[\text{2) CH}_3\text{O}^-\text{Na}^+]{\text{1) Cl}_2, \text{H}_2\text{O}} Ph-CH-CH-Ph$$

1,2-Diphenylethylene　　　　　　　　　　　　　　2,3-Diphenyloxirane

Suppose that the starting alkene is *trans*-1,2-diphenylethylene.
(a) What is the configuration of the oxirane formed in each sequence?

In each case, the oxirane has the *trans* configuration.

(b) Does the oxirane formed in either sequence rotate the plane of polarized light? Explain.

Neither product will rotate the plane of polarized light. The *trans* isomer can exist as a pair of enantiomers. In each reaction, the product formed is a racemic mixture and will not be optically active.

Problem 11.30 Complete these reactions. *Hint:* each reaction shows a stereoselectivity typical of nucleophilic opening of the epoxide ring.

A racemic mixture is observed in both cases.

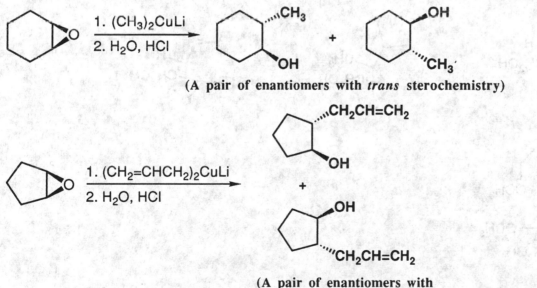

(A pair of enantiomers with *trans* sterochemistry)

(A pair of enantiomers with
trans stereochemistry)

Problem 11.31 One of the most useful organic reactions discovered in the last 20 years is the titanium- catalyzed asymmetric epoxidation of allylic alcohols developed by Professor Barry Sharpless and coworkers (see K. B. Sharpless et al., Pure and Appl. Chem., **55**, 589 (1983)). The reagent combination consists of Ti[(OCH(CH$_3$)$_2$]$_4$, a hydroperoxide, and a chiral molecule such as (+)-diethyl tartrate.

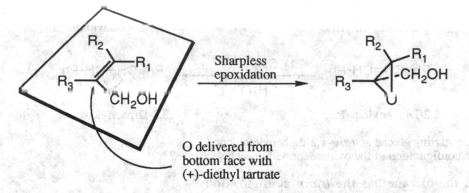

Two new stereocenters are created in the product. If (+)-diethyl tartrate is used in the reaction, the product arises by delivery of O from the hydroperoxide to the bottom face of the molecule, and the product epoxide is the stereoisomer shown. If (-)-diethyl tartrate is used, the product is the enantiomer of the stereoisomer shown. Draw the expected products of Sharpless epoxidation of the following allylic alcohols using (+)-diethyl tartrate.

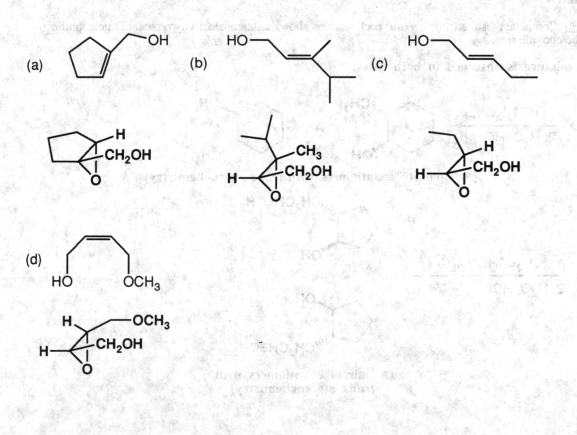

Problem 11.32 The following chiral epoxide is an intermediate in the synthesis of the insect pheromone frontalin. How can this epoxide be prepared from an allylic alcohol precursor, using the Sharpless epoxidation reaction described in Problem 11.31?

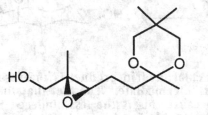

A close inspection indicates that this epoxide has the oxygen atom on the opposite face compared to the ones in problem 11.31 that utilized the (+)-diethyl tartrate. Thus the desired product is produced from the appropriate alkene using (-)-diethyl tartrate along with the rest of the Sharpless epoxidation reagents.

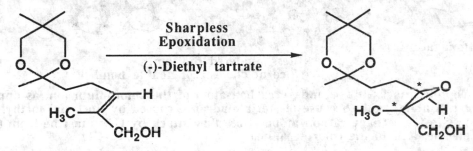

Problem 11.33 Human white cells produce an enzyme called myeloperoxidase. This enzyme catalyzes the reaction between hydrogen peroxide and chloride ion to produce hypochlorous acid, HOCl, which reacts as if it is Cl⁺OH⁻. When attacked by white cells, cholesterol reacts to give the chlorohydrin shown below as the major product

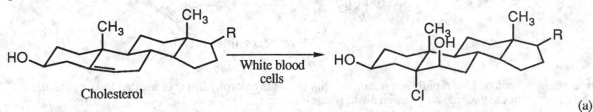

(a)

Propose a mechanism for this reaction. Account for both the regioselectivity and the stereoselectivity.

The mechanism involves reaction of the double bond of cholesterol with HOCl to produce a chloronium ion intermediate, that is attacked by HO⁻ to produce the chlorohydrin.

Step 1:

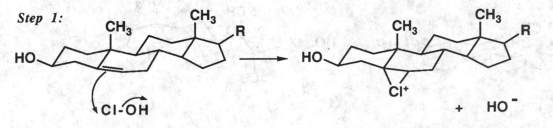

Step 2:

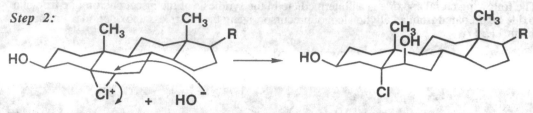

The *trans* stereochemistry of the chlorohydrin product is the result of the backside attack that is required by the chloronium ion intermediate. The fact that the chlorine atom ends up on the bottom face of the molecule is because this is the less-hindered face of the double bond. The axial methyl group adjacent to the double bond provides a steric barrier to reaction on the top face.

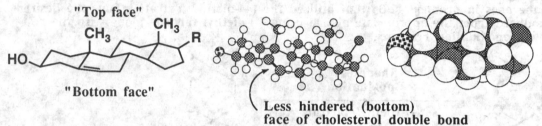

Less hindered (bottom)
face of cholesterol double bond

The HO⁻ nucleophile attacks the secondary carbon atom of the chloronium ion, as opposed to the tertiary carbon atom, again because of steric hindrance caused by the axial methyl group preventing access to the tertiary carbon atom. Note that attack by HO⁻ must be from the top face of the molecule, anti to the chlorine atom.

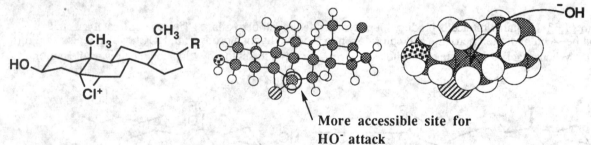

More accessible site for
HO⁻ attack

(b) On standing or (much more rapidly) on treatment with base, the chlorohydrin is converted to an epoxide. Show the structure of the epoxide and a mechanism for its formation.

The OH group of the chlorohydrin is deprotonated by the base, allowing for epoxide formation as chloride anion departs. The backside attack of the oxygen atom dictates that the epoxide product will have the oxygen atom on what we are calling the top face of the molecule.

Step 1:

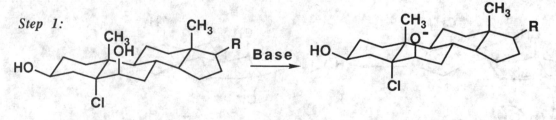

Step 2:

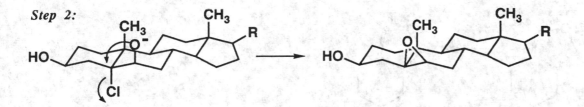

Problem 11.34 Propose a mechanism for the following acid-catalyzed rearrangement.

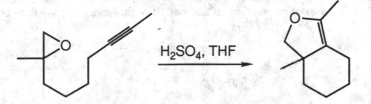

The mechanism of this rearrangement involves initial protonation of the epoxide oxygen atom, making the epoxide electrophilic enough to react with the weakly nucleophilic pi bond of the adjacent triple bond. The triple bond reacts at the more reactive, more highly substituted tertiary side of the protonated epoxide to give the six-membered ring intermediate.

Step 1:

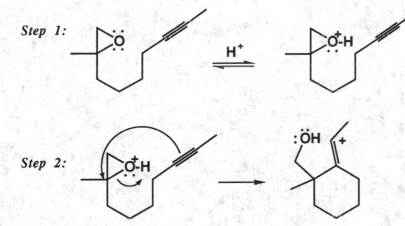

Attack of the alcohol upon the carbocation of this intermediate followed by loss of a proton completes the reaction.

Step 3:

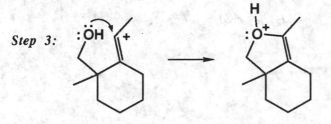

Step 4:

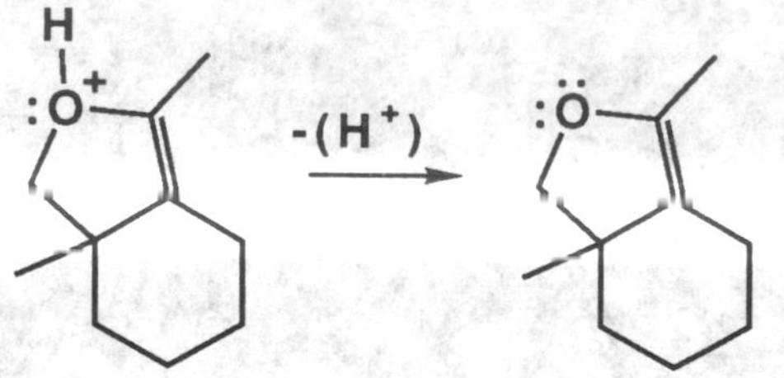

Synthesis

Problem 11.35 Show reagents and experimental conditions to synthesize the following compounds from 1-propanol. Any derivative of 1-propanol prepared in an earlier part of this problem may then be used for a later synthesis.

(a) Propanal

$$CH_3CH_2CH_2OH + PCC \longrightarrow CH_3CH_2\overset{O}{\overset{\|}{C}}H$$

(b) Propanoic acid

$$CH_3CH_2CH_2OH \xrightarrow{H_2CrO_4} CH_3CH_2\overset{O}{\overset{\|}{C}}OH$$

(c) Propene

$$CH_3CH_2CH_2OH \xrightarrow[\text{heat}]{H_3PO_4} CH_3CH=CH_2$$

(d) 2-Propanol

$$CH_3CH=CH_2 \xrightarrow[H_2SO_4]{H_2O} CH_3\overset{OH}{\overset{|}{C}}HCH_3$$
From (c)

(e) 2-Bromopropane

$$\overset{OH}{\overset{|}{CH_3CHCH_3}} \quad \text{or} \quad H_2C=CHCH_3 + HBr \xrightarrow{heat} CH_3\overset{Br}{\overset{|}{C}}HCH_3$$
From (d)   From (c)

(f) 1-Chloropropane

$$CH_3CH_2CH_2OH \xrightarrow[\text{Pyridine}]{SOCl_2} CH_3CH_2CH_2Cl + SO_2 + HCl$$

(g) 1,2-Dibromopropane

$$CH_3CH=CH_2 \xrightarrow[CCl_4]{Br_2} CH_3\overset{Br}{\overset{|}{C}}HCH_2Br$$
From (c)

(h) Propyne

$$CH_3\overset{Br}{\overset{|}{C}}HCH_2Br \xrightarrow{2NaNH_2} HC\equiv CCH_3$$
From (g)

(i) 2-Propanone

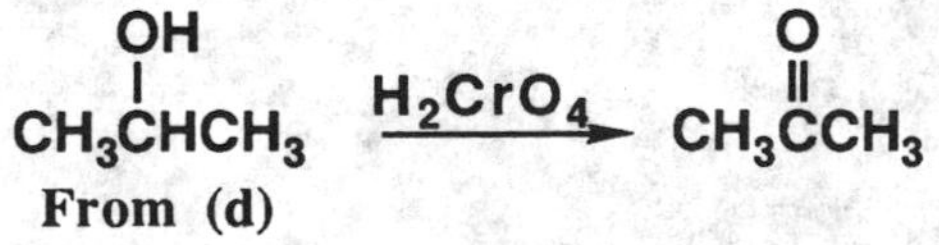

$$\underset{\text{From (d)}}{CH_3\overset{\overset{\displaystyle OH}{|}}{C}HCH_3} \xrightarrow{H_2CrO_4} CH_3\overset{\overset{\displaystyle O}{\|}}{C}CH_3$$

(j) 1-Chloro-2-propanol

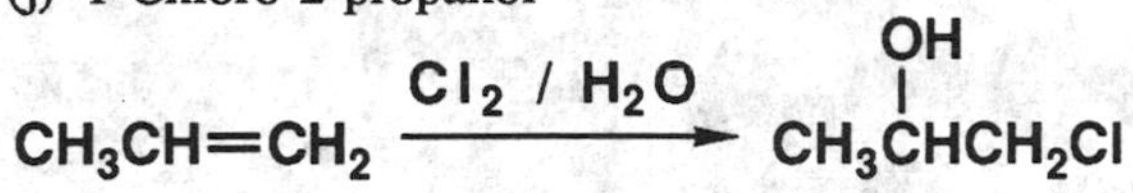

$$CH_3CH{=}CH_2 \xrightarrow{Cl_2\ /\ H_2O} CH_3\overset{\overset{\displaystyle OH}{|}}{C}HCH_2Cl$$

(k) Methyloxirane

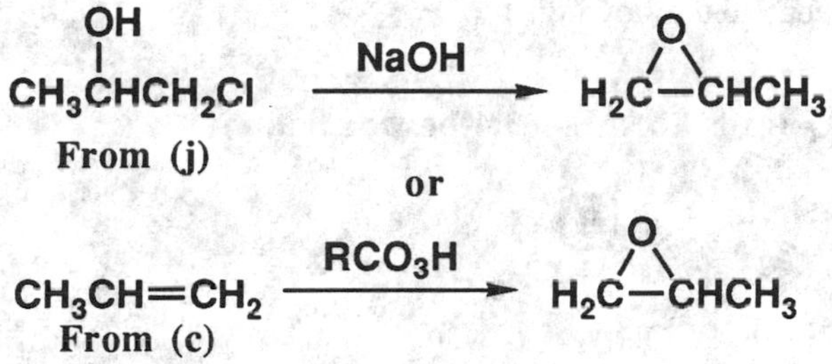

$$\underset{\text{From (j)}}{CH_3\overset{\overset{\displaystyle OH}{|}}{C}HCH_2Cl} \xrightarrow{NaOH} H_2C\overset{\displaystyle O}{\overset{\displaystyle \diagup\diagdown}{-}}CHCH_3$$

or

$$\underset{\text{From (c)}}{CH_3CH{=}CH_2} \xrightarrow{RCO_3H} H_2C\overset{\displaystyle O}{\overset{\displaystyle \diagup\diagdown}{-}}CHCH_3$$

(l) Dipropyl ether

$$\underset{\text{From (f)}}{CH_3CH_2CH_2Cl} + \underset{\substack{\text{From 1-propanol}\\\text{and Na}}}{CH_3CH_2CH_2O^-Na^+} \longrightarrow CH_3CH_2CH_2OCH_2CH_2CH_3$$

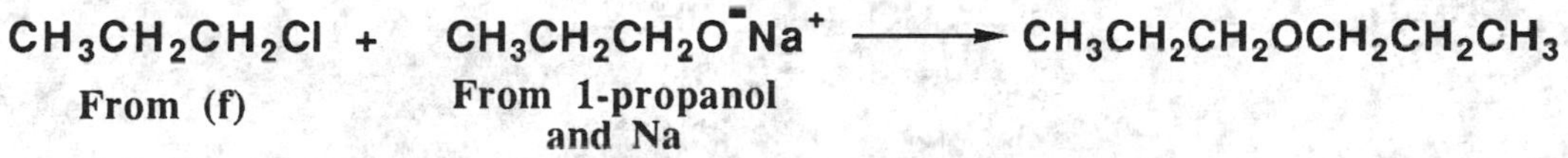

(m) Isopropyl propyl ether

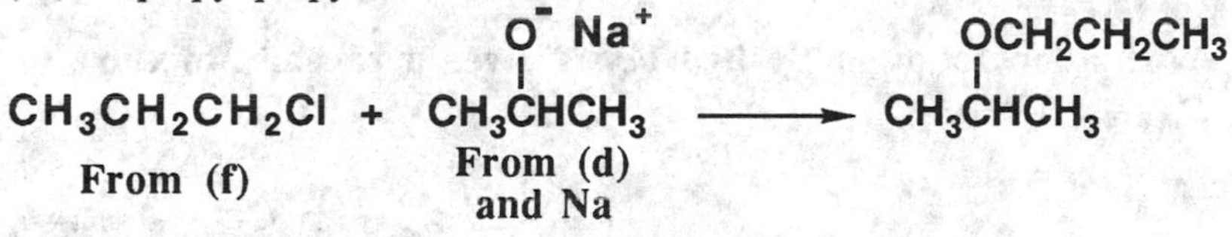

$$\underset{\text{From (f)}}{CH_3CH_2CH_2Cl} + \underset{\substack{\text{From (d)}\\\text{and Na}}}{CH_3\overset{\overset{\displaystyle O^-Na^+}{|}}{C}HCH_3} \longrightarrow CH_3\overset{\overset{\displaystyle OCH_2CH_2CH_3}{|}}{C}HCH_3$$

(n) 1-Mercapto-2-propanol

$$\underset{\text{From (k)}}{H_2C\overset{\displaystyle O}{\overset{\displaystyle \diagup\diagdown}{-}}CHCH_3} \xrightarrow[H_2O]{HS^-Na^+} CH_3\overset{\overset{\displaystyle OH}{|}}{C}HCH_2SH$$

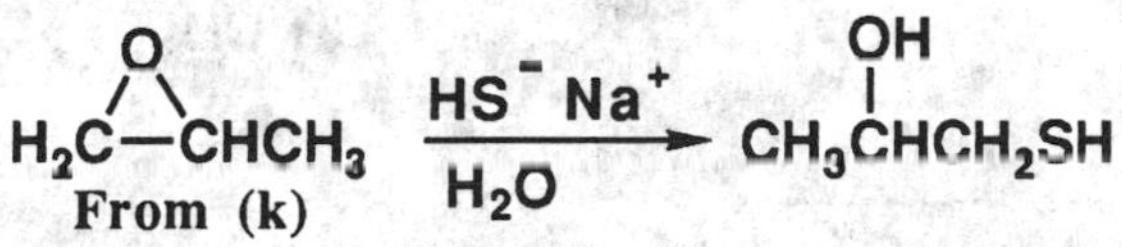

(o) 1-Amino-2-propanol

$$\underset{\text{From (k)}}{H_2C\overset{\displaystyle O}{\overset{\displaystyle \diagup\diagdown}{-}}CHCH_3} \xrightarrow{NH_3} CH_3\overset{\overset{\displaystyle OH}{|}}{C}HCH_2NH_2$$

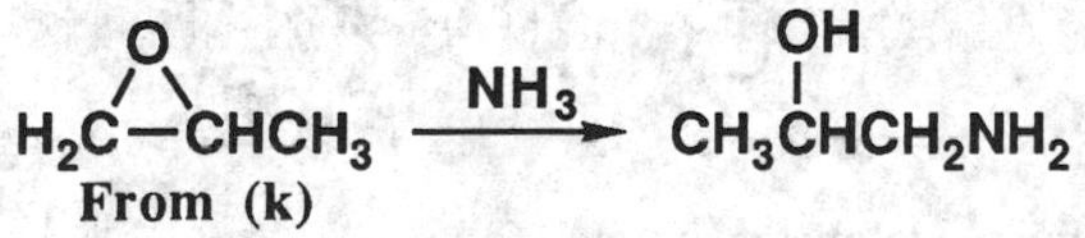

(p) 1,2-Propanediol

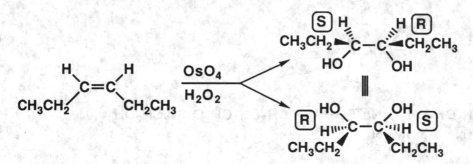

$$CH_3CH{=}CH_2 \xrightarrow[H_2O_2]{OsO_4} \overset{HO\quad OH}{CH_3CH{-}CH_2}$$

From (c)

or

$$\underset{\text{From (k)}}{\overset{O}{H_2C{-}CHCH_3}} \xrightarrow[H_3O^+]{HO^-\ or} \overset{HO\quad OH}{CH_3CH{-}CH_2}$$

Note that the products in g, j, k, n, o, and p will be racemic mixtures.

<u>Problem 11.36</u> Starting with *cis*-3-hexene, show how to prepare the following:
(a) Meso 3,4-hexanediol

The syn geometry of addition observed with OsO$_4$ leads to meso 3,4-hexanediol.

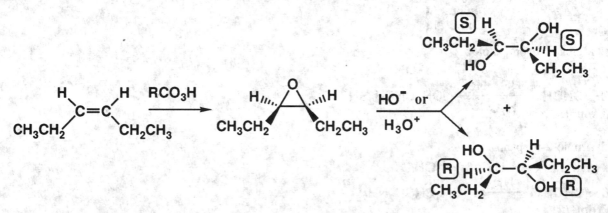

(b) Racemic 3,4-hexanediol

The anti geometry produced via epoxidation followed by hydrolysis gives a racemic mixture of enantiomers.

<u>Problem 11.37</u> Show reagents and experimental conditions to convert cycloheptene to the following. Any compound made in an earlier part of this problem may be used as an intermediate in any following conversion.

(a)

Oxidation of cycloheptene with a peroxyacid such as trifluoroperacetic acid.

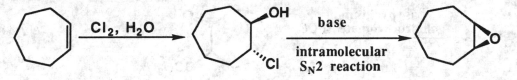

As an alternative, addition of HOCl to cycloheptene to form a chlorohydrin followed by treatment of the chlorohydrin with strong base.

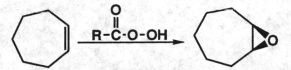

(b)

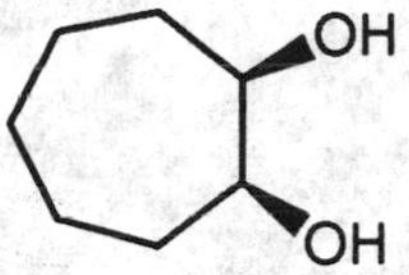

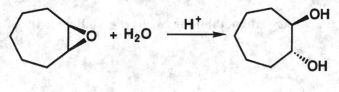

Oxidation of cycloheptene with osmium tetroxide in the presence of hydrogen peroxide.

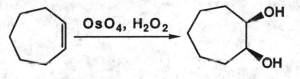

(c)

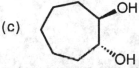

Acid-catalyzed hydrolysis of the epoxide from part (a).

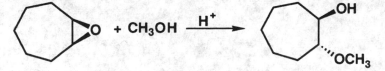

(d)

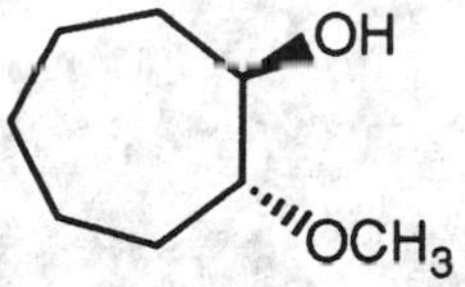

Treatment of the epoxide from part (a) with methanol in the presence of an acid catalyst, or alternatively, treatment with methanol in the presence of sodium methoxide.

(e)

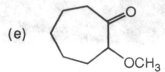

Oxidation of the secondary alcohol of part (d) using chromium trioxide in pyridine or as pyridinium chlorochromate or H_2CrO_4.

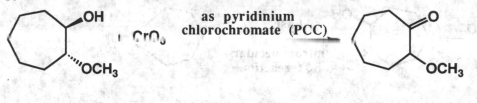

(f)

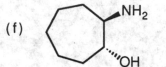

Treatment of the epoxide from part (a) with ammonia.

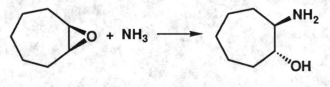

(g)

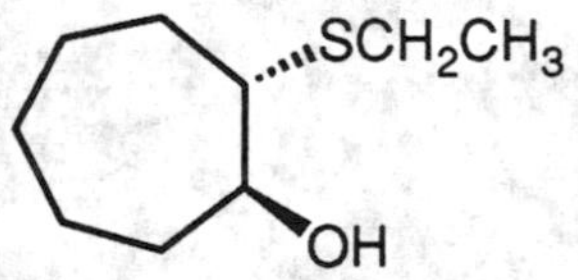

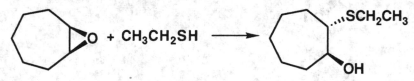

Treatment of the epoxide from part (a) with either ethanethiol or with the sodium salt of ethanethiol.

(h)

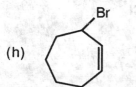

Allylic bromination of cycloheptene with *N*-bromosuccinimide (NBS) or with bromine at high temperature. In each case, reaction is by a radical chain mechanism.

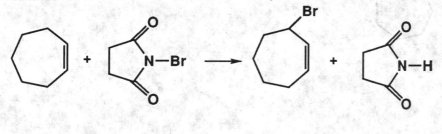

(i)

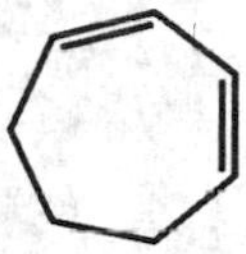

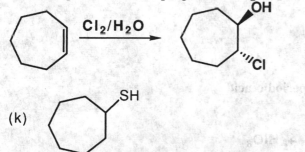

Dehydrohalogenation of product (h) with a strong base such.

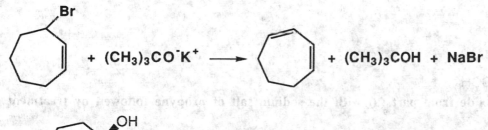

(j)

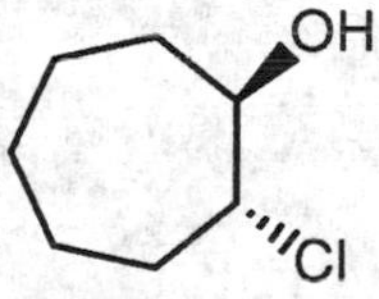

This compound can be prepared using Cl_2 in H_2O.

(k)

As a first step, hydration of cycloheptene to cycloheptanol. Hydration can be accomplished by (1) acid-catalyzed hydration, (2) oxymercuration followed by reduction with sodium borohydride, or (3) hydroboration followed by oxidation of the organoborane intermediate with alkaline hydrogen peroxide. Treatment of cycloheptanol with thionyl chloride and then sodium hydrosulfide gives the product.

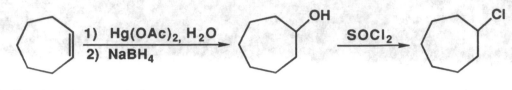

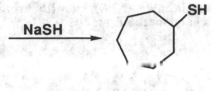

(l)

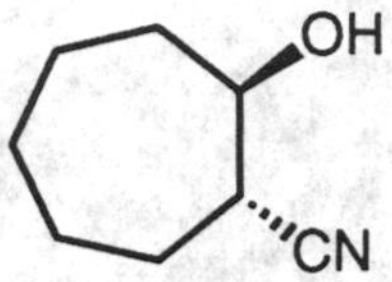

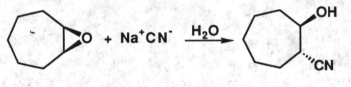

Treatment of the epoxide from part (a) with aqueous sodium cyanide.

(m)

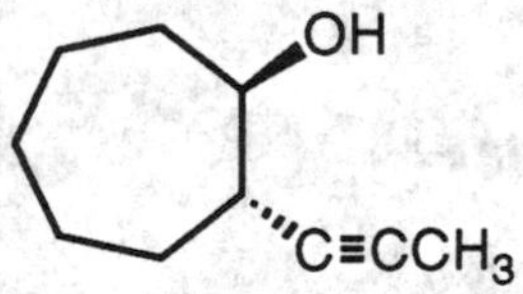

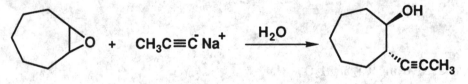

Treatment of the epoxide from part (a) with the sodium salt of propyne followed by treatment with H_2O.

(n) $\overset{O}{\overset{\|}{H C}}(CH_2)_5 \overset{O}{\overset{\|}{C H}}$

Oxidation of the glycol from part (b) or (c) with periodic acid

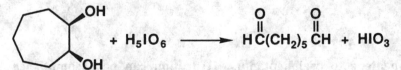

Alternatively, oxidation of cycloheptene with ozone followed by treatment of the ozonide with dimethyl sulfide.

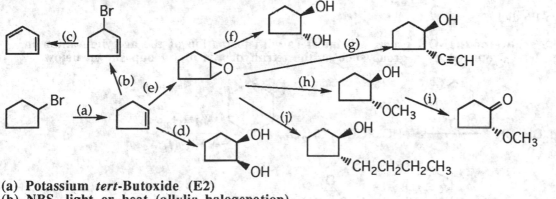

Problem 11.38 Show reagents to bring about each reaction:

(a) Potassium *tert*-Butoxide (E2)
(b) NBS, light or heat (allylic halogenation)
(c) Potassium *tert*-Butoxide (E2)
(d) OsO₄ / H₂O₂
(e) RCO₃H
(f) H₃O⁺ or HO⁻
(g) 1. HC≡C⁻Na⁺ 2. H₂O
(h) CH₃O⁻Na⁺ / CH₃OH
(i) PCC
(j) 1. [(CH₃CH₂CH₂CH₂)₂Cu]Li 2. H₂O

Problem 11.39 Propose a synthesis of the following alcohol from styrene and 1-chloro-3-methyl-2-butene.

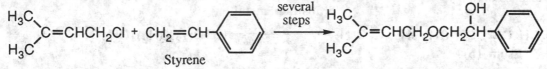

The 1-chloro-3-methyl-2-butene is converted to the alcohol in base then the alkoxide by reaction with Na metal. The styrene is converted to the epoxide, then reacted with the alkoxide followed by an aqueous workup to give the final product. Notice that the last step requires reaction of the nucleophile at the less-hindered carbon of the epoxide, so an acid-catalyzed step could not be used.

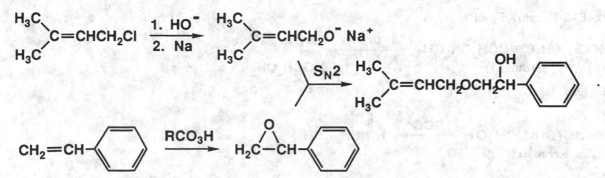

Problem 11.40 Starting with acetylene and ethylene oxide as the only sources of carbon atoms, show how to prepare these compounds.
(a) 3-Butyn-1-ol

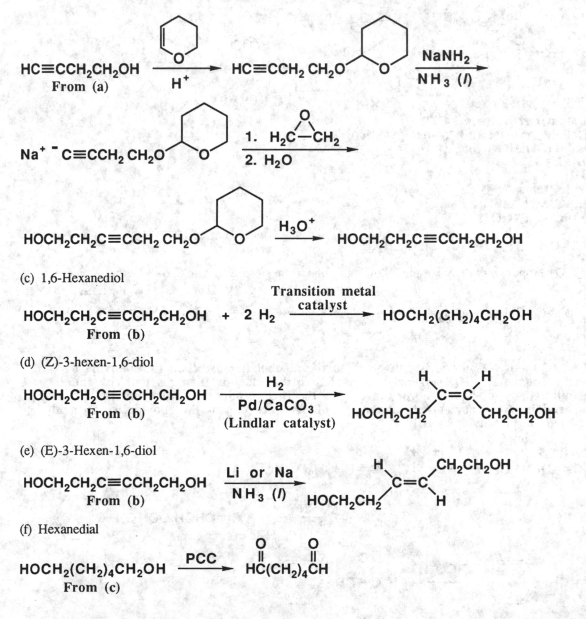

$$HC\equiv CH \xrightarrow[NH_3\ (l)]{NaNH_2} HC\equiv C^-Na^+ \xrightarrow[2.\ H_2O]{1.\ H_2C\overset{O}{\overbrace{\quad}}CH_2} HC\equiv CCH_2CH_2OH$$

(b) 3-Hexyn-1,6-diol

Because the acidity of the -OH group will interfere with formation of the acetylide anion in base, it must be protected by a group such as the tetrahydropyranyl group shown below.

$$HC\equiv CCH_2CH_2OH \xrightarrow[H^+]{} HC\equiv CCH_2CH_2O\!-\!\overset{O}{\bigcirc} \xrightarrow[NH_3\ (l)]{NaNH_2}$$
From (a)

$$Na^+\ ^-C\equiv CCH_2CH_2O\!-\!\overset{O}{\bigcirc} \xrightarrow[2.\ H_2O]{1.\ H_2C\overset{O}{\overbrace{\quad}}CH_2}$$

$$HOCH_2CH_2C\equiv CCH_2CH_2O\!-\!\overset{O}{\bigcirc} \xrightarrow{H_3O^+} HOCH_2CH_2C\equiv CCH_2CH_2OH$$

(c) 1,6-Hexanediol

$$HOCH_2CH_2C\equiv CCH_2CH_2OH\ +\ 2\ H_2 \xrightarrow[catalyst]{Transition\ metal} HOCH_2(CH_2)_4CH_2OH$$
From (b)

(d) (Z)-3-hexen-1,6-diol

$$HOCH_2CH_2C\equiv CCH_2CH_2OH \xrightarrow[\substack{Pd/CaCO_3 \\ (Lindlar\ catalyst)}]{H_2}$$
From (b)

with product drawn as:

$$\underset{HOCH_2CH_2}{\overset{H}{\diagup}}C=C\underset{CH_2CH_2OH}{\overset{H}{\diagdown}}$$

(e) (E)-3-Hexen-1,6-diol

$$HOCH_2CH_2C\equiv CCH_2CH_2OH \xrightarrow[NH_3\ (l)]{Li\ or\ Na}$$
From (b)

with product drawn as:

$$\underset{HOCH_2CH_2}{\overset{H}{\diagup}}C=C\underset{H}{\overset{CH_2CH_2OH}{\diagdown}}$$

(f) Hexanedial

$$HOCH_2(CH_2)_4CH_2OH \xrightarrow{PCC} \overset{O}{\overset{\|}{H C}}(CH_2)_4\overset{O}{\overset{\|}{C H}}$$
From (c)

Problem 11.41 Following are the steps in the industrial synthesis of glycerin. Provide structures for all intermediates and describe the type of mechanism by which each is formed.

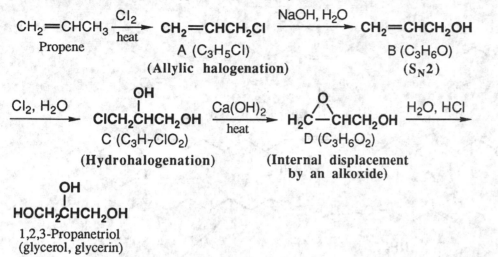

$$CH_2=CHCH_3 \xrightarrow[\text{heat}]{Cl_2} CH_2=CHCH_2Cl \xrightarrow{NaOH, H_2O} CH_2=CHCH_2OH$$

Propene

A (C₃H₅Cl)

(Allylic halogenation)

B (C₃H₆O)

(Sₙ2)

$$\xrightarrow{Cl_2, H_2O} ClCH_2\overset{\underset{\displaystyle OH}{|}}{CH}CH_2OH \xrightarrow[\text{heat}]{Ca(OH)_2} H_2C\overset{O}{\overset{\diagdown\diagup}{-}}CHCH_2OH \xrightarrow{H_2O, HCl}$$

C (C₃H₇ClO₂)

(Hydrohalogenation)

D (C₃H₆O₂)

(Internal displacement
by an alkoxide)

$$HOCH_2\overset{\underset{\displaystyle OH}{|}}{CH}CH_2OH$$

1,2,3-Propanetriol
(glycerol, glycerin)

(Acid-catalyzed
epoxide hydrolysis)

Problem 11.42 The following is a retrosynthetic scheme for the preparation of *trans*-2-allylcyclohexanol. Show reagents to bring about the synthesis of this compound from cyclohexane.

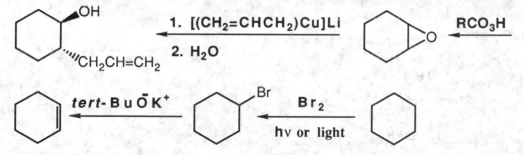

Problem 11.43 Gossyplure, the sex pheromone of the pink bollworm, is the acetic ester of 7,11-hexadecadien-1-ol. The active pheromone has the Z configuration at the C7-C8 double bond and is a mixture of E,Z isomers at the C11-C12 double bond. Shown here is the Z,E isomer.

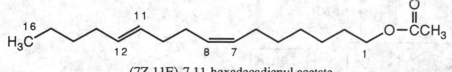

(7Z,11E)-7,11-hexadecadienyl acetate

Following is a retrosynthetic analysis for (7Z,11E)-7,11-hexadecadien-1-ol, which then led to a successful synthesis of gossyplure.

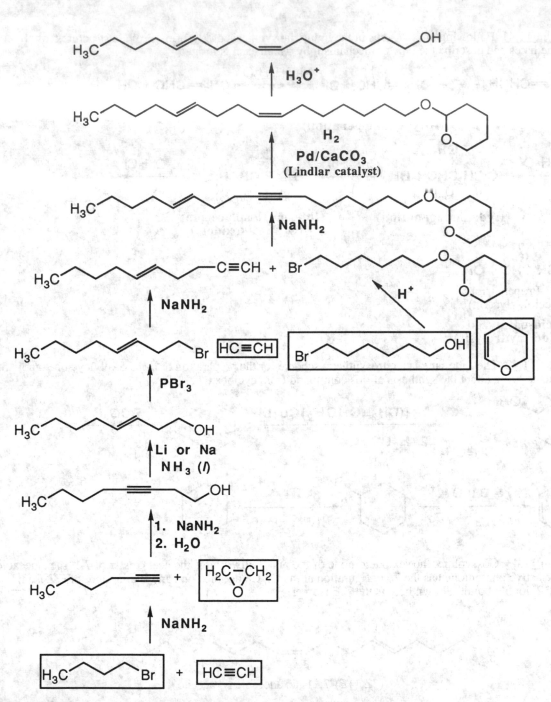

(a) Suggest reagents and experimental conditions for each step in this synthesis.
(b) Why is it necessary to protect the -OH group of 6-bromohexanol?

A hydroxyl group is considerably more acidic than an alkyne. Thus, left unprotected, the hydroxyl group would be deprotonated by NaNH₂, creating anucleophilic alkoxide anion that would displace bromine to give an ether.

(c) How might you modify this synthesis to prepare the 7Z,11Z isomer of gossyplure?

Lindlar's catalyst would be used in place of the Li or Na in NH₃ for the reduction of the 3-octyn-2-ol.

CHAPTER 12: MASS SPECTROMETRY

12.0 OVERVIEW
- Today, instrumental methods are used to determine molecular structure. **Mass spectrometry** is an important instrumental method used routinely by organic chemists. ✳
- In a **mass spectrometer**, electrons are removed from atoms or molecules to produce a stream of positive ions, which is accelerated in an electric field then passed through a magnetic field. The extent of curvature of the path of individual ions depends on the ratio of **mass-to-charge** (m/z) of the ion. Differences in curvature are measured. These measurements are then used to determine the mass of the ions, thereby providing important structural information. ✳

12.1 A MASS SPECTROMETER
- In a **mass spectrometer**:
  - The neutral atoms or molecules are converted to a beam of positively-charged ions in the **ionization chamber**. The atoms or molecules are bombarded with a stream of **high energy electrons**, and the resulting collisions result in loss of electrons from the sample atoms or molecules to produce the positive ions.

    The species formed by removal of a single electron from a molecule are called **molecular ions**. These species are actually **radical cations**, since they contain both an odd number of electrons and a positive charge.
  - Once molecular ions are formed, they are transformed into a rapidly traveling ion beam by a positively-charged **repeller plate** and negatively-charged **accelerator plates**.
  - The ion beam is focused by focusing slits and passed into the **mass analyzer**. The mass analyzer has a magnetic field, so the accelerated ions are deflected along a circular path. Ions with larger mass are deflected less than those with smaller mass. By varying the accelerating voltage or the strength of the magnetic field, ions can be focused on a detector where the ion current is detected and counted.
- A **mass spectrum** is a scan of relative ion abundance versus mass-to-charge ratio, and is normally plotted as a bar graph. The tallest peak in the mass spectrum is called the **base peak**.

12.2 FEATURES OF A MASS SPECTRUM
- **Resolution** is an important operating characteristic of a mass spectrometer that refers to how well it separates ions of differing mass. **Low resolution** mass spectrometers are capable of separating ions differing by **nominal mass** (ions that differ by one or more mass unit), and **high resolution** mass spectrometers are capable of separating ions differing in mass by as little as 0.0001 **atomic mass units (amu)**.
- Most common elements found in organic molecules (H, C, N, O, S, Cl, and Br) are found as a **mixture of isotopes**, which occur in characteristic ratios. Of particular interest are chlorine and bromine because they each have two predominant isotopes. Chlorine is found in nature as ^{35}Cl and ^{37}Cl in a 75.77 to 24.23 ratio, and bromine is found as ^{79}Br and ^{81}Br in a 50.69 to 49.31 ratio. Common elements that occur in nature only as single isotopes include F, P, and I.
- The electrons used in the ionization chamber are of such high energy that additional reactions can occur with the molecular ion that is initially formed. A common type of reaction is **fragmentation**, that is breaking up of the molecular ion into smaller fragments. Some of these fragments can break up further into even smaller pieces.✳
  - Cation radicals usually break into a cation fragment and a radical fragment. Only the cation is observed in the mass spectrum, because a charge is required for the fragment to be accelerated into the mass analyzer. Uncharged radical fragments are not detected in the mass spectrum. *[This would be a good time to review the definitions of a cation, a radical, and a cation radical.]*
  - When it comes to ion fragmentation, more stable cations are formed in preference to less stable cations. Thus, the probability of fragmentation to form a new carbocation increases in the order $CH_3 < 1° < 2° < 3° =$ allylic.

12.3 INTERPRETING MASS SPECTRA
- When a chemist interprets a mass spectrum, identification of the compounds comes primarily from the mass of the molecular ion, and on the appearance of (M+1) and (M+2) peaks as evidence for the presence of heteroatoms such as Br and Cl. Important structural information can be obtained by analysis of the fragmentation patterns in mass spectra, even though they can be quite complicated (many peaks). Chemists rarely attempt a total analysis of a complicated mass spectrum, only certain key peaks are analyzed. This is in contrast to NMR spectra described in the next chapter, in which every peak is usually scrutinized. ✳
- Different types of molecules give characteristic fragmentation patterns, and these will be discussed in turn.

- **Alkanes**
  - Straight chain hydrocarbons fragment by breaking carbon-carbon bonds to form a homologous series of cations differing by 14 amu (a CH_2 group). Fragmentation tends to occur in the middle of unbranched chains and the most stable carbocation tends to be formed in preference to the more stable radical. Recall that the initially formed molecular ion fragments to give a cation and a radical, but only the cation is detected in the mass spectrum.
  - Branched hydrocarbons fragment to form secondary and tertiary cations.
  - Cycloalkanes fragment by losing side chains and also ethylene via fragmentation of the ring.
- **Alkenes**
  - Alkenes show a strong molecular ion peak due to removal of one electron from the pi bond.
  - Alkenes also fragment readily to form stable allylic cations.
  - Cyclohexenes fragment in a characteristic pattern that is a reverse Diels-Alder reaction.
- **Alkynes**
  - Alkynes show a strong molecular ion peak due to removal of one electron from a pi bond.
  - In general, alkynes fragment in ways that are analogous to alkenes.
  - The propargyl cation or substituted propargyl cations are usually prominent peaks in the mass spectra of alkynes.
- **Alcohols**
  - The molecular ion peak for primary and secondary alcohols is usually small, and usually not detectable for tertiary alcohols.
  - A common fragmentation pattern for alcohols is the loss of water to give a peak corresponding to the molecular ion minus 18.
  - Another common pattern is loss of an alkyl group from the carbon bearing the -OH group to from an oxonium ion and an alkyl radical. The oxonium ion is resonance stabilized.
- **Ethers**
  - Ethers can fragment via cleavage of the carbon-carbon bond adjacent to the ether oxygen atom to give an oxonium ion.
  - The oxonium ion can fragment further by migration of a hydrogen atom beta to the oxygen, resulting in elimination of an alkene.

CHAPTER 12
Solutions to the Problems

Problem 12.1 Calculate the nominal mass of each ion. Unless otherwise indicated, use the mass of the most abundant isotope of each element.

(a) $\left[CH_3Br\right]^{+}$ (b) $\left[CH_3{}^{81}Br\right]^{+}$ (c) $\left[{}^{13}CH_3Br\right]^{+}$

    94   **96**   **95**

Problem 12.2 Propose a structural formula for the cation of m/z 41 observed in the mass spectrum of methylcyclopentane.

The cation of m/z 41 must have a molecular formula of C_3H_5. This corresponds to the stable allyl cation:

$$CH_2{=}CH{-}CH_2^{+}$$

Problem 12.3 The low-resolution mass spectrum of 2-pentanol shows 15 peaks. Account for the formation of the peaks at m/z 73, 70, 55, 45, 43, and, 41. (*Hint:* Consider (1) the loss of water to form an alkene and then fragmentations that the resulting alkene might undergo, and (2) the fragmentation of bonds to the carbon bearing the -OH group.

2-Pentanol has a molecular formula of $C_5H_{12}O$ and thus a nominal mass of 88.

$$\overset{\displaystyle OH}{\overset{\displaystyle |}{CH_3\text{-}CH_2\text{-}CH_2{-}CH{-}CH_3}}$$

2-Pentanol

Loss of a methyl radical would leave a fragment of mass 73:

$$\overset{\displaystyle {}^{+}OH}{\overset{\displaystyle \|}{CH_3\text{-}CH_2\text{-}CH_2{-}CH}}$$

Loss of water results in a fragment of mass 70:

$$\left[\, CH_3\text{-}CH_2\text{-}CH_2{-}CH{=}CH_2 \,\right]^{+\cdot} \qquad \left[\, CH_3\text{-}CH_2\text{-}CH{=}CH{-}CH_3 \,\right]^{+\cdot}$$

The above alkene cation radicals could then lose a methyl radical to give the following fragments of mass 55:

$$\left[\, CH_2\text{-}CH_2{-}CH{=}CH_2 \,\right]^{+} \left[\, CH_2\text{-}CH{=}CH{-}CH_3 \,\right]^{+}$$

An alkyl radical could break off from the alcohol to give a fragment of mass 45:

$$\overset{\displaystyle {}^{+}OH}{\overset{\displaystyle \|}{CH{-}CH_3}}$$

The original alcohol cation radical could have fragmented in such as way as to generate the following cation with mass 43:

$$CH_3\text{-}CH_2\text{-}CH_2^{+}$$

The allyl cation has a mass of 41:

$$CH_2=CH—CH_2^+$$

<u>Problem 12.4</u> Draw acceptable Lewis structures for the molecular ion (radical cation) formed from the following molecules when each is bombarded by high-energy electrons in a mass spectrometer.

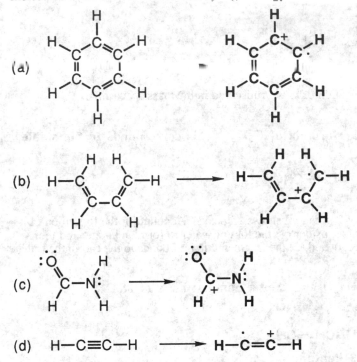

(a)

(b)

(c)

(d) H—C≡C—H ⟶ H—Ċ=Ċ⁺—H

<u>Problem 12.5</u> Some organic molecules can add a single electron to form unstable species called radical anions. A radical anion possesses both a negative charge and an unpaired electron. For example:

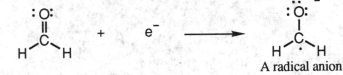

A radical anion

Draw an acceptable Lewis structure for a radical anion formed from the following molecules:

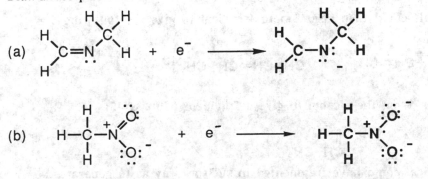

Problem 12.6 The molecular ion for compounds containing only C, H, and O is always at an even mass-to-charge value. Why is this so? What can you say about mass-to-charge ratio of ions that arise from fragmentation of one bond in the molecular ion? From fragmentation of two bonds in the molecular ion?

Stable organic molecules composed of C, H, and O with filled valences will always have an even number of hydrogen atoms because carbon and oxygen make an even number of bonds (4 and 2, respectively). Try making some molecular formulas for yourself to verify this. This fact, combined with the even-numbered atomic weights of carbon and oxygen, gaurantee an even mass-to-charge ratio of molecular ions. If one bond is broken in the molecular ion, then a cation and a radical are formed. The cation is the only species observed in the mass spectrum, and this will have an odd mass-to-charge ratio. If the molecular ion fragments so that two bonds are broken, then a new radical cation is formed, and the new radical cation will have an even mass-to-charge ratio.

Problem 12.7 For which compounds containing a heteroatom (an atom other than carbon or hydrogen) does a molecular ion have an even-numbered mass and for which does it have an odd-numbered mass?
(a) A chloroalkane of molecular formula $C_nH_{2n+1}Cl$

The molecular ion has an even-numbered mass. The C and H atoms will add up to an odd-numbered mass, since there will be an odd number of hydrogen atoms. The chlorine provides an additional odd-numbered mass as either of the two most abundant isotopes (35 and 37, respectively).

(b) A bromoalkane of molecular formula $C_nH_{2n+1}Br$

The molecular ion has an even-numbered mass. Again, the C and H atoms add up to an odd-numbered mass, and bromine provides an additional odd-numbered mass as either of the two most abundant isotopes (79 and 81, respectively).

(c) An alcohol of molecular formula $C_nH_{2n+1}OH$

The molecular ion has an even-numbered mass. There is an even number of hydrogen atoms, so the C and H atoms add up to an even-numbered mass. The most abundant isotope of oxygen (16) is also even, so the entire molecular ion has an even-numbered mass.

(d) A primary amine of molecular formula $C_nH_{2n+1}NH_2$.

The molecular ion has an odd-numbered mass. These compounds have an odd number of hydrogens, so the C and H atoms add up to an odd-numbered mass. The nitrogen contributes an even-numbered mass (14) so the molecular ion will have an odd-numbered mass.

(e) A thiol of molecular formula $C_nH_{2n+1}SH$.

The molecular ion has an even-numbered mass. These compounds have an even number of hydrogens, so the C and H atoms add up to an even-numbered mass. The sulfur contributes an even-numbered mass (32 or 34) so the entire molecular ion has an even-numbered mass.

Problem 12.8 The so-called nitrogen rule states that if a compound has an odd number of nitrogen atoms, the value of m/z for its molecular ion will be an odd number. Why is this so?

Nitrogen atoms have an even-numbered mass, but one lone pair of electrons in the neutral state. Thus, they make an odd number of bonds, namely three, to other atoms. As a result, compounds with with an odd number of nitrogens in the neutral state will have an odd-numbered m/z ratio for the molecular ion.

Problem 12.9 Both $C_6H_{10}O$ and C_7H_{14} have the same nominal mass, namely 98. Show how these compounds can be distinguished by the m/z ratio of their molecular ions in high resolution mass spectrometry.

The values in Table 12.1 can be used to calculate precise masses for each molecular ion.
$C_6H_{10}O$: 6(12.0000) + 10(1.00783) + 15.9949 = <u>98.0732</u>
C_7H_{14}: 7(12.0000) + 14(1.00783) = <u>98.1096</u>
A high resolution mass spectrum will distinguish these compounds based on this small difference in the expected m/z ratio of the molecular ions.

Problem 12.10 Show how the compounds of molecular formula C_6H_9N and C_5H_5NO can be distinguished by the m/z ratio of their molecular ions in high resolution mass spectrometry.

The values in Table 12.1 can be used to calculate precise masses for each molecular ion.
C_6H_9N: 6(12.0000) + 9(1.00783) + 14.0031 = <u>95.0736</u>
C_5H_5NO: 5(12.0000) + 5(1.00783) + 14.0031 + 15.9949= <u>95.0372</u>
A high resolution mass spectrum will distinguish these compounds based on this small difference in the expected m/z ratio of the molecular ions.

Problem 12.11 What rule would you expect for the m/z values of fragment ions resulting from the cleavage of one bond in a compound with an odd number of nitrogen atoms?

As stated in Problem 12.8, molecules having an odd number of nitrogen atoms have an odd-numbered m/z ratio for the molecular ion. Cleavage of one bond of a molecular ion would generate a cation and a radical, only the cation of which would be observed in the mass spectrum. If the cation fragment retained an odd number of nitrogen atoms, then the fragment would have an even m/z ratio. If the cation fragment contained an even number of nitrogen atoms (or zero), then the fragment would have an odd m/z ratio.

Problem 12.12 In a natural sample of ethane, what is the probability that:
(a) One carbon in an ethane molecule is ^{13}C?

The relative abundance of ^{13}C listed in Table 12.1 is 1.11 atoms of ^{13}C for every 100 atoms of ^{12}C. This corresponds to a probability that any one atom is a ^{13}C atom of 1.11/(100 + 1.11) = 0.0110. There are two ways to have this situation since either of the carbon atoms could be the ^{13}C so the final answer is 2(0.0110) = <u>0.0220</u>. Converting to percent by multiplying by 100 means that there is a <u>2.20%</u> chance that one of the carbon atoms in ethane is ^{13}C.

(b) Both carbons in an ethane molecule are ^{13}C?

The probability that both of the two carbon atoms in ethane is ^{13}C is $(0.0110)^2$ = <u>1.21 x 10^{-4}</u>. Converting to percent by multiplying by 100 means that there is a <u>0.012%</u> chance that both of the carbon atoms in ethane is ^{13}C.

(c) Two hydrogens in an ethane molecule are replaced by deuterium atoms?

The relative abundance of ^2H listed in Table 12.1 is 0.016 atoms of ^2H for every 100 atoms of ^1H. This corresponds to a probability of 0.016/(100 + 0.016) = 1.6 x 10^{-4} that any given H atom is a ^2H atom. The probability for having two ^2H $(1.6 \times 10^{-4})^2$ = 2.56 x 10^{-8}. Converting to percent by multiplying by 100 means that there is a <u>2.56 x 10^{-6}%</u> chance that two of the hydrogen atoms in ethane are ^2H.

Problem 12.13 The molecular ions of both $C_5H_{10}S$ and $C_6H_{14}O$ appear at m/z 102 in low-resolution mass spectrometry. Show how determination of the correct molecular formula can be made from the appearance and relative intensity of the (M + 2) peak of each compound.

As shown in Table 12.1, ^{16}O occurs in greater than 99.7% abundance, so no (M + 2) peak will be observable for the case of $C_6H_{14}O$. On the other hand, sulfur has one isotope, ^{34}S, that is 95% abundant and another isotope 2 amu higher, namely ^{32}S, that has an abundance of 4.2%.

Thus, if the low-resolution mass spectrum has an (M + 2) peak that is 4.2% the height of the molecular ion peak, the compound must be $C_5H_{10}S$.

Problem 12.14 In Section 12.3, we saw several examples of fragmentation of molecular ions to give resonance-stabilized cations. Make a list of these resonance-stabilized cations and write important contributing structures of each. Estimate the relative importance of the contributing structures in each set.

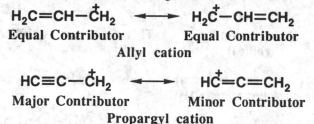

The major contributor of the propargyl cation has the positive charge on the less-electronegative sp^2 hybridized carbon atom.

The more stable contributor of the oxonium ion is predicted to be the one that has the positive charge on oxygen due to having filled valence shells on both carbon and oxygen.

Problem 12.15 Carboxylic acids often give strong fragment ions at m/z (M - 17). What is the likely structure of these cations, and how might they be formed? Show by drawing contributing structures that these cations are stabilized by resonance.

Loss of 17 results from losing the -OH group of the carboxylic acid. This produces an acylium ion:

These ions are probably derived from fragmentation of the molecular ion as follows:

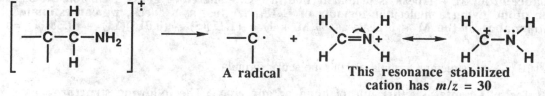

Problem 12.16 For primary amines with no branching on the carbon bearing the nitrogen, the base peak occurs at m/z 30. What cation does this peak represent and how is it formed? Show by drawing contributing structures that this cation is stabilized by resonance.

<u>Problem 12.17</u> The base peak in the mass spectrum of propanone (acetone) occurs at *m/z* 43. What cation does this peak represent?

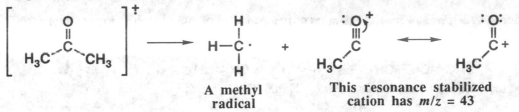

A methyl radical   This resonance stabilized cation has *m/z* = 43

<u>Problem 12.18</u> A characteristic peak in the mass spectrum of most aldehydes occurs at *m/z* 29. What cation does this peak represent? (No, it is not an ethyl cation, $CH_3CH_2^+$.)

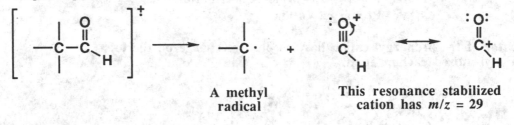

A methyl radical   This resonance stabilized cation has *m/z* = 29

Interpretation of Mass Spectra
<u>Problem 12.19</u> Predict the relative intensities of the M and M + 2 peaks for

The following are based on the values in Table 12.1.

(a) CH_3CH_2Cl (b) CH_3CH_2Br

   100 : 32.5   **100 : 98**

(c) $BrCH_2CH_2Br$

We will assume that the M + 2 peak is due to the ^{81}Br isotope. To get a mass of M + 2 there must be only one ^{81}Br present. Using the formula given in the text: %(M + 2) = (98)(2) = 196. Thus 100 : 196 are the expected relative intensities.

(d) CH_3CH_2SH

   100 : 4.40

<u>Problem 12.20</u> The mass spectrum of Compound B shows the molecular ion at *m/z* 85, an M + 1 peak at *m/z* 86 of approximately 6% abundance relative to M, and an M + 2 peak at *m/z* 87 of less than 0.1% abundance relative to M.
(a) Propose a molecular formula for Compound B. *Hint:* Remember the nitrogen rule.

The odd mass indicates an odd number of nitrogen atoms. Assuming only C, H, and N atoms, the 6% abundance of an M + 1 peak is primarily due to ^{13}C. Therefore guess 5 carbon atoms and 1 nitrogen atom giving a molecular formula of $C_5H_{11}N$. Just as a check, we can estimate the relative abundance of the M + 1 peak as %(M + 1) = (1.11)(5) + (0.016)(11) + (0.38)(1) = 6.1%.

(b) Draw at least 10 possible structural formulas for this molecular formula.

This molecular formula contains either one pi bond or one ring. The following structures are representative examples out of the large number of possible stable molecules that are consistent with this formula.

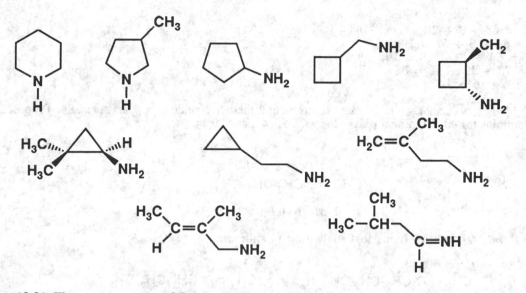

Problem 12.21 The mass spectrum of Compound E, a colorless liquid, shows these peaks in its mass spectrum. The base peak is at m/z 43. Other peaks are given in relative abundance to the base peak. Determine the molecular formula of Compound E and propose a structural formula for it. *Hint:* Calculate the ratio of the M to M + 2 peak.

m/z	Relative Abundance
43	100 (base)
78	23.6 (M)
79	1.00
80	7.55
81	0.25

The ratio of the M to M + 2 peaks is 23.6/7.55 = 3.13 to 1. This is approximately the same ratio as seen with ^{35}Cl to ^{37}Cl, so there must be a chlorine atom in the molecule. The base peak at 43 corresponds to either a propyl cation or isopropyl cation. Therefore, the molecular formula for compound E must be C_3H_7Cl. The two structures consistent with this formula are:

$$CH_3CH_2CH_2Cl$$

$$\overset{\displaystyle Cl}{\underset{\displaystyle}{CH_3CHCH_3}}$$

Problem 12.22 Write molecular formulas for the five possible molecular ions of m/z 88 containing the elements C, H, N, and O. (*Note:* Because the value of the mass of this set of molecular ions is an even number, members of the set must have either no nitrogen atoms or an even number of nitrogen atoms.)

$C_3H_4O_3$ = 36 + 4 + 48 = 88
$C_3H_8N_2O$ = 36 + 8 + 28 + 16 = 88
$C_4H_8O_2$ = 48 + 8 + 32 = 88
$C_4H_{12}N_2$ = 48 + 12 + 28 = 88
$C_5H_{12}O$ = 60 + 12 + 16 = 88

Problem 12.23 Write molecular formulas for the five possible molecular ions of m/z 100 containing only the elements C, H, N, and O.

Because of the nitrogen rule, there must be an even number of nitrogen atoms or no nitrogen atoms in the molecular formulas.

$C_4H_8N_2O = 48 + 8 + 28 + 16 = 100$
$C_5H_8O_2 = 60 + 8 + 32 = 100$
$C_5H_{12}N_2 = 60 + 12 + 28 = 100$
$C_6H_{12}O = 72 + 12 + 16 = 88$
$C_7H_{16} = 84 + 16 = 100$

Problem 12.24 The molecular ion in the mass spectrum of 2-methyl-1-pentene appears at m/z 84. Propose structural formulas for the prominent peaks at m/z 69, 55, 41, and 29.

$$CH_2{=}\overset{\overset{\displaystyle CH_3}{|}}{C}{-}CH_2{-}CH_2{-}CH_3$$

2-Methyl-1-pentene

The peak at m/z 69 results from loss of a methyl radical:

$$CH_2{=}\overset{+}{C}{-}CH_2{-}CH_2{-}CH_3 \quad \text{or} \quad CH_2{=}\overset{\overset{\displaystyle CH_3}{|}}{C}{-}CH_2{-}CH_2{-}CH_2{}^{+}$$

The peak at m/z 55 results from loss of an ethyl group to generate an allylic cation:

$$CH_2{=}\overset{\overset{\displaystyle CH_3}{|}}{C}{-}\underset{+}{CH_2}$$

The peak at m/z 41 corresponds to the allyl cation:

$$CH_2{=}CH{-}\underset{+}{CH_2}$$

The peak at m/z 29 corresponds to the ethyl cation:

$$CH_3{-}\underset{+}{CH_2}$$

Problem 12.25 Following is the mass spectrum of 1,2-dichloroethane.
(a) Account for the appearance of an (M + 2) peak with approximately two-thirds the intensity of the molecular ion peak.

The relative abundance of ^{37}Cl is 32.5 (Table 12.1) so it makes sense that the M + 2 peak observed in 1,2-dichloroethane is 2(32.5) = 65%.

(b) Predict the intensity of the (M + 4) peak.

The probability of having a ^{37}Cl at both Cl atoms is $(32.5/100)^2$ = 0.105, which corresponds to 10.5% of the molecular ion peak.

(c) Propose structural formulas for the cations of m/z 64, 63, 62, 51, 49, 27, and 26.

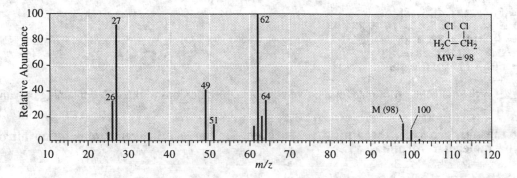

$$CH_2Cl\text{-}CH_2Cl$$
1,2-Dichloroethane

The peaks at m/z 64, 63, and 62 correspond to the following structures:

$$\left[\,^{37}Cl—CH{=}CH_2 \right]^{\overset{+}{\cdot}} \qquad \left[Cl—^{13}CH{=}CH_2 \right]^{\overset{+}{\cdot}} \text{ or } \left[Cl—CH{=}^{13}CH_2 \right]^{\overset{+}{\cdot}}$$
$$m/z = 64 \qquad\qquad\qquad\qquad m/z = 63$$

$$\left[\,^{35}Cl—CH{=}CH_2 \right]^{\overset{+}{\cdot}}$$
$$m/z = 62$$

The peaks at m/z 51 and 49 correspond to the following structures:

$$^{37}Cl—CH_2^{+} \qquad\qquad ^{35}Cl—CH_2^{+}$$
$$m/z = 51 \qquad\qquad m/z = 49$$

The peaks at m/z 27 and 26 correspond to the following structures:

$$^{+}CH{=}CH_2 \qquad \left[HC{\equiv}CH \right]^{\overset{+}{\cdot}}$$
$$m/z = 27 \qquad m/z = 26$$

<u>Problem 12.26</u> Following is the mass spectrum of 1-bromobutane.
(a) Account for the appearance of the (M + 2) peak of approximately 95% of the intensity of the molecular ion peak.

As detailed in Table 12.1, the most abundant isotope of bromine has a mass of 79 amu. Bromine has another isotope with a mass of 81 amu that has a natural abundance of 49.31%. The (M+ 2) peak results from the presence of a ^{81}Br atom in the 1-bromobutane molecule.

(b) Propose structural formulas for the cations of m/z 57, 41, and 29.

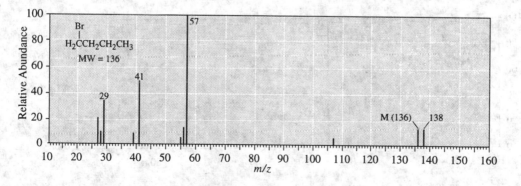

The peak at m/z 57 results from loss of a bromine atom:

$$\underset{+}{CH_2\text{-}CH_2\text{-}CH_2\text{-}CH_3}$$

The peak at m/z 41 corresponds to the allyl cation:

$$\underset{+}{CH_2{=}CH\text{-}CH_2}$$

The peak at *m/z* **29** corresponds to the ethyl cation:

$$CH_3\text{-}CH_2$$
$$+$$

Problem 12.27 Following is the mass spectrum of bromocyclopentane. The molecular ion *m/z* 148 is of such low intensity that it does not appear in this spectrum. Assign structural formulas for the cations of *m/z* 69 and 41.

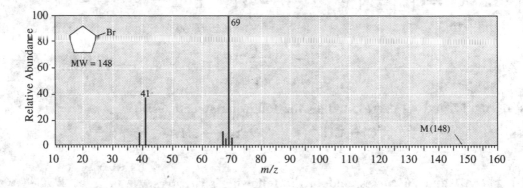

The peak at *m/z* **69** corresponds to the loss of the bromine atom to give the cyclopentyl cation:

The peak at *m/z* **41** corresponds to the allyl cation:

$$CH_2\!=\!CH\text{-}CH_2$$
$$+$$

Problem 12.28 Following is the mass spectrum of 3-methyl-2-butanol. The molecular ion *m/z* 88 does not appear in this spectrum. Propose structural formulas for the cations of *m/z* 45, 43, and 41.

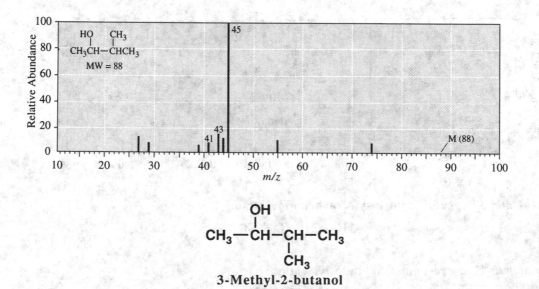

$$OH$$
$$CH_3\text{—}CH\text{—}CH\text{—}CH_3$$
$$CH_3$$

3-Methyl-2-butanol

The peak at *m/z* 45 corresponds to the following resonance stabilized cation:

$$\overset{\displaystyle +OH}{\underset{}{CH_3-\overset{\|}{CH}}}$$

The peak at *m/z* 43 corresponds to the isopropyl cation:

$$\overset{+}{CH}-CH_3 \atop \underset{}{CH_3}$$

The peak at *m/z* 41 corresponds to the allyl cation:

$$CH_2=CH\text{-}CH_2 \atop +$$

Problem 12.29 The following is the mass spectrum of a compound A, C_3H_8O. Compound A is infinitely soluble in water, undergoes reaction with sodium metal with the evolution of a gas, and undergoes reaction with thionyl chloride to give a water-insoluble chloroalkane. Propose a structural formula for compound A, and write equations for each of its reactions.

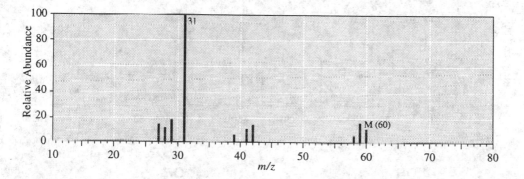

Compound A is 1-propanol.

$$CH_3\text{-}CH_2\text{-}CH_2OH$$

Compound A

In the mass spectrum, the peak at *m/z* 31 could only have come from 1-propanol due to the following fragmentation.

$$CH_3\text{-}CH_2-\overset{\displaystyle OH}{\underset{}{CH_2}} \longrightarrow CH_3CH_2 \cdot \; + \; \overset{\displaystyle +OH}{\underset{}{CH_2}}$$
$$m/z \; 31$$

The reaction with sodium liberates molecular hydrogen:

$$2 \; CH_3\text{-}CH_2\text{-}CH_2OH \; + \; 2Na \longrightarrow 2 \; CH_3\text{-}CH_2\text{-}O^- \; Na^+ \; + \; H_2$$

The reaction with thionyl chloride produces 1-chloropropane, SO_2, and H-Cl.

$$CH_3\text{-}CH_2\text{-}CH_2OH \; + \; SOCl_2 \longrightarrow CH_3\text{-}CH_2\text{-}CH_2Cl \; + \; SO_2 \; + \; H\text{-}Cl$$

Problem 12.30 Following are mass spectra for the constitutional isomers 2-pentanol and 2-methyl-2-butanol. Assign each isomer its correct spectrum.

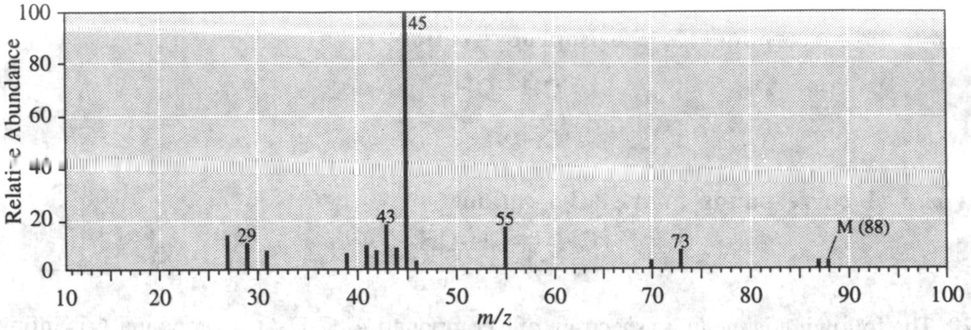

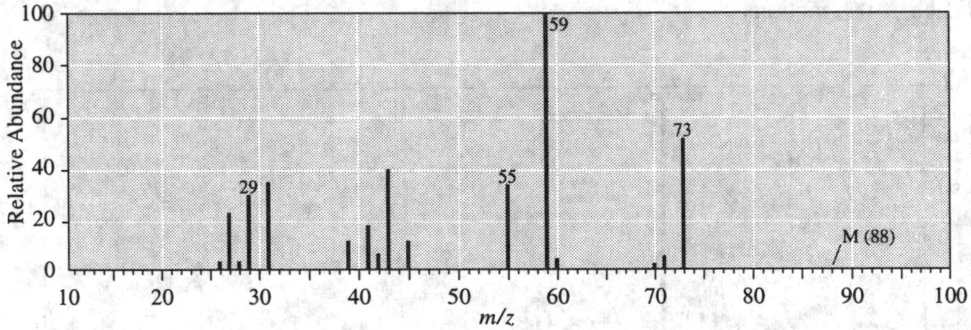

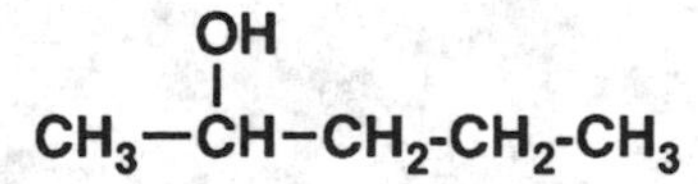

$$
\begin{array}{cc}
\text{OH} & \text{OH} \\
| & | \\
CH_3-CH-CH_2-CH_2-CH_3 & CH_3-C-CH_2-CH_3 \\
& | \\
& CH_3 \\
\text{2-Pentanol} & \text{2-Methyl-2-butanol}
\end{array}
$$

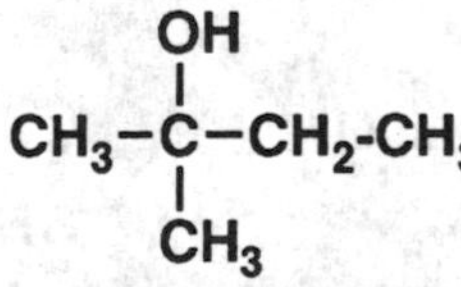

The main difference in the above spectra is the base peak at 45 in the first spectrum that is only about 10% of the base peak in the second spectrum. This peak corresponds to the following resonance stabilized cation:

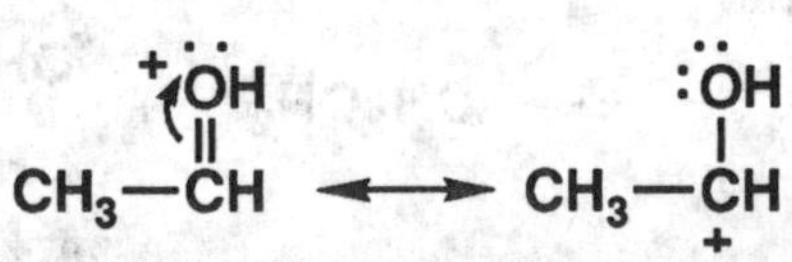

This species can be readily produced by loss of the propyl group from 2-pentanol. It is unlikely that 2-methyl-2-butanol could produce such a cation, thus the upper mass spectrum is that of 2-pentanol and the lower mass spectrum is that of 2-methyl-2-butanol. Also, the base peak in the second spectrum has a m/z of 59 that corresponds to the following fragmentation.

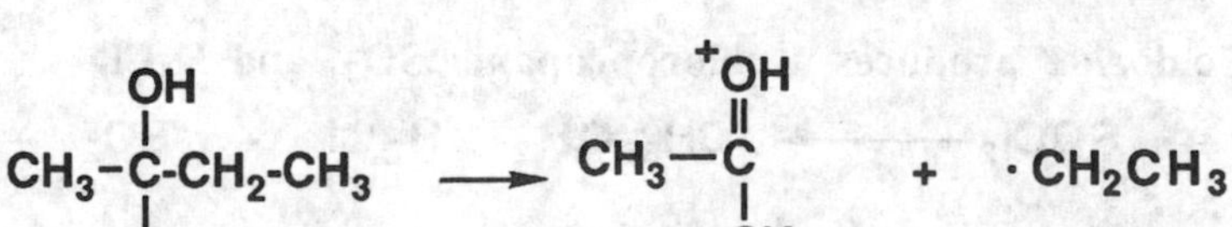

Problem 12.31 Water-^{18}O, in which enrichment is 10 atom % ^{18}O, is available commercially. Also available, at a considerably higher price, is water-^{18}O in which enrichment is 97 atom % ^{18}O. The oxygen-18 label can be transferred from water to acetone by establishing the following acid-catalyzed equilibrium. (We discuss the chemistry of this equilibration in Chapter 15).

$$CH_3-\overset{O}{\overset{||}{C}}-CH_3 \;+\; H_2^{18}O \;\overset{H^+}{\rightleftharpoons}\; CH_3-\overset{^{18}O}{\overset{||}{C}}-CH_3 \;+\; H_2O$$

Suppose 5.00 g of water-^{18}O, 97 atom % ^{18}O, is mixed with 11.6 g of acetone-^{16}O in the presence of an acid catalyst until equilibrium is established. Assume for the purposes of this problem that the value of K_{eq}, the equilibrium constant for this reaction, is 1.00.
(a) Calculate the percent atom enrichment in acetone recovered after equilibration.

$$\frac{5.00 \text{ gm}}{18 \text{ gm/mol}} = 0.28 \text{ mol } H_2O \qquad \frac{11.6 \text{ gm}}{58 \text{ gm/mol}} = 0.20 \text{ mol acetone}$$

After equilibrium has been established, the following amounts of material will be present, written in terms of the amount of ^{18}O-containing acetone (as "X").

$$\left[H_2^{18}O\right] = (0.28)(0.97) - X \qquad \left[H_2O\right] = (0.28)(0.03) + X$$

$$\left[CH_3-\overset{O}{\overset{||}{C}}-CH_3\right] = 0.2 - X \qquad \left[CH_3-\overset{^{18}O}{\overset{||}{C}}-CH_3\right] = X$$

$$K_{eq} = 1.00 = \frac{\left[CH_3-\overset{^{18}O}{\overset{||}{C}}-CH_3\right]\left[H_2O\right]}{\left[CH_3-\overset{O}{\overset{||}{C}}-CH_3\right]\left[H_2^{18}O\right]}$$

so at equilibrium
$$\left[CH_3-\overset{O}{\overset{||}{C}}-CH_3\right]\left[H_2^{18}O\right] = \left[CH_3-\overset{^{18}O}{\overset{||}{C}}-CH_3\right]\left[H_2O\right]$$

Substituting the values into the equation gives:
$$(0.20 - X)(0.27 - X) = (X)(0.01 + X)$$
$$0.054 - 0.47 X + X^2 = 0.01 X + X^2$$
The X^2 terms cancel, so we are left with:
$$0.054 - 0.47 X = 0.08 X$$
$$0.054 = 0.48 X$$
$$0.113 \text{ mol} = X$$
Thus, the atom percent (%) enrichment is (0.113 mol) / (0.20 mol)
$$\boxed{= 57 \text{ atom } \%}$$

(b) Show how the ratio of (M + 2)/M can be used to verify the atom percent enrichment.

A 57 atom percent enrichment means that the ratio of $(M^{+}+ 2)/M^{+}$ will be (0.57 / 0.43). This ratio measured in the mass spectrum would thus verify the atom percent enrichment.

<u>Problem 12.32</u> Because of the sensitivity of mass spectrometry, it is often used to detect the presence of drugs in blood, urine, or other biological fluids.   Tetrahydrocannabinol (nominal mass 314), a component of marijuana, exhibits two strong fragment ions at *m/z* 246 and 231 (the base peak). What is the likely structure of each ion?

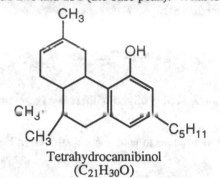

Tetrahydrocannibinol
($C_{21}H_{30}O$)

The peak at *m/z* 246 is probably the result of a reverse Diels-Alder reaction as described in section 12.4B. In this case, the cyclohexene ring in the top left of the structure is split from the rest of the molecule to leave the radical cation shown below. The peak at *m/z* 231 results from loss of a methyl group from the *m/z* 246 radical cation to give a tertiary cation as shown.

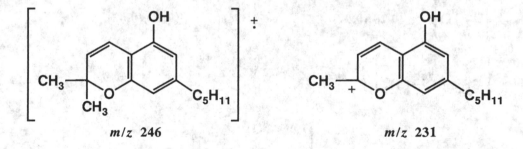

m/z 246   *m/z* 231

<u>Problem 12.33</u> Electrospray mass spectrometry is a recently developed technique for looking at large molecules with a mass spectrometer. In this technique, molecular ions, each associated with one or more H^+ ions, are prepared under mild conditions in the mass spectrometer. As an example, a protein (P) with a molecular weight of 11,812 gives clusters of the type $(P + 8H)^{8+}$, $(P + 7H)^{7+}$, and $(P + 6H)^{6+}$. At what mass-to-charge values do these three clusters appear in the mass spectrum? (See B. Ganem, Y.-T. Li and J. D. Henion, *J. Am. Chem. Soc.,* **113**, 6294 (1991)).

The key to this question is to notice that these ions have multiple charges. Since ions are recorded in a mass spectrum according to their mass divided by their total charge, the mass-to-charge values for these ions are calculated by dividing their total mass (11,812 + the number of protons) by their total charge (8, 7, and 6, respectively). The final answers are thus (11,812 + 8)/8 = <u>1477.5 *m/z*.</u>, (11,812 + 7)/7 = <u>1688.4 *m/z*.</u>, and (11,812 + 6)/6 = <u>1969.7 *m/z*.</u>

Problem 12.34 Occasionally, weak, broad peaks are observed in a mass spectrum. These often have fractional mass-to-charge values, for example 46.3 or 30.2. These are called metastable ion peaks and arise when an ion fragments after exiting the ionization chamber and while it is passing into the analyzer region of the mass spectrometer. The observed mass m^* of a metastable ion depends on the mass of the precursor ion (m_1) and the product ion (m_2) according to the following equation.

$$m^* = \frac{(m_2)^2}{m_1}$$

An ion of mass 59, for example, that fragments into an ion of mass 41 while passing into the analyzing chamber gives a metastable peak at $(41)^2/(59)$ or m/z 28.49. Metastable ions link two peaks together, as for example the peaks at m/z 59 and 41 and thus are useful in analyzing proposed fragmentation patterns. What is the m/z ratio of the metastable ion resulting from the fragmentation of $(CO_2CH_3)^+$ to $(OCH_3)^+$?

In this case, the ion with mass 45 $(CO_2CH_3)^+$ fragments to the ion with mass 29 $(OCH_3)^+$. As a result, the metastable ion shows up in the mass spectrum as $(29)^2/45$ = <u>16.3 m/z</u>.

CHAPTER 13: NUCLEAR MAGNETIC RESONANCE SPECTROSCOPY

OVERVIEW
• **Nuclear magnetic resonance (NMR)** spectroscopy was developed in the early 1960's, and is now the most important technique for the determination of molecular structure. *[NMR is a very complicated type of spectroscopy, because there are so many difficult concepts involved. The best way to learn about NMR is to read through the concepts, then work through as many spectra as possible.]* ✳

13.1 ELECTROMAGNETIC RADIATION
• **Electromagnetic radiation** can be described like a wave, in terms of its wavelength and frequency. ✳
   - **Wavelength** is the distance between any two identical points on a wave and is given the symbol λ **(lambda)** and usually expressed in **meters (m)**.
   - **Frequency** is the number of full cycles of a wave that pass a given point in a fixed period of time. Frequency is given the symbol ν **(nu)** and is usually expressed in **hertz (Hz)** and given the units **sec^{-1}**. For example, 1 Hz corresponds to one cycle per second and 1 MHz corresponds to 10^6 Hz.
   - Wavelength and frequency are related to each other, and one can be calculated from the other using the expression $\nu = c/\lambda$. Here **c** is the **speed of light**, equal to **3.00×10^8 m/sec**.
• **Electromagnetic radiation** can also be described as a particle, and the particle is called a **photon**. ✳
   - The **energy (E)** measured in **kcal** of one mol of photons is related to wavelength and frequency according to the following relationships: $E=h\nu=hc/\lambda$. Here **h** is **Planck's constant** and is equal to **9.537×10^{-14} kcal-sec-mol^{-1}**.

13.2 MOLECULAR SPECTROSCOPY
• In general, an atom or molecule can be made to **undergo a transition** from energy state E_1 to a higher energy state E_2 by irradiating it with electromagnetic radiation corresponding to the energy difference between states E_1 and E_2. During this process, the atom or molecule **absorbs the energy** of the electromagnetic radiation. When the atom or molecule **returns to the ground state** E_1, an equivalent amount of **energy is emitted**.✳
• **Molecular spectroscopy** involves measuring the frequencies of electromagnetic radiation that are absorbed or emitted by a molecule, then correlating the observed patterns with the details of molecular structure.
   - Certain regions of the electromagnetic spectrum are particularly interesting to the chemist because they represent the energies involved in transitions between important types of molecular energy levels. In particular, **radiofrequency electromagnetic radiation** corresponds to transitions between **nuclear spin** energy levels, **infrared electromagnetic radiation** corresponds to transitions between **vibrational levels of chemical bonds** and **ultraviolet-visible electromagnetic radiation** corresponds to **electronic energy** levels of pi and nonbonding electrons. *[It may prove helpful to review how these different regions of electromagnetic radiation fit into the entire spectrum shown in Table 13.1.]*
   - The nuclear spin transitions caused by absorbed radiofrequency electromagnetic radiation form the basis for **nuclear magnetic resonance (NMR) spectroscopy.** ✳

13.3 NUCLEAR SPIN STATES
• Like electrons, certain nuclei have spin. That is, they behave as if they are spinning on an axis and thus have an associated **magnetic moment**. This is only true for nuclei that have **spin quantum numbers** that are not zero. *[It might help to review the concept of spin quantum numbers in a General Chemistry text.]*
   - Both ^{1}H and ^{13}C nuclei have spin quantum numbers of 1/2, so they have two allowed spin states with values of +1/2 and -1/2. These are the nuclei that are most often studied by NMR spectroscopy, but other nuclei, such as ^{31}P and ^{15}N, can also be used because they too have spin quantum numbers that are not zero. ✳
   - Some common nuclei such as ^{12}C and ^{16}O have a spin quantum number of zero, so they are not observed by NMR.

13.4 ORIENTATION OF NUCLEAR MAGNETIC SPINS IN AN APPLIED MAGNETIC FIELD
• Ordinarily, nuclear spins are oriented in a completely random fashion. However, in an **applied magnetic field (B_o), nuclei with non-zero spin interact with the applied field.** Recall that the nuclear spin produces it own magnetic moment, so this interaction is between the magnetic moments of the nuclei with the

applied magnetic field. For ^{1}H and ^{13}C nuclei, there are two allowed orientations in the field. By convention, **nuclei that have spin +1/2 are aligned with the applied field** and are in the **lower energy state**, and **nuclei with spin -1/2 are aligned against the applied field** and are in the **higher energy state**. ✳

• The **difference** in **energy** between the **nuclear spin states increases** with **increasing applied field strength**. ✳ Nevertheless, these energy differences are small compared to other types of energy levels such as vibrational and electronic energy levels in molecules.

13.5 NUCLEAR MAGNETIC "RESONANCE"

• When the nuclei are placed in an applied magnetic field, a majority of spins are aligned with the field and are thus in the lower spin state. **Electromagnetic radiation** can **cause a transition from the lower spin state to the higher spin state**. An NMR spectrum is a plot of how much and of which energies are absorbed by a molecule as its atoms undergo these transitions from the lower to the higher nuclear spin state in an applied magnetic field. ✳

  - The amount of energy required depends on the strength of the applied field and the type of nuclei being used, but this energy corresponds to electromagnetic radiation somewhere in the radiofrequency range. For example, in an applied magnetic field of strength 7.05 Tesla (T), the energy between the spin states of ^{1}H is around 0.0286 cal/mol, corresponding to electromagnetic energy of approximately 300 MHz (300,000,000 Hz).

  - The electromagnetic energy is absorbed because the nuclei **precess** in the applied magnetic field. That is, the nuclei that are lined up with or against the applied magnetic field precess just like a spinning top or gyroscope precesses and traces out a cone in the earth's gravitational field. The **rate of precession** can be expressed as a frequency in Hertz. **When the precessing nuclei are irradiated with electromagnetic radiation of the same frequency as the rate of precession, then the two frequencies couple, energy is absorbed, and the nuclear spin flips from +1/2 to -1/2.** ✳

  - **Resonance** is the term used to describe the absorption of electromagnetic radiation when the precession frequency and the electromagnetic radiation frequencies couple. That is why the entire process is called nuclear magnetic resonance. The absorption of energy by a molecule is measured and plotted to give the NMR spectrum.

• The different types of atomic nuclei in a molecule, for example ^{13}C or ^{1}H, do not absorb energy at the same frequency. If they did, NMR would not be a useful probe of molecular structure. The fact is that there are usually different local chemical environments in a molecule that change the resonance frequencies of the different nuclei. By looking at the **different resonance frequencies measured in the NMR spectrum**, information is obtained about the **different chemical environments in the molecule**, leading to an understanding **of the molecular structure**. ✳

  - Atomic nuclei from different elements resonate at different frequencies in the same applied field. For example, in the presence of an applied field of 7.05 T, ^{1}H nuclei resonate at about 300 MHz, while ^{13}C nuclei resonate at about 75 MHz.

  - The different nuclei in a molecule are in different chemical environments, if they are surrounded by electrons to varying degrees. The electrons themselves have spin and thereby create their own **local magnetic fields**. These local magnetic fields **shield the nucleus** from the applied magnetic fields. **The greater the electron density around a nucleus**, the **greater the shielding**. This means that at constant magnetic field strength, **the greater the shielding of a nucleus, the lower the frequency of electromagnetic radiation required to bring about a spin flip**. Another way to look at it is from the point of view of constant electromagnetic radiation frequency. At constant frequency of electromagnetic radiation, **the greater the shielding of a nucleus, the higher the magnetic field strength required to bring the nucleus into resonance**. ✳ *[It is important to be able to think about shielding and spin flipping from the point of view of constant frequency of electromagnetic radiation as well as at constant magnetic field strength.]*

  - The local magnetic fields are small compared to the applied magnetic fields. The differences in resonance frequencies for different nuclei caused by the local magnetic fields are usually on the order of 1×10^{-6} times as large as the original resonance frequencies. In other words, different ^{1}H nuclei in the same molecule have resonance frequencies that are different by an amount on the order of **parts per million (ppm)**. Thus, ppm is a convenient measurement unit for NMR spectroscopy. For example, a difference of 100 Hz is 1 ppm of 100 MHz, and a difference of 300 Hz is 1 ppm of 300 MHz.

  - In order to increase the precision of resonance frequency measurements, a reference compound is used. By convention, the ^{1}H resonance in **tetramethylsilane (TMS)** is used as the reference against which the frequency of other ^{1}H resonances are measured. Similarly, the ^{13}C resonance in TMS is used as the reference against which the frequency of other ^{13}C resonances are measured.

- A unit called **chemical shift** is used to standardize reporting of NMR data. **Chemical shift** (δ) is the frequency shift from TMS divided by the operating radiofrequency of the spectrometer. **Chemical shift is reported in** the units **ppm**.

$$\delta = \frac{\text{shift in frequency from TMS (Hz)}}{\text{frequency of spectrometer (Hz)}}$$

13.6 AN NMR SPECTROMETER
• The essential features of an NMR spectrometer are a powerful magnet, a radiofrequency generator, a radiofrequency detector, and a sample tube.
   Newer machines are called **Fourier transform NMR (FT-NMR) spectrometers.** In FT-NMR machines, the magnetic field strength is held constant, and the sample is irradiated with a short pulse of radiofrequency energy that flips all of the spins at once. The NMR spectrum is determined as the spins return to their equilibrium states. A mathematical process called Fourier transformation is used to convert the intensity versus time information produced in the experiment into the intensity versus frequency information that is actually plotted. Two advantages of FT-NMR are that it takes much less time to run a scan, and multiple scans can be averaged to produce more intense signals from a dilute sample.

13.7 EQUIVALENT HYDROGENS
• **Equivalent hydrogens** have the same chemical environment within a molecule and **have identical chemical shifts.** For example, all of the hydrogens in dimethyl ether (CH_3-O-CH_3) are equivalent, so there is only one signal in the ^{1}H-NMR spectrum of this compound.
• Hydrogens that are not equivalent give rise to different signals with different chemical shifts. For example, there are two signals in the ^{1}H-NMR spectrum of ethyl bromide (CH_3-CH_2-Br).

13.8 SIGNAL AREAS
• **The area under each signal** is **proportional to the number of hydrogens** giving rise to that signal in an ^{1}H-NMR spectrum. All modern NMR spectrometers can integrate the area under each signal. Please note that ^{13}C signals cannot be integrated accurately (Section 13.13) in ^{13}C-NMR spectra.

13.9 CHEMICAL SHIFT
• Each type of **equivalent hydrogen** within a molecule has only a **limited range of δ values**, and thus the value of the chemical shift for a signal in a ^{1}H-NMR spectrum gives valuable information about the type of hydrogen giving rise to that absorption. For example, the three hydrogens on methyl groups bonded to sp^3 hybridized carbons resonate near δ 1.0 ppm, while the three hydrogens on methyl groups bonded to an sp^2 hybridized carbonyl carbon atom resonate near δ 2.0 ppm. ✳
• A signal is considered **downfield** if it is shifted toward the left (weaker applied field) on the chart paper, and it is **upfield** if it is shifted to the right (stronger applied field).
• The chemical shift of a particular hydrogen depends primarily on the following three factors, but the influence of these factors on chemical shift falls off very quickly with distance.
   - **Electronegativity of adjacent atoms.** The greater the electronegativity of atoms near a particular hydrogen, the greater the chemical shift because the hydrogen is deshielded.
   - **Hybridization of adjacent atoms.** Hydrogens attached to an sp^3 hybridized carbon typically resonate at δ 0.8 to 1.7 ppm, hydrogens attached to an sp^2 hybridized carbon typically resonate at δ 4.6 to 5.7 ppm, and hydrogens attached to an sp hybridized carbon typically resonate at δ 2.0 to 3.0 ppm. These shifts are caused by a combination of the percent s character of the carbon orbital taking part in the C-H bond, and magnetic induction through the pi system.
   - **Magnetic induction in a pi system.** Pi electrons induce a magnetic field that can either increase or decrease the chemical shift of adjacent hydrogens in alkenes and alkynes, respectively. Carbonyl group pi electrons increase the chemical shift of aldehyde hydrogens all the way to near δ 10 ppm.

13.10 SIGNAL SPILITTING AND THE (n+1) RULE
• Signals can be split into several peaks and this phenomenon is called **spin-spin or signal splitting**. In **signal splitting**, the ^{1}H-NMR signal from one set of equivalent hydrogens is split by the influence of neighboring nonequivalent hydrogens. ✳
   - If a hydrogen has a set of **n** nonequivalent hydrogens on the same or adjacent atoms, its NMR signal will be split into **(n+1)** peaks. The nuclei of all adjacent hydrogens couple, but it is only between nonequivalent hydrogens that the coupling results in signal splitting. For example, in ethyl bromide (CH_3-CH_2-Br) the CH_3- signal is split into three peaks by the two hydrogens on the adjacent -CH_2- group, and the -CH_2-

signal is split into four peaks by the three hydrogens on the adjacent CH_3- group. *[Understanding signal splitting is absolutely essential for the interpretation of ^1H-NMR spectra, so it is critical that these ideas are understood before going on.]*

13.11 THE ORIGINS OF SIGNAL SPLITTING
• The chemical shift of a given hydrogen is influenced by the magnetic field derived from the spin of an adjacent nonequivalent nucleus. When this happens, the nuclei are said to be **coupled**. For coupling to be important, the nuclei must usually be no more than three bonds away from each other. In other words, hydrogens on adjacent carbon atoms can show significant coupling.
   - The magnetic field derived from the nuclear spin of one nucleus is what influences the adjacent nucleus. For example, if there are two coupled hydrogens that we call H_a and H_b, then H_a has an equal probability of finding the nucleus of H_b in the +1/2 or the -1/2 spin state. Because the magnetic fields of the two different spin states are different, H_a could feel either of two total magnetic fields. (The total magnetic field felt by a given nucleus such as that in H_a is a sum of the applied field plus contributions from electrons and nearby nuclear spin magnetic fields such as that from H_b.) The bottom line is that H_a is split into two ^1H-NMR peaks because the spin state of H_b could be either +1/2 or -1/2. ✳
   - H_b is also coupled to H_a to the same extent, so it is also split into two peaks.
   - When a hydrogen nucleus is coupled to three equivalent hydrogens such as those in a methyl group, the nucleus can feel (n+1) or 4 different magnetic fields. This is because each of the three hydrogen nuclei in the methyl group can be in the +1/2 or -1/2 spin state, and these can add up in four different ways (+1/2,+1/2,+1/2; +1/2,+1/2,-1/2, etc.), in a 1:2:2:1 ratio.

13.12 COUPLING CONSTANTS (J)
• A **coupling constant (J)** is the distance, in ppm, between adjacent peaks in a multiplet. In other words, **coupling constants** are a quantitative measure of the shielding/deshielding influence of induced magnetic fields from adjacent nuclei. ✳
   Coupling constants are usually in the range of a few Hz. Because they are a property of the molecule, they are independent of applied field strength. For this reason, spectra recorded at higher field strength are easier to interpret because the signals (many of which are multiplets due to signal splitting) are separated by more Hz and the individual peaks are less likely to overlap. *[This is a very important yet complicated idea, and it is worth understanding this before going on.]*
   - When two nuclei couple, the coupling constant is the same for both of their signals. Using the same example discussed in Section 13.11, both H_a and H_b will be doublets. Furthermore, the distance between the two peaks in each of these two doublets will be the same number of Hz, and this is the coupling constant for this interaction. ✳
   - When a single signal is coupled to two or more different sets of nuclei, then both sets of interactions lead to splitting according to the (n+1) rule. This can lead to some very complex multiplets. For example, in a molecule with a hydrogen that is coupled to two nonequivalent adjacent hydrogen atoms, the signal is split into 2 x 2 or 4 peaks. Such a signal is referred to as a doublet of doublets.

13.13 ^{13}C-NMR
• Carbon-12 (^{12}C) is the most abundant natural isotope of carbon (98.89%), but it is not seen in an NMR spectrum because its nucleus has only one allowed spin state. On the other hand, **carbon-13 (^{13}C)** (natural abundance of 1.11%) has two allowed nuclear spin states and it **can be detected by NMR**. ✳
• The NMR signals from ^{13}C are only about 10^{-4} times as strong as those from ^1H-NMR. This is because of the relatively low abundance of ^{13}C and the relatively small magnetic moment of the ^{13}C nuclei. For this reason, only the modern FT-NMR machines are able to measure routinely ^{13}C-NMR spectra.
• ^1H nuclei couple to the ^{13}C nuclei. Unfortunately, this can make the spectra very difficult to interpret. As a result, ^{13}C spectra are usually measured in the **hydrogen-decoupled mode** in which the sample is irradiated such that all hydrogens are in the same spin state and thus signal splitting is prevented. The ^{13}C spectra can then be measured without interference from complex signals due to ^1H-^{13}C signal splitting.
• As stated above, ^{13}C nuclei are very rare, so ordinarily only ^{12}C carbon atoms are adjacent to a given ^{13}C nucleus in a molecule. Thus, ^{13}C-^{13}C coupling and/or signal splitting are usually not observed.
• Because of the way ^{13}C nuclei return to equilibrium states, ^{13}C signals cannot be integrated accurately in ^{13}C-NMR spectra.
• ^{13}C-NMR spectra provide important structural information, because each different carbon atom in a molecule gives rise to a different signal. Thus, by simply looking at a ^{13}C-NMR spectra, a chemist can tell how many different types of carbon atoms are in a molecule.

- Most important, the observed chemical shift of a ^{13}C-NMR signal provides important information. For example; sp^3 alkyl carbon atoms, sp^2 atoms in an alkene, or sp^2 carbonyl carbon atoms all have characteristic chemical shifts. Please see table 13.10 for a detailed list of ^{13}C chemical shifts. ✻

13.14 THE DEPT METHOD

- **Distortionless enhancement by polarization transfer (DEPT)** is an advanced ^{13}C NMR technique that provides information about **the number of hydrogen atoms attached to a given carbon atom.** This information is useful in determining structure and assigning spectra. ✻
- DEPT spectra are obtained using a complex set of pulse sequences in both the ^1H and ^{13}C ranges with the result that the ^{13}C signal for CH_3, CH_2, and CH groups have different phases. One type of pulse sequence records only CH_3 signals (as positive peaks), another pulse sequence records only CH_2 signals, but as negative peaks, and a third pulse sequence records only CH signals, but as positive peaks. Still another pulse sequence records all three types of signals, but the phasing of the CH_3 and CH signals are positive while the phasing of the CH_2 signals is negative. Carbon atoms that do not have any hydrogens attached do not give signals in DEPT spectra.

13.15 INTERPRETATION OF NMR SPECTRA

- Different types of molecules have **characteristic chemical shifts and splitting patterns** in NMR spectra. The following is a list of these characteristics for a number of important types of molecules.
  - **Alkanes**: All hydrogens in alkanes are in similar chemical environments and fall within a narrow range of chemical shifts. ^1H-NMR chemical shifts for alkanes fall in the range of δ 0.9-1.5. ^{13}C-NMR chemical shifts for alkanes fall in the range of δ 0 to 60.
  - **Alkenes**: The chemical shifts of vinylic hydrogens fall in the range δ 4.6-5.7. ^{13}C-NMR chemical shifts for the sp^2 carbon atoms of alkenes fall in the range of δ 100-150. Coupling constants are generally larger for *trans* vinylic hydrogens (~16 Hz) than for *cis* vinylic hydrogens (~8 Hz) because the *trans* hydrogens are more strongly coupled to each other. Notice how this usually allows a chemist to distinguish between *cis* and *trans* alkenes.
  - **Alcohols**: The hydrogen of the OH group is variable and appears in the range δ 2.0-6.0. This hydrogen is rarely split by signal splitting because of the phenomenon of fast exchange. The hydrogens attached to the carbon adjacent to the OH group are deshielded by the inductive effect of the electron withdrawing nature of the oxygen atom. These hydrogen signals appear at δ 3.5-4.5.
  - **Ethers**: Hydrogens attached to carbon atoms adjacent to an ether oxygen atom usually fall in the range δ 3.3-4.0.

13.16 SOLVING NMR PROBLEMS

- Before analyzing the NMR spectrum of a given molecule, it is helpful to analyze the molecular formula. This could be determined by elemental analysis or mass spectrometry. An important piece of information contained in the molecular formula is the **index of hydrogen deficiency**. The **index of hydrogen deficiency** is the **number of rings and/or pi bonds in a molecule.** This is determined by comparing the number of hydrogens in the molecular formula of a compound of unknown structure with the number of hydrogens in a **reference compound**, a compound with the same number of carbon atoms that contains no rings or pi bonds. In particular, the **index of hydrogen deficiency** is defined according to the following formula: ✻

$$\text{index of hydrogen deficiency} = \frac{\text{\# of hydrogen atoms}_{(reference)} - \text{\# of hydrogens}_{(molecule of interest)}}{2}$$

  - For reference compounds that contain only C and H atoms, the molecular formula is C_nH_{2n+2}.
  - For each atom of F, Cl, Br, or I subtract one hydrogen.
  - No correction is necessary for O, S, or Se.
  - For each atom of N or P add one hydrogen.
- **After the index of hydrogen deficiency has been determined,** the following steps should be followed when solving a spectral problem. *[Practice is the best and possibly only way to become good at this.]* ✻
  - **Count the number of signals** to determine how many different types of hydrogens are present.
  - **Examine the pattern of chemical shifts** and correlate them with the known characteristic chemical shifts for different types of hydrogen atoms.
  - **Analyze the integration of each signal,** to see how many hydrogen atoms of each type are present.
  - **Analyze the signal splitting patterns.** This is usually the most difficult task by far, but also the most informative.
  - **Write the formula** that is consistent with all of the above information.

CHAPTER 13
Solutions to the Problems

Problem 13.1 Calculate the energy of red light (680 nm) in kilocalories per mol. Which form of radiation carries more energy, infrared radiation of wavelength 2.50 μm or red light of wavelength 680 nm?

Combining the two equations given in the text gives:

$$E = h\nu = h\left(\frac{c}{\lambda}\right)$$

Plugging in the appropriate values gives the desired answer:

$$E = \frac{(9.537 \times 10^{-14} \text{ kcal-sec-mol}^{-1})(3.00 \times 10^8 \text{ m-sec}^{-1})}{680 \times 10^{-9} \text{ m}} = \boxed{42.1 \text{ kcal-mol}^{-1}}$$

Notice how the units canceled to give the final answer in kcal-mol^{-1}. As can be seen from the equations, the longer the wavelength, the lower the energy, thus red light carries more energy.

Problem 13.2 Calculate the ratio of nuclei in the higher spin state to those in the lower spin state, N_h/N_l, for ^{13}C at 25°C in an applied field strength of 7.05 T. The difference in energy between the higher and lower nuclear spin states in this applied field is approximately 0.00715 cal/mol.

The important equation relates the change in energy of two spin states to their equilibrium concentrations:

$$\Delta E = -2.303RT\log\frac{N_h}{N_l}$$

Rearranging this expression in terms of N_h/N_l gives:

$$\log\frac{N_h}{N_l} = \frac{-\Delta E}{2.303RT}$$

Substituting in the appropriate values for R (1.987 cal·deg^{-1}·mol^{-1}), T (298 K) and ΔE (0.00715 cal·mol^{-1}) gives:

$$\log\frac{N_h}{N_l} = \frac{-0.00715 \text{ cal·mol}^{-1}}{2.303(1.987 \text{ cal·deg}^{-1}\text{·mol}^{-1})(298 \text{ deg})} = -5.243 \times 10^{-6}$$

$$\frac{N_h}{N_l} = \boxed{0.9999879 = \frac{1.0000000}{1.0000121}}$$

Problem 13.3 State the number of sets of equivalent hydrogens in each compound and the number of hydrogens in each set.

Numbers have been added to the carbon atoms of the structures to aid in referring to specific hydrogens. Use the "test atom" approach if you have trouble understanding the answers.

(a)
$$\overset{4}{\underset{}{\overset{\textcircled{c}}{CH_3}}}$$
$$\overset{1\textcircled{a}}{CH_3}-\overset{2\textcircled{b}}{CH_2}-\overset{3|}{\underset{\textcircled{d}}{CH}}-\overset{5\textcircled{b}}{CH_2}-\overset{6\textcircled{a}}{CH_3}$$

There are four sets of equivalent hydrogens. Set a: 6 hydrogens from the methyl groups of carbon atoms 1 and 6. Set b: 4 hydrogens from the -CH$_2$- groups of carbon atoms 2 and 5. Set c: 3 hydrogens from the methyl group of carbon atom 4. Set d: 1 hydrogen from the -CH- group of carbon atom 3.

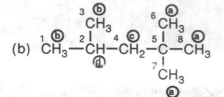

(b)

There are four sets of equivalent hydrogens. **Set a:** 9 hydrogens from the methyl groups of carbon atoms 6, 7, and 8. **Set b:** 6 hydrogens from the methyl groups of carbon atoms 1 and 3. **Set c:** 2 hydrogens from the -CH_2- group of carbon atom 4. **Set d:** 1 hydrogen from the -CH- group of carbon atom 2.

Problem 13.4 Each compound gives only one signal in its ^{1}H-NMR spectrum. Propose a structural formula for each compound.

In order for these molecules to give a single absorption peak, each of the hydrogen nuclei must be in an identical environment. This will only occur in symmetrical molecules.

(a) C_3H_6O (b) C_5H_{10} (c) C_5H_{12}

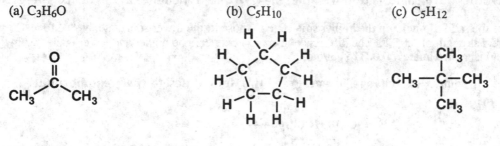

(d) $C_4H_6Cl_4$

CH_3-CCl_2-CCl_2-CH_3

Problem 13.5 The line of integration of the two signals in the ^{1}H-NMR spectrum of a ketone of molecular formula $C_7H_{14}O$ rises 62 and 10 chart divisions, respectively. Calculate the number of hydrogens giving rise to each signal, and propose a structural formula for this ketone.

The ratio of signals is approximately 6:1, which corresponds to a 12:2 ratio of hydrogens. Thus, the larger signal represents 12 hydrogens and the smaller signal represents 2 hydrogens. A structure consistent with this assignment is 2,4-dimethyl-3-pentanone as shown below:

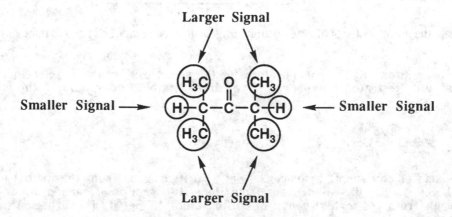

<u>Problem 13.6</u> Following are two constitutional isomers of molecular formula $C_4H_8O_2$.

$$\underset{(1)}{CH_3CH_2O\overset{\overset{\displaystyle O}{\|}}{C}CH_3} \qquad \underset{(2)}{CH_3CH_2\overset{\overset{\displaystyle O}{\|}}{C}OCH_3}$$

(a) Predict the number of signals in the ^{1}H-NMR spectrum of each isomer.

Each isomer will have three signals.

(b) Predict the ratio of areas of the signals in each spectrum.

In each spectrum, the ratio of areas of the three signals will be 3:2:3.

(c) Show how to distinguish between these isomers on the basis of chemical shift.

The -CH₃ singlet signals will be diagnostic. Isomer (1) is the only one with a -CH₃ attached to the carbonyl carbon atom, while isomer (2) is the only one with a -CH₃ attached to an ester oxygen atom. Therefore, the spectrum of isomer (1) will be the one with a -CH₃ singlet at δ 2.1-2.3, and the spectrum of isomer (2) will be the one with a -CH₃ singlet at δ 3.7-3.9.

<u>Problem 13.7</u> Following are pairs of constitutional isomers. Predict the number of signals in the ^{1}H-NMR spectrum of each isomer and the splitting pattern of each signal.

(a) $$CH_3OCH_2\overset{\overset{\displaystyle O}{\|}}{C}CH_3 \quad \text{and} \quad CH_3CH_2\overset{\overset{\displaystyle O}{\|}}{C}OCH_3$$

The molecule on the left will have three signals that are all singlets, and the molecule on the right will have three signals with splitting patterns as indicated.

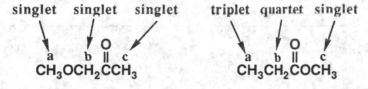

(b) $$CH_3\overset{\overset{\displaystyle Cl}{|}}{\underset{\underset{\displaystyle Cl}{|}}{C}}CH_3 \quad \text{and} \quad ClCH_2CH_2CH_2Cl$$

The molecule on the left will have one signal and the molecule on the right will have two signals with splitting patterns as indicated.

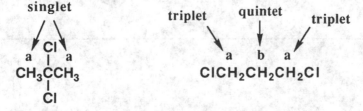

Problem 13.8 Explain how to distinguish between the members of each pair of constitutional isomers based on the number of signals in the proton-decoupled ^{13}C-NMR spectrum of each member.

(a) 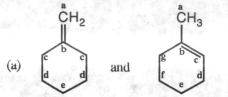 and

These molecules can be distinguished because they have different numbers of nonequivalent carbon nuclei and thus will have different numbers of ^{13}C-NMR signals. Different signals are indicated by different letters on the above structures. The molecule on the left has higher symmetry and will have 5 signals corresponding to the carbon atoms labeled as a - e, while the molecule on the right has less symmetry and will have 7 signals corresponding to the carbon atoms labeled as a - g.

(b) 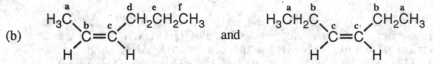 and

These molecules can also be distinguished because they have different numbers of nonequivalent carbon nuclei and thus will have different numbers of ^{13}C-NMR signals. Different signals are indicated by different letters on the above structures. The molecule on the right has higher symmetry and will only have 3 signals corresponding to the carbon atoms labeled as a - c, while the molecule on the left has less symmetry and will have 6 signals corresponding to the carbon atoms labeled as a - f.

Problem 13.9 Assign all signals in the ^{13}C-NMR spectrum of 4-methyl-2-pentanone.

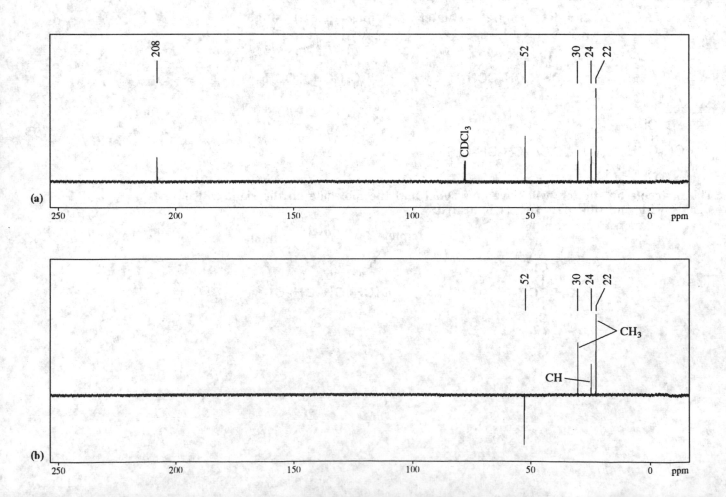

The appropriate peaks have been assigned using the DEPT spectra to distinguish the CH_3, CH_2 and CH groups. The different CH_3 groups were distinguished by the fact that the CH_3 group adjacent to the carbonyl group is less shielded (higher chemical shift) than the other two methyl groups, which are equivalent. The carbonyl group carbon atom has a partial positive charge, thus explaining the relative lack of shielding observed for the adjacent carbon nuclei.

4-Methyl-2-pentanone

Problem 13.10 Calculate the index of hydrogen deficiency of cyclohexene, C_6H_{10}, and account for this deficiency by reference to its structural formula.

The molecular formula for cyclohexene is C_6H_{10}. The molecular formula for the reference compound with 6 carbon atoms is C_6H_{14}. Thus the index of hydrogen deficiency is (14-10)/2 or 2. This makes sense since cyclohexene has one ring and one pi bond.

Cyclohexene

Problem 13.11 The index of hydrogen deficiency of niacin is 5. Account for this index of hydrogen deficiency by reference to the structural formula of niacin.

Nicotinamide
(Niacin)

The index of hydrogen deficiency of niacin is 5 because there are four pi bonds and one ring in the structure.

Index of Hydrogen Deficiency
Problem 13.12 Complete the following table:

Class of Compound	General Molecular Formula	Index of Hydrogen Deficiency	Reason for Hydrogen Deficiency
alkane	C_nH_{2n+2}	0	(reference hydrocarbon)
alkene	C_nH_{2n}	1	one pi bond
alkyne	C_nH_{2n-2}	2	two pi bonds
alkadiene	C_nH_{2n-2}	2	two pi bonds
cycloalkane	C_nH_{2n}	1	one ring
cycloalkene	C_nH_{2n-2}	2	one ring and one pi bond
bicycloalkane	C_nH_{2n-2}	2	two rings

Problem 13.13 Calculate the index of hydrogen deficiency of these compounds:
(a) Aspirin, $C_9H_8O_4$ (b) Ascorbic acid (vitamin C), $C_6H_8O_6$

(20-8)/2 = 6   (14-8)/2 = 3

(c) Pyridine, C_5H_5N (d) Urea, CH_4N_2O

(13-5)/2 = 4 (nitrogen correction)   (6-4)/2 = 1 (nitrogen correction)

(e) Cholesterol, $C_{27}H_{46}O$ (f) Dopamine, $C_8H_{11}NO_2$

(56-46)/2 = 5   (19-11)/2 = 4 (nitrogen correction)

Interpretation of ^{1}H-NMR and ^{13}C-NMR Spectra
Problem 13.14 Complete the following table. Which nucleus requires the least energy to flip its spin at this applied field? Which nucleus requires the most energy?

Nucleus	Applied field (T)	Radiofrequency (MHz)	Energy (cal/mol)
^{1}H	7.05	300	2.86×10^{-2}
^{13}C	7.05	75.5	7.20×10^{-3}
^{19}F	7.05	282	2.69×10^{-2}

Based on the entries in the table, the ^{13}C requires the least energy to flip its spin and the ^{1}H requires the most.

Problem 13.15 The natural abundance of ^{13}C is only 1.1%. Furthermore, its sensitivity in NMR spectroscopy (a measure of the energy difference between a spin aligned with or against an external magnetic field) is only 1.6% that of ^{1}H. What are the relative signal intensities expected for the ^{1}H-NMR and ^{13}C-NMR spectra of the same sample of $Si(CH_3)_4$?

A given ^{13}C signal is (0.011)(0.016) = 0.000176 as strong as a given ^{1}H signal. There are three times as many H atoms as C atoms in $Si(CH_3)_4$, so overall the ratio of H to C signals is 1 : (0.000176/3) = 1 : 0.000059. (Note that this is the same as 17,000 to 1)

Problem 13.16 Following are structural formulas for three constitutional isomers of molecular formula $C_7H_{16}O$ and three sets of ^{13}C-NMR spectral data. Assign each constitutional isomer its correct spectral data.

(a) $CH_3CH_2CH_2CH_2CH_2CH_2CH_2OH$

	Spectrum 1	Spectrum 2	Spectrum 3
	74.66	70.97	62.93
	30.54	43.74	32.79
	7.73	29.21	31.86
		26.60	29.14
		23.27	25.75
		14.09	22.63
			14.08

(b) $CH_3\overset{\displaystyle OH}{\underset{\displaystyle CH_3}{C}}CH_2CH_2CH_2CH_3$

(c) $CH_3CH_2\overset{\displaystyle OH}{\underset{\displaystyle CH_2CH_3}{C}}CH_2CH_3$

These constitutional isomers are most readily distinguished by the number of sets of nonequivalent carbon atoms and thus different ^{13}C signals. Using the following analysis, it can be seen that compound (a) has 7 sets of nonequivalent carbon atoms corresponding to Spectrum 3, compound (b) has 6 sets of nonequivalent carbon atoms corresponding to

Spectrum 2, and compound (c) has 3 sets of nonequivalent carbon atoms corresponding to Spectrum 1.

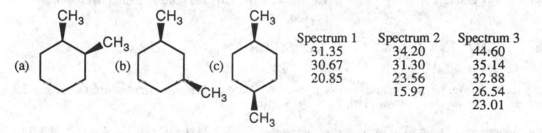

Problem 13.17 Following are structural formulas for the *cis* isomers of 1,2-, 1,3-, and 1,4-dimethylcyclohexane and three sets of ^{13}C-NMR spectral data. Assign each constitutional isomer its correct spectral data.

Spectrum 1	Spectrum 2	Spectrum 3
31.35	34.20	44.60
30.67	31.30	35.14
20.85	23.56	32.88
	15.97	26.54
		23.01

These constitutional isomers are most readily distinguished by the number of sets of nonequivalent carbon atoms and thus different ^{13}C signals. Using the analysis shown below, it can be seen that compound (a) has 4 sets of nonequivalent carbon atoms corresponding to Spectrum 2, compound (b) has 5 sets of nonequivalent carbon atoms corresponding to Spectrum 3, and compound (c) has 3 sets of nonequivalent carbon atoms corresponding to Spectrum 1. The different sets of equivalent carbon atoms are indicated by the letters.

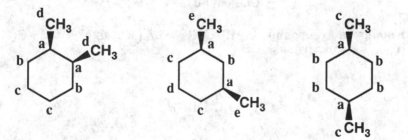

Problem 13.18 Following are structural formulas, dipole moments, and ^{1}H-NMR chemical shifts for acetonitrile, fluoromethane, and chloromethane.

$CH_3-C\equiv N$	CH_3-F	CH_3-Cl
Acetonitrile	Fluoromethane	Chloromethane
3.92 D	1.85 D	1.87 D
δ 1.97	δ 4.26	δ 3.05

(a) How do you account for the fact that the dipole moments of fluoromethane and chloromethane are almost identical even though fluorine is considerably more electronegative than chlorine?

Recall that dipole moment is proportional to the partial charge times the distance of that charge separation. Fluorine is a much smaller atom than chlorine, so it makes shorter bonds leading to relatively short charge separation distances. Thus, the differences in bond lengths almost lengths almost exactly offsets the differences in electronegativities between fluorine and chlorine and the dipole moments come out almost the same.

(b) How do you account for the fact that the dipole moment of acetonitrile is considerably greater than that of either fluoromethane or chloromethane?

Again, the key is distance. The acetonitrile has partial charge distributed over more atoms and thus a larger distance than fluoromethane or chloromethane.

(c) How do you account for the fact that the chemical shift of the methyl hydrogens in acetonitrile is considerably less than that for either fluoromethane or chloromethane? (*Hint:* Consider the magnetic induction in the pi bonds of acetonitrile.)

A magnetic field is induced in the pi system of the nitrile that is against the applied field, thus decreasing the chemical shift.

<u>Problem 13.19</u> Following are three compounds of molecular formula $C_4H_8O_2$, and three ^{1}H-NMR spectra. Assign each compound its correct spectrum and assign all signals to their corresponding hydrogens.

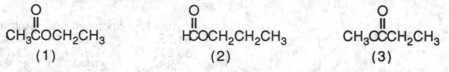

$$CH_3COCH_2CH_3$$
(1)

$$HCOCH_2CH_2CH_3$$
(2)

$$CH_3OCCH_2CH_3$$
(3)

For the spectral interpretations in the rest of this chapter the chemical shift (δ) is shown on the structure adjacent to the appropriate hydrogen atom.

Spectrum A corresponds to compound 2: ^{1}H-NMR δ 8.1 (1H, singlet, H-C(O)-), 4.2 (2H, triplet, -O-CH$_2$-), 1.7 (2H, multiplet; a doublet of triplets, -CH$_2$-), 1.0 (3H, triplet, -CH$_3$).

$$\overset{8.1}{\underset{HC-O-CH_2CH_2CH_3}{\overset{O}{\parallel}}}$$
                4.2   1.7   1.0
(2)
Compound A

Spectrum B corresponds to compound 3: ^{1}H-NMR δ 3.7 (3H, singlet, CH$_3$-C(O)-), 2.3 (2H, quartet, -C(O)-CH$_2$-), 1.2 (3H, triplet, -CH$_3$).

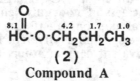

$$CH_3-O-CCH_2CH_3$$
3.7   2.3 1.2
(3)
Compound B

Spectrum C corresponds to compound 1: ^{1}H-NMR δ 4.1 (2H, quartet, -O-CH$_2$-), 2.0 (3H, singlet, CH$_3$-C-), 1.2 (3H, triplet, -CH$_3$).

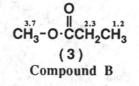

$$CH_3C-O-CH_2CH_3$$
2.0   4.1 1.2
(1)
Compound C

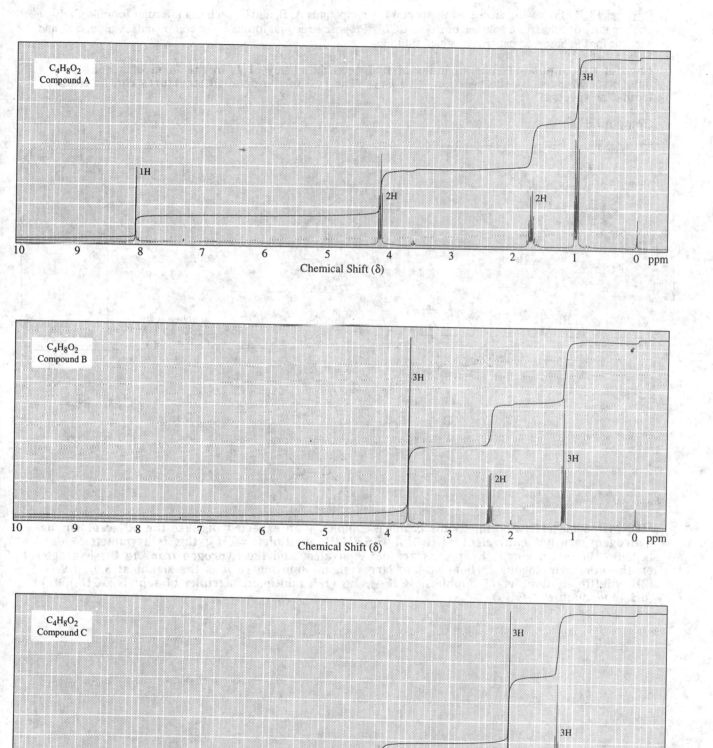

<u>Problem 13.20</u> Following are ^{1}H-NMR spectra for compounds A, B, and C, each of molecular formula C_6H_{12}. Each readily decolorizes a solution of Br_2 in CCl_4. Propose structural formulas for compounds A, B, and C, and account for the observed patterns of signal splitting.

Each of the compounds has an index of hydrogen deficiency of 1, in the form of a double bond as evidenced by the reaction with Br_2. The rest of the detailed structures can be deduced from the spectra.

Compound A:

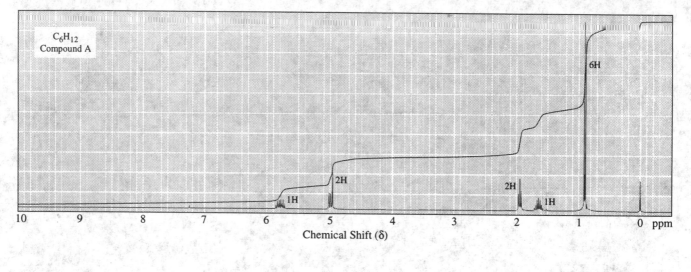

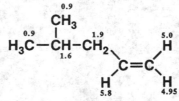

^{1}H-NMR δ 5.8 (1H, multiplet; this is more complex than expected because the adjacent vinylic hydrogens are not equivalent, -CH=), 4.95-5.0 (2H, multiplet, =CH$_2$; this is asymmetric because these two vinylic hydrogens are not equivalent and the hydrogen *trans* to the hydrogen on the other vinylogous carbon has the larger signal splitting so it is the signal at 5.0), 1.9 (2H, multiplet; doublet of doublets, -CH$_2$-), 1.6 (1H, multiplet; a triplet of septets, -CH-), 0.9 (6H, one doublet, -CH$_3$).

Compound B:

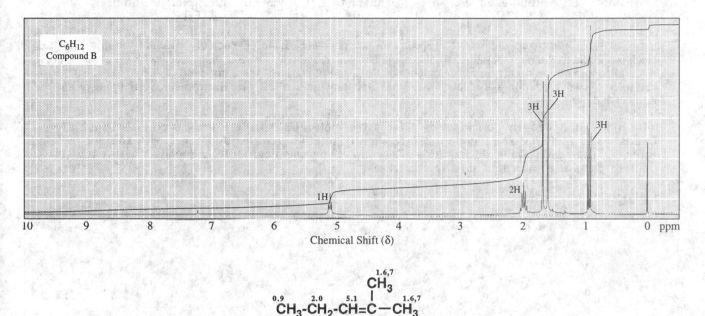

^{1}H-NMR δ 5.1 (1H, triplet, -CH=), 2.0 (2H, multiplet; a doublet of quartets, -CH$_2$-), 1.6 and 1.7 (6H, two singlets, =C(CH$_3$)$_2$), 0.9 (3H, triplet, -CH$_3$)

Compound C:

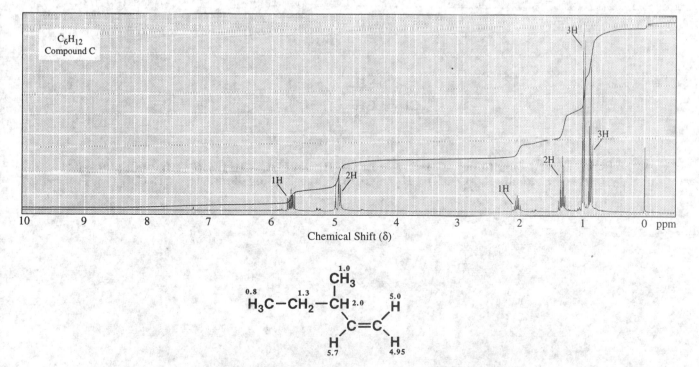

^{1}H-NMR δ 5.7 (1H, multiplet; this is more complex than expected because the adjacent vinylic hydrogens are not equivalent, -CH=), 4.9-5.0 (2H, multiplet, =CH$_2$; this is asymmetric because these two vinylic hydrogens are not equivalent and the hydrogen *trans* to the hydrogen on the other vinylogous carbon has the larger signal splitting so it is the signal at 5.0), 2.0 (1H, multiplet; a doublet of a triplet of a quartet, -CH-), 1.3 (2H, multiplet; a doublet of a quartet, -CH$_2$-), 1.0 (3H, doublet,-CH-CH$_3$), 0.8 (3H, triplet, -CH$_2$-CH$_3$)

<u>Problem 13.21</u> Following are ^{1}H-NMR spectra for compounds D, E, and F, each of molecular formula C$_5$H$_{12}$O. Each is a liquid at room temperature, slightly soluble in water, and reacts with sodium metal with the evolution of a gas. Propose structural formulas of compounds D, E, and F. (Hints: For compound D, the signal at δ 0.9 results from two overlapping doublets. The signal for compound E at δ 0.9 results from the overlapping of a doublet and a triplet.)

The index of hydrogen deficiency is 0 for these molecules, so there are no rings or double bonds. The fact that the compounds are slightly soluble in water and react with sodium metal indicates that each molecule has an -OH group. The chemical shifts associated with each set of hydrogens are indicated on the structures.

Compound D:

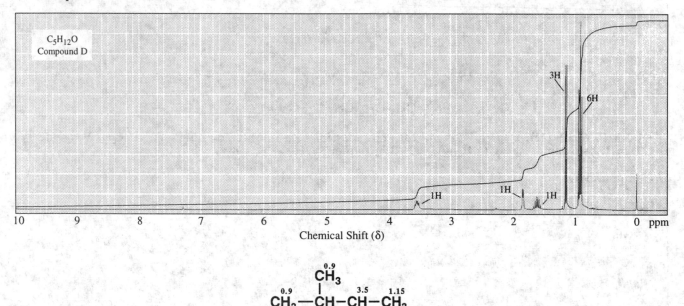

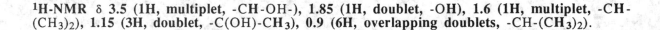

^{1}**H-NMR** δ 3.5 (1H, multiplet, -CH-OH-), 1.85 (1H, doublet, -OH), 1.6 (1H, multiplet, -CH-(CH$_3$)$_2$), 1.15 (3H, doublet, -C(OH)-CH$_3$), 0.9 (6H, overlapping doublets, -CH-(CH$_3$)$_2$).

Compound E:

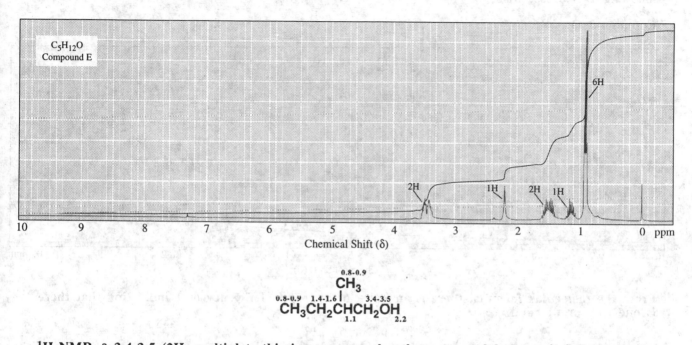

$$\overset{0.8-0.9}{CH_3}$$

$$\underset{1.1}{\overset{0.8-0.9}{CH_3}}\underset{}{\overset{1.4-1.6}{CH_2}}\overset{}{CH}\overset{3.4-3.5}{CH_2}\overset{2.2}{OH}$$

¹H-NMR δ 3.4-3.5 (2H, multiplet; this is more complex than expected because it is adjacent to a stereocenter, -CH₂-OH), 2.2 (1H, broad triplet, -OH), 1.4-1.6 (2H, multiplet; this is more complex than expected because it is adjacent to a stereocenter, CH₃-CH₂-), 1.1 (1H, multiplet, -CH-), 0.8-0.9 (6H, broad multiplet, both -CH₃ groups).

Compound F:

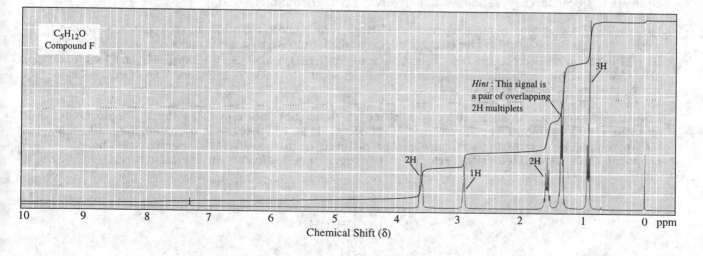

$$\underset{0.9}{CH_3}\underset{1.4}{CH_2}\underset{1.4}{CH_2}\underset{1.55}{CH_2}\underset{3.6}{CH_2}\underset{2.9}{OH}$$

¹H-NMR δ 3.6 (2H, broad multiplet, -CH₂-OH), 2.9 (1H, broad peak, -OH), 1.55 (2H, multiplet; a triplet of triplets, -CH₂-CH₂-OH), 1.4 (4H, multiplet, CH₃-CH₂-CH₂- and CH₃-CH₂-CH₂-), 0.9 (3H, triplet, -CH₃)

<u>Problem 13.22</u> Propose a structural formula for compound G, molecular formula C_3H_6O, consistent with the following ^{1}H-NMR spectrum:

Compound G:

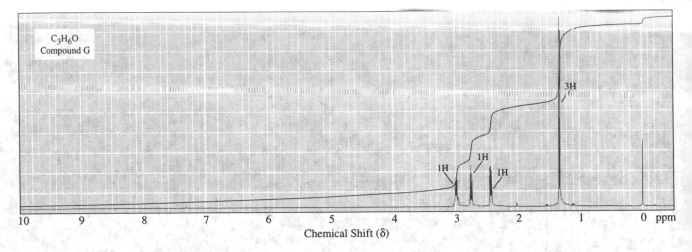

From the molecular formula, there is an index of hydrogen deficiency of 1 indicating that there is one ring or pi bond.

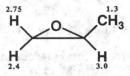

^{1}**H-NMR δ 3.0 (1H, multiplet, H$_c$), 2.4 and 2.75 (2H, multiplets; these hydrogens are not equivalent because they cannot rotate freely), 1.3 (3H, doublet, -CH$_3$)**

<u>Problem 13.23</u> Compound H, molecular formula $C_6H_{14}O$, readily undergoes acid-catalyzed dehydration when warmed with phosphoric acid to give compound I, molecular formula C_6H_{12}, as the major organic product. The ^{1}H-NMR spectrum of compound H shows signals at δ 0.90 (t, 6H), 1.12 (s, 3H), 1.38 (s, 1H), and 1.48 (q, 4H). The ^{13}C-NMR spectrum of compound H shows signals at 72.98, 33.72, 25.85, and 8.16. Deduce the structural formulas of compounds H and I.

From the molecular formula, there is a hydrogen deficiency index of 0, so there are no rings or pi bonds in compound E. From the ^{13}C-NMR peak at 72.98 there is a carbon bonded to an -OH group. The rest of the structure can be deduced from the ^{1}H-NMR spectrum. The chemical shifts associated with each set of hydrogens are indicated on the structure.

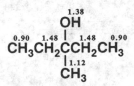

^{1}**H-NMR δ 1.48 (4H, quartet, -CH$_2$), 1.38 (1H, singlet, -OH), 1.12 (3H, singlet, -C(OH)-CH$_3$), 0.90 (6H, triplet, both -CH$_2$-CH$_3$ groups)**

Dehydration of compound H gives the following alkene as compound I:

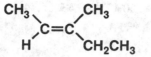

Problem 13.24 Compound J, molecular formula $C_5H_{10}O$, readily decolorizes Br_2 in CCl_4, and is converted by H_2/Ni into compound K, molecular formula $C_5H_{12}O$. Following is the ^1H-NMR spectrum of compound J. The ^{13}C-NMR spectrum of compound J shows signals at 146.12, 110.75, 71.05, and 29.38. Deduce the structural formulas of compounds J and K.

Compound J:

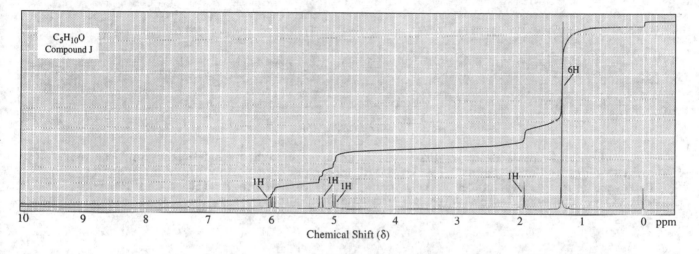

From the reaction with Br_2 and H_2/Ni it is clear the compound J has a carbon-carbon double bond. These conclusions are supported by the ^{13}C-NMR signals corresponding to the sp^2 carbons at δ 146.12 and 110.75. There is also a carbon bonded to an -OH group judging from the signal at δ 71.05. The rest of the structure is deduced from the ^1H-NMR spectrum.

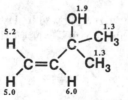

^1H-NMR δ 6.0 (1H, doublet of doublets), 4.9 and 5.2, (2H, two doublets; the vinylic hydrogens are not equivalent and the hydrogen *trans* to the hydrogen on the other vinylogous carbon has the larger signal splitting so it is the signal at 5.2), 1.9 (1H, singlet, -OH), 1.3 (6H, singlet, C-(CH$_3$)$_2$)

Upon hydrogenation, compound J is reduced to the alcohol shown below as compound K:

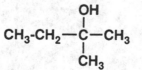

Problem 13.25 Following is the ^{1}H-NMR spectrum of compound L, molecular formula C_7H_{12}. Compound L reacts with bromine in carbon tetrachloride to give a compound of molecular formula $C_7H_{12}Br_2$. The ^{13}C-NMR spectrum of compound L shows signals at 150.12, 106.43, 35.44, 28.36, and 26.36. Deduce the structural formula of compound L.

Compound L:

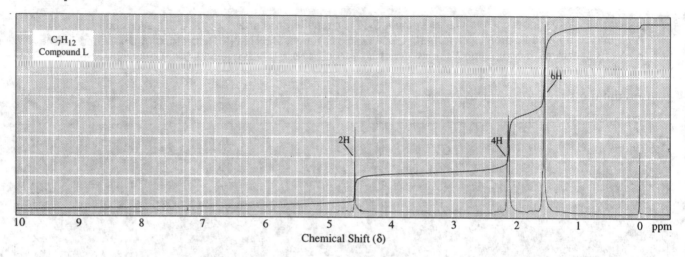

The molecular formula indicates that there is an index of hydrogen deficiency of 2, so there are two rings and/or pi bonds. The ^{13}C-NMR indicates there is only one double bond because there are only two resonances corresponding to sp^2 carbon atoms (150.12 and 106.43). Therefore, compound L must have one ring and one pi bond.

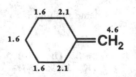

^{1}H-NMR δ 4.6 (2H, singlet, =CH$_2$), 2.1 (4H, broad peak, two -CH$_2$- groups), 1.6 (6H, broad peak, three -CH$_2$- groups)

Problem 13.26 Treatment of compound M with BH$_3$ followed by H$_2$O$_2$/NaOH gives compound N. Following are ^{1}H-NMR spectra for compounds M and N along with ^{13}C-NMR spectral data. From this information, deduce structural formulas for compounds M and N

$$C_7H_{12} \xrightarrow[\text{2) H}_2\text{O}_2\text{, NaOH}]{\text{1) BH}_3} C_7H_{14}O$$

(M)   (N)

^{13}C-NMR	
(M)	(N)
132.38	72.71
32.12	37.59
29.14	28.13
27.45	22.68

Compound M:

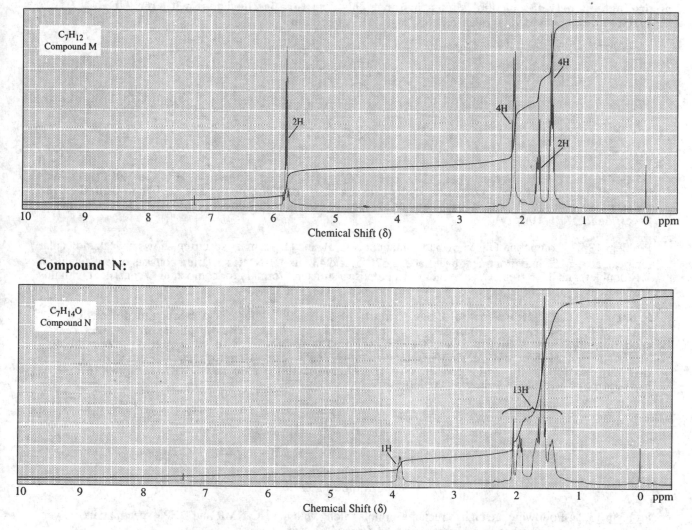

C_7H_{12}
Compound M

Chemical Shift (δ)

Compound N:

$C_7H_{14}O$
Compound N

Chemical Shift (δ)

The molecular formula for M indicates that it has an index of hydrogen deficiency of 2 so that is has two rings and/or pi bonds. The ^{13}C-NMR spectral data shows that there is an sp^2 carbon atom (132.38). Since there must be two sp^2 carbon atoms to make a pi bond, then the molecule must be symmetric so that both sp^2 carbon atoms are equivalent. This also explains why there are so few other ^{13}C-NMR signals. Since there is presumably only one pi bond, then there must be one ring in the molecule. The rest of the structure can be deduced from the ^{1}H-NMR spectrum.

^{1}H-NMR δ 5.8 (2H, triplet, both =CH-), 2.1 (4H, multiplet; doublet of triplets, two -CH$_2$- groups), 1.7 (2H, quintet, the unique -CH$_2$- group), 1.5 (4H, multiplet; triplet of triplets, two -CH$_2$- groups).

Given the structural formula for M, it is clear that compound N would be the hydroboration product, namely the alcohol shown below. This structure is consistent with the ¹³C-NMR spectral data provided as well as the ¹H-NMR spectrum.

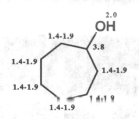

¹H-NMR δ 3.8 (1H, broad peak, -C(OH)H-), 2.0 (1H, sharp singlet, -OH), 1.4-1.9 (12H, broad multiplets, all the remaining hydrogens on the ring. The peaks are so broad and the patterns so complex because this ring does not have a double bond to hold it rigid, so it has a great deal of flexibility)

Problem 13.27 Compound O is known to contain only C, H, and O. Its mass spectrum shows a weak molecular ion peak at m/z 102 and prominent peaks at m/z 87, 45, and 43. Its ¹H-NMR spectrum consists of two signals: δ 1.1 (doublet) and δ 3.6 (septet) in the ratio 6:1. Propose a structural formula for compound O consistent with this information.

From $m/z = 102$, deduce a molecular formula of $C_6H_{14}O$. There is a hydrogen deficiency index of 0, so there are no rings or pi bonds in compound O. The simplicity of the ¹H-NMR spectrum indicates a highly level of symmetry in the molecule, with each methyl group being attached to a carbon with a single hydrogen atom. The only structure consistent with all of this information is the following ether. The chemical shifts associated with each set of hydrogens are indicated on the structure.

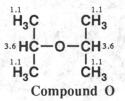

Compound O

Problem 13.28 Following are eight structural formulas along with the ¹³C-NMR and DEPT spectral data. Given this data, assign each carbon in each compound its correct ¹³C chemical shift.

(a)
$$\overset{13.40}{CH_3}\overset{43.22}{CH_2}\overset{}{\underset{26.46}{CH_2}}\overset{\overset{Br}{|}}{\underset{51.55}{CH}}\overset{21.00}{CH_3}$$

¹³C	DEPT
51.55	CH
43.22	CH₂
26.46	CH₂
21.00	CH₃
13.40	CH₃

(b)
$$\overset{}{\underset{12.23}{CH_3}}\overset{30.56}{CH_2}\overset{\overset{\overset{22.47}{CH_3}}{|}}{\underset{147.70}{C}}=\overset{108.33}{CH_2}$$

¹³C	DEPT
147.70	–
108.33	CH₂
30.56	CH₂
22.47	CH₃
12.23	CH₃

(c)
$$\overset{115.26}{CH_2}=\overset{137.81}{CH}\overset{43.35}{CH_2}\overset{\overset{\overset{22.26}{CH_3}}{|}}{\underset{28.12}{CH}}\overset{22.26}{CH_3}$$

¹³C	DEPT
137.81	CH
115.26	CH₂
43.35	CH₂
28.12	CH
22.26	CH₃

(d)

$$\overset{28.72}{CH_3}$$
$$\overset{28.72}{CH_3}\overset{49.02}{\underset{33.15}{C}}CH_2Br$$
$$\overset{28.72}{CH_3}$$

^{13}C	DEPT
49.02	CH_2
33.15	–
28.72	CH_3

(e)

$$\overset{35.1}{CH_3}\overset{35.1}{CH_2}\overset{O}{\underset{207.8}{C}}CH_2CH_3$$
$$\underset{7.5}{}\underset{7.5}{}$$

^{13}C	DEPT
207.8	–
35.1	CH_2
7.5	CH_3

(f)

$$\overset{37.6}{CH_3}\overset{}{CH_2}\overset{O}{\underset{208.7}{C}}\overset{30.1}{CH_3}$$
$$\underset{9.2}{}$$

^{13}C	DEPT
208.7	–
37.6	CH_2
30.1	CH_3
9.2	CH_3

(g)

$$\overset{33.94}{CH_3}\overset{}{\underset{19.01}{\underset{CH_3}{\underset{19.01}{C}}}}H\overset{O}{\underset{51.50}{\underset{177.48}{C}}}OCH_3$$

^{13}C	DEPT
177.48	–
51.50	CH_3
33.94	CH
19.01	CH_3

(h)

$$\overset{24.45}{CH_3}\overset{O}{\underset{171.17}{C}}\overset{63.12}{CH_2}\overset{}{\underset{37.21}{CH_2}}\overset{21.02}{\underset{CH_3}{\underset{21.02}{C}}}H\underset{25.05}{CH}CH_3$$

^{13}C	DEPT
171.17	–
63.12	CH_2
37.21	CH_2
25.05	CH
24.45	CH_3
21.02	CH_3

Problem 13.29   Write structural formulas for the following compounds:
(a) $C_2H_4Br_2$: δ 2.5 (d, 3H) and 5.9 (q, 1H)

$$\overset{2.5}{CH_3}-\overset{5.9}{CH}Br_2$$

(b) $C_4H_8Cl_2$: δ 1.60 (d, 3H), 2.15 (m, 2H), 3.72 (t, 2H), and 4.27 (m, 1H)

$$\overset{1.6}{CH_3}-\overset{4.27}{CH}Cl-\overset{2.15}{CH_2}-\overset{3.72}{CH_2}Cl$$

(c) $C_5H_8Br_4$: δ 3.6 (s, 8H)

$$\overset{3.6}{CH_2}Br$$
$$\overset{3.6}{CH_2}Br-\underset{\underset{CH_2Br}{\overset{3.6}{|}}}{\overset{|}{C}}-\overset{3.6}{CH_2}Br$$

(d) C_4H_8O: δ 1.0 (t, 3H), 2.1 (s, 3H), and 2.4 (q, 2H)

$$\overset{1.0}{CH_3}\overset{2.4}{CH_2}-\overset{O}{\underset{}{C}}-\overset{2.1}{CH_3}$$

(e) $C_4H_8O_2$: δ 1.2 (t, 3H), 2.1 (s, 3H) and 4.1 (quartet, 2H); contains an ester group

$$\overset{2.1}{CH_3}-\overset{O}{\underset{}{C}}-O-\overset{4.1}{CH_2}-\overset{1.2}{CH_3}$$

(f) $C_4H_8O_2$: δ 1.2 (t, 3H), 2.3 (quartet, 2H) and 3.6 (s, 3H); contains an ester group

$$\overset{1.2}{CH_3}-\overset{2.3}{CH_2}-\overset{O}{\overset{\|}{C}}-O-\overset{3.6}{CH_3}$$

(g) C_4H_9Br: δ 1.1 (d, 6H), 1.9 (m, 1H), and 3.4 (d, 2H)

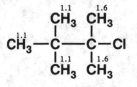

(h) $C_6H_{12}O_2$: δ 1.5 (s, 9H) and 2.0 (s, 3H)

$$\overset{2.0}{CH_3}-\overset{O}{\overset{\|}{C}}-O-\overset{\overset{1.5}{CH_3}}{\underset{\underset{1.5}{CH_3}}{\overset{|}{\underset{|}{C}}}}-\overset{1.5}{CH_3}$$

(i) $C_7H_{14}O$: δ 0.9 (t, 6H), 1.6 (sextet, 4H), and 2.4 (t, 4H)

$$\overset{0.9}{CH_3}-\overset{1.6}{CH_2}-\overset{2.4}{CH_2}-\overset{O}{\overset{\|}{C}}-\overset{2.4}{CH_2}-\overset{1.6}{CH_2}-\overset{0.9}{CH_3}$$

(j) $C_5H_{10}O_2$: δ 1.2 (d, 6H), 2.0 (s, 3H) and 5.0 (septet, 1H)

$$\overset{2.0}{CH_3}-\overset{O}{\overset{\|}{C}}-O-\overset{\overset{5.0}{H}}{\underset{\underset{1.2}{CH_3}}{\overset{|}{\underset{|}{C}}}}-\overset{1.2}{CH_3}$$

(k) $C_5H_{11}Br$: δ 1.1 (s, 9H) and 3.2 (s, 2H)

$$\overset{1.1}{CH_3}-\overset{\overset{1.1}{CH_3}}{\underset{\underset{1.1}{CH_3}}{\overset{|}{\underset{|}{C}}}}-\overset{3.2}{CH_2}Br$$

(l) $C_7H_{15}Cl$: δ 1.1 (s, 9H) and 1.6 (s, 6H)

$$\overset{1.1}{CH_3}-\overset{\overset{1.1}{CH_3}}{\underset{\underset{1.1}{CH_3}}{\overset{|}{\underset{|}{C}}}}-\overset{\overset{1.6}{CH_3}}{\underset{\underset{1.6}{CH_3}}{\overset{|}{\underset{|}{C}}}}-Cl$$

Problem 13.30 The percent s character of carbon participating in a C-H bond can be established by measuring the ^{13}C-^{1}H coupling constant and using the relationship

$$\text{percent s character} = 0.2\ J(^{13}C\text{-}^{1}H)$$

The ^{13}C-^{1}H coupling constant observed for methane, for example, is 125 Hz, which gives 25% s character, the value expected for an sp^3 hybridized carbon atom.
(a) Calculate the expected ^{13}C-^{1}H coupling constant in ethylene and acetylene.

For ethylene and acetylene the carbon atoms are sp^2 and sp hybridized and thus 33% and 50% s character, respectively. Using the above equation gives coupling constants of 165 Hz and 250 Hz, respectively.

(b) In cyclopropane, the ^{13}C-^{1}H coupling constant is 160 Hz. What is the hybridization of carbon in cyclopropane?

The carbon atoms in cyclopropane are (0.2)(160) = 32% s character. This corresponds roughly to an sp^2 hybridized carbon atom.

CHAPTER 14: INFRARED AND ULTRAVIOLET-VISIBLE SPECTROSCOPY

14.0 OVERVIEW
• Infrared and ultraviolet-visible spectroscopy give information about functional groups. **Infrared spectroscopy provides information about molecular vibrations**, while **ultraviolet-visible spectroscopy provides information about electronic transitions of pi and non-bonding electrons**. ✳
  - The main use of infrared spectroscopy is to determine the presence or absence of certain functional groups in a molecule.
  - The main use of ultraviolet-visible spectroscopy is to study molecules with double bonds, especially conjugated double bonds. Ultraviolet-visible spectroscopy is also used to analyze aromatic molecules and molecules with non-bonding electrons.

14.1 INFRARED SPECTROSCOPY
• The **vibrational infrared** region of electromagnetic radiation has wavelengths that extend from 2.5 μm to 25 μm. For the purposes of **infrared spectroscopy (IR spectroscopy)** it is useful to describe infrared radiation in terms of a unit called the **wavenumber** that is defined by the following:

$$\text{wavenumber} = \frac{10,000}{\lambda} \qquad \text{where } \lambda \text{ is wavelength measured in micrometers}$$

Wavenumbers are reported in the units of **cm^{-1}**.
• **Atoms** joined by covalent bonds can **vibrate in a quantized fashion**, that is only specific vibrational energy levels are allowed. The **energy of these vibrations** corresponds to that of the **vibrational infrared region**. Therefore, absorption of infrared radiation of the appropriate wavelength results in a vibrationally excited state. ✳
• **Infrared radiation is absorbed** if the **frequency** of radiation **matches** that of an **allowed vibrational transition**. Furthermore, the **bond(s)** undergoing the vibrational transition **must have a dipole moment**, and **absorption** of radiation must result in a **change of** that **dipole moment**. The **greater** this **change in dipole moment**, the **more intense** the **absorption**. **Infrared active** vibrations meet the above criteria. Note that symmetrical bonds such as those in homonuclear diatomics (Br_2, O_2, etc.) do not absorb infrared radiation, because they do not have a dipole moment. ✳
• **Nonlinear molecules** (molecules with branches, etc.) that have **n** atoms will have **3n-6 allowed fundamental vibrations**. The simplest vibrations are **stretching** and **bending**. Stretching vibrations can be symmetric or asymmetric. Bending vibrations can be relatively complicated motions such as scissoring, rocking, wagging, or twisting. ✳
• Useful information may be obtained from a simplified calculation that is based upon analyzing one bond at a time in a molecule, and thus ignoring the other bonds. The two atoms are assumed to be two masses on a spring, so a form of **Hooke's law** is used:

$$\begin{array}{ll} \text{frequency of vibration} = \dfrac{1}{2\pi c} \sqrt{\dfrac{NK}{\mu}} & \mu = \dfrac{m_1 \times m_2}{m_1 + m_2} = \text{reduced mass} \\ \text{(in } cm^{-1}) & \end{array}$$

Here **c** is the **speed of light** (2.998×10^{10} cm/sec), **N** is **Avogadro's number** (6.022×10^{23} atoms/mol), **K** is the **force constant of the bond** in dynes/cm, and μ is the **reduced mass** calculated as shown above using the mass of each atom calculated as grams per atom. The mass in grams per atom is calculated by dividing the mass of the element in grams per mol by Avogadro's number.
• An **infrared spectrophotometer** is the instrument that measures which frequencies of infrared radiation are absorbed by a given sample.
• The output is recorded in the form of a chart called an **infrared spectrum**. The **horizontal axis** of the chart is in wavelength plotted as **wavenumbers (cm^{-1})**. The **vertical axis** of the chart is **percent transmittance**, with **100% transmittance at the top** and **0% transmittance at the bottom**. Remember that **100% transmittance corresponds to no absorption**. Thus, absorption peaks are actually recorded as inverted signals that start at the top, come down to a point corresponding to maximum absorption, then go back to the top. *[Understanding this now will save a lot of confusion when it comes to interpretation of IR spectra.]* ✳
  - IR spectra are recorded from samples that are usually neat if they are liquids, or compacted wafers mixed with KBr if they are solids.

- Most **IR spectra are very complex.** This is because besides the fundamental vibration absorptions discussed above, there are **overtones** and **coupling peaks. Overtones** are **higher frequency harmonic vibrations** of fundamental vibrations that occur at integral multiples. For example, an absorption at 600 cm^{-1} can have weaker overtone peaks at 1200 cm^{-1}, 1800 cm^{-1}, and 2400 cm^{-1}. **Coupling peaks** result from the **coupling of two vibrations** by addition or subtraction in certain allowed combinations.
- **Characteristic absorptions** for different functional groups are recorded in **correlation tables.** The **intensity** of a particular absorption is referred to as being **strong (s), medium (m),** or **weak (w).**
- In general, most attention is paid to the region from 3500 cm^{-1} to 1000 cm^{-1} of an infrared spectrum, because most functional groups have characteristic absorptions in this area. Absorptions in the 1000 cm^{-1} to 400 cm^{-1} are much more complex and difficult to analyze, so this area is referred to as the **fingerprint region.**

14.2 INTERPRETING INFRARED SPECTRA

- **Alkanes** have **C-H stretching** vibrations near **2850-3000 cm^{-1}** and **methylene bending** at **1450 cm^{-1}**.
- **Alkenes** have a vinylic **C-H stretch** near **3000 cm^{-1}**. The **C=C stretching** near **1600-1680 cm^{-1}** is often **weak** and may not be visible.
- **Alkynes** with a **hydrogen on the triple bonded carbon** have a **C-H stretching** vibration near **3300 cm^{-1}** that is usually sharp and strong. The **C≡C stretching** occurs at **2150 cm^{-1}**.
- **Alcohols** have an **OH stretch** that **depends on the amount of hydrogen bonding** present in the sample. Normally, alcohol samples show **extensive hydrogen bonding** and a **broad absorption** near **3200-3650 cm^{-1}**. The **C-O stretch** occurs at **1050 cm^{-1}, 1100 cm^{-1}, and 1150 cm^{-1}** for **primary, secondary, and tertiary alcohols,** respectively.
- **Ethers** have a **strong C-O stretch** near **1000 to 1150 cm^{-1}**.
- **Carbonyl**-containing species such as aldehydes, ketones, carboxylic acids, and carboxylic acid derivatives also have **very strong** and **characteristic IR absorptions** in **1630 cm^{-1}** to **1810 cm^{-1}** region. As a result, infrared spectroscopy is especially useful for characterizing these species.

14.3 ULTRAVIOLET-VISIBLE SPECTROSCOPY

- **Absorption of ultraviolet-visible region electromagnetic radiation** corresponds to **transitions** in molecules **between electronic energy levels of pi and nonbonding electrons.** As a result, ultraviolet-visible spectroscopy is most useful for only those molecules with pi or nonbonding electrons. ✳
- The vacuum ultraviolet region refers to wavelengths below 200 nm and this region is not generally used. The near ultraviolet extends from 200 to 400 nm, and the visible region extends from 400 nm (violet light) to 700 nm (red light).
- **Ultraviolet and visible spectra** are **recorded** as a chart of **wavelength** along the **horizontal axis** and **absorbance (A)** on the **vertical axis.** Absorbance is a surprisingly complicated measure, and is defined as:

$$\text{Absorbance (A)} = \log \frac{I_0}{I}$$

Here I_0 is the **intensity of radiation incident** on the sample, and I is the **intensity of radiation leaving** the sample. ✳

- The **relationship between absorbance, concentration, and length of sample** is known as the **Beer-Lambert law:**

$$\text{Absorbance (A)} = \varepsilon c l$$

Here ε is a proportionality constant called the **molar absorptivity, c** is the **concentration of the solute** in mol/liter and l is the **length of the cell** in cm.
- This relationship means that absorbance is linearly dependent on concentration.
- The value of ε ranges from 0 to 10^6 for different molecules, and absorptions with ε greater than 10^4 are considered high-intensity absorptions.
- **Absorption** of radiation in the **near ultraviolet-visible** spectrum results in a **transition of electrons** from a **lower energy occupied molecular orbital** to a **higher energy unoccupied molecular orbital.** The energy of this radiation can cause **$\pi \rightarrow \pi^*$ transitions** and **$n \rightarrow \pi^*$ transitions,** involving interactions with pi and nonbonding electrons, respectively. Important examples of these include the excitation of pi electrons in molecules with conjugated pi systems, and excitation of the nonbonding electrons on the oxygen atom of carbonyl groups. On the other hand, **ultraviolet-visible radiation is not of sufficient energy to affect electrons in sigma bonding molecular orbitals.**

between the various vibrations and rotations are superposed on the electronic absorptions. These are so closely spaced, that they are not resolved but appear as a broad band. This phenomenon emphasizes that ultraviolet-visible radiation is of higher energy than infrared radiation, and the corresponding absorption of ultraviolet-visible radiation involves higher energy transitions as well.

- The **more double bonds** that are conjugated to each other, the **smaller the energy spacing** between the **highest occupied** π **orbital**, and the **lowest unoccupied** π^* **orbital**. Thus, **the greater the number of the conjugated double bonds, the lower the wavelength (more toward the visible region) of the observed absorption**.

CHAPTER 14
Solutions to the Problems

Problem 14.1 Which is higher in energy?
(a) Infrared radiation of wavelength 1715 cm^{-1} or 2800 cm^{-1}?

The higher the wavenumber, the higher the energy. As a result, 2800 cm^{-1} is higher energy than 1705 cm^{-1}.

(b) Microwave radiation of frequency 300 MHz or 60 Hz?

Energy is directly proportional to frequency, so 300 MHz is higher energy than 60 Hz.

Problem 14.2 Without doing the calculation, which member of each pair do you expect to occur at the higher wavenumber?
(a) C=O or C=C stretching?

We must assume that C=O and C=C have similar force constants, since no actual values were provided. Using this assumption, the C=C bond vibrational absorption is predicted to occur at a higher wavenumber. The atomic weight of C is lower than O, and the frequency of vibration is predicted to be higher for bonds with atoms of lower atomic weight.

(b) C=O or C-O stretching?

Double bonds have higher force constants than single bonds, so the C=O bond will have a stretching frequency that occurs at a higher wavenumber than C-O.

(c) C≡C and C=C stretching?

Triple bonds have higher force constants than double bonds, so the C≡C bond will have a stretching frequency that occurs at a higher wavenumber than C=C.

(d) C-H or C-Cl stretching?

Assuming that C-H and C-Cl have similar force constants, then the C-H will have an absorbance at a higher wavenumber because the atomic weight of H is much smaller than Cl.

Problem 14.3 A compound shows strong, very broad IR absorption in the region 3300-3600 cm^{-1} and strong absorption at 1715 cm^{-1}. What functional groups account for these absorptions?

The very broad IR absorption centered at 3300-3600 cm^{-1} corresponds to C-H stretching and the strong absorption at 1715 cm^{-1} corresponds to a carbonyl C=O stretch.

Problem 14.4 Propanoic acid and methyl ethanoate are constitutional isomers. Show how to distinguish between them by IR spectroscopy.

$$CH_3CH_2COH$$
Propanoic acid

$$CH_3COCH_3$$
Methyl ethanoate
(Methyl acetate)

The key difference between these constitutional isomers is the OH group of propanoic acid that is not present in methyl ethanoate. The -OH group will show a strong, broad O-H stretch near 3300-3650 cm^{-1} in the IR spectrum of propanoic acid that will be absent in the spectrum of methyl ethanoate. Note that both spectra will have C-H stretching absorptions near 3000

cm^{-1}, C-O stretching absorptions at 1100 cm^{-1}, and a strong C=O stretching absorption near 1715 cm^{-1}.

Problem 14.5 Wavelengths in ultraviolet-visible spectroscopy are commonly expressed in nanometers; wavelengths in infrared spectroscopy are commonly expressed in micrometers. Carry out the following conversions:
(a) 2.5 μm to nanometers. **2.5 μm is equal to 2500 nm.**
(b) 200 nm to micrometers. **200 nm is equal to 0.2 μm.**

Problem 14.6 The visible spectrum of the tetraterpene β-carotene (C$_{40}$H$_{56}$, MW 536.89, the orange pigment in carrots) dissolved in hexane shows intense absorption maxima at 463 nm and 494 nm, both in the blue-green region. Because light of these wavelengths is absorbed by β-carotene, we perceive the color of this compound as that of the complement to blue-green, namely, red-orange.

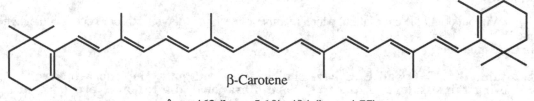

β-Carotene

λ_{max} 463 (log ε 5.10): 494 (log ε 4.77)

Calculate the concentration in milligrams per milliliter of β-carotene that gives an absorbance of 1.8 at λ_{max} 463.

$$c = 1.8/(1.00 \text{ cm})(10^{5.10}) = 1.43 \times 10^{-5} \text{ mol/liter}$$

The molecular weight of β-carotene, C$_{40}$H$_{56}$, is (12 x 40) + (1 x 56) = 536 g/mol. Concentration in units of milligrams per milliliter is equal to the value of concentration of grams per liter, so:

$$c = (1.43 \times 10^{-5} \text{ mol/liter})(536 \text{ g/mol}) = 7.7 \times 10^{-3} \text{ g/liter}$$

$$\boxed{= 7.7 \times 10^{-3} \text{ milligram / milliliter.}}$$

Problem 14.7 In molecular mechanics calculations, the energy required to stretch or compress a bond is given by $E_b = k_b (r - r_o)^2$ where k_b is a constant for a given type of bond, r_o is the equilibrium bond length, and r is the length of the bond in the stretched or compressed state. Values of these parameters for some common types of bonds are shown in the table:

Bond type	k_b (kcal/mol·nm^2)	r_o (nm)	$r_{5\%}$ (nm)	$r_{10\%}$ (nm)	$E_{5\%}$ stretch kcal/mol	$E_{10\%}$ stretch kcal/mol
C=O	58.0×10^3	0.123	0.0062	0.0123	2.23	8.78
C(sp^3)-C(sp^3)	20.0×10^3	0.153	0.0077	0.015	1.19	4.68
C(sp^3)-H	30.0×10^3	0.108	0.0054	0.010	0.875	3.50
O-H	45.0×10^3	0.096	0.0048	0.010	1.04	4.15

How much energy is required to stretch each type of bond by 5% of its length? By 10% of its length?

The answers in the table were calculated using the equation given in the question.

<u>Problem 14.8</u> In molecular mechanics calculations, the energy required to bend a bond is given by $E_q = k_q (q - q_0)^2$ where k_q is a constant for a given type of bond, q_0 is the equilibrium bond angle, and q is the angle of the bond in its bent state. Values of these parameters for some common types of bonds are shown in the table.

Bond type	k_q (kcal/mol·radians^2)	q_0 (degrees)	q_0 (radians)	$E_{5\% \text{ bend}}$ kcal/mol	$E_{10\% \text{ bend}}$ kcal/mol
C(sp^3)-C=O	86	121.6	2.12	0.966	3.86
C(sp^3)-C(sp^3)-C(sp^3)	70	109.5	1.91	0.638	2.55
H-C(sp^3)-H	40	109.5	1.91	0.365	1.46
C(sp^3)-C(sp^3)-O	50	109.5	1.91	0.456	1.82

How much energy is required to bend each type of bond by 5% ? By 10% ?

The answers are included in the table. The angle must be converted from degrees to radians (radians = degrees x 0.01745), since the bending constant is given in radians.

<u>Problem 14.9</u> Given your answers to the two previous problems, is it easier to bend bonds or to stretch bonds?

As can be seen in the tables, it is easier (takes less energy) to bend bonds than to stretch them.

Infrared Spectra
<u>Problem 14.10</u> Following are infrared spectra of methylenecyclopentane and 2,3-dimethyl-2-butene. Assign each compound its correct spectrum.

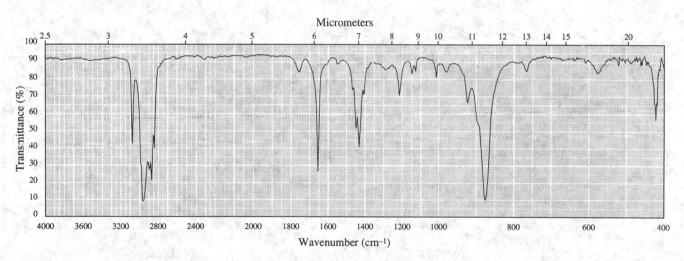

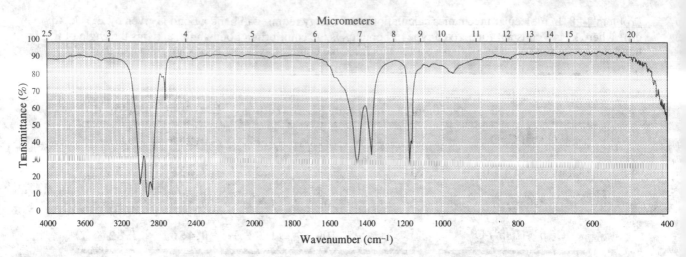

Both molecules have several C-H bonds and thus both spectra have C-H stretches and C-H bending vibrations at 2900 cm^{-1} and 1450 cm^{-1}, respectively. The alkene in methylenecyclopentane is unsymmetrical and therefore has a permanent dipole, so the C=C stretching will have a prominent band at 1654 cm^{-1} as seen in the first spectrum. On the other hand, 2,3-dimethyl-2-butene has a symmetrically substituted carbon-carbon double bond with no permanent dipole, so no C=C stretching should be prominent. In addition, the four methyl groups of 2,3-dimethyl-2-butene should give a prominent CH$_3$ bending band at 1375 cm^{-1} as is seen in the second spectrum.

Thus, the first spectrum corresponds to methylenecyclopentane and the second spectrum corresponds to 2,3-dimethyl-2-butene.

Problem 14.11 Following are infrared spectra of nonane and 1-hexanol. Assign each compound its correct spectrum.

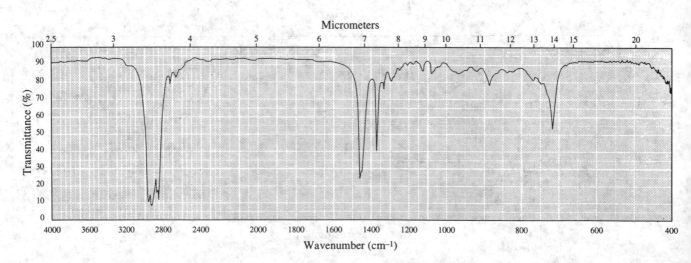

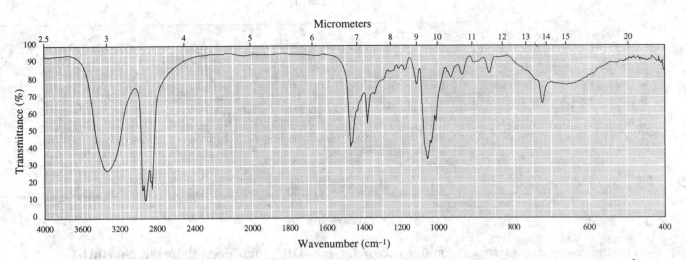

Both compounds have C-H bonds, so both spectra have C-H stretches and bends at 2900 cm⁻¹ and 1450 cm⁻¹, respectively. Furthermore, both compounds have methyl groups so both spectra have methyl bending vibrations at 1375 cm⁻¹. On the other hand, the 1-hexanol has an OH group, that will give rise to an O-H and C-O stretching vibrations at 3340 cm⁻¹ and 1050 cm⁻¹, respectively.

These two features are in the second spectra, so the second spectra must correspond to the 1-hexanol and the first spectra must correspond to nonane.

Problem 14.12 Following are infrared spectra of 2-methyl-1-butanol and *tert*-butyl methyl ether. Assign each compound its correct spectra.

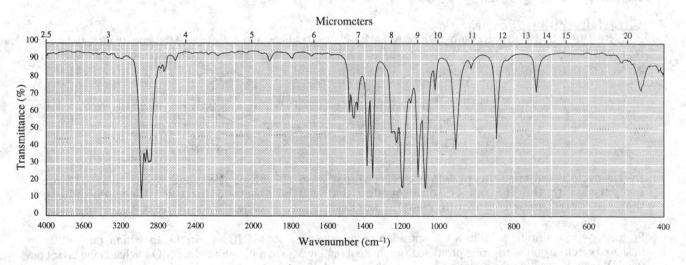

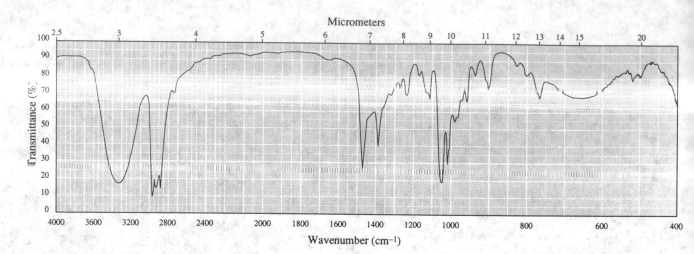

The molecules are extremely similar except for the -OH group present in the 2-methyl-1-butanol. The characteristic O-H stretch is present at 3625 cm^{-1} in the second spectrum.

> This verifies that the second spectrum corresponds to 2-methyl-1-butanol. Therefore, the first spectrum corresponds to *tert*-butyl methyl ether.

Problem 14.13 The IR C≡C stretching absorption in symmetrical alkynes is usually absent. Why is this so?

For a bond vibration to have a corresponding absorption of high intensity in the IR, that bond must have a dipole moment, and absorption of energy must result in a large change in that dipole moment. For symmetrical alkynes, there is no dipole moment associated with the C≡C bond, and thus no large change in dipole moment associated with bond stretching. Thus, the IR band is usually absent.

Ultraviolet-Visible Spectra
Problem 14.14 Show how to distinguish between 1,3-cyclohexadiene and 1,4-cyclohexadiene by ultraviolet spectroscopy.

1,3-Cyclohexadiene   1,4-Cyclohexadiene

In 1,3-cyclohexadiene, the pi bonds are conjugated so the absorption will occur near 217 nm compared with the 165 nm or so absorption that will be observed for the 1,4-cyclohexadiene that has only isolated (unconjugated) pi bonds.

Problem 14.15 Pyridine exhibits a UV transition of the type n --> π* at 270 nm. In this transition, one of the unshared electrons on nitrogen is promoted from a nonbonding MO to a pi antibonding MO. What is the effect on this UV peak if pyridine is protonated?

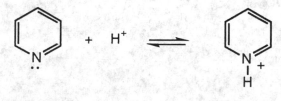

Pyridine   Pyridinium ion

When the pyridium ion is protonated, the lone pair of electrons on nitrogen is tied up in a bond to hydrogen. As a result, it is lower in energy compared with the unprotonated form, so it takes more energy to promote one of these electrons into the π^* orbital. Therefore, protonation shifts the absorbance peak to lower wavelength (higher energy).

Problem 14.16 The weight of proteins or nucleic acids in solution is commonly determined by UV spectroscopy using the Beer-Lambert law. For example, the ε of double-stranded DNA at 260 nm is 6670 $M^{-1}cm^{-1}$. The formula weight of the repeating unit in DNA (650) can be used as the molecular weight. What is the weight of DNA in 2.0 mL of aqueous buffer if the absorbance, measured in a 1-cm cuvette, is 0.75?

According to the Beer-Lambert law:

$$0.75 = (6670)(1)(x)$$

where x equals the unknown concentration of DNA in mol per liter. Rearranging gives:

$$x = 0.75/(6670)(1) = 1.12 \times 10^{-4} \text{ mol/liter}$$

The molecular weight used is that of a single base pair of DNA, namely 650 grams/mol. Furthermore, there is a total of 2.0 mL of solution, so the total weight of double stranded DNA in the sample is:

$$(1.12 \times 10^{-4} \text{ mol/liter})(650 \text{ grams/mol})(2.0 \times 10^{-3} \text{ liter}) =$$

$$\boxed{1.46 \times 10^{-4} \text{ gram}}$$

Problem 14.17 A sample of adenosine triphosphate (ATP) (MW 507, ε = 14,700 $M^{-1}cm^{-1}$ at 257 nm) is dissolved in 5.0 mL of buffer. A 250-μL aliquot is removed and placed in a 1-cm cuvette with sufficient buffer to give a total volume of 2.0 mL. The absorbance of the sample at 257 nm is 1.15. Calculate the weight of ATP in the original 5.0-mL sample.

The concentration of sample in the cuvette can be determined by using the Beer-Lambert law as follows:

$$1.15 = (14,700)(1.0 \text{ cm})(x \text{ mol/liter})$$
$$x = 1.15/14,700 = 7.82 \times 10^{-5} \text{ mol/liter}$$

The sample measured in the cuvette is actually diluted 0.250 in 2.0 or 1 in 8 compared to the original unknown sample, so the concentration of ATP in the original unknown sample is:

$$(8)(7.82 \times 10^{-5}) = 6.26 \times 10^{-4} \text{ mol/liter}$$

Since there is a total of 5.0 mL or 5×10^{-3} liter and ATP has a MW = 507 grams/mol then the weight of ATP in the original sample can be calculated as:

$$6.26 \times 10^{-4} \text{ mol/liter} = (x \text{ grams})/(506 \text{ grams/mol})(5 \times 10^{-3} \text{ liters})$$
$$x = (6.26 \times 10^{-4})(5 \times 10^{-3} \text{ liters})(506) =$$
$$\boxed{1.59 \times 10^{-3} \text{ grams} = 1.59 \text{ mg}}$$

Problem 14.18 Biochemical molecules are frequently sold by optical density (OD) units, where one OD unit is the amount of compound that gives an absorbance of 1.0 at its UV maximum in 1.0 mL of solvent in a 1-cm cuvette. If the cost of 10.0 OD units of a DNA polymer, ε = 6600 $M^{-1}cm^{-1}$ at 262 nm, is $51, what is the cost per gram of this biochemical?

According to the information given, 10.0 OD units is equal to the following amount of compound in 1.0 mL (1×10^{-3} liters). Recall from problem 14.17 that the molecular weight used for DNA is 650 g/mol:

$$10.0 = (6600)(1 \text{ cm})(1/650 \text{ g/mol})(x \text{ grams}/1 \times 10^{-3} \text{ liters})$$
$$x = (650 \text{ g/mol})(10.0)(1 \times 10^{-3} \text{ liters})/(6600) = 9.8 \times 10^{-4} \text{ grams}$$

Converting to dollars per gram gives:

$$(x \text{ dollars})/\text{gram} = (\$51)/(9.8 \times 10^{-4} \text{ gram})$$
$$\boxed{= \$52,000 \text{ !}}$$

UV-visible Spectroscopy: optical density (OD)

Problem 14.19 The Beer-Lambert law applies to IR spectroscopy as well as to UV. Whereas ultraviolet spectra are a plot of absorbance (A) versus wavelength, IR spectra are a plot of percent transmittance (%T) versus wavenumber. Absorbance and percent transmittance are related in the following way:

$$\text{Absorbance (A)} = \log \frac{T_o}{T}$$

where T_o is the baseline (100%) transmittance; and T is the transmittance at the peak.
(a) What is the absorbance of a peak with 10% transmittance in an IR spectrum?

$$\textbf{Absorbance} = \log (100\%/10\%) = \log (10) = 1.0$$

(b) If the concentration of this sample is halved, how does the absorbance and percent transmittance change?

If the concentration of the sample is halved, the absorbance is halved as well. On the other hand, the change in transmittance is harder to calculate and is given by:

$$0.5 = \log (T_o/T) \text{ so } 10.0^{0.5} = T_o/T = 3.16$$
$$\text{thus } T = T_o/3.16 \text{ so it is decreased by 31\%}$$

Combined Spectral Problems

Problem 14.20 Compound A, a hydrocarbon, bp 81°C, decolorizes a solution of bromine in carbon tetrachloride. Following are its mass spectrum, ^{1}H-NMR spectrum, and infrared spectrum. Compound A is transparent to ultraviolet-visible radiation.

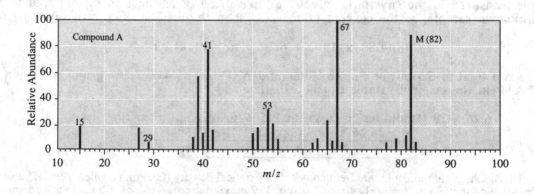

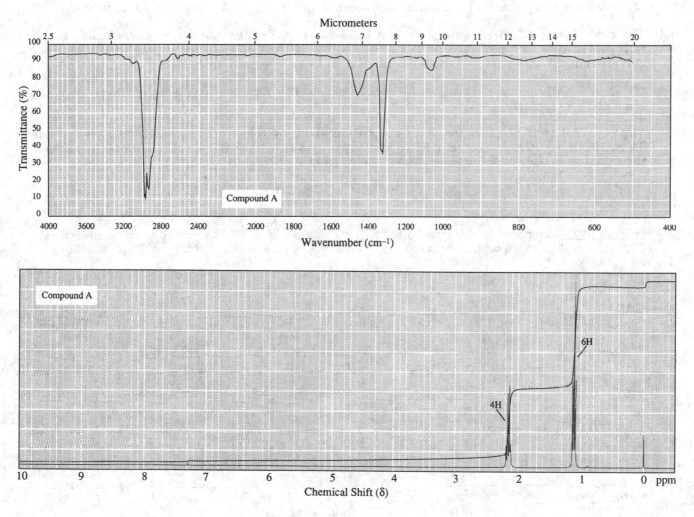

Compound A

Compound A

(a) What is the molecular formula of compound A?

Based on the mass spectrum, compound A has a MW of 82. This is only possible for a hydrocarbon with a molecular formula of C_6H_{10}.

(b) From its molecular formula, calculate the index of hydrogen deficiency of compound A. How many rings are possible for this compounds? How many pi bonds?

The index of hydrogen deficiency for compound A is 2. Compound A could have 2 rings, 2 pi bonds, 1 ring and 1 double bond, or 1 triple bond.

(c) Propose a structural formula for compound A consistent with the spectral information.

Because compound A decolorizes the bromine solution, it must have at least 1 pi bond. Furthermore, the ¹H-NMR shows only two types of hydrogens and the IR does not show any alkene or alkyne stretching vibrations. Thus, compound B is a highly symmetrical compound, and must be 3-hexyne.

$$\overset{1.1}{C}H_3\overset{2.15}{C}H_2C \equiv C\overset{2.15}{C}H_2\overset{1.1}{C}H_3$$

3-Hexyne

(d) Account for the presence of peaks in the mass spectrum of compound A at *m/z* 67, 53, 41, 29, and 15.

The peak at *m/z* 67 corresponds to [CH₃CH₂C≡CCH₂]⁺, the peak at *m/z* 53 corresponds to [CH₃CH₂C≡C]⁺, the peak at *m/z* 41 corresponds to [CH₂=CH-CH₂]⁺, the peak at *m/z* 29 corresponds to [CH₃CH₂]⁺, and the peak at *m/z* 15 corresponds to [CH₃]⁺.

(e) The ¹³C-NMR spectrum of compound A shows peaks at δ 12.4, 14.4, and 80.9. Assign each carbon in compound A its appropriate ¹³C chemical shift.

The chemical shift becomes larger as the carbon atoms get closer to the center of the molecule. Thus, the CH₃ carbons correspond to the signal at δ 12.4, the CH₂ carbons correspond to the signal at δ 14.4, and the -C≡ carbons correspond to the signal at δ 80.9.

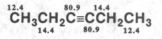

Problem 14.21 Compound B is a liquid, bp 122°C. Following are its ¹H-NMR spectrum and infrared spectrum. Compound B shows a molecular ion peak at *m/z* 136 and an M+2 peak of almost equal intensity at *m/z* 138.

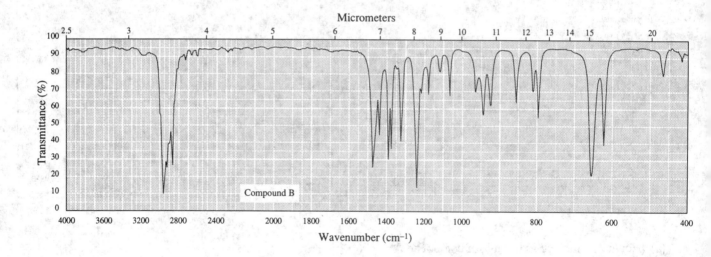

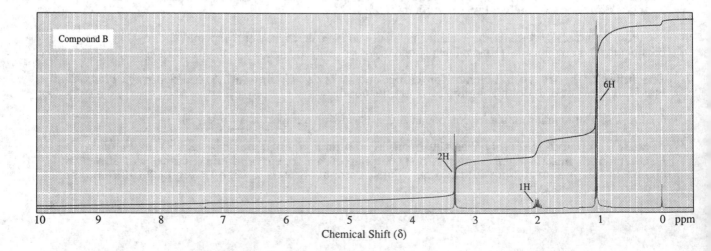

(a) From this information, deduce the structural formula of compound B.

From the mass spectrum, the MW is 136. The IR shows no alkenes or alcohols. The ^{1}H-NMR shows that there are three types of hydrogens in a 2:1:6 ratio. These ratios along with the observed splitting pattern correspond to a $(CH_3)_2CH-CH_2-$ group. This piece has a molecular weight of 57 leaving 79 in the formula. Since bromine corresponds to 79, this means the spectra corresponds to isobutyl bromide.

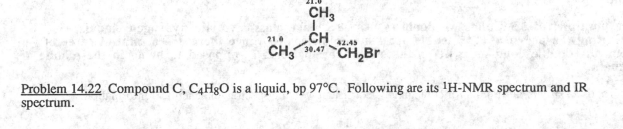

Isobutyl bromide

(b) Account for the presence of peaks in its mass spectrum at m/z 138, 136, 123, 121, 43, and 41.

The peak at m/z 138 corresponds to $C_4H_9{}^{81}Br$, the peak at m/z 136 corresponds to $C_4H_9{}^{79}Br$, the peak at m/z 123 corresponds to loss of a methyl group $[CH_3-CH-CH_2-Br]^+$, the peak at m/z 121 corresponds to $[CH_2=C-CH_2-Br]^+$, the peak at m/z 43 corresponds to $[CH_3-CH-CH_3)]^+$, and the peak at m/z 41 corresponds to $[CH_2=CH-CH_2]^+$.

(c) The DEPT ^{13}C-NMR spectrum of compound B shows signals at δ 21.0 (CH_3), 30.47 (CH), and 42.45 (CH_2). Assign each carbon in compound B its appropriate ^{13}C chemical shift.

The chemical shifts of the carbon atom signals increase as they get closer to the bromine atom. Thus, the methyl groups correspond to the signal at δ 21.0, the CH carbon atom corresponds to the signal at δ 30.47, and the $-CH_2-Br$ carbon atom corresponds to the signal at δ 42.45.

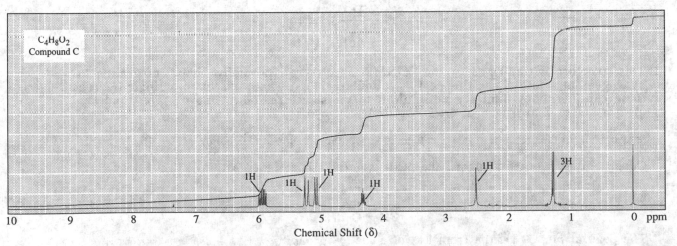

Problem 14.22 Compound C, C_4H_8O is a liquid, bp 97°C. Following are its ^{1}H-NMR spectrum and IR spectrum.

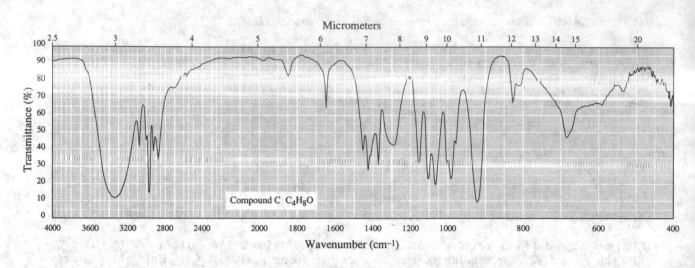

(a) What is the index of hydrogen deficiency of compound C? How many pi bonds can it contain?

The index of hydrogen deficiency is 1, so compound C must contain 1 ring or 1 pi bond.

(b) What information can you get from the infrared spectrum about the oxygen-containing functional group?

The broad peak at 3400 cm⁻¹ indicates the oxygen atom is contained in an -OH group.

(c) What information can you learn from the ¹H-NMR spectrum about the presence or absence of vinylic hydrogens.

The multiplet at δ 5.9 and two doublets near δ 5.1 are classic vinylic hydrogen signals, indicating the presence of a double bond in the molecule. Since there are a total of three of them, the double bond must be on the end of the molecule as opposed to being in the middle of a chain.

(d) Propose a structural formula for compound C consistent with the spectral and chemical information.

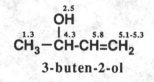

3-buten-2-ol

This structure is confirmed by the ¹H-NMR since there is a methyl signal integrating to 3 hydrogens at δ 1.3. Since this peak is a doublet, it must be adjacent to a single hydrogen, and these considerations limit the possibilities to the structure shown above. The singlet integrating to 1 hydrogen at δ 2.5 is the alcohol hydrogen.

(e) Account for the presence of peaks in the mass spectrum of compound C at m/z 71, 57, 45, 27, and 15.

Consistent with other alcohols, the peak at *m/z* 71 is assigned as:

$$\overset{+}{O}H$$
$$\|$$
$$CH_3-C-CH=CH_2$$

The peak at *m/z* 57 results from loss of a methyl group:

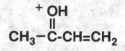

The peak at m/z 45 results from cleavage on the other side of the alcohol:

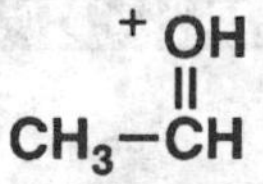

$$\overset{\overset{+ \text{ OH}}{\|}}{CH_3-CH}$$

The peak at m/z 27 is the other part of that fragmentation, namely $[CH_2=CH]^+$ and the peak at m/z 15 is the methyl group $[CH_3]^+$.

(f) Account for the splitting patterns of single hydrogens at δ 5.1, 5.3, and 5.8. (*Hint:* Review the ^{1}H-NMR spectrum of vinyl acetate (Figure 13.19)..

The two terminal hydrogens on the alkene are in different environments, so they both couple to the =CH- hydrogen. This same =CH- hydrogen is also split by the adjacent hydrogen on the alcoholic carbon atom. Thus, the signal at δ 5.8, the =CH- hydrogen, is actually a doublet of doublet of doublets. The coupling between the two terminal hydrogens is very small and not observed in this spectrum. Thus, each of the terminal hydrogens gives rise to a doublet, split by the =CH- hydrogen. Recall that the coupling between *trans* vinylic hydrogens is stronger than between *cis* hydrogens, so the terminal vinylic hydrogens are assigned as on the structure below.

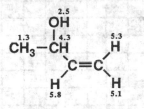

(g) The ^{13}C-NMR spectrum of compound C shows peaks at δ 23.1, 68.92, 113.5, and 143.3. Assign these peaks to the appropriate carbon on compound C.

The signal at δ 23.1 corresponds to the methyl carbon atom, the signal at δ 68.92 corresponds to the carbon with the -OH group attached, the signal at δ 113.5 corresponds to the terminal sp^2 carbon (=CH$_2$), and the signal at δ 143.3 corresponds to the sp^2 carbon atom (-CH=) that is adjacent to the carbon with the -OH group attached.

$$\overset{\overset{\text{OH}}{|}}{\underset{68.92 \quad\quad 113.5}{\overset{23.1 \quad\quad\quad 143.3}{CH_3-CH-CH=CH_2}}}$$

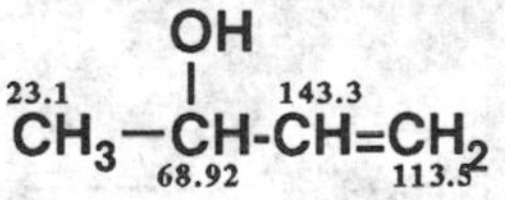

CHAPTER 15: ALDEHYDES AND KETONES

SUMMARY OF REACTIONS

Starting Material ＼ Product	Acetals	Alcohols		Aldehydes, Ketones	Alkanes	Alkenes	Carboxylic Acids	Cyanohydrins	α-Deuterated Aldehydes, Ketones	1,1-Diols	1,3-Dithianes	Enamines	Enol	Enolates	Esters	α-Halogenated Aldehydes, Ketones	Hemiacetals	Hydrazones	α-Hydroxy Alkynes	Imines	Ketones
Aldehydes							15A 15.13A*														
Aldehydes, Ketones		15B 15.14A	15C 15.14B		15D 15.14C			15E 15.6D	15F 15.12B	15G 15.8A						15H 15.11A	15I 15.12C	15J 15.10B			
Aldehydes, Ketones Alcohols	15K 15.8B																15L 15.8B				
Aldehydes, Ketones Amines												15M 15.10A								15N 15.10A	
Aldehydes, Ketones Grignard Reagents		15O 15.6A																			
Aldehydes, Ketones Organolithium Reagents		15P 15.6B																			
Aldehydes, Ketones Phosphonium Ylides						15Q 15.7															
Aldehydes, Ketones Terminal Alkynes																			15R 15.6C		
Aldehydes, Ketones Thiols											15S 15.9										
1,3-Dithianes		15T 15.9																			
1,3-Dithianes Alkyl Halides																					15U 15.9
Keto													15V 15.11B								
Ketones															15W 15.13B						
Methyl Ketones							15X 15.12														

*Section in book that describes reaction.

REACTION 15A: OXIDATION OF AN ALDEHYDE TO A CARBOXYLIC ACID (Section 15.13A)

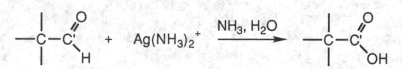

- **Aldehydes** are easily **oxidized to carboxylic acids** by various oxidizing agents including the **Tollens' reagent** shown above, as well as $KMnO_4$, $K_2Cr_2O_7$, H_2O_2, and O_2. The Tollens' reagent is especially interesting because a silver mirror can deposit on the side of the flask as the reaction is carried out. ✳
- Note that aldehydes will react with molecular oxygen to give carboxylic acids, so liquid aldehydes are usually stored under an inert atmosphere, such as pure nitrogen, for long term-storage.

REACTION 15B: CATALYTIC REDUCTION (Section 15.14B)

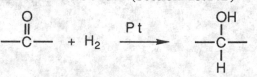

- **Aldehydes and ketones** are easily **reduced to alcohols** in the presence of **hydrogen** and catalysts such as **platinum, palladium, nickel, ruthenium or a copper-chromium complex.** ✳
- Unfortunately, other functional groups that may be present in a molecule, such as double and triple bonds, may also be reduced under these conditions. As a result, the metal hydride reductions described below are more common.

REACTION 15C: METAL HYDRIDE REDUCTIONS (Section 15.14C)

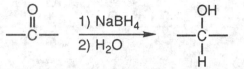

- **Aldehydes** are **reduced** by **sodium borohydride** (NaBH$_4$) in methanol, ethanol, or mixtures of these solvents with water to form the corresponding **alcohols**. Water is required to convert the initially formed tetraalkoxyborate into the product alcohol and boric acid salts.∗
- Lithium aluminum hydride (LiAlH$_4$) also reduces aldehydes and ketones to alcohols. LiAlH$_4$ reacts violently with water and other protic solvents, so the first step of the reactions is carried out in aprotic solvents, such as diethyl ether or tetrahydrofuran. Once the tetraalkoxy aluminate is formed, it is reacted with water in a second step to generate the product alcohol. **LiAlH$_4$ also reduces carboxylic acids** and their derivatives, whereas **NaBH$_4$ does not**. Neither reagent will reduce carbon-carbon double bonds.
- The mechanism of the reaction involves an initial transfer of a nucleophilic hydride ion from the metal reagent to the electrophilic carbon atom of the carbonyl group. A negatively-charged alkoxide species (negative charge on the oxygen atom) that makes an ionic bond to the metal is the product of this step. This process repeats itself three more times until all of the reducing equivalents are used up. The resulting tetraalkoxy metal species are then hydrolyzed by water to form the product alcohols.

REACTION 15D: REDUCTION OF A CARBONYL GROUP TO A METHYLENE GROUP (Section 15.14C)

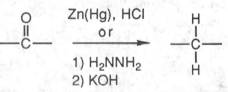

- **Aldehydes and ketones** can be reduced to the **corresponding methylene compounds** by either of two reactions. The first, the **Clemmensen reduction,** uses amalgamated zinc, Zn(Hg), in concentrated HCl. The second reaction, the **Wolff-Kishner reduction**, uses treatment with hydrazine to form a hydrazone (Reaction 15J, Section 15.10B), that is then converted to the methylene compound by reaction with potassium hydroxide. ∗
- Notice that the Clemmensen reduction uses acid, while the Wolff-Kishner reduction uses base. Thus, the Clemmensen reduction can be used with molecules that have other groups sensitive to base but not acid, while the Wolff-Kishner reduction can be used with molecules that have other groups sensitive to acid but not base.

REACTION 15E: REACTION WITH HCN TO FORM CYANOHYDRINS (Section 15.6)

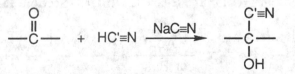

- **Aldehydes and ketones** react with **HCN** to form **cyanohydrins**. Equilibrium favors cyanohydrin formation for aldehydes and most ketones, except for sterically hindered aliphatic or aryl ketones. ∗
- The mechanism of cyanohydrin formation involves an initial attack of the nucleophilic cyanide ion on the electrophilic carbonyl carbon atom. The resulting tetrahedral carbonyl addition intermediate is protonated to give the final product.
- The main value of cyanohydrins is that they can be converted via dehydration into α,β-unsaturated nitriles, or via acid-catalyzed hydrolysis to α-hydroxycarboxylic acids, or converted to α-aminoalcohols using catalytic hydrogenation.

REACTION 15F: DEUTERIUM EXCHANGE AT AN α-CARBON (Section 15.12B)

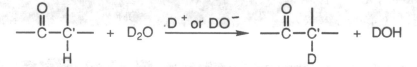

- **Hydrogen atoms** on the α-**carbon atoms of aldehydes** and **ketones** can be **exchanged with deuterium** in the presence of **acid or base catalysts**. ✳
- The mechanism involves acid- or base-catalyzed enol formation, followed by incorporation of the deuterium as the enol form converts back to the keto form
- This is a useful method from preparing deuterated samples, including the deuterated solvents like acetone-d_6 used as an NMR solvent.

REACTION 15G: HYDRATION (Section 15.8A)

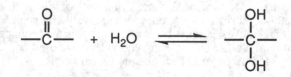

- **Aldehydes and ketones** react with **water** to form **1,1-diols**, also referred to as geminal diols.✳
- The position of equilibrium greatly favors the carbonyl form rather than the diol form for most ketones, and all but the most simple aldehydes such as formaldehyde.

REACTION 15H: ACIDITY OF THE α-HYDROGEN; ENOLATE FORMATION (Section 15.11A)

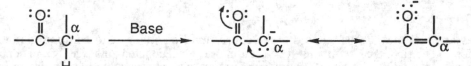

- **Hydrogen atoms** on the α-**carbon atoms of aldehydes and ketones** are relatively acidic ($pK_a \sim 18$) for a C-H bond. This is because the resulting **anion, called an enolate, is resonance stabilized** as shown by the two resonance forms drawn above. ✳
- Enolates are important because they are strong nucleophiles, capable of reacting with electrophiles including other carbonyl groups thereby forming carbon-carbon bonds.
- Enolates are involved in the mechanism of base-promoted reactions such as halogenation at an α-carbon (Reaction 15I, Section 15.12C) and the haloform reaction (Reaction 15X, Section 15.12C).

REACTION 15I: HALOGENATION AT THE α-CARBON (Section 15.12C)

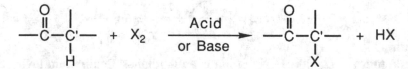

- **Hydrogen atoms** on the α-**carbon atoms of aldehydes and ketones** can be **exchanged for a halogen atom** in the presence of **acid or base**. ✳
- The **mechanism in acid** involves **acid-catalyzed enol formation**, followed by **reaction of X_2** with the enol double bond.
- **In base**, the mechanism involves **formation of the enolate anion**, that **reacts with X_2 to give the product**. Since one mol of base is **consumed** for each mol of starting material used, the reaction is **base-promoted** (as opposed to base-catalyzed). The resulting α-**halogenated aldehyde or ketone** is even **easier to deprotonate** than the starting aldehyde or ketone, so **successive base-promoted halogenation** becomes **more and more rapid**. Thus, as opposed to acid-catalyzed α-halogentaion, the

base-promoted reaction is not generally synthetically useful because it cannot be reliably stopped after a single substitution.

REACTION 15J: ADDITION OF HYDRAZINE AND ITS DERIVATIVES (Section 15.10B)

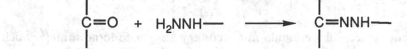

- **Hydrazine (NH_2NH_2)** and its derivatives react with **aldehydes and ketones** to produce **hydrazones.**∗

REACTION 15K: ADDITION OF ALCOHOLS TO FORM ACETALS (Section 15.8B)

$$O = \overset{|}{\underset{|}{C}} \ + \ 2 \ \overset{|}{\underset{|}{-C'-OH}} \ \overset{H^+}{\rightleftharpoons} \ \overset{|}{\underset{|}{-C'}} - O - \overset{|}{\underset{|}{C}} - O - \overset{|}{\underset{|}{C'}} - \ + \ H_2O$$

- **Aldehydes and ketones** react with **alcohols** in the presence of an **acid catalyst** to generate an **acetal**. The process is reversible, so adding excess alcohol drives the reaction toward acetal formation, adding excess water drives the reaction the other way. ∗
- The reaction is initiated by reversible protonation of the carbonyl oxygen atom. This makes the carbonyl group electrophilic enough to react with the nucleophilic oxygen atom of an alcohol, and following loss of a proton, a hemiacetal is formed. Next, the oxygen atom of the remaining -OH group is protonated to generate a positively-charged oxonium ion that then leaves to produce a highly electrophilic cation. Another nucleophilic alcohol molecule reacts with this cation then loses a proton to generate the final acetal product. Acetals are generally more stable than hemiacetals, so acetals are usually the product isolated from the reaction.
- Diols such as ethylene glycol, produce cyclic acetals, that can be used as base-stable protecting groups for the carbonyl groups of aldehydes and ketones.

REACTION 15L: ADDITION OF ALCOHOLS TO FORM HEMIACETALS (Section 15.8B)

$$O = \overset{|}{\underset{|}{C}} \ + \ \overset{|}{\underset{|}{-C'-OH}} \ \overset{H^+}{\rightleftharpoons} \ HO - \overset{|}{\underset{|}{C}} - O - \overset{|}{\underset{|}{C'}} -$$

- **Aldehydes and ketones** react with **alcohols** in the presence of **acid catalysts** to generate **hemiacetals**. Usually, the reaction immediately continues with addition of another molecule of alcohol to form the more stable acetal (Reaction 15K; Section 15.8B), and the hemiacetal cannot be isolated.
- The reaction is initiated by reversible protonation of the carbonyl oxygen atom. This makes the carbonyl group electrophilic enough to react with the nucleophilic oxygen atom of an alcohol, and following loss of a proton, a hemiacetal is formed.
- Hemiacetal formation is highly favored when the -OH and carbonyl group are part of the same molecule and can form a hemiacetal with a five- or six-membered ring.

REACTION 15M: ADDITION OF SECONDARY AMINES: FORMATION OF ENAMINES (Section 15.10A)

- **Aldehydes and ketones** react with **secondary amines** to form **enamines**. ∗
- The mechanism involves an initial attack of the nucleophilic nitrogen atom on the electrophilic carbonyl carbon atom. The proton on the nitrogen is then transferred to the oxygen to create a tetrahedral carbonyl addition intermediate. Acid-catalyzed dehydration leads to the enamine.
- Enamines are important because they are useful for certain synthetic reactions.

REACTION 15N: ADDITION OF AMMONIA AND ITS DERIVATIVES: FORMATION OF IMINES (Section 15.10A)

$$O{=}\overset{|}{\underset{|}{C}} \ + \ H_2N{-} \ \overset{H^+}{\rightleftharpoons} \ \overset{|}{C}{=}N{-} \ + \ H_2O$$

- **Aldehydes and ketones** react with **ammonia and primary amines** to form **imines**, which are also called **Schiff bases.** ✳
- The mechanism involves an initial attack of the nucleophilic nitrogen atom on the electrophilic carbonyl carbon atom. The proton on the nitrogen is then transferred to the oxygen to create a tetrahedral carbonyl addition intermediate. Acid-catalyzed dehydration involving loss of another proton on the nitrogen atom leads to the imine.

REACTION 15O: REACTION WITH GRIGNARD REAGENTS (Section 15.6A)

$$O{=}\overset{|}{\underset{|}{C}} \ + \ {-}\overset{|}{\underset{|}{C}}'{-}MgX \ \longrightarrow \ MgX^+O^-{-}\overset{|}{\underset{|}{C}}{-}\overset{|}{\underset{|}{C}}'{-} \ \overset{H_2O}{\longrightarrow} \ HO{-}\overset{|}{\underset{|}{C}}{-}\overset{|}{\underset{|}{C}}'{-}$$

- **Formaldehyde, other aldehydes,** and **ketones** react with **Grignard reagents** to produce **primary, secondary, and tertiary alcohols,** respectively. ✳
- The mechanism of the reaction involves an initial attack of the nucleophilic Grignard reagent on the electrophilic carbonyl carbon atom. The magnesium alkoxide is protonated after water is added to the reaction to generate the product alcohol.
- Grignard reagents **react with CO$_2$** to give **carboxylic acids.**
- Grignard reactions are important because they involve the formation of carbon-carbon single bonds.

REACTION 15P: REACTION WITH ORGANOLITHIUM REAGENTS (Section 15.6B)

$$O{=}\overset{|}{\underset{|}{C}} \ + \ {-}\overset{|}{\underset{|}{C}}'{-}Li \ \longrightarrow \ Li^+O^-{-}\overset{|}{\underset{|}{C}}{-}\overset{|}{\underset{|}{C}}'{-} \ \overset{HCl, H_2O}{\longrightarrow} \ HO{-}\overset{|}{\underset{|}{C}}{-}\overset{|}{\underset{|}{C}}'{-}$$

- **Organolithium** reagents react with **aldehydes and ketones** to produce **alcohols.** ✳
- The mechanism of the reaction involves an initial attack of the organolithium reagent on the electrophilic carbonyl carbon atom. The lithium alkoxide is protonated after water is added to the reaction to generate the product alcohol.
- In general, organolithium reagents are more reactive than Grignard reagents and give higher yields.

REACTION 15Q: THE WITTIG REACTION (Section 15.7)

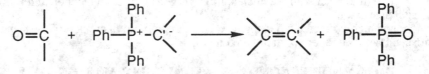

- **Aldehydes and ketones** react with **alkyltriphenylphosphonium ylides** to produce **alkenes** and **triphenylphosphine oxide,** a reaction that is referred to as a **Wittig reaction.** ✳
- The mechanism of the reaction involves cycloaddition of the alkyltriphenylphosphonium ylide with the aldehyde or ketone carbonyl to form a four-membered ring oxaphosphetane intermediate. Upon warming, the oxaphosphetane intermediate decomposes to give the alkene and triphenylphosphine oxide.
- This reaction is important because it involves the formation of a carbon-carbon double bond from two different fragments.
- The alkyltriphenylphosphonium ylide is produced by an S$_N$2 reaction between the nucleophilic alkyltriphenylphosphine and a primary or secondary alkyl halide to give a triphenylphosphonium halide. This is deprotonated with a strong base such as butyllithium, NaH, or NaNH$_2$ to give the alkyltriphenylphosphonium ylide.

REACTION 15R: REACTION WITH ALKALI METAL SALTS OF TERMINAL ALKYNES (Section 15.6C)

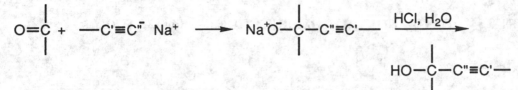

- **Aldehydes and ketones** react with the **alkali metal salts of terminal alkynes** to produce α-**hydroxyalkynes.** *
- The mechanism of the reaction involves attack of the nucleophilic alkyne anion on the electrophilic carbonyl carbon atom to produce a tetrahedral carbonyl addition intermediate that is protonated to give the α-hydroxyalkyne.
- The α-hydroxyalkynes can be reacted further to produce α-hydroxyketones or α-hydroxyaldehydes.

REACTION 15S: ADDITION OF SULFUR NUCLEOPHILES: FORMATION OF 1,3-DITHIANES (Section 15.9)

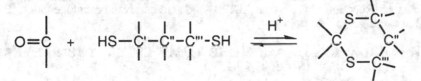

- **Aldehydes and ketones** react with **1,3-propanedithiol** to produce **cyclic thioacetals** called **1,3-dithianes.** *
- The mechanism is analogous to that of acetal formation (Reaction 15K, Section 15.8B).

REACTION 15T: HYDROLYSIS OF THIOACETALS (Section 15.9)

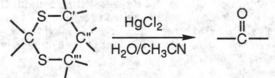

- **1,3-Dithianes** are **hydrolyzed** by reaction with **HgCl₂ in aqueous acetonitrile** to produce a carbonyl group, in the form of an aldehyde to ketone. *
- Since the 1,3-dithianes are produced from aldehydes and ketones (Reaction 15S, Section 15.9) and do not react with nucleophiles like carbonyl groups, 1,3-dithianes can be used as protecting groups for aldehydes and ketones i.e. with Grignard reagents.

REACTION 15U: ALKYLATION OF ALDEHYDE 1,3-DITHIANE ANIONS (Section 15.9)

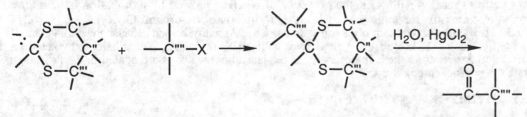

- Aldehyde 1,3-dithiane anions are produced by reaction of aldehyde 1,3-dithianes with the strong base butyllithium.
- The **anions of aldehyde 1,3-dithianes** are strong nucleophiles, and **react with primary alkyl halides** via an S_N2 mechanism to produce ketone 1,3-dithianes that are converted to the final product **ketones** by hydrolysis using aqueous HgCl₂. *
- Using this scheme, an aldehyde can be converted into a ketone.

- An aldehyde 1,3-dithiane anion also reacts as a nucleophile with the carbonyl group of aldehydes and ketones to give hydroxy ketones following removal of the 1,3-dithiane with aqueous $HgCl_2$ (Reaction 15.T, Section 15.9).

REACTION 15V: KETO-ENOL TAUTOMERIZATION (Section 15.11B)

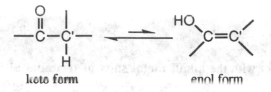

keto form   enol form

- The **keto forms of aldehydes and ketones** are in **equilibrium with the enol form**. This interconversion is called **keto-enol tautomerization**. ✳
- For most simple aldehydes and ketones, the equilibrium lies far to side of the keto form.
- Interconversion of these two forms is catalyzed by acids and bases. Remember that catalysis only increases the rate at which the two forms interconvert; even in the presence of a catalyst the keto form still predominates to the same extent.
- The enol form is important to keep in mind because it has unique reactivities that are exploited in some transformations, despite the fact that it is present in only small amounts at any one time. For example, the α-deuterium exchange (Reaction 15F, Section 15.12B) and acid-catalyzed α-halogenation (Reaction 15I, Section 15.12C) can involve the enol form.

REACTION 15W: OXIDATION OF A KETONE TO AN ESTER: THE BAEYER-VILLIGER OXIDATION (Section 15.13B)

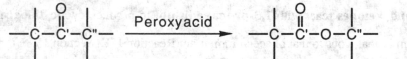

- A **ketone** can be oxidized to an **ester** by **peroxyacids** such as **trifluoroperoxyacetic acid** or **peroxyacetic acid** in a reaction referred to as the **Baeyer-Villiger oxidation**. ✳
- With unsymmetrical ketones, the observed migratory aptitude is:

    3° alkyl > 2° alkyl > benzylic = aryl > 1° alkyl > methyl

  For example, a 3° alkyl group will end up bonded to the ester oxygen atom in preference to a 2° alkyl group, if both were present in the original ketone.

REACTION 15X: THE HALOFORM REACTION (Section 15.12)

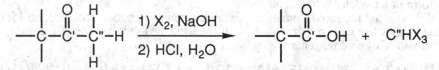

- **Methyl ketones** reacted with X_2 in the **presence of NaOH** followed by protonation in acid are converted into **carboxylic acids** accompanied by formation of **haloform** (CHX_3).✳
- The mechanism of the reaction involves formation of a trihalomethylketone because the enolate anion formed on the methyl side of the ketone reacts three times with the X_2 (halogen). The trihalomethylketone is then hydrolyzed in base to form the haloform and carboxylate anion, which is protonated in the last step to generate the product carboxylic acid.

SUMMARY OF IMPORTANT CONCEPTS

15.0 OVERVIEW
• The **carbonyl group** is one of the most important functional groups in organic chemistry, and is found in aldehydes, ketones, carboxylic acids, and their functional derivatives. ✳

15.1 STRUCTURE AND BONDING

- The **characteristic functional group** of aldehydes and ketones is the **carbonyl group**. In **aldehydes**, the carbonyl carbon atom is bonded to **at least one hydrogen atom**. In **ketones**, the carbonyl carbon atom is bonded to **two other carbon atoms** of alkyl or aryl groups.
- The carbonyl group is made up of a **carbon atom and oxygen atom**, both in the sp^2 **hybridization state**. The **carbonyl double bond** is composed of **one sigma bond** formed by **overlap of sp^2 orbitals** of **carbon and oxygen**, and **one pi bond** formed by overlap of **parallel 2p orbitals** on these atoms. **Two lone pairs** on oxygen reside in sp^2 **hybridized orbitals**. *

15.2 NOMENCLATURE

- According to **IUPAC nomenclature:**
  - An **aldehyde** is named by changing the suffix **e** to **al**. Unsaturated aldehydes are designated with the infixes **en** or **yn** for carbon-carbon double and triple bonds, respectively. The number of the aldehyde carbonyl carbon atom is always 1. For cyclic molecules in which the -CHO group is attached to the ring, the suffix **carbaldehyde** is used.
  - A **ketone** is named by changing the suffix **e** to **one**. The parent chain is chosen as the longest chain that contains the carbonyl group, and is numbered so that the carbonyl carbon atom has the lowest possible number.
  - The IUPAC nomenclature retains the common names benzaldehyde, and cinnamaldehyde for aldehydes, as well as acetone, acetophenone, and benzophenone for ketones.
  - When a complex molecule contains more than one functional group, an **order of precedence** is used to determine the name. Table 15.1 lists these functional groups in order of precedence that ranges from carboxylic acids at the high end to thiols at the low end. Each functional group has a suffix that is used if it is the highest ranking functional group in the molecule, and a prefix if it is not the highest ranking functional group. For example, aldehydes and ketones use the suffixes **al** and **one** if they are the highest ranking functional groups present, but the prefix of **oxo** in both cases if there are higher ranking functional groups present. The location of the aldehyde or ketone carbonyl group is indicated by the appropriate number.
- The **common name** for an **aldehyde** is analogous to that for the corresponding carboxylic acid, except the **ic** or **oic** suffix is replaced by the suffix **aldehyde**. The common names for ketones are derived by naming the two alkyl or aryl groups attached to the carbonyl group, followed by the word **ketone**.

15.3 PHYSICAL PROPERTIES

- The most **important electronic feature of carbonyl groups** is **their polarity**, with a **partial negative charge** on the more electronegative oxygen atom, and a corresponding **partial positive charge** on the **carbon atom**. *
  - The **polar nature of carbonyl compounds** means that they **interact via dipole-dipole interactions**, and thus have **higher boiling points** than corresponding hydrocarbons. Similarly, the lower-molecular-weight aldehydes and ketones are more soluble in water than the corresponding hydrocarbons. Dipole-dipole interactions are not as strong as hydrogen bonds, so **aldehydes and ketones** have **lower boiling points** and **lower water solubilities** than the **corresponding alcohols**.

15.4 SPECTROSCOPIC PROPERTIES

- In **mass spectrometry, aliphatic aldehydes** characteristically give fragments that result from cleavage of one of the groups bonded to the carbonyl carbon atom (α-**cleavage**), cleavage of a C-C bond that is adjacent to the carbonyl group (β-**cleavage**), and aldehydes with at least one γ-hydrogen atom undergo fragmentation by the **McClafferty rearrangement**. Aliphatic ketones characteristically give fragments that result from α-**cleavage** and the **McClafferty rearrangement**.
- ^{1}**H-NMR** is especially important for identifying aldehydes since the **aldehydic hydrogen atom** produces a **sharp singlet** that has a very characteristic chemical shift of δ **9.5** to δ **10.1**. This is a relatively empty region of most ^{1}H-NMR spectra, so this signal can usually be identified clearly. Signals for **hydrogens** on carbon atoms α **to a carbonyl group** of an aldehyde or ketone usually appear at δ **2.2** to δ **2.6**. The signal for **aldehyde and ketone carbonyl carbon** atoms appear at characteristic chemical shifts of δ **180** and δ **210**, respectively, in ^{13}C-NMR spectra. *
- In **infrared spectroscopy**, the **carbonyl group of aldehydes and ketones** shows a very **strong absorption between 1630 and 1810 cm^{-1}** that is the result of a **stretching vibration**. This signal is in a relatively empty region of the spectrum, so it is often used as a diagnostic indication for the presence of a carbonyl group. Furthermore, the exact location of the carbonyl stretching absorption is very sensitive to the local environment within the molecule, so structural information can be obtained by analyzing exactly where it occurs.*

• Ordinarily, **carbonyl groups** show only **weak absorptions** in the **ultraviolet region** due to the n→π* electronic transition. On the other hand, **carbonyl groups conjugated** to one or more carbon-carbon double bonds show a much **more intense absorption** due to a π→π* **transition**. The position of the transition also shifts to longer wavelengths with increasing conjugation.

15.5 REACTIONS

• **Carbonyl compounds react in several characteristic ways.** They are useful functional groups because a large number of different types of molecules react with carbonyl groups, **including several different reactions** that lead to **formation of new carbon-carbon bonds**. Carbon-carbon bond forming reactions are important because they allow the synthesis of larger, more complex carbon skeletons from smaller ones. *[A secret to learning carbonyl chemistry is to learn the characteristic reaction mechanisms, then figure out which reagents react by which reaction mechanism. In this way, the large number of reactions encountered in this chapter can be effectively considered in groups, thus simplifying greatly the task of learning. The following discussion is intended to help group the reactions, the details of which are described in the "Summary of Reactions" section.]* *

- As discussed above, the **carbonyl carbon atom possesses a partial positive charge**. Therefore, the first characteristic reaction of carbonyl groups is that they **react with nucleophiles** at the carbonyl carbon atom. The pi bond to oxygen is broken in this attack, leading to a negative charge on oxygen. This intermediate is called a tetrahedral carbonyl addition intermediate. The tetrahedral carbonyl addition intermediate is usually protonated to give an alcohol that is either the final product or sometimes reacts further to give additional substitution.

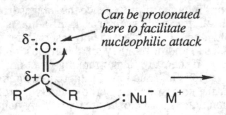

Tetrahedral carbonyl
addition intermediate

- A list of nucleophiles that react with carbonyl groups in this way includes **carbon nucleophiles** (Grignard reagents, organolithium reagents, anions of terminal alkynes, Wittig reagents, and cyanide anions), **oxygen nucleophiles** (water and alcohols), **sulfur nucleophiles** (thiols), and **nitrogen nucleophiles** (amines, hydrazines). *

15.6 ADDITION OF CARBON NUCLEOPHILES

• The **organometallic reagents** that act as carbon nucleophiles in these reactions are the **Grignard** and **organolithium reagents**. In both these types of reagents, a **carbon atom** is **bonded** to a more **electropositive metal atom**. Thus, the bond is polar covalent and polarized with a **partial negative charge on the carbon atom**. Because of this, it **behaves as a carbon nucleophile** in the reactions with carbonyl groups. *

• **Grignard reagents and organolithium reagents** are prepared from **alkyl halides** by reaction with the free metal.

• Because **Grignard and organolithium reagents** are also **highly basic**, they react with any species that has even moderately acidic protons such as water or alcohols. For this reason, when a Grignard or organolithium reagent reacts with a carbonyl group, a metal alkoxide is actually the product formed. The more acidic **water is added** to the reaction mixture in a second step to protonate the alkoxide and **generate the product alcohol**.

• Note that Grignard reagents cannot ordinarily be made from alkyl halides that also contain a carbonyl group, since such a Grignard reagent would immediately react with itself to make unusable mixtures. On the other hand, an aldehyde or ketone can be turned into a cyclic acetal by treatment with ethylene glycol and acid in the absence of water. The acetal will not react with Grignard reagents. At the end of the synthesis, the acetal can be converted back into the aldehyde or ketone be treatment with aqueous acid. The acetal is serving as a **protecting group** in this capacity, protecting the carbonyl function from unwanted reactions.

15.7 THE WITTIG REACTION
• **Alkylphosphonium ylides** react as carbon nucleophiles. The final product is an **alkene (Wittig reaction)**. ✳

15.8 ADDITION OF OXYGEN NUCLEOPHILES
• **Water and alcohols** react as oxygen nucleophiles, leading to 1,1-diols (also called gem diols) and acetals. That latter are produced via the hemiacetal intermediates. Note that acetals can be cyclic if a diol is used as starting material. ✳

15.9 ADDITION OF SULFUR NUCLEOPHILES
• **Thiols** react as nucleophiles with carbonyl groups. Perhaps the most useful example is the reaction of **1,3 dithiols** such as 1,3-propanedithiol that react to produce **1,3 dithianes**. ✳
- 1,3-Dithianes produced from aldehydes can be deprotonated to give anions that can act as carbon nucleophiles to react with electrophiles such as alkyl halides. This is yet another important way to make carbon-carbon bonds.
- 1,3-Dithianes can be hydrolyzed in the presence of $HgCl_2$ to give back the original aldehyde or ketone.

15.10 ADDITION OF NITROGEN NUCLEOPHILES
• **Nitrogen nucleophiles** (ammonia, 1° amines, and hydrazines) react as nucleophiles to produce molecules that have **carbon-nitrogen double bonds**. **Secondary amines** give **enamines**.

15.11 KETO-ENOL EQUILIBRIA
• A carbonyl group is usually described as a **keto form**, but it is actually **in equilibrium** with a very small amount of the **enol form**.

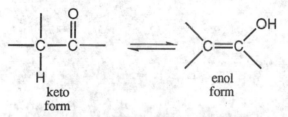

keto
form

enol
form

• Both **acid and base catalyze** the **interconversion** of the **keto** and **enol forms**. Remember that a catalyst does not change the position of an equilibrium, just the speed at which the species equilibrate. The enol form reacts as an alkene, so certain reactions take place with the small amount of enol that is present at any one time. For example, α-**deuteration** and α-**halogenation** involve **reactions** with the **enol form**. *[Note that the keto-enol equilibrium involves two different molecules, not contributing structures of a resonance hybrid. The atomic nucleii are in different positions in the keto and enol forms.]* ✳

15.12 REACTIONS AT THE α-CARBON
• The **hydrogens** on the carbon atoms adjacent to a carbonyl group (the so-called α-**hydrogens**) are **relatively acidic** (pK_a ` 18). The **anion produced** upon **deprotonation** of an α-hydrogen is called an **enolate ion**. The α-hydrogens are relatively acidic for a C-H bond because **enolate ions** are relatively **stable due to resonance stabilization**. The negative charge of the enolate ion is shared by the carbonyl oxygen and α-carbon atoms as predicted by the two most important contributing resonance structures. In addition, the electron-withdrawing carbonyl group weakens the C-H bond via an inductive effect. ✳

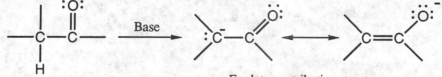

Enolate contributing structures

[Note that although the two enolate contributing structures have certain features that resemble keto and enol forms, the keto and enol forms are equilibrating molecules, while the enolate contributing structures together represent only a single resonance hybrid.]

- **Enolate ions** are strong **nucleophiles**, reacting primarily at the α-carbon atom with other electrophiles. This is interesting because aldehydes and ketones are usually thought of as acting like electrophiles, capable of reacting with other nucleophiles. By turning them into enolate ions they are converted into nucleophiles. ✳

15.13 OXIDATION

- Aldehydes can be oxidized to carboxylic acids. Ketones can only be oxidized under special conditions because ketones do not have any hydrogen atoms attached to the carbonyl carbon atom. Treatment of ketones with peroxyacids leads to esters (Baeyer-Villiger reaction).

15.14 REDUCTION

- Aldehydes and ketones are **reduced** by **hydride reagents** ($LiAlH_4$ or $NaBH_4$) to primary and secondary **alcohols**, respectively.
- The **carbonyl function** of **aldehydes** and **ketones** can also be reduced to a **methylene group** using two different methods; the Clemmensen and Wolff-Kishner reductions. These two reactions take place under completely different conditions, and the reaction chosen for a particular molecule might depend on what other function groups are present. For example, a Wolff-Kishner reduction, not a Clemmensen reduction, is appropriate for a compound that has other functional groups that will be destroyed by acid.

CHAPTER 15
Solutions to the Problems

<u>Problem 15.1</u> Write IUPAC names for these compounds. Specify configuration in (c).

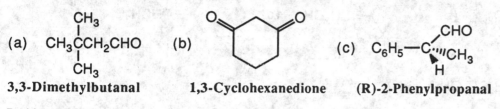

(a) 3,3-Dimethylbutanal   (b) 1,3-Cyclohexanedione   (c) (R)-2-Phenylpropanal

<u>Problem 15.2</u> Write structural formulas for all aldehydes of molecular formula $C_6H_{12}O$ and give each its IUPAC name. Which of these aldehydes are chiral?

$$CH_3-CH_2-CH_2-CH_2-CH_2-\overset{O}{\overset{\|}{C}}H$$

Hexanal

$$CH_3-\overset{CH_3}{\underset{|}{CH}}-CH_2-CH_2-\overset{O}{\overset{\|}{C}}H$$

4-Methylpentanal

$$CH_3-CH_2-\overset{*}{\underset{\underset{CH_3}{|}}{CH}}-CH_2-\overset{O}{\overset{\|}{C}}H$$

3-Methylpentanal
(chiral)

$$CH_3-CH_2-CH_2-\overset{*}{\underset{\underset{CH_3}{|}}{CH}}-\overset{O}{\overset{\|}{C}}H$$

2-Methylpentanal
(chiral)

$$CH_3-CH_2-\underset{\underset{\underset{CH_3}{|}}{\overset{|}{CH_2}}}{CH}-\overset{O}{\overset{\|}{C}}H$$

2-Ethylbutanal

$$CH_3-\overset{*}{\underset{\underset{CH_3}{|}}{CH}}-\underset{\underset{CH_3}{|}}{CH}-\overset{O}{\overset{\|}{C}}H$$

2,3-Dimethylbutanal
(chiral)

$$CH_3-\underset{\underset{CH_3}{|}}{\overset{\overset{CH_3}{|}}{C}}-CH_2-\overset{O}{\overset{\|}{C}}H$$

3,3-Dimethylbutanal

$$CH_3-CH_2-\underset{\underset{CH_3}{|}}{\overset{\overset{CH_3}{|}}{C}}-\overset{O}{\overset{\|}{C}}H$$

2,2-Dimethylbutanal

<u>Problem 15.3</u> Write IUPAC names for these compounds.

(a) $HOCH_2\overset{O}{\overset{\|}{C}}CH_2OH$ (b) 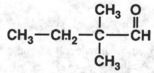   (c) $H_2NCH_2CH_2CH_2\overset{O}{\overset{\|}{C}}H$

1,3-Dihydroxy-2-propanone   **1,2-Cyclohexanedione**   **4-Aminobutanal**

Problem 15.4 Show how these three products can be synthesized from the same Grignard reagent; that is, treating the Grignard reagent with one compound gives (a), with another compound gives (b), and so on. (*Hint:* Grignard reagents react with ethylene oxide in the same manner as Gilman reagents, Section 11.10B.)

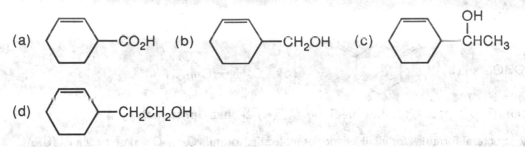

(d) ![] —CH₂CH₂OH

All of the products can be obtained using **2-cyclohexenylmagnesium bromide** and the other reactants shown.

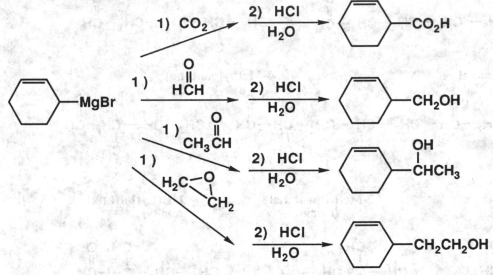

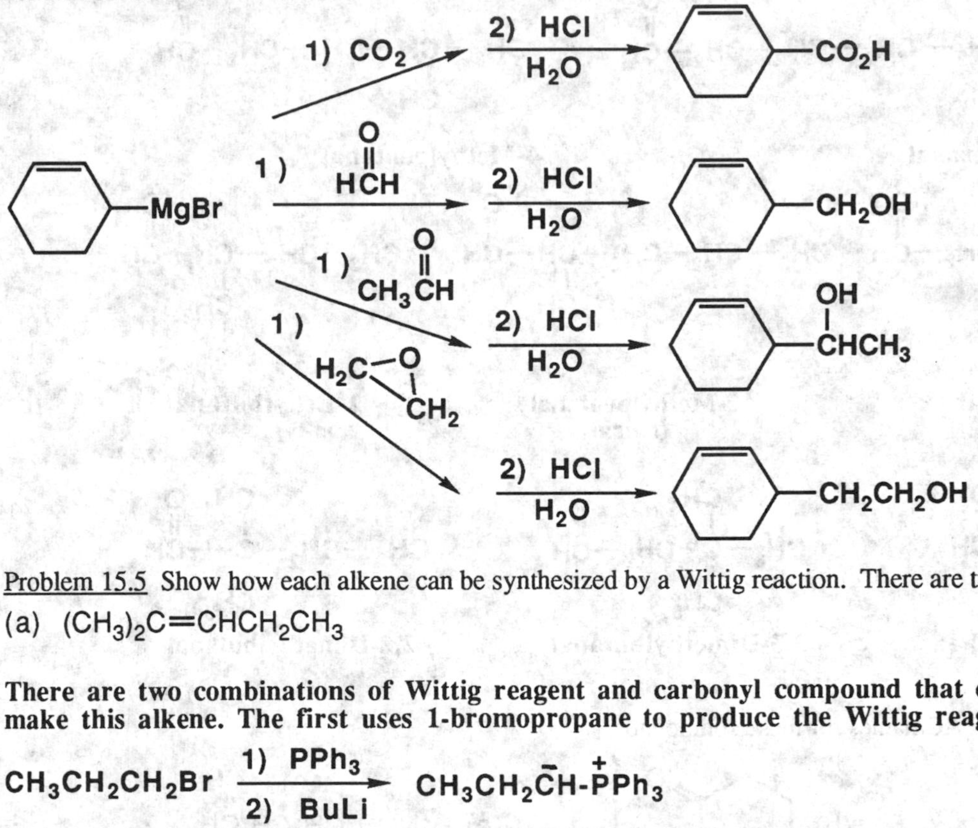

Problem 15.5 Show how each alkene can be synthesized by a Wittig reaction. There are two routes to each.

(a) $(CH_3)_2C{=}CHCH_2CH_3$

There are two combinations of Wittig reagent and carbonyl compound that could be used to make this alkene. The first uses 1-bromopropane to produce the Wittig reagent.

$$CH_3CH_2CH_2Br \xrightarrow[\text{2) BuLi}]{\text{1) PPh}_3} CH_3CH_2\bar{C}H\text{-}\overset{+}{P}Ph_3$$

$$CH_3CH_2\bar{C}H\text{-}\overset{+}{P}Ph_3 + H_3C\overset{\displaystyle O}{\overset{\|}{C}}CH_3 \longrightarrow (CH_3)_2C{=}CHCH_2CH_3$$

The other involves the Wittig reagent prepared from 2-bromopropane

$$CH_3CHBrCH_3 \xrightarrow[\text{2) BuLi}]{\text{1) PPh}_3} (CH_3)_2\bar{C}\text{-}\overset{+}{P}Ph_3$$

$$(CH_3)_2\bar{C}\text{-}\overset{+}{P}Ph_3 + CH_3CH_2\overset{\displaystyle O}{\overset{\|}{C}}H \longrightarrow (CH_3)_2C{=}CHCH_2CH_3$$

(b) $(CH_3)_3C$—⬡$=CH_2$

There are also two combinations of Wittig reagent and carbonyl compound that could be used to make this alkene. The first uses 1-bromo-4-*tert*-butylcyclohexane to produce the Wittig reagent.

$(CH_3)_3C$—⬡—Br $\xrightarrow[\text{2) BuLi}]{\text{1) PPh}_3}$ $(CH_3)_3C$—⬡—$\overset{-}{C}$—$\overset{+}{P}Ph_3$

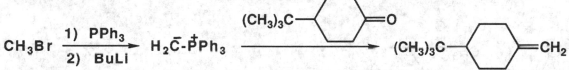

$(CH_3)_3C$—⬡—$\overset{-}{C}$—$\overset{+}{P}Ph_3$ + $\overset{O}{\overset{\|}{H\overset{}{C}H}}$ ⟶ $(CH_3)_3C$—⬡$=CH_2$

The other involves the Wittig reagent prepared from bromomethane

$CH_3Br \xrightarrow[\text{2) BuLi}]{\text{1) PPh}_3} H_2\overset{-}{C}\text{-}\overset{+}{P}Ph_3$ $\xrightarrow{(CH_3)_3C—⬡=O}$ $(CH_3)_3C$—⬡$=CH_2$

In the above examples, alkyl bromides were used. In fact, alkyl chlorides and iodides could have also been utilized.

<u>Problem 15.6</u> Hydrolysis of an acetal in aqueous acid forms an aldehyde or ketone and two molecules of alcohol. Following are structural formulas for three acetals. Draw the structural formulas for the products of hydrolysis of each in aqueous acid.

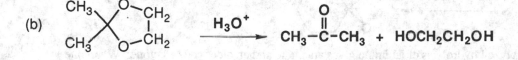

(a) CH_3O—⬡—$\overset{OCH_3}{\underset{|}{C}HOCH_3}$ $\xrightarrow{H_3O^+}$ CH_3O—⬡—$\overset{O}{\overset{\|}{C}}$-H + $2CH_3OH$

(b) $\overset{CH_3}{\underset{CH_3}{}}$⟩$\overset{O—CH_2}{\underset{O—CH_2}{}}$ $\xrightarrow{H_3O^+}$ CH_3-$\overset{O}{\overset{\|}{C}}$-$CH_3$ + $HOCH_2CH_2OH$

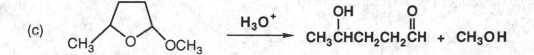

(c) [structure] $\xrightarrow{H_3O^+}$ $CH_3\overset{OH}{\underset{|}{C}}HCH_2CH_2\overset{O}{\overset{\|}{C}}H$ + CH_3OH

Problem 15.7 Show how to convert the given starting material into the indicated product using a 1,3-dithiane as an intermediate.

(a)

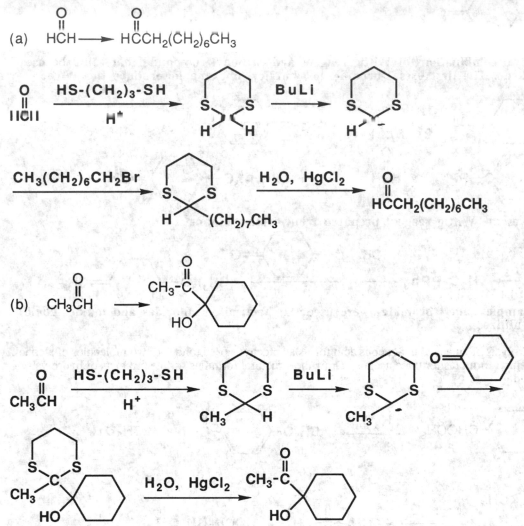

(b)

Problem 15.8 Acid-catalyzed hydrolysis of an imine gives an amine and an aldehyde or ketone. When one equivalent of acid is used, the amine is converted to an ammonium salt. Write structural formulas for the products of hydrolysis of the following imines using one equivalent of HCl:

(a) CH_3O-⟨benzene ring⟩$-CH=NCH_2CH_3 + H_2O \xrightarrow{H^+}$

Hydrolysis of (a) gives an aldehyde and a primary amine.

CH_3O-⟨benzene ring⟩$-\overset{\overset{O}{\|}}{C}H + H_2NCH_2CH_3$

(b) 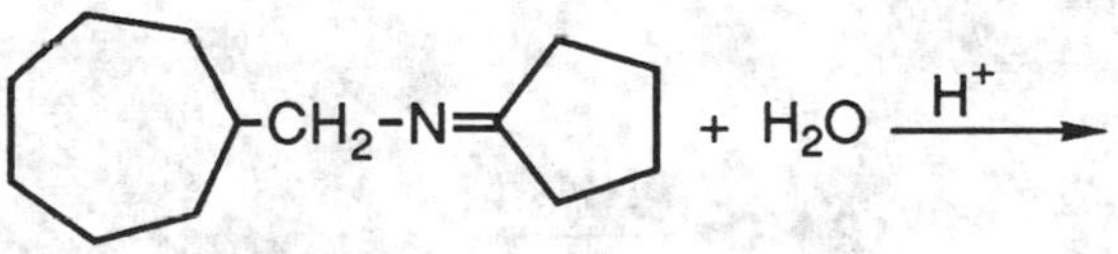 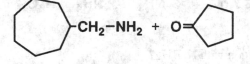 $+ \; H_2O \; \xrightarrow{\; H^+ \;}$

Hydrolysis of (b) gives a ketone and a primary amine.

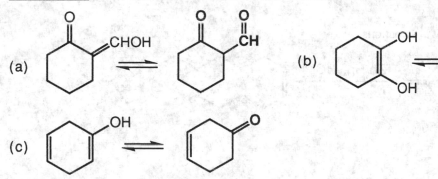

Problem 15.9 Predict the position of the following equilibrium.

Like all acid-base reactions, equilibrium favors formation of the weaker acid/weaker base. Since ammonia, with pK_a of 33, is a much weaker acid than acetophenone (pK_a ~20) the position of equilibrium will be strongly to the right.

Problem 15.10 Draw the structural formula for the keto form of each enol.

(a)

(b)

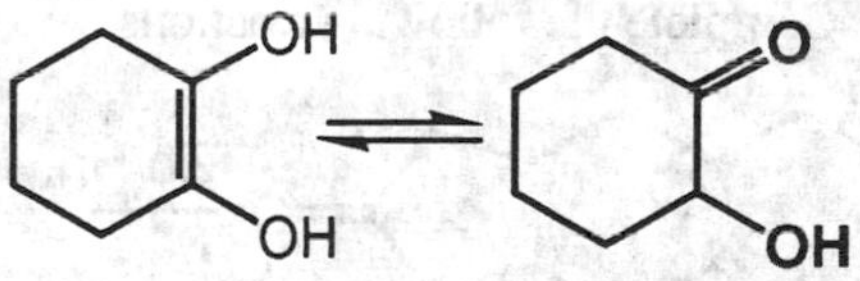

(c)

Problem 15.11 Complete these oxidations.

(a) Hexanedial + H_2O_2 $\longrightarrow$

$$\underset{HC}{\overset{O}{\|}}(CH_2)_4\underset{CH}{\overset{O}{\|}} \; + \; 2 \; H_2O_2 \longrightarrow \underset{HOC}{\overset{O}{\|}}(CH_2)_4\underset{COH}{\overset{O}{\|}} \; + \; 2 \; H_2O$$

(b) 3-Phenylpropanal + Tollens' reagent ⟶

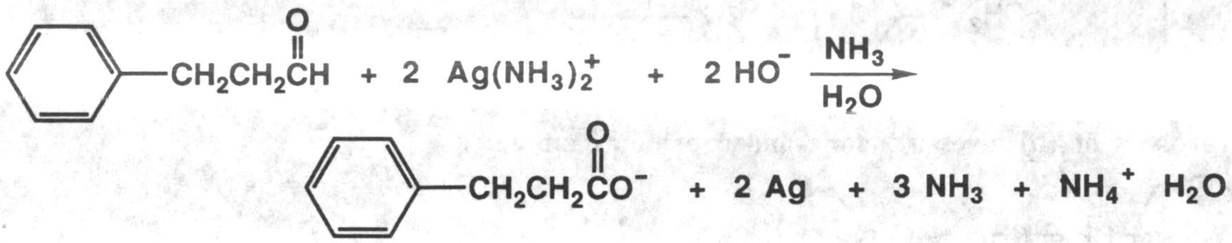

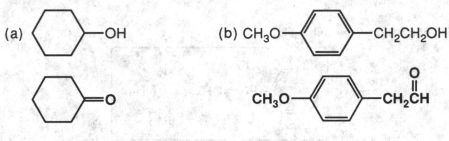

Problem 15.12 What aldehyde or ketone gives these alcohols on reduction with $NaBH_4$?

(a)

(b)

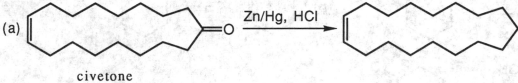

(c)

Problem 15.13 Complete the following reactions:

(a)

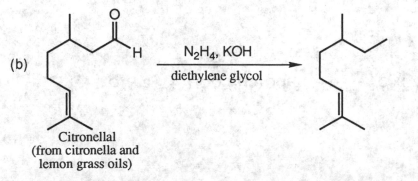

civetone
(from the civet cat; used
in perfumery)

The above reaction is an example of a Clemmensen reduction.

(b)

Citronellal
(from citronella and
lemon grass oils)

The above reaction is an example of a Wolff-Kishner reduction.

Structure and Nomenclature
Problem 15.14 Name these compounds.

(a) $(CH_3CH_2CH_2)_2C=O$

4-Heptanone
(dipropyl ketone)

(b)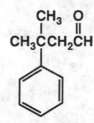

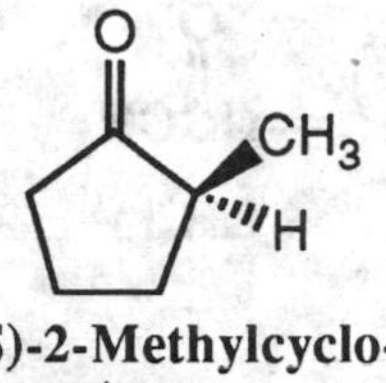

(S)-2-Methylcyclo-
pentanone

(c)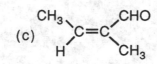

(Z)-2-Methyl-2-butenal

(d)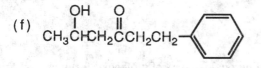

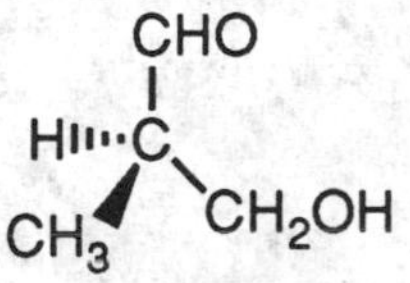

(R)-3-Hydroxy-2-methyl-
propanal

(e) 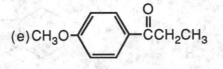 CH_3O—⟨ ⟩—CCH_2CH_3

1-(4-Methoxyphenyl)-1-propanone

(f) 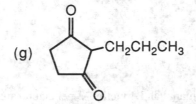 $CH_3CHCH_2CCH_2CH_2$—⟨ ⟩

5-Hydroxy-1-phenyl-3-hexanone

(g) $CH_2CH_2CH_3$

2-Propyl-1,3-cyclopentanedione

(h) 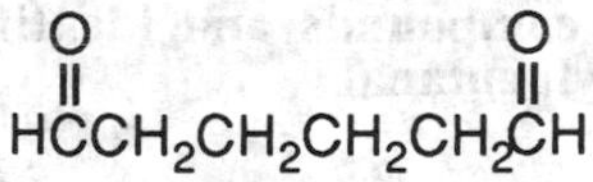 $HCCH_2CH_2CH_2CH_2CH$

Hexanedial
(Adipaldehyde)

(i) 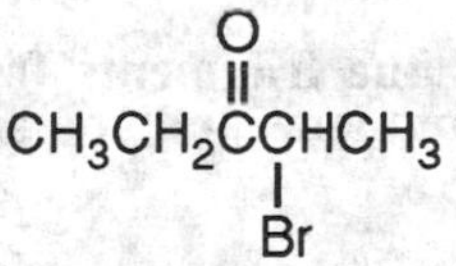 $CH_3CH_2CCHCH_3$
            $|$
            Br

2-Bromo-3-pentanone

Problem 15.15 Draw structural formulas for these compounds.
(a) 1-Chloro-2-propanone

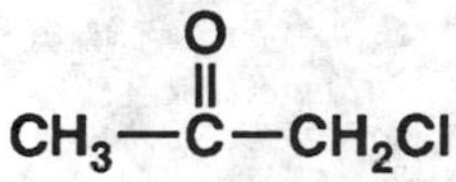

CH_3—C—CH_2Cl
        $\|$
        O

(b) 3-Hydroxybutanal

CH_3CHCH_2CH
      $|$ $\|$
      OH O

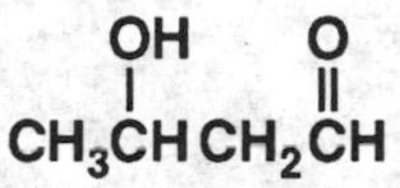

(c) 4-Hydroxy-4-methyl-2-pentanone

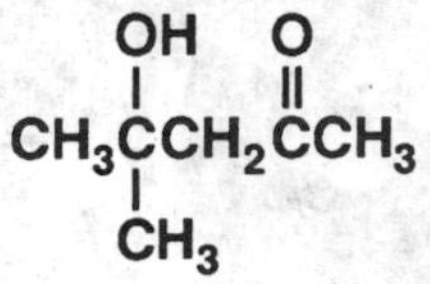

$CH_3CCH_2CCH_3$
      $|$ $\|$
      OH O
      $|$
      CH_3

(d) 3-Methyl-3-phenylbutanal

CH_3CCH_2CH
      $|$ $\|$
      CH_3 O
      $|$
      ⟨ ⟩

(e) 1,3-Cyclohexanedione

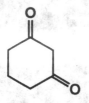

(f) 3-Methyl-3-butene-2-one

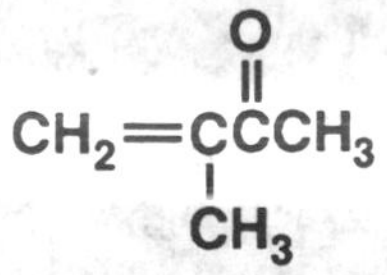

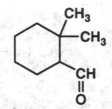

$$CH_2=\overset{\overset{\displaystyle O}{\|}}{\underset{\underset{\displaystyle CH_3}{|}}{C}}CCH_3$$

(g) 5-Oxohexanal

$$CH_3\overset{\overset{\displaystyle O}{\|}}{C}CH_2CH_2CH_2\overset{\overset{\displaystyle O}{\|}}{C}H$$

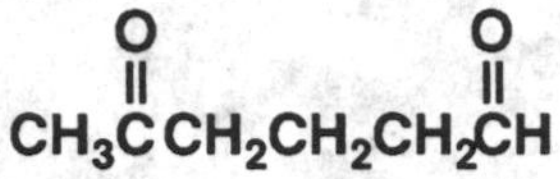

(h) 2,2-Dimethylcyclohexanecarbaldehyde

(i) 3-Oxobutanoic acid

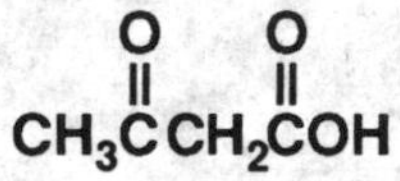

$$CH_3\overset{\overset{\displaystyle O}{\|}}{C}CH_2\overset{\overset{\displaystyle O}{\|}}{C}OH$$

Spectroscopy
Problem 15.16 2-Methylpentanal and 4-methyl-2-pentanone are constitutional isomers of molecular formula $C_6H_{12}O$. Each shows a molecular ion peak in its mass spectrum at m/z 100. Spectrum A shows significant peaks at m/z 85, 58, 57, 43, and 42. Spectrum B shows significant peaks at m/z 71, 58, 57, 43, and 29. Assign each compound its correct spectrum.

The following table describes the fragments from the two different compounds and identifies spectrum A as 4-methyl-2-pentanone and the spectrum B as 2-methylpentanal

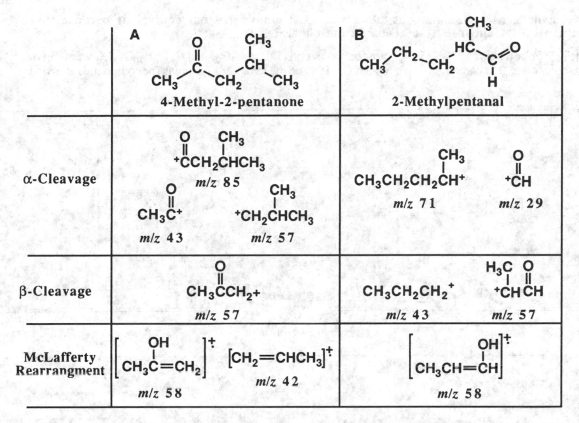

	A 4-Methyl-2-pentanone	B 2-Methylpentanal
α-Cleavage	$^+$CCH$_2$CHCH$_3$ m/z 85 CH$_3$C$^+$ m/z 43 $^+$CH$_2$CHCH$_3$ m/z 57	CH$_3$CH$_2$CH$_2$CH$^+$ m/z 71 $^+$CH m/z 29
β-Cleavage	CH$_3$CCH$_2$+ m/z 57	CH$_3$CH$_2$CH$_2$$^+$ m/z 43 $^+$CHCH m/z 57
McLafferty Rearrangment	[CH$_3$C=CH$_2$]$^+$ m/z 58 [CH$_2$=CHCH$_3$]$^+$ m/z 42	[CH$_3$CH=CH]$^+$ m/z 58

Problem 15.17 The infrared spectrum of Compound A, C$_6$H$_{12}$O, shows a strong, sharp peak at 1724 cm^{-1}. From this information and its ^{1}H-NMR spectrum, deduce the structure of compound A.

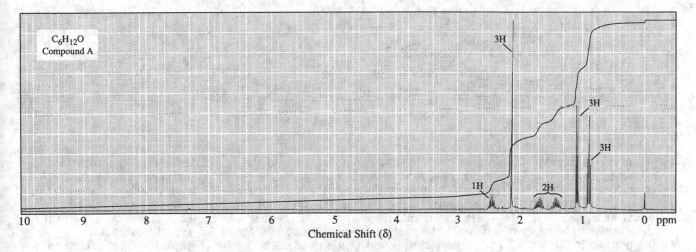

Compound A has an aldehyde or ketone function judging from the peak at 1724 cm^{-1} in the IR spectrum. The rest of the structure can be deduced from the ^{1}H-NMR, with assignments as listed below.

$$\underset{0.89}{CH_3}-\underset{1.42, 1.68}{CH_2}-\underset{\underset{\underset{1.08}{CH_3}}{|}}{\underset{2.45}{CH}}-\underset{O}{\overset{O}{C}}-\underset{2.1}{CH_3}$$

The -CH$_2$- hydrogens with signals at δ 1.42 and δ 1.68 above are not equivalent because they are adjacent to a tetrahedral stereocenter in the molecule.

<u>Problem 15.18</u> Following are ^{1}H-NMR spectra for compounds B, C$_6$H$_{12}$O$_2$, and C, C$_6$H$_{10}$O. On warming in dilute acid, compound B is converted to compound C. Deduce the structural formulas for compounds B and C.

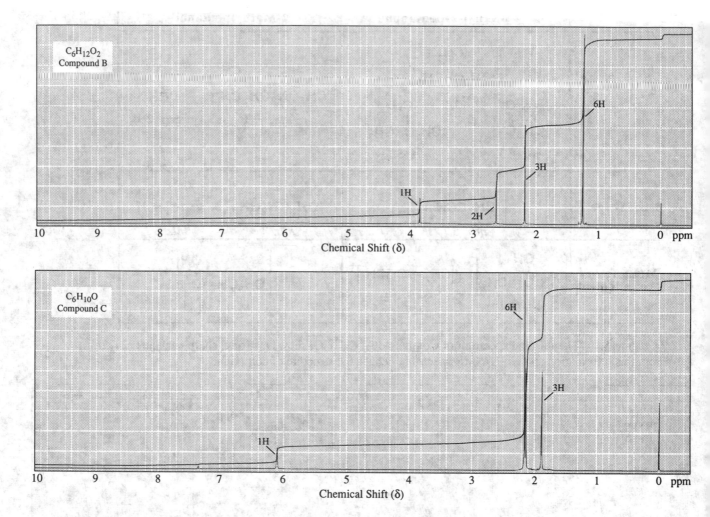

From the molecular formula it is clear that compound B undergoes acid-catalyzed dehydration to create compound C. Thus, compound B must have an -OH group. Furthermore, because its index of hydrogen deficiency is one, compound B must have one ring or pi bond. However, compound C has an index of hydrogen deficiency of two, so it must have two pi bonds or rings. The ^{1}H-NMR spectrum of compound B shows all singlets. Especially helpful are the methyl group resonances; the singlet integrating to 6H at δ 1.22 and the singlet integrating to 3H at δ 2.18. This latter signal is assigned as a methyl ketone by comparison to the previous problem. The other two methyl groups are equivalent. There is the -OH hydrogen at δ 3.85 and a -CH$_2$- resonance at δ 2.62. The only structure consistent with these signals is 4-hydroxy-4-methyl-2-pentanone.

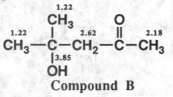

Compound B

Upon dehydration, Compound B would be converted to 4-methyl-3-pentene-2-one.

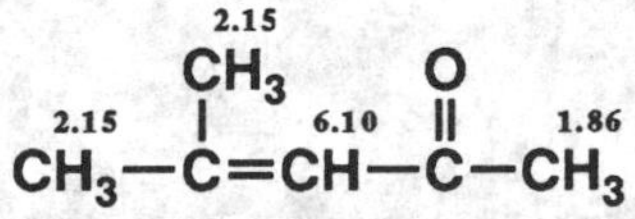

$$\overset{2.15}{CH_3}-\overset{\underset{|}{\overset{2.15}{CH_3}}}{C}=\overset{6.10}{CH}-\overset{\overset{O}{\|}}{C}-\overset{1.86}{CH_3}$$

Compound C

The structure of 4-methyl-3-pentene-2-one is entirely consistent with the ^{1}H-NMR spectrum, especially the presence of the signal at δ 6.10 (-CH=C) and the methyl singlets at δ 1.86 (integrating to 3H) and δ 2.15 (integrating to 6H).

Addition of Carbon Nucleophiles

Problem 15.19 Draw structural formulas for the product formed by treating each compound with propylmagnesium bromide followed by aqueous HCl.

The products after acid hydrolysis are given in bold.

(a) CH_2O

$CH_3CH_2CH_2CH_2OH$

(b) 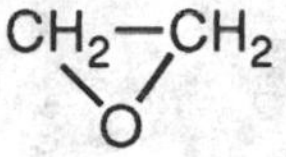 $\underset{\overset{\diagdown}{O}}{CH_2-CH_2}$

$\overset{\qquad\qquad\qquad\overset{OH}{|}}{CH_3CH_2CH_2CH_2CH_2}$

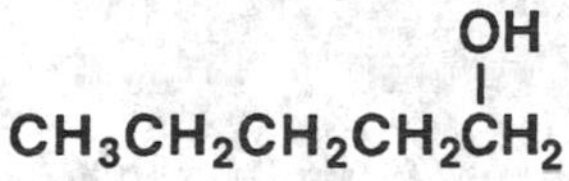

(c) 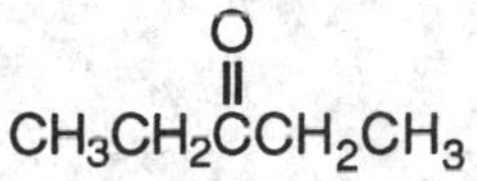 $\overset{\overset{O}{\|}}{CH_3CH_2CCH_2CH_3}$

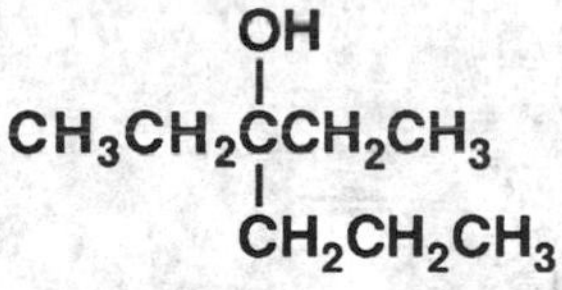

$\overset{\overset{OH}{|}}{CH_3CH_2CCH_2CH_3}$
$\underset{CH_2CH_2CH_3}{|}$

(d)

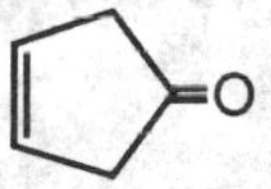

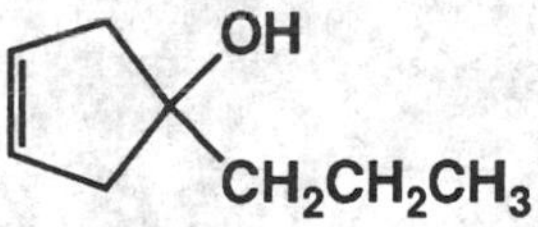

(e) CO_2

$\overset{\overset{O}{\|}}{CH_3CH_2CH_2COH}$

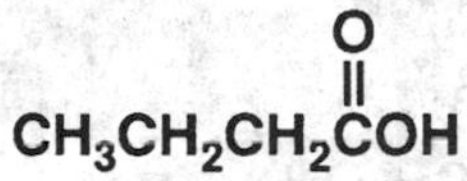

<u>Problem 15.20</u> Suggest a synthesis for the following alcohols starting from an aldehyde or ketone and an appropriate Grignard reagent. Below each target molecule is the number of combinations of Grignard reagent and aldehyde or ketone that might be used.

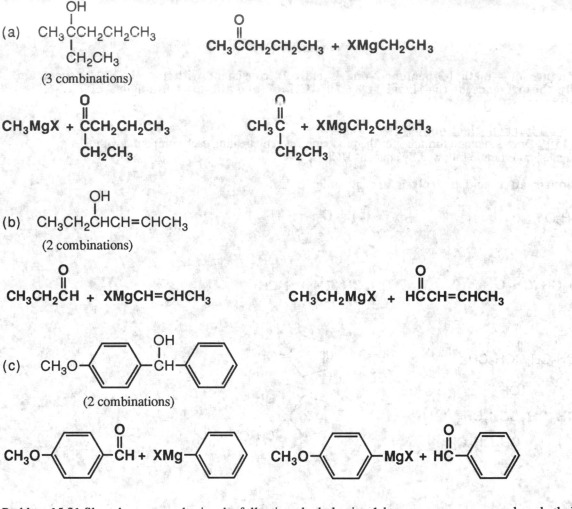

<u>Problem 15.21</u> Show how to synthesize the following alcohol using 1-bromopropane, propanal, and ethylene oxide as the only sources of carbon atoms. It can be done using each compound only once. (*Hint:* Do one Grignard reaction to form an alcohol, convert the alcohol to an alkyl halide, and then do a second Grignard reaction.)

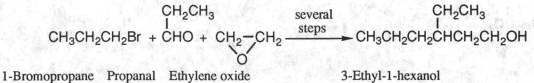

1-Bromopropane   Propanal   Ethylene oxide   3-Ethyl-1-hexanol

This synthesis is divided into two stages. In the first stage, 1-bromopropane is treated with magnesium to form a Grignard reagent and then with propanal followed by hydrolysis in aqueous acid to give 3-hexanol.

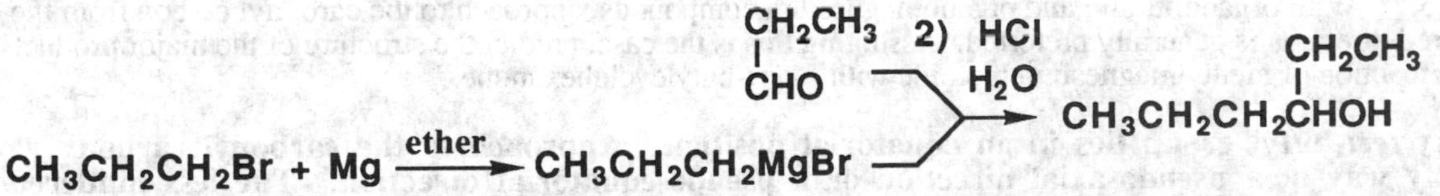

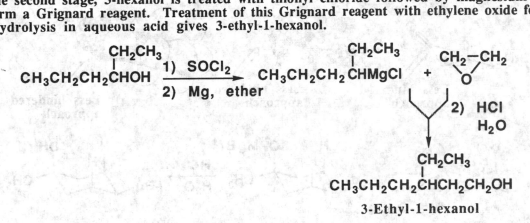

$$CH_3CH_2CH_2Br + Mg \xrightarrow{\text{ether}} CH_3CH_2CH_2MgBr$$

3-Hexanol

In the second stage, 3-hexanol is treated with thionyl chloride followed by magnesium in ether to form a Grignard reagent. Treatment of this Grignard reagent with ethylene oxide followed by hydrolysis in aqueous acid gives 3-ethyl-1-hexanol.

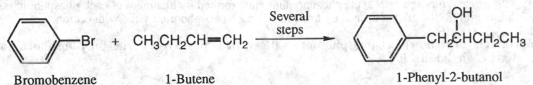

3-Ethyl-1-hexanol

Problem 15.22 1-Phenyl-2-butanol is used in perfumery. Show how to synthesize this alcohol from bromobenzene, 1-butene, and any necessary inorganic reagents.

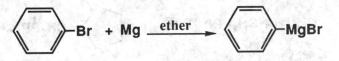

Bromobenzene   1-Butene   1-Phenyl-2-butanol

(a) Bromobenzene is treated with magnesium in diethyl ether to form phenylmagnesium bromide, in preparation for part (c).

(b) Treatment of 1-butene with a peroxycarboxylic acid gives 1,2-epoxybutane.

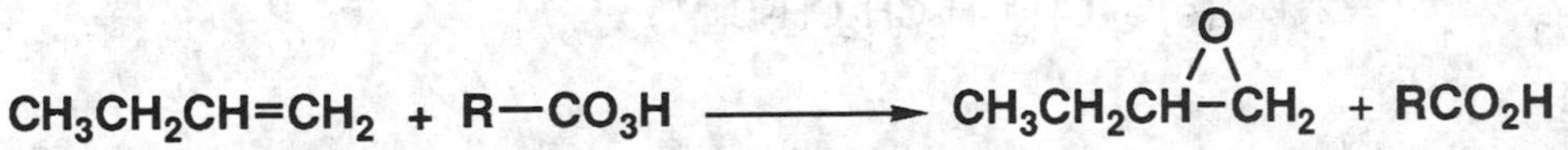

$$CH_3CH_2CH{=}CH_2 + R{-}CO_3H \longrightarrow CH_3CH_2CH{-}CH_2 + RCO_2H$$

(c) Treatment of phenylmagnesium bromide with 1,2-epoxybutane followed by hydrolysis in aqueous acid gives 1-phenyl-2-butanol.

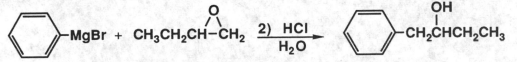

<u>Problem 15.23</u> With organolithium and organomagnesium compounds, approach to the carbonyl carbon from the less hindered direction is generally preferred. Assuming this is the case, predict the structure of the major product formed by reaction of methylmagnesium bromide with 4-*tert*-butylcyclohexanone.

The bulky *tert*-butyl group lies in an equatorial position. Approach to the carbonyl carbon may be by way of a pseudo-axial direction or a pseudo-equatorial direction. The less hindered approach is from the pseudo-equatorial direction which then places the incoming group equatorial and the -OH axial.

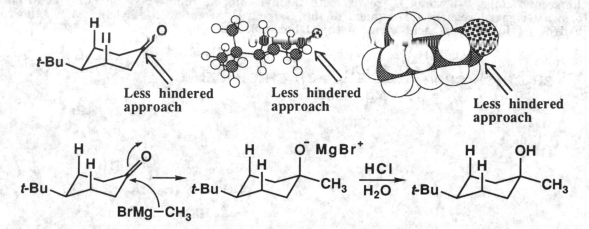

Less hindered approach   Less hindered approach   Less hindered approach

Wittig Reaction
<u>Problem 15.24</u> Draw structural formulas for (1) the triphenylphosphonium salt formed by treatment of each haloalkane with triphenylphosphine, (2) the phosphonium ylide formed by treatment of each phosphonium salt with butyllithium, and (3) the alkene formed by treatment of each phosphonium ylide with acetone.

(1) The products are shown as phosphonium salts with a positive charge on phosphorus and a negative charge on halide ion.

(a) $(CH_3)_2CHBr$ (b) $CH_2=CHCH_2Br$ (c)   cyclopentyl–CH_2Cl

$CH_3-\overset{+}{\underset{CH_3}{CH}}-PPh_3\ Br^-$ $CH_2=CHCH_2\overset{+}{P}Ph_3\ Br^-$ cyclopentyl–$CH_2\overset{+}{P}Ph_3\ Cl^-$

(d) $ClCH_2\overset{O}{\overset{\|}{C}}OCH_2CH_3$ (e) $CH_3CH_2CH_2CH_2Br$

$Cl^-\ Ph_3\overset{+}{P}CH_2\overset{O}{\overset{\|}{C}}OCH_2CH_3$ $CH_3CH_2CH_2CH_2\overset{+}{P}Ph_3Br^-$

(f)

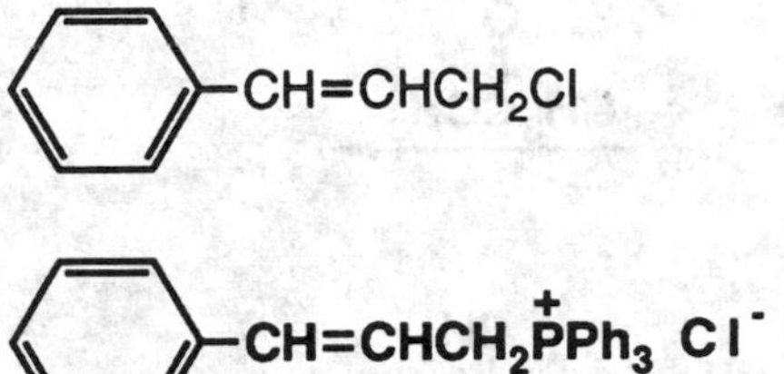

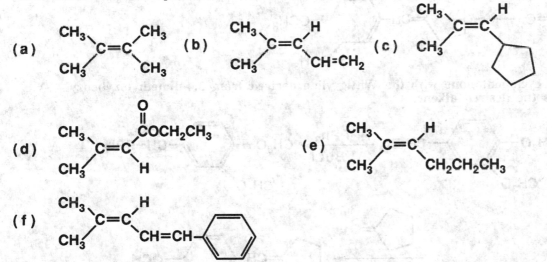

$C_6H_5-CH=CHCH_2Cl$

$C_6H_5-CH=CHCH_2\overset{+}{P}Ph_3 \ Cl^-$

The phosphonium ylides formed in part (2) are as follows:

(a) $CH_3-\overset{..\ -}{\underset{\overset{|}{CH_3}}{C}}-\overset{+}{P}Ph_3$

(b) $CH_2=CH\overset{..\ -}{C}H\overset{+}{P}Ph_3$

(c) 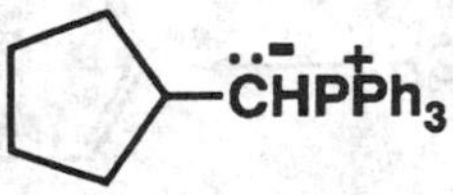 (cyclopentyl)$-\overset{..\ -}{C}H\overset{+}{P}Ph_3$

(d) $Ph_3\overset{+}{P}\overset{..\ -}{C}H\overset{O}{\overset{||}{C}}OCH_2CH_3$

(e) $CH_3CH_2CH_2\overset{..\ -}{C}H\overset{+}{P}Ph_3$

(f) 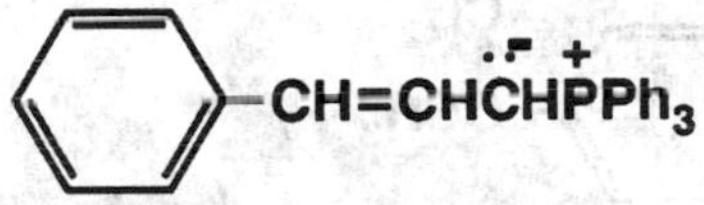 (phenyl)$-CH=CH\overset{..\ -}{C}H\overset{+}{P}Ph_3$

The alkenes formed in part (3) are as follows:

(a) $\underset{CH_3}{\overset{CH_3}{>}}C=C\underset{CH_3}{\overset{CH_3}{<}}$

(b) $\underset{CH_3}{\overset{CH_3}{>}}C=C\underset{CH=CH_2}{\overset{H}{<}}$

(c) $\underset{CH_3}{\overset{CH_3}{>}}C=C\underset{(cyclopentyl)}{\overset{H}{<}}$

(d) $\underset{CH_3}{\overset{CH_3}{>}}C=C\underset{H}{\overset{COCH_2CH_3}{<}}$ (with O on the carbonyl)

(e) $\underset{CH_3}{\overset{CH_3}{>}}C=C\underset{CH_2CH_2CH_3}{\overset{H}{<}}$

(f) $\underset{CH_3}{\overset{CH_3}{>}}C=C\underset{CH=CH-C_6H_5}{\overset{H}{<}}$

Problem 15.25 Show how to bring about the following conversions using a Wittig reaction.

(a) $CH_3\overset{O}{\overset{||}{C}}CH_3 \longrightarrow CH_3\overset{CH_3}{\overset{|}{C}}=CH(CH_2)_3CH_3$

Start with 1-halopentane and, by treatment with triphenylphosphine followed by butyllithium, convert it to a Wittig ylide. Treatment of this ylide with acetone gives the desired alkene.

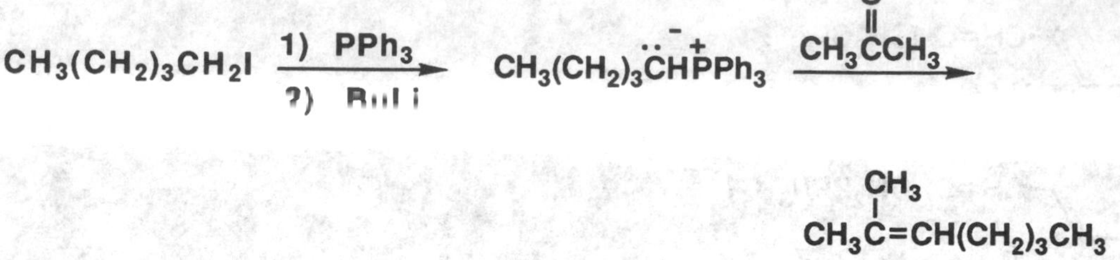

(b) 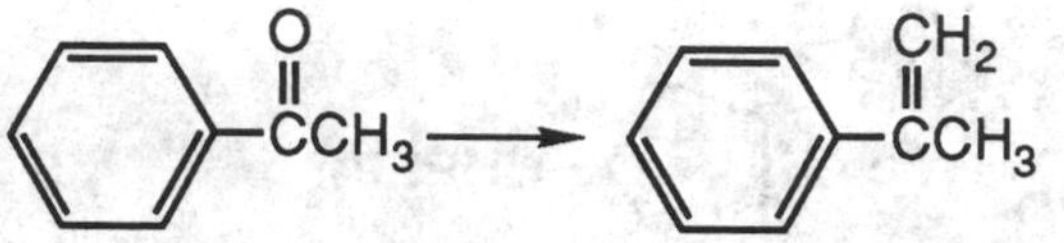

Treatment of acetophenone with the Wittig ylide derived from methyl iodide gives the desired alkene.

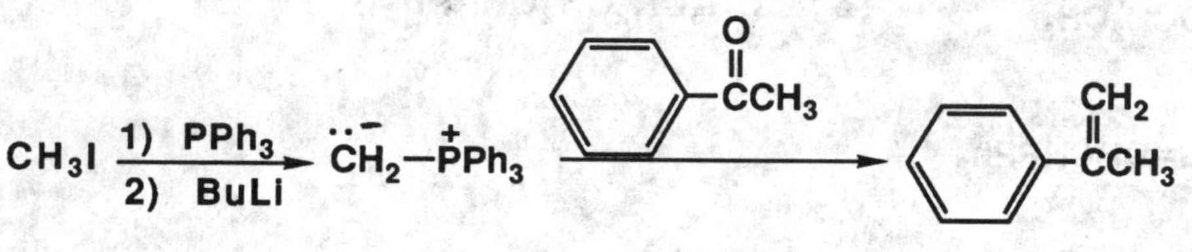

(c) 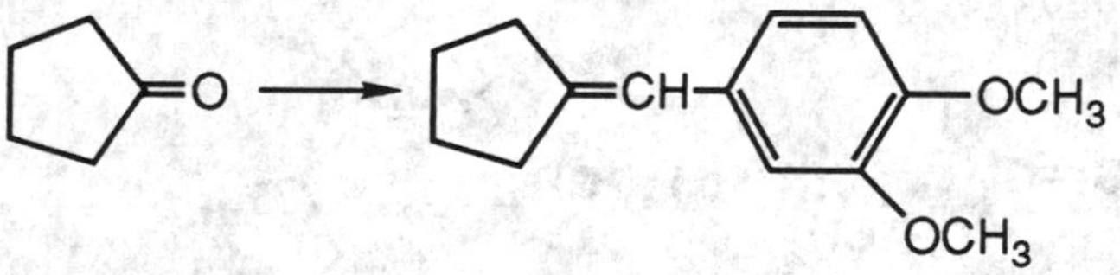

Treatment of cyclopentanone with the Wittig ylide derived from 3,4-dimethyloxybenzyl bromide gives the desired alkene.

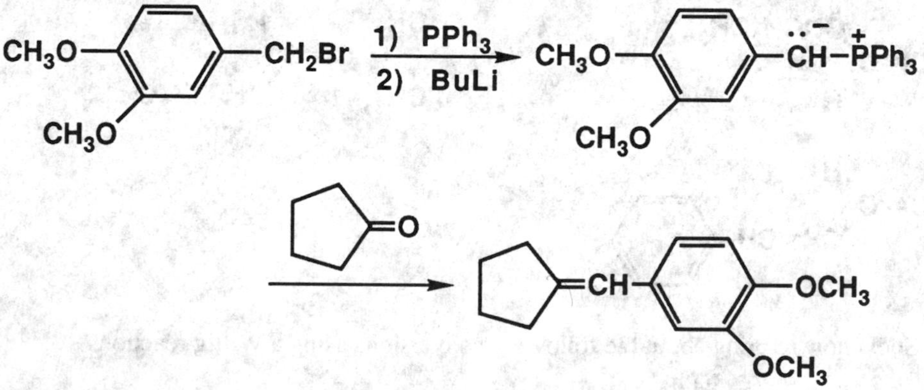

Problem 15.26 The Wittig reaction can be used for the synthesis of conjugated dienes, as, for example, 1-phenyl-1,3-pentadiene. Propose two sets of reagents that might be combined in a Wittig reaction to give this conjugated diene.

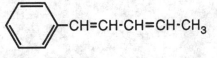

1-Phenyl-1,3-pentadiene

Allylic halides can be used as starting materials for preparation of Wittig reagents. Below are combinations of allylic halides and aldehydes that can be used to prepare the desired diene.

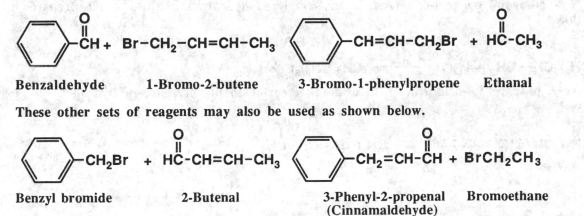

Benzaldehyde   1-Bromo-2-butene   3-Bromo-1-phenylpropene   Ethanal

These other sets of reagents may also be used as shown below.

Benzyl bromide   2-Butenal   3-Phenyl-2-propenal   Bromoethane
**   (Cinnamaldehyde)**

Problem 15.27 Show how to convert heptanal into the following:

(a) $CH_3(CH_2)_5CH=CH_2$

Prepare 1-octene by treatment of heptanal with the Wittig reagent derived from methyl iodide.

$$CH_3(CH_2)_5\overset{\overset{O}{\|}}{C}H + \overset{..\,-}{C}H_2\!-\!\overset{+}{P}Ph_3 \longrightarrow CH_3(CH_2)_5CH=CH_2 + Ph_3P=O$$

(b) 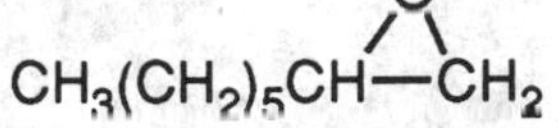 $CH_3(CH_2)_5CH\!-\!CH_2$ (epoxide)

Treat 1-octene, from part (a), with a peroxycarboxylic acid.

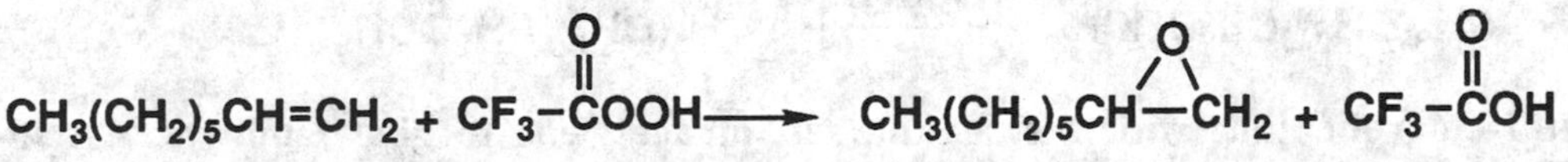

$$CH_3(CH_2)_5CH=CH_2 + CF_3\!-\!\overset{\overset{O}{\|}}{C}OOH \longrightarrow CH_3(CH_2)_5CH\!-\!CH_2 + CF_3\!-\!\overset{\overset{O}{\|}}{C}OH$$

(c) $CH_3(CH_2)_5CH=CH-CH=CH_2$

Prepare 1,3-decadiene by treatment of heptanal with the Wittig reagent derived from 3-chloropropene (allyl chloride).

$$CH_3(CH_2)_5CH \overset{O}{\parallel} \quad + \quad Ph_3\overset{+}{P}-\overset{..}{\overset{-}{C}}H-CH=CH_2 \longrightarrow CH_3(CH_2)_5CH=CH-CH=CH_2$$

$$+$$

$$Ph_3P=O$$

(d) $$CH_3(CH_2)_5\overset{HO}{\underset{|}{C}}H-\overset{OH}{\underset{|}{C}}H_2$$

Prepare 1,2-octanediol by acid-catalyzed or base-catalyzed hydrolysis of the epoxide prepared in part (b).

$$CH_3(CH_2)_5CH\overset{O}{\overbrace{}}CH_2 + H_2O \xrightarrow{H^+} CH_3(CH_2)_5\overset{HO}{\underset{|}{C}}H-\overset{OH}{\underset{|}{C}}H_2$$

Alternatively, prepare 1,2-octanediol with osmium tetroxide in the presence of hydrogen peroxide.

$$CH_3(CH_2)_5CH=CH_2 \xrightarrow[H_2O_2]{OsO_4} CH_3(CH_2)_5\overset{HO}{\underset{|}{C}}H-\overset{OH}{\underset{|}{C}}H_2$$

(e) $$CH_3(CH_2)_5CH_2\overset{O}{\overset{\parallel}{C}}CH_3$$

The key is to devise a method for adding a two-carbon fragment to the aldehyde carbon of heptanal in such a way that there is either an oxygen-containing functional group on carbon-2 of the newly formed nine-carbon chain or a group that can be converted to an oxygen-containing functional group. A way to do this is to convert heptanal to 1-chloroheptane, then, to a Grignard reagent followed by reaction with acetaldehyde, to a nine-carbon chain with an alcohol on carbon-2 of the chain. Oxidation of this alcohol gives the desired ketone.

$$CH_3(CH_2)_5CH \overset{O}{\parallel} \xrightarrow[\text{2) SOCl}_2]{\text{1) NaBH}_4} CH_3(CH_2)_5CH_2Cl \xrightarrow[\substack{\text{2) CH}_3\text{CHO} \\ \text{3) H}_2\text{O, HCl}}]{\text{1) Mg, ether}}$$

$$CH_3(CH_2)_5CH_2\overset{OH}{\underset{|}{C}}HCH_3 \xrightarrow{H_2CrO_4} CH_3(CH_2)_5CH_2\overset{O}{\overset{\parallel}{C}}CH_3$$

Alternatively, treat 1-chloroheptane with the sodium salt of acetylene to give 1-nonyne. Acid-catalyzed hydration of this alkyne gives 2-nonanone.

$$CH_3(CH_2)_5CH_2Cl \xrightarrow{HC\equiv C^-Na^+} CH_3(CH_2)_5CH_2C\equiv CH$$

$$\xrightarrow[H_2SO_4, HgSO_4]{H_2O} CH_3(CH_2)_5CH_2\overset{O}{\overset{\parallel}{C}}CH_3$$

(f) $CH_3(CH_2)_5CH_2CH_2CH_2OH$

Catalytic reduction of 1-nonyne, an intermediate in part (c), over a Lindlar catalyst gives 1-nonene. Hydroboration of 1-nonene followed by oxidation in alkaline hydrogen peroxide gives 1-nonanol.

$$CH_3(CH_2)_6C\equiv CH \xrightarrow[\text{Lindlar catalyst}]{H_2} CH_3(CH_2)_6CH=CH_2$$

$$\xrightarrow[\text{2) } H_2O_2, \text{ NaOH}]{\text{1) } BH_3} CH_3(CH_2)_6CH_2CH_2OH$$

Alternatively, treatment of 1-chloroheptane from part (a) with magnesium in diethyl ether followed by reaction of the Grignard reagent with ethylene oxide and then hydrolysis of the resulting magnesium alkoxide gives 1-nonanol.

$$CH_3(CH_2)_6Cl \xrightarrow[\text{3) } H_2O, \text{ HCl}]{\begin{array}{c}\text{1) Mg, ether}\\ \text{2) } CH_2\text{—}CH_2 \end{array}} CH_3(CH_2)_6CH_2CH_2OH$$

<u>Problem 15.28</u> Wittig reactions with the following α-chloroethers can be used for the synthesis of aldehydes and ketones:

$$\begin{array}{ccc} & & CH_3 \\ & & | \\ ClCH_2OCH_3 & & ClCHOCH_3 \\ (A) & & (B) \end{array}$$

(a) Draw the structure of the triphenylphosphonium salt and Wittig reagent formed from each chloroether.

$$Ph_3^+P\text{-}\overset{..}{\overset{-}{C}}HOCH_3 \qquad\qquad Ph_3^+P\text{-}\overset{..}{\overset{-}{C}}OCH_3$$
$$\qquad\qquad\qquad\qquad\qquad | $$
$$\qquad\qquad\qquad\qquad\qquad CH_3$$

(b) Draw the structural formula of the product formed by treatment of each Wittig reagent with cyclopentanone. Note that the functional group is an enol ether, or, alternatively, a vinyl ether.

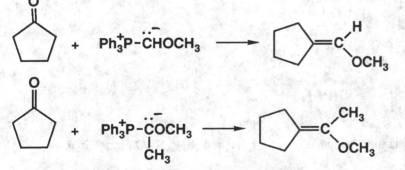

(c) Draw the structural formula of the product formed on acid-catalyzed hydrolysis of each enol ether from part (b).

The acid catalyzed hydrolysis leads initially to an enol that tautomerizes to give an aldehyde and methyl ketone for (A) and (B), respectively.

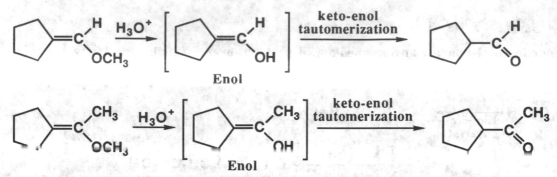

keto-enol tautomerization

Enol

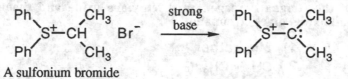

keto-enol tautomerization

Enol

<u>Problem 15.29</u> It is possible to generate sulfur ylides in a manner similar to that used to produce phosphonium ylides. For example, treating a sulfonium salt with a strong base gives the sulfur ylide.

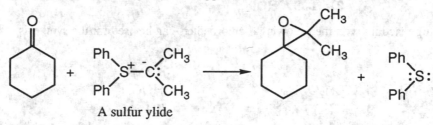

strong base

A sulfonium bromide

Sulfur ylides react with ketones to give epoxides. Suggest a mechanism for this reaction.

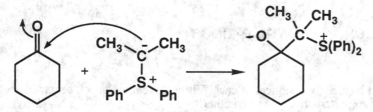

A sulfur ylide

Step 1:

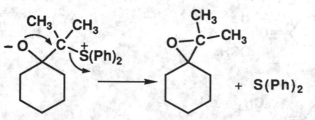

Step 2:

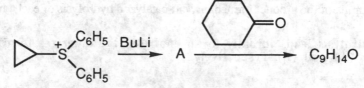

<u>Problem 15.30</u> Propose a structural formula for Compound A and for the product, $C_9H_{14}O$ formed in this reaction sequence.

BuLi   A $C_9H_{14}O$

The product of this transformation is an epoxide with both a three-membered and a six-membered ring attached via so-called "spiro" attachments.

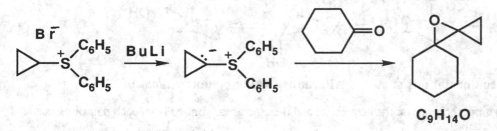

$C_9H_{14}O$

Addition of Oxygen Nucleophiles

Problem 15.31 5-Hydroxyhexanal forms a six-membered cyclic hemiacetal, which predominates at equilibrium in aqueous solution.

$$CH_3CHCH_2CH_2CH_2CH \overset{H^+}{\rightleftharpoons} \text{A cyclic hemiacetal}$$

(with a C=O at the right end and OH below the first carbon)

(a) Draw a structural formula for this cyclic hemiacetal.

5-Hydroxyhexanal forms a six-membered cyclic hemiacetal.

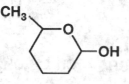

(b) How many stereoisomers are possible for 5-hydroxyhexanal?

Two stereoisomers are possible for 5-hydroxyhexanal; a pair of enantiomers.

(c) How many stereoisomers are possible for this cyclic hemiacetal?

Four stereoisomers are possible for the cyclic hemiacetal; two pair of enantiomers. Following are planar hexagon formulas for each pair of enantiomers of the cyclic hemiacetal.

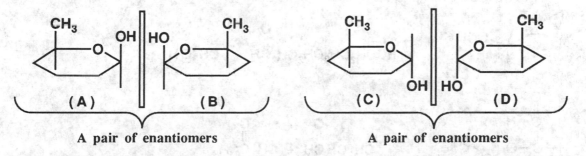

(d) Draw alternative chair conformations for each stereoisomer of the cyclic hemiacetal and label groups axial or equatorial. Also predict which of the alternative chair conformations for each stereoisomer is the more stable.

Alternative chair conformations are drawn for (A), one of the *cis* enantiomers, and for (C), one of the *trans* enantiomers. Methyl and hydroxyl groups are 1.74 kcal/mol and 0.95 kcal/mol more stable in the equatorial position, respectively. For (A), the diequatorial chair is the more stable by 1.74 + 0.95 = 2.69 kcal/mol. For (C), the chair with the methyl group is equatorial (structure on the left) is 1.74 - 0.95 = 0.79 kcal/mol moe stable.

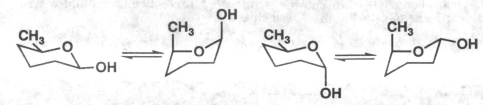

Alternative chair conformations of A   **Alternative chair conformations of C**

Problem 15.32 Draw structural formulas for the hemiacetal and then the acetal formed from each pair of reactants in the presence of an acid catalyst.

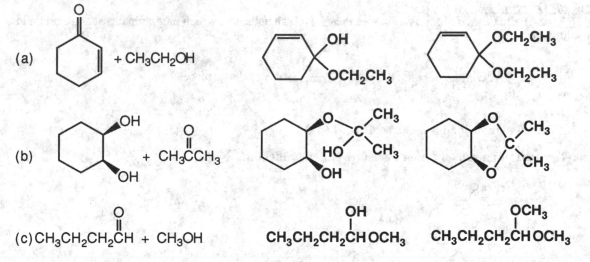

(a)   + CH_3CH_2OH

(b)   + $CH_3\overset{O}{\overset{\|}{C}}CH_3$

(c) $CH_3CH_2CH_2\overset{O}{\overset{\|}{C}}H$ + CH_3OH

$CH_3CH_2CH_2\overset{OH}{\overset{|}{C}H}OCH_3$ $CH_3CH_2CH_2\overset{OCH_3}{\overset{|}{C}H}OCH_3$

Problem 15.33 Draw structural formulas for the products of hydrolysis of the following acetals.

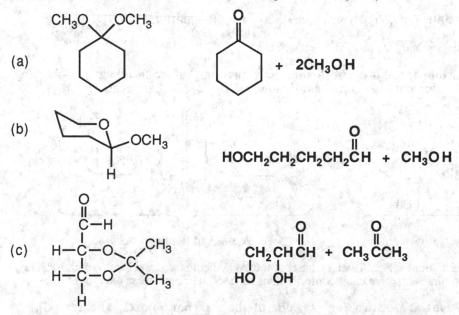

(a)   + $2CH_3OH$

(b) $HOCH_2CH_2CH_2CH_2\overset{O}{\overset{\|}{C}}H$ + CH_3OH

(c) $\underset{HO\quad OH}{CH_2CHCH}$ + $CH_3\overset{O}{\overset{\|}{C}}CH_3$

Problem 15.34 Propose a mechanism to account for the formation of a cyclic acetal from 4-hydroxypentanal and one equivalent of methanol. If the carbonyl oxygen of 4-hydroxypentanal is enriched with oxygen-18, do you predict that the oxygen label appears in the cyclic acetal or in the water?

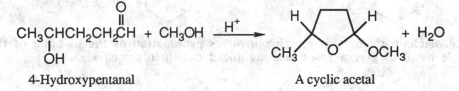

4-Hydroxypentanal   A cyclic acetal

Propose formation of a hemiacetal followed by protonation of the hemiacetal -OH and loss of water to form a resonance-stabilized cation. Then propose a Lewis acid-base reaction between this cation and methanol followed by loss of a proton to give the product. If the carbonyl group of 4-hydroxypentanal is enriched with oxygen-18, the oxygen-18 label appears in the water.

Step 1:

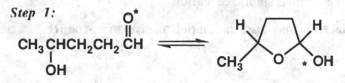

Step 2:

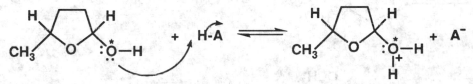

derived from the oxygen of the carbonyl group

Step 3:

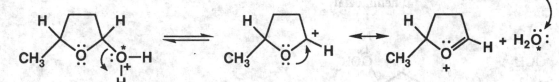

A resonance-stabilized cation

Step 4:

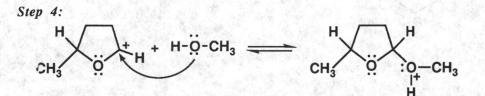

Step 5:

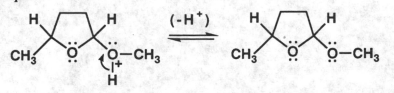

Problem 15.35 Propose a mechanism for this acid-catalyzed hydrolysis. (*Hint:* Review Problem 15.28.)

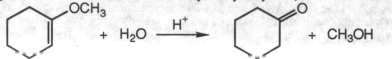

A reasonable mechanism for this reaction involves protonation of the pi bond of the carbon-carbon double bond to give a resonance stabilized cation intermediate.

Step 1:

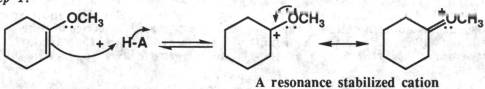

A resonance stabilized cation

Water reacts with the cation to form a hemiacetal, that loses methanol to give the product ketone.

Step 2:

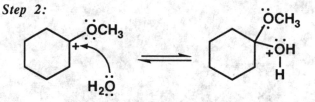

Step 3:

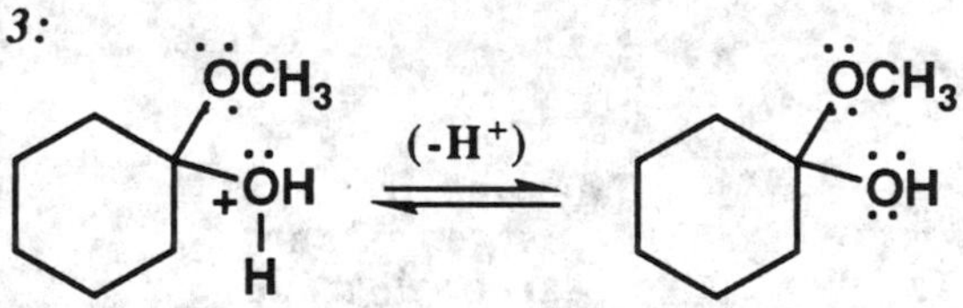

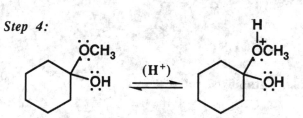

A hemiacetal

Step 4:

Step 5:

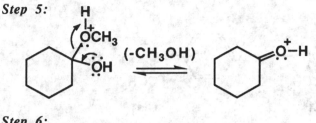

(-CH₃OH)

Step 6:

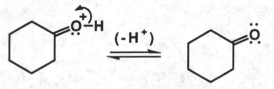

(-H⁺)

Problem 15.36 All rearrangements we have discussed so far involve generation of a positively-charged electron-deficient carbon atom (a carbocation) followed by a 1,2-shift of an atom or group of atoms from an adjacent carbon to the carbocation. A mechanism that can be written for the following reaction also involves generation of an electron-deficient atom (in this case an oxygen) followed by a 1,2-shift from an adjacent carbon to the electron-deficient atom.

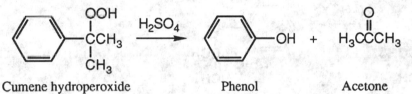

Cumene hydroperoxide   Phenol   Acetone

Propose a mechanism for the acid-catalyzed rearrangement of cumene hydroperoxide to phenol and acetone based on the previous rationale. In completing a mechanism, you will want to review the characteristic structural features of a hemiacetal (Section Problem 15.8B) and its equilibration with a ketone by loss of an alcohol.

A reasonable mechanism for this reaction involves protonation of the terminal oxygen atom of the hydroperoxide group followed by a 1,2 shift of the benzene ring. Attack of the resulting carbocation by water and loss of a proton yields a hemiacetal, that decomposes to give phenol and acetone.

Step 1:

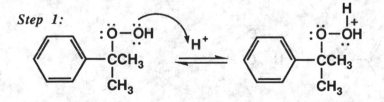

Step 2:

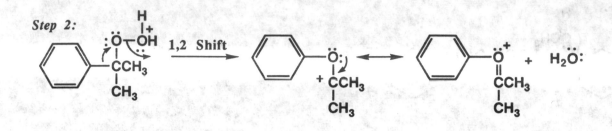

Step 3:

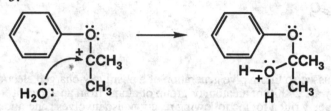

Step 4:

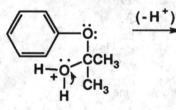

A hemiacetal

Step 5:

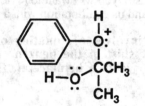

Step 6:

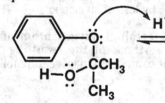

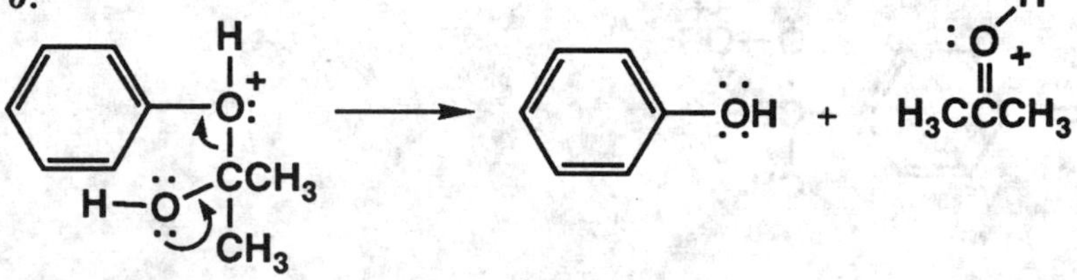

Step 7:

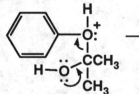

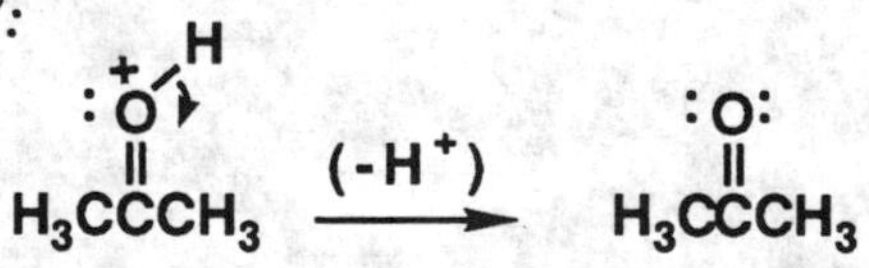

Problem 15.37 In Section 11.6A we saw that ethers, such as diethyl ether and tetrahydrofuran, are quite resistant to the action of dilute acids and require hot concentrated HI or HBr for cleavage. Acetals, however, in which two ether groups are linked to the same carbon, undergo hydrolysis readily even in dilute aqueous acid. How do you account for this marked difference in chemical reactivity toward dilute aqueous acid between ethers and acetals?

The first step of the cleavage reactions in acid for both ethers and acetals is protonation of an oxygen to form an oxonium ion. For acetals, the following step is cleavage of a carbon-oxygen bond to form a resonance-stabilized cation. For ethers, similar cleavage occurs, but the cation formed has no comparable resonance stabilization. Therefore, it is the resonance-stabilization of the cation intermediate formed during the cleavage of acetals that lowers the activation energy for their hydrolysis much below that for the hydrolysis of ethers.

Problem 15.38 Draw a structural formula for the magnesium alkoxide formed in the following Grignard reaction and then the product formed on hydrolysis of the magnesium alkoxide with aqueous acid.

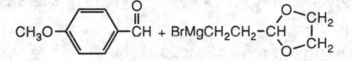

Hydrolysis of the magnesium alkoxide in aqueous acid converts the alkoxide to an alcohol and causes hydrolysis of the acetal to an aldehyde.

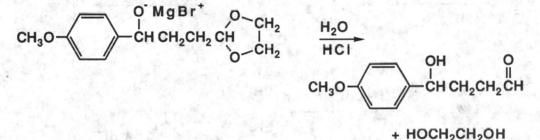

Problem 15.39 Show how to bring about the following conversion:

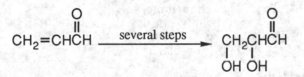

The most convenient way to convert an alkene to a glycol is to oxidize the alkene with osmium tetroxide in the presence of hydrogen peroxide. These conditions, however, will also oxidize an aldehyde to a carboxylic acid. Therefore, it is necessary to first protect the aldehyde by transformation to an acetal. In the following answer, ethylene glycol is the protecting agent.

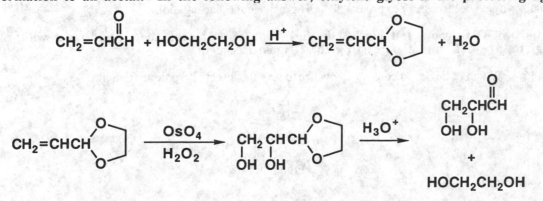

Problem 15.40 Multistriatin is a component of the aggregating pheromone of the European elm bark beetle, the insect vector of Dutch elm disease. In one laboratory synthesis of this molecule, (Z)-2-butene-1,4-diol was used as a starting material. Because of the requirements of subsequent steps, it was necessary to protect the two -OH groups of this molecule, which was done by making a cyclic acetal with acetone. The cyclic acetal was then treated with 3-chloroperoxybenzoic acid followed by methylmagnesium iodide and acid hydrolysis to give a compound of molecular formula $C_5H_{12}O_3$.

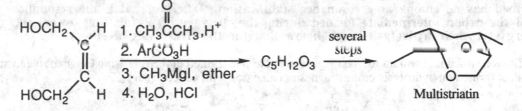

(Z)-2-Butene-1,4-diol

(a) Draw structural formulas for the products of Steps 1, 2, 3 and 4.

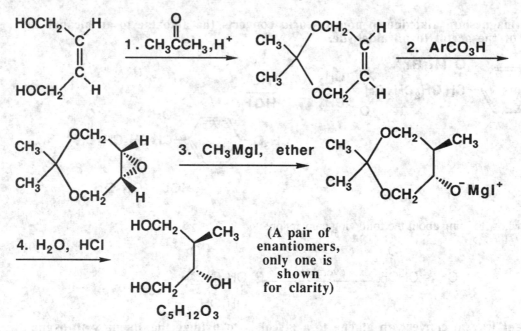

Note that the products of reaction 3 and 4 give racemic mixtures of enantiomers. Only one enantiomer was shown above for clarity. Reaction 2 gives a meso compound.

(b) Explain why it was necessary to protect the -OH groups in the starting diol.

Were the two -OH groups not blocked by formation of an acetal, they would react with CH_3MgI to give CH_4 and thus destroy the Grignard reagent.

Problem 15.41 Which of these molecules will cyclize to give the insect pheromone frontalin?

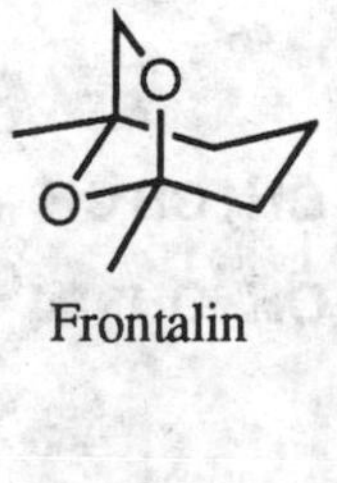

Frontalin

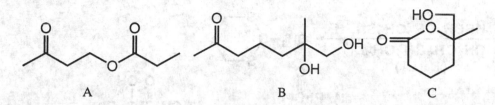

A   B   C

The answer is B. Using models may help with this answer.

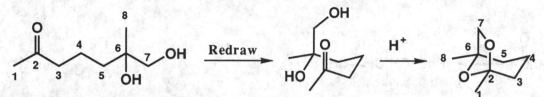

Addition of Sulfur Nucleophiles

<u>Problem 15.42</u> Draw a structural formula for the product of reaction of benzaldehyde with the following dithiols in the presence of an acid catalyst.
(a) 1,2-ethanedithiol   (b) 1,3-propanedithiol

Each compound forms a cyclic dithioacetal with benzaldehyde.

<u>Problem 15.43</u> Draw a structural formula for the product formed by treating each of these compounds with (1) the lithium salt of the 1,3-dithiane derived from acetaldehyde and then (2) H_2O, $HgCl_2$.

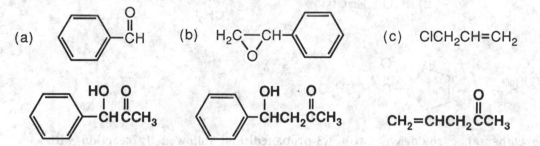

In reaction of the 1,3-dithiane of acetaldehyde with butyllithium, the carbonyl carbon is in effect converted to a carbanion that then adds to the carbonyl carbon of benzaldehyde in (a), brings about an S_N2 opening of the epoxide in (b), and an S_N2 displacement of chlorine in (c). Hydrolysis in aqueous mercuric chloride regenerates the carbonyl group of acetaldehyde.

<u>Problem 15.44</u> Show how to bring about the following conversions using a 1,3-dithiane:

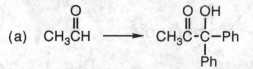

First convert acetaldehyde to a 1,3-dithiane and then to a lithium salt with butyllithium. Nucleophilic addition of this carbanion to the carbonyl group of benzophenone followed by hydrolysis of the 1,3-dithiane gives the desired ketoalcohol.

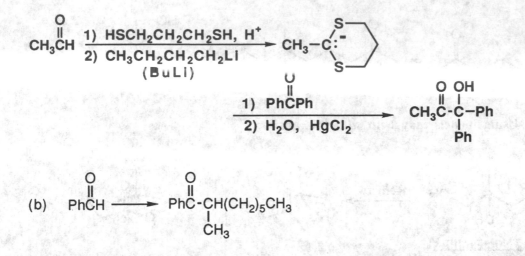

(b)

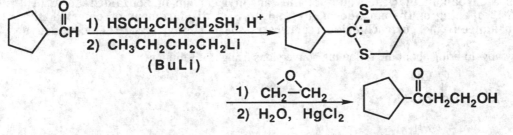

Convert benzaldehyde to a 1,3-dithiane and then to a lithium salt. Treatment of this nucleophile with 2-iodooctane via an S_N2 pathway followed by hydrolysis gives the desired product.

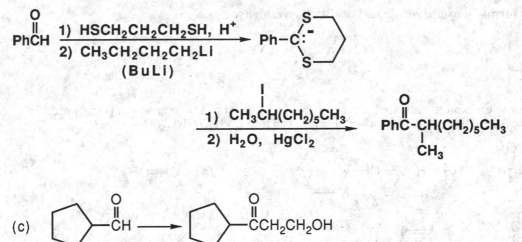

(c)

Treatment of cyclopentanecarbaldehyde with 1,3-propanedithiol followed by reaction with butyllithium gives a carbanion. Treatment of this carbanion with ethylene oxide followed by hydrolysis in aqueous mercuric chloride gives the product.

Addition of Nitrogen Nucleophiles

Problem 15.45 Draw structural formulas for the products of the following acid-catalyzed reactions:

(a) Phenylacetaldehyde + hydrazine ----->

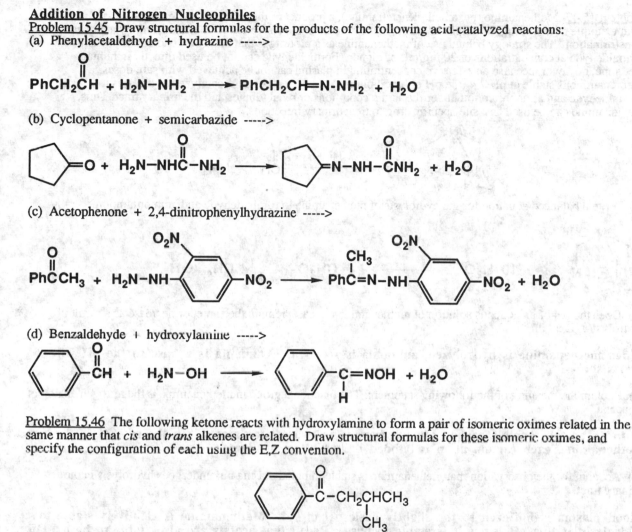

(b) Cyclopentanone + semicarbazide ----->

(c) Acetophenone + 2,4-dinitrophenylhydrazine ----->

(d) Benzaldehyde + hydroxylamine ----->

Problem 15.46 The following ketone reacts with hydroxylamine to form a pair of isomeric oximes related in the same manner that *cis* and *trans* alkenes are related. Draw structural formulas for these isomeric oximes, and specify the configuration of each using the E,Z convention.

The stereoisomers have a *cis* and *trans* relationship of groups about the carbon-nitrogen double bond. The configuration of these oximes is shown here using the E-Z system. Note that an unshared pair of electrons has a lower priority than hydrogen

Problem 15.47 Methenamine (hexamethylenetetramine), a product of the reaction of formaldehyde and ammonia, is an example of a prodrug, a compound that is inactive itself but is converted to an active drug by a biochemical transformation. The strategy behind use of methenamine as a prodrug is that nearly all bacteria are sensitive to formaldehyde at concentrations of 20 mg/mL or higher. Formaldehyde cannot be used directly in human medicine, however, because an effective concentration in plasma cannot be achieved with safe doses. Methenamine is stable at pH 7.4 (the pH of blood plasma) but undergoes acid-catalyzed hydrolysis to formaldehyde and ammonium ion under the acidic conditions of renal tubules and the urinary tract. Thus, methenamine can be used as a site-specific drug to treat urinary infections.

$$\text{methenamine} + H_2O \xrightarrow{HCl} HCH(=O) + NH_4^+$$

(a) Write a balanced equation for the hydrolysis of methenamine to formaldehyde and ammonium ion.

$$\text{methenamine} + 10\ H_2O \longrightarrow 6\ CH_2O + 4\ NH_4^+\ OH^-$$

(b) Does the pH of an aqueous solution of methenamine increase, remain the same, or decrease as a result of hydrolysis? Explain.

When methenamine is hydrolyzed, ammonia is released. Ammonia is a base so the pH will increase.

(c) Explain the meaning of the following statement: The functional group in methenamine is the nitrogen analog of an acetal.

With an acetal, a single carbon atom is bonded to two oxygen atoms. In the case of methenamine, each carbon atom is bonded to two nitrogen atoms.

(d) Account for the observation that methenamine is stable in blood plasma but undergoes hydrolysis in the urinary tract.

Blood plasma is buffered to the slightly basic pH of 7.4. Methenamine is relatively stable to hydrolysis at this pH, since it is stable to base. Recall that acetals are also stable to base. On the other hand, both methenamine and acetals are readily hydrolyzed at acidic pH. The urinary tract is more acidic, so the methenamine is hydrolyzed much more rapidly there.

(e) Propose a mechanism for the acid-catalyzed hydrolysis of methenamine to formaldehyde and ammonium ion.

The entire mechanism for the hydrolysis of methenamine has many steps. The general mechanism involves protonation of the nitrogen atom, followed by formation of an iminium ion that is attacked by water as a nucleophile. This mechanism is highly analogous to that for acid-catalyzed hydrolysis of acetals.

Step 1:

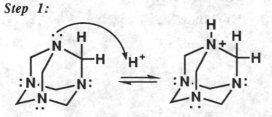

Step 2:

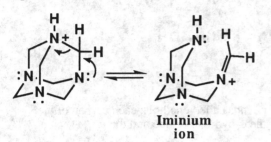

**Iminium
ion**

Step 3:

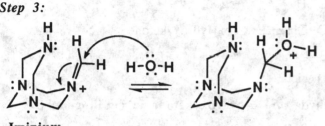

**Iminium
ion**

Step 4:

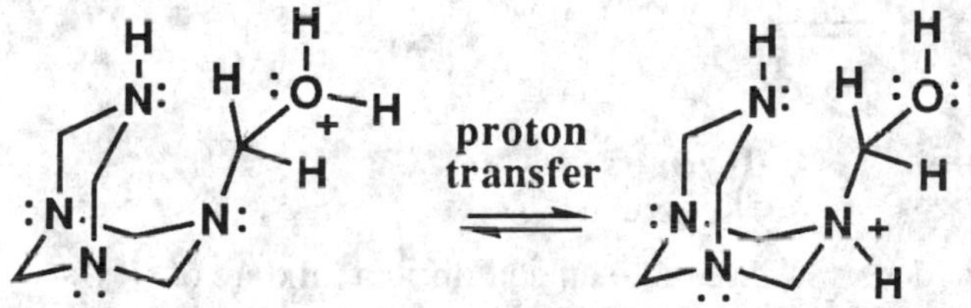

proton
transfer

Step 5:

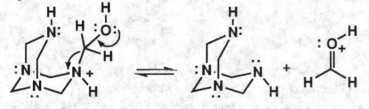

Step 6:

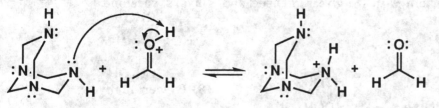

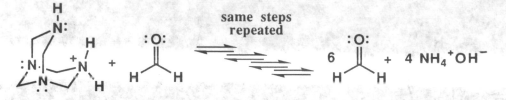

Keto-Enol Tautomerism

<u>Problem 15.48</u> The following molecule belongs to a class of compounds called enediols; each carbon of the double bond carries an OH group. Draw structural formulas for the α-hydroxyketone and the α-hydroxyaldehyde with which this enediol is in equilibrium.

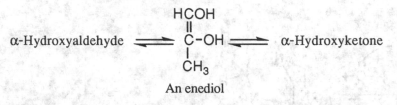

α-Hydroxyaldehyde ⇌ [enediol] ⇌ α-Hydroxyketone

An enediol

Following are formulas for the α-hydroxyaldehyde and α-hydroxyketone in equilibrium by way of the enediol intermediate.

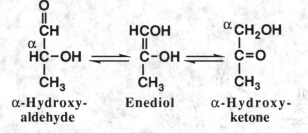

α-Hydroxy-aldehyde   Enediol   α-Hydroxy-ketone

<u>Problem 15.49</u> In dilute aqueous base, (R)-glyceraldehyde is converted into an equilibrium mixture of (R,S)-glyceraldehyde and dihydroxyacetone. Propose a mechanism for this isomerization.

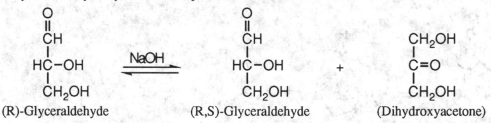

(R)-Glyceraldehyde   (R,S)-Glyceraldehyde   (Dihydroxyacetone)

The key is keto-enol tautomerism. In the presence of base, (R)-glyceraldehyde undergoes base-catalyzed keto-enol tautomerism to form an enediol in which carbon-2 is achiral. This enediol is in turn in equilibrium with (S)-glyceraldehyde and dihydroxyacetone.

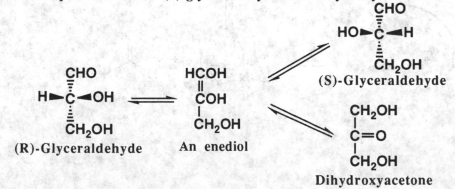

(R)-Glyceraldehyde   An enediol   (S)-Glyceraldehyde

Dihydroxyacetone

Problem 15.50 When *cis*-2-decalone is dissolved in ether containing a trace of HCl, an equilibrium is established with *trans*-2-decalone. The latter ketone predominates in the equilibrium mixture. Propose a mechanism for this isomerization and account for the fact that the *trans* isomer predominates at equilibrium.

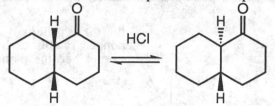

The *cis*-2-decalone and *trans*-2-decalone are equilibrating via the enol intermediate.

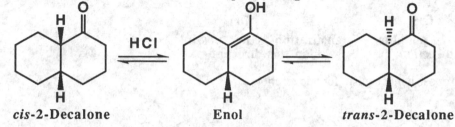

      cis-2-Decalone   Enol   *trans*-2-Decalone

The *trans*-2-decalone is more stable because all of the carbons of the ring fusions are in the more stable equatorial arrangement. In *cis*-2-decalone, there is one equatorial and one less-favored axial arrangement at the ring fusion.

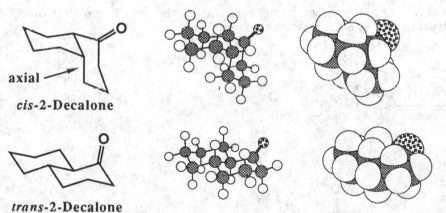

axial

cis-2-Decalone

trans-2-Decalone

Reactions at the α-Carbon

Problem 15.51 The following bicyclic ketone has two α-carbons and three α-hydrogens. When this molecule is treated with D_2O in the presence of an acid catalyst, only two α-hydrogens exchange with deuterium. The α-hydrogen at the bridgehead does not exchange. How do you account for the fact that two α-hydrogens do exchange but the third does not ? You will find it helpful to build a model of this molecule and of the enols by which exchange of α-hydrogens occurs.

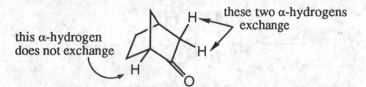

this α-hydrogen
does not exchange

these two α-hydrogens
exchange

Exchange of α-hydrogens is through keto-enol tautomerism and an enol intermediate. The key to this problem centers on the possibility of placing a carbon-carbon double bond between carbons 2-3 of the bicyclic ring and between carbons 1-2 of the ring. Enolization in the first direction is possible with the result that the two α-hydrogens on carbon-3 exchange.

Enolization toward the bridgehead carbon is not possible because of the geometry of the
bicyclo[2.2.1]heptane ring. The energy required to force the bridgehead carbon and the three
other carbons attached to it into a planar conformation with bond angles of 120° is
prohibitively high.

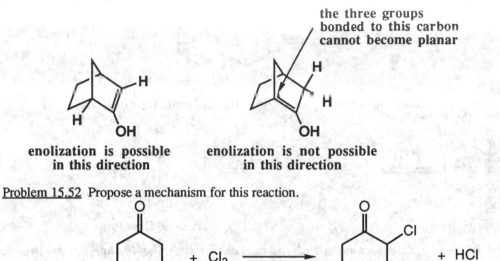

the three groups
bonded to this carbon
cannot become planar

enolization is possible   enolization is not possible
in this direction   in this direction

Problem 15.52 Propose a mechanism for this reaction.

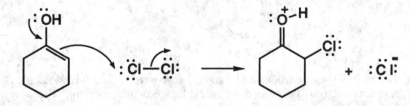

A reasonable mechanism for this reaction involves tautomerization to the enol form, that then
reacts with the Cl₂. A final proton transfer completes the reaction.

Step 1:

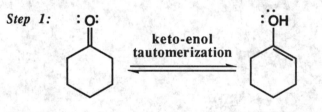

Step 2:

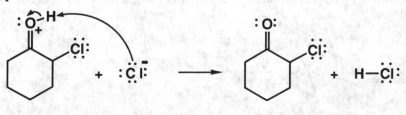

Step 3:

<u>Problem 15.53</u> The base-promoted rearrangement of an α-haloketone to a carboxylic acid, known as the Favorskii rearrangement, is illustrated by the conversion of 2-chlorocyclohexanone to cyclopentanecarboxylic acid. It is proposed that NaOH first converts the α-haloketone to the substituted cyclopropanone shown in brackets, and then to the sodium salt of cyclopentanecarboxylic acid.

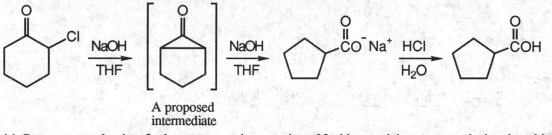

A proposed
intermediate

(a) Propose a mechanism for base-promoted conversion of 2-chlorocyclohexanone to the bracketed intermediate.

A reasonable mechanism involves formation of an enolate anion that then displaces chloride in an intramolecular step that produces the three-membered ring intermediate.

Step 1:

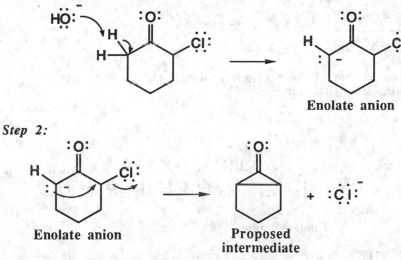

Enolate anion

Step 2:

Enolate anion   Proposed
                                 intermediate

(b) Propose a mechanism for base-promoted conversion of the bracketed intermediate to sodium cyclopentanecarboxylate. (*Hint:* Begin by adding hydroxide ion to the carbonyl carbon to form a tetrahedral carbonyl addition intermediate.)

As stated in the hint, a reasonable mechanism can be written that starts with formation of a tetrahedral carbonyl addition intermediate, followed by collapse of the intermediate and breaking of one of the cyclopropane ring bonds. A proton transfer completes the reaction.

Step 1:

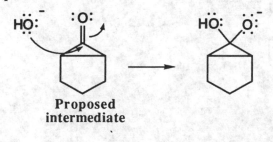

Proposed
intermediate

Step 2:

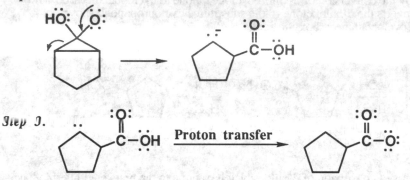

Step 3.   **Proton transfer**

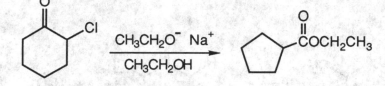

<u>Problem 15.54</u> If the Favorskii rearrangement 2-chlorocyclohexanone is carried out using sodium ethoxide in ethanol, the product is ethyl cyclopentanecarboxylate. Propose a mechanism for this reaction.

$$CH_3CH_2O^- \; Na^+$$
$$CH_3CH_2OH$$

This mechanism is entirely analogous to that of the previous problem, except this time ethoxide is the base and nucleophile, not hydroxide.

Step 1:

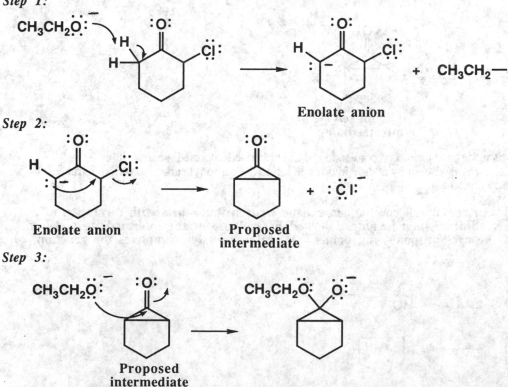

Step 4:

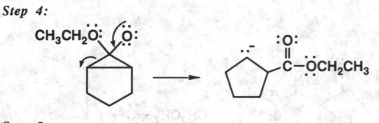

Step 5:

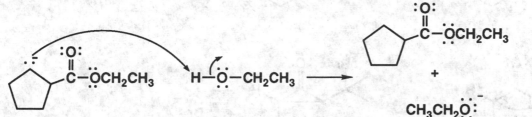

<u>Problem 15.55</u> (R)-(+)-Pulegone, readily available from the pennyroyal oils (Merck Index, 12th Ed., #8124), is an important enantiopure building block for organic syntheses. Propose a mechanism for each step in this transformation of pulegone. (*Hint:* The first stages of the mechanism for the second reaction are essentially identical to those of the Favorskii rearrangement.)

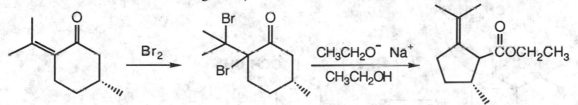

The first step involves bromination of an alkene via a bromonium ion intermediate.
Step 1:

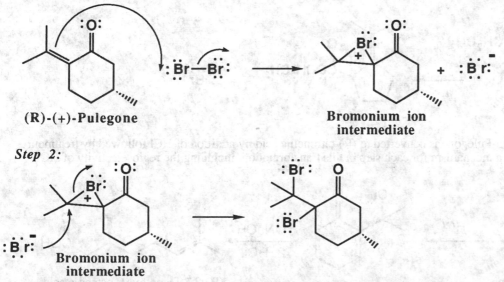

Step 2:

A Favorskii rearrangement then occurs via the proposed cyclopropanone intermediate. The novel feature in this case is shown in step 6 wherein decomposition of the tetrahedral carbonyl addition intermediate is accompanied by displacement of bromide and formation of the carbon-carbon double bond.

Step 3:

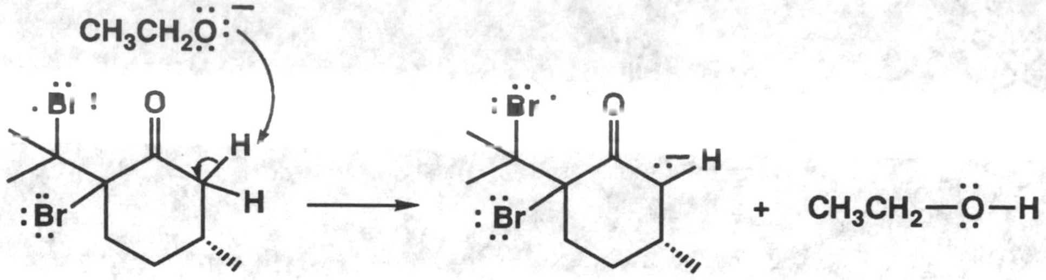

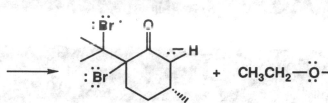

Step 4:

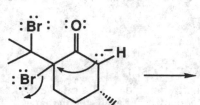

Step 5:

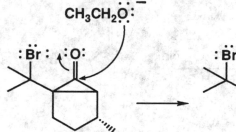

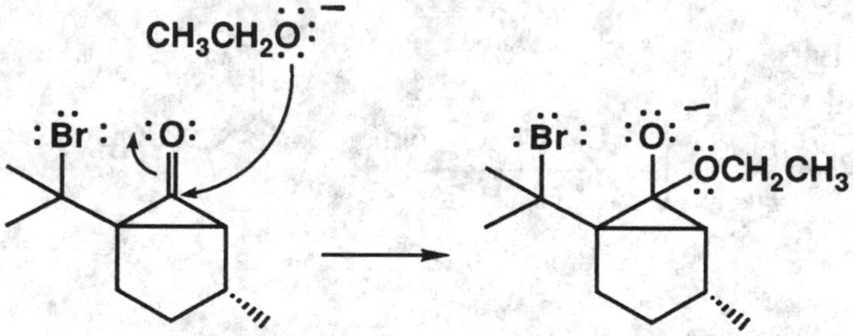

Step 6:

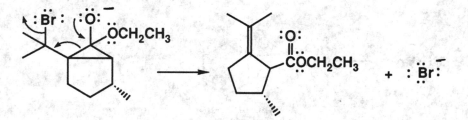

<u>Problem 15.56</u> (R)-(+)-Pulegone is converted to (R)-citronellic acid by addition of HCl followed by treatment with NaOH. Propose a mechanism for each step in this transformation, including the regioselectivity of HCl addition.

(R)-(+)-Pulegone

(R)-3,7-Dimethyl-6-octenoic acid
(R)-Citronellic acid

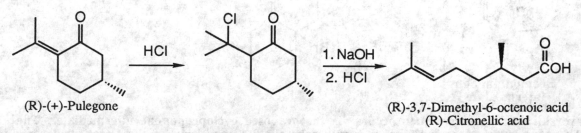

The first step of the process involves protonation of the carbon-carbon double bond to create the tertiary carbocation shown below. Chloride reacts with this carbocation to give the

chlorinated intermediate. This regiochemistry of hydrochlorination reflects the most stable carbocation that is possible. The other possible carbocation produced upon protonation would also have been a tertiary carbocation, but one that is α to the carbonyl group. Carbonyl groups are electron withdrawing and thus destabilizing to an adjacent carbocation.

Step 1:

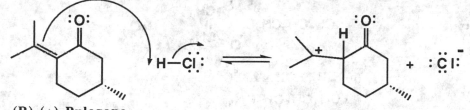

(R)-(+)-Pulegone

Step 2:

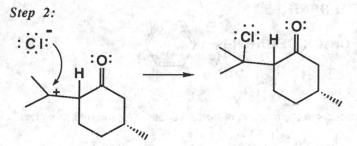

Hydroxide attacks the carbonyl group to give a tetrahedral carbonyl addition intermediate that then decomposes in step 4 to break one bond of the six-membered ring while displacing chloride to create a new carbon-carbon double bond.

Step 3:

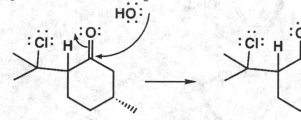

Step 4:

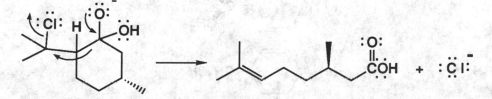

Oxidation/Reduction of Aldehydes and Ketones
Problem 15.57 Draw structural formulas for the products formed by treatment of butanal with the following reagents.

(a) LiAlH₄ followed by H₂O

  CH₃CH₂CH₂CH₂OH

(b) NaBH₄ in CH₃OH/H₂O

  CH₃CH₂CH₂CH₂OH

(c) H_2/Pt

$CH_3CH_2CH_2CH_2OH$

(d) $Ag(NH_3)_2{}^+$ in NH_3/H_2O

$CH_3CH_2CH_2CO_2{}^-NH_4{}^+$

(e) H_2CrO_4, heat

$CH_3CH_2CH_2CO_2H$

(f) $HOCH_2CH_2OH$, HCl

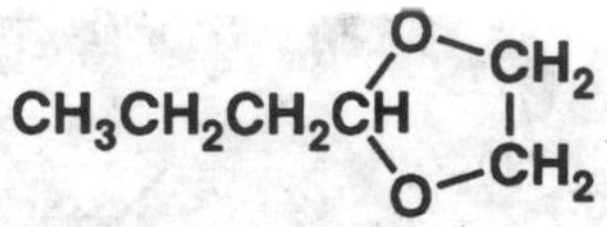

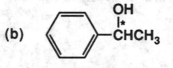

$+ H_2O$

(g) Zn(Hg)/HCl

$CH_3CH_2CH_2CH_3$

(h) N_2H_4, KOH at 250°C

$CH_3CH_2CH_2CH_3$

(i) $C_6H_5NH_2$

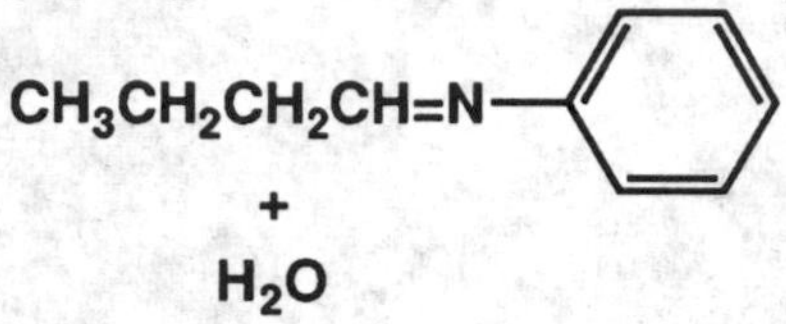

$CH_3CH_2CH_2CH=N$—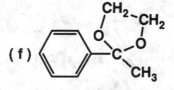

+

H_2O

(j) $C_6H_5NHNH_2$

$CH_3CH_2CH_2CH=N-N$—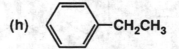
                              H

+

H_2O

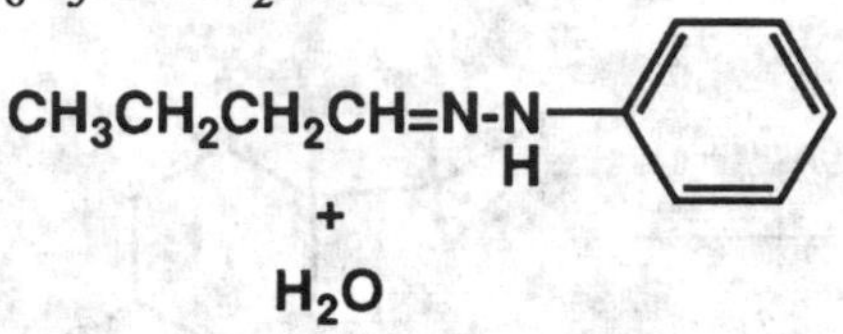

<u>Problem 15.58</u> Draw structural formulas for the products of the reaction of acetophenone with the reagents given in Problem 15.57.

Following are structural formulas for each product. Note that in parts (a), (b), and (c) a new stereocenter is created in the reaction so the product will actually be a racemic mixture.

(a)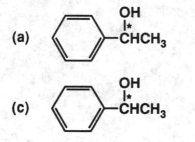

(b)

(c)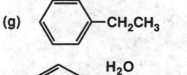

(d) No reaction

(e)   No reaction

(f)

(g)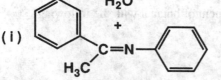

(h)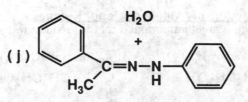

(i)

(j)

Synthesis

Problem 15.59 Starting with cyclohexanone, show how to prepare these compounds. In addition to the given starting material, use any other organic or inorganic reagents as necessary.

(a) Cyclohexanol

This transformation can be accomplished with any of three different sets of reagents:

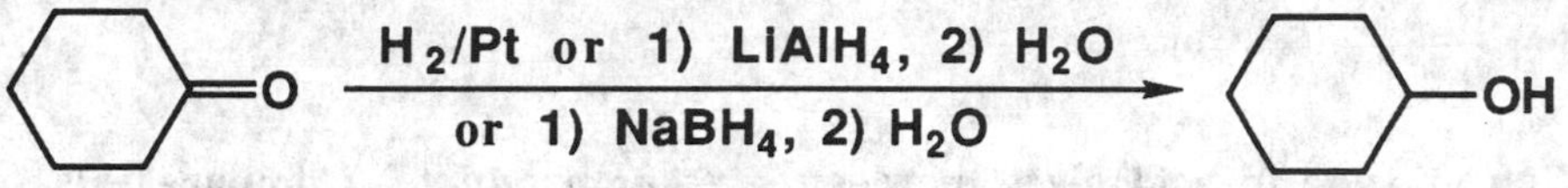

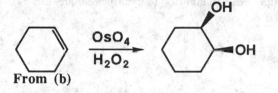

(b) Cyclohexene

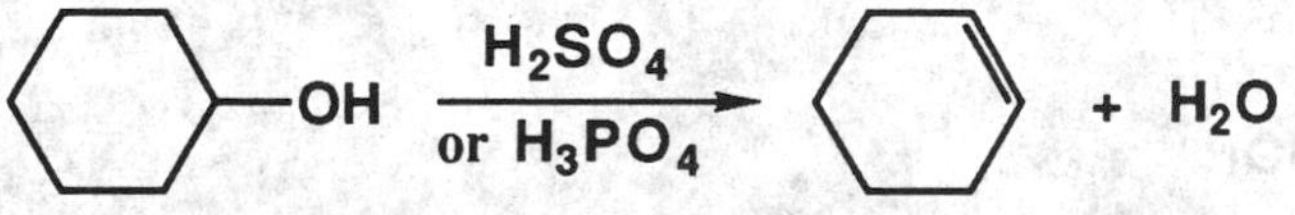

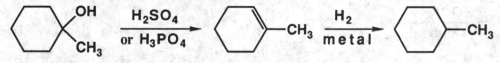

From (a)

(c) *cis*-1,2-Cyclohexanediol

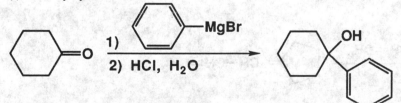

From (b)

(d) 1-Methylcyclohexanol

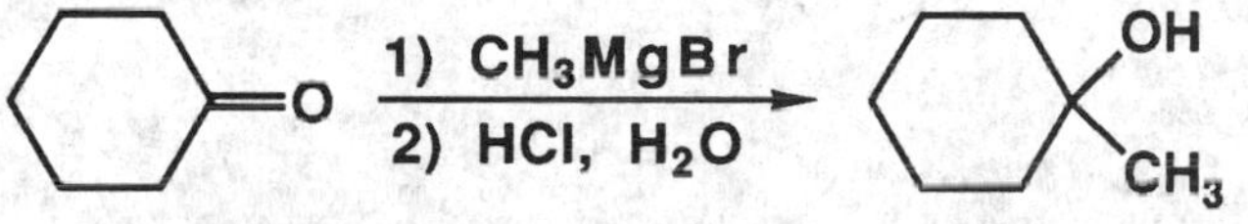

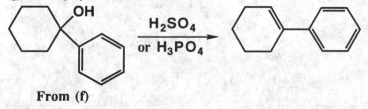

(e) 1-Methylcyclohexane

From (d)

(f) 1-Phenylcyclohexanol

(g) 1-Phenylcyclohexene

From (f)

(h) Cyclohexene oxide

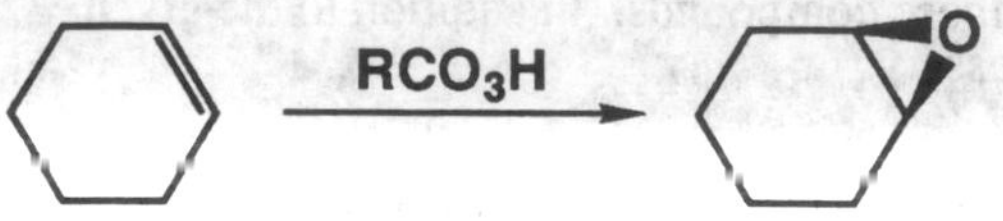

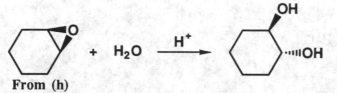

From (b)

(i) *trans*-1,2-Cyclohexanediol

Recall that ring-opening of an epoxide in acid gives the desired *trans* product. Compare this to part (c) of this problem in which the *cis* product is desired.

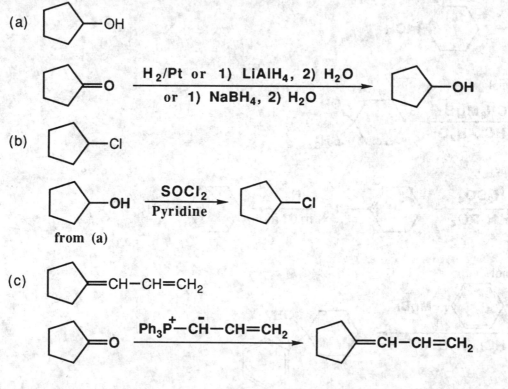

From (h)

Problem 15.60 Show how to convert cyclopentanone to these compounds. In addition to cyclopentanone, use other organic or inorganic reagents as necessary.

(a)

$\overset{\text{H}_2/\text{Pt or 1) LiAlH}_4,\ 2)\ \text{H}_2\text{O}}{\underset{\text{or 1) NaBH}_4,\ 2)\ \text{H}_2\text{O}}{\longrightarrow}}$

(b)

$\overset{\text{SOCl}_2}{\underset{\text{Pyridine}}{\longrightarrow}}$

from (a)

(c)

$\overset{\text{Ph}_3\overset{+}{\text{P}}-\overset{-}{\text{CH}}-\text{CH}=\text{CH}_2}{\longrightarrow}$

(d)

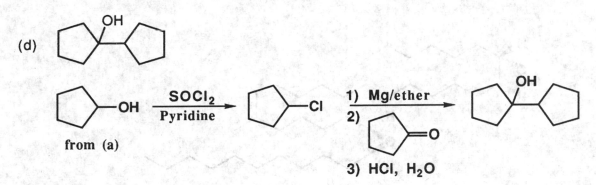

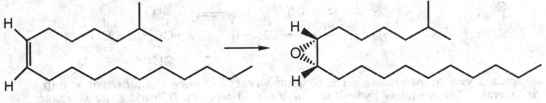

from (a)

1) Mg/ether
2)
3) HCl, H₂O

Problem 15.61 Disparlure is a sex attractant of the gypsy moth (*Porthetria dispar*). It has been synthesized in the laboratory from the following (Z)-alkene.

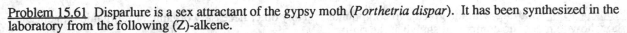

(Z)-2-methyl-7-octadecene   Disparlure

(a) Propose two sets of reagents that might be combined in a Wittig reaction to give the indicated (Z)-alkene. Note that at least for simple phosphonium ylides, the product of a Wittig reaction is a (Z)-alkene.

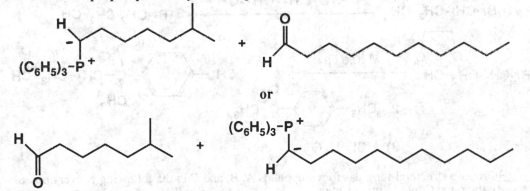

$(C_6H_5)_3$–P⁺   +

or

$(C_6H_5)_3$–P⁺

+

(b) How might the (Z)-alkene be converted to disparlure?

The (Z)-alkene is converted to disparlure by reaction with a peracid.

(c) How many stereoisomers are possible for disparlure? How many are formed in the sequence you have chosen?

Disparlure has two stereocenters, so there are a total of 4 stereoisomers possible. Starting with the (Z)-alkene means that only (Z) epoxides are formed. Thus, there will only be two stereoisomers formed as shown below.

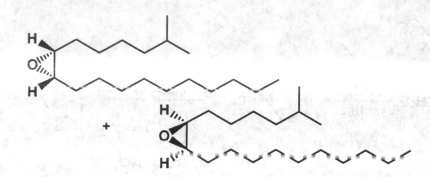

Problem 15.62 Starting with the given two compounds and any other necessary organic or inorganic reagents, show how to make the compound shown at the right.

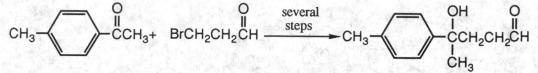

The desired compound can be produced via a Grignard reaction, once the aldehyde group is protected as an acetal. The acetal is hydrolyzed in the acid workup following the Grignard reaction.

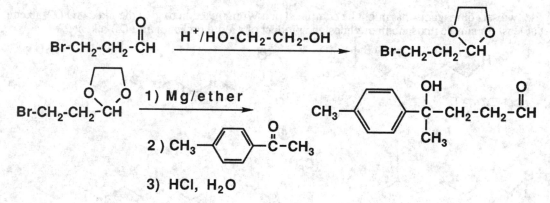

Problem 15.63 Propose structural formulas for compounds A, B, and C in the following conversion. Show also how to prepare compound C by a Wittig reaction.

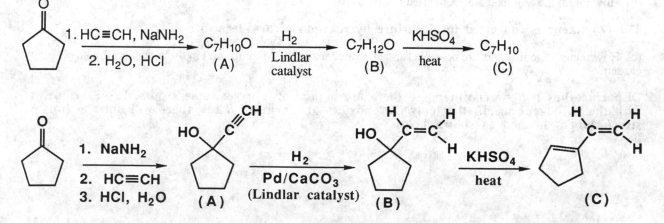

Compound C could also be prepared by the following Wittig reaction:

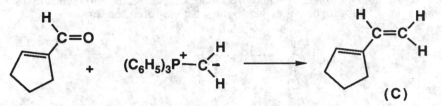

Problem 15.64 Following is a retrosynthetic scheme for the synthesis of *cis*-3-penten-2-ol. Write a synthesis for this compound from the indicated starting materials.

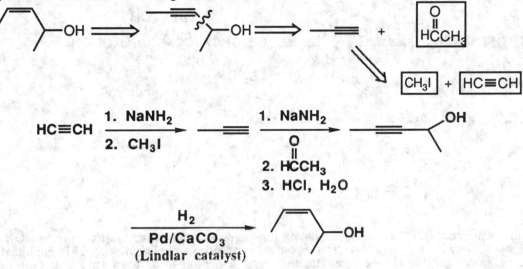

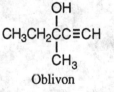

Problem 15.65 Following is the structural formula of the tranquilizer Oblivon (meparfynol). Propose a synthesis for this molecule starting with acetylene and a ketone.

$$CH_3CH_2\overset{\overset{\displaystyle OH}{|}}{\underset{\underset{\displaystyle CH_3}{|}}{C}}C\equiv CH$$

Oblivon

Oblivon is prepared by reaction of sodium acetylide with 2-butanone followed by hydrolysis of the resulting sodium alkoxide in aqueous acid.

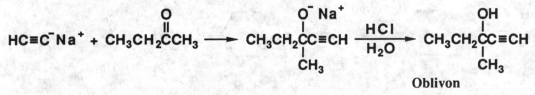

<u>Problem 15.66</u> Following is the structural formula of surfynol, a defoaming surfactant. Describe the synthesis of this molecule from acetylene and a ketone. You may wish to look up the word "surfactant" in a science reference book and find out what surfactants are, how they work, and what they are used for.

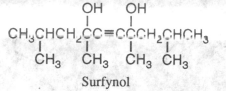

Surfynol

Surfynol (surfactant containing alkyne and alcohol functional groups) is synthesized by reaction of the disodium salt of acetylene with two mol of 4-methyl-2-pentanone.

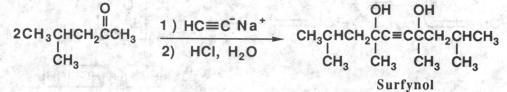

Surfynol

<u>Problem 15.67</u> Propose a mechanism for this isomerization.

A reasonable mechanism for this transformation involves initial protonation of the -OH group, followed by loss of water to generate a resonance stabilized carbocation. This carbocation is attacked by water to give an intermediate that loses a proton to give an enol that tautomerizes to produce the final product.

Step 1:

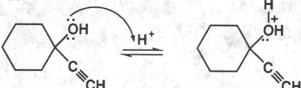

Step 2:

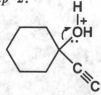 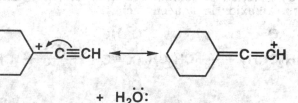

Step 3:

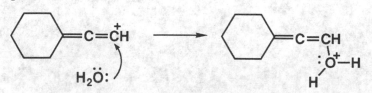

Step 4:

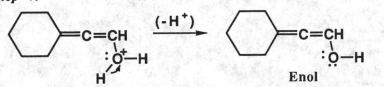

Step 5:

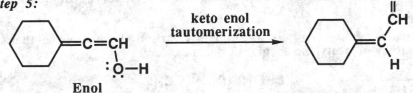

Problem 15.68 Propose a mechanism for this isomerization.

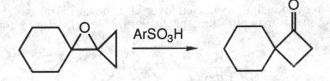

A reasonable mechanism for this reaction involves an initial protonation of the epoxide, leading to epoxide ring opening to form a tertiary carbocation. This undergoes a rearrangement to create a resonance stabilized cation that contains the less strained four-membered ring instead of the original, more highly strained three-membered ring. Deprotonation gives the final product.

*Step 1:*

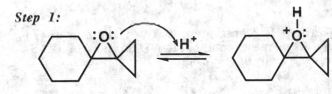

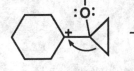*Step 2:*

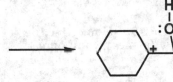

Step 3:

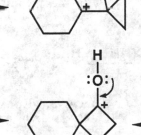

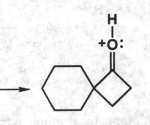

Step 4:

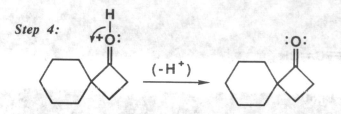

Problem 15.69 Starting with acetylene and 1-bromobutane as the only sources of carbon atoms, show how to synthesize the following:

All molecules in this problem are decanes. The way to build this carbon skeleton using the compounds given is by dialkylation of acetylene to give 5-decyne.

$$HC{\equiv}CH \xrightarrow[\text{2) } 2CH_3CH_2CH_2CH_2Br]{\text{1) } 2NaNH_2} CH_3(CH_2)_3C{\equiv}C(CH_2)_3CH_3$$

5-Decyne

Catalytic reduction of 5-decyne using hydrogen over a Lindlar catalyst gives (Z)-5-decene. Chemical reduction of 5-decyne using sodium or lithium metal in liquid ammonia gives (E)-5-decene.

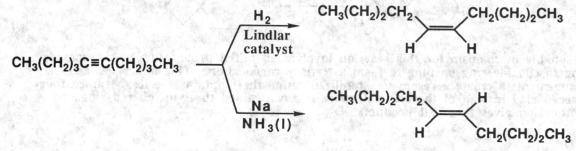

In the following formulas, the $CH_3(CH_2)_2CH_2$- group is represented as R-. Oxidation of (Z)-5-decene with a peracid gives *cis*-5,6-epoxydecane as a meso compound. Oxidation of (E)-5-decene by a peracid gives *trans*-5,6-epoxydecane as a pair of enantiomers.

(Z)-5-Decene **cis-5,6-Epoxydecane (a meso compound)**

(E)-5-Decene ***trans*-5,6-Epoxydecane (as a pair of enantiomers)**

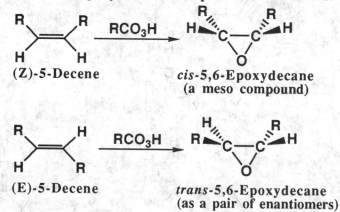

(a) Meso-5,6-decanediol

Oxidation of (Z)-5-decene by osmium tetroxide in the presence of hydrogen peroxide gives meso-5,6-decanediol.

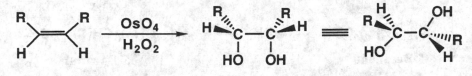

(Z)-5-Decene   **meso-5,6-Decanediol**

Alternatively, acid-catalyzed hydrolysis of the *trans*-epoxide also gives meso-5,6-decanediol.

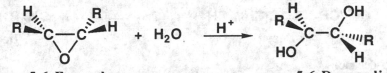

trans-5,6-Epoxydecane   **meso-5,6-Decanediol**

(b) Racemic 5,6-decanediol

Oxidation of (E)-5-decene by osmium tetroxide in the presence of hydrogen peroxide gives racemic 5,6-decanediol.

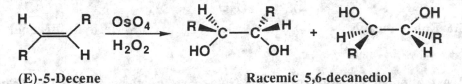

(E)-5-Decene   **Racemic 5,6-decanediol**

Alternatively, acid-catalyzed hydrolysis of the *cis*-epoxide also gives racemic 5,6-decanediol.

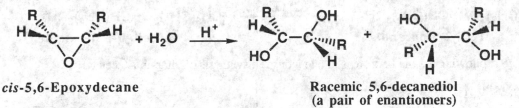

cis-5,6-Epoxydecane   **Racemic 5,6-decanediol**
                                               (a pair of enantiomers)

(c) 5-Decanone

Acid-catalyzed hydration of 5-decyne gives 5-decanone.

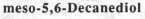

$$CH_3(CH_2)_3C{\equiv}C(CH_2)_3CH_3 \quad + \quad H_2O \quad \xrightarrow[\text{HgSO}_4]{\text{H}_2\text{SO}_4} \quad CH_3(CH_2)_3\overset{\overset{\displaystyle O}{\|}}{C}CH_2(CH_2)_3CH_3$$

5-Decyne   **5-Decanone**

(d) 5,6-Epoxydecane

See the beginning of this solution for methods to prepare both the *cis*-epoxide and the *trans*-epoxide.

(e) 5-Decanol

Catalytic reduction of 5-decanone using hydrogen over a transition metal catalyst, or chemical reduction using $NaBH_4$ or $LiAlH_4$ gives 5-decanol.

$$CH_3(CH_2)_3\overset{O}{\overset{\|}{C}}(CH_2)_4CH_3 \ + \ NaBH_4 \longrightarrow CH_3(CH_2)_3\overset{OH}{\overset{|}{CH}}(CH_2)_4CH_3$$

5-Decanone   5-Decanol

(f) Decane

Catalytic reduction of 5-decyne using hydrogen over a transition metal catalyst gives decane.

$$CH_3(CH_2)_3C\equiv C(CH_2)_3CH_3 \ + \ 2H_2 \xrightarrow{\ Pt\ } CH_3(CH_2)_8CH_3$$

5-Decyne   Decane

(g) 6-Methyl-5-decanol

Treatment of either the *cis*-epoxide or the *trans*-epoxide from part (d) with methylmagnesium iodide gives 6-methyl-5-decanol.

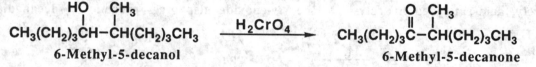

5,6-Epoxydecane   6-Methyl-5-decanol

(h) 6-Methyl-5-decanone

Chromic acid oxidation of 6-methyl-5-decanol from part (g) gives 6-methyl-5-decanone.

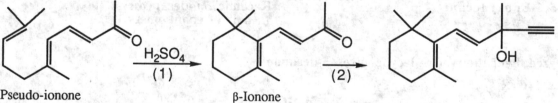

6-Methyl-5-decanol   6-Methyl-5-decanone

Problem 15.70 Following are the final steps in one industrial synthesis of vitamin A acetate:

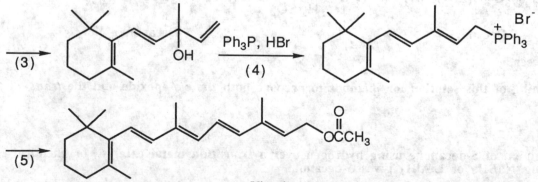

Vitamin A acetate

(a) Propose a mechanism for the acid-catalyzed cyclization in Step 1.

Step 1:

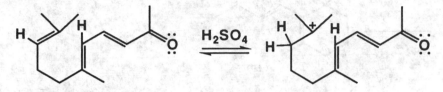

Step 2:

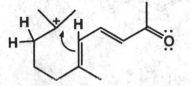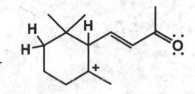

Step 3:

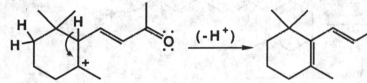

(b) Propose reagents to bring about Step 2.

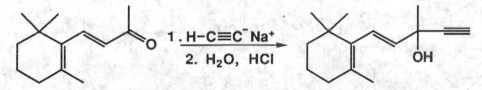

(c) Propose reagents to bring about Step 3

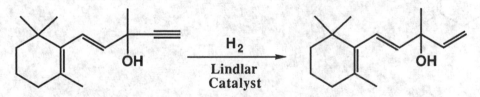

(d) Propose a mechanism for formation of the phosphonium salt in Step 4.

Step 1:

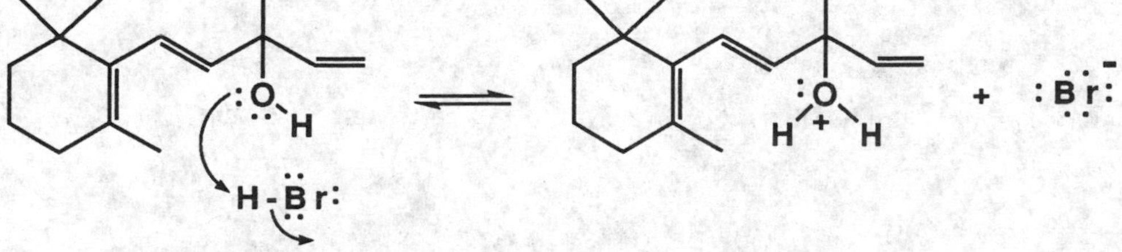

Step 2:

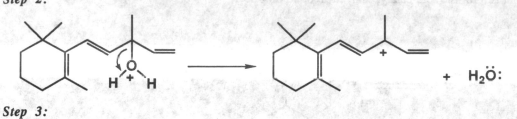

Step 3:

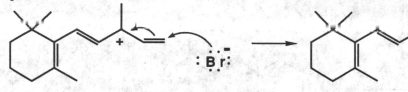

Step 4:

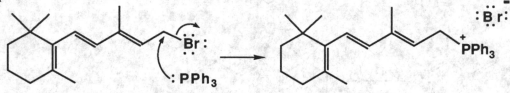

(e) Show how Step 5 can be completed by a Wittig reaction.

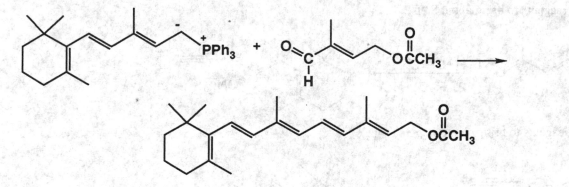

CHAPTER 16: CARBOXYLIC ACIDS

SUMMARY OF REACTIONS

Starting Material \ Product →	Acid Halides	Alcohols	Carboxyates	Carboxylic Acids	Esters	Ketones	Methyl Esters
Carboxylic Acids	**16A** 16.9*	**16B** 16.7	**16C** 16.4B				**16D** 16.8C
Carboxylic Acids Alcohols					**16E** 16.8A		
β-Dicarboxylic Acids				**16F** 16.10B			
β-Ketoacids				**16G** 16.10A			

*Section in book that describes reaction.

REACTION 16A: CONVERSION TO ACID HALIDES (Section 16.9)

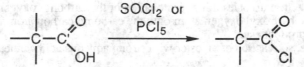

- **Carboxylic acids** react with either **thionyl chloride** or **phosphorus pentachloride** to yield acid chlorides. ✳
- The mechanism for the reaction with thionyl chloride involves an initial formation of an alkyl chlorosulfite intermediate, followed by its decomposition to SO_2 and the acid chloride.

REACTION 16B: REDUCTION BY LITHIIUM ALUMINUM HYDRIDE (Section 16.7)

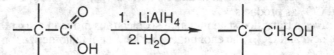

- **Carboxylic acids** react with **lithium aluminum hydride** to give **primary alcohols**.✳
- The reaction involves delivery of two hydride ions from the $LiAlH_4$ to the carbonyl group, while the hydroxyl hydrogen is derived from water in the work-up.
- $NaBH_4$ cannot reduce carboxylic acids to primary alcohols, so this reagent can be used to reduce selectively other functional groups such as aldehydes or ketones in the presence carboxylic acid groups.
- Catalytic hydrogenation does not reduce carboxylic acids.

REACTION 16C: REACTION OF CARBOXYLIC ACIDS WITH BASES (Section 16.4A)

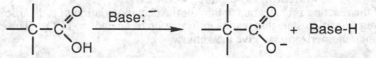

- **Carboxylic acids** are **relatively acidic**, so they react with **bases** such as NaOH, KOH, or NH_3 to give a **carboxylate anion** and protonated base. ✳
- Carboxylate anions are highly water soluble, and this is often exploited in the isolation of carboxylic acids.

REACTION 16D: REACTION WITH DIAZOMETHANE (Section 16.8C)

- **Carboxylic acids** react with **diazomethane** to form **methyl esters.** ∗
- This is a very clean reaction, but unfortunately diazomethane must be used with care because it is poisonous and potentially explosive.
- The mechanism of this reaction involves an initial proton transfer from the carboxylic acid to diazomethane, followed by nucleophilic attack of the carboxylate and loss of N_2 as the leaving group to generate the product methyl ester.

REACTION 16E: FISCHER ESTERIFICATION (Section 16.8B)

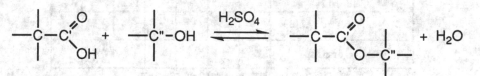

- **Carboxylic acids** react with **excess alcohol** in the presence **sulfuric acid** under anhydrous conditions to generate **esters.** ∗
- The mechanism of the reaction involves initial protonation of the carbonyl oxygen atom, followed by nucleophilic attack of the hydroxyl oxygen atom onto the carbonyl carbon atom, transfer of a proton, and loss of H_2O to give the product ester.
- The reaction is reversible, so reaction of an ester with strong aqueous acid can lead to hydrolysis of the ester.

REACTION 16F: DECARBOXYLATION OF β–DICARBOXYLIC ACIDS (Section 16.10B)

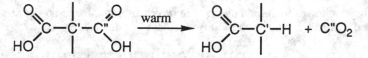

- When a **β-dicarboxylic acid** (malonic acid and its derivatives) is **heated**, it can **decarboxylate** to give a carboxylic acid and CO_2 as products. ∗
- The mechanism involves a transition state in which six electrons move in a six-membered ring to give the enol of a carboxylic acid and CO_2. The enol undergoes keto-enol tautomerization to give the product carboxylic acid.
- This reaction is important because malonic acid derivatives are easily produced.

REACTION 16G: DECARBOXYLATION OF β-KETOACIDS (Section 16.10A)

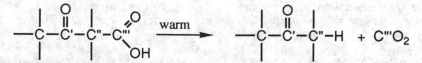

- **β-Ketoacids** lose CO_2 upon **heating** to produce **ketones.** ∗
- The mechanism of the reaction is analogous to that for the decarboxylation of β-dicarboxylic acids, involving a six-membered ring transition state that results in formation of an enol and CO_2. The enol rapidly undergoes keto-enol tautomerization to give the ketone.

SUMMARY OF IMPORTANT CONCEPTS

16.0 OVERVIEW
- **Carboxylic acids** are acidic, and they can be converted to a number of important derivatives such as acid chlorides, anhydrides, esters, and amides. ✱

16.1 STRUCTURE
- The functional group of carboxylic acids is the **carboxyl group**, which contains a **carbonyl group** that has an **-OH group attached** to the **carbonyl carbon atom.** ✱

16.2 NOMENCLATURE
- According to the **IUPAC system**, a carboxylic acid is named by **dropping** the **e** from the name of the longest chain that contains the carboxyl group, and replacing it with the suffix **oic acid**. The carboxyl group takes precedence over most other functional groups.
  - Dicarboxylic acids are named the same as above, except the final **e** is retained and the suffix **dioic acid** is used. Higher carboxylic acids are named with the suffixes **tricarboxylic acid, tetracarboxylic acid**, etc.
  - Aromatic carboxylic acids are named as derivatives of benzoic acid.
- Many **common names** of carboxylic acids are still used, some of which are listed in Table 16.2 of the text. When these common names are used, then the Greek letters $\alpha, \beta, \gamma, \delta$, etc. are used to locate substituents.

16.3 PHYSICAL PROPERTIES
- In the liquid and solid state, **carboxylic acids exist in dimeric, hydrogen-bonded structures.** In these dimeric structures, the hydrogen atom of each carboxyl group is hydrogen bonded to the carbonyl oxygen atom of its partner. ✱
  - Because of these dimeric structures, carboxylic acids have higher boiling points than analogous alcohols or aldehydes.
  - Carboxylic acids can readily take part in hydrogen bonding with water molecules, so they are also more soluble in water than analogous alcohols and aldehydes.

16.4 ACIDITY
- Unsubstituted **carboxylic acids** are relatively **acidic**, with pK_a values in the **range of 4-5.** ✱
  - In general, an anion is generated along with a proton when an acid dissociates, and **the more stable the anion, the stronger the acid.** That is, the relative **acidity** of carboxylic acids can be **attributed to the stability of the carboxylate anion** that is produced upon deprotonation. The **carboxylate anion is stabilized** by **delocalization** of the **negative charge** onto both **oxygen atoms.**
  - Another factor contributing to the acidity of the carboxylic acid is the electron withdrawing capability of the carbonyl group via an inductive effect. **The electrons in the -OH bond are polarized away from hydrogen, thereby weakening the bond and increasing acidity.**
  - **Electron withdrawing groups**, such as **halogen atoms** adjacent to the carboxylic group functions, **increase acid strength**. This effect can be quite large in certain cases.

16.5 SPECTROSCOPIC PROPERTIES
- The **mass spectrum** of carboxylic acids usually contains at least some of the **molecular ion** peak, but α-**cleavage** of the carboxyl group is common.
- In **^{1}H-NMR** spectra, the hydrogens α **to the carboxyl group** give rise to signals in the **δ 2.0 to 2.5 range**, while the **acidic carboxyl hydrogen** gives a signal between δ **10 and 13.**
- In infrared spectra, the carboxyl group gives rise to two characteristic absorptions; a **carbonyl stretch** between **1700 cm^{-1} and 1725 cm^{-1}**, and an **-OH stretch** between **2400 cm^{-1} and 3400 cm^{-1}.**

CHAPTER 16
Solutions to the Problems

Problem 16.1 Each of these compounds has a well-recognized common name. A derivative of glyceric acid is an intermediate in glycolysis. Maleic acid is an intermediate in the tricarboxylic acid (TCA) cycle. Mevalonic acid is an intermediate in the biosynthesis of steroids. Write the IUPAC name for each compound. Be certain to specify configuration.

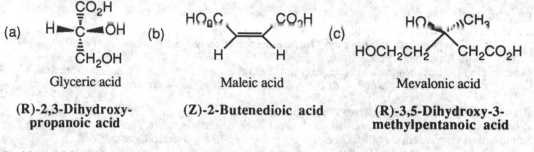

(a) Glyceric acid

(R)-2,3-Dihydroxy-propanoic acid

(b) Maleic acid

(Z)-2-Butenedioic acid

(c) Mevalonic acid

(R)-3,5-Dihydroxy-3-methylpentanoic acid

Problem 16.2 Which is the stronger acid in each pair?

(a) CH_3CO_2H or $\boxed{CH_3SO_3H}$

Acetic acid   Methanesulfonic acid

$pK_a = 4.8$ $pK_a = -1.8$

(b) $\boxed{CH_3CCO_2H}$ or $CH_3CH_2CO_2H$

2-Oxopropanoic acid   Propanoic acid
(Pyruvic acid)

$pK_a = 2.4$ $pK_a = 4.8$

In (a), the third oxygen atom on sulfur increases acidity by both inductive and resonance effects compared to the carboxylic acid. In (b), the electron withdrawing carbonyl group in the 2 position of pyruvic acid increases acidity due to inductive effects.

Problem 16.3 Write equations for the reaction of each acid in Example 16.3 with ammonia and name the carboxylate salt formed.

(a) $CH_3(CH_2)_2CO_2H$ + NH_3 ⟶ $CH_3(CH_2)_2CO_2^-$ NH_4^+

Butanoic acid   Ammonium butanoate

(b) $CH_3\overset{OH}{\underset{|}{C}}HCO_2H$ + NH_3 ⟶ $CH_3\overset{OH}{\underset{|}{C}}HCO_2^-$ NH_4^+

2-Hydroxy-propanoic acid
(Lactic acid)

Ammonium 2-hydroxy-propanoate
(Ammonium lactate)

Problem 16.4 Complete these Fischer esterifications.

(a)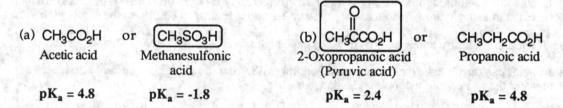

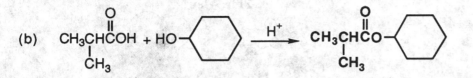

(b)

Problem 16.5 Complete the following equations.

(a)

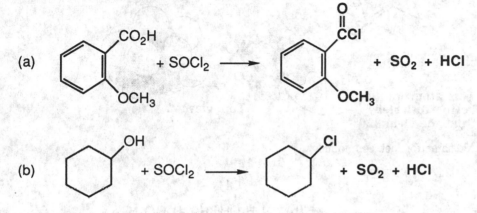

(b)

Problem 16.6 How might you account for the observation that the following β-ketoacid is stable to thermal decarboxylation. It can be heated for extended periods of time at temperatures above its melting point without noticeable decomposition.

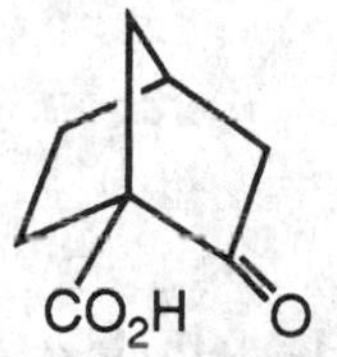

2-Oxobicyclo[2.2.1]heptane-1-carboxylic acid

The mechanism we have proposed for decarboxylation of a β-ketoacid involves formation of an enol intermediate and equilibrium of the enol, via keto-enol tautomerism, with the keto-form. Given the geometry of a bicyclo[2.2.1]alkane, it is not possible to have a carbon-carbon double bond to a bridgehead carbon because of the prohibitively high angle strain. Therefore, because the enol intermediate cannot be formed, this β-ketoacid does not undergo thermal decarboxylation. Note that this argument is similar to that presented in the answer to Problem 15.51.

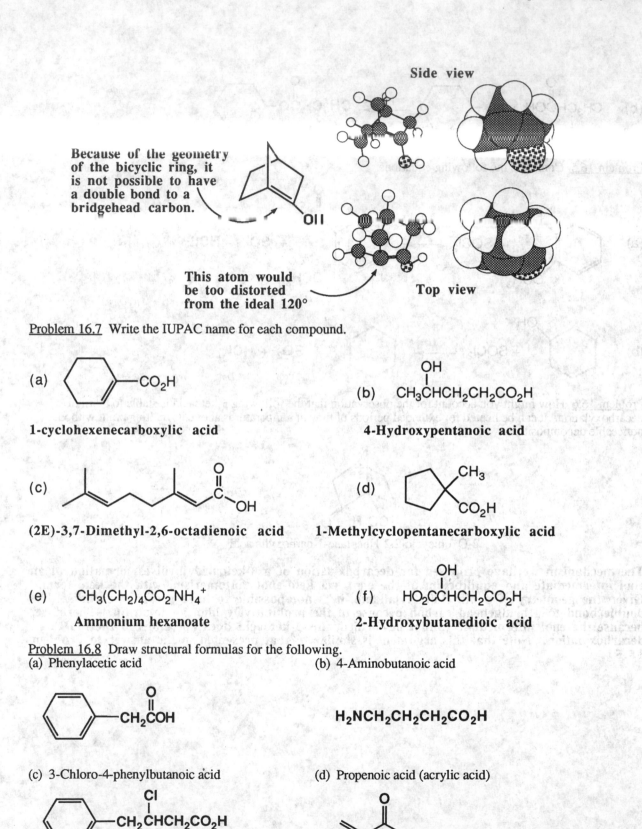

Side view

Because of the geometry
of the bicyclic ring, it
is not possible to have
a double bond to a
bridgehead carbon.

This atom would
be too distorted
from the ideal 120°

Top view

Problem 16.7 Write the IUPAC name for each compound.

(a) ⬡—CO₂H

1-cyclohexenecarboxylic acid

(b)   OH
     CH₃CHCH₂CH₂CO₂H

4-Hydroxypentanoic acid

(c)

(2E)-3,7-Dimethyl-2,6-octadienoic acid

(d)

1-Methylcyclopentanecarboxylic acid

(e)   CH₃(CH₂)₄CO₂⁻NH₄⁺

Ammonium hexanoate

(f)   OH
       HO₂CCHCH₂CO₂H

2-Hydroxybutanedioic acid

Problem 16.8 Draw structural formulas for the following.
(a) Phenylacetic acid   (b) 4-Aminobutanoic acid

              O
              ‖
   —CH₂COH   H₂NCH₂CH₂CH₂CO₂H

(c) 3-Chloro-4-phenylbutanoic acid   (d) Propenoic acid (acrylic acid)

              Cl
              |
   —CH₂CHCH₂CO₂H

(e) (Z)-3-Hexenedioic acid

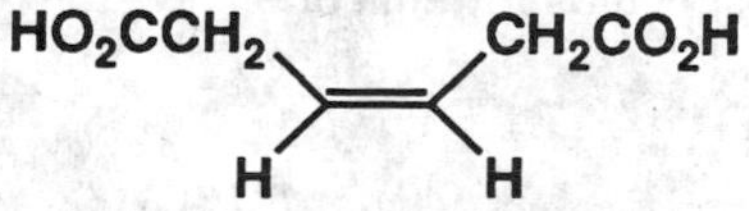

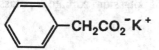

(f) 2-Pentynoic acid

$$CH_3CH_2C\equiv CCO_2H$$

(g) Potassium phenylacetate

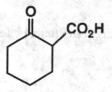

(h) Sodium oxalate

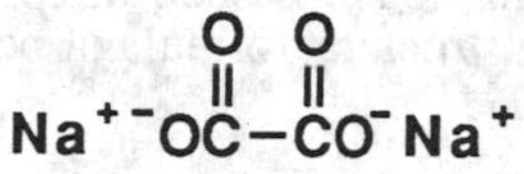

(i) 2-Oxocyclohexanecarboxylic acid

(j) 2,2-Dimethylpropanoic acid

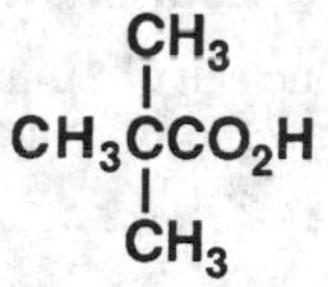

$$CH_3\overset{\overset{\displaystyle CH_3}{|}}{\underset{\underset{\displaystyle CH_3}{|}}{C}}CO_2H$$

Problem 16.9 Megatomoic acid, the sex attractant of the female black carpet beetle, has the structure:

$$CH_3(CH_2)_7CH=CHCH=CHCH_2CO_2H$$

Megatomoic acid

(a) What is its IUPAC name?

Its IUPAC name is 3,5-tetradecadienoic acid.

(b) State the number of stereoisomers possible for this compound.

Four stereoisomers are possible; each double bond can have either an E or Z (*trans* or *cis*) configuration.

Problem 16.10 Draw structural formulas for these salts.
(a) Sodium benzoate

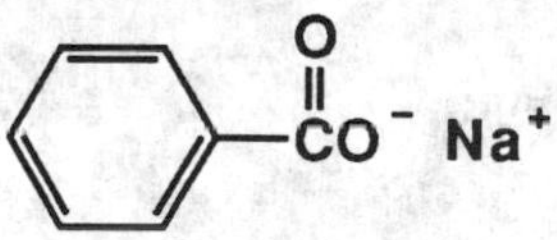

(b) Lithium acetate

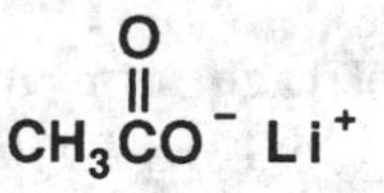

$$CH_3\overset{\overset{\displaystyle O}{\|}}{C}O^-\ Li^+$$

(c) Ammonium acetate

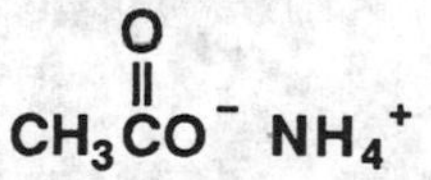

$$CH_3\overset{\overset{\displaystyle O}{\|}}{C}O^-\ NH_4^+$$

(d) Disodium adipate

$$Na^+\ ^-O\overset{\overset{\displaystyle O}{\|}}{C}(CH_2)_4\overset{\overset{\displaystyle O}{\|}}{C}O^-\ Na^+$$

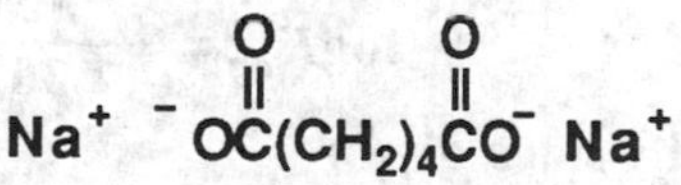

(e) Sodium salicylate

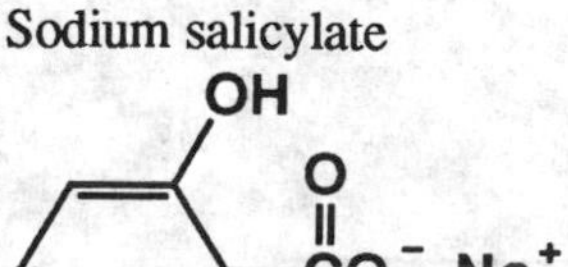

(f) Calcium butanoate

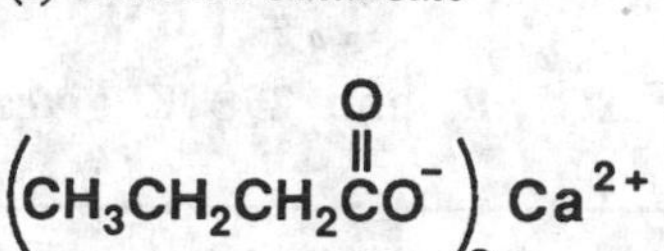

$$\left(CH_3CH_2CH_2\overset{\overset{\displaystyle O}{\|}}{C}O^-\right)_2 Ca^{2+}$$

<u>Problem 16.11</u> The monopotassium salt of oxalic acid is present in certain leafy vegetables, including rhubarb. Both oxalic acid and its salts are poisonous in high concentrations. Draw the structural formula of monopotassium oxalate.

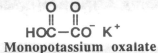

Monopotassium oxalate

<u>Problem 16.12</u> Potassium sorbate is added as a preservative to certain foods to prevent bacteria and molds from causing food spoilage and to extend the foods' shelf life. The IUPAC name of potassium sorbate is potassium (E,E)-2,4-hexadienoate. Draw the structural formula of potassium sorbate.

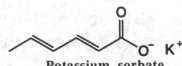

Potassium sorbate

<u>Problem 16.13</u> Zinc 10-undecenoate, the zinc salt of 10-undecenoic acid, is used to treat certain fungal infections, particularly *tinea pedis* (athlete's foot). Draw the structural formula of this zinc salt.

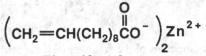

Zinc 10-undecenoate

<u>Problem 16.14</u> On a cyclohexane ring, an axial carboxyl group has a conformational energy of +1.4 kcal/mol (5.9 kJ/mol) relative to an equatorial carboxyl group. Consider the equilibrium for the alternative chair conformations of *trans*-1,4-cyclohexanedicarboxylic acid. Draw the less stable chair conformation on the left of the equilibrium arrows and the more stable chair on the right. Calculate $\Delta G°$ for the equilibrium as written and calculate the ratio of more stable chair to less stable chair at 25°C.

As written, the conformation on the right is (2 x 1.4) = 2.8 kcal/mol more stable than the conformation on the left.

At equilibrium the relative amounts of each form are given by the equation:

$$\Delta G° = -RT\ln K_{eq}$$

Here K_{eq} refers to the ratio of the alternative chair conformations. Converting to base 10 and rearranging gives

$$\log K_{eq} = \frac{-\Delta G°}{(2.303)RT}$$

Taking the antilog of both sides gives:

$$K_{eq} = 10^{\left(\frac{-\Delta G°}{(2.303)RT}\right)}$$

Plugging in the values for $\Delta G°$, R, and 298 K gives theanswer:

$$K_{eq} = 10^{\left(\frac{-(-2.8 \ \text{kcal/mol})}{(2.303)(1.987 \ \times \ 10^{-3} \ \text{kcal/mol K})(298 \ \text{K})}\right)} = 10^{2.05} = 1.12 \ \times \ 10^2$$

Physical Properties

Problem 16.15 Arrange the compounds in each set in order of increasing boiling point:

The better the hydrogen bond capability, the higher the boiling point.

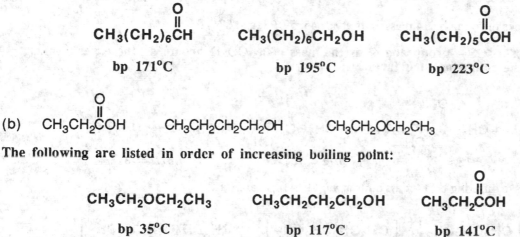

(a) $CH_3(CH_2)_5COH$ $CH_3(CH_2)_6CH$ $CH_3(CH_2)_6CH_2OH$

The following are listed in order of increasing boiling point:

$CH_3(CH_2)_6CH$ $CH_3(CH_2)_6CH_2OH$ $CH_3(CH_2)_5COH$

bp 171°C   bp 195°C   bp 223°C

(b) CH_3CH_2COH $CH_3CH_2CH_2CH_2OH$ $CH_3CH_2OCH_2CH_3$

The following are listed in order of increasing boiling point:

$CH_3CH_2OCH_2CH_3$ $CH_3CH_2CH_2CH_2OH$ CH_3CH_2COH

bp 35°C   bp 117°C   bp 141°C

Problem 16.16 Acetic acid has a boiling point of 118°C, whereas its methyl ester has a boiling point of 57°C. Account for the fact that the boiling point of acetic acid is higher than that of its methyl ester, even though acetic acid has a lower molecular weight.

Acetic acid can make strong hydrogen bonds, but the methyl ester lacks a hydrogen bonding hydrogen atom. Thus, acetic acid will have the much higher boiling point compared to the methyl ester.

Spectroscopy

Problem 16.17 Account for the presence of peaks at m/z 135 and 107 in the mass spectrum of 4-methoxybenzoic acid (*p*-anisic acid).

The peak at m/z 135 results from loss of the OH group by α-cleavage, and the peak at m/z 107 results from decarboxylation, also by α-cleavage.

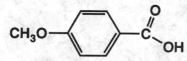

4-methoxybenzoic acid $m/z = 135$ $m/z = 107$
(*p*-anisic acid)   (an acylium ion)
$m/z = 152$

<u>Problem 16.18</u> Account for the presence of the following peaks in the mass spectrum of hexanoic acid.
(a) m/z 60.

The peak at m/z 60 is the result of a McLafferty rearrangement.

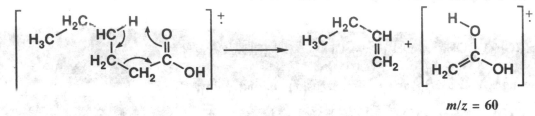

$m/z = 60$

(b) A series of peaks differing by 14 mass units at m/z 45, 59, 73, and 87.

14 is the mass of a $-CH_2-$ group and 45 is the mass of a $-CO_2H$ group, so these peaks must correspond to the following structures.

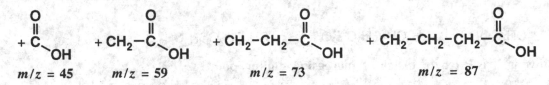

$m/z = 45$ $m/z = 59$ $m/z = 73$ $m/z = 87$

(c) A series of peaks differing by 14 mass units at m/z 29, 43, 57, and 71.

$$\left[CH_3CH_2 \right]^{\ddot{+}} \quad \left[CH_3CH_2CH_2 \right]^{+} \quad \left[CH_3CH_2CH_2CH_2 \right]^{\ddot{+}} \quad \left[CH_3CH_2CH_2CH_2CH_2 \right]^{\ddot{+}}$$

$m/z = 29$ $m/z = 43$ $m/z = 57$ $m/z = 71$

<u>Problem 16.19</u> Given here are ^{1}H-NMR and ^{13}C-NMR spectral data for ten compounds. Each compound shows strong absorption between 1720 and 1700 cm^{-1}, and strong, broad absorption over the region 2500 - 3500 cm^{-1}. Propose a structural formula for each compound.

(a) $C_5H_{10}O_2$

^{1}H-NMR	^{13}C-NMR
0.94 (t, 3H)	180.71
1.39 (m, 2H)	33.89
1.62 (m, 2H)	26.76
2.35 (t, 2H)	22.21
12.0 (s, 1H)	13.69

$$\overset{0.94}{CH_3}\overset{1.39}{CH_2}\overset{1.62}{CH_2}\overset{2.35}{CH_2}\overset{12.0}{CO_2H}$$

(b) $C_6H_{12}O_2$

^{1}H-NMR	^{13}C-NMR
1.08 (s, 9H)	179.29
2.23 (s, 2H)	47.82
12.1 (s, 1H)	30.62
	29.57

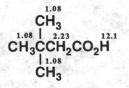

(c) $C_5H_8O_4$

^{1}H-NMR	^{13}C-NMR
0.93 (t, 3H)	170.94
1.80 (m, 2H)	53.28
3.10 (t, 1H)	21.90
12.7 (s, 2H)	11.81

$$\overset{12.7}{HO_2C}\overset{3.10}{CH}\overset{12.7}{CO_2H}$$
$$\underset{\underset{0.93}{CH_2CH_3}}{|_{1.80}}$$

(d) $C_5H_8O_4$

^{1}H-NMR	^{13}C-NMR
1.29 (s, 6H)	174.01
12.8 (s, 2H)	48.77
	22.56

$$\overset{1.29}{CH_3}$$
$$\overset{12.8}{HO_2C}\overset{|}{C}\overset{12.8}{CO_2H}$$
$$\underset{\underset{1.29}{CH_3}}{|}$$

(e) $C_4H_6O_2$

^{1}H-NMR	^{13}C-NMR
1.91 (d, 3H)	172.26
5.86 (d, 1H)	147.53
7.10 (m, 1H)	122.24
12.4 (s, 1H)	18.11

$$\overset{1.91}{CH_3}\overset{7.10}{CH}=\overset{5.86}{CH}\overset{12.4}{CO_2H} \ .$$

(f) $C_3H_4Cl_2O_2$

^{1}H-NMR	^{13}C-NMR
2.34 (s, 3H)	171.82
11.3 (s, 1H)	79.36
	34.02

$$\overset{Cl}{\underset{\underset{Cl}{|}}{\overset{|}{\underset{2.34}{CH_3}C}}}\overset{11.3}{CO_2H}$$

(g) $C_5H_8Cl_2O_2$

^{1}H-NMR	^{13}C-NMR
1.42 (s, 6H)	180.15
6.10 (s, 1H)	77.78
12.4 (s, 1H)	51.88
	20.71

$$\overset{1.42}{CH_3}$$
$$\overset{6.10}{Cl_2C}\overset{|}{H}C\overset{12.4}{CO_2H}$$
$$\underset{\underset{1.42}{CH_3}}{|}$$

(h) $C_5H_9BrO_2$

^{1}H-NMR	^{13}C-NMR
0.97 (t, 3H)	176.36
1.50 (m, 2H)	45.08
2.05 (m, 2H)	36.49
4.25 (t, 1H)	20.48
12.1 (s, 1H)	13.24

$$\overset{Br}{}$$
$$\overset{0.97}{CH_3}\overset{1.50}{CH_2}\overset{2.05}{CH_2}\overset{|}{\overset{4.25}{CH}}\overset{12.1}{CO_2H}$$

(i) $C_4H_8O_3$

^{1}H-NMR	^{13}C-NMR
2.62 (t, 2H)	177.33
3.38 (s, 3H)	67.55
3.68 (t, 2H)	58.72
11.5 (s, 1H)	34.75

$$\overset{3.38}{CH_3}O\overset{3.68}{CH_2}\overset{2.62}{CH_2}\overset{11.5}{CO_2H}$$

(j) $C_6H_{10}O_3$

^{1}H-NMR	^{13}C-NMR
1.90 (m, 2H)	208.44
2.16 (s, 3H)	179.08
2.40 (t, 2H)	42.29
2.55 (t, 2H)	32.93
11.4 (s, 1H)	29.91
	18.84

$$\overset{O}{\overset{||}{}}$$
$$\overset{2.16}{CH_3}C\overset{2.40}{CH_2}\overset{1.90}{CH_2}\overset{2.55}{CH_2}\overset{11.4}{CO_2H}$$

Preparation of Carboxylic Acids
<u>Problem 16.20</u> Complete these reactions:

(a)

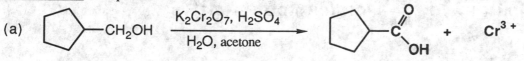

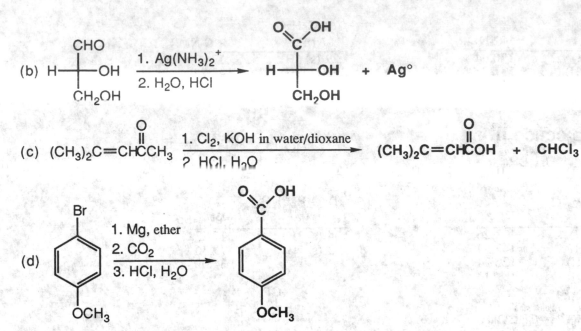

(b) H—[CHO / OH / CH₂OH] $\xrightarrow{\text{1. Ag(NH}_3)_2{}^+ \quad 2.\ H_2O,\ HCl}$ H—[CO₂H / OH / CH₂OH] + **Ag°**

(c) $(CH_3)_2C{=}CHCCH_3$ $\xrightarrow{\text{1. Cl}_2,\ \text{KOH in water/dioxane} \quad 2.\ HCl,\ H_2O}$ $(CH_3)_2C{=}CHCOH$ + $CHCl_3$

(d) [aromatic ring with Br, 1. Mg, ether; 2. CO₂; 3. HCl, H₂O, OCH₃] → [aromatic ring with CO₂H, OCH₃]

Problem 16.21 Show how to bring about each conversion in good yield.

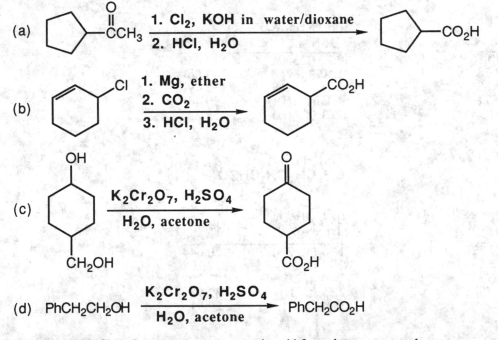

(a) [cyclopentane-C(=O)CH₃] $\xrightarrow{\text{1. Cl}_2,\ \text{KOH in water/dioxane} \quad 2.\ HCl,\ H_2O}$ [cyclopentane-CO₂H]

(b) [cyclohexene with Cl] $\xrightarrow{\text{1. Mg, ether} \quad 2.\ CO_2 \quad 3.\ HCl,\ H_2O}$ [cyclohexene with CO₂H]

(c) [cyclohexane with OH and CH₂OH] $\xrightarrow{\text{K}_2\text{Cr}_2\text{O}_7,\ \text{H}_2\text{SO}_4 \quad H_2O,\ \text{acetone}}$ [cyclohexanone with CO₂H]

(d) $PhCH_2CH_2OH$ $\xrightarrow{\text{K}_2\text{Cr}_2\text{O}_7,\ \text{H}_2\text{SO}_4 \quad H_2O,\ \text{acetone}}$ $PhCH_2CO_2H$

Problem 16.22 Show how to prepare pentanoic acid from these compounds:
(a) 1-Pentanol

Oxidation of 1-pentanol by chromic acid gives pentanoic acid.

$$CH_3(CH_2)_3CH_2OH + H_2CrO_4 \xrightarrow[\text{heat}]{H^+} CH_3(CH_2)_3CO_2H + Cr^{3+}$$

(b) Pentanal

Oxidation of pentanal by chromic acid, Tollens' solution, Benedict's solution, or molecular oxygen gives pentanoic acid.

$$CH_3(CH_2)_3\overset{O}{\overset{\|}{C}}H + O_2 \longrightarrow CH_3(CH_2)_3\overset{O}{\overset{\|}{C}}OH$$

(c) 1-Pentene

Hydroboration/oxidation of 1-pentene gives 1-pentanol. Oxidation of 1-pentanol as in part (a) gives pentanoic acid.

$$CH_3CH_2CH_2CH=CH_2 \xrightarrow[\text{2) } H_2O_2, \text{ NaOH}]{\text{1) } BH_3 \cdot THF} CH_3(CH_2)_3CH_2OH \text{ then as in part (a)}$$

(d) 1-Butanol

Conversion of 1-butanol to 1-chlorobutane, then to butyllithium or butylmagnesium bromide followed by carbonation and then acidification gives pentanoic acid.

$$CH_3(CH_2)_3OH \xrightarrow[\text{2) } \text{Mg,ether}]{\text{1) } SOCl_2} CH_3(CH_2)_3MgBr \xrightarrow[\text{2) } HCl,H_2O]{\text{1) } CO_2} CH_3(CH_2)_3CO_2H$$

(e) 1-Bromopropane

Treatment of 1-bromopropane with magnesium in ether followed by treatment of the Grignard reagent with ethylene oxide gives 1-pentanol. Oxidation of 1-pentanol as in part (a) gives pentanoic acid.

$$CH_3CH_2CH_2Br \xrightarrow{\text{Mg, ether}} CH_3CH_2CH_2MgBr \xrightarrow[\text{2) } HCl, H_2O]{\text{1) } H_2C\overset{O}{-}CH_2}$$

$$CH_3(CH_2)_3CH_2OH \xrightarrow{\text{Oxidize as in part (a)}} CH_3(CH_2)_3CO_2H$$

(f) 2-Hexanone

Haloform reaction of 2-hexanone gives pentanoic acid and a haloform.

$$CH_3(CH_2)_3\overset{O}{\overset{\|}{C}}CH_3 \xrightarrow[\text{2) } HCl, H_2O]{\text{1) } Br_2, \text{ NaOH}} CH_3(CH_2)_3\overset{O}{\overset{\|}{C}}OH + CHBr_3$$

(g) 1-Hexene

Acid-catalyzed hydration or oxymercuration/reduction of 1-hexene gives 2-hexanol. Oxidation of 2-hexanol as in part (f) gives pentanoic acid.

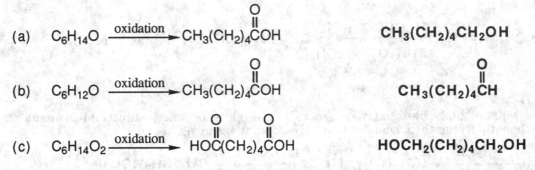

$$CH_3(CH_2)_3CH=CH_2 + H_2O \xrightarrow{H_2SO_4} CH_3(CH_2)_3\overset{\underset{|}{OH}}{C}HCH_3$$

$$\xrightarrow[\text{2) HCl, H}_2\text{O}]{\text{1) Br}_2,\ \text{NaOH}} CH_3(CH_2)_3\overset{\overset{O}{\|}}{C}OH$$

Problem 16.23 Draw the structural formula of a compound of the given molecular formula that, on oxidation by potassium dichromate in aqueous sulfuric acid, gives the carboxylic acid or dicarboxylic acid shown.

(a) $C_6H_{14}O \xrightarrow{\text{oxidation}} CH_3(CH_2)_4\overset{\overset{O}{\|}}{C}OH$ $CH_3(CH_2)_4CH_2OH$

(b) $C_6H_{12}O \xrightarrow{\text{oxidation}} CH_3(CH_2)_4\overset{\overset{O}{\|}}{C}OH$ $CH_3(CH_2)_4\overset{\overset{O}{\|}}{C}H$

(c) $C_6H_{14}O_2 \xrightarrow{\text{oxidation}} HO\overset{\overset{O}{\|}}{C}(CH_2)_4\overset{\overset{O}{\|}}{C}OH$ $HOCH_2(CH_2)_4CH_2OH$

Problem 16.24 Show the reagents and experimental conditions necessary to bring about each conversion in good yield. Shown over each reaction arrow is the number of steps (not including workup in aqueous acid) required for each conversion.

(a)
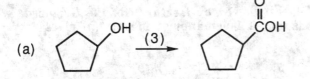

The most convenient way to add a carbon atom in the form of a carboxyl group is carbonation of an organolithium or organomagnesium compound.

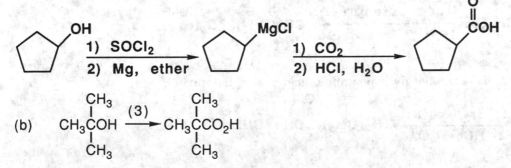

Use the same set of reactions in this part as you used in part (a), except since the starting material is a tertiary alcohol, HCl is used in place of $SOCl_2$.

$$CH_3\overset{\underset{|}{CH_3}}{\underset{\underset{|}{CH_3}}{C}}OH \xrightarrow{HCl} CH_3\overset{\underset{|}{CH_3}}{\underset{\underset{|}{CH_3}}{C}}Cl \xrightarrow[\substack{\text{2) CO}_2 \\ \text{3) HCl, H}_2\text{O}}]{\text{1) Mg,ether}} CH_3\overset{\underset{|}{CH_3}}{\underset{\underset{|}{CH_3}}{C}}CO_2H$$

(c) $CH_3\underset{\underset{CH_3}{|}}{\overset{\overset{CH_3}{|}}{C}}OH \xrightarrow{(3)} CH_3\underset{\underset{CH_3}{|}}{C}HCO_2H$

This conversion is best accomplished by acid-catalyzed dehydration of the tertiary alcohol to an alkene followed by hydroboration/oxidation. These two reactions in sequence shift the -OH group from the tertiary carbon to a primary carbon. Oxidation of this primary alcohol with chromic acid gives 2-methylpropanoic acid.

$CH_3\underset{\underset{CH_3}{|}}{\overset{\overset{CH_3}{|}}{C}}OH \xrightarrow[(-H_2O)]{H_2SO_4} CH_3\overset{\overset{CH_3}{|}}{C}=CH_2 \xrightarrow[2)\ H_2O_2,\ NaOH]{1)\ BH_3 \cdot THF} CH_3\overset{\overset{CH_3}{|}}{C}HCH_2OH$

$\xrightarrow[H_2O,\ acetone]{K_2Cr_2O_7,\ H_2SO_4} CH_3\overset{\overset{CH_3}{|}}{C}HCO_2H$

(d) $CH_3\underset{\underset{CH_3}{|}}{\overset{\overset{CH_3}{|}}{C}}OH \xrightarrow{(5)} CH_3\overset{\overset{CH_3}{|}}{C}HCH_2CO_2H$

Repeat the first set of steps as in part (c) to convert *tert*-butyl alcohol to isobutyl alcohol. Then use the sequence of steps as in part (a) and (b) to convert this alcohol to a carboxylic acid containing one more carbon atom.

$CH_3\overset{\overset{CH_3}{|}}{C}HCH_2OH \xrightarrow[2)\ Mg,\ ether]{1)\ SOCl_2} CH_3\overset{\overset{CH_3}{|}}{C}HCH_2Cl \xrightarrow[2)\ HCl,\ H_2O]{1)\ CO_2} CH_3\overset{\overset{CH_3}{|}}{C}HCH_2CO_2H$

From part (c)

(e) $CH_3CH{=}CHCH_3 \xrightarrow{(3)} CH_3CH{=}CHCH_2CO_2H$

This problem is analogous to part (a) except an allylic halogenation of the alkene starting material is the first step.

$CH_3CH{=}CHCH_3 \xrightarrow[Light\ or\ peroxides]{NBS} CH_3CH{=}CHCH_2Br$

$\xrightarrow[3)\ HCl,\ H_2O]{\substack{1)\ Mg,\ ether \\ 2)\ CO_2}} CH_3CH{=}CHCH_2CO_2H$

Problem 16.25 Succinic acid can be synthesized by the following series of reactions from acetylene. Show the reagents and experimental conditions necessary to carry out this synthesis in good yield.

$HC{\equiv}CH \longrightarrow HOCH_2C{\equiv}CCH_2OH \longrightarrow HOCH_2CH_2CH_2CH_2OH \longrightarrow HO\overset{O}{\overset{\|}{C}}CH_2CH_2\overset{O}{\overset{\|}{C}}OH$

Acetylene   2-Butyne-1,4-diol   1,4-Butanediol   Butanedioic acid
                                                                                   (Succinic acid)

Two one-carbon fragments in the form of formaldehyde are added to the carbon skeleton of acetylene. To bring about this double addition, acetylene is treated with two mol of sodamide or sodium hydride followed by two mol of formaldehyde. Acidification of the resulting disodium salt gives 2-butyne-1,4-diol.

$$HC\equiv CH + 2NaNH_2 \longrightarrow Na^{+-}C\equiv C^-Na^+ + 2NH_3$$

Acetylene   Sodamide   Sodium acetylide

$$Na^{+-}C\equiv C^-Na^+ + 2H\overset{\overset{\displaystyle O}{\|}}{-C}-H \longrightarrow Na^{+-}OCH_2C\equiv CCH_2O^-Na^+$$

Formaldehyde

$$Na^{+-}OCH_2C\equiv CCH_2O^-Na^+ + 2HCl \longrightarrow HOCH_2C\equiv CCH_2OH$$

2-Butyne-1,4-diol

Catalytic reduction of the carbon-carbon triple bond in 2-butyne-1,4-diol over a transition metal catalyst gives 1,4-butanediol.

$$HOCH_2C\equiv CCH_2OH + 2H_2 \xrightarrow[\substack{\text{or other}\\\text{transition metal}\\\text{catalyst}}]{Pt} HOCH_2CH_2CH_2CH_2OH$$

2-Butyne-1,4-diol   **1,4-Butanediol**

Oxidation of 1,4-butanediol by chromic acid gives succinic acid.

$$HOCH_2CH_2CH_2CH_2OH + H_2CrO_4 \longrightarrow HO\overset{\overset{\displaystyle O}{\|}}{C}CH_2CH_2\overset{\overset{\displaystyle O}{\|}}{C}OH$$

1,4-Butanediol   **Butanedioic acid**
                                                                    (Succinic acid)

<u>Problem 16.26</u> The reaction of an α–diketone with concentrated sodium or potassium hydroxide to give the salt of an α–hydroxyacid is given the general name benzil-benzilic acid rearrangement. It is illustrated by the conversion of benzil to sodium benzilate and then to benzilic acid.

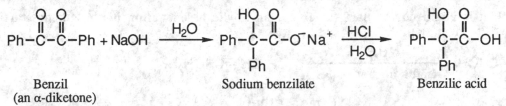

Benzil   **Sodium benzilate**   **Benzilic acid**
(an α-diketone)

Propose a mechanism for this base-catalyzed rearrangement of benzil to sodium benzilate.

Addition of hydroxide ion to one of the carbonyl groups to form a tetrahedral carbonyl addition intermediate followed by collapse of this intermediate with simultaneous regeneration of the carbonyl group and migration of a phenyl group gives sodium benzilate.

Step 1:

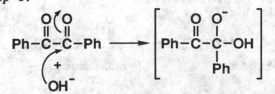

Tetrahedral carbonyl
addition intermediate

Step 2:

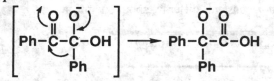

Step 3:

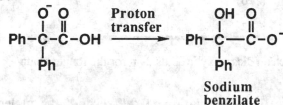

Sodium
benzilate

Acidity of Carboxylic Acids

Problem 16.27 Select the stronger acid in each set.
(a) Phenol (pK_a 9.95) and benzoic acid (pK_a 4.17)

Recall that pK_a is the negative $\log_{10}$ of K_a. The smaller the pK_a, the stronger the acid, so benzoic acid is the stronger acid.

(b) Lactic acid (K_a 8.4 x 10^{-4}) and ascorbic acid (K_a 7.9 x 10^{-5})

The larger the value of K_a, the stronger the acid, so lactic acid is the stronger acid.

Problem 16.28 Assign the acid in each set its appropriate pK_a.

(a) 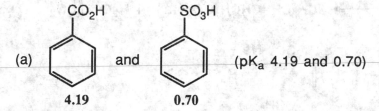 and (pK$_a$ 4.19 and 0.70)

4.19   0.70

The third oxygen on sulfur increases acidity compared to an analogous carboxylic acid.

(b) CH$_3$CCH$_2$CO$_2$H and CH$_3$CH$_2$CCO$_2$H   (pK$_a$ 3.58 and 2.49)

3.58   2.49

The electron–withdrawing carbonyl group increases acidity through inductive effects, so the closer the carbony group to the carboxylic acid, the greater the effect and the stronger the acid.

(c) CH$_3$CH$_2$CO$_2$H   and   N≡CCH$_2$CO$_2$H   (pK$_a$ 4.78 and 2.45)

4.78   2.45

The nitrile is electron–withdrawing, so it increases acidity through inductive effects.

<u>Problem 16.29</u> Low-molecular-weight dicarboxylic acids normally exhibit two different pK_a values. Ionization of the first carboxyl group is easier than the second. This effect diminishes with molecular size and, for adipic acid and longer chain dicarboxylic acids, the two acid ionization constants differ by about one pK unit.

Dicarboxylic acid	Structural formula	pK_{a1}	pK_{a2}
oxalic	HO_2CCO_2H	1.23	4.19
malonic	$HO_2CCH_2CO_2H$	2.83	5.69
succinic	$HO_2C(CH_2)_2CO_2H$	4.16	5.61
glutaric	$HO_2C(CH_2)_3CO_2H$	4.31	5.41
adipic	$HO_2C(CH_2)_4CO_2H$	4.43	5.41

Why do the two pK_a values differ more for the shorter chain dicarboxylic acids than for the longer chain dicarboxylic acids?

For these dicarboxylic acids, in going from the first dissociation to the second, the molecule is changing from a monoanion to a dianion. Electrostatic repulsion hinders formation of two negative charges in nearby regions of space. Thus, the second pK_a values are higher than the first, and the effect is more pronounced the shorter the chain between the two carboxyl groups.

<u>Problem 16.30</u> Complete the following acid-base reactions:

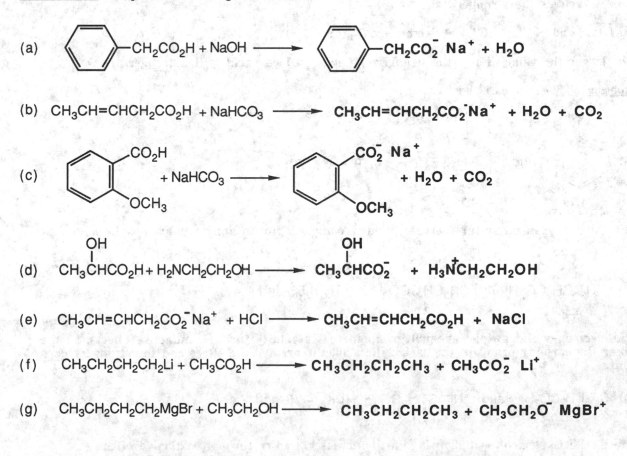

Problem 16.31 The normal pH range for blood plasma is 7.35 - 7.45. Under these conditions, would you expect the carboxyl group of lactic acid (pK_a 3.07) to exist primarily as a carboxyl group or as a carboxylate anion? Explain.

Recall from the definition of K_a that:

$$K_a = \frac{[A^-][H^+]}{[H-A]} \qquad \text{so dividing both sides by } [H^+] \text{ gives} \qquad \frac{K_a}{[H^+]} = \frac{[A^-]}{[H-A]}$$

Here, $[H^+]$ is concentration of H^+, $[H-A]$ is concentration of protonated acid (lactic acid in this case) and $[A^-]$ is the concentration of deprotonated acid (lactic acid carboxylate anion in this case). Therefore, if the ratio of $K_a / [H^+]$ is greater than 1, $[A^-]$ will be the predominant form, and if the ratio of $K_a / [H^+]$ is less than 1, then $[H-A]$ will be the predominant form. Recall that $pH = -\log_{10}[H^+]$, so a pH of 7.4 corresponds to a $[H^+]$ of $10^{-(pH)} = 10^{-(7.4)} = 4.0 \times 10^{-8}$. Similarly, $pK_a = -\log_{10} K_a$, so for lactic acid $K_a = 10^{-(K_a)} = 10^{-(3.07)} = 8.5 \times 10^{-4}$. Using these numbers:

$$\frac{[A^-]}{[H-A]} = \frac{K_a}{[H^+]} = \frac{8.5 \times 10^{-4}}{4.0 \times 10^{-8}} = 2.1 \times 10^4$$

Therefore, lactic acid will exist primarily as the carboxylate anion in blood plasma.

Problem 16.32 The K_{a1} of ascorbic acid (Section 24.6) is 7.94×10^{-5}. Would you expect ascorbic acid dissolved in blood plasma to exist primarily as ascorbic acid or as ascorbate anion? Explain.

Using the same reasoning described in the answer to Problem 16.31:

$$\frac{[A^-]}{[H-A]} = \frac{K_a}{[H^+]} = \frac{7.9 \times 10^{-5}}{4.0 \times 10^{-8}} = 2.0 \times 10^3$$

Therefore, ascorbic acid will exist primarily as the ascorbate anion in blood plasma.

Problem 16.33 Excess ascorbic acid is excreted in the urine, the pH of which is normally in the range 4.8 - 8.4. What form of ascorbic acid would you expect to be present in urine of pH 8.4, free ascorbic acid or ascorbate anion? Explain.

At pH 8.4, $[H^+] = 4.0 \times 10^{-9}$, therefore using the same reasoning as described in the answer to Problem 16.31 and 16.32:

$$\frac{[A^-]}{[H-A]} = \frac{K_a}{[H^+]} = \frac{7.9 \times 10^{-5}}{4.0 \times 10^{-9}} = 2.0 \times 10^4$$

Ascorbic acid will exist primarily as the ascorbate anion in urine of pH 8.4.

Reactions of Carboxylic Acids

Problem 16.34 Give the expected organic products when $PhCH_2CO_2H$, phenylacetic acid, is treated with each of these reagents.

(a) $SOCl_2$

$$\underset{\displaystyle PhCH_2\overset{\displaystyle O}{\overset{\|}{C}}OH}{} + SOCl_2 \longrightarrow \underset{\displaystyle PhCH_2\overset{\displaystyle O}{\overset{\|}{C}}Cl}{} + SO_2 + HCl$$

(b) $NaHCO_3$, H_2O

$$\underset{\displaystyle PhCH_2\overset{\displaystyle O}{\overset{\|}{C}}OH}{} + NaHCO_3 \longrightarrow \underset{\displaystyle PhCH_2\overset{\displaystyle O}{\overset{\|}{C}}O^-Na^+}{} + H_2O + CO_2$$

(c) NaOH, H_2O

$$\underset{\substack{\| \\ O}}{PhCH_2COH} + NaOH \longrightarrow \underset{\substack{\| \\ O}}{PhCH_2COO^-Na^+} + H_2O$$

(d) CH_3MgBr (1 equivalent)

$$\underset{\substack{\| \\ O}}{PhCH_2COH} + CH_3MgBr \longrightarrow \underset{\substack{\| \\ O}}{PhCH_2COMgBr'} + CH_4$$

(e) $LiAlH_4$ followed by H_2O

$$\underset{\substack{\| \\ O}}{PhCH_2COH} \xrightarrow[\text{2) } H_2O]{\text{1) } LiAlH_4} PhCH_2CH_2OH$$

(f) CH_2N_2

$$\underset{\substack{\| \\ O}}{PhCH_2COH} + CH_2N_2 \longrightarrow \underset{\substack{\| \\ O}}{PhCH_2COCH_3} + N_2$$

(g) $CH_3OH + H_2SO_4$ (catalyst)

$$\underset{\substack{\| \\ O}}{PhCH_2COH} + CH_3OH \xrightarrow{H_2SO_4} \underset{\substack{\| \\ O}}{PhCH_2COCH_3} + H_2O$$

Problem 16.35 Show how to convert *trans*-3-phenyl-2-propenoic acid (cinnamic acid) to these compounds.

(a)
$$\underset{C_6H_5}{\overset{H}{\diagdown}}C=C\underset{H}{\overset{C-OH \ (=O)}{\diagdown}} \xrightarrow[\text{2) } H_2O]{\text{1) } LiAlH_4} \underset{C_6H_5}{\overset{H}{\diagdown}}C=C\underset{H}{\overset{CH_2OH}{\diagdown}}$$

(b)
$$\underset{C_6H_5}{\overset{H}{\diagdown}}C=C\underset{H}{\overset{C-OH \ (=O)}{\diagdown}} \xrightarrow[\substack{Pt \quad 25°C \\ 2 \text{ atm}}]{H_2} C_6H_5CH_2CH_2CO_2H$$

(c)
$$\underset{C_6H_5}{\overset{H}{\diagdown}}C=C\underset{H}{\overset{C-OH \ (=O)}{\diagdown}} \xrightarrow[\text{2) } H_2O]{\text{1) } LiAlH_4} \xrightarrow[\substack{Pt \quad 25°C \\ 2 \text{ atm}}]{H_2} C_6H_5CH_2CH_2CH_2OH$$

Problem 16.36 Show how to convert 3-oxobutanoic acid (acetoacetic acid) to these compounds.

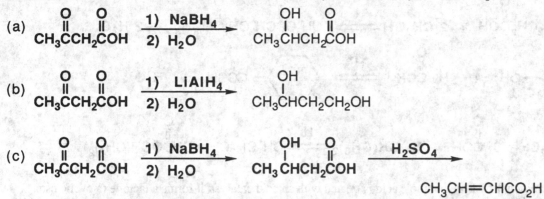

(a) CH_3CCH_2COH (with two C=O) $\xrightarrow[\text{2) } H_2O]{\text{1) NaBH}_4}$ CH_3CHCH_2COH (with OH and C=O)

(b) CH_3CCH_2COH (with two C=O) $\xrightarrow[\text{2) } H_2O]{\text{1) LiAlH}_4}$ $CH_3CHCH_2CH_2OH$ (with OH)

(c) CH_3CCH_2COH (with two C=O) $\xrightarrow[\text{2) } H_2O]{\text{1) NaBH}_4}$ CH_3CHCH_2COH (with OH and C=O) $\xrightarrow{H_2SO_4}$

$CH_3CH=CHCO_2H$

Problem 16.37 Complete these examples of Fischer esterification. Assume alcohol is present in excess.

(a) $CH_3CO_2H + HOCH_2CH_2CH(CH_3)_2 \underset{}{\overset{H^+}{\rightleftharpoons}} CH_3COCH_2CH_2CH(CH_3)_2 + H_2O$

(b) [benzene ring with two CO_2H groups ortho] $+ 2 CH_3OH \overset{H^+}{\rightleftharpoons}$ [benzene ring with two CO_2CH_3 groups ortho] $+ 2 H_2O$

(c) $HO_2C(CH_2)_2CO_2H + 2 CH_3CH_2OH \overset{H^+}{\rightleftharpoons} CH_3CH_2OC(CH_2)_2COCH_2CH_3 + 2 H_2O$

Problem 16.38 Benzocaine, a topical anesthetic, is prepared by treatment of 4-aminobenzoic acid with ethanol in the presence of an acid catalyst followed by neutralization. Draw the structural formula of benzocaine.

H_2N—[benzene ring]—$CO_2H + CH_3CH_2OH \xrightarrow[\text{2) Mild base to deprotonate amino group}]{\text{1) } H_2SO_4}$

4-Aminobenzoic acid

H_2N—[benzene ring]—$COCH_2CH_3 + H_2O$

Benzocaine
(a topical anesthetic)

Problem 16.39 From what carboxylic acid and what alcohol is each ester derived?

(a) HO—[cyclohexane ring]—$OH + 2 CH_3COH \overset{H^+}{\rightleftharpoons} CH_3CO$—[cyclohexane ring]—$OCCH_3$

$+ 2 H_2O$

(b) $\overset{O}{\overset{\|}{HOCCH_2CH_2COH}}$ + 2 CH$_3$OH $\rightleftharpoons$ $\overset{O}{\overset{\|}{CH_3OCCH_2CH_2COCH_3}}$ + 2 H$_2$O

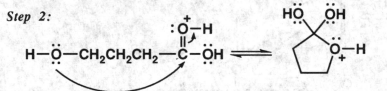

(c) $\langle\text{cyclohexyl}\rangle$—OH + $\overset{O}{\overset{\|}{HOCCH_3}}$ $\overset{H^+}{\rightleftharpoons}$ $\langle\text{cyclohexyl}\rangle$—$\overset{O}{\overset{\|}{OCCH_3}}$ + H$_2$O

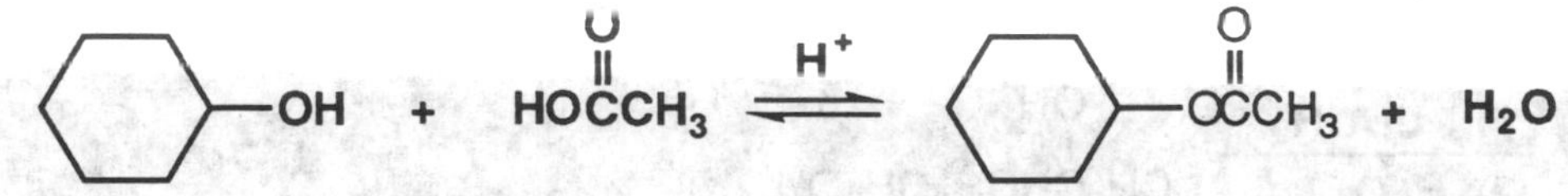

(d) CH$_3$CH$_2$CH=CHCOH + HOCH(CH$_3$)$_2$ $\overset{H^+}{\rightleftharpoons}$ CH$_3$CH$_2$CH=CHCOCH(CH$_3$)$_2$

+ H$_2$O

<u>Problem 16.40</u> When 4-hydroxybutanoic acid is treated with an acid catalyst, it forms a lactone (a cyclic ester). Draw the structural formula of this lactone and propose a mechanism for its formation.

Step 1:

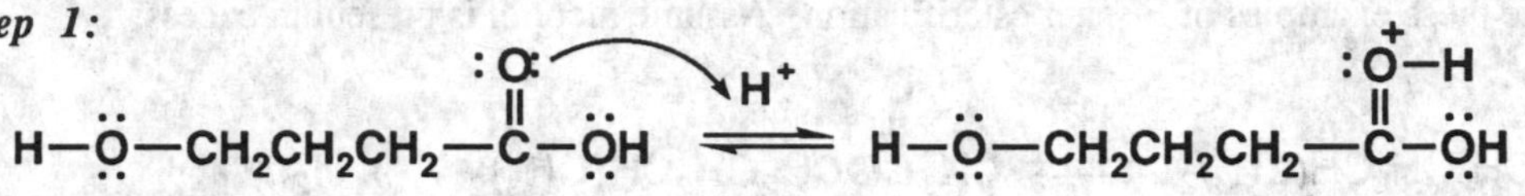

H—Ö—CH$_2$CH$_2$CH$_2$—C—ÖH $\rightleftharpoons$ H—Ö—CH$_2$CH$_2$CH$_2$—C—ÖH

4-Hydroxybutanoic acid

Step 2:

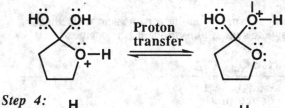

Step 3:

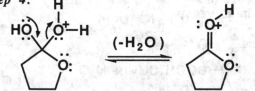

Proton transfer

Step 4:

(-H$_2$O)

Step 5:

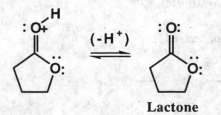

(-H$^+$)

Lactone

Problem 16.41 Fischer esterification cannot be used to prepare *tert*-butyl esters. Instead, carboxylic acids are treated with 2-methylpropene and an acidic catalyst to generate them.
(a) Why does the Fischer esterification fail for the synthesis of *tert*-butyl esters?

The Fischer esterification does not work for at least three reasons. First, the *t*-butyl alcohol is not very nucleophilic because of steric hindrance. Second, the *t*-butyl alcohol dehydrates in acid. Third, any *t*-butyl ester that forms falls apart to give the carboxylic acid and *t*-butyl cation in acid.

Step 1:

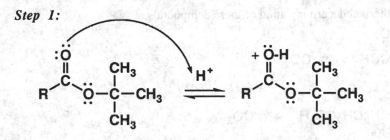

Step 2:

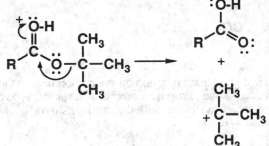

(b) Propose a mechanism for the 2-methylpropene method.

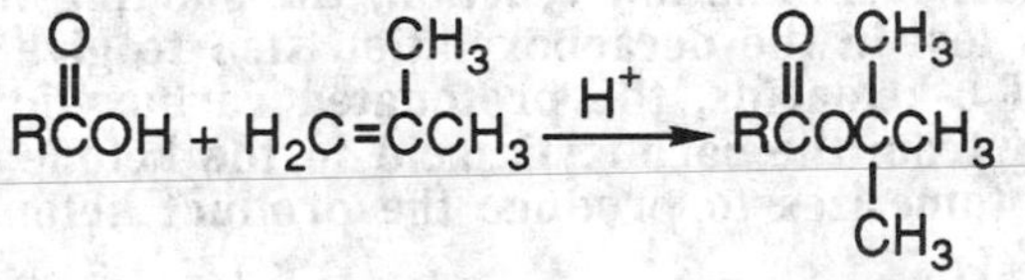

$$\underset{\text{2-Methylpropene}}{\underset{\text{(Isobutylene)}}{}} \; RCOH + H_2C=CCH_3 \; \xrightarrow{\text{H}^+} \; RCOCCH_3 \quad \text{(a } tert\text{-butyl ester)}$$

Step 1:

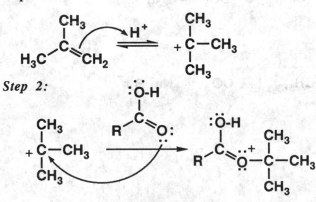

Step 2:

Step 3:

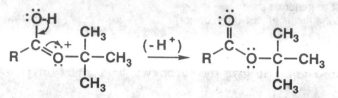

Problem 16.42 Draw the product formed on thermal decarboxylation of these compounds.

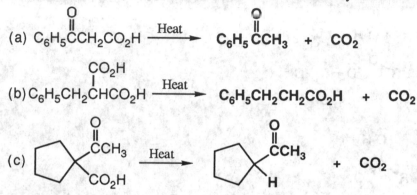

Problem 16.43 When heated, carboxylic salts in which there is a good leaving group on the carbon beta to the carboxylate group undergo decarboxylation/elimination to give an alkene. Propose a mechanism for this type of decarboxylation/elimination. Compare the mechanism of these decarboxylations with the mechanism for decarboxylation of β-ketoacids; in what way(s) are the mechanisms similar?

The mechanisms are similar in that, in both cases, CO_2 is generated and a carbon-carbon bond is broken. As can be seen in the mechanisms shown below, that is where the similarity ends. In the case of the decarboxylation/elimination reaction, the starting material is the carboxylate anion and a bromide anion is lost in the decarboxylation step to give the product alkene. In the case of decarboxylation of β-ketoacids, the protonated carboxylic acid group is usually used. A proton is transferred from the carboxylic acid to the ketone to give an enol in the decarboxylation step, that tautomerizes to produce the product ketone.

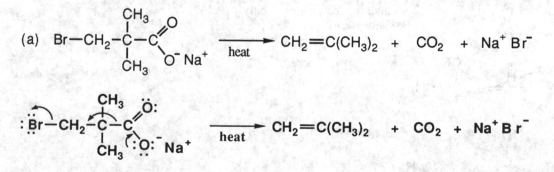

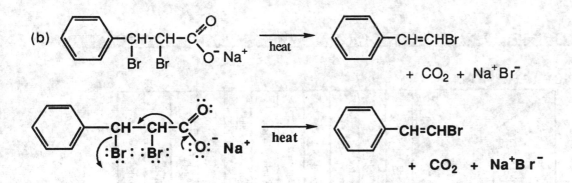

CHAPTER 17: FUNCTIONAL DERIVATIVES OF CARBOXYLIC ACIDS

SUMMARY OF REACTIONS

Starting Material ↓ / Product →	Acid Anhydrides	Alcohols	Alcohols Aldehydes	Amines	Amines	Carboxylic Acids	Carboxylic Acids alcohols	Carboxylic Acids Amines	Esters	Ketones	Primary Amines
Acid Anhydrides				17A 17.7B*		17B 17.5B			17C 17.6B		
Acid Chlorides				17D 17.7A		17E 17.5A			17F 17.6A		
Acid Chlorides Carboxylates	17G 17.8										
Acid Chlorides Lithium Diorganocopper Reagents										17H 17.9C	
Amides					17I 17.11B			17J 17.5D			
Esters		17K 17.11A	17L 17.11A	17M 17.7C			17N 17.5C		17O 17.6C		
Esters and Grignard Reagents or Organolithium Reagents		17P 17.9A									
Nitriles						17Q 17.5E					17R 17.11C
Primary Amides											17S 20.12

*Section in book that describes reaction.

REACTION 17A: REACTION OF ACID ANHYDRIDES WITH AMMONIA AND AMINES (Section 17.7B)

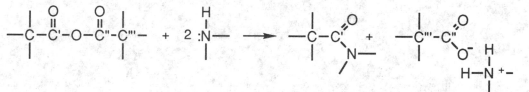

- **Acid anhydrides** react with **ammonia, primary amines,** and **secondary amines** to produce **amides**. ✳
- Two mol of amine are usually used in the reaction; one mol to form the amide and the other mol to neutralize the carboxylic acid that is also produced.
- For this reaction as well as the others mentioned in this chapter, the acid anhydride will usually be symmetric, that is, derived from two mol of the same carboxylic acid.

REACTION 17B: HYDROLYSIS OF ACID ANHYDRIDES (Section 17.5B)

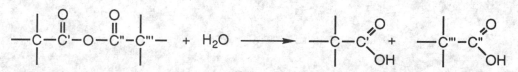

- **Lower molecular weight acid anhydrides** react with **water** to from **two mol of carboxylic acid.**✳
- Like the other reactions that involve attack by nucleophiles on acid anhydrides, the mechanism of the reaction begins when the nucleophile reacts with the carbonyl carbon atom to form a tetrahedral carbonyl addition intermediate, that collapses to product by elimination of a carboxylate leaving group.

REACTION 17C: REACTION OF ACID ANHYDRIDES WITH ALCOHOLS (Section 17.6B)

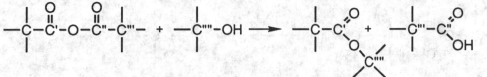

- **Acid anhydrides** also react with **alcohols** to produce **one mol** of **ester** and **one mol** of **carboxylic acid.** ✳

REACTION 17D: REACTION OF ACID CHLORIDES WITH AMMONIA AND AMINES (Section 17.7A)

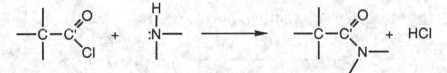

- **Acid chlorides** react with **ammonia, primary amines,** and **secondary amines** to produce **amides.**✳
- Like the other reactions that involve attack by nucleophiles on acid chlorides, the mechanism of the reaction begins when the nucleophile reacts with the carbonyl carbon atom to form a tetrahedral carbonyl addition intermediate, that collapses to product by elimination of a chloride leaving group.

REACTION 17E: HYDROLYSIS OF ACID CHLORIDES (Section 17.5A)

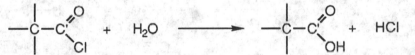

- Lower molecular weight **acid chlorides** are so reactive to nucleophiles that they are **readily hydrolyzed** to **carboxylic acids** by water. ✳
- The higher molecular weight acid chlorides are less soluble in water, so they react more slowly.

REACTION 17F: REACTION OF ACID CHLORIDES WITH ALCOHOLS (Section 17.6A)

- **Acid chlorides** react with **alcohols** to produce **esters.** ✳

REACTION 17G: REACTION OF ACID CHLORIDES WITH CARBOXYLIC ACID SALTS (Section 17.8)

- **Acid chlorides** react with **carboxylate salts** to produce **acid anhydrides.** ✳

REACTION 17H: REACTION OF ACID CHLORIDES WITH LITHIUM DIORGANOCOPPER (GILMAN) REAGENTS (Section 17.9C)

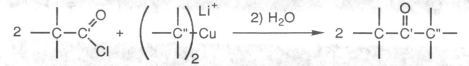

- **Acid chlorides** react with **lithium diorganocopper (Gilman) reagents** to produce **ketones**. ✳
- **Lithium diorganocopper (Gilman)** reagents are less nucleophilic than Grignard reagents or organolithium reagents. Thus, lithium diorganocopper reagents attack the highly reactive acid chlorides, but they are not reactive enough to react with the product ketones like Grignard and organolithium reagents do. The reaction stops at the ketone stage.
- **Lithium diorganocopper** reagents are usually prepared by reacting alkyllithium reagents with CuCl.

REACTION 17I: REDUCTION OF AMIDES (Section 17.11B)

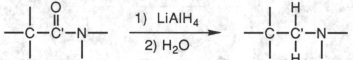

- **Amides** are **reduced by LiAlH$_4$** to produce **primary, secondary,** and **tertiary amines**, depending on the substitution of the original amide. ✳

REACTION 17J: HYDROLYSIS OF AMIDES (Section 17.5D)

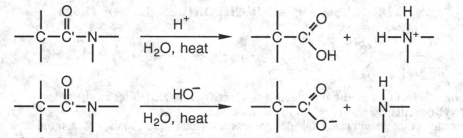

- **Amides** are **hydrolyzed** in the presence of **one mol acid** or **one mol hydroxide** to produce a **carboxylic acid** and a protonated **ammonium ion or protonated amine**, or a **carboxylate salt** and an **amine**, respectively. ✳
- In the acid-promoted reaction, the mechanism involves an initial protonation of the carbonyl oxygen atom that facilitates nucleophilic attack of water. A proton transfer gives a tetrahedral carbonyl addition intermediate, that collapses upon protonation and departure of the nitrogen atom. The ammonia or amine that leaves is finally protonated, thus completing the reaction.
- In the base-promoted reaction, the nucleophilic hydroxide attacks the carbonyl carbon atom to produce a tetrahedral carbonyl addition intermediate. This collapses in a step that involves protonation by water and departure of the nitrogen-containing group to give the products.

REACTION 17K: REDUCTION OF ESTERS (Section 17.11A)

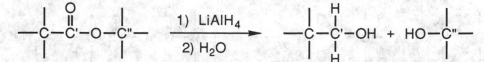

- **Esters** are **reduced by LiAlH$_4$** to produce **two mol of alcohols;** one derived from the carboxylic acid portion of the ester, and the other is from the alcohol portion. ✳
- NaBH$_4$ reacts much more slowly with esters, so it is possible to use this reagent to reduce an aldehyde or ketone group without reducing an ester in the same molecule.

REACTION 17L: REDUCTION OF ESTERS WITH DIBALH (Section 17.11A)

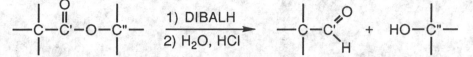

- **Esters** are **reduced by DIBALH** (a modified hydride reducing agent) at low temperature to produce **an aldehyde** derived from the carboxyl group and **an alcohol.** ✳

REACTION 17M: REACTION OF ESTERS WITH AMMONIA AND AMINES (Section 17.7C)

- **Esters** react very slowly with **ammonia, primary amines,** and **secondary amines** to produce **amides.** ✳

REACTION 17N: HYDROLYSIS OF ESTERS (Section 17.5C)

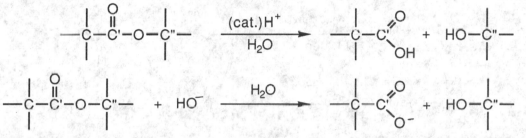

- **Esters** are **hydrolyzed** in the presence of a **catalytic amount of acid** or **one mol of hydroxide** to produce a **carboxylic acid** and **alcohol,** or **carboxylate salt** and **alcohol,** respectively. ✳
- In the acid-catalyzed reaction, the mechanism involves an initial protonation of the carbonyl oxygen atom that facilitates nucleophilic attack of water. A proton transfer gives a tetrahedral carbonyl addition intermediate, that can collapse upon departure of the alkoxide and a proton. The alkoxide is protonated to give an alcohol, thus completing the reaction. Note that the acid is not consumed in the reaction, so it is catalytic. Furthermore, this reaction is just the reverse of the Fischer esterification reaction.
- In the base-promoted reaction, the nucleophilic hydroxide attacks the carbonyl carbon atom to produce a tetrahedral carbonyl addition intermediate, that collapses due to the departure of alkoxide that is protonated by water to give the products.

REACTION 17O: REACTION OF ESTERS WITH ALCOHOLS: TRANS-ESTERIFICATION
(Section 17.6C)

- **Esters** react with **alcohols** in the presence of **anhydrous acid** in a process called **transesterification.**✳
- Transesterification is an equilibrium process, the position of which is determined by the experimental conditions. For example, carrying out the reaction in the presence of a large excess of alcohol reagent, or removal of the departing alcohol product from the reaction mixture as it is liberated, will drive the reaction to completion.

REACTION 17P: REACTION OF ESTERS WITH GRIGNARD OR ORGANO-LITHIUM REAGENTS (Section 17.9A)

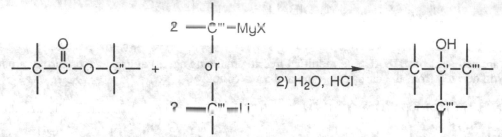

- Most **esters** react with two mol of **Grignard or organolithium reagent** to produce a **tertiary alcohol**. An exception to this is that a **secondary alcohol** is produced from **esters of formic acid**. ✳
- The mechanism involves an initial reaction of the Grignard or organolithium reagent with the ester to produce a ketone, that itself reacts with another Grignard or organolithium reagent to produce a magnesium or lithium alkoxide salt that is protonated in an acid work-up to give the product alcohol.

REACTION 17Q: HYDROLYSIS OF NITRILES (Section 17.5E)

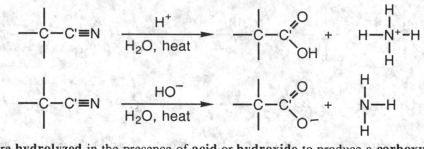

- **Nitriles** are **hydrolyzed** in the presence of **acid** or **hydroxide** to produce a **carboxylic acid** and **ammonium ion**, or **carboxylate salt** and **ammonia**, respectively. ✳
- The reaction in acid involves an initial protonation of the nitrile nitrogen atom, that facilitates nucleophilic attack of a water molecule to produce an imidic acid that undergoes keto-enol tautomerization to give an amide. The amide is then hydrolyzed according to the same mechanism described under Reaction 17J, Section 17.4D. It is possible to stop at the primary amide stage of the reaction if sulfuric acid and one mol of water are used in the reaction.

REACTION 17R: REDUCTION OF NITRILES (Section 17.11C)

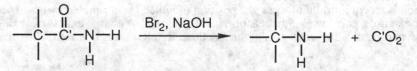

- **Nitriles** are reduced by **LiAlH$_4$** to produce **primary amines**. ✳

REACTION 17S: THE HOFMANN REARRANGEMENT OF PRIMARY AMIDES (Section 17.12)

- The **Hofmann rearrangement** involves the conversion of a **primary amide** to a **primary amine**, with loss of the original **carbonyl carbon atom** as CO_2. ✳

- The mechanism involves initial deprotonation of the amide to produce an amide anion that reacts with bromine to produce an *N*-bromoamide. The second amide hydrogen is then removed by base, followed by migration of the alkyl group to nitrogen, and departure of the bromine. The resulting isocyanate reacts with base to produce a carbamic acid, that undergoes spontaneous decarboxylation to give the product amine and CO_2.

SUMMARY OF IMPORTANT CONCEPTS

17.0 OVERVIEW
• **Nitriles**, as well as the derivatives of carboxylic acids, namely **acid chlorides**, **acid anhydrides**, **esters**, and **amides** all exhibit similar types of chemistry. This chapter covers two characteristic reactions of these groups. ✱

17.1 STRUCTURE AND NOMENCLATURE
• The functional group of **acid halides** is a **halogen atom, usually chlorine, attached** to the **carbon atom** of a **carbonyl group**. Acid halides are named the same as a carboxylic acid, but the **ic acid** suffix is replaced by **yl halide**. ✱
• The functional group of **acid anhydrides** is **two acyl groups attached** to an oxygen atom. Acid anhydrides are usually symmetrical, that is both acyl groups are the same, although unsymmetrical acid anhydrides can also be prepared and used. Acid anhydrides are named by adding the word **anhydride** to the name of the corresponding carboxylic acid. Cyclic anhydrides may be formed from dicarboxylic acids. ✱
• The functional group of **esters** is an **-OR** group bonded to the **carbon atom** of a **carbonyl group**. The R group may be an alkyl or aryl group. Esters are named by listing the name of the R group first, followed by the name of the carboxylic acid, except the suffix **ic acid** is replaced by the suffix **ate**. **Cyclic esters**, derived from molecules with both an alcohol and carboxylic acid function, are called **lactones**. ✱
• The functional group of **amides** is a **trivalent nitrogen atom attached** to the **carbon atom** of a **carbonyl group**. Primary, secondary, and tertiary amides have two, one, or zero hydrogen atoms attached to the nitrogen atom, respectively. Amides are named the same as a carboxylic acid, but the **oic acid** suffix of an IUPAC name or **ic acid** suffix of a common name is replaced by **amide**. If there is an R group on the nitrogen atom, it is listed first and designated with *N*-. Cyclic amides are called **lactams**. **Imides** contain a nitrogen atom attached to two acyl groups. ✱
• The functional group of **nitriles** is a **nitrogen atom triple bonded** to a **carbon atom**. Although nitriles do not contain a carbonyl group, they undergo many of the reactions characteristic of carboxylic acid derivatives. The IUPAC names are given as an alkanenitrile, while common names replace the **ic** or **oic acid** suffix of a corresponding carboxylic acid with **onitrile**. ✱

17.3 SPECTROSCOPIC PROPERTIES
• **In mass spectra**, **esters** and **amides** usually produce a **molecular ion peak**. In addition, α-**cleavage** and **McLafferty rearrangements** are common.
• In ^{1}**H-NMR spectra**, the signals due to the **hydrogens** α to a **carboxyl group** occur in the range δ **2.0 to 2.6**. The **H** of a **carboxyl group** gives a signal in the range δ **10 to 13**. The signals for the **hydrogens on carbon atoms attached** to the **oxygen atom** of esters occur in the range δ **3.6 to 4.1**.
• **Infrared spectroscopy** is especially **useful** with **carboxylic acid derivatives**, and the characteristic absorptions are listed in Table 17.1.

17.4 CHARACTERISTIC REACTIONS
• The different **groups (chloride, carboxyl, alkoxyl, amino) attached** to the **carbonyl groups** of the carboxylic acid derivatives can be **considered** as **leaving groups** of varying ability. Thus, when the **carbonyl group is attacked by a nucleophile**, the **tetrahedral carbonyl addition intermediate** that is formed can collapse by **expulsion** of the **leaving group** to **regenerate** a **carbonyl group** and give the final product. This process is referred to as **nucleophilic acyl substitution**. ✱ *[Variations of this mechanism operate in a large number of different reactions. It is helpful to distinguish these different reactions by keeping track of 1)the nature of the nucleophile and 2) proton transfers for each of the reactions described throughout the chapter.]*

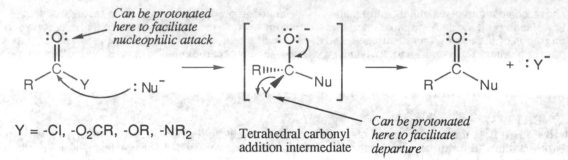

$Y = -Cl, -O_2CR, -OR, -NR_2$

Tetrahedral carbonyl
addition intermediate

- In general, the **chloride ion** of acid chlorides is the **best leaving group**, followed by the carboxylate of acid anhydrides, the alkoxy group of esters, and the amino group of amides, respectively. Note that when sufficient acid is present, the **leaving group can be protonated** as it departs. This deprotonation can assist in the departure of the more sluggish alkoxyl and amino leaving groups in esters and amides, respectively.
- Because of the relative leaving group abilities listed above, the **acid chlorides** are the **most reactive** class of carboxylic acid derivatives; followed by anhydrides, esters, and amides, respectively.
- **Strong acid** can **protonate the carbonyl oxygen atom**, resulting in an oxonium species that is a much better electrophile. Thus, acid catalysis can be used to facilitate nucleophilic attack of the less reactive carboxylic acid derivatives such as esters, nitriles, and amides.

CHAPTER 17
Solutions to the Problems

__Problem 17.1__ Draw structural formulas for the following compounds.

(a) *N*-Cyclohexylacetamide.

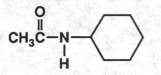

(b) *sec*-Butyl methanoate

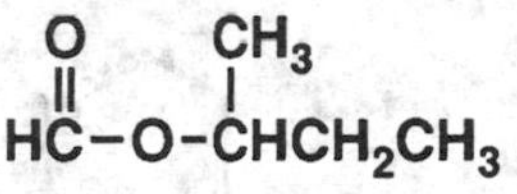

O   CH₃
‖ |
HC–O–CHCH₂CH₃

(c) Cyclobutyl butanoate

CH₃CH₂CH₂CO–▢

(d) 1-(2-Octyl)succinimide

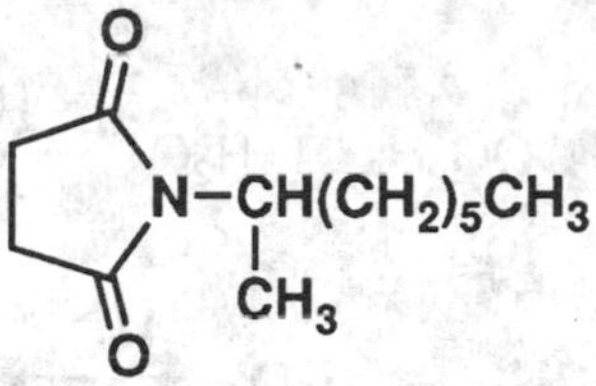

N–CH(CH₂)₅CH₃
          |
          CH₃

(e) Diethyl adipate

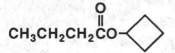

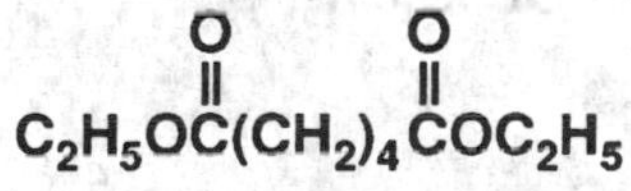

$C_2H_5OC(CH_2)_4COC_2H_5$

(f) 2-Aminopropanamide

O
‖
CH₃—CH—CNH₂
      |
      NH₂

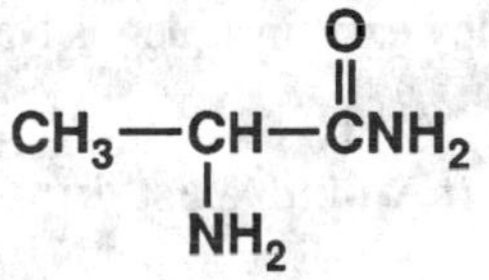

__Problem 17.2__ Will phthalimide dissolve in aqueous sodium bicarbonate?

In order to dissolve in water, the phthalimide must be deprotonated to give the negatively charged phthalimide anion. The questions becomes whether sodium bicarbonate is a strong enough base to deprotonate phthalimide according to the following equilibrium:

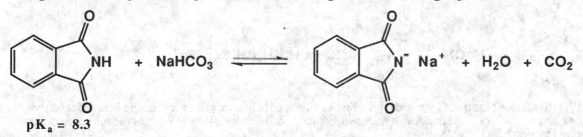

$pK_a = 8.3$

The pK_a of phthalimide is 8.3, while the pKa of carbonic acid H_2CO_3, which dissociates into H_2O and CO_2, is 6.4. Since equilibria in acid-base reactions favor formation of the weaker acid (phthalimide), the equilibrium will be to the left and the phthalimide will not dissolve.

Problem 17.3 Complete and balance equations for the hydrolysis of each ester in aqueous solution. Show each product as it is ionized under the indicated experimental conditions.

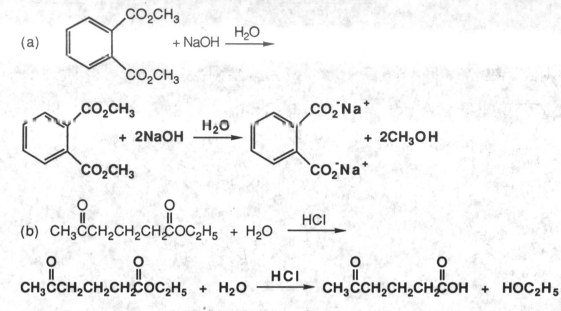

(a)

(b) $CH_3\overset{O}{\overset{\|}{C}}CH_2CH_2CH_2\overset{O}{\overset{\|}{C}}OC_2H_5$ + H_2O $\xrightarrow{\text{HCl}}$

$CH_3\overset{O}{\overset{\|}{C}}CH_2CH_2CH_2\overset{O}{\overset{\|}{C}}OC_2H_5$ + H_2O $\xrightarrow{\text{HCl}}$ $CH_3\overset{O}{\overset{\|}{C}}CH_2CH_2CH_2\overset{O}{\overset{\|}{C}}OH$ + HOC_2H_5

Problem 17.4 Complete equations for the hydrolysis of the amides in Example 17.4 in concentrated aqueous NaOH. Show all products as they exist in aqueous NaOH, and the number of mol of NaOH required for hydrolysis of each amide.

Each product is shown as it would exist in aqueous NaOH.

(a) $CH_3\overset{O}{\overset{\|}{C}}-N-CH_3$ + NaOH $\xrightarrow{\text{H}_2\text{O}}$ $CH_3\overset{O}{\overset{\|}{C}}O^-Na^+$ + $(CH_3)_2NH$
                 |
                 CH_3

(b) [structure] + NaOH $\xrightarrow{\text{H}_2\text{O}}$ $H_2NCH_2CH_2CH_2CH_2\overset{O}{\overset{\|}{C}}O^-\ Na^+$

Problem 17.5 Synthesis of nitriles by nucleophilic displacement of halide from an alkyl halide is practical only with primary and secondary alkyl halides. It fails with tertiary alkyl halides. Why? What is the major product of the following reaction?

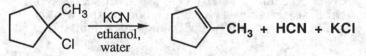

Cyanide ion is both a base and a nucleophile. With a tertiary halide and moderate base, moderate nucleophile such as CN⁻, E2 is the principal reaction. Thus, the major product in this instance is 1-methylcyclopentene.

Problem 17.6 Complete the following transesterification reaction. The stoichiometry is given in the equation.

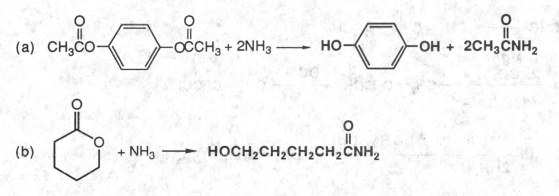

Problem 17.7 Complete and balance equations for the following reactions. The stoichiometry of each reaction is given in the equation.

Each is an example of aminolysis of an ester.

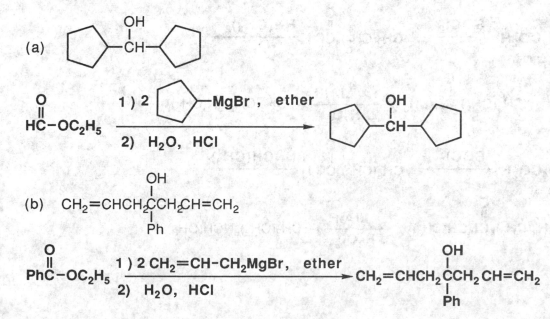

Problem 17.8 Show how to prepare these alcohols by treatment of an ester with a Grignard reagent.

Each can be prepared by treatment of an ester with two mol of an organomagnesium reagent. In these solutions, the ester chosen is the ethyl ester.

(a)

(b)

Problem 17.9 Show how to bring about these conversions in good yield.

Both of these conversions can be brought about using a lithium diorganocopper reagent.

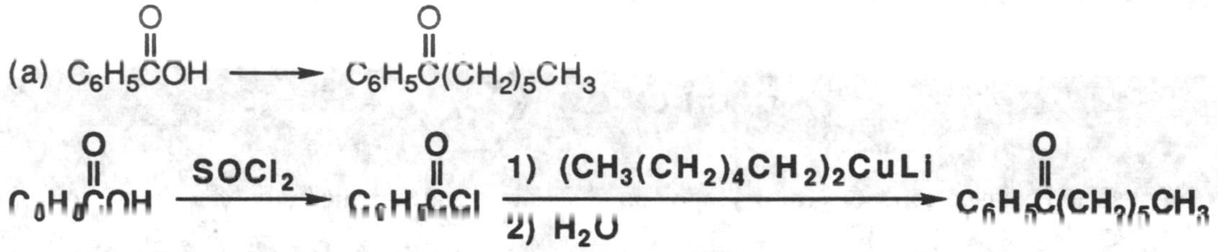

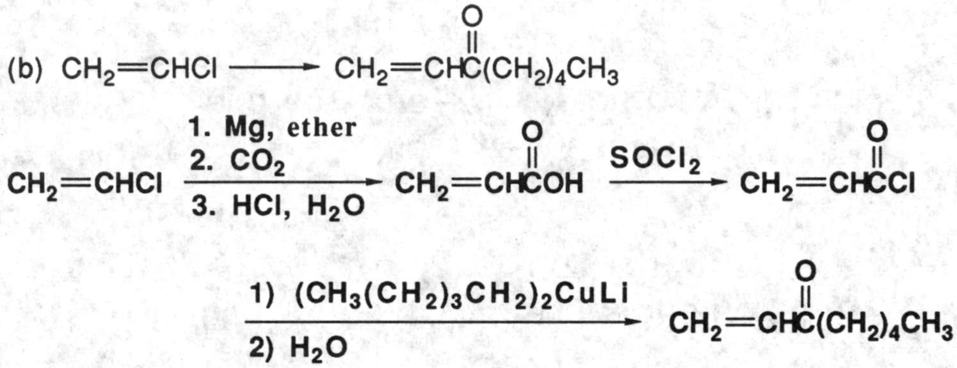

(a) $C_6H_5\overset{O}{\overset{\|}{C}}OH \longrightarrow C_6H_5\overset{O}{\overset{\|}{C}}(CH_2)_5CH_3$

$C_6H_5\overset{O}{\overset{\|}{C}}OH \xrightarrow{SOCl_2} C_6H_5\overset{O}{\overset{\|}{C}}Cl \xrightarrow[\text{2) } H_2O]{\text{1) } (CH_3(CH_2)_4CH_2)_2CuLi} C_6H_5\overset{O}{\overset{\|}{C}}(CH_2)_5CH_3$

(b) $CH_2{=}CHCl \longrightarrow CH_2{=}CH\overset{O}{\overset{\|}{C}}(CH_2)_4CH_3$

$CH_2{=}CHCl \xrightarrow[\substack{\text{1. Mg, ether}\\ \text{2. } CO_2 \\ \text{3. HCl, } H_2O}]{} CH_2{=}CH\overset{O}{\overset{\|}{C}}OH \xrightarrow{SOCl_2} CH_2{=}CH\overset{O}{\overset{\|}{C}}Cl$

$\xrightarrow[\text{2) } H_2O]{\text{1) } (CH_3(CH_2)_3CH_2)_2CuLi} CH_2{=}CH\overset{O}{\overset{\|}{C}}(CH_2)_4CH_3$

Problem 17.10 Show how to convert hexanoic acid to each amine in good yield.

In both cases, the conversions can be carried out through formation of an amide, followed by reduction to give the desired amine.

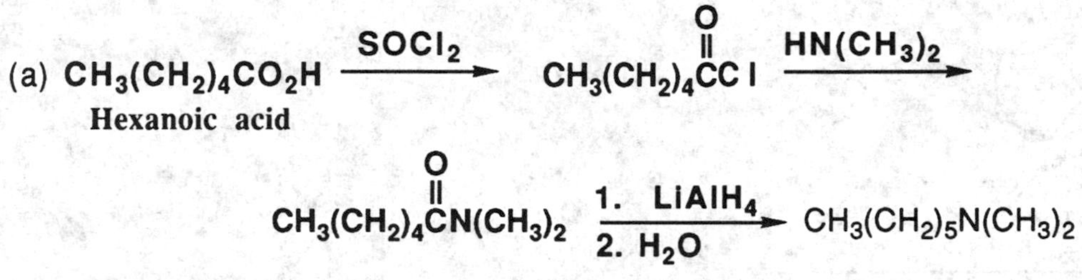

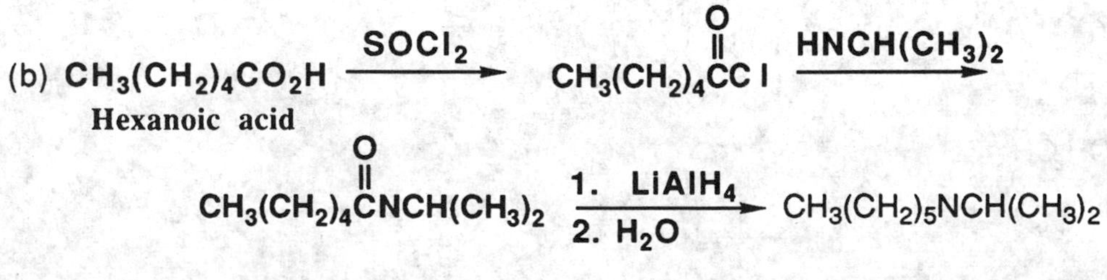

(a) $CH_3(CH_2)_4CO_2H \xrightarrow{SOCl_2} CH_3(CH_2)_4\overset{O}{\overset{\|}{C}}Cl \xrightarrow{HN(CH_3)_2}$
   Hexanoic acid

$CH_3(CH_2)_4\overset{O}{\overset{\|}{C}}N(CH_3)_2 \xrightarrow[\text{2. } H_2O]{\text{1. } LiAlH_4} CH_3(CH_2)_5N(CH_3)_2$

(b) $CH_3(CH_2)_4CO_2H \xrightarrow{SOCl_2} CH_3(CH_2)_4\overset{O}{\overset{\|}{C}}Cl \xrightarrow{HNCH(CH_3)_2}$
   Hexanoic acid

$CH_3(CH_2)_4\overset{O}{\overset{\|}{C}}NCH(CH_3)_2 \xrightarrow[\text{2. } H_2O]{\text{1. } LiAlH_4} CH_3(CH_2)_5NCH(CH_3)_2$

Problem 17.11 Show how to convert (R)-2-phenylpropanoic acid to these compounds:

(a)

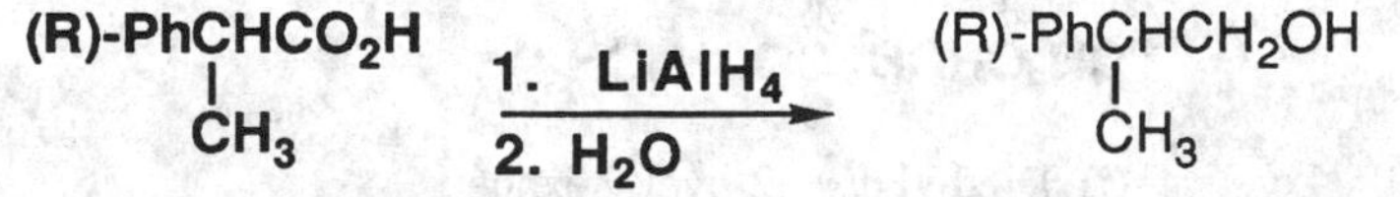

(R)-PhCHCO₂H
|
CH₃
(R)-2-Phenyl-
propanoic acid

$\xrightarrow{\text{1. LiAlH}_4 \atop \text{2. H}_2\text{O}}$

(R)-PhCHCH₂OH
|
CH₃
(R)-2-Phenyl-1-propanol

(b)

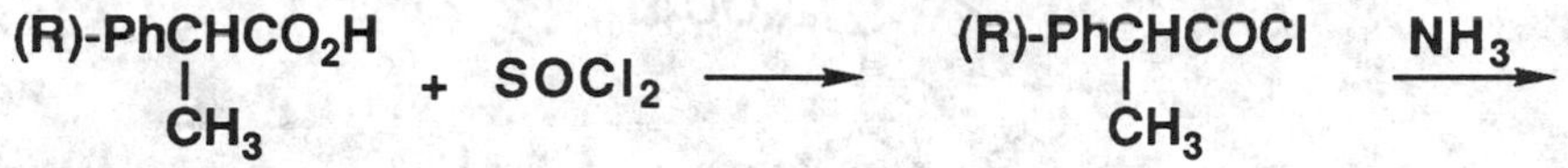

(R)-PhCHCO₂H
|
CH₃
(R)-2-Phenyl-
propanoic acid

+ SOCl₂ ⟶

(R)-PhCHCOCl
|
CH₃

$\xrightarrow{\text{NH}_3}$

(R)-PhCHCONH₂
|
CH₃

$\xrightarrow{\text{1. LiAlH}_4 \atop \text{2. H}_2\text{O}}$

(R)-PhCHCH₂NH₂
|
CH₃

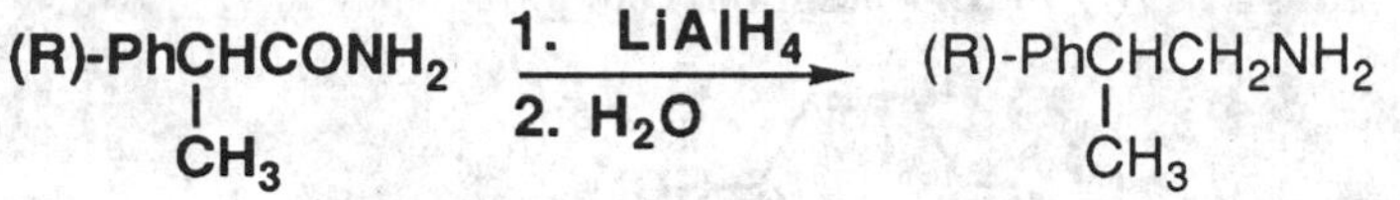

(R)-2-Phenyl-1-propanamine

Problem 17.12 Show how to convert phenylacetic acid into the following in good yield.

(a)

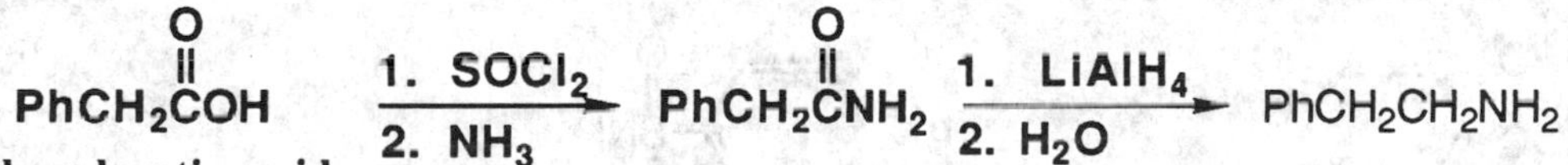

O
‖
PhCH₂COH
Phenylacetic acid

$\xrightarrow{\text{1. SOCl}_2 \atop \text{2. NH}_3}$

O
‖
PhCH₂CNH₂

$\xrightarrow{\text{1. LiAlH}_4 \atop \text{2. H}_2\text{O}}$

PhCH₂CH₂NH₂

(b)

O
‖
PhCH₂CNH₂
From part (a)

$\xrightarrow{\text{Br}_2/\text{NaOH}}$

PhCH₂NH₂

PROBLEMS
Structure and Nomenclature
Problem 17.13 Draw structural formulas for these compounds:
(a) Dimethyl carbonate

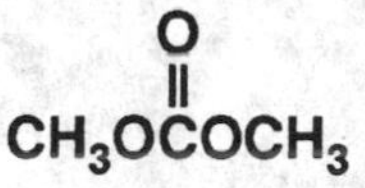

O
‖
CH₃OCOCH₃

(b) Benzonitrile

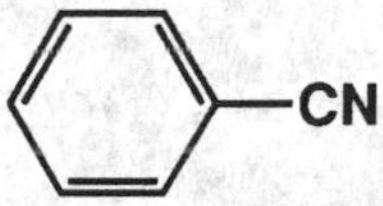

(c) Isopropyl 3-methylhexanoate

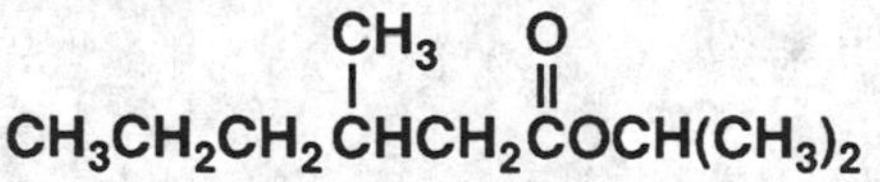

CH₃   O
| ‖
CH₃CH₂CH₂CHCH₂COCH(CH₃)₂

(d) Diethyl oxalate

O O
‖ ‖
CH₃CH₂OC COCH₂CH₃

(e) Ethyl (Z)-2-pentenoate

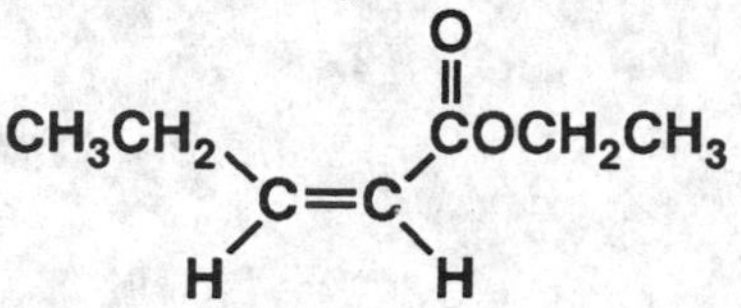

        O
        ‖
CH₃CH₂   COCH₂CH₃
     \\   /
      C=C
     /   \\
    H   H

(f) Butanoic anhydride

(CH₃CH₂CH₂CO)₂O

(g) Dodecanamide

$$CH_3(CH_2)_{10}\overset{\displaystyle O}{\overset{\|}{C}}NH_2$$

(h) Ethyl 3-hydroxybutanoate

$$CH_3\overset{\displaystyle OH}{\overset{|}{C}}HCH_2\overset{\displaystyle O}{\overset{\|}{C}}OCH_2CH_3$$

(i) Octanoyl chloride

$$CH_3(CH_2)_6\overset{\displaystyle O}{\overset{\|}{C}}Cl$$

(j) Diethyl *cis* 1,2-cyclohexane dicarboxylate

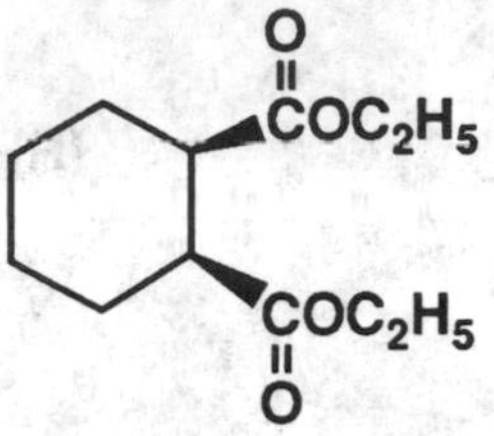

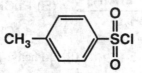

(k) Methanesulfonyl chloride

$$CH_3-\overset{\displaystyle O}{\underset{\displaystyle O}{\overset{\|}{\underset{\|}{S}}}}-Cl$$

(l) *p*-Toluenesulfonyl chloride

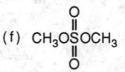

Problem 17.14 Write the IUPAC name for each compound.

(a)

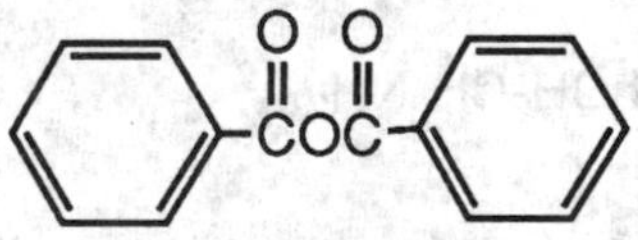

Benzoic anhydride

(b)

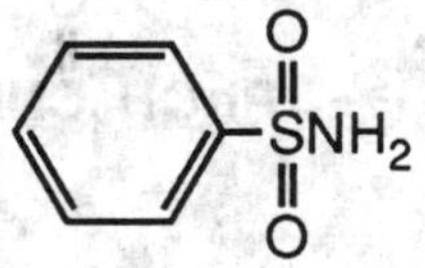

Benzenesulfonamide

(c) $CH_3(CH_2)_4\overset{\displaystyle O}{\overset{\|}{C}}NHCH_3$

N-**Methylhexanamide**

(d) $CH_3(CH_2)_6\overset{\displaystyle O}{\overset{\|}{C}}NH_2$

Octanamide

(e) $CH_2(CO_2CH_2CH_3)_2$

**Diethyl propanedioate
(Diethyl malonate)**

(f) $CH_3O\overset{\displaystyle O}{\underset{\displaystyle O}{\overset{\|}{\underset{\|}{S}}}}OCH_3$

Dimethyl sulfate

(g) $PhCH_2\overset{\displaystyle O}{\overset{\|}{C}}\overset{\displaystyle }{\underset{\displaystyle CH_3}{\overset{\displaystyle }{\underset{|}{C}}}}H\overset{\displaystyle O}{\overset{\|}{C}}OCH_3$

**Methyl 2-methyl-3-oxo-
4-phenylbutanoate**

(h) $Cl\overset{\displaystyle O}{\overset{\|}{C}}(CH_2)_4\overset{\displaystyle O}{\overset{\|}{C}}Cl$

**Hexanedioyl chloride
(Adipoyl chloride)**

(i) $CH_3(CH_2)_5CN$

Heptanenitrile

Physical Properties

<u>Problem 17.15</u> Both the melting point and boiling point of acetamide are higher than those of its *N,N*-dimethyl derivative.

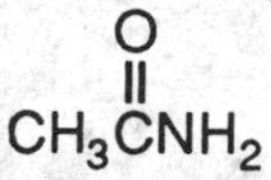

O
‖
CH_3CNH_2
Acetamide
mp 82.3°C, bp 221°C

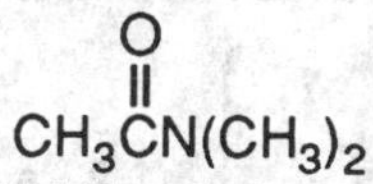

O
‖
$CH_3CN(CH_3)_2$
N,N-dimethylacetamide
mp -20°C, bp 165°C

How do you account for these differences?

Acetamide has two amide N-H atoms that can take part in hydrogen bonding with carbonyl oxygen atoms, while *N,N*-dimethylacetamide does not. Thus, due to this ability to hydrogen bond, acetamide molecules can associate with each other more strongly, leading to higher melting and boiling points compared to *N,N*-dimethylacetamide. Additionally, in the case of acetamide, overlap of unhybridized 2p orbitals of the carbonyl group and the amide nitrogen results in delocalization and resonance stabilization of the molecule. For this overlap of 2p orbitals to be possible, the six atom system of the amide group must be planar; only when these atoms lie in a plane can the three 2p orbitals be parallel to each other and thereby overlap. One of the graphic ways in which we indicate this overlap and electron delocalization is by drawing contributing structures such as the ones shown below. These contributing structures are possible only if the six-atom system of the amide bond is planar. With two groups larger than hydrogen on the amide nitrogen, there is steric hindrance to planarity with the result that the dipolar contributing structure shown makes less contribution to the hybrid, the *N,N*-disubstituted amide is less polar, there is less intermolecular association, and consequently a lower melting point and a lower boiling point.

non-bonded interaction between these sets of groups hinders achievement of planarity

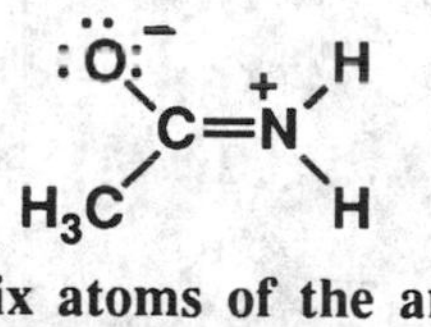

the six atoms of the amide
bond must be planar in this
contributing structure

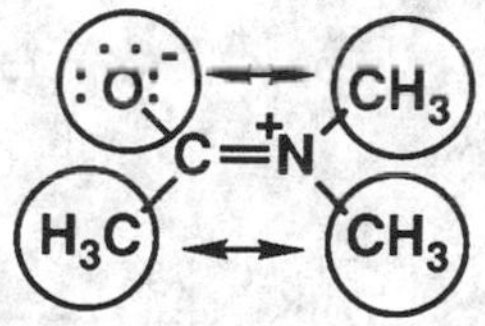

Spectroscopy

<u>Problem 17.16</u> All methyl esters of long-chain aliphatic acids (for example, methyl tetradecanoate, $C_{13}H_{27}CO_2CH_3$), show significant fragment ions at *m/z* 74, 59, and 31. What are the structures of these ions and how do they form?

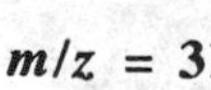

$^+OCH_3$

m/z = 31
(by α–cleavage)

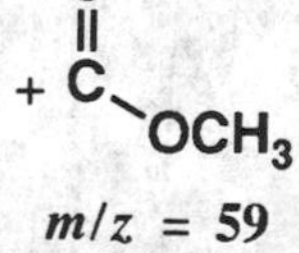

O
‖
$+ C$
OCH_3

m/z = 59
(by α–cleavage)

$$\left[\begin{array}{c} H \\ O \\ C \\ H_2C \quad OCH_3 \end{array} \right]^{+\cdot}$$

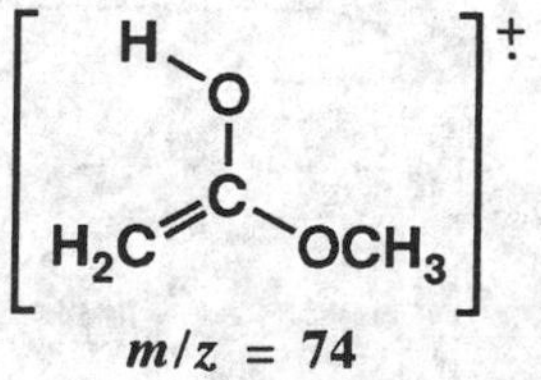

m/z = 74

(by McLafferty
rearrangement)

Problem 17.17 The two hydrogens of primary amides typically have separate ^{1}H-NMR resonances, as illustrated by the separate signals for the two amide hydrogens of propanamide. Furthermore, each methyl group of the *N*,*N*-dimethylformamide has a separate resonance. How do you account for these observations?

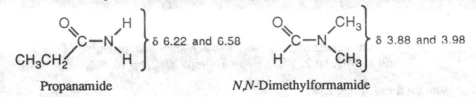

Propanamide   *N*,*N*-Dimethylformamide

Hint: Consider the covalent bonding in a resonance-stabilized amide group and the likely orientation in space of the six atoms of the amide group.

Amide resonance causes a rotation barrier around the C(O)-N bond that prevents interconversion of the two nitrogen substituents on the NMR time scale. For this reason, the two hydrogens and two methyl groups, respectively, are in different chemical environments and thus give different signals.

Problem 17.18 Propose a structural formula for compound A, $C_7H_{14}O_2$, consistent with its ^{1}H-NMR and infrared spectra.

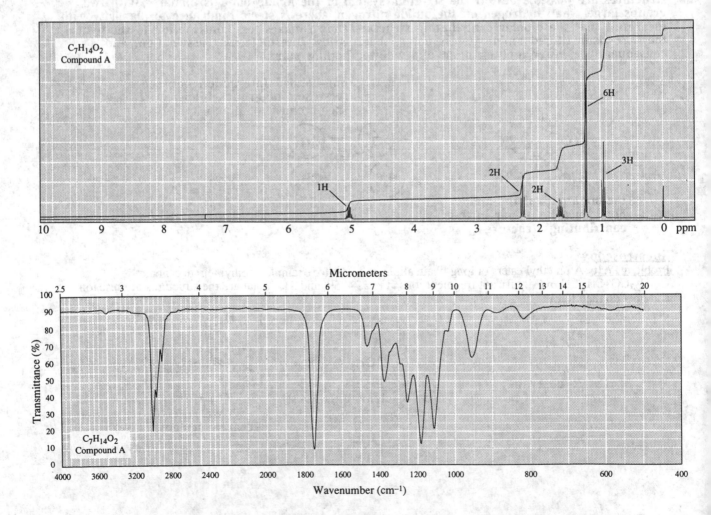

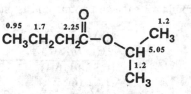

Isopropyl butanoate

<u>Problem 17.19</u> Propose a structural formula for compound B, $C_6H_{13}NO$, consistent with its ^{1}H-NMR and infrared spectra.

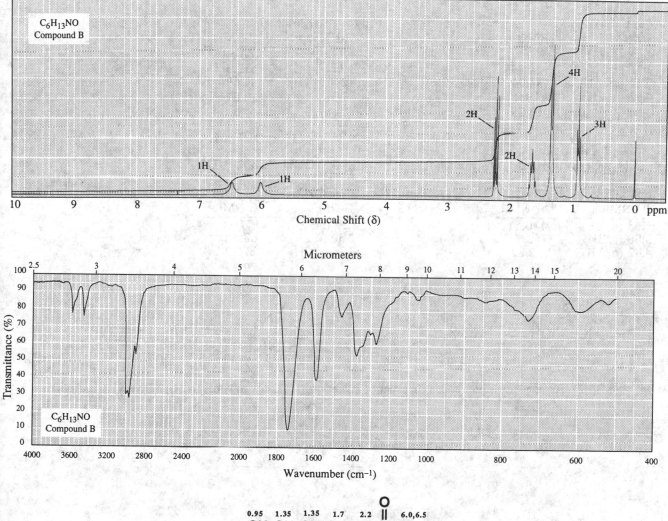

Hexanamide

<u>Problem 17.20</u> Propose a structural formula for each compound consistent with its ^{1}H-NMR and ^{13}C-NMR spectra.

(a) $C_5H_{10}O_2$

^{1}H-NMR	^{13}C-NMR
0.96 (d, 6H)	161.11
1.96 (m, 1H)	70.01
3.95 (d, 2H)	27.71
8.08 (s, 1H)	19.00

$$\overset{0.96}{}\overset{1.96\ 3.95}{}\overset{O}{\overset{\|}{C}}\overset{8.08}{}$$
$$(CH_3)_2CHCH_2COH$$

(b) $C_7H_{14}O_2$

^{1}H-NMR	^{13}C-NMR
0.92 (d, 6H)	171.15
1.52 (m, 2H)	63.12
1.70 (m, 1H)	37.31
2.09 (s, 3H)	25.05
4.10 (t, 2H)	22.45
	21.06

$$\overset{2.09}{}\overset{O}{\overset{\|}{}}\overset{4.10\ \ 1.52\ \ 1.70\ \ 0.92}{}$$
$$CH_3COCH_2CH_2CH(CH_3)_2$$

(c) $C_6H_{12}O_2$

^{1}H-NMR	^{13}C-NMR
1.18 (d, 6H)	177.16
1.26 (t, 3H)	60.17
2.51 (m, 1H)	34.04
4.13 (q, 2H)	19.0
	14.25

$$\overset{1.18}{}\overset{2.51}{}\overset{O}{\overset{\|}{}}\overset{4.13\ \ 1.26}{}$$
$$(CH_3)_2CHCOCH_2CH_3$$

(d) $C_7H_{12}O_4$

^{1}H-NMR	^{13}C-NMR
1.28 (t, 6H)	166.52
3.36 (s, 2H)	61.43
4.21 (q, 4H)	41.69
	14.07

$$\overset{1.28\ \ 4.21}{}\overset{O}{\overset{\|}{}}\overset{3.36}{}\overset{O}{\overset{\|}{}}\overset{4.21\ \ 1.28}{}$$
$$CH_3CH_2OCCH_2COCH_2CH_3$$

(e) $C_4H_7ClO_2$

^{1}H-NMR	^{13}C-NMR
1.68 (d, 3H)	170.51
3.80 (s, 3H)	52.92
4.42 (q, 1H)	52.32
	21.52

$$\overset{1.68\ \ 4.42}{}\overset{O}{\overset{\|}{}}\overset{3.80}{}$$
$$CH_3CHClCOCH_3$$

(f) $C_4H_6O_2$

^{1}H-NMR	^{13}C-NMR
2.29 (m, 2H)	177.81
2.50 (t, 2H)	68.58
4.36 (t, 2H)	27.79
	22.17

$$\begin{array}{c} O \\ \| \\ C \\ 2.50\quad\diagup\quad\diagdown \\ H_2C\qquad\quad O \\ |\qquad\quad\ | \\ H_2C-CH_2 \\ 2.29\qquad 4.36 \end{array}$$

<u>Problem 17.21</u> Arrange these compounds in order of decreasing reactivity toward nucleophilic acyl substitution.

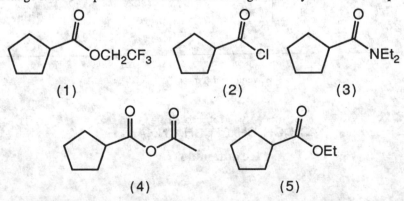

(1)   (2)   (3)

(4)   (5)

The order of reactivity of carboxylic acid derivatives is acid chlorides > anhydrides > esters > amides. In addition, the fluorine atoms of (1) inductively stabilize the negative charge of

2,2,2-triflouroethoxide making it a better leaving group than ethoxide, so (1) is more reactive than the other ester compound, (5).

The order of reactivity, listed from most to least reactive is:

| (2) > (4) > (1) > (5) > (3) |

Problem 17.22 Write the structural formula of the principal product formed when benzoyl chloride is treated with the following reagents:

(a)   Cyclohexanol

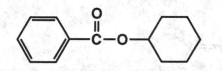

(b) $CH_3CH_2CH_2CH_2OH$, pyridine

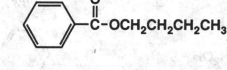

(c) $CH_3CH_2CH_2CH_2SH$, pyridine

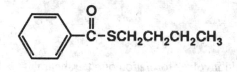

(d) $CH_3CH_2CH_2CH_2NH_2$ (two equivalents)

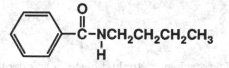

(e) $CH_3CH_2CH_2CO_2^- Na^+$

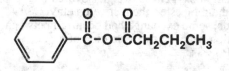

(f) $(CH_3)_2CuLi$, then H_3O^+

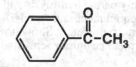

(g) CH_3O—⟨⟩—NH_2, pyridine

(h) C_6H_5MgBr (two equivalents)

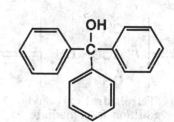

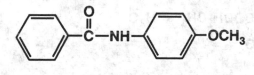

Problem 17.23 Write the structural formula of the principal product formed when ethyl benzoate is treated with the following reagents:

(a) H_2O, NaOH, heat

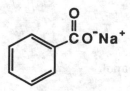

(b) H_2O, H_2SO_4, heat

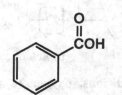

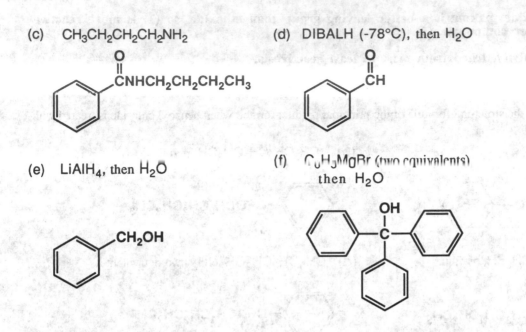

(c)   CH₃CH₂CH₂CH₂NH₂

(d)   DIBALH (-78°C), then H₂O

(e)   LiAlH₄, then H₂O

(f)   C₆H₅MgBr (two equivalents)
       then H₂O

Problem 17.24 The mechanism for hydrolysis of an ester in aqueous acid involves formation of a tetrahedral carbonyl addition intermediate. Evidence in support of this mechanism comes from an experiment designed by Myron Bender. He first prepared ethyl benzoate enriched with oxygen-18 in the carbonyl oxygen and then carried out acid-catalyzed hydrolysis of the ester in water containing no enrichment in oxygen-18. He discovered that if he stopped the experiment after only partial hydrolysis and isolated the remaining ester, the recovered ethyl benzoate had lost a portion of its enrichment in oxygen-18. In other words, some exchange had occurred between oxygen-18 of the ester and oxygen-16 of water. Show how this observation bears on the formation of a tetrahedral carbonyl addition intermediate during acid-catalyzed ester hydrolysis.

This observation can only be explained by proposing reversible formation of a tetrahedral carbonyl addition intermediate, and the assumption that exchange of oxygen-18 in this manner is more rapid than collapse of the intermediate to give benzoic acid and ethanol.

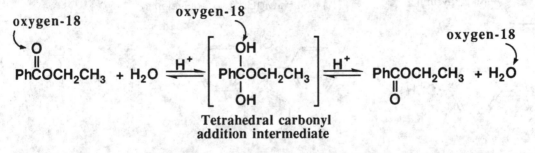

Tetrahedral carbonyl
addition intermediate

Problem 17.25 Predict the distribution of oxygen-18 in the products obtained from hydrolysis of ethyl benzoate labeled in the ethoxy oxygen under the following conditions:

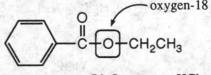

(a) In aqueous NaOH.   (b) In aqueous HCl.

Under each set of experimental conditions, oxygen-18 will appear in ethanol.

Problem 17.26 Write the structural formula of the principal product formed when benzamide is treated with the following reagents:

(a) H_2O, HCl, heat

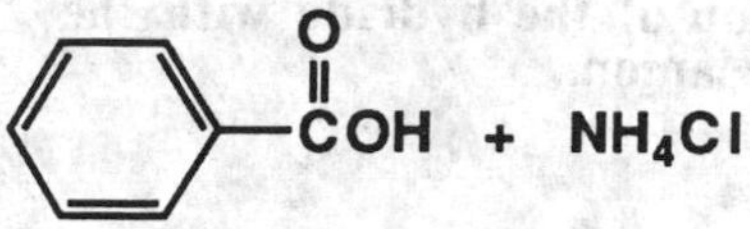

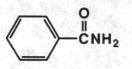

⬡—COH + NH_4Cl

(b) NaOH, H_2O, heat

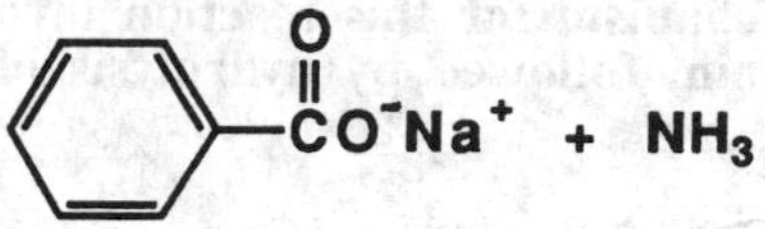

⬡—$\overset{O}{\overset{\|}{C}}$O⁻Na⁺ + NH_3

(c) $LiAlH_4$, then H_2O

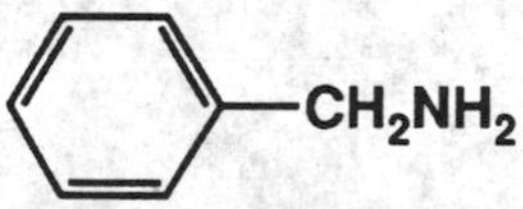

⬡—CH_2NH_2

(d) Br_2, NaOH, heat

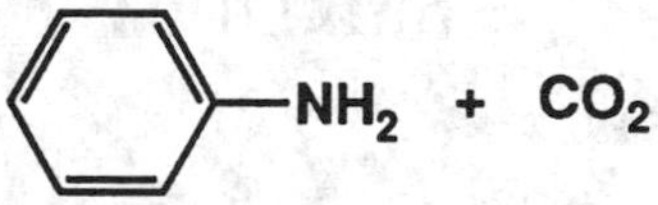

⬡—NH_2 + CO_2

Problem 17.27 Write the structural formula of the principal product formed when benzonitrile is treated with the following reagents.

(a) H_2O (one equivalent), H_2SO_4, heat

⬡—$\overset{O}{\overset{\|}{C}}NH_2$

(b) H_2O (excess), H_2SO_4, heat

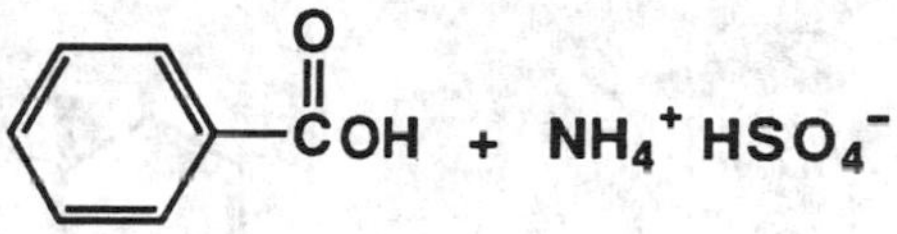

⬡—$\overset{O}{\overset{\|}{C}}OH$ + $NH_4^+ HSO_4^-$

(c) NaOH, H_2O, heat

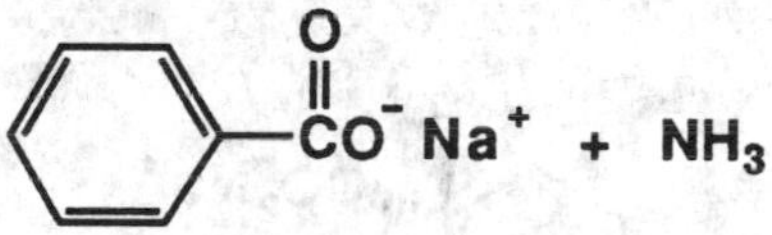

⬡—$\overset{O}{\overset{\|}{C}}O^-$ Na⁺ + NH_3

(d) $LiAlH_4$, then H_2O

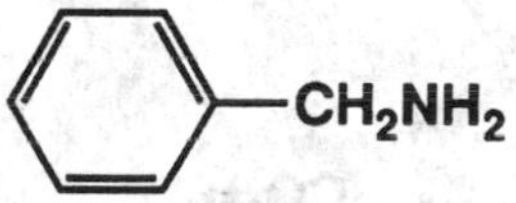

⬡—CH_2NH_2

Problem 17.28 Show the product expected when the following unsaturated δ-ketoester is treated with each reagent.

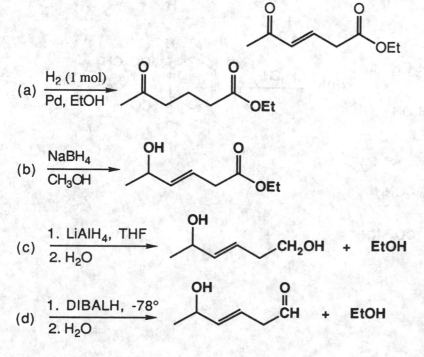

(a) $\dfrac{H_2\ (1\ mol)}{Pd,\ EtOH}$

(b) $\dfrac{NaBH_4}{CH_3OH}$

(c) $\dfrac{1.\ LiAlH_4,\ THF}{2.\ H_2O}$ + EtOH

(d) $\dfrac{1.\ DIBALH,\ -78°}{2.\ H_2O}$ + EtOH

Problem 17.29 The reagent diisobutylaluminum hydride (DIBALH) reduces esters to aldehydes. When nitriles are treated with DIBALH, followed by mild acid hydrolysis, the product is also an aldehyde. Give a mechanism for this reaction.

A reasonable mechanism for this reaction involves an initial reaction of the hydride with the nitrile carbon atom, followed by hydrolysis of the resulting imine anion.

Step 1:

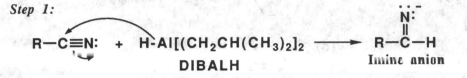

DIBALH
Imine anion

Step 2:

H3O+
Several steps

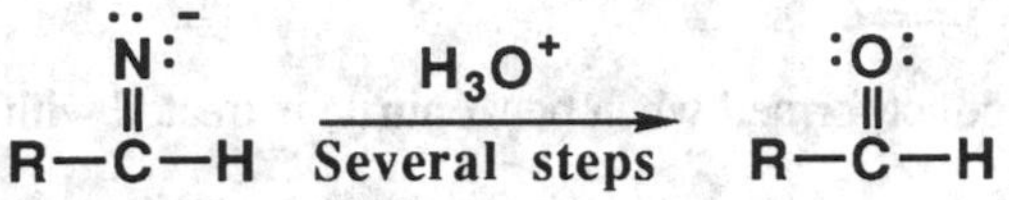

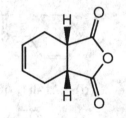

Problem 17.30 Show the product of treatment of this anhydride with each reagent.

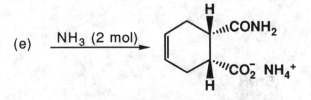

(a)

H2O, HCl

(b)

H2O, NaOH

(c)

1. LiAlH4
2. H2O

(d)

CH3OH

(e)

NH3 (2 mol)

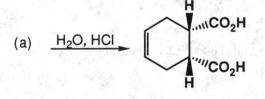

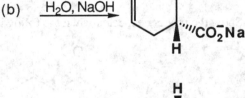

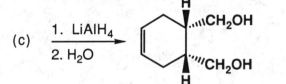

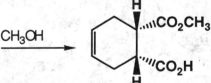

<u>Problem 17.31</u> The analgesic acetaminophen is synthesized by treating 4-aminophenol with one equivalent of acetic anhydride. Write an equation for the formation of acetaminophen. (*Hint:* An -NH$_2$ group is a better nucleophile than an -OH group.)

Note how in the following reaction scheme the acylation occurs at the more nucleophilic amino group rather than the less nucleophilic hydroxyl group of 4-aminophenol.

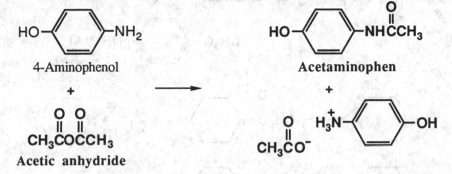

<u>Problem 17.32</u> Treatment of choline with acetic anhydride gives acetylcholine, a neurotransmitter. Write an equation for the formation of acetylcholine.

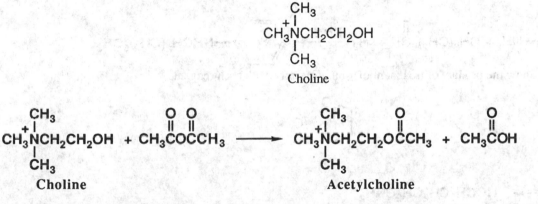

<u>Problem 17.33</u> Nicotinic acid, more commonly named niacin, is one of the B vitamins. Show how nicotinic acid can be converted to (a) ethyl nicotinate and then to (b) nicotinamide.

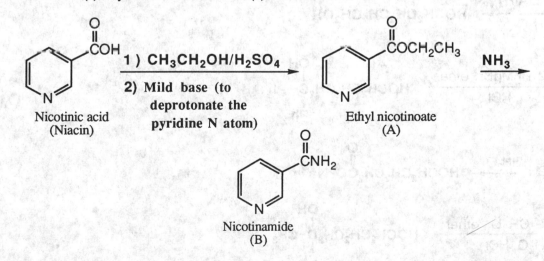

Problem 17.34 Complete these reactions.

(a) 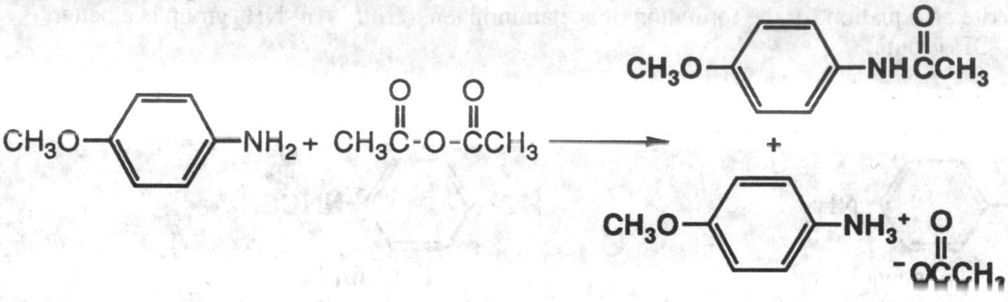 CH_3O—⟨benzene ring⟩—NH_2 + $CH_3\overset{O}{\underset{\|}{C}}$-O-$\overset{O}{\underset{\|}{C}}CH_3$ ⟶ CH_3O—⟨benzene ring⟩—$NH\overset{O}{\underset{\|}{C}}CH_3$

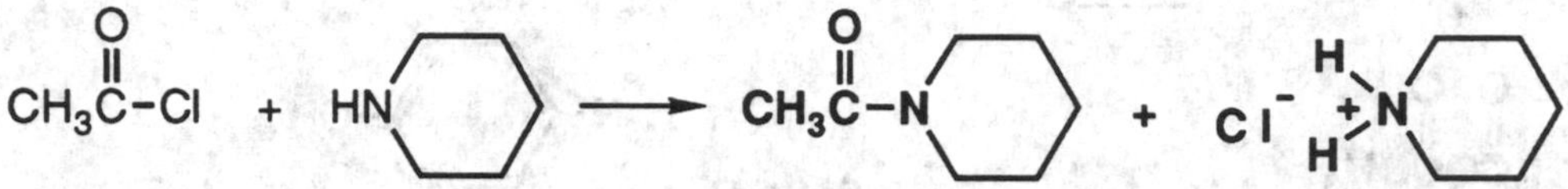

+

CH_3O—⟨benzene ring⟩—NH_3^+ $\overset{O}{\underset{\|}{^-OCCH_3}}$

(b) $CH_3\overset{O}{\underset{\|}{C}}$-Cl + HN⟨piperidine ring⟩ ⟶ $CH_3\overset{O}{\underset{\|}{C}}$-N⟨piperidine ring⟩ + Cl^- $\overset{H}{\underset{H}{\overset{+}{N}}}$⟨piperidine ring⟩

(c) $CH_3\overset{O}{\underset{\|}{C}}OCH_3$ + HN⟨piperidine ring⟩ ⟶ $CH_3\overset{O}{\underset{\|}{C}}$-N⟨piperidine ring⟩ + CH_3OH

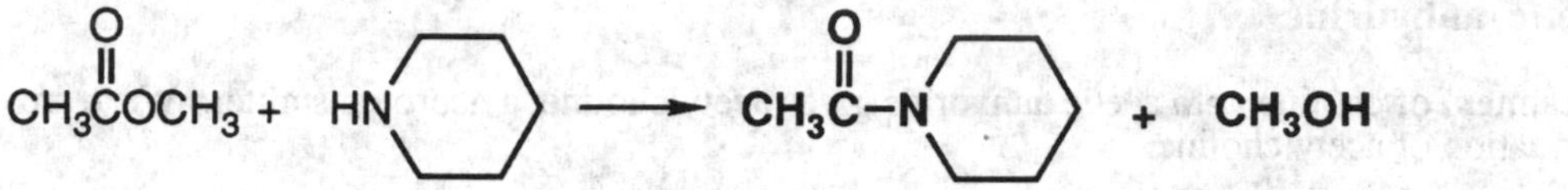

(d) 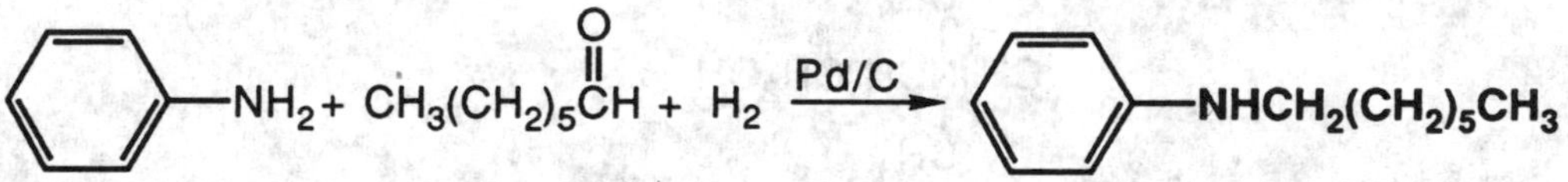 ⟨benzene ring⟩—NH_2 + $CH_3(CH_2)_5\overset{O}{\underset{\|}{C}}H$ + H_2 $\xrightarrow{Pd/C}$ ⟨benzene ring⟩—$NHCH_2(CH_2)_5CH_3$

Problem 17.35 Show the product of treatment of γ-butyrolactone with each reagent.

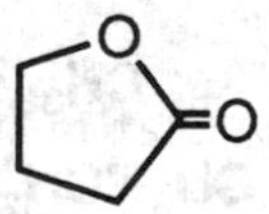

(a) NH_3 ⟶ $HOCH_2CH_2CH_2\overset{O}{\underset{\|}{C}}NH_2$

(b) $\dfrac{1.\ LiAlH_4}{2.\ H_2O}$ $HOCH_2CH_2CH_2CH_2OH$

(c) $\dfrac{1.\ 2\ PhMgBr,\ ether}{2.\ H_2O,\ HCl}$ $HOCH_2CH_2CH_2\overset{OH}{\underset{Ph}{\overset{|}{\underset{|}{C}}}}-Ph$

(d) NaOH $\xrightarrow{H_2O}$ $HOCH_2CH_2CH_2\overset{O}{\underset{\|}{C}}O^-\ Na^+$

(e) $\dfrac{1.\ 2\ CH_3Li,\ ether}{2.\ H_2O,\ HCl}$ $HOCH_2CH_2CH_2\overset{OH}{\underset{CH_3}{\overset{|}{\underset{|}{C}}}}-CH_3$

(f) $\dfrac{\text{1. DIBALH, ether, -78°C}}{\text{2. } H_2O, HCl}$ ⟶ $HOCH_2CH_2CH_2\overset{\overset{\displaystyle O}{\|}}{C}H$

Problem 17.36 Show the product of treatment of the following γ-lactam with each reagent.

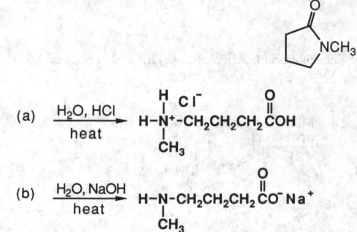

(a) $\dfrac{H_2O, HCl}{\text{heat}}$ ⟶ $\overset{\overset{\displaystyle H}{}}{\underset{\underset{\displaystyle CH_3}{}}{H-N^+}}\,Cl^-\,-CH_2CH_2CH_2\overset{\overset{\displaystyle O}{\|}}{C}OH$

(b) $\dfrac{H_2O, NaOH}{\text{heat}}$ ⟶ $\underset{\underset{\displaystyle CH_3}{}}{H-N}-CH_2CH_2CH_2\overset{\overset{\displaystyle O}{\|}}{C}O^-\,Na^+$

(c) $\dfrac{\text{1. } LiAlH_4}{\text{2. } H_2O}$ ⟶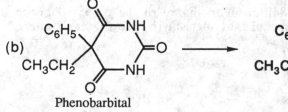

Problem 17.37 Draw structural formulas for the products of complete hydrolysis of meprobamate, phenobarbital, and pentobarbital in hot aqueous acid. Meprobamate is a tranquilizer prescribed under 58 different trade names, including Equanil and Miltown. Phenobarbital is a long-acting sedative, hypnotic, and anticonvulsant. Luminal is one of over a dozen names under which it is prescribed. Pentobarbital is a short-acting sedative, hypnotic, and anticonvulsant. Nembutal is one of several trade names under which it is prescribed. (*Hint:* Remember that when heated, β-dicarboxylic acids and β-ketoacids undergo decarboxylation.)

(a) $H_2N\overset{\overset{\displaystyle O}{\|}}{C}OCH_2\underset{\underset{\displaystyle CH_2CH_2CH_3}{\overset{\overset{\displaystyle CH_3}{|}}{|}}}{C}CH_2O\overset{\overset{\displaystyle O}{\|}}{C}NH_2$ ⟶ $NH_4^+ + CO_2 + HOCH_2\underset{\underset{\displaystyle CH_2CH_2CH_3}{\overset{\overset{\displaystyle CH_3}{|}}{|}}}{C}CH_2OH + NH_4^+ + CO_2$

Meprobamate

(b)

$$\begin{array}{c}\text{phenobarbital ring structure}\end{array}$$ ⟶ $\underset{\underset{\displaystyle CH_3CH_2}{}}{\overset{\overset{\displaystyle C_6H_5}{}}{}}CH\overset{\overset{\displaystyle O}{\|}}{C}OH + CO_2 + 2NH_4^+ + CO_2$

Phenobarbital

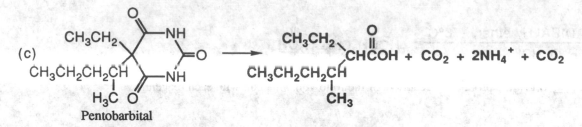

(c)

Pentobarbital

The Hofmann Rearrangement

Problem 17.38 Following are steps in a synthesis of anthranilic acid from phthalic anhydride. Describe how you could bring about each step.

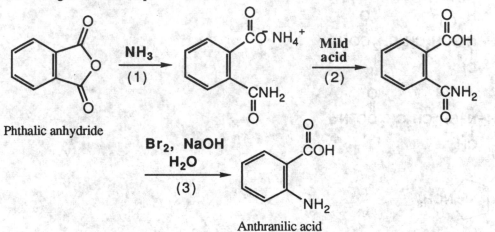

Phthalic anhydride

Anthranilic acid

Problem 17.39 Hofmann rearrangements of lower-molecular-weight primary amides can be brought about using bromine in aqueous NaOH. Primary amides larger than about seven or eight carbon atoms are not sufficiently soluble in aqueous solution to react. Instead, they are dissolved in methanol or ethanol, and the corresponding sodium alkoxide is used as the base. Under these conditions, the isocyanate intermediate reacts with the alcohol to form a carbamate.

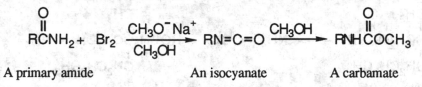

A primary amide   An isocyanate   A carbamate

Propose a mechanism for the reaction of an isocyanate with methanol to form a methyl carbamate.

A reasonable mechanism for this reaction involves addition of methanol to the carbonyl group to give an enol or a carbamic ester, followed by keto-enol tautomerism to give the carbamic ester product.

Step 1:

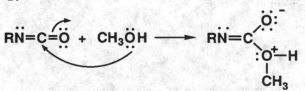

Step 2:

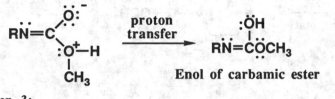

Enol of carbamic ester

Step 3:

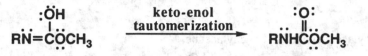

Enol of carbamic ester   **A carbamic ester**

Synthesis

Problem 17.40 *N,N*-Diethyl *m*-toluamide (DEET) is the active ingredient in several common insect repellents. DEET can be synthesized in two steps: (1) treatment of 3-methylbenzoic acid (*m*-toluic acid) with thionyl chloride to form an acid chloride followed by (2) treatment of the acid chloride with diethylamine. Write equations for each step in this synthesis of DEET.

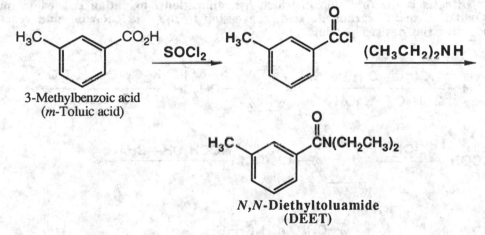

N,N-**Diethyltoluamide**
(DEET)

Problem 17.41 Following is the structural formula of isoniazid, a drug used to treat tuberculosis. It is estimated that one-third of the world's population is infected with tuberculosis, which results in approximately 3 million TB-related deaths per year. Show how to prepare isoniazid from pyridine 4 carboxylic acid. (See The Merck Index, 12th ed., # 5203.)

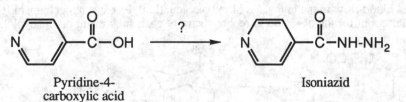

Pyridine-4-   Isoniazid
carboxylic acid

Isoniazid can be prepared by reaction of hydrazine (NH₂-NH₂) with a suitable carboxylic acid derivative, such as an ester.

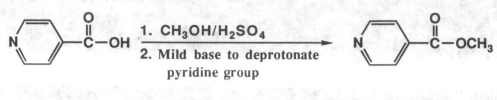

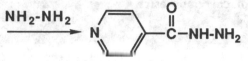

Problem 17.42 Show how to convert phenylacetylene to allyl phenylacetate.

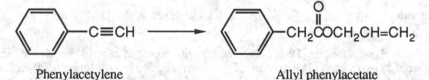

Phenylacetylene   Allyl phenylacetate

The key to this synthesis is the first step in which hydroboration / oxidation is used to make the aldehyde. Oxidation to the carboxylic acid, conversion to the acid chloride, and reaction with allyl alcohol give the desired product.

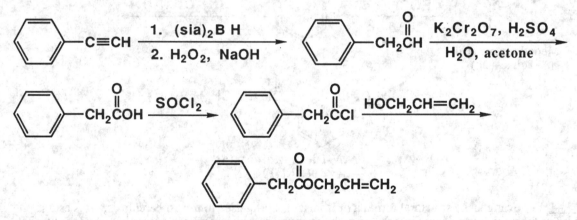

Problem 17.43 A step in a synthesis of PGE$_1$ (prostaglandin E$_1$, alprostadil) is reaction of a trisubstituted cyclohexene with bromine to form a bromolactone. Alprostadil is used as a temporary therapy for infants born with congenital heart defects that restrict pulmonary blood flow. It brings about dilation of the ductus arteriosus, which in turn increases blood flow in the lungs and blood oxygenation. Propose a mechanism for formation of this bromolactone and account for the observed stereochemistry of each substituent on the cyclohexane ring.

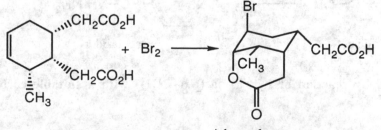

A bromolactone

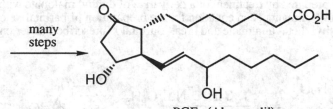

PGE₁ (Alprostadil)

The mechanism likely involves formation of a bridged bromonium ion that is attacked by the axial carboxylic acid group. It is the axial carboxylic acid group that is in the proper location to attack the bromonium ion as shown:

Step 1:

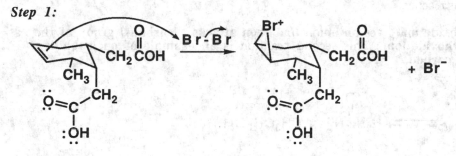

Step 2:

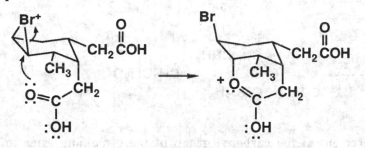

Step 3:

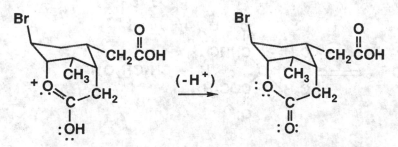

The stereochemistry is fixed by the bicyclic structure. The bromine atom is axial, while the -CH₂CO₂H and methyl group are equatorial.

<u>Problem 17.44</u> Barbiturates are prepared by treatment of a derivative of diethyl malonate with urea in the presence
of sodium ethoxide as a catalyst. Following is an equation for the preparation of barbital, a long-duration
hypnotic and sedative, from diethyl diethylmalonate and urea. Barbital is prescribed under one of a dozen or more
trade names.

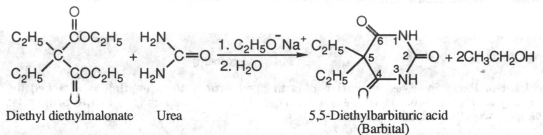

$$\text{Diethyl diethylmalonate} \qquad \text{Urea} \qquad\qquad\qquad \begin{array}{c}\text{5,5-Diethylbarbituric acid}\\ \text{(Barbital)}\end{array}$$

(a) Propose a mechanism for this reaction.

**Treatment of urea with ethoxide ion gives an anion that then attacks a carbonyl group of the
malonic ester to displace ethoxide ion. This second reaction is an example of nucleophilic
displacement at a carbonyl carbon.**

Step 1:

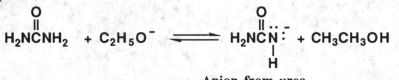

Anion from urea

Step 2:

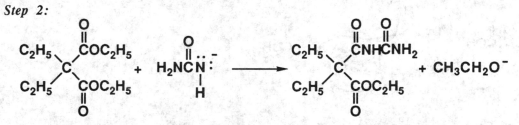

**This is followed by nucleophilic displacement at the carbonyl group of the remaining ester to
complete formation of the six-membered ring.**

Step 3:

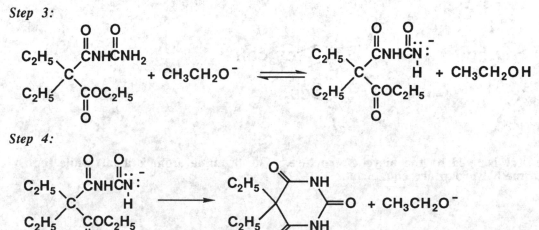

Step 4:

(b) The pK$_a$ of barbital is 7.4. Which is the most acidic hydrogen in this molecule and how do you account for its acidity?

The most acidic hydrogen is the imide hydrogen. Acidity results from the inductive effects of the adjacent carbonyl groups and stabilization of the deprotonated anion by resonance interaction with the carbonyl groups. Following are three contributing structures for the barbiturate anion.

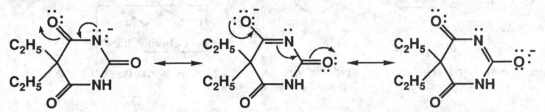

The two contributing structures that place the negative charge on the more electronegative oxygen atom make the greater contribution to the resonance hybrid.

Problem 17.45 The following compound is one of a group of β-chloroamines, many of which have antitumor activity. Describe a synthesis of this compound from anthranilic acid and ethylene oxide. [*Hint:* To see how the seven-membered ring might be formed, review the chemistry of the nitrogen mustards (Section 8.5)].

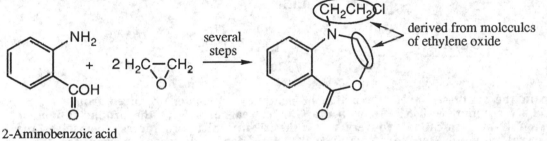

2-Aminobenzoic acid
  (Anthranilic acid)

Reaction of the nucleophilic amine group with two molecules of ethylene oxide gives the diol intermediate that is converted to the dichloride by treatment with HCl. The dichloride intermediate is actually a nitrogen mustard derivative that will react with the -OH group of the carboxyl function to give the desired lactone, presumably via the three-membered ring intermediate characteristic of nitrogen mustards (Section 8.5).

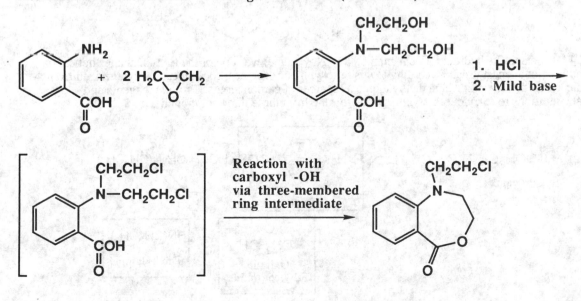

Problem 17.46 Following is a retrosynthetic scheme for the synthesis of 5-nonanone from 1-bromobutane as the only organic starting material. Show reagents and experimental conditions to bring about this synthesis.

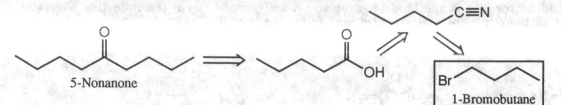

5-Nonanone   1-Bromobutane

1-Bromobutane is converted to pentanoic acid via the nitrile, which is hydrolyzed to the carboxylic acid.

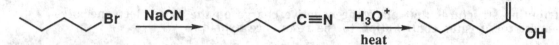

1-Bromobutane is also converted to a Gilman reagent via the organolithium species.

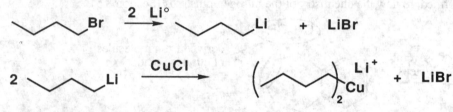

The key step in the synthesis involves reaction of pentanoyl chloride, produced from pentanoic acid by reaction with $SOCl_2$, with the Gilman reagent to give the product 5-nonone. This is an example of a so-called "convergent synthesis" in which two pieces, themselves the product of several synthetic steps, are combined in a late step in the reaction sequence.

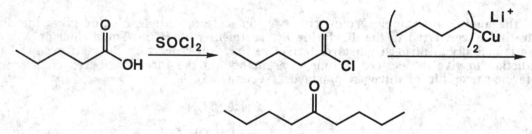

Problem 17.47 Procaine (its hydrochloride is marketed as Novocaine) was one of the first local anesthetics for infiltration and regional anesthesia. According to this retrosynthetic scheme, procaine can be synthesized from 4-aminobenzoic acid, ethylene oxide, and diethylamine as sources of carbon atoms. Provide reagents and experimental conditions to carry out the synthesis of procaine from these three compounds.

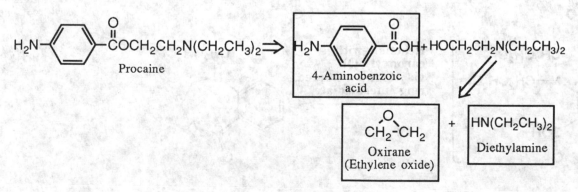

A reasonable synthetic scheme is shown below. First, the amine is reacted with ethylene oxide to create the corresponding β-aminoalcohol:

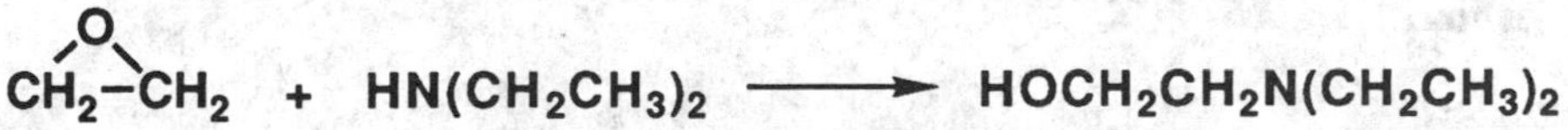

$$CH_2\text{--}CH_2 + HN(CH_2CH_3)_2 \longrightarrow HOCH_2CH_2N(CH_2CH_3)_2$$

Next, p-aminobenzoic acid is converted to procaine by Fischer esterification.

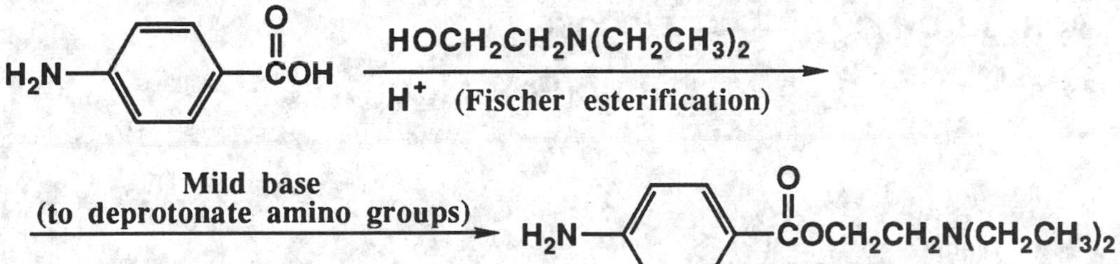

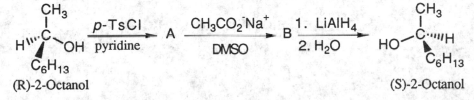

Procaine

Problem 17.48   The following sequence of steps converts (R)-2-octanol to (S)-2-octanol. Propose structural formulas for intermediates A and B, specify the configuration of each, and account for the inversion of configuration in this sequence.

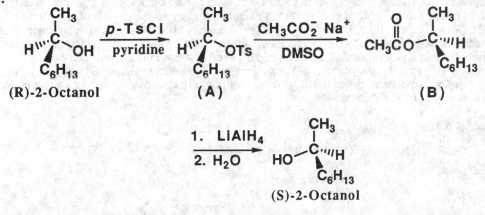

(R)-2-Octanol   (S)-2-Octanol

Formation of the tosylate ester (A) occurs with retention of configuration; reaction takes place on oxygen of the alcohol and does not involve the tetrahedral stereocenter. Nucleophilic substitution to form (B) occurs by an S_N2 pathway with inversion of configuration at the stereocenter undergoing reaction. Reduction of the ester with LAH occurs at the carbonyl carbon, not at the stereocenter. These relationships are shown in the following structural formulas.

(R)-2-Octanol   (A)   (B)

(S)-2-Octanol

Problem 17.49   Reaction of a primary or secondary amine with diethyl carbonate under controlled conditions gives a carbamic ester. Propose a mechanism for this reaction.

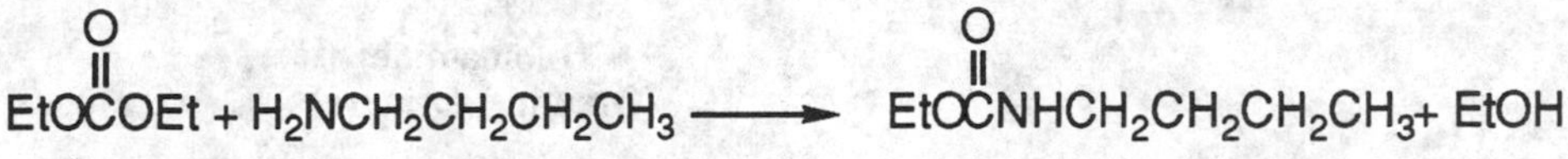

$$EtOCOEt + H_2NCH_2CH_2CH_2CH_3 \longrightarrow EtOCNHCH_2CH_2CH_2CH_3 + EtOH$$

Diethyl carbonate   Butylamine   Ethyl N-butylcarbamate

A reasonable mechanism for this reaction involves initial attack by the nucleophilic nitrogen atom of butylamine on the carbonyl carbon atom of diethyl carbonate to create a tetrahedral carbonyl addition intermediate that leads to the product carbamate after a proton transfer then elimination of ethoxide.

Step 1:

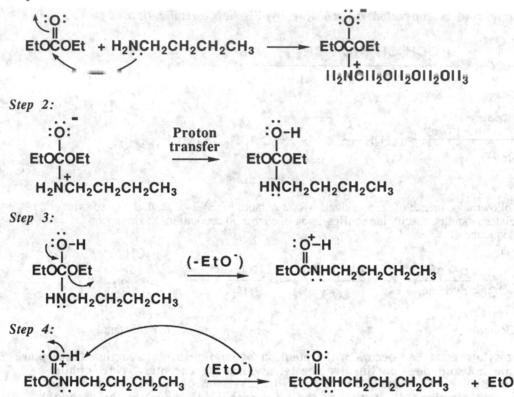

Step 2:

Step 3:

Step 4:

Problem 17.50 Several sulfonylureas, a class of compounds containing $RSO_2NHCONHR$, are useful drugs as orally active replacements for injected insulin in patients with adult-onset diabetes. It was discovered in 1942 that certain members of this class cause hypoglycemia in laboratory animals. Clinical trials of tolbutamide were begun in the early 1950s and since that time, more than 20 sulfonylureas have been introduced into clinical medicine. The sulfonylureas decrease blood glucose concentrations by stimulating β cells of the pancreas to release insulin and by increasing the sensitivity of insulin receptors in peripheral tissues to insulin stimulation.

Tolbutamide is synthesized by the reaction of the sodium salt of *p*-toluenesulfonamide and ethyl *N*-butylcarbamate (see the previous problem for the synthesis of this carbamic ester). Propose a mechanism for the following step in the synthesis of tolbutamide:

Sodium salt of
p-toluenesulfonamide

A carbamic ester

Tolbutamide
(Oramide, Orinase)

A reasonable mechanism for this reaction involves initial attack by the nucleophilic nitrogen atom of the sulfonamide sodium salt on the carbonyl carbon atom of the carbamic ester to create a tetrahedral carbonyl addition intermediate that eliminates ethoxide to give the final product.

Step 1:

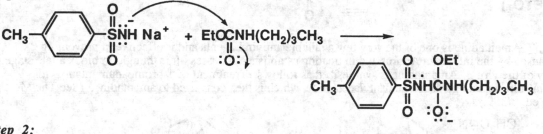

Step 2:

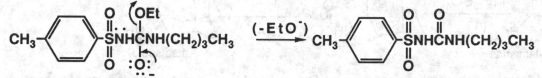

<u>Problem 17.51</u> Following are structural formulas for two more widely used sulfonylurea hypoglycemic agents. Show how each might be synthesized by converting an appropriate amine to a carbamic ester and then treating the carbamate with the sodium salt of a substituted benzenesulfonamide.

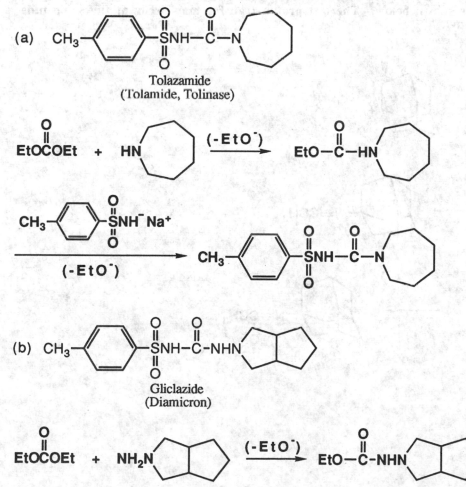

(a) Tolazamide (Tolamide, Tolinase)

(b) Gliclazide (Diamicron)

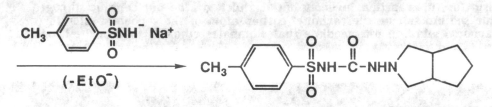

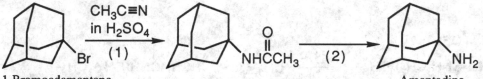

<u>Problem 17.52</u> Amantadine is one of the very few available antiviral agents and is effective in preventing infections caused by the influenza A virus and in treating established illnesses. It is thought to block a late stage in the assembly of the virus. Amantadine is synthesized as follows. Treatment of 1-bromoadamantane with acetonitrile in sulfuric acid gives *N*-adamantylacetamide, which is then converted to amantadine. (See The Merck Index, 12th ed., #389.)

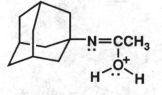

1-Bromoadamantane

Amantadine

(a) Propose a mechanism for the transformation in Step 1. Consider the possibility of forming an adamantyl cation under these experimental conditions and then the manner in which this carbocation might undergo reaction with acetonitrile.

A reasonable mechanism is shown below. The last proton transfer may occur at the same time as formation of the carbonyl.

Step 1:

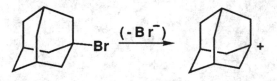

Step 2:

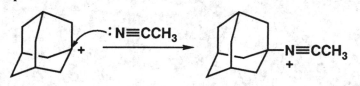

Step 3:

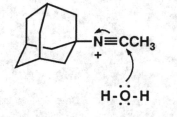

Step 4:

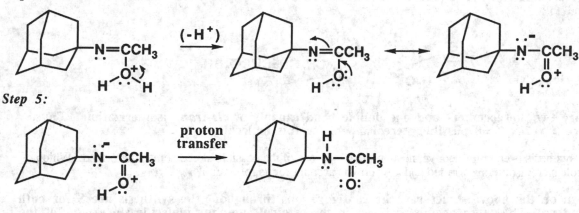

Step 5:

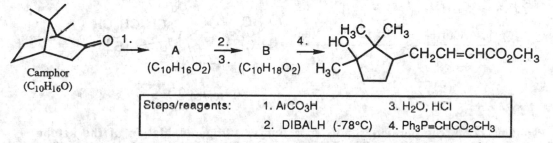

(b) Describe experimental conditions to bring about Step 2.

Step 2 is a simple hydrolysis carried with HCl and H_2O.

<u>Problem 17.53</u> The following four-step sequence converts natural camphor, an enantiomerically pure compound, to the product shown on the right. The configuration of camphor is shown, but the configuration of the product is not shown.

Camphor
($C_{10}H_{16}O$)

A
($C_{10}H_{16}O_2$)

B
($C_{10}H_{18}O_2$)

$CH_2CH=CHCO_2CH_3$

Steps/reagents:	1. AlCO₃H	3. H_2O, HCl
	2. DIBALH (-78°C)	4. $Ph_3P=CHCO_2CH_3$

(a) Propose structural formulas for intermediates A and B. Also specify the configuration at all stereocenters in A and B.

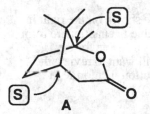

A
($C_{10}H_{16}O_2$)

This is an example of a Baeyer-Villiger oxidation. Note how the tertiary carbon atom ends up adjacent to the ester oxygen atom.

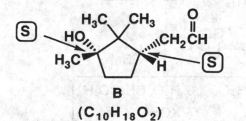

B
($C_{10}H_{18}O_2$)

The ester is reduced to an aldehyde with DIBALH and an acidic workup.

(b) Label all stereocenters in the product of this sequence and tell how many stereoisomers are possible for this compound?

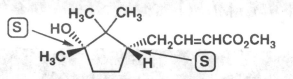

There are two stereocenters and one double bond capable of *cis-trans* isomerization. Thus, there are $2 \times 2 \times 2 = 8$ possible stereoisomers for this molecule.

(c) This synthesis is enantioselective, meaning that only one of the possible stereoisomers is formed. Which of the possible stereoisomers is formed and how do you account for its formation?

As shown on the above structures, the configurations throughout the synthesis are S for both stereocenters. This stereochemistry is set in the original camphor molecule since none of the four reactions involves a change or loss of stereochemistry at these stereocenters.

Problem 17.54 When natural camphor is treated first with peroxybenzoic acid and then with lithium aluminum hydride in diethyl ether, there is formed an optically active compound of molecular formula $C_{10}H_{20}O_2$. Propose a structural formula for this compound and show configurations at all of its stereocenters.

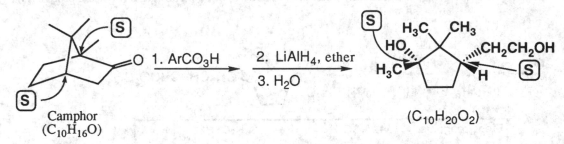

As shown in Problem 17.54, the peracid results in a Baeyer-Villiger oxidation of the ketone function to give a lactone intermediate. The LiAlH₄ simultaneously cleaves and reduces the lactone to the diol shown. The stereochemistry of the two stereocenters in the product is set in the original camphor molecule so that both have the S configuration.

Problem 17.55 In a series of seven steps, (S)-malic acid is converted to the bromoepoxide shown on the right in 50% overall yield. (S)-Malic acid occurs in apples and many other fruits. It is also available in enantiopure form from microbiological fermentation. This synthesis is enantioselective: of the stereoisomers possible for the bromoepoxide, only one is formed. In thinking about the chemistry of these steps, you will want to review the use of dihydropyran as an -OH protecting group (Section 11.7) and the use the p-toluenesulfonyl chloride to convert the -OH, a poor leaving group, into a tosylate, a good leaving group (Section 9.6D).

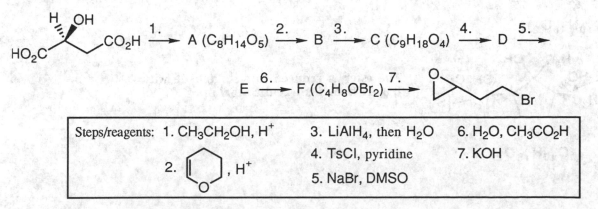

Steps/reagents: 1. CH₃CH₂OH, H⁺   3. LiAlH₄, then H₂O   6. H₂O, CH₃CO₂H
                 2. [dihydropyran], H⁺   4. TsCl, pyridine   7. KOH
                                            5. NaBr, DMSO

(a) Propose structural formulas for intermediates A through F. Also specify configuration at each stereocenter.

Transformation 2. deserves special comment. This reaction produces a tetrahydropyranyl ether (intermediate B) that serves as an -OH protecting group. Note how a new stereocenter is created as shown, but that it will be a mixture of stereoisomers (R,S). This center has no influence over the stereochemistry of the product since the tetrahydropyranyl ether is cleaved in acid during the second to last step of the sequence.

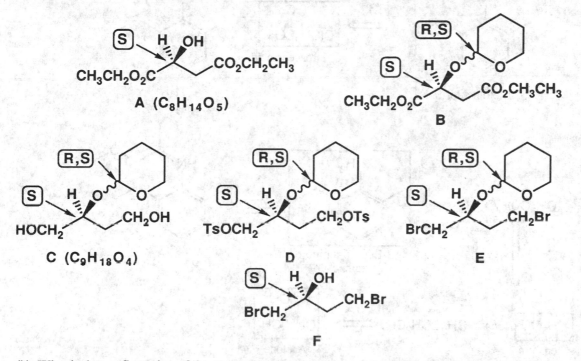

(b) What is the configuration of the stereocenter in the bromoepoxide and how do you account for the stereoselectivity of this seven-step conversion?

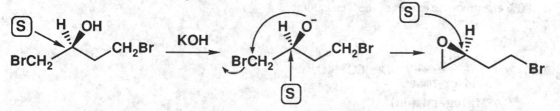

As shown on the above structures, the configuration at the 2 position during the synthesis is S. This stereochemistry is set in the starting material and maintained throughout the synthetic sequence, since none of the reactions involves a change or loss of stereochemistry at this position.

Problem 17.56 Following is a retrosynthetic analysis for the synthesis of the herbicide (S)-Metolachlor. Ciba synthesizes approximately 10,000 tons of this agrochemical per year. According to this analysis, starting materials are 2-ethyl-6-methylaniline, chloroacetic acid, acetone, and methanol.

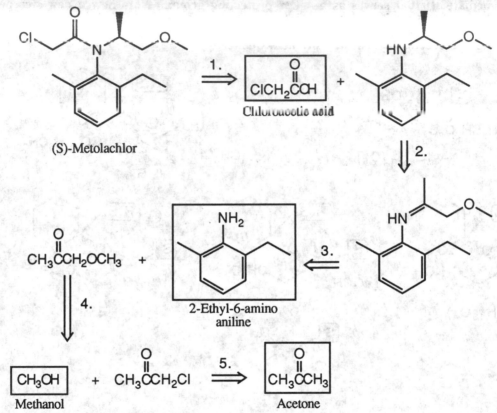

Show reagents and experimental conditions for the conversion of these three organic starting materials to Metolachlor. Note that your synthesis will most likely give a racemic mixture. The chiral catalyst used by Ciba for Step 2 gives 80% enantiomeric excess of the S enantiomer.

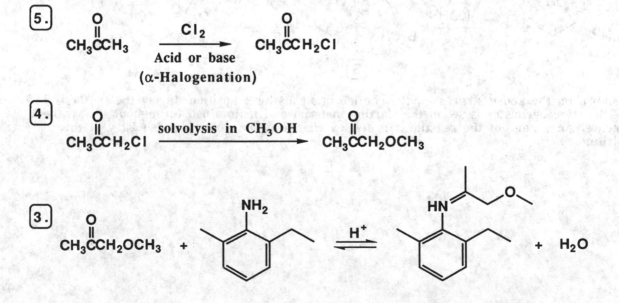

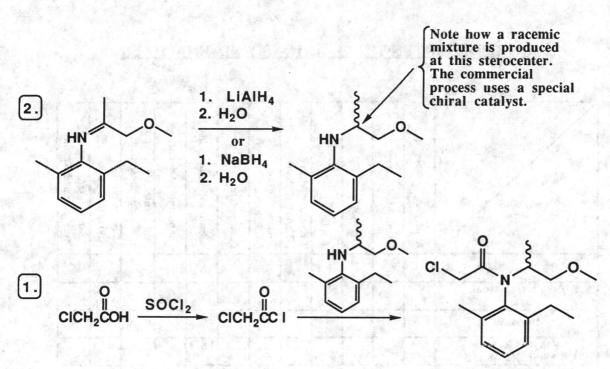

Note how a racemic mixture is produced at this sterocenter. The commercial process uses a special chiral catalyst.

(racemic)-Metolachlor

CHAPTER 18: ENOLATE ANIONS AND ENAMINES

SUMMARY OF REACTIONS

Starting Material ↓ \ Product →	Aldehydes, Ketones	Cyclic β-Ketoesters	Enamines	Enolates	β-Hydroxyaldehydes β-Hydroxyketones	β-Ketoaldehydes β-Diketones	β-Ketoesters	Substituted Acetic Acids	Substituted Acetones	Substituted Carbonyl Compounds, etc.	Substituted Cyclohexenones	α,β-Unsaturated Aldehydes and Ketones
Aldehydes, Ketones					18A 18.1*							
Aldehydes, Ketones, Esters				18B 18.1								
Diesters		18C 18.3B										
Diethyl Malonate Alkyl Halides								18D 18.7				
Enamines Acyl Halides						18E 18.5B						
Enamines Alkyl Halides	18F 18.5A											
Enolates Aldehydes, Ketones				18G 18.2B								
Enolates Conjugated Carbonyls											18H 18.8	18I 18.8A
Esters							18J 18.3A,C					
Ethyl Acetoacetate Alkyl Halides									18K 18.6			
Gilman Reagents Conjugated Carbonyls										18L 18.8B		
β-Hydroxyaldehydes β-Hydroxyketones												18M 18.1
Secondary Amines Aldehydes, Ketones			18N 18.5									

*Section in book that describes reaction.

REACTION 18A: THE ALDOL REACTION (Section 18.1)

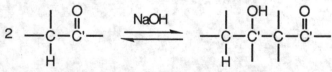

- In the aldol reaction, two aldehyde and/or ketone molecules react to form a β-hydroxyaldehyde or β-hydroxyketone. ✳
- The base-catalyzed mechanism involves the nucleophilic attack by an enolate anion of one aldehyde or ketone on the electrophilic carbonyl carbon atom of another aldehyde or ketone to produce a tetrahedral carbonyl addition intermediate that is protonated to yield the final product.
- The acid-catalyzed mechanism involves reaction between the enol form of one aldehyde or ketone and the highly electrophilic protonated form of another aldehyde or ketone.

- This is an important reaction because it represents a relatively easy way to make a carbon-carbon bond, and thus create a larger, more complex molecule from simpler ones. ✳
- Usually, the two molecules of the same aldehyde or ketone are used to produce the aldol product.
- If attempts are made to react two different aldehydes or ketones, a mixture of aldol products results from the various combinations of enolates and carbonyls that can react. However, these mixed aldol reactions are more successful if one of the aldehydes or ketones cannot form an enolate (such as formaldehyde), thus limiting the possible reactions and ensuring that fewer products will be produced.
- The equilibrium constant for the aldol product formation is usually not large. However, product yields can be improved by allowing the initially formed β-hydroxyaldehyde or ketone to eliminate water, thereby generating the more stable α,β-unsaturated aldehyde or ketone (Reaction 18M, Section 18.1).

REACTION 18B: FORMATION OF ENOLATE ANIONS (Section 18.1)

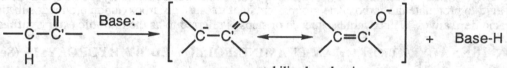

- The **α-hydrogen** of carbonyl **compounds is weakly acidic**, so carbonyl compounds can be converted to resonance-stabilized **enolate anions** by reaction with strong **base**. ✳
- Enolate anions are important because they are intermediates in reactions such as the aldol and Claisen condensations, and they also take part as nucleophiles in S_N2 reactions.
- For ketones with non-equivalent α-hydrogens, different enolates can be formed by reaction with strong bases such as lithium diisopropylamide (LDA). The predominant product formed depends on the conditions.

    Under conditions of **thermodynamic control**, for example when the ketone is in slight excess over the base, the ketone and enolate are in equilibrium so that the more stable (more highly substituted) enolate predominates.

    Under conditions of **kinetic control**, for example when there is a slight excess of base, the enolate is formed quantitatively and there is no equilibrium established. Therefore, the enolate that forms faster predominates. This is usually the less substituted enolate since in this case there is less steric hindrance around the α-hydrogen that must react with the base to create the enolate.

REACTION 18C: THE DIECKMANN CONDENSATION (Section 18.3B)

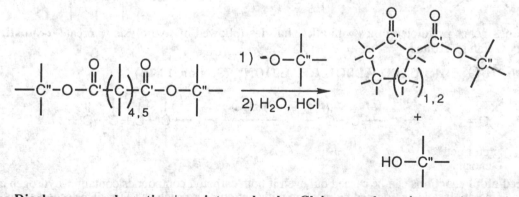

- The **Dieckmann condensation** is an **intramolecular Claisen condensation** reaction carried out on an appropriate diester molecule to produce a **five- or six-membered ring**. ✳
- The reaction mechanism involves initial deprotonation by the base to form an anion on the α-carbon of one ester, that then attacks the other carbonyl carbon atom to produce a tetrahedral carbonyl addition intermediate. This collapses when the alkoxide departs. The resulting β-keto ester is deprotonated under these basic conditions, but then protonated by the acid work-up to produce the final product.

REACTION 18D: MALONIC ESTER SYNTHESIS (Section 18.7)

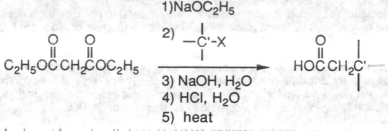

- **Esters of malonic acid** react with base to create **enolate anions** that can react with electrophiles such as **alkyl halides**. Treatment of the resulting compound with **aqueous hydroxide** to hydrolyze the esters, **acid** to protonate the carboxylate groups, and **heat** to cause decarboxylation produces substituted acetic acid derivatives. This overall sequence of steps is referred to as the **malonic ester synthesis**. ✳

REACTION 18E: ACYLATION OF AN ENAMINE FOLLOWED BY HYDROLYSIS (Section 18.5B)

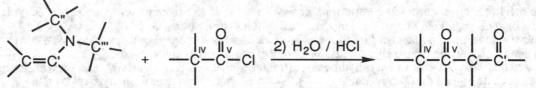

- **Enamines react as nucleophiles** with **acyl halides** followed by hydrolysis to create β-**ketoaldehydes** and β-**diketones**. ✳

REACTION 18F: ALKYLATION OF AN ENAMINE FOLLOWED BY HYDROLYSIS (Section 18.5A)

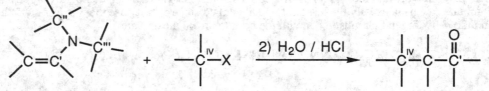

- **Enamines react as nucleophiles** with **alkyl halides** followed by hydrolysis to create α-**substituted aldehydes** and **ketones**. ✳

REACTION 18G: DIRECTED ALDOL REACTIONS (Section 18.2B)

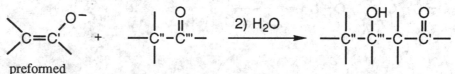

preformed
enolate

- **Directed aldol reactions** can be carried out even if both carbonyl compounds contain α-hydrogen atoms if the desired **enolate is preformed** by reaction with a strong base such as LDA. This scheme can prevent the many unwanted products that would have formed by simply adding a base such as NaOH to a mixture of the two carbonyl compounds. ✳

REACTION 18H: THE MICHAEL REACTION (Section 18.8)

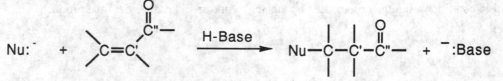

- **Nucleophiles** such as enolate anions can react via **conjugate addition** with **electrophilic alkenes** in a reaction referred to as a **Michael reaction**. If the alkene is conjugated to a carbonyl group, then the product of the reaction is a substituted carbonyl compound. The Michael reaction can also take place with α,β-unsubstituted nitriles and nitro compounds. ✳
- In the Michael reaction with an alkene conjugated to a carbonyl species, the reaction mechanism involves an attack of the nucleophile onto the electrophilic alkene to produce a resonance-stabilized enolate anion. Protonation of this enolate anion creates the enol, that rearranges to the more stable keto product.

REACTION 18I: THE ROBINSON ANNULATION (Section 18.8A)

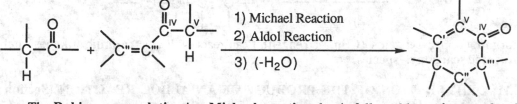

1) Michael Reaction
2) Aldol Reaction
3) (-H₂O)

- The **Robinson annulation** is a **Michael reaction** that is followed by an **intramolecular aldol reaction**. **Dehydration** of the aldol product leads to **substituted cyclohexenones**. ✳

REACTION 18J: THE CLAISEN CONDENSATION (Section 18.3 A,C)

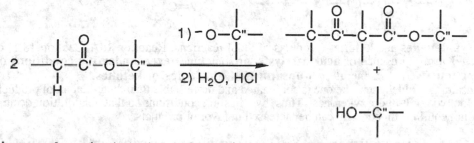

1) ⁻O—C″—
2) H₂O, HCl

- A **Claisen condensation** is carried out by treating an **ester** with **base**, to produce a β-ketoester. ✳
- The reaction mechanism involves initial deprotonation by the base to form an anion on the α-carbon of one ester molecule, that then attacks the carbonyl carbon atom of another ester molecule to produce a tetrahedral carbonyl addition intermediate. This collapses when the alkoxide departs. The resulting β-ketoester is deprotonated under these basic conditions, but then protonated by the acid work-up to produce the final product.
- Usually, the reaction takes place between two molecules of the same ester. However, crossed Claisen condensations are possible if there is an appreciable difference in the reactivity between the two esters. For example, crossed Claisen condensations can be carried out if one of the esters cannot form an enolate because it does not have any hydrogens on the α-carbon atom.
- The Claisen condensation is important because it allows for the synthesis of a carbon-carbon bond under relatively mild conditions.

REACTION 18K: THE ACETOACETIC ESTER SYNTHESIS (Section 18.6)

1) NaOC₂H₅

2) —C′-X

CH₃CCH₂COC₂H₅ ⟶ CH₃CCH₂C′—

3) NaOH, H₂O
4) HCl, H₂O
5) heat

- **Esters of acetoacetic acid** react with **base to create enolate anions** that can **react with electrophiles** such as **alkyl halides**. Treatment of the resulting compound with **aqueous hydroxide** to hydrolyze the ester, **acid** to protonate the carboxylate groups, and **heat** to cause decarboxylation produces substituted methyl ketone derivatives. This overall sequence of steps is referred to as the **acetoacetic ester synthesis**. ✳

REACTION 18L: ADDITION OF GILMAN REAGENTS (Section 18.8B)

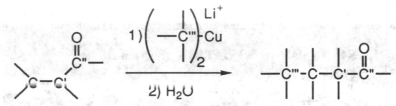

- **Lithium diorganocopper reagents (Gilman reagents)** can react with **electrophilic alkenes such as** α,β-**unsaturated carbonyl species**. ✳

REACTION 18M: DEHYDRATION OF THE PRODUCT OF AN ALDOL REACTION (Section 18.1)

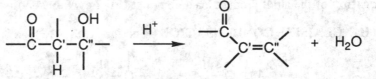

- β-**Hydroxyaldehydes** and **ketones** (products of aldol reactions; Reaction 18A, Section 18.1) can be **dehydrated** in the presence of **acid catalysts**, or sometimes **even under the conditions of the aldol reaction itself**, to generate α,β-**unsaturated aldehydes** or **ketones**. ✳
- The α,β-unsaturated aldehyde or ketone is conjugated and more stable than the initial aldol products; the β-hydroxyaldehyde or β-hydroxyketones. Thus, by adjusting conditions so that dehydration occurs the usually small equilibrium constant can be increased in favor of products.

REACTION 18N: PREPARATION OF ENAMINES FROM SECONDARY AMINES AND ALDEHYDES OR KETONES (Section 18.3)

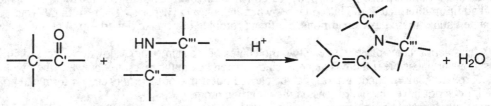

- **Enamines** are prepared by the reaction of **secondary amines** with **aldehydes** or **ketones**. It is removal of water that drives the reaction to completion. ✳
- Enamines are important because they can behave as enolate equivalents in reactions with strong electrophiles such as alkyl halides and acyl chlorides (reactions 18C and 18D).
- Morpholine and pyrrolidine are two commonly used secondary amines for the formation of enamines.

CHAPTER 18
Solutions to the Problems

Problem 18.1 Draw the product of the base-catalyzed aldol reaction of these compounds.
(a) Phenylacetaldehyde, $C_6H_5CH_2CHO$

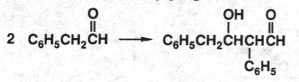

$$2 \ C_6H_5CH_2\overset{\overset{\displaystyle O}{\|}}{C}H \longrightarrow C_6H_5CH_2\underset{\underset{\displaystyle C_6H_5}{|}}{C}H\overset{OH}{C}H\overset{\overset{\displaystyle O}{\|}}{C}H$$

(b) Cyclopentanone

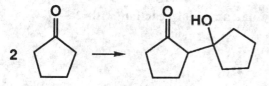

Problem 18.2 Draw the product of dehydration of each aldol product in Problem 18.1.

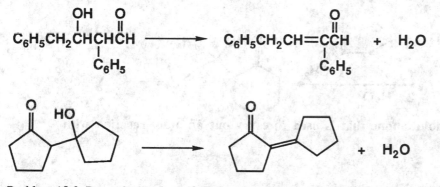

$$C_6H_5CH_2\underset{\underset{\displaystyle C_6H_5}{|}}{C}H\overset{OH}{C}H\overset{\overset{\displaystyle O}{\|}}{C}H \longrightarrow C_6H_5CH_2CH=\underset{\underset{\displaystyle C_6H_5}{|}}{C}\overset{\overset{\displaystyle O}{\|}}{C}H \ + \ H_2O$$

Problem 18.3 Draw the product of the base-catalyzed crossed aldol reaction between benzaldehyde and 3-pentanone and the product formed by its dehydration.

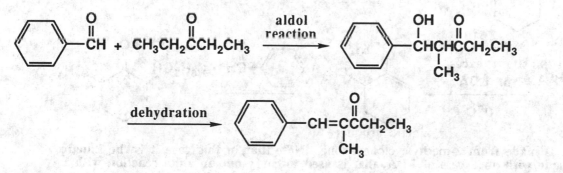

Problem 18.4 Show how you might prepare these compounds by directed aldol reactions.

(a)

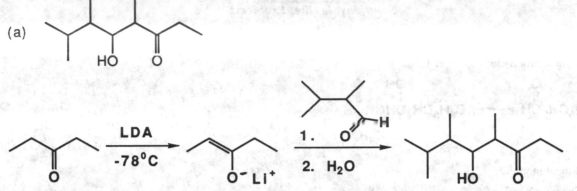

An enolate is made from 3-pentanone, and used to carry out an aldol reaction with 2,3-dimethylbutanal.

(b)

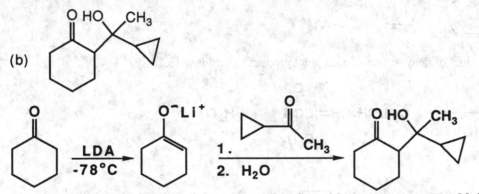

An enolate is made from cyclohexanone that is used to carry out an aldol reaction with cyclopropyl methyl ketone.

(c)

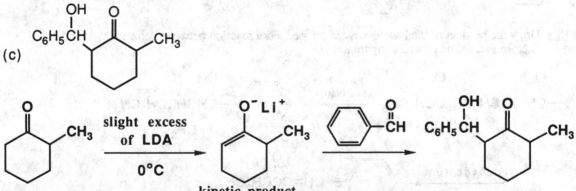

An enolate is made from 2-methylcyclohexanone. Note that, in this case, it is the kinetic product, made with an excess of LDA, that is used to carry out an aldol reaction with benzaldehyde.

Problem 18.5 Show the product of Claisen condensation of ethyl 3-methylbutanoate in the presence of sodium ethoxide followed by acidification with aqueous HCl.

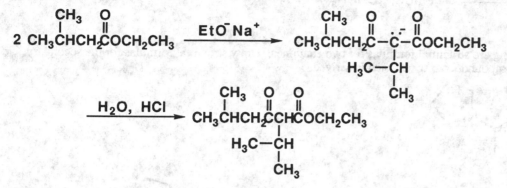

Problem 18.6 Complete the equation for this crossed Claisen condensation:

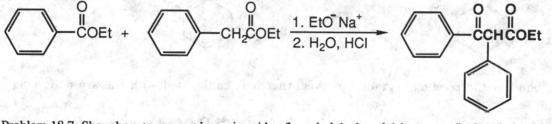

Problem 18.7 Show how to convert benzoic acid to 3-methyl-1-phenyl-1-butanone (isobutyl phenyl ketone) by the following synthetic strategies, each of which uses a different type of reaction to form the new carbon-carbon bond to the carbonyl group of benzoic acid.

$$
\underset{\text{Benzoic acid}}{\overset{\displaystyle O}{\underset{\displaystyle \|}{Ph\overset{}{C}OH}}} \quad \xrightarrow{\quad ? \quad} \quad \underset{\text{3-Methyl-1-phenyl-1-butanone}}{\overset{\displaystyle O \quad CH_3}{Ph\overset{}{C}CH_2\overset{}{C}HCH_3}}
$$

(a) A lithium diorganocopper (Gilman) reagent

Treatment of benzoic acid with thionyl chloride gives benzoyl chloride. Treatment of benzoyl chloride with lithium diisobutylcopper followed by hydrolysis in aqueous acid gives the desired product.

$$
\underset{}{\overset{O}{\underset{\|}{PhCOH}}} \xrightarrow{SOCl_2} \underset{}{\overset{O}{\underset{\|}{PhCCl}}} \xrightarrow[\text{2) } H_2O, \ HCl]{\text{1) } [(CH_3)_2CHCH_2]_2CuLi} \underset{}{\overset{O \quad CH_3}{PhCCH_2\overset{}{C}HCH_3}}
$$

(b) A Claisen condensation

Crossed Claisen condensation of ethyl benzoate and ethyl 3-methylbutanoate gives a β-ketoester. Saponification of the ester followed by acidification gives the β-ketoacid. Heating causes decarboxylation and gives the desired product.

$$
\underset{}{\overset{O}{\underset{\|}{PhCOC_2H_5}}} + \underset{\underset{CO_2C_2H_5}{|}}{\overset{CH_3}{CH_2\overset{}{C}HCH_3}} \xrightarrow[\text{2) } H_2O, \ HCl]{\text{1) } C_2H_5O^-Na^+} \underset{\underset{CO_2C_2H_5}{|}}{\overset{O \quad CH_3}{Ph\overset{}{C}CH\overset{}{C}HCH_3}} \xrightarrow[\text{2) } H_2O, \ HCl]{\text{1) } NaOH, \ H_2O}
$$

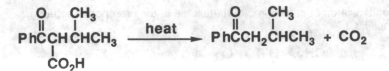

Problem 18.8 Following are structural formulas for two enamines. Draw structural formulas for the secondary amine and carbonyl compound from which each is derived.

(a)

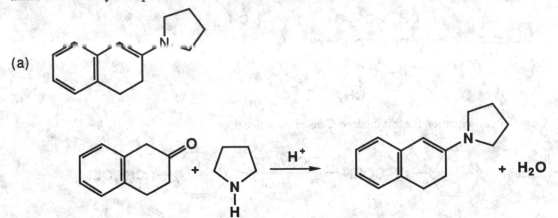

The double bond of the product enamine forms so that it is conjugated with the aromatic ring.

(b)

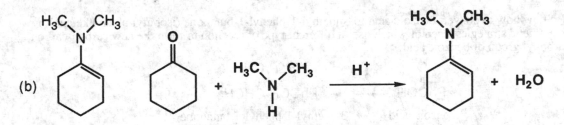

Problem 18.9 Write a mechanism for the hydrolysis of the following iminium chloride in aqueous HCl:

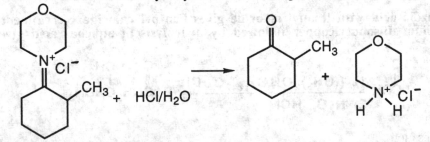

The positively-charged iminium ion is attacked by water acting as a nucleophile. A proton is transferred to the nitrogen, followed by departure of the amine and formation of the carbon-oxygen double bond. The reaction is completed with a proton transfer step to produce the product ketone and protonated morpholine. Under the acidic conditions in this reaction, an enamine is not formed, so hydrolysis predominates.

Step 1:

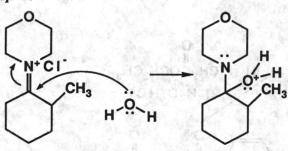

Step 2:

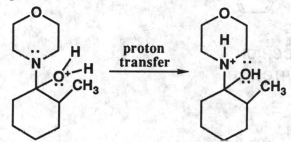

Step 3:

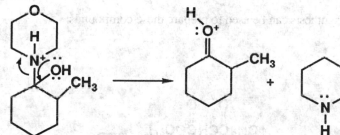

Step 4:

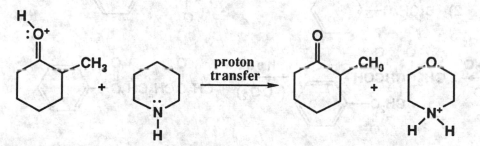

Problem 18.10 Show how to use alkylation or acylation of an enamine to convert acetophenone to the following compounds.

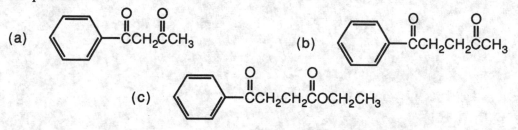

(a)

(b)

(c)

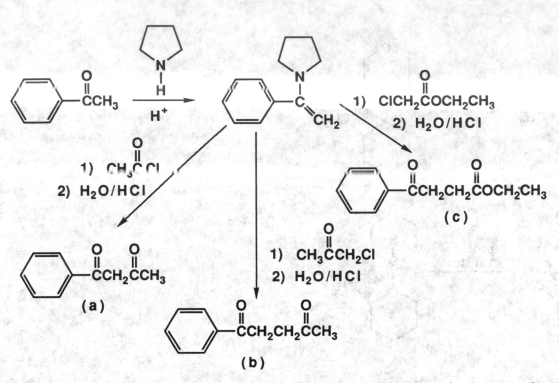

Problem 18.11 Show how the acetoacetic ester synthesis can be used to prepare these compounds:

(a) $CH_3\overset{O}{\overset{\|}{C}}CH_2CH_2\overset{O}{\overset{\|}{C}}$—⟨phenyl⟩

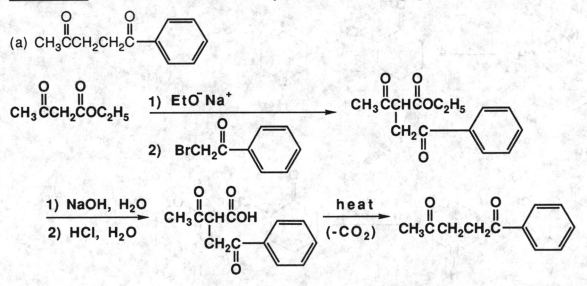

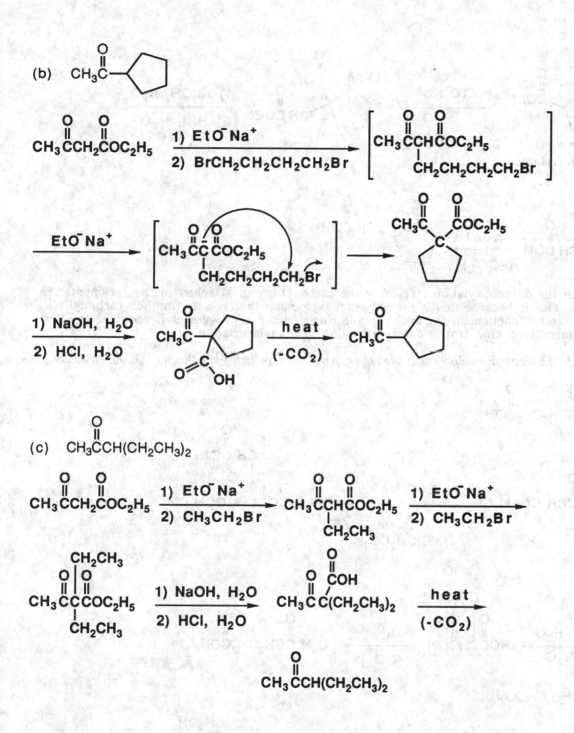

Problem 18.12 Show how to convert ethyl 2-oxocyclopentanecarboxylate to this compound.

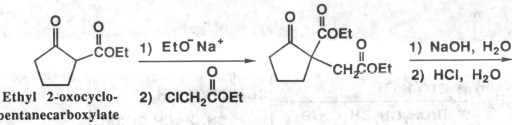

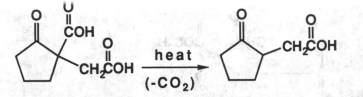

Note that, in the decarboxylation step, only the carboxyl group attached to the cyclopentane ring is lost. This is because this is the only carboxyl group that is β to another carbonyl, namely the ketone function on the cyclopentane ring. This β arrangement is required to allow for the six-membered ring transition state leading to decarboxylation.

Problem 18.13 Show how the malonic ester synthesis can be used to prepare the following substituted acetic acids.

(a) $C_6H_5CCH_2CH_2COH$ (with two C=O groups)

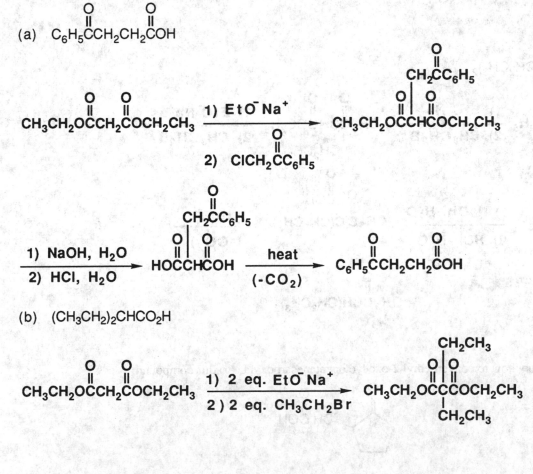

(b) $(CH_3CH_2)_2CHCO_2H$

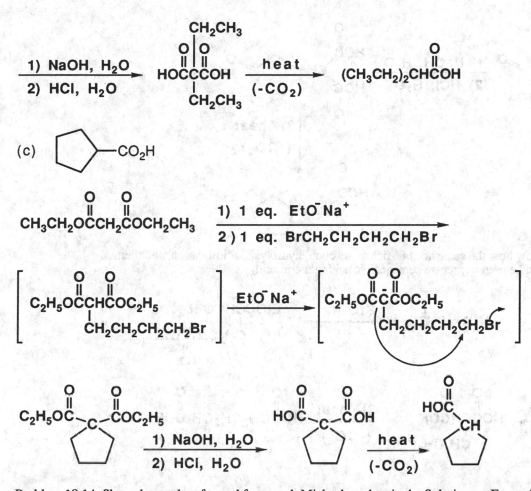

Problem 18.14 Show the product formed from each Michael product in the Solution to Example 18.14 after (1) hydrolysis in aqueous NaOH, (2) acidification, and (3) thermal decarboxylation of each β-ketoacid or β-dicarboxylic acid. These reactions illustrate the usefulness of the Michael reaction for the synthesis of 1,5-dicarbonyl compounds.

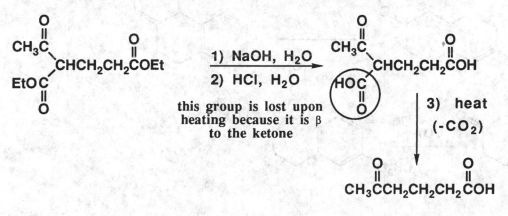

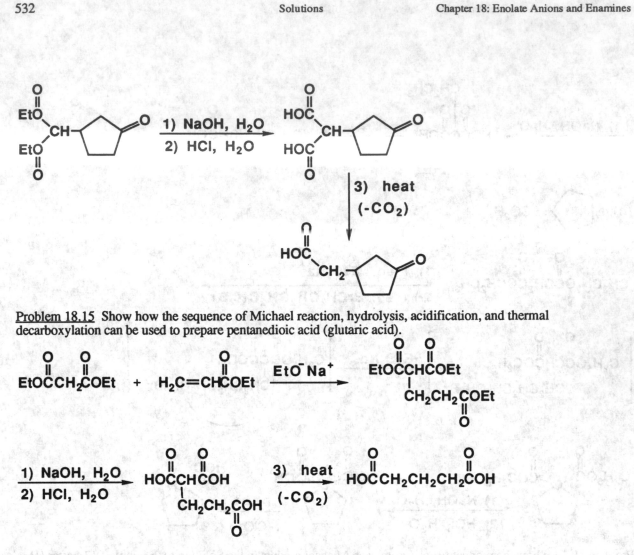

Problem 18.15 Show how the sequence of Michael reaction, hydrolysis, acidification, and thermal decarboxylation can be used to prepare pentanedioic acid (glutaric acid).

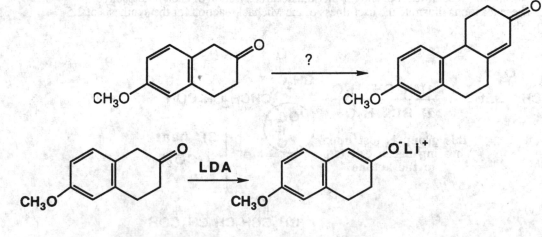

Problem 18.16 Show how to bring about the following conversion.

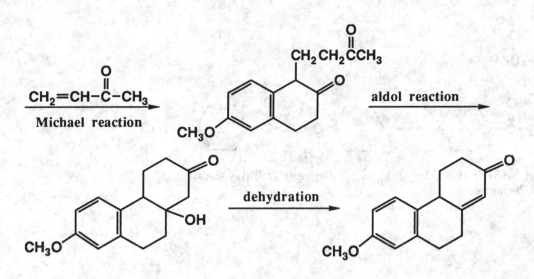

Problem 18.17 Propose two syntheses of 4-phenyl-2-pentanone, each involving conjugate addition of a lithium diorganocopper reagent.

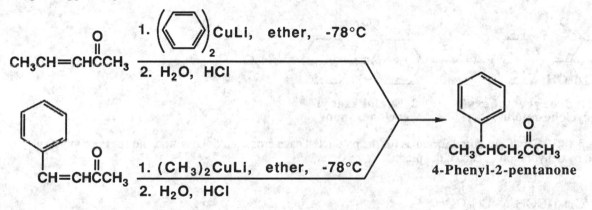

PROBLEMS
The Aldol Reaction
Problem 18.18 Draw structural formulas for the product of the aldol reaction of each compound and for the α,β-unsaturated aldehyde or ketone formed from dehydration of each aldol product.

(a) $CH_3CH_2\overset{\overset{\displaystyle O}{\|}}{C}H$

$$CH_3CH_2\underset{\underset{\displaystyle CH_3}{|}}{\overset{\overset{\displaystyle OH}{|}}{C}H}CH\overset{\overset{\displaystyle O}{\|}}{C}H \longrightarrow CH_3CH_2CH=\underset{\underset{\displaystyle CH_3}{|}}{C}\overset{\overset{\displaystyle O}{\|}}{C}H + H_2O$$

3-Hydroxy-2-methylpentanal   **2-Methyl-2-pentenal**

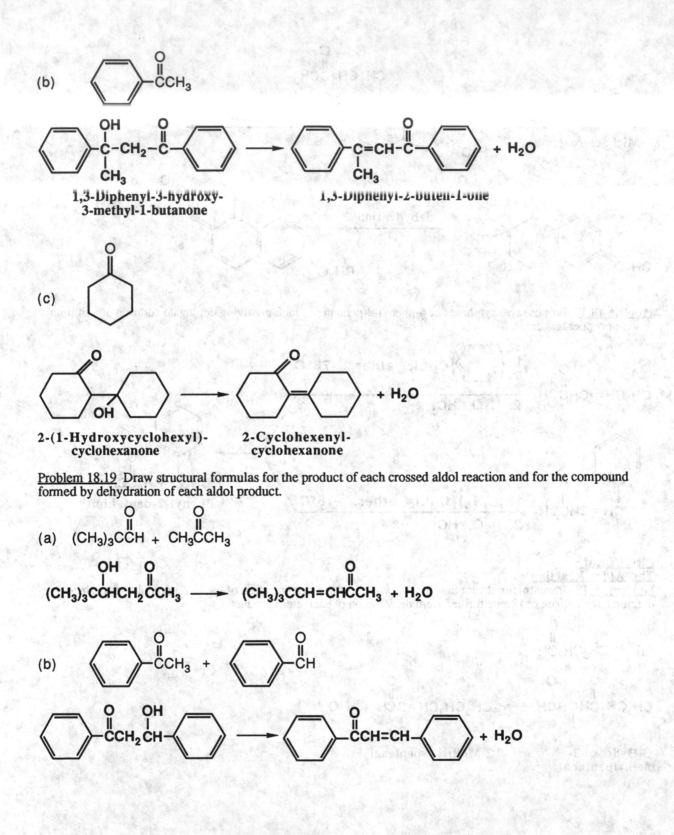

(b)

1,3-Diphenyl-3-hydroxy-
3-methyl-1-butanone

1,3-Diphenyl-2-buten-1-one

(c)

2-(1-Hydroxycyclohexyl)-
cyclohexanone

2-Cyclohexenyl-
cyclohexanone

__Problem 18.19__ Draw structural formulas for the product of each crossed aldol reaction and for the compound formed by dehydration of each aldol product.

(a) $(CH_3)_3CCH$ + CH_3CCH_3

$(CH_3)_3CCHCH_2CCH_3$ $\longrightarrow$ $(CH_3)_3CCH=CHCCH_3$ + H_2O

(b)

(c)

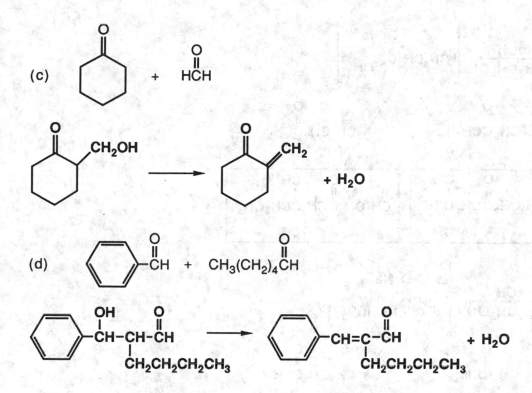

(d)

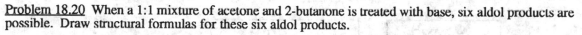

Problem 18.20 When a 1:1 mixture of acetone and 2-butanone is treated with base, six aldol products are possible. Draw structural formulas for these six aldol products.

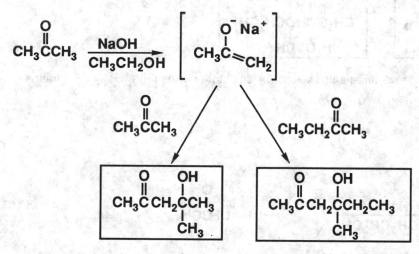

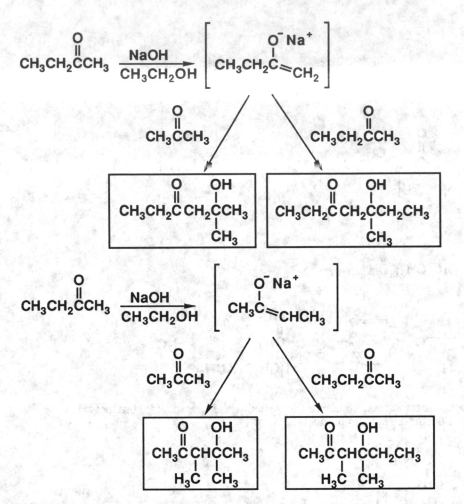

Problem 18.21 Show how to prepare these α,β-unsaturated ketones by an aldol reaction followed by dehydration of the aldol product.

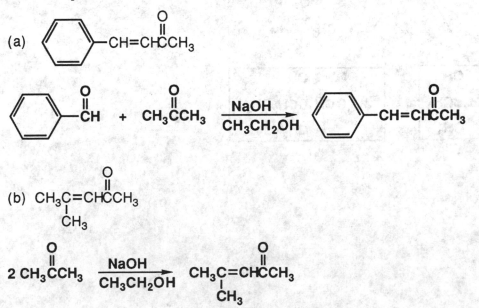

Problem 18.22 Show how to prepare these α,β-unsaturated aldehydes by an aldol reaction followed by dehydration of the aldol product.

(a)

$$\text{C}_6\text{H}_5\!-\!\text{CH}\!=\!\text{CHCH} \; (=\!\text{O})$$

$$\text{C}_6\text{H}_5\!-\!\overset{\text{O}}{\overset{\|}{\text{CH}}} \;+\; \text{CH}_3\overset{\text{O}}{\overset{\|}{\text{CH}}} \xrightarrow[\text{CH}_3\text{CH}_2\text{OH}]{\text{NaOH}} \text{C}_6\text{H}_5\!-\!\text{CH}\!=\!\text{CHCH}(=\!\text{O})$$

Note that in the reaction above, some CH₃CH=CHCHO may form, but the product shown should predominate because of conjugation with the aromatic ring.

(b) $\text{C}_7\text{H}_{15}\text{CH}\!=\!\text{CHCH}(=\!\text{O})$
    $|$
    C_6H_{13}

$$2\;\text{C}_7\text{H}_{15}\overset{\text{O}}{\overset{\|}{\text{CH}}} \xrightarrow[\text{CH}_3\text{CH}_2\text{OH}]{\text{NaOH}} \text{C}_7\text{H}_{15}\text{CH}\!=\!\overset{\text{O}}{\overset{\|}{\text{CHCH}}}$$
$$|$$
$$\text{C}_6\text{H}_{13}$$

Problem 18.23 When treated with base, the following compound undergoes an intramolecular aldol reaction to give a product containing a ring (yield 78%). Propose a structural formula for this product.

$$\text{CH}_3\text{CH}_2\text{CH}\!=\!\text{CHCH}_2\text{CH}_2\overset{\text{O}}{\overset{\|}{\text{C}}}\text{CH}_2\text{CH}_2\overset{\text{O}}{\overset{\|}{\text{CH}}} \xrightarrow[\substack{\text{aldol} \\ \text{reaction}}]{\text{base}} \text{C}_{10}\text{H}_{14}\text{O} \;+\; \text{H}_2\text{O}$$

Analyze this problem in the following way. There are three α-carbons which might form an anion and then condense with either of the other carbonyl groups. Two of these condensations lead to three-membered rings and, therefore, are not feasible. The third anion leads to formation of a five-membered ring, so this is the pathway followed.

aldol condensation by
this α-carbon gives
a five–membered ring

$$\text{CH}_3\text{CH}_2\text{CH}\!=\!\text{CHCH}_2\text{CH}_2\overset{\text{O}}{\overset{\|}{\text{C}}}\text{CH}_2\text{CH}_2\overset{\text{O}}{\overset{\|}{\text{CH}}} \longrightarrow \text{CH}_3\text{CH}_2\text{CH}\!=\!\text{CHCH}_2\!-\!\text{(cyclopentenone)}$$

Problem 18.24 Cyclohexene can be converted to 1-cyclopentenecarbaldehyde by the following series of reactions. Propose a structural formula for each intermediate compound.

$$\text{(cyclohexene)} \xrightarrow[\text{H}_2\text{O}_2]{\text{OsO}_4} \text{C}_6\text{H}_{12}\text{O}_2 \xrightarrow{\text{HIO}_4} \text{C}_6\text{H}_{10}\text{O}_2 \xrightarrow{\text{base}} \text{(cyclopentene-CHO)}$$

1-Cyclopentenecarbaldehyde

Oxidation of cyclohexene by osmium tetroxide gives a *cis*-glycol, which is then oxidized by periodic acid to hexanedial. Base-catalyzed aldol reaction of this dialdehyde followed by dehydration of the aldol product gives 1-cyclopentenecarbaldehyde.

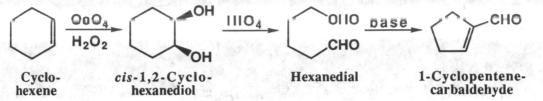

Cyclo-   *cis*-1,2-Cyclo-   Hexanedial   1-Cyclopentene-
hexene   hexanediol   carbaldehyde

<u>Problem 18.25</u> Propose a structural formula for each lettered compound.

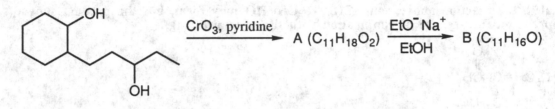

Oxidation of the starting material by CrO_3 gives the diketone, compound A. Base-catalyzed aldol reaction of this diketone could potentially give cyclic products with either a four-membered or six-membered ring. The six-membered ring is much more stable, so this product predominates. Dehydration of the aldol product gives compound B

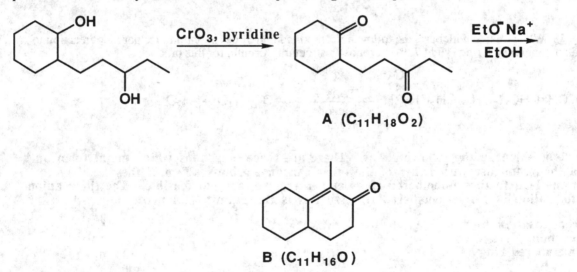

A $(C_{11}H_{18}O_2)$

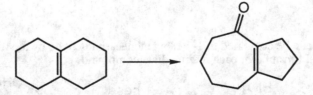

B $(C_{11}H_{16}O)$

<u>Problem 18.26</u> How might you bring about the following conversions?

The alkene is oxidized to a diketone by OsO_4/H_2O_2 followed by treatment with HIO_4. Sodium ethoxide or other base catalyzes an intramolecular aldol reaction of the diketone to give a β-hydroxyketone. Dehydration of this aldol product under the conditions of the aldol reaction gives the desired α,β-unsaturated ketone.

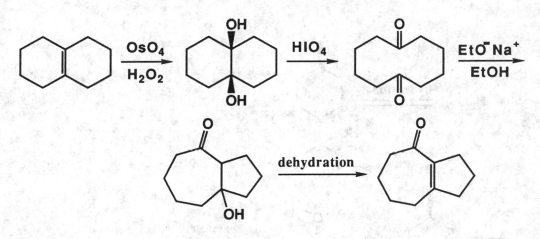

Problem 18.27 Pulegone, $C_{10}H_{16}O$, a compound from oil of pennyroyal, has a pleasant odor midway between peppermint and camphor. (See The Merck Index, 12th ed., # 8124.) Treatment of pulegone with steam produces acetone and 3-methylcyclohexanone.

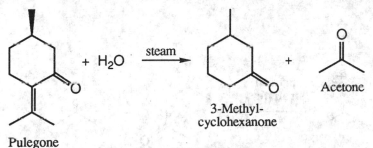

Pulegone

3-Methyl-cyclohexanone

Acetone

(a) Natural pulegone has the configuration shown. Assign an R,S configuration to its stereocenter.

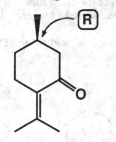

(b) Propose a mechanism for the steam hydrolysis of pulegone to the compounds shown.

An aldol reaction is reversible, and this steam hydrolysis of pulegone is formally the reverse of an aldol reaction. Thus, a reasonable mechanism is exactly the reverse of what would be written for an aldol reaction between acetone and 3-methylcyclohexanone.

Step 1:

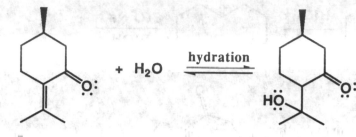

    + H₂O $\xrightarrow{\text{hydration}}$

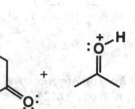

Step 2:

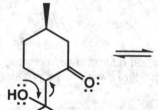

Step 3:

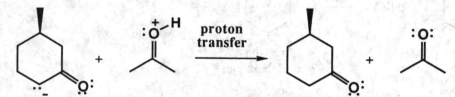

    $\xrightarrow{\substack{\textbf{proton} \\ \textbf{transfer}}}$

(c) In what way does this steam hydrolysis affect the configuration of the stereocenter in pulegone? Assign an R,S configuration to the 3-methylcyclohexanone formed in this reaction.

This reaction mechanism does not involve the stereocenter; it remains in the R configuration in the 3-methylcyclohexanone product.

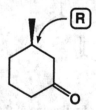

Problem 18.28 Propose a mechanism for this acid-catalyzed aldol reaction and the dehydration of the resulting aldol product.

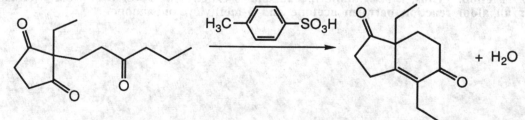

     + H₂O

A reasonable mechanism for this acid-catalyzed aldol reaction involves attack of the enol shown on the protonated carbonyl group, followed by acid-catalyzed dehydration. Note how the first two steps could occur in either order.

Step 1:

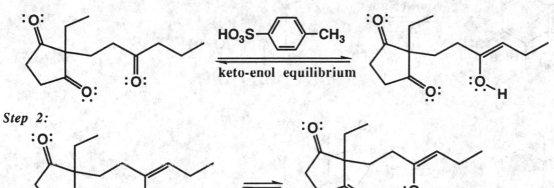

keto-enol equilibrium

Step 2:

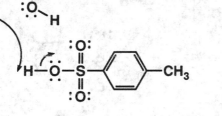

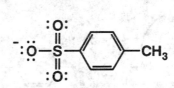

Step 3:

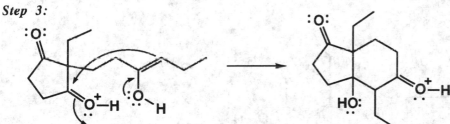

Step 4:

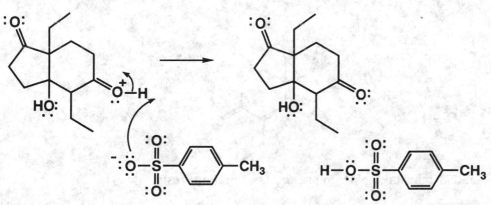

Step 5:

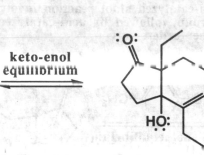

keto-enol
equilibrium

Step 6:

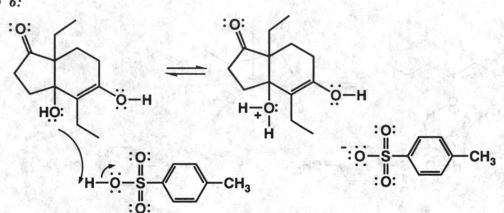

Step 7:

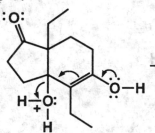

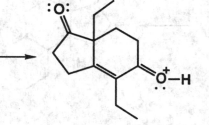

Step 8:

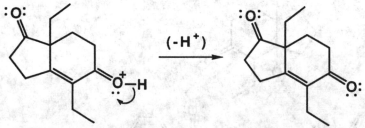

Note that steps 7 and 8 may take place more or less simultaneously.

Directed Aldol Reactions
Problem 18.29 In Section 18.2B, it was stated that four possible aldol products are formed when phenylacetaldehyde and acetone are mixed in the presence of base. Draw structural formulas for each of these aldol products.

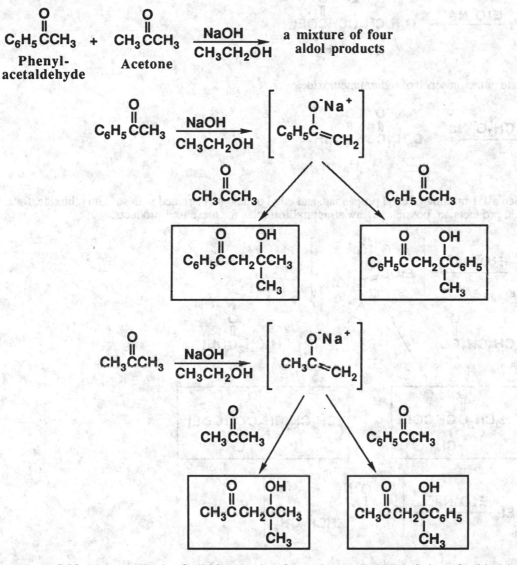

Problem 18.30 In the synthesis of a lithium enolate from a ketone and LDA, is it preferable (a) to add a solution of LDA to a solution of the ketone, or (b) to add a solution of the ketone to a solution of LDA, or (c) to conclude that the order in which the solutions are mixed makes no difference? Explain your answer.

It is important to (b), add a solution of the ketone to the LDA. This ensures that excess base is available to convert all of the ketone to the enolate form as quickly as possible during the addition. This prevents any ketone from being available to take part in an unwanted aldol reaction.

The Claisen Condensation

Problem 18.31 Show the product of Claisen condensation of these esters.
(a) Ethyl phenylacetate in the presence of sodium ethoxide.

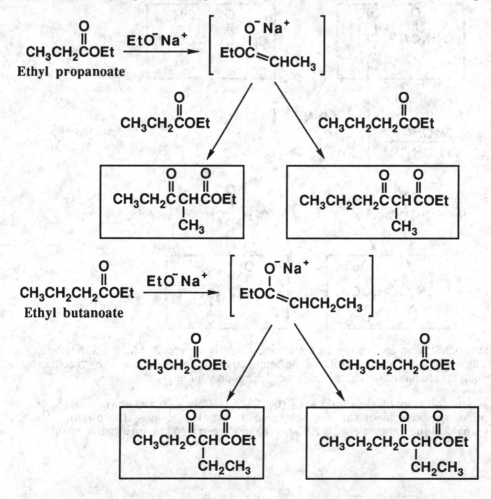

$$C_6H_5CH_2\overset{\overset{\displaystyle O}{\|}}{C}OEt \xrightarrow{\text{EtO}^- \text{Na}^+} C_6H_5CH_2\overset{\overset{\displaystyle O}{\|}}{C}\underset{\underset{\displaystyle C_6H_5}{|}}{C}H\overset{\overset{\displaystyle O}{\|}}{C}OEt$$

Ethyl phenylacetate

(b) Methyl hexanoate in the presence of sodium methoxide.

$$C_5H_{11}\overset{\overset{\displaystyle O}{\|}}{C}OCH_3 \xrightarrow{\text{CH}_3\text{O}^- \text{Na}^+} C_5H_{11}\overset{\overset{\displaystyle O}{\|}}{C}\underset{\underset{\displaystyle C_4H_9}{|}}{C}H\overset{\overset{\displaystyle O}{\|}}{C}OCH_3$$

Methyl hexanoate

Problem 18.32 When a 1:1 mixture of ethyl propanoate and ethyl butanoate is treated with sodium ethoxide, four Claisen condensation products are possible. Draw structural formulas for these four products.

$$CH_3CH_2\overset{\overset{\displaystyle O}{\|}}{C}OEt \xrightarrow{\text{EtO}^- \text{Na}^+} \left[EtO\underset{\underset{\displaystyle O^- \text{Na}^+}{|}}{C}{=}CHCH_3 \right]$$

Ethyl propanoate

$$CH_3CH_2\overset{\overset{\displaystyle O}{\|}}{C}OEt \qquad\qquad CH_3CH_2CH_2\overset{\overset{\displaystyle O}{\|}}{C}OEt$$

$$\boxed{CH_3CH_2\overset{\overset{\displaystyle O}{\|}}{C}\underset{\underset{\displaystyle CH_3}{|}}{C}H\overset{\overset{\displaystyle O}{\|}}{C}OEt} \qquad \boxed{CH_3CH_2CH_2\overset{\overset{\displaystyle O}{\|}}{C}\underset{\underset{\displaystyle CH_3}{|}}{C}H\overset{\overset{\displaystyle O}{\|}}{C}OEt}$$

$$CH_3CH_2CH_2\overset{\overset{\displaystyle O}{\|}}{C}OEt \xrightarrow{\text{EtO}^- \text{Na}^+} \left[EtO\underset{\underset{\displaystyle O^- \text{Na}^+}{|}}{C}{=}CHCH_2CH_3 \right]$$

Ethyl butanoate

$$CH_3CH_2\overset{\overset{\displaystyle O}{\|}}{C}OEt \qquad\qquad CH_3CH_2CH_2\overset{\overset{\displaystyle O}{\|}}{C}OEt$$

$$\boxed{CH_3CH_2\overset{\overset{\displaystyle O}{\|}}{C}\underset{\underset{\displaystyle CH_2CH_3}{|}}{C}H\overset{\overset{\displaystyle O}{\|}}{C}OEt} \qquad \boxed{CH_3CH_2CH_2\overset{\overset{\displaystyle O}{\|}}{C}\underset{\underset{\displaystyle CH_2CH_3}{|}}{C}H\overset{\overset{\displaystyle O}{\|}}{C}OEt}$$

Problem 18.33 Draw structural formulas for the β-ketoesters formed by Claisen condensation of ethyl propanoate with each ester:

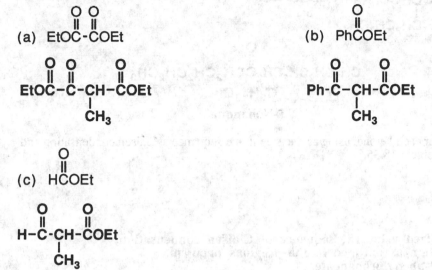

(a) EtOC-COEt

EtOC-C-CH-COEt
|
CH₃

(b) PhCOEt

Ph-C-CH-COEt
|
CH₃

(c) HCOEt

H-C-CH-COEt
|
CH₃

Problem 18.34 Draw a structural formula for the product of saponification, acidification, and decarboxylation of each β-ketoester formed in Problem 18.33.

Following are the structures for the products of saponification and decarboxylation of each β-ketoester from the previous problem.

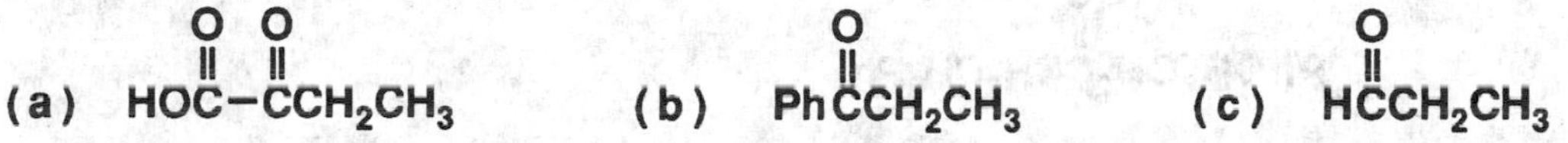

(a) $HOC-CCH_2CH_3$ (b) $PhCCH_2CH_3$ (c) $HCCH_2CH_3$

Problem 18.35 The Claisen condensation can be used as one step in the synthesis of ketones, as illustrated by this reaction sequence. Propose structural formulas for compounds A, B, and the ketone formed in this sequence.

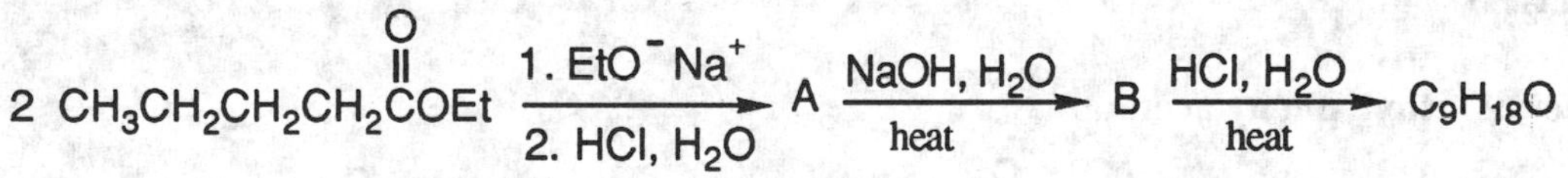

$$2\ CH_3CH_2CH_2CH_2COEt \xrightarrow[\text{2. HCl, H}_2\text{O}]{\text{1. EtO}^-\text{Na}^+} A \xrightarrow[\text{heat}]{\text{NaOH, H}_2\text{O}} B \xrightarrow[\text{heat}]{\text{HCl, H}_2\text{O}} C_9H_{18}O$$

Compound (A) is a β-ketoester, compound (B) is the sodium salt of a β-ketoacid, and the final ketone is 5-nonanone.

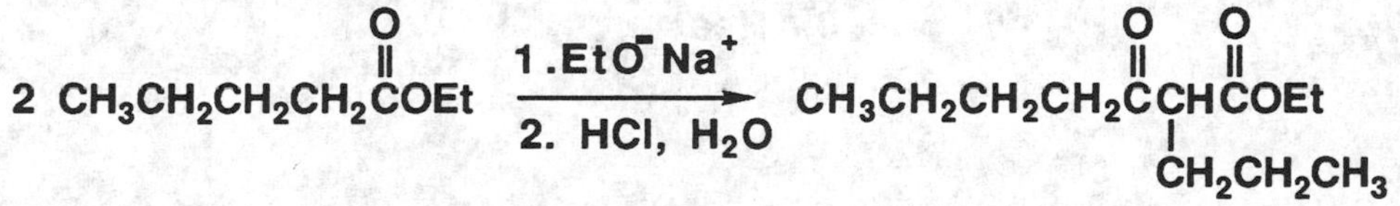

$$2\ CH_3CH_2CH_2CH_2COEt \xrightarrow[\text{2. HCl, H}_2\text{O}]{\text{1. EtO}^-\text{Na}^+} CH_3CH_2CH_2CH_2CCHCOEt$$

CH₂CH₂CH₃

A

$$\xrightarrow[\text{heat}]{\text{NaOH, H}_2\text{O}} \quad \underset{\underset{\text{CH}_2\text{CH}_2\text{CH}_3}{|}}{\text{CH}_3\text{CH}_2\text{CH}_2\text{CH}_2\overset{\overset{\text{O}}{\|}}{\text{C}}\overset{\overset{\text{O}}{\|}}{\text{CHCO}^-} \text{ Na}^+} \quad \xrightarrow[\text{heat}]{\text{HCl, H}_2\text{O}}$$

B

$$\text{CH}_3\text{CH}_2\text{CH}_2\text{CH}_2\overset{\overset{\text{O}}{\|}}{\text{C}}\text{CH}_2\text{CH}_2\text{CH}_2\text{CH}_3$$

$$\text{C}_9\text{H}_{18}\text{O}$$

5-Nonanone

Problem 18.36 Propose a synthesis for each ketone, using as one step in the sequence a Claisen condensation and the reaction sequence illustrated in Problem 18.35.

(a) $\text{PhCH}_2\text{CH}_2\overset{\overset{\text{O}}{\|}}{\text{C}}\text{CH}_2\text{CH}_2\text{Ph}$

Each target molecule is synthesized using the sequence of Claisen condensation, saponification, and decarboxylation as outlined in the previous problem.
(a) The starting ester is ethyl 3-phenylpropanoate.

$$2\;\text{PhCH}_2\text{CH}_2\overset{\overset{\text{O}}{\|}}{\text{C}}\text{OEt} \quad \xrightarrow[\text{C}_2\text{H}_5\text{OH}]{\text{EtO}^-\text{Na}^+} \quad \underset{\underset{\text{CH}_2\text{Ph}}{|}}{\text{PhCH}_2\text{CH}_2\overset{\overset{\text{O}}{\|}}{\text{C}}\overset{\overset{\text{O}}{\|}}{\text{CHC}}\text{OEt}} \quad \xrightarrow{\begin{array}{l}\text{1. NaOH, H}_2\text{O}\\ \text{2. HCl, H}_2\text{O}\\ \text{3. Heat}\end{array}}$$

$$\text{PhCH}_2\text{CH}_2\overset{\overset{\text{O}}{\|}}{\text{C}}\text{CH}_2\text{CH}_2\text{Ph}$$

(b) $\text{PhCH}_2\overset{\overset{\text{O}}{\|}}{\text{C}}\text{CH}_2\text{Ph}$

The starting ester is ethyl phenylacetate.

$$2\;\text{PhCH}_2\overset{\overset{\text{O}}{\|}}{\text{C}}\text{OEt} \quad \xrightarrow[\text{C}_2\text{H}_5\text{OH}]{\text{EtO}^-\text{Na}^+} \quad \underset{\underset{\text{Ph}}{|}}{\text{PhCH}_2\overset{\overset{\text{O}}{\|}}{\text{C}}\overset{\overset{\text{O}}{\|}}{\text{CHC}}\text{OEt}} \quad \xrightarrow{\begin{array}{l}\text{1. NaOH, H}_2\text{O}\\ \text{2. HCl, H}_2\text{O}\\ \text{3. Heat}\end{array}} \quad \text{PhCH}_2\overset{\overset{\text{O}}{\|}}{\text{C}}\text{CH}_2\text{Ph}$$

(c)

The starting diester is derived from a *cis*-4,5-disubstituted cyclohexene.

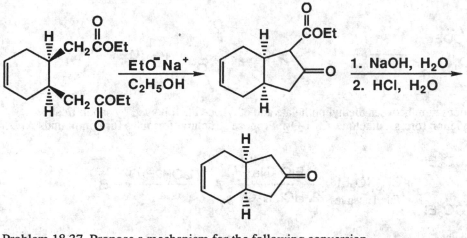

<u>Problem 18.37</u> Propose a mechanism for the following conversion.

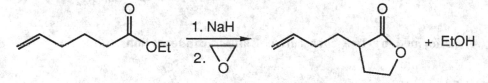

A reasonable mechanism for this conversion involves an initial reaction of the ester with the strong base NaH to give an enolate, that attacks the ethylene oxide.

Step 1:

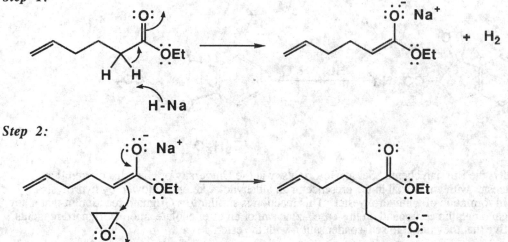

Step 2:

Step 3: The alkoxide intermediate then undergoes an intramolecular reaction with the ester function to give the product.

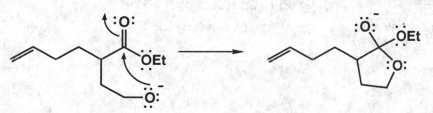

Step 4:

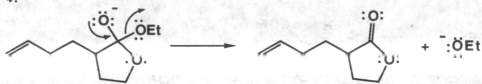

Problem 18.38 Claisen condensation between diethyl phthalate and ethyl acetate followed by saponification, acidification, and decarboxylation forms a diketone, $C_9H_6O_2$. Propose structural formulas for compounds A, B, and the diketone.

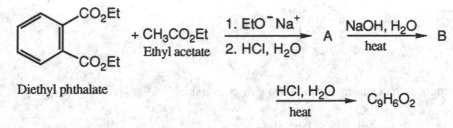

Diethyl phthalate

Compound (A) is formed by two consecutive Claisen condensations.

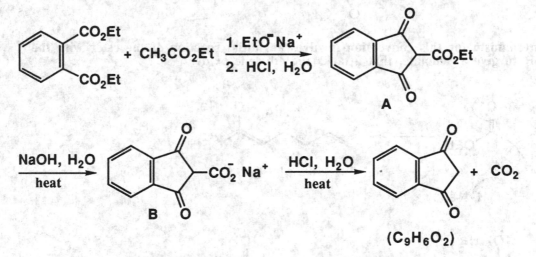

Problem 18.39 In 1887, the Russian chemist Sergei Reformatsky at the University of Kiev discovered that treatment of an α-haloester with zinc metal in the presence of an aldehyde or ketone followed by hydrolysis in aqueous acid results in formation of a β-hydroxyester. This reaction is similar to a Grignard reaction in that a key intermediate is an organometallic compound, in this case a zinc salt of an ester enolate anion. Grignard reagents, however, are so reactive that they undergo self-condensation with the ester.

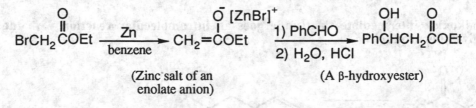

Show how a Reformatsky reaction can be used to synthesize the following compounds from an aldehyde or
ketone and an α-haloester:

(a)

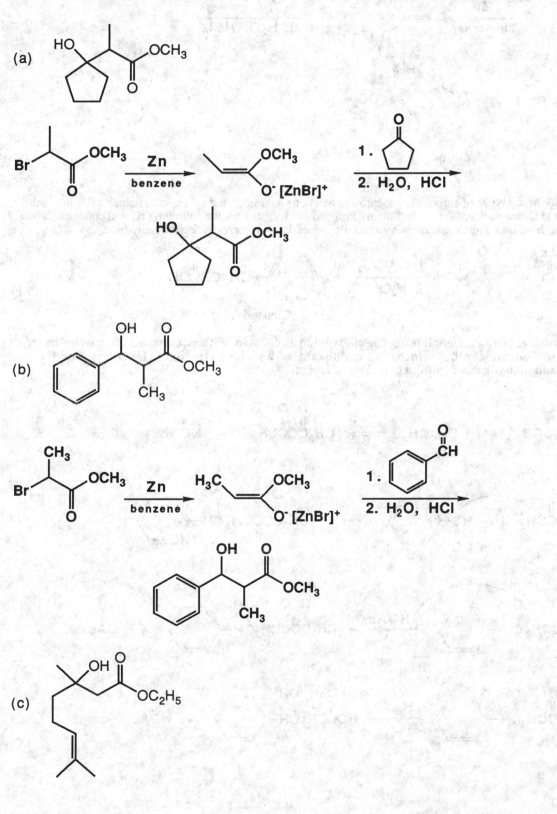

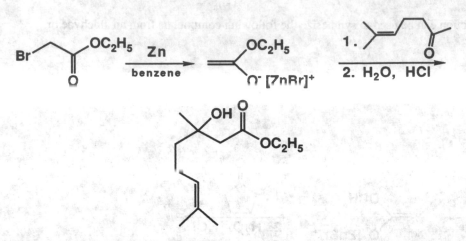

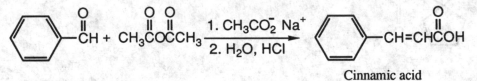

Problem 18.40 Many types of carbonyl condensation reactions have acquired specialized names, after the 19th century organic chemists who first studied them. Propose mechanisms for the following named condensations.
(a) Perkin condensation: Condensation of an aromatic aldehyde with a carboxylic acid anhydride.

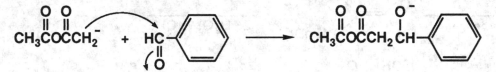

Cinnamic acid

In the Perkin reaction, an enolate of the anhydride is formed with acetate acting as the base. This enolate then attacks benzaldehyde reminiscent of an aldol reaction. Hydrolysis of the anhydride and dehydration complete the mechanism.

Step 1:

$$CH_3COCCH_3 + {}^-O_2CCH_3 \rightleftharpoons CH_3COCCH_2^- + HOAc$$

Step 2:

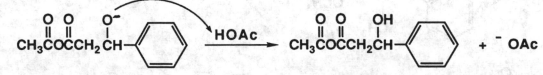

Step 3:

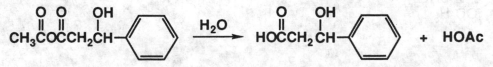

Step 4:

Step 5:

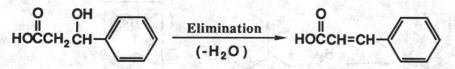

(b) Darzens condensation: Condensation of an α-haloester with a ketone or an aromatic aldehyde.

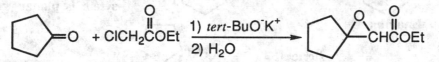

In the Darzens condensation, the enolate of the α-haloester is formed, then this attacks the ketone carbonyl group reminiscent of an aldol reaction.

Step 1:

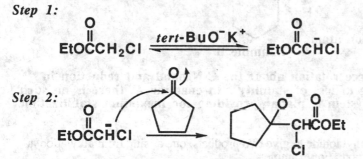

Step 2:

The oxyanion displaces chloride via backside attack to create an expoxide.

Step 3:

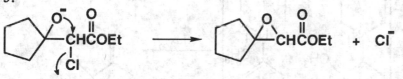

Enamines
Problem 18.41 When 2-methylcyclohexanone is treated with pyrrolidine, two isomeric enamines are formed. Why is enamine A with the less substituted double bond the thermodynamically favored product? You will find it helpful to build models.

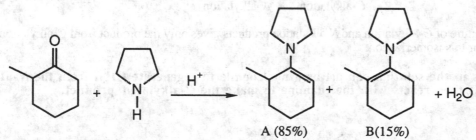

Remember that in the six atom system of an enamine, the nitrogen is sp² hybridized, so that there is overlap of the 2p orbital of nitrogen atom with the 2p orbitals of the carbon-carbon double bond. This overlap allows electron delocalization and thus stabilization of the enamine.

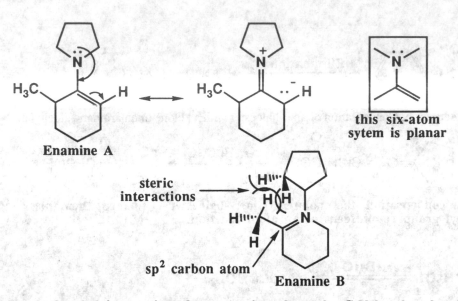

Enamine A

this six-atom
sytem is planar

steric
interactions

sp² carbon atom

Enamine B

In enamine B, non-bonded interactions force rotation about the C-N bond and reduction in planarity of the enamine system, resulting in loss of stability. In enamine A, there is no such non-bonded interaction, so the enamine system is planar providing for maximum stabilization. Thus, enamine A predominates.

<u>Problem 18.42</u> Enamines normally react with methyl iodide to give two products: one arising from alkylation at nitrogen, the second arising from alkylation at carbon. For example:

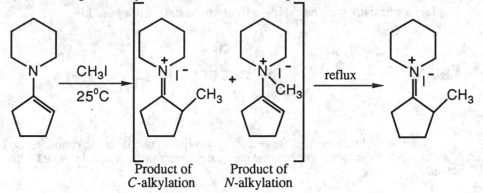

Product of
C-alkylation

Product of
N-alkylation

Heating the mixture of *C*-alkylation and *N*-alkylation products gives only the product from *C*-alkylation. Propose a mechanism for this isomerization.

The I⁻ that is in the solution can act as a nucleophile to regenerate CH₃I from the *N*-alkylated product. This then reacts with the enamine to make the *C*-alkylated product.

Step 1:

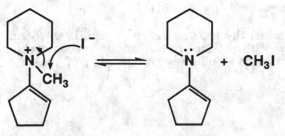

Step 2:

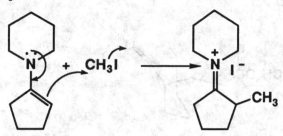

<u>Problem 18.43</u> Propose a mechanism for the following conversion.

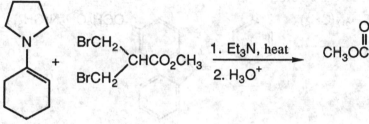

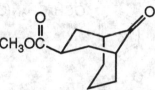

Step 1:

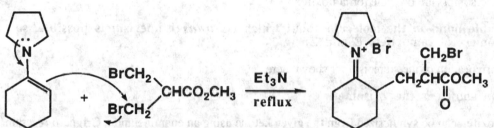

Step 2:

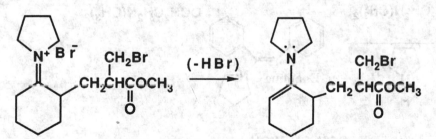

Step 3:

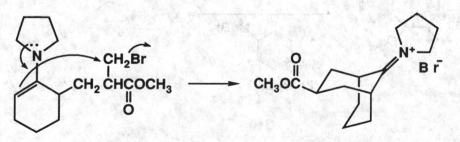

Step 4:

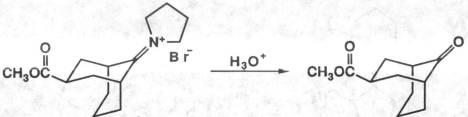

Problem 10.11 Many tumors of the breast are estrogen dependent. Drugs that interfere with estrogen binding have antitumor activity and may even help prevent tumor occurrence. A widely used antiestrogen drug is tamoxifen. (See The Merck Index, 12th ed., # 9216.)

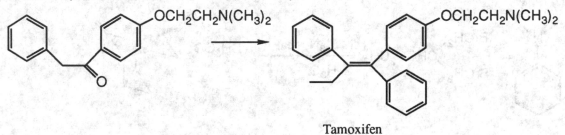

Tamoxifen

(a) How many stereoisomers are possible for tamoxifen?

There is one double bond in the molecule about which *cis-trans* isomerism is possible, so there are two isomers of tamoxifen, E and Z.

(b) Specify the configuration of the stereoisomer shown here.

The configuration shown is the Z isomer.

(c) Show how tamoxifen can be synthesized from the given ketone using an enamine and a Grignard reaction.

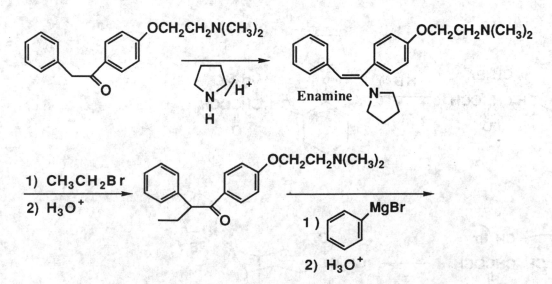

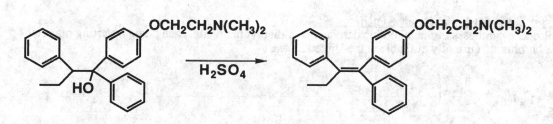

Problem 18.45 Propose a mechanism for the following reaction.

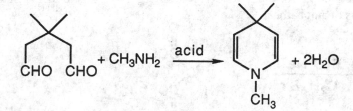

Step 1:

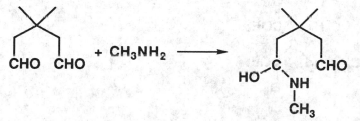

Step 2:

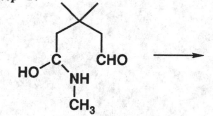 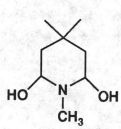

Step 3:

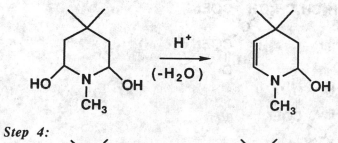

Step 4:

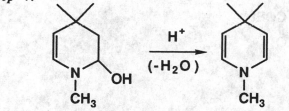

Acetoacetic Ester and Malonic Ester Syntheses

<u>Problem 18.46</u> Propose syntheses of the following derivatives of diethyl malonate, each being a starting material for synthesis of a barbiturate currently available in the United States.

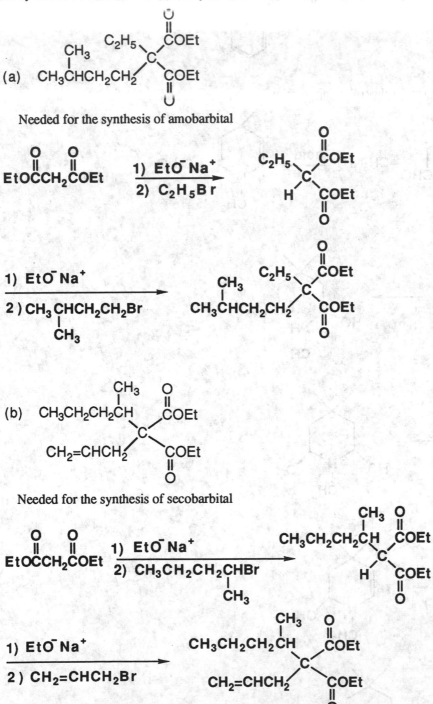

(a)

Needed for the synthesis of amobarbital

(b)

Needed for the synthesis of secobarbital

Problem 18.47 2-Propylpentanoic acid (valproic acid) is an effective drug for treatment of several types of epilepsy, particularly absence seizures, which are generalized epileptic seizures characterized by brief and abrupt loss of consciousness. (See The Merck Index, 12th ed., #10049.) Propose a synthesis of valproic acid starting with diethyl malonate.

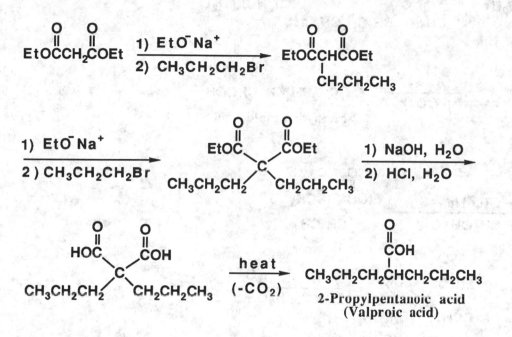

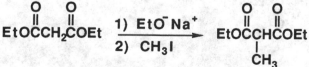

Problem 18.48 Show how to synthesize the following compounds using either the malonic ester synthesis or the acetoacetic ester synthesis.
a) 4-Phenyl-2-butanone

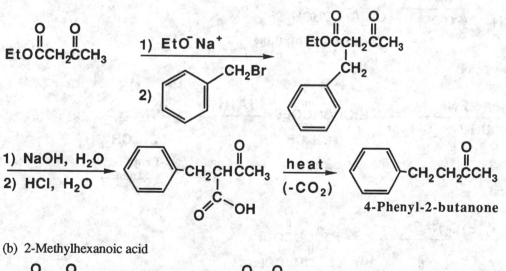

(b) 2-Methylhexanoic acid

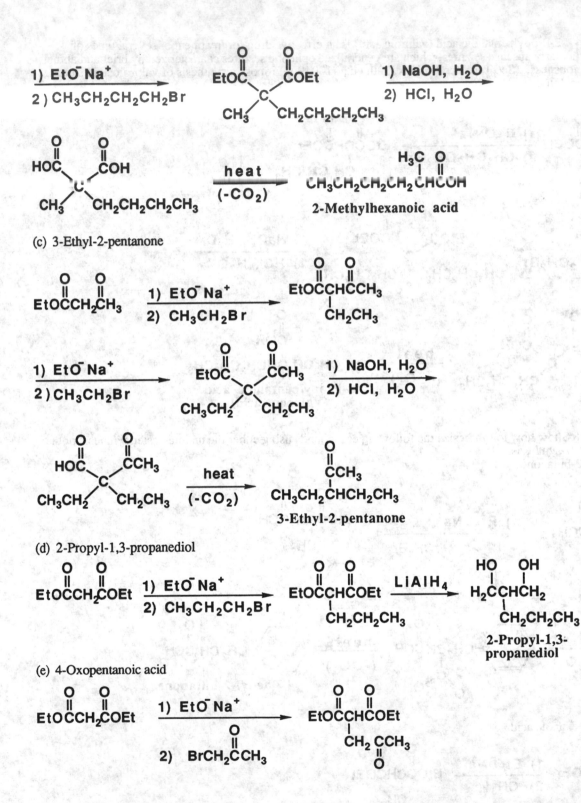

(c) 3-Ethyl-2-pentanone

(d) 2-Propyl-1,3-propanediol

(e) 4-Oxopentanoic acid

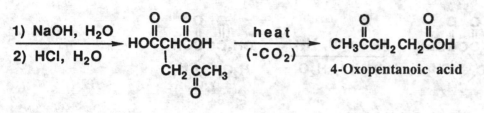

(f) 3-Benzyl-5-hexen-2-one

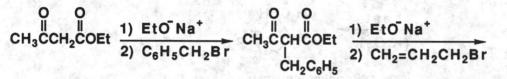

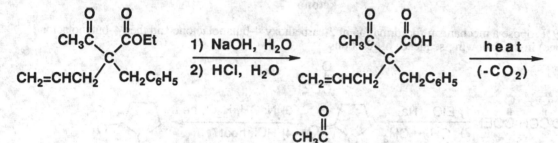

3-Benzyl-5-hexen-2-one

(g) Cyclopropanecarboxylic acid

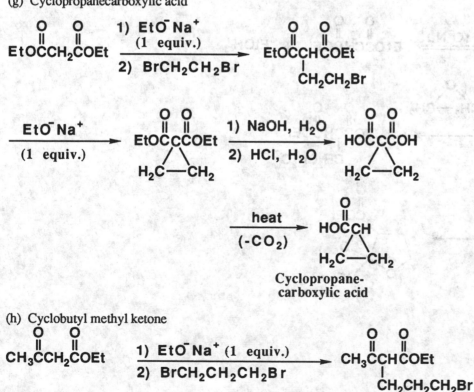

Cyclopropane-
carboxylic acid

(h) Cyclobutyl methyl ketone

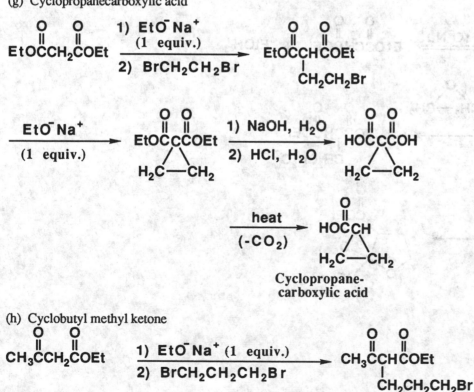

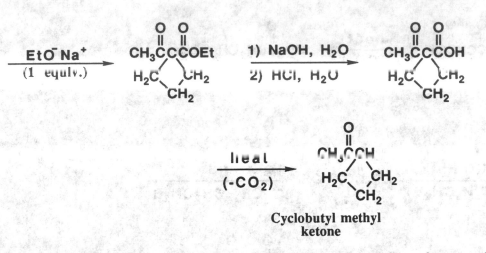

Problem 18.49 Propose a mechanism for formation of 2-carbethoxy-4-butanolactone and then 4-butanolactone (γ-butyrolactone) in the following sequence of reactions.

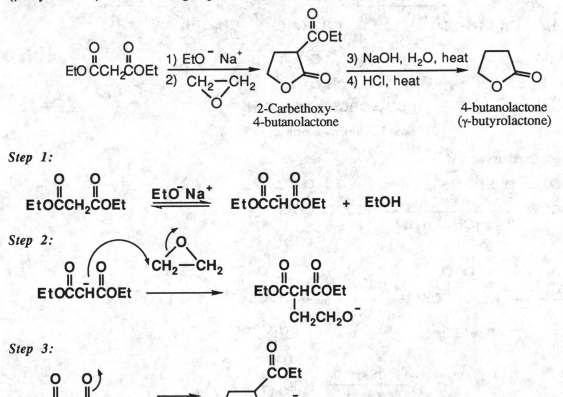

Step 4:

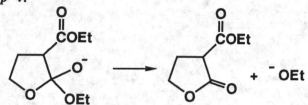

Step 5:

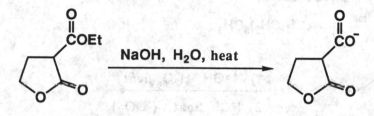

NaOH, H$_2$O, heat

Step 6:

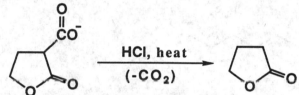

HCl, heat
(-CO$_2$)

Problem 18.50 Show how the scheme for formation of 4-butanolactone in Problem 18.49 can be used to synthesize lactones (a) and (b). Each has a peach odor and is used in perfumery. As sources of carbon atoms for these syntheses, use diethyl malonate, ethylene oxide, 1-bromoheptane, and 1-nonene.

(a)

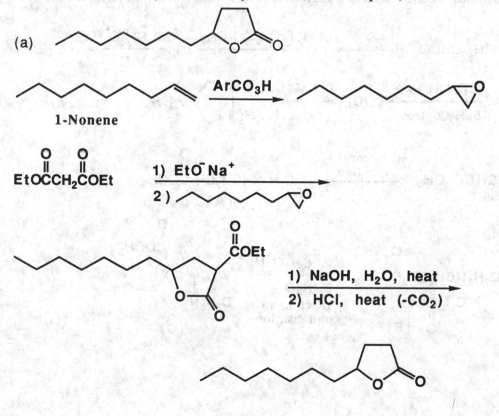

1-Nonene

ArCO$_3$H

EtOCCH$_2$COEt

1) EtO$^-$ Na$^+$

2)

1) NaOH, H$_2$O, heat
2) HCl, heat (-CO$_2$)

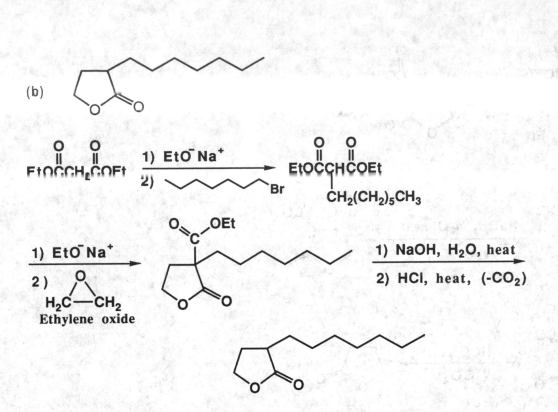

Michael Reactions

Problem 18.51 The following synthetic route is used to prepare an intermediate in the total synthesis of the anticholinergic drug benzilonium bromide. Write structural formulas for intermediates(A), (B), (C), and (D) in this synthesis.

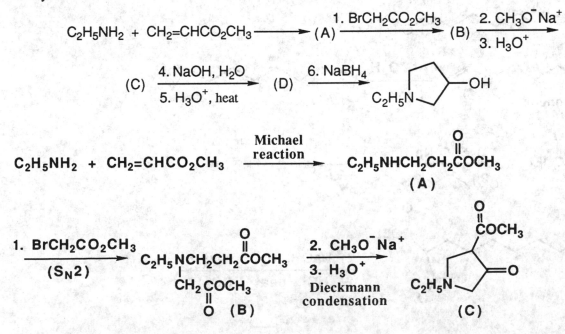

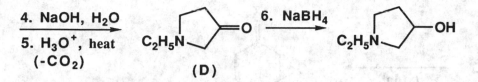

$$\frac{\text{4. NaOH, H}_2\text{O}}{\text{5. H}_3\text{O}^+, \text{ heat}}$$
$$(-\text{CO}_2)$$

(D)

6. NaBH$_4$

Problem 18.52 Propose a mechanism for formation of the bracketed intermediate, and for the bicyclic ketone formed in the following reaction sequence.

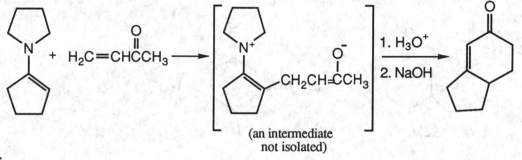

Step 1:

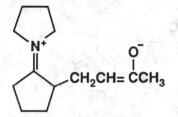

Michael reaction

(an intermediate not isolated)

Step 2:

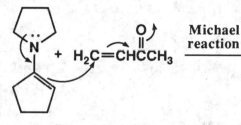

1. H$_3$O$^+$

(an intermediate not isolated)

Step 3:

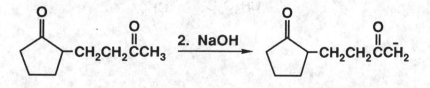

2. NaOH

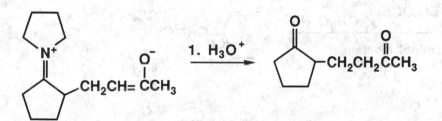

Step 4:

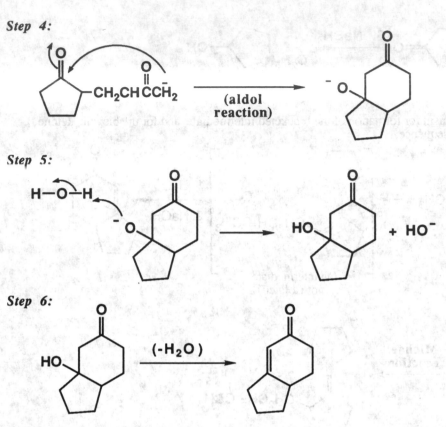

Step 5:

Step 6:

Synthesis
Problem 18.53 Show experimental conditions by which to carry out the following synthesis starting with benzaldehyde and methyl acetoacetate.

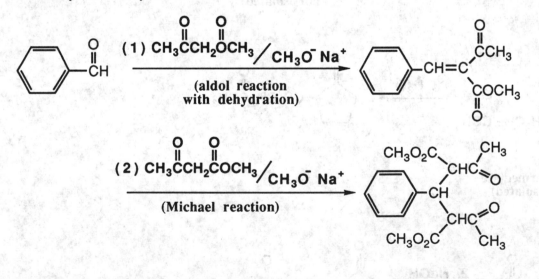

Problem 18.54 Nifedipine (Procardia, Adalat) belongs to a class of drugs called calcium channel blockers and is effective in the treatment of various types of angina, including that induced by exercise. (See The Merck Index, 12 ed., # 6617.) Show how nifedipine can be synthesized from 2-nitrobenzaldehyde, methyl acetoacetate, and ammonia. (*Hint:* Review the chemistry of your answers to Problems 18.45 and 18.53, and then combine that chemistry to solve this problem.)

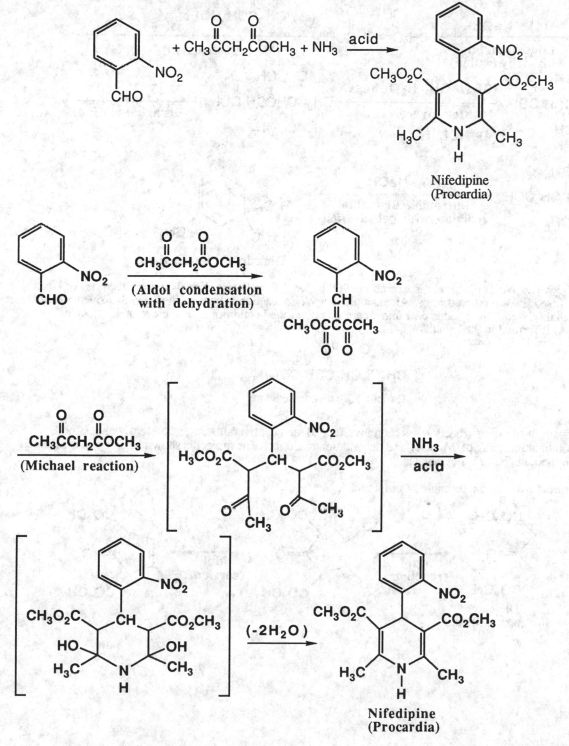

Nifedipine
(Procardia)

<u>Problem 18.55</u> The compound 3,5,5-trimethyl-2-cyclohexenone can be synthesized using acetone and ethyl acetoacetate as sources of carbon atoms. New carbon-carbon bonds in this synthesis are formed by a combination of aldol reactions and Michael reactions. Show reagents and conditions by which this synthesis might be accomplished.

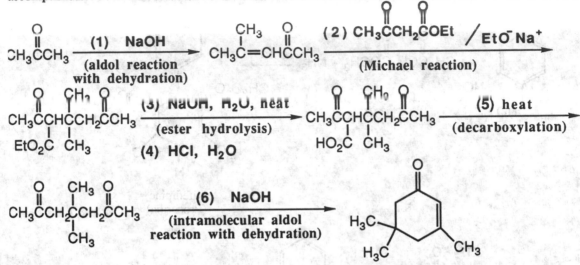

3,5,5-Trimethyl-2-cyclohexanone

<u>Problem 18.56</u> The Weiss reaction, discovered in 1968 by Dr. Ulrich Weiss at the National Institutes of Health, is a route to fused five-membered rings. An example of a Weiss reaction is treating dimethyl 3-oxopentanedioate with ethanedial (glyoxal) in aqueous base under carefully controlled conditions. The bicyclo[3.3.0]octane derivative (A) is formed in 90% yield.

$$CH_3OCCH_2CCH_2COCH_3$$

Dimethyl 3-oxopentanedioate

The mechanism of the Weiss reaction has been investigated, and the overall steps, as presently understood, involve a combination of aldol, Michael, and dehydration reactions. The molecule shown in brackets is assumed to be an intermediate, but it is not isolated.

Propose a mechanism for the formation Compound A.

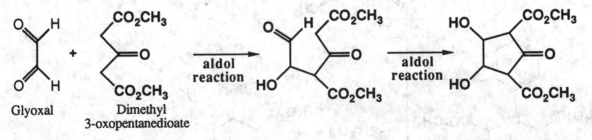

Glyoxal   Dimethyl
            3-oxopentanedioate

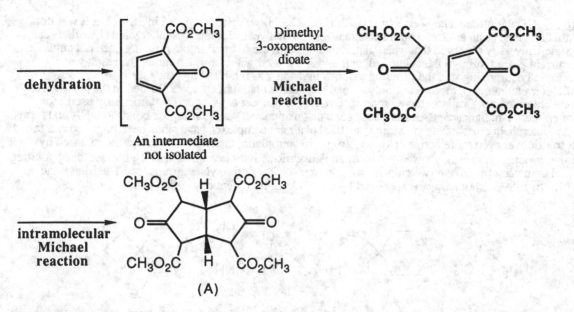

An intermediate
not isolated

(A)

Problem 18.57 The following β-diketone (A) can be synthesized from cyclopentanone and an acid chloride using an enamine reaction.

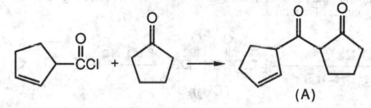

(A)

(a) Propose a synthesis of the starting acid chloride from cyclopentene.

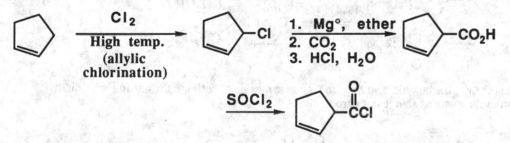

(b) Show the steps in the synthesis of compound A using a morpholine enamine.

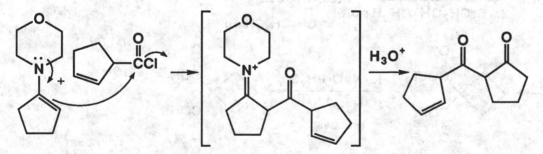

<u>Problem 18.58</u>   Cisplatin (See The Merck Index, 12th ed., # 2378) was first prepared in 1844, but it was not until 1964 that its value as an anticancer drug was realized. In that year, Barnett Rosenberg and coworkers at Michigan State University observed that when platinum electrodes are inserted into a growing bacterial culture and an electric current passed through the culture, all cell division ceased within 1 to 2 hours. The result was surprising. Equally surprising was their finding that cell division was inhibited by *cis*-diamminedichloroplatinum(II), more commonly named cisplatin, a platinum complex formed in the presence of ammonia and chloride ion. Cisplatin has a broad spectrum of anticancer activity and is particularly useful for treatment of epithelial malignancies. Evidence suggests that platinum(II) in the complex bonds to DNA and forms intrachain and interchain cross linkages. More than 1000 platinum complexes have since been prepared and tested in attempts to discover even more active cytotoxic drugs. In spiroplatin, the two NH_3 groups are replaced by primary amino groups. This drug showed excellent antileukemic activity in animal models, but was disappointing in human trials. In carboplatin, the two chloride ions are replaced by carboxylate groups (see The Merck Index, 12 ed., 1870). In 1989, carboplatin was approved by the FDA for treatment of ovarian cancers.

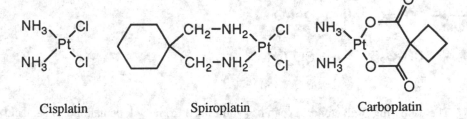

Cisplatin   Spiroplatin   Carboplatin

(a) Devise a synthesis for the diamine required in the synthesis of spiroplatin starting with diethyl malonate and 1,5-dibromopentane as the sources of carbon atoms.

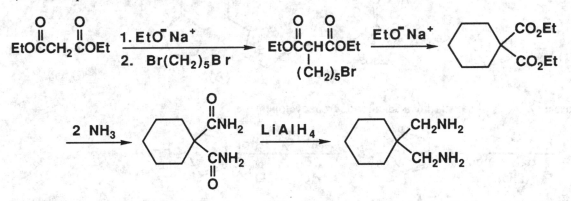

(b) Devise a synthesis for the dicarboxylic acid required in the synthesis of carboplatin starting with diethyl malonate and 1,3-dibromopropane as sources of carbon atoms.

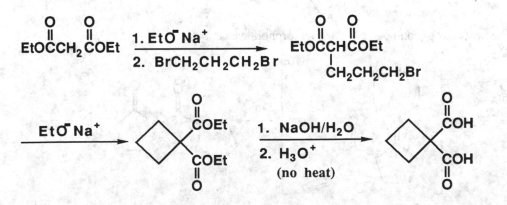

<u>Problem 18.59</u> Oxanamide is a mild sedative belonging to a class of molecules called oxanamides (it contains an **oxirane** (epoxide) group and an **amide** group). (See The Merck Index, 12th ed., #7053. As seen in this retrosynthetic scheme, the source of carbon atoms for the synthesis of oxanamide is butanal.

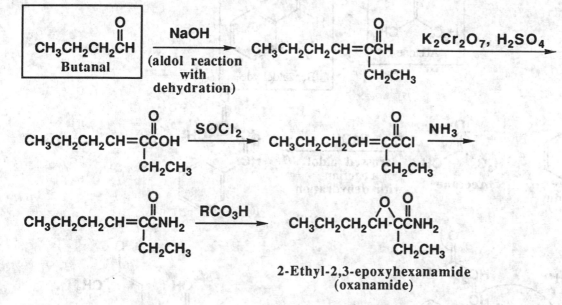

(a) Show reagents and experimental conditions by which oxanamide can be synthesized from butanal.

(b) How many stereocenters are there in oxanamide ? How many stereoisomers are possible for this compound?

There are two stereocenters in oxanamide (marked with the *) so there are 2 x 2 = 4 possible stereoisomers.

Problem 18.60 The widely used anticoagulant warfarin (see the Chemistry in Action box "From Moldy Clover to a Blood Thinner") is synthesized from 4-hydroxycoumarin, benzaldehyde, and acetone as shown in this retrosynthesis. Show how warfarin is synthesized from these reagents.

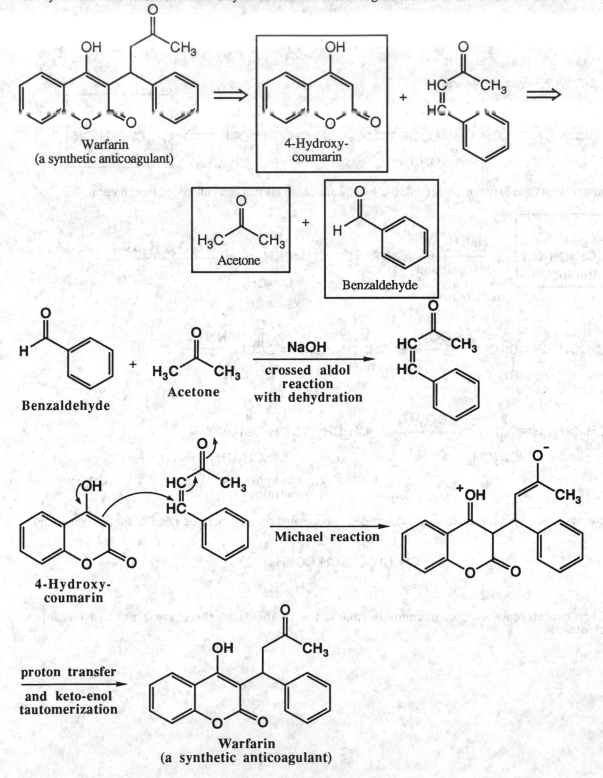

CHAPTER 19: AROMATICS 1: BENZENE AND ITS DERIVATIVES

SUMMARY OF REACTIONS

Starting Material \ Product →	Alkyl Aryl Ethers	Benzoic Acids	Cyclohexanes	1-Haloalkylbenzenes	Hydroquinones	Methyl benzenes Alcohols	Phenols, Alkyl Halides	Phenoxides	Quinones	Salicylates
Alkylbenzenes		**19A** 19.6A*		**19B** 19.6B						
Aryl Alkyl Ethers							**19C** 19.5D			
Benzenes		**19D** 19.1C								
Benzyl ethers					**19E** 19.6C					
Phenols								**19F** 19.5B,C		
Phenols, Catechols Hydroquinones									**19G** 15.5C	
Phenoxides										**19H** 19.5E
Phenoxides Alkyl Halides	**19I** 19.5D									
Quinones				**19J** 19.5F						

*Section in book that describes reaction.

REACTION 19A: OXIDATION AT A BENZYLIC POSITION (Section 19.6A)

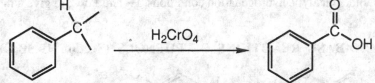

- Compounds with at least **one benzylic hydrogen** react with H_2CrO_4 to produce **benzoic acid**. Notice that other groups attached to the benzylic carbon atom are removed in the process. ✳

REACTION 19B: HALOGENATION AT A BENZYLIC POSITION (Section 19.6B)

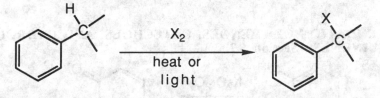

- **Benzylic hydrogens** can be **replaced by bromine or chlorine** in the presence of light or heat. ✳
- In compounds such as toluene, more that one of the benzylic hydrogens can be replaced when excess halogen in used.

- The reaction occurs via a radical chain mechanism initiated when the X_2 is converted to two X· radicals. An X· radical then abstracts the benzylic hydrogen to create a resonance stabilized benzylic radical that reacts with another molecule of X_2 to give the halogenated product and a new X· radical that continues the chain reaction.
- In molecules with alkyl groups attached to the benzylic carbon, the reaction is selective for replacement of the benzylic hydrogen because the benzylic radical is more stable than the other possible radicals.
- Bromine is more selective than chlorine in these reactions, because the transition state for bromination is reached later. The later transition state means more radical character, so the relative stability of the benzylic radical becomes more important.
- *N*-bromosuccinimide (NBS) in the presence of a radical initiator such as a peroxides will also lead to benzylic halogenation.

REACTION 19C: CLEAVAGE OF ALKYL ARYL ETHERS WITH H-X (Section 19.5D)

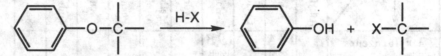

- **Alkyl aryl ethers** are **cleaved by H-X** to form an alkyl halide and a phenol. ✳

REACTION 19D: HYDROGENATION OF BENZENE WITH Ni AND HIGH PRESSURES OF H_2 (Section 19.1C)

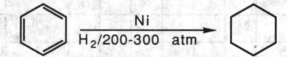

- **Benzene** can be converted into **cyclohexane** by **hydrogenation** over a **Ni catalyst**. Because benzene is aromatic, it reacts very slowly under normal conditions. Hydrogen pressures of several hundred atmospheres are usually used. ✳

REACTION 19E: HYDROGENOLYSIS OF BENZYL ETHERS (Section 19.6C)

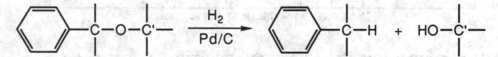

- **Benzyl ethers** react under catalytic hydrogenation conditions, H_2 and Pd/C, to give a methylbenzene and alcohol. ✳

REACTION 19F: ACID-BASE REACTIONS OF PHENOLS (Section 19.5B,C)

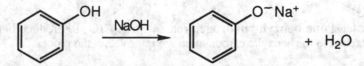

- **Phenols** are **weak acids** (pK_a **9-10**) that react with strong bases such as NaOH to form water-soluble salts. ✳

REACTION 19G: OXIDATION OF PHENOLS, CATECHOLS, AND HYDROQUINONES TO QUINONES (Section 19.5F)

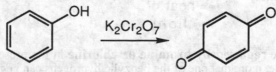

- **Phenols** are **susceptible to oxidation** because of the electron-donating -OH group on the ring. As a result, they react with strong oxidizing agents like potassium dichromate to give 1,4-benzoquinones. *
- **Catechols** (1,2-benzenediols) and **hydroquinones** (1,4-benzenediols) are also oxidized to quinones under these conditions.

REACTION 19H: KOLBE SYNTHESIS: CARBOXYLATION OF PHENOLS (Section 19.5F)

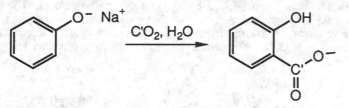

- **Phenoxides** react with CO_2 under **pressure** to give a **salicylate** product. *
- The mechanism of the reaction involves nucleophilic attack of the phenoxide onto the CO_2 carbon atom to give a cyclohexadienone anion, that undergoes keto-enol tautomerization to give the product salicylate.
- This reaction is important because can it be used in the industrial synthesis of salicylic acid, from which aspirin is synthesized.

REACTION 19I: FORMATION OF ALKYL ARYL ETHERS FROM PHENOXIDES AND ALKYL HALIDES (Section 19.5D)

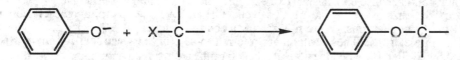

- The **Williamson ether synthesis** can be used to **prepare** certain **alkyl aryl ethers**. In these cases, the weakly nucleophilic phenol must be converted to the much more nucleophilic phenoxide ion that then reacts with an alkyl halide. Ether synthesis is often accomplished using phase transfer catalysis. *
- Note that the reverse reagents cannot be used, since aryl halides are not reactive enough to react with alkoxides.

REACTION 19J: REDUCTION OF QUINONES TO HYDROQUINONES (Section 19.5F)

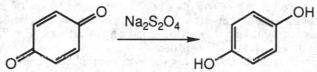

- **Quinones** can be **reduced to hydroquinones** by reducing agents such as sodium dithionite in neutral or alkaline solutions. *

SUMMARY OF IMPORTANT CONCEPTS

19.0 OVERVIEW
• Benzene and its derivatives have marked distinctions from other types of unsaturated compounds, and they are broadly classified as **aromatic**. They have remarkable stability that makes them unreactive toward reagents that attack other species such as alkenes and alkynes. **Aromaticity** is the term used to describe this special stability of benzene and its derivatives, and the term **arene** refers to aromatic hydrocarbons. *

19.1 THE STRUCTURE OF BENZENE
• An important development in the determination of benzene's structure was Kekulé's proposal that benzene is composed of six carbon atoms in a ring, with one hydrogen atom attached to each carbon. *
• The six carbon atoms of the ring are equivalent, and the carbon-carbon bond lengths are all intermediate between a single and double bond. Thus, it is not accurate to think of benzene as simply having alternating single and double bonds that are static, because this would predict alternating longer and shorter carbon-carbon bonds.

- The concepts of **hybridization of atomic orbitals** and **resonance** provide an accurate picture of the bonding in benzene. ✳
    - Each carbon atom of the ring is sp² hybridized.
    - Each carbon atom of the ring makes sigma bonds by sp²-sp² overlap with the two adjacent carbon atoms and sp²-1s overlap with a hydrogen atom.
    - Each carbon atom also has a single unhybridized 2p orbital containing one electron. These six 2p orbitals overlap to form a continuous pi system that extends over all six carbon atoms. The electron density of this pi system thus lies in two bagel-shaped regions, one above and one below the plane of the ring.
    - Benzene can be represented as a resonance hybrid of two contributing structures in which the locations of the double bonds are reversed. Alternatively, benzene is represented as a hexagon with a circle drawn on the inside.

    - **Resonance energy** is the difference in energy between a resonance hybrid and the most stable hypothetical contributing structure in which electron density is localized on particular atoms and on particular bonds. The resonance energy for benzene is large, namely 36.0 kcal/mol. What this means is that the pi system of benzene is extremely stable, and dramatically less reactive than would be expected for a normal alkene under conditions like catalytic hydrogenation (reaction 19E, Section 19.1C).
- **Molecular orbitals of benzene.** The six 2p orbitals give a set of six pi molecular orbitals. These molecular orbitals are arranged in a 1:2:2:1 pattern with respect to energy. The six pi electrons fill the three pi bonding molecular orbitals, all of which are at lower energy than the six isolated 2p orbitals. Thus, the molecular orbital approach also explains the extra stability of benzene and its derivatives. ✳
- To be **aromatic** like benzene a molecule must be cyclic, planar, have one 2p orbital on each atom of the ring, and contain (**4n + 2**) pi electrons, where **n** is zero or a positive integer (1, 2, 3, 4, 5, . . .). That is, a total of 2, 6, 10, 14, . . . pi electrons. These criteria are referred to as **Hückel's criteria for aromaticity**, named after the chemist who first described them. Note that all of the atoms in an aromatic ring must be sp² hybridized, so there cannot be any -CH₂- groups in the ring. ✳
- An **antiaromatic hydrocarbon** is the same as an aromatic hydrocarbon described above in that they are cyclic, planar, and have one 2p orbital on each atom of the ring. On the other hand, antiaromatic hydrocarbons are different because they have **4n** pi electrons. ✳
    - Unlike aromatic hydrocarbons that have extremely stable pi systems, antiaromatic hydrocarbons are less stable than an acyclic analog with the same number of pi electrons.
    - This instability can be explained for antiaromatic hydrocarbons such as cyclobutadiene by using molecular orbital theory. The four 2p orbitals of the pi system form four molecular orbitals in a 1:2:1 pattern. The four pi electrons fill these orbitals to give one filled bonding pi molecular orbital and two half-filled degenerate nonbonding molecular orbitals. It is the presence of the two unpaired electrons that makes cyclobutadiene so reactive and unstable relative to aromatic hydrocarbons. See Figure 19.6 for the molecular orbital energy diagram of cyclobutadiene.
- Some larger structures like cyclooctatetraene are only stable in a non-planar geometry. In this case, there are alternating double and single bonds, and as a result, there are two different carbon-carbon bonds lengths observed by experiment corresponding to single and double bonds, respectively. On the other hand, certain large rings can adopt a planar geometry (see annulene below) so aromaticity is possible with monocyclic systems larger than benzene.

- In order to predict the pattern of molecular orbitals to be found on a molecular orbital energy diagram, it is helpful to use the **inscribed polygon (Frost) method**. Here the shape of the polygon being analyzed (for example, a hexagon for benzene) is drawn in a ring with one point down, and the relative energies of the molecular orbitals are indicated by the points of the polygon. A horizontal line is drawn through the center of the figure. Bonding molecular orbitals are below the line, nonbonding molecular orbitals (if any) are on the line, and antibonding molecular orbitals are above the line.
- An **annulene** is a planar, cyclic hydrocarbon with a continuous overlapping pi system. Thus, cyclobutadiene and benzene are annulenes, namely [4]annulene and [6]annulene, respectively. Annulenes can be much larger, such as [14]annulene and [18]annulene. Annulenes that have (4n+2) pi electrons are aromatic as long as the ring can accommodate a planar structure. For example, [14] annulene and [18]annulene are aromatic.

[10]annulene is not aromatic because it is too small to adopt a planar geometry due to nonbonded steric interactions between hydrogen atoms on the interior of the ring. ✳

- A **heterocyclic compound** is one that contains more than one kind of atom in a ring. Certain heterocycles can be aromatic if the Hückel criteria are met. Nature is filled with aromatic heterocycles such as indoles, purines, and pyrimidines.

- An important parameter to keep track of in aromatic heterocycles is whether lone pairs of electrons are part of the aromatic pi system or not. ✳
  - For example, in pyridine (C_5H_5N), the lone pair of electrons on nitrogen lies in an sp^2 orbital that is perpendicular to the six 2p orbitals of the aromatic 6 pi electron system. Thus, the lone pair of electrons on the nitrogen is not part of the aromatic pi system, and is free to take part in interactions with other species.
  - On the other hand, in compounds such as pyrrole (C_4H_5N), the lone pair of electrons on nitrogen is part of the pi system to allow for a total of 6 pi electrons and aromaticity. Thus, this lone pair of electrons is not as available to take part in interactions with other species. *[This is an important but subtle point, so drawing the appropriate structures may be helpful.]*

- Cyclic hydrocarbon ions can also be aromatic. For example, the **cyclopropenyl cation** has 2 pi electrons, is planar, and all of the carbon atoms are sp^2 hybridized. Thus, this species satisfies the Hückel criteria for aromaticity. Other aromatic hydrocarbon ions include the **cyclopentadienyl anion** and **cycloheptatrienyl cation**. These ionic species are, of course, not anywhere near as stable as benzene or other neutral aromatic compounds, but they are highly stabilized compared to other nonaromatic hydrocarbon cations or anions. ✳

19.3 NOMENCLATURE OF AROMATIC COMPOUNDS

- The **IUPAC system retains certain common names** for several of the simpler benzene derivatives including **toluene, cumene, styrene, xylene, phenol, aniline, benzoic acid,** and **anisole**.

- In more complicated molecules, the benzene ring is named as a substituent on a parent chain.
  - The C_6H_5 group is given the name **phenyl**. For example, the IUPAC name for $C_6H_5CH_2CH_2OH$ is 2-phenylethanol.
  - The $C_6H_5CH_2$- group is given the name **benzyl**. These compounds are derivatives of toluene $C_6H_5CH_3$. *[The terms phenyl and benzyl are often confused by students. Make sure you know when each should be used.]*

- For benzene rings with **two substituents**, the three possible constitutional isomers are named as **ortho** (1,2 substitution), **meta** (1,3 substitution), and **para** (1,4 substitution). These are abbreviated as *o*, *m*, and *p*, respectively. It is also acceptable to name these species with numbers as locators. When one of the substituents has a special name (if NH_2 is present, the molecule is an aniline, etc.) then the molecule is named after that parent molecule. For example, 3-chloroaniline and *m*-chloroaniline are both acceptable names for the same molecule. If neither group imparts a special name, then the substituents are listed in alphabetical order. For example, 1-chloro-4-ethylbenzene and *p*-chloroethylbenzene are acceptable names for the same molecule.✳

- **Polynuclear aromatic hydrocarbons** (PAH) contain more than one benzene ring, each pair of which shares two carbon atoms. For example, naphthalene is two benzene rings fused together and anthracene is three benzene rings fused together in a linear fashion. Other common PAH's include phenanthrene, pyrene, coronene, and benzo[a]pyrene. Benzo[a]pyrene has been especially well-studied because it is a potent carcinogen. ✳

19.4 SPECTROSCOPIC PROPERTIES

- The **mass spectra** of aromatic hydrocarbons generally have a **strong molecular ion peak**. Alkyl-substituted aromatic hydrocarbons also contain fragments derived from **cleavage at the benzylic carbon**. Interestingly, there is some evidence that the benzyl cation (*m/z* 91) is not the compound observed in the mass spectrum, but rather a **rearrangement occurs** to generate the **more stable tropylium cation** (*m/z* 91).

- **Hydrogens** attached to benzene rings come into resonance at about δ **6.5 to** δ **8.5** in the **^{1}H-NMR spectrum**. These signals are this far downfield because of the **induced ring current** present in aromatic pi systems.
  - The **ring current** in the pi system of an aromatic compound is **induced by the applied magnetic field** when the plane of the aromatic ring is perpendicular to that of the applied magnetic field. The pi electrons are induced to circulate around the aromatic ring, which in turn, sets up a local magnetic field that reinforces the applied field on the outside of the ring. As a result, the aromatic hydrogen atoms come into resonance at a lower applied field (larger chemical shift). Note that **hydrogens on the inside of a large aromatic annulene**, such as, those in [18]annulene, exhibit the opposite effect and these signals **appear at negative values** relative to TMS.
  - With multiple substituents, the aromatic hydrogen signals may be difficult to interpret due to complex splitting patterns. One splitting pattern that is easy to recognize is the so-called **para pattern** that is a **pair of doublets** resulting from **1,4 disubstitution**.

- In ^{13}C-NMR, carbon atoms of aromatic rings give signals in the range of δ **110 to** δ **160**. The substitution patterns of aromatic rings can usually be discerned by simply counting the aromatic signals in ^{13}C-NMR. Note that alkene carbons come into resonance in the same region as the aromatic carbons, so care must be taken when interpreting spectra for molecules that contain both types of carbon atoms.
- In **infrared spectra**, the C-H hydrogen stretching vibration shows up as a moderate peak near **3030 cm^{-1}**. There are C=C **stretching** peaks at **1600 cm^{-1}** and **1450 cm^{-1}**, and a strong **out-of-plane C-H bend** in the region **690 cm^{-1} to 900 cm^{-1}**.
- Molecules with aromatic rings show **strong absorptions** in **ultraviolet-visible absorption spectra** as a result of π to π* transitions. There are **usually two peaks**, the first **high intensity peak** is **near 205 nm**, and a second, **less intense peak** near **270 nm**.

19.5 PHENOLS
- The functional group of **phenols** is a **hydroxyl group attached to a benzene ring**.
- **Phenols are more acidic** than alkyl alcohols, because of **weakening of the O-H bond by the inductive effect of the sp^2 hybridized atoms of the aromatic ring. In addition,** the **phenoxide anion is more stable** than an alkoxide anion. This **increased stability** is due to **resonance stabilization of the phenoxide anion.** In other words, upon deprotonation of a phenol, the resulting phenoxide is more stable, because the phenoxide can be considered a resonance hybrid of three contributing structures that delocalize the negative charge onto three different carbon atoms of the ring. This charge delocalization means that no one atom must absorb the entire negative charge, and the delocalized anion is thus more stable. There is no similar resonance stabilization possible for alkoxide anions. ✳

  - **Substituents on the ring** can have a **dramatic influence** on the **acidity** of phenols.
    The **inductive effect** is due to **electron polarization** caused by **differences in relative electronegativities of bonded atoms.** Atoms or groups that are **more electronegative than the sp^2 carbon atoms of the ring** are said to be **electron-withdrawing**, while atoms or groups that are **less electronegative than the sp^2 carbon atoms of the ring** are said to be **electron-releasing**. Electron-withdrawing groups help weaken the O-H bond, while also stabilizing the phenoxide anion by absorbing some of the negative charge. Electron-releasing groups destabilize a phenoxide anion by dumping even more electron density into the ring. The bottom line is that **phenols with electron-withdrawing groups like fluorine atoms are more acidic** than phenol, while **phenols with electron-releasing groups are less acidic** than simple phenol. ✳
    Another important effect is the **resonance effect.** Certain groups, especially at the ortho and para positions, also stabilize phenoxide anions because the **negative charge can be further delocalized** into the group. For example, an ortho or para nitro (-NO$_2$) group increases phenol acidity in part due to a resonance effect, since the phenoxide negative charge can be distributed partially onto the nitro group oxygen atoms. ✳

CHAPTER 19
Solutions to the Problems

<u>Problem 19.1</u> Construct a Frost circle for a planar eight-membered ring with one 2p orbital on each atom of the ring and show the relative energies of its eight pi molecular orbitals. Which are bonding, which are antibonding, and which are nonbonding?

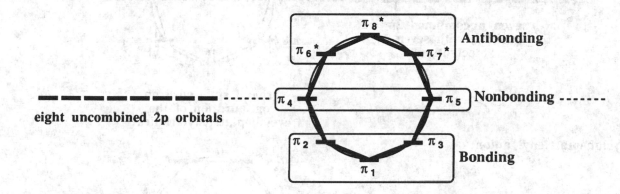

eight uncombined 2p orbitals

<u>Problem 19.2</u> Describe the ground-state electron configuration of the cyclopentadienyl cation and radical. Assuming each species is planar, would you expect it to be aromatic or antiaromatic?

Cyclopentadienyl cation:

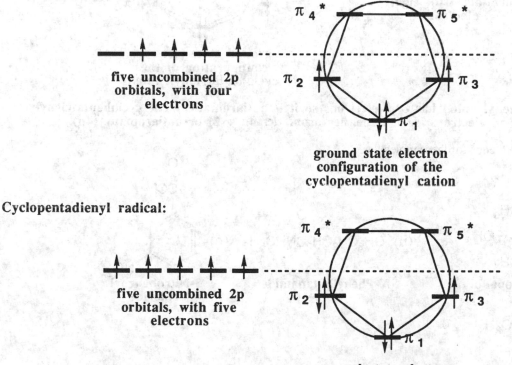

five uncombined 2p orbitals, with four electrons

ground state electron configuration of the cyclopentadienyl cation

Cyclopentadienyl radical:

five uncombined 2p orbitals, with five electrons

ground state electron configuration of the cyclopentadienyl anion

The cyclopentadienyl cation has 4 pi electrons so it is antiaromatic. The cyclopentadienyl radical has five pi electrons, so it is neither aromatic (4n + 2) or antiaromatic (4n).

Problem 19.3 Describe the ground-state electron configuration of the cycloheptatrienyl radical and anion. Assuming each species is planar, would you expect it to be aromatic or antiaromatic?

Cycloheptatrienyl radical:

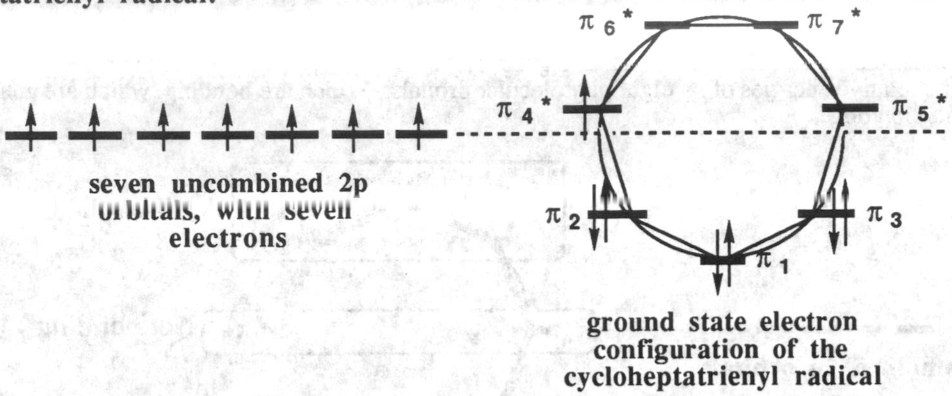

π_6^* π_7^*

π_4^* π_5^*

seven uncombined 2p orbitals, with seven electrons

π_2 π_3

π_1

ground state electron configuration of the cycloheptatrienyl radical

Cycloheptatrienyl anion:

π_6^* π_7^*

π_4^* π_5^*

seven uncombined 2p orbitals, with eight electrons

π_2 π_3

π_1

ground state electron configuration of the cycloheptatrienyl anion

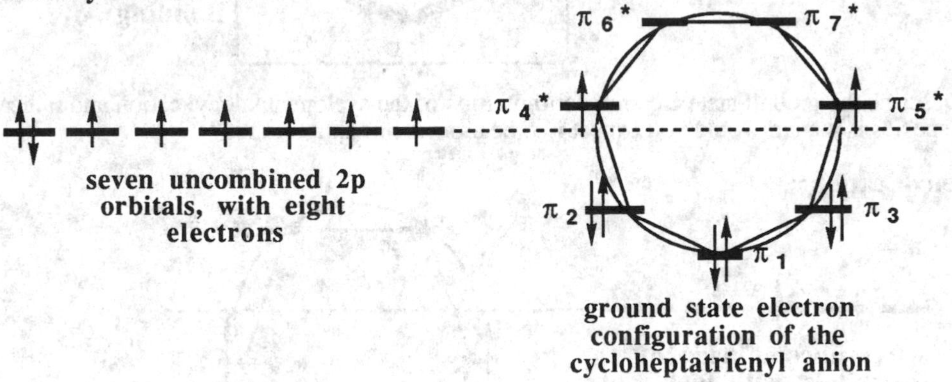

The cycloheptatrienyl anion has 8 pi electrons so it is antiaromatic. The cycloheptatrienyl radical has seven pi electrons, so it is neither aromatic ($4n + 2$) or antiaromatic ($4n$).

Problem 19.4 Write names for these molecules.

(a)

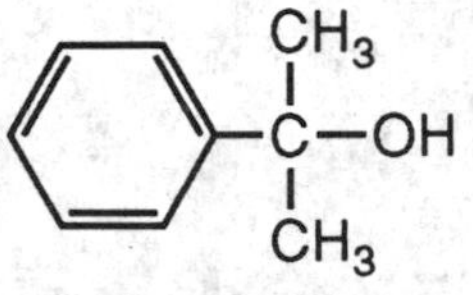

CH_3

C—OH

CH_3

2-Phenyl-2-propanol

(b) $CH_3CH_2CH_2\overset{\displaystyle O}{\overset{\displaystyle \|}{C}}NHC_6H_5$

***N*-Phenylbutanamide**

(c)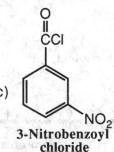

3-Nitrobenzoyl chloride

(d)

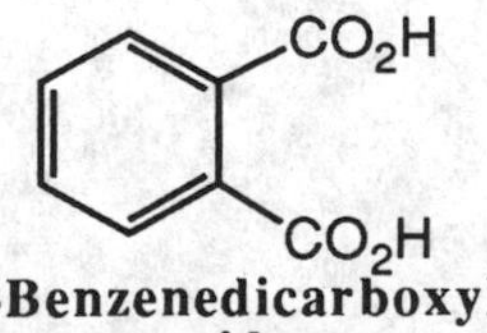

CO_2H

CO_2H

1,2-Benzenedicarboxylic acid (Phthalic acid)

Problem 19.5 Which compound gives a signal in the ^{1}H-NMR spectrum at a lower applied field (with a larger chemical shift): furan or cyclopentadiene? Explain.

Furan. Furan is an aromatic compound. Hydrogens on this aromatic ring are deshielded by the induced ring current and appear farther downfield compared to those of the nonaromatic cyclopentadiene.

Problem 19.6 Arrange these compounds in order of increasing acidity: 2,4-dichlorophenol, phenol, cyclohexanol.

The following compounds are ranked from least to most acidic:

$$\text{Cyclohexanol} \quad < \quad \text{Phenol} \quad < \quad \text{2,4-Dichlorophenol}$$

A good way to predict relative acidities between related compounds is to keep track of the anionic conjugate bases produced upon deprotonation. In general, the more stable the conjugate base anion, the stronger the acid. Anions become increasingly stabilized as the negative charge is more delocalized around the molecule. Thus, phenol is more acidic than an aliphatic alcohol like cyclohexanol, because resonance involving the aromatic ring of phenol leads to increased charge delocalization and thus stabilization of the phenoxide anion compared to the cyclohexylalkoxide anion. The electronegative chlorine atoms of 2,4-dichlorophenol withdraw electron density from the aromatic ring and thus help to stabilize the 2,4-dichlorophenoxide anion even further than what is seen with phenoxide. The inductive effect also operates to increase acidity by weakening the O-H bond. The sp^2 atoms of phenol are more electron-withdrawing than the sp^3 atoms of cyclohexanol, so the phenol O-H bond is weaker. The electron-withdrawing Cl atoms of 2,4-dichlorophenol weaken the O-H bond even further than what is seen with phenol.

Problem 19.7 Predict the products resulting from vigorous oxidation of these compounds by $K_2Cr_2O_7$ in aqueous H_2SO_4.

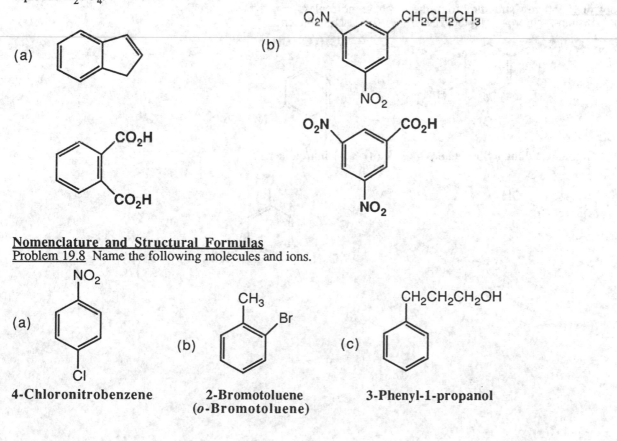

Nomenclature and Structural Formulas
Problem 19.8 Name the following molecules and ions.

(a)

(b)

(c)

4-Chloronitrobenzene   **2-Bromotoluene**   **3-Phenyl-1-propanol**
                             (*o*-**Bromotoluene**)

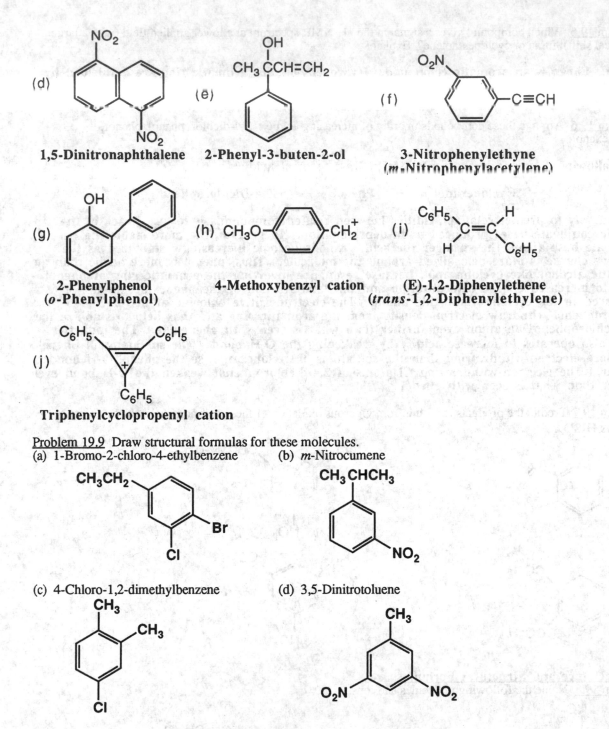

(d) 1,5-Dinitronaphthalene

(e) 2-Phenyl-3-buten-2-ol

(f) 3-Nitrophenylethyne
(*m*-Nitrophenylacetylene)

(g) 2-Phenylphenol
(*o*-Phenylphenol)

(h) 4-Methoxybenzyl cation

(i) (E)-1,2-Diphenylethene
(*trans*-1,2-Diphenylethylene)

(j) Triphenylcyclopropenyl cation

Problem 19.9 Draw structural formulas for these molecules.
(a) 1-Bromo-2-chloro-4-ethylbenzene   (b) *m*-Nitrocumene

(c) 4-Chloro-1,2-dimethylbenzene   (d) 3,5-Dinitrotoluene

(e) 2,4,6-Trinitrotoluene

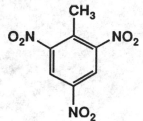

(f) 4-Phenyl-2-pentanol

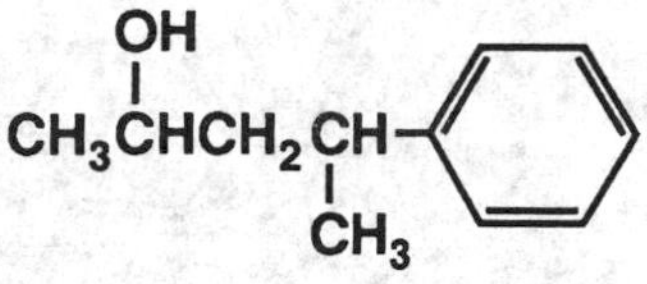

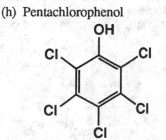

(g) *p*-Cresol

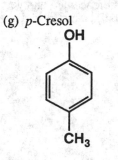

(h) Pentachlorophenol

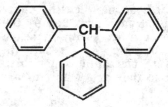

(i) 1-Phenylcyclopropanol

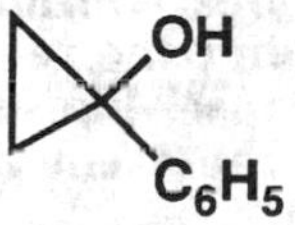

(j) Triphenylmethane

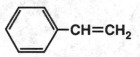

(k) Phenylethylene

$-CH=CH_2$

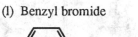

(l) Benzyl bromide

$-CH_2Br$

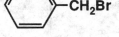

(m) 1-Phenyl-1-butyne

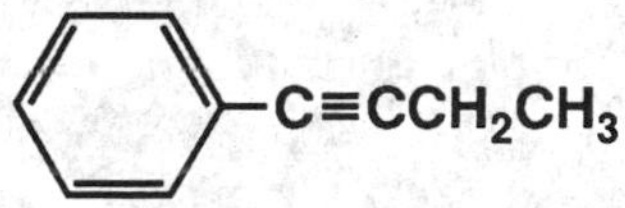

$-C\equiv CCH_2CH_3$

(n) 3-Phenyl-2-propen-1-ol

$-CH=CHCH_2OH$

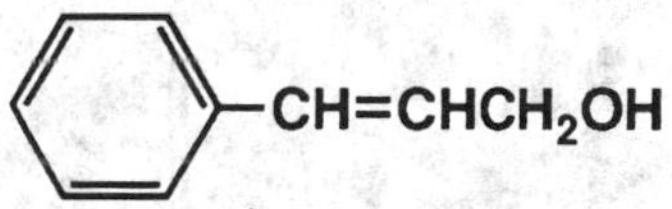

<u>Problem 19.10</u> Draw structural formulas for these molecules.
(a) 1-Nitronaphthalene

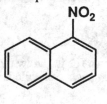

(b) 1,6-Dichloronaphthalene

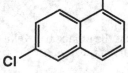

(c) 9-Bromoanthracene   (d) 2-Methylphenanthrene

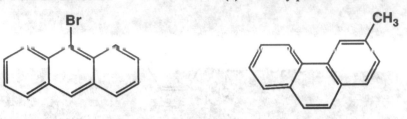

Problem 19.11 Molecules of 6,6'-dinitrobiphenyl-2,2'-dicarboxylic acid have no tetrahedral stereocenter, and yet they can be resolved to a pair of enantiomers. Account for this chirality. *Hint:* It will help to build a model and study the possibility of rotation about the single bond joining the two benzene rings.

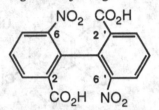

6,6'-Dinitrobiphenyl-2,2'-dicarboxylic acid

The key here is that the central bond between the benzene rings cannot rotate freely at room temperature due to the steric hindrance provided by the nitro and carboxyl groups. In other words, these groups run into each other as the molecule attempts to rotate around the central bond, so it is prevented from rotating. As a result, the molecule is chiral because, like a propeller, there are two different orientations possible. The two orientations represent non-superimposable mirror images so they are a pair of enantiomers.

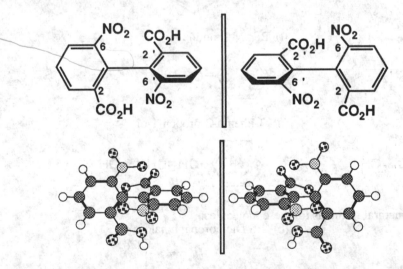

Resonance in Aromatic Compounds
Problem 19.12 Following each name is the number of Kekulé structures that can be drawn for it. Draw these Kekulé structures, and show, using curved arrows how the first contributing structure for each molecule is converted to the second and so forth.
(a) Naphthalene (3)

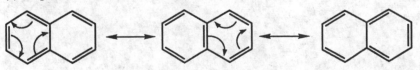

(b) Phenanthrene (5)

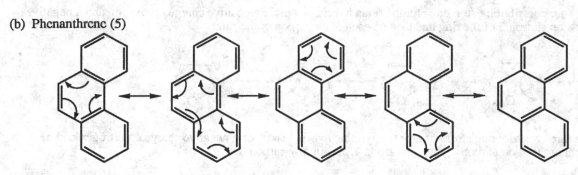

<u>Problem 19.13</u> Each molecule can be drawn as a hybrid of five contributing structures: two Kekulé structures and three that involve creation and separation of unlike charges. For (a) and (b), the creation and separation of unlike charges places a positive charge on the substituent and a negative charge on the ring. For (c), a positive charge is placed on the ring and an additional negative charge is placed on the $-NO_2$ group. Draw these five contributing structures for each molecule.

(a) Chlorobenzene

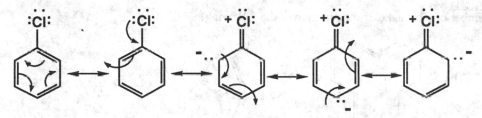

(b) Phenol

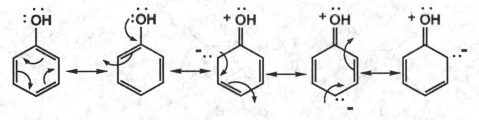

(c) Nitrobenzene

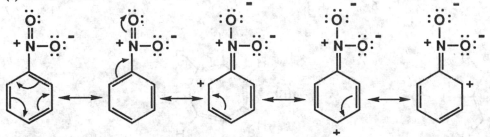

<u>Problem 19.14</u> Following are structural formulas for furan and pyridine.

Furan

Pyridine

(a) Write four contributing structures for the furan hybrid that place a positive charge on oxygen and a negative charge first on carbon 3 of the ring and then on each other carbon of the ring.

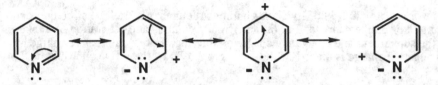

(b) Write three contributing structures for the pyridine hybrid that place a negative charge on nitrogen and a positive charge first on carbon 2, then on carbon 4 and finally carbon 6.

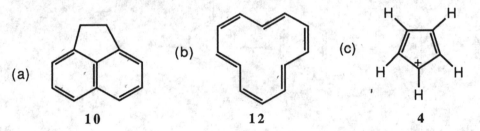

The Concept of Aromaticity
Problem 19.15 State the number of p orbital electrons in each of the following.

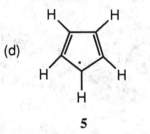

(a)

10

(b)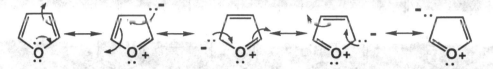

12

(c)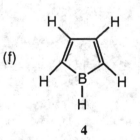

4

(d)

5

(e)

6

(f)

4

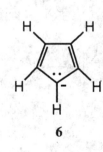

(g)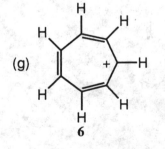

6

(h)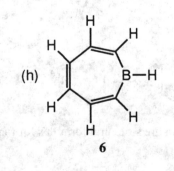

6

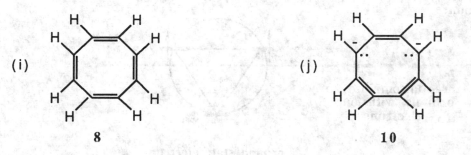

(i)   (j)

8   10

<u>Problem 19.16</u> Which of the molecules and ions given in the previous problem are aromatic according to the Hückel criteria? Which, if planar, would be antiaromatic?

The following molecules are aromatic because they have 4n + 2 π electrons: a, e, g, h, and j.

The following molecules would be antiaromatic if planar because they have 4n π electrons: b, c, f, and i.

<u>Problem 19.17</u> Construct MO energy diagrams for the cyclopropenyl cation, radical, and anion. Which of these species is aromatic according to the Hückel criteria?

Cyclopropenyl cation:

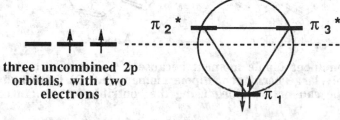

three uncombined 2p
orbitals, with two
electrons

ground state electron
configuration of the
cyclopropenyl cation

Cyclopropenyl radical:

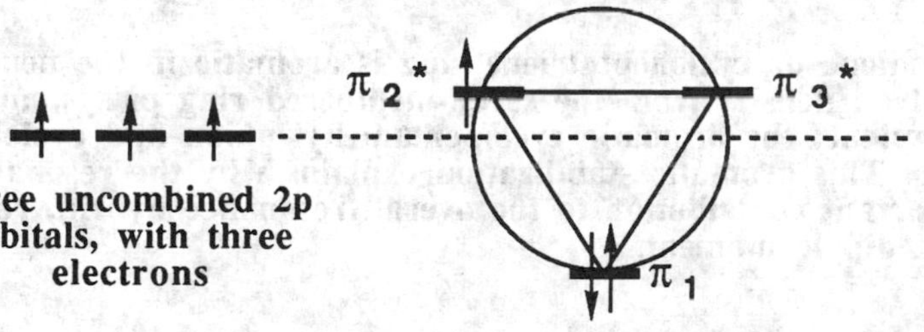

three uncombined 2p
orbitals, with three
electrons

ground state electron
configuration of the
cyclopropenyl radical

Cyclopropenyl anion:

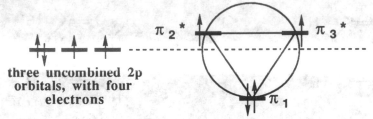

three uncombined 2p
orbitals, with four
electrons

ground state electron
configuration of the
cyclopropenyl anion

Of these species, the only one that satisfies the Hückel criteria is the first one, namely the
cyclopropenyl cation with 2 pi electrons, a Hückel number (4n + 2, here n = 0).

Problem 19.18 Naphthalene and azulene are constitutional isomers of molecular formula $C_{10}H_8$. Naphthalene is
a colorless solid with a dipole moment of zero. Azulene is a solid with an intense blue color and a dipole moment
of 1.0 D. Account for the difference in dipole moments of these constitutional isomers.

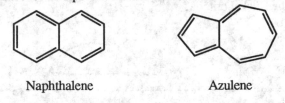

Naphthalene   Azulene

Naphthalene has no permanent dipole moment because it possesses a high degree of symmetry.
Azulene has a remarkably large permanent dipole moment (1.0 D) for a hydrocarbon. The
dipole moment of azulene can be explained using the contributing structure shown below on
the right:

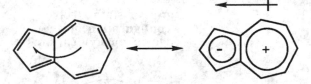

Neither a cyclopentadiene or cycloheptatriene ring is aromatic in the neutral state. However,
by transferring electron density from the seven-membered ring pi system to the five-membered
ring pi system of azulene, the aromatic cyclopentadienyl anion and cycloheptatrienyl cation pi
systems are formed. This aromatic stabilization explains why the resonance structure on the
right makes an important contribution to the overall resonance hybrid, resulting in the large
observed permanent dipole moment.

Spectroscopy
Problem 19.19 Compound A, molecular formula C_9H_{12}, shows prominent peaks in its mass spectrum at m/z 120
and 105. Compound B, also molecular formula C_9H_{12}, shows prominent peaks at m/z 120 and 91. On vigorous
oxidation with chromic acid, both compounds give benzoic acid. From this information, deduce the structural
formulas of compounds A and B.

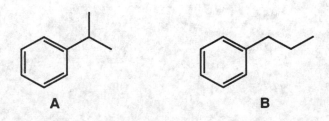

A   B

The compounds can be identified by the major fragments in the mass spectrum, both of which are benzylic cations. For compound A, the peak at *m/z* 105 corresponds to the cation shown below:

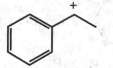

Peak at *m/z* 105 for compound A

For compound B, the peak at *m/z* 91 corresponds to the tropylium ion that was produced by rearrangement of the benzyl cation (Section 15.4)

Peak at *m/z* 91 for compound B

Both of the compounds will produce benzoic acid upon oxidation with chromic acid.

Problem 19.20 Compound C shows a molecular ion at *m/z* 148, and other prominent peaks at *m/z* 105, and 77. Following are its infrared and ^{1}H-NMR spectra.

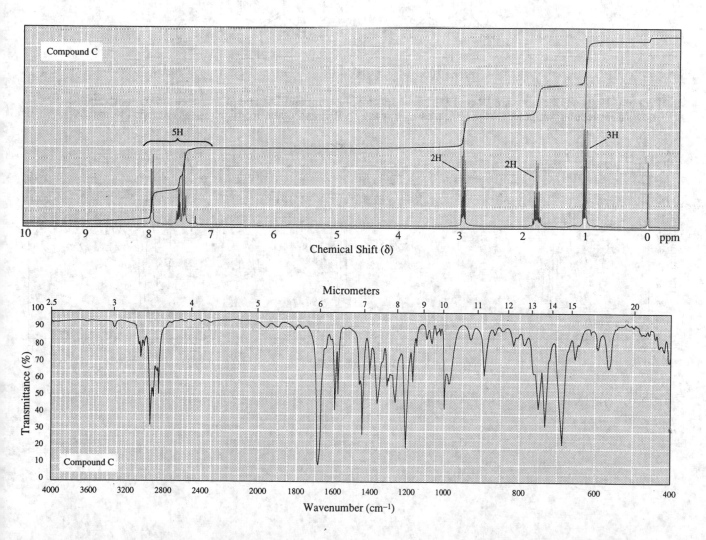

(a) Deduce the structural formula of Compound C.

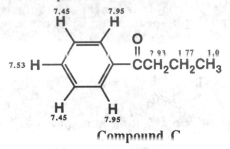

Compound C

This compound, $C_{10}H_{12}O$, has the correct molecular formula of 148. This compound also has a carbonyl corresponding to the peak at 1680 cm^{-1} in the IR spectrum and the correct pattern of hydrogens to explain the ^1H-NMR spectrum.

(b) Account for the appearance of peaks in its mass spectrum at m/z 105 and 77.

The peaks at m/z 105 and 77 correspond to the following fragments produced by α-cleavage (Section 17.4) on either side of the carbonyl group.

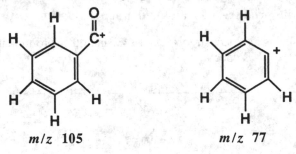

Problem 19.21   Following are IR and ^1H-NMR spectra of compound D. The mass spectrum of compound D shows a molecular ion peak at m/z 136, a base peak at m/z 107, and other prominent peaks at m/z 118 and 59.

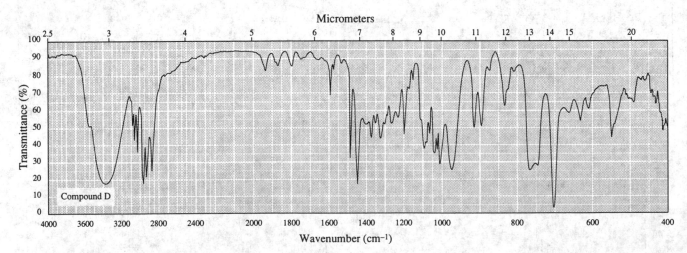

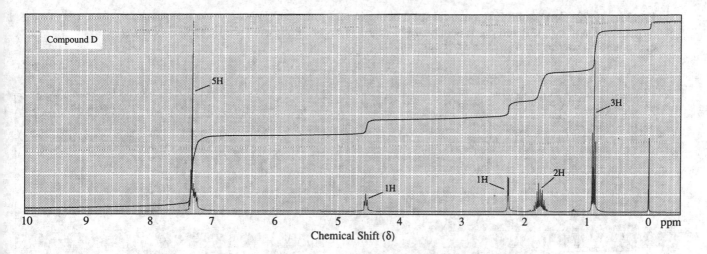

(a) Propose a structural formula for compound D based on this information.

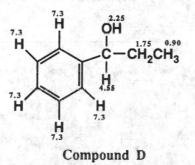

Compound D

This compound, $C_9H_{12}O$, has the correct molecular weight of 136. The hydroxyl group gives the broad peak at 3300 cm^{-1} in the IR spectrum and the pattern of hydrogens present explains the ^{1}H-NMR spectrum.

(b) Propose structural formulas for ions in the mass spectrum at *m/z* 118, 107 and 59.

The peaks at *m/z* 118, 107 and 77 correspond to the following fragments:

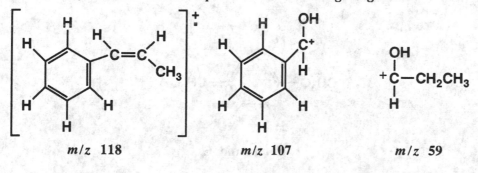

Problem 19.22 Compound E is a neutral solid of molecular formula $C_8H_{10}O_2$. Its mass spectrum shows a molecular ion at *m/z* 138 and prominent peaks at M-1 and M-17. Following are IR and ^{1}H-NMR spectra of compound E. Deduce the structure of compound E.

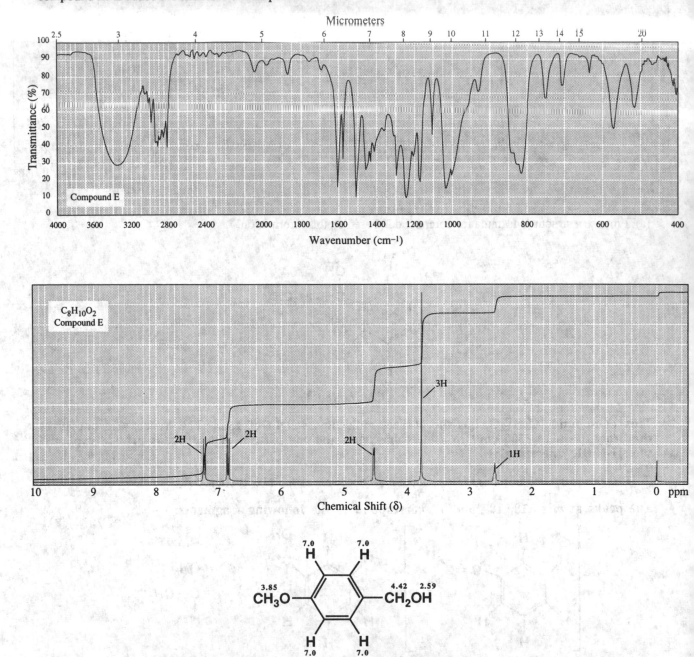

Compound E

A molecule of molecular formula $C_8H_{10}O_2$ has an index of hydrogen deficiency of 4, which is accounted for by the three double bonds and ring of the benzene ring of compound E. Compound E also has an alcohol function corresponding to the broad peak at 3350 cm^{-1} in the IR spectrum, and the correct pattern of hydrogens to explain the ^{1}H-NMR spectrum. Notice especially the distinctive *para* pattern near δ 7.0 and the methyl group signal at δ 3.85.

Problem 19.23 Following are ^{1}H-NMR and ^{13}C-NMR spectral data for compound F, $C_{12}H_{16}O$. From this information, deduce the structure of compound F.

^{1}H-NMR	^{13}C-NMR	
0.83 (d, 6H)	207.82	50.88
2.11 (m, 1H)	134.24	50.57
2.30 (d, 2H)	129.36	24.43
3.64 (s, 2H)	128.60	22.48
7.2-7.4 (m, 5H)	126.86	

The signal in the ^{13}C-NMR spectrum at δ 207.82 indicates the presence of a carbonyl group. The signals around δ 130 indicate there is an aromatic ring. The doublet in the ^{1}H-NMR at δ 0.83 that integrates to 6H indicates two methyl groups adjacent to a -CH- group. There are also two -CH$_2$- groups; one that is not adjacent to other hydrogens (the singlet at δ 3.64) and one next to a -CH- group (the doublet at δ 2.30). The multiplet at δ 2.11 must be this -CH- group that is also adjacent to the two methyl groups. Five aromatic hydrogens are found in the complex set of signals at δ 7.2-7.4. The only structure that is consistent with all of these facts is 4-methyl-1-phenyl-2-pentanone.

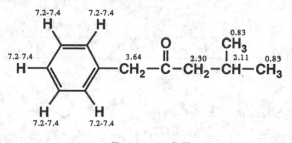

Compound F

A molecule of molecular formula $C_{12}H_{16}O$ has an index of hydrogen deficiency of 5, which is accounted for by the three double bonds and ring of the benzene ring plus the pi bond of the carbonyl group in compound F.

Problem 19.24 Following are ^{1}H-NMR and ^{13}C-NMR spectral data for compound G, $C_{10}H_{10}O$. From this information, deduce the structure of compound G.

^{1}H-NMR	^{13}C-NMR	
2.50 (t, 2H)	210.19	126.82
3.05 (t, 2H)	136.64	126.75
3.58 (s, 2H)	133.25	45.02
7.1-7.3 (m, 4H)	128.14	38.11
	127.75	28.34

The signals at δ 210.19 in the ^{13}C-NMR indicate the presence of a carbonyl species. The six signals between δ 126 and δ 137 indicate that there is a phenyl ring and the three signals between δ 28 and δ 45 indicate there are three more sp^3 carbon atoms. The signals between δ 7.1 and δ 7.3 in the ^{1}H-NMR integrate to 4H so the phenyl ring must have two hydrogens replaced by other atoms. The three signals between δ 2.6 and δ 3.6 integrate to 2H each so these must be three -CH$_2$- groups. Furthermore, the splitting pattern indicates that two of these are adjacent to each other (the two triplets) while one is not adjacent to any carbons with hydrogen atoms attached. The only structure fully consistent with these spectra is β-tetralone.

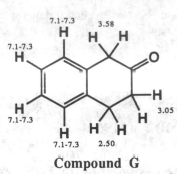

Compound G

A molecule of molecular formula $C_{10}H_{10}O$ has an index of hydrogen deficiency of 6, which is accounted for by the three double bonds and ring of the benzene ring, the pi bond of the carbonyl group, and the cyclohexane ring in compound G.

<u>Problem 19.25</u> Compound H, $C_8H_6O_3$, gives a precipitate when treated with hydroxylamine in aqueous ethanol, and a silver mirror when treated with Tollens' solution. Following is its ^{1}H-NMR spectrum. Deduce the structure of compound H.

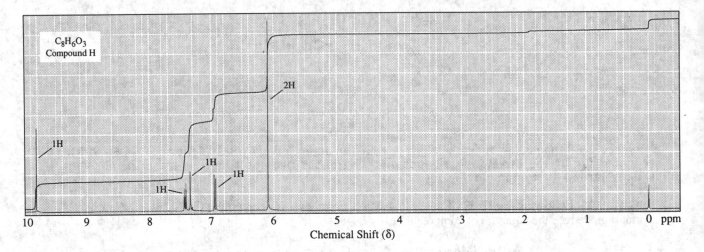

The hydroxylamine reaction indicates the presence of a carbonyl species, and the positive Tollens test confirms the presence of an aldehyde. The aldehyde is also indicated by the singlet in the ^{1}H-NMR at δ 9.8. The splitting pattern of the aromatic hydrogen signals between δ 6.95 and 7.5 indicates that the phenyl ring only has three hydrogens, two are adjacent to each other (the two doublets), while the third is not adjacent to any hydrogens. The singlet at δ 6.1, integrating to 2H, indicates the presence of a -CH$_2$- group bound to two very electronegative atoms (oxygen). Given the molecular formula, the only structure that agrees with this information is piperonal.

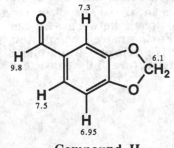

Compound H

A molecule of molecular formula $C_8H_6O_3$ has an index of hydrogen deficiency of 6, which is accounted for by the three double bonds and ring of the benzene ring, the pi bond of the carbonyl group, and the five-membered ring in compound H.

Problem 19.26 Compound I, $C_{11}H_{14}O_2$, is insoluble in water, aqueous acid, and aqueous $NaHCO_3$ but dissolves readily in 10% Na_2CO_3 and 10% NaOH. When these alkaline solutions are acidified with 10% HCl, compound I is recovered unchanged. Given this information and its ^{1}H-NMR spectrum, deduce the structure of compound I.

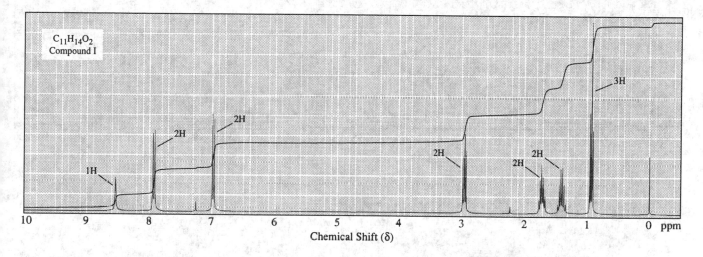

Compound I dissolves in Na_2CO_3 or NaOH so it must be deprotonated by these bases. This behavior is expected for a weakly acidic species like a phenol. The ^{1}H-NMR is also consistent with the presence of a phenol since there is a broad singlet at δ 8.54 that integrates to 1H. There is no aldehyde hydrogen signal and the molecule does not undergo an aldol reaction since it is recovered unchanged from alkaline solution, therefore there is no aldehyde in the molecule. The two doublets at δ 6.9 and δ 7.9, each integrating to 2H, indicate a 1,4 disubstituted phenyl ring. The signals between δ 0.9 and δ 3.0 indicate a -CH_2-CH_2-CH_2-CH_3 group. Finally, the molecular formula indicates an index of hydrogen deficiency of 5. The phenyl ring accounts for 4 so there must be one other π bond (or ring) in the molecule. The only structure consistent with the molecular formula and the spectrum is 1-(4-hydroxyphenyl)-1-pentanone (4-hydroxy-valerophenone).

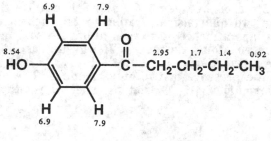

Compound I

Problem 19.27 Propose a structural formula for compound J, $C_{11}H_{14}O_3$, consistent with its ^{1}H-NMR and infrared spectra.

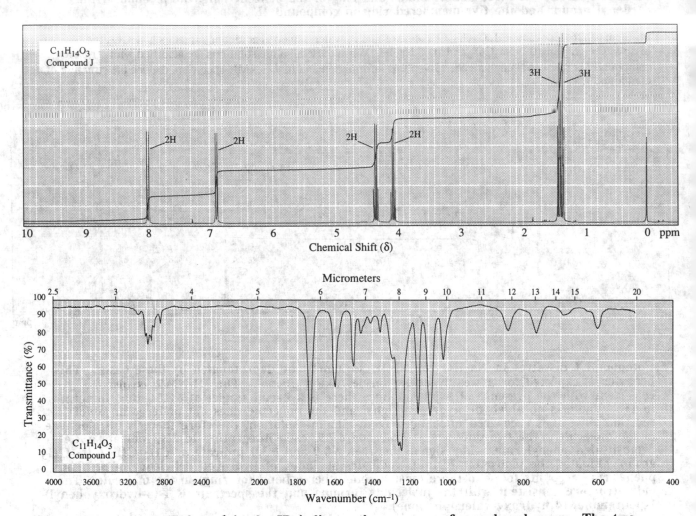

The strong peak at 1720 cm^{-1} in the IR indicates the presence of a carbonyl group. The two sets of doublets integrating to 2H each between δ 6.9 and 8.0 indicate the presence of a 1,4 disubstituted benzene ring. The two quartets integrating to 2H each at δ 4.35 and δ 4.05 indicate there are two -CH$_2$- groups attached to oxygen atoms as well methyl groups. The molecular formula has three oxygen atoms, so the -OCH$_2$- groups are likely part of one ester and one ether. The two triplets integrating to a total of 6H at δ 1.4 are the signals from the two methyl groups, confirming the presence of two ethyl groups. The only structure that is consistent with the molecular formula $C_{11}H_{14}O_3$ and the spectra is ethyl 4-ethoxybenzoate.

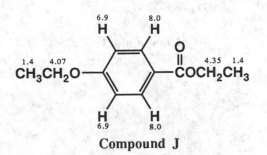

Compound J

A molecule of molecular formula $C_{11}H_{14}O_3$ has an index of hydrogen deficiency of 5, which is accounted for by the three double bonds and ring of the benzene ring plus the pi bond of the carbonyl group in compound J.

<u>Problem 19.28</u> Propose a structural formula for the analgesic phenacetin, molecular formula $C_{10}H_{13}NO_2$, based on its ^{1}H-NMR spectrum.

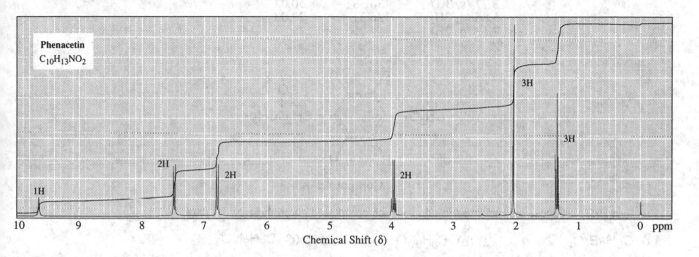

This structure is not only consistent with the molecular formula, but also with the ^{1}H-NMR spectrum. The characteristic two doublets centered at δ 6.8 and 7.5 indicate the presence of a 1,4-disubstituted benzene ring. The singlet at δ 9.65 integrating to 1H indicates a primary amide, and the singlet integrating to 3H at δ 2.05 indicates this is an acetamide. Finally, the typical ethyl splitting pattern for the signals at δ 1.32 and δ 3.95 indicates the presence of an ethyl group. These signals are shifted so far downfield that they must be part of an ethoxy group.

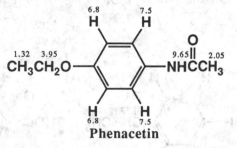

Phenacetin

A molecule of molecular formula $C_{10}H_{13}NO_2$ has an index of hydrogen deficiency of 5, which is accounted for by the three double bonds and ring of the benzene ring plus the pi bond of the carbonyl group in phenacetin.

Problem 19.29 Compound K, $C_{10}H_{12}O_2$, is insoluble in water, 10% NaOH, and 10% HCl. Given this information and the following ^{1}H-NMR and ^{13}C-NMR spectral information, deduce the structural formula of Compound K.

^{1}H-NMR	^{13}C-NMR	
2.10 (s, 3H)	206.51	114.17
3.61 (s, 2H)	158.67	55.21
3.77 (s, 3H)	130.33	50.07
6.86 (d, 2H)	126.31	29.03
7.12 (d, 2H)		

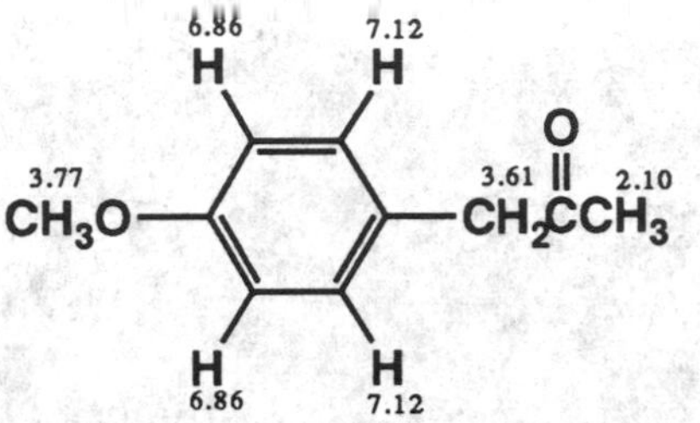

Compound K

Problem 19.30 Propose a structural formula for each compound given these NMR data.

(a) $C_9H_9BrO_2$

^{1}H-NMR	^{13}C-NMR
1.39 (t, 3H)	165.73
4.38 (q, 2H)	131.56
7.57 (d, 2H)	131.01
7.90 (d, 2H)	129.84
	127.81
	61.18
	14.18

(b) C_8H_9NO

^{1}H-NMR	^{13}C-NMR
2.06 (s, 3H)	168.14
7.01 (t, 1H)	139.24
7.30 (m, 2H)	128.51
7.59 (d, 2H)	122.83
9.90 (s, 1H)	118.90
	23.93

(c) $C_9H_9NO_3$

^{1}H-NMR	^{13}C-NMR
2.10(s, 3H)	168.74
7.72 (d, 2H)	166.85
7.91 (d, 2H)	143.23
10.3 (s, 1H)	130.28
12.7 (s, 1H)	124.80
	118.09
	24.09

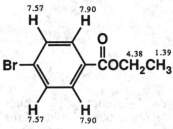

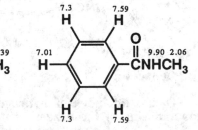

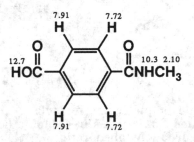

Problem 19.31 Given here are ^{1}H-NMR and ^{13}C-NMR spectral data for two compounds. Each shows strong, sharp absorption between 1700 and 1720 cm^{-1}, and strong, broad absorption over the region 2500 - 3000 cm^{-1}. Propose a structural formula for each compound.

(a) $C_{10}H_{12}O_3$

^{1}H-NMR	^{13}C-NMR
2.49 (t, 2H)	173.89
2.80 (t, 2H)	157.57
3.72 (s, 3H)	132.62
6.78 (d, 2H)	128.99
7.11 (d, 2H)	113.55
12.4 (s 1H)	54.84
	35.75
	29.20

(b) $C_{10}H_{10}O_2$

^{1}H-NMR	^{13}C-NMR
2.34(s, 3H)	167.82
6.38 (d, 1H)	143.82
7.18 (d, 1H)	139.96
7.44 (d, 2H)	131.45
7.56 (d, 2H)	129.37
12.0 (s 1H)	127.83
	111.89
	21.13

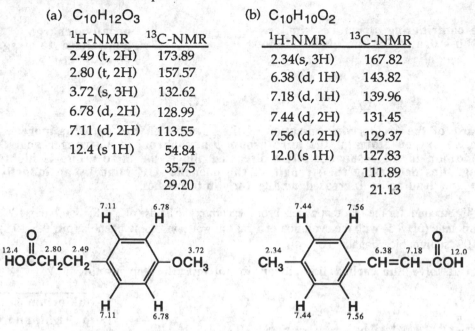

Acidity of Phenols
Problem 19.32 Account for the fact that p-nitrophenol is a stronger acid than phenol. Consider both the resonance and inductive effects of the nitro group.

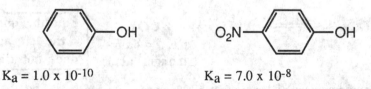

$$K_a = 1.0 \times 10^{-10} \qquad\qquad K_a = 7.0 \times 10^{-8}$$

As seen from the acid ionization constants, p-nitrophenol is the stronger acid. To account for this fact, draw contributing structures for each anion and compare the degree of delocalization of negative charge (i.e., the resonance stabilization of each anion). Phenoxide ion is a resonance hybrid of five important contributing structures, two of which place the negative charge on the phenoxide oxygen, and three of which place the negative charge on the atoms of the ring.

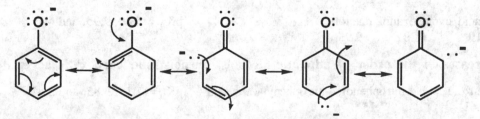

The p-nitrophenoxide ion is a hybrid of six important contributing structures. In addition to the five similar to those drawn above for the phenoxide ion, there is a sixth that places the negative charge on the oxygen atoms of the p-nitro group.

**five contributing structures similar
to those drawn for the phenoxide ion** ⟷ **negative charge
delocalized to oxygens
of nitro group**

Thus, because of the greater delocalization of the negative charge onto the more
electronegative oxygen atoms of the nitro group, *p*-nitrophenol is a stronger acid than phenol.
     In addition to the resonance effect discussed above, the nitro group is highly electron-
withdrawing, thus decreasing the strength of the phenolic -OH bond by an inductive
mechanism and leading to increased acidity for 4-nitrophenol.

Problem 19.33 Account for the fact that water-insoluble carboxylic acids (pK_a 4-5) dissolve in 10% aqueous
sodium bicarbonate (pH 8.5) with the evolution of a gas but water-insoluble phenols (pK_a 9.5-10.5) do not
dissolve in 10% sodium bicarbonate.

In order to dissolve, the carboxylic acid or phenol must be deprotonated.

$$R-\overset{\overset{\displaystyle O}{\|}}{C}OH + HCO_3^- \rightleftharpoons R-\overset{\overset{\displaystyle O}{\|}}{C}O^- + H_2CO_3$$

pK_a = 4-5   pK_a = 6.36
(stronger acid)   (weaker acid)

| Equilibrium lies to the right, and it dissolves |

$$Ar-OH + HCO_3^- \rightleftharpoons Ar-O^- + H_2CO_3$$

pK_a = 10   pK_a = 6.36
(weaker acid)   (stronger acid)

| Equilibrium lies to the left, and it does not dissolve |

Problem 19.34 Match each compound with its appropriate pK_a value.
(a) 4-Nitrobenzoic acid, benzoic acid, 4-chlorobenzoic acid   pK_a = 4.19, 3.98, and 3.41
   pK_a 3.41   4.19   3.98

Electron-withdrawing groups increase acidity through a combination of resonance and
inductive effects. The nitro group is more withdrawing than the chloro group, explaining the
observed trend.

(b) Benzoic acid, cyclohexanol, phenol   pK_a = 18.0, 9.95, and 4.19
   pK_a 4.19   18.0   9.95

Acidity increases in the order of aliphatic alcohols, phenols, and carboxylic acids.

(c) 4-Nitrobenzoic acid, 4-nitrophenol, 4-nitrophenylacetic acid   pK_a = 7.15, 3.85, and 3.41
   pK_a 3.41   7.15   3.85

From part (a) it is clear that 4-nitrobenzoic acid has a pK_a of 3.41, meaning that the other
carboxylic acid, 4-nitrophenylacetic acid, must be slightly less acidic with a pK_a of 3.85.
This makes sense since the electron-withdrawing nitro group is farther away from the
carboxylic acid group in 4-nitrophenylacetic acid compared to 4-nitrobenzoic acid. The 4-
nitrophenol is significantly less acidic with a pK_a of 7.15.

Problem 19.35 Arrange the molecules and ions in each set in order of increasing acidity (from least acidic to most acidic).

(a)

To arrange these in order of increasing acidity, refer to Table 4.2. For those compounds not listed in Table 3.2, estimate pK$_a$ using values for compounds that are given in the table.

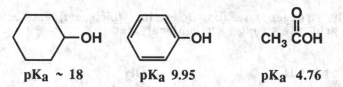

pK$_a$ ~ 18   pK$_a$ 9.95   pK$_a$ 4.76

(b)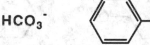

HCO_3^- H_2O

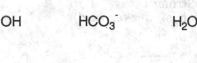

H_2O HCO_3^-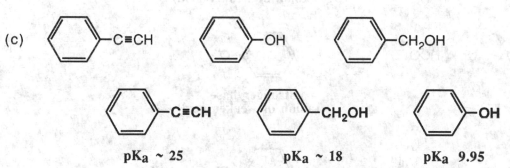

pK$_a$ 15.7   pK$_a$ 10.33   pK$_a$ 9.95

(c)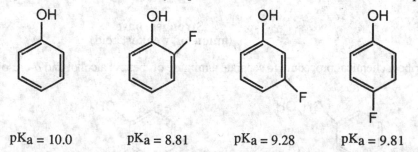

—C≡CH   —OH   —CH$_2$OH

—C≡CH   —CH$_2$OH   —OH

pK$_a$ ~ 25   pK$_a$ ~ 18   pK$_a$ 9.95

Problem 19.36 Explain the trends in the acidity of phenol and the monofluoro derivatives of phenol:

pK$_a$ = 10.0   pK$_a$ = 8.81   pK$_a$ = 9.28   pK$_a$ = 9.81

The electronegative fluoro substituent increases the acidity of the phenol through an inductive effect, so all of the monofluoro isomers of phenol are more acidic than phenol itself. Because this is an inductive effect, the closer the fluorine atom to the phenolic OH group, the stronger the effect and the greater the acidity. Thus, the *ortho* fluoro derivative is most acidic, followed by the *meta* fluoro derivative, then finally the *para* fluoro derivative.

Problem 19.37 You wish to determine the inductive effects of a series of functional groups, for example Cl, Br, CN, CO_2H, and C_6H_5. Is it best to use a series of ortho-, meta-, or para-substituted phenols? Explain your answer.

The question to be addressed involves inductive effects only. It would be best to use the derivatives with the substituents in the *meta* position, because this would minimize any contributions from resonance effects. The resonance of effects are maximal when substituents are in the *ortho* and *para* positions.

Problem 19.38 From each pair, select the stronger base.

To estimate which is the stronger base, first determine which conjugate acid is the weaker acid. The weaker the acid, the stronger its conjugate base.

(a) 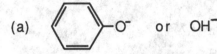   or   OH⁻   **OH⁻**

                                                          Stronger base
                                                     (anion of weaker acid)

(b) 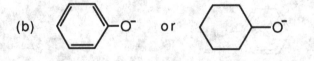   or

                                                          Stronger base
                                                     (anion of weaker acid)

(c) 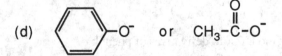   or HCO_3^-

                                                          Stronger base
                                                     (anion of weaker acid)

(d)   or $CH_3-\overset{O}{\overset{\|}{C}}-O^-$

                                                          Stronger base
                                                     (anion of weaker acid)

Problem 19.39 Describe a chemical procedure to separate a mixture of benzyl alcohol and *o*-cresol and recover each in pure form.

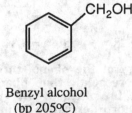
Benzyl alcohol
(bp 205°C)

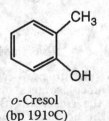
o-Cresol
(bp 191°C)

Following is a flow chart for an experimental method for separating these two compounds. Separation is based on the facts that each is insoluble in water, soluble in diethyl ether, and

that *o*-cresol reacts with 10% NaOH to form a water-soluble phenoxide salt while benzyl alcohol does not.

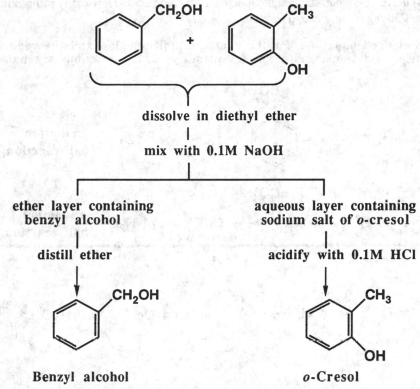

Problem 19.40 The compound 2-hydroxypyridine, a derivative of pyridine, is in equilibrium with 2-pyridone. 2-Hydroxypyridine is aromatic. Does 2-pyridone have comparable aromatic character? Explain.

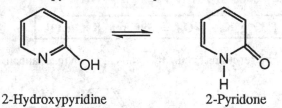

2-Hydroxypyridine   2-Pyridone

2-Pyridones have aromatic character because of the contribution from the contributing structure shown on the right.

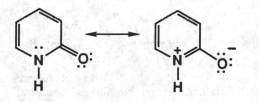

Reactions at the Benzylic Position

<u>Problem 19.41</u> Write a balanced equation for the oxidation of *p*-xylene to 1,4-benzenedicarboxylic acid (terephthalic acid) using potassium dichromate in aqueous sulfuric acid. How many milligrams of $K_2Cr_2O_7$ is required to oxidize 250 mg of *p*-xylene to terephthalic acid?

Following are balanced equations for the oxidation half-reaction and the reduction half-reaction. Because oxidation takes place in aqueous acid, the reactions are balanced with H_2O and H^+.

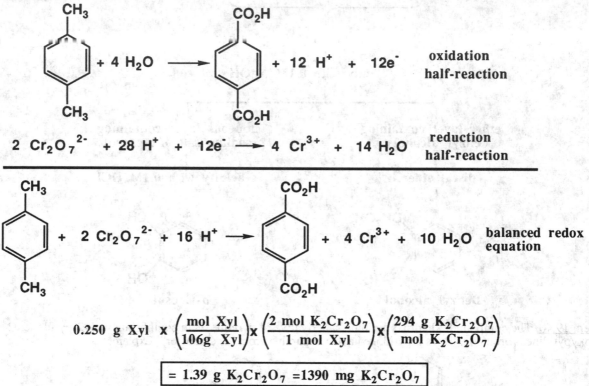

$$2\ Cr_2O_7{}^{2-} + 28\ H^+ + 12e^- \longrightarrow 4\ Cr^{3+} + 14\ H_2O \qquad \text{reduction half-reaction}$$

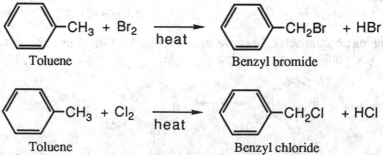

$$0.250\ \text{g Xyl} \times \left(\frac{\text{mol Xyl}}{106\text{g Xyl}}\right) \times \left(\frac{2\ \text{mol } K_2Cr_2O_7}{1\ \text{mol Xyl}}\right) \times \left(\frac{294\ \text{g } K_2Cr_2O_7}{\text{mol } K_2Cr_2O_7}\right)$$

$$\boxed{= 1.39\ \text{g } K_2Cr_2O_7 = 1390\ \text{mg } K_2Cr_2O_7}$$

<u>Problem 19.42</u> Each of the following reactions occurs by a radical chain mechanism. (Consult Appendix 3 for bond dissociation energies.)

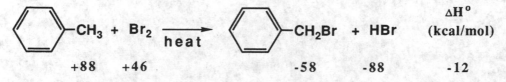

(a) Calculate the heat of reaction, $\Delta H°$, in kilocalories per mol for each reaction.

Following are the calculations of the enthalpies of reaction.

					$\Delta H°$
					(kcal/mol)

C₆H₅–CH₃ + Br₂ →(heat) C₆H₅–CH₂Br + HBr

| +88 | +46 | | −58 | −88 | −12 |

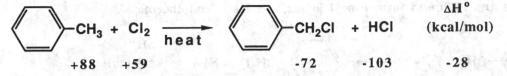

+88   +59   -72   -103   -28

(b) Write a pair of chain propagation steps for each mechanism, and show that the net result of the chain propagation steps is the observed reaction.
(c) Calculate $\Delta H°$ for each chain propagation step, and show that the sum for each pair of chain propagation steps is identical with the $\Delta H°$ value calculated in part (a).

Shown below are pairs of chain propagation steps for each reaction. Each pair adds up to the observed reaction and the observed enthalpy of reaction. This answers both (b) and (c).

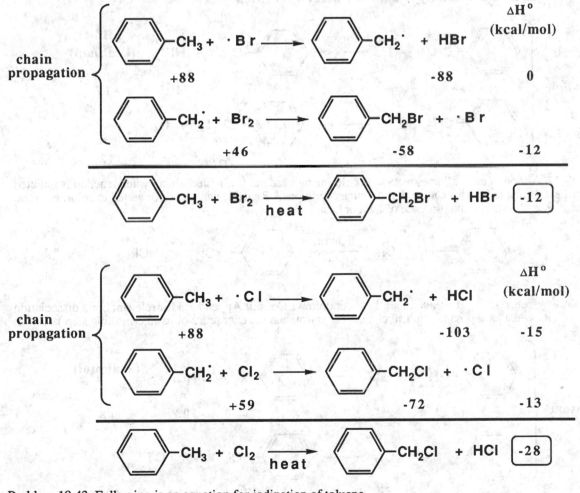

Problem 19.43 Following is an equation for iodination of toluene.

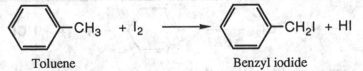

Toluene   Benzyl iodide

This reaction does not take place. All that happens under experimental conditions for the formation of radicals is initiation to form iodine radicals, I•, followed by termination to reform I_2. How do you account for these observations?

Reaction of toluene and iodine to form benzyl iodide and HI is endothermic.

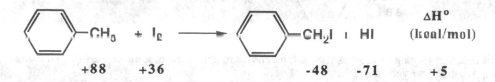

| | +88 | +36 | | -48 | -71 | +5 |

Using values for bond dissociation energies, calculate the enthalpy for each of the most likely chain propagation steps. Abstraction of hydrogen by an iodine radical (an iodine atom) is endothermic by 17 kcal/mol. The activation energy for this step is approximately a few kcal/mol greater than 17 kcal/mol. Given this large activation energy and the fact that the overall reaction is endothermic, it does not occur as written.

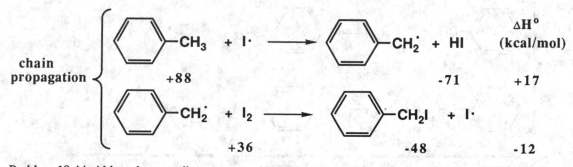

Problem 19.44 Although most alkanes react with chlorine by a radical chain mechanism when reaction is initiated by light or heat, benzene fails to react under the same conditions. Benzene cannot be converted to chlorobenzene by treatment with chlorine in the presence of light or heat.

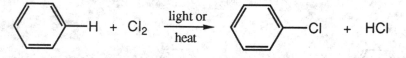

(a) Explain why benzene fails to react under these conditions. Consult Appendix 3 for relevant bond dissociation energies. (*Hint:* consider a possible radical chain mechanism, and the energetics of its rate-limiting step.)

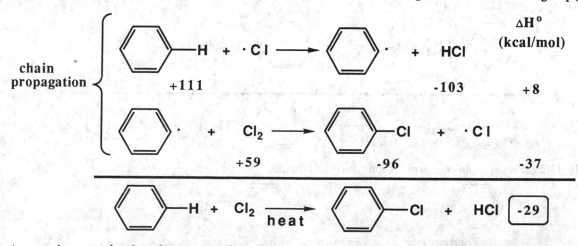

As can be seen in the above equations, the rate-limiting step is the abstraction of a benzene hydrogen atom by a chlorine radical. This process is endothermic by 8 kcal/mol because the C-H bonds are relatively strong (111 kcal/mol). Therefore, even though the entire process is exothermic by 29 kcal/mol, the reaction does not proceed because of the endothermic rate-limiting step.

(b) Explain why the bond dissociation energy of a C-H bond in benzene is significantly greater than that in alkanes. (*Hint:* Think of the orbitals forming the C-H bond in alkanes compared with those in benzene.)

The benzene carbon atoms are sp^2 hybridized, so the C-H bonds are derived from sp^2 hybridized orbitals. Compared with alkanes that use sp^3 hybridized orbitals, the sp^2 orbitals of benzene have greater s character and thus the electrons are held closer to the nucleus. This has the effect of increasing C-H bond strength in benzene relative to alkanes.

Problem 19.45 Following is an equation for hydroperoxidation of cumene.

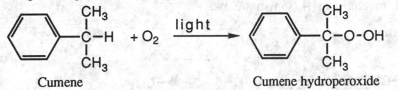

Cumene   Cumene hydroperoxide

Propose a radical chain mechanism for this reaction. Assume that initiation is by an unspecified radical, R•.

The stability of the benzyl radical, especially with the added methyl groups, facilitates the reaction with the radical initiator according to the following mechanism.

Step 1: **Inititiation**

Step 2: **Propagation**

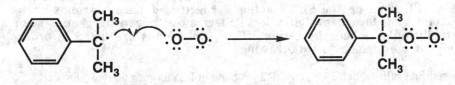

Step 3: **2nd Propagation Step**

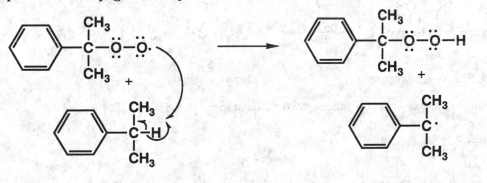

Problem 19.46 Para-substituted benzyl halides undergo reaction with methanol by an S_N1 mechanism to give a benzyl ether. Account for the following order of reactivity under these conditions.

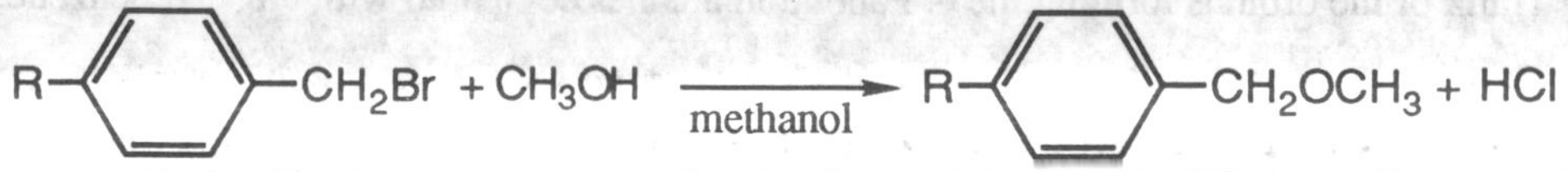

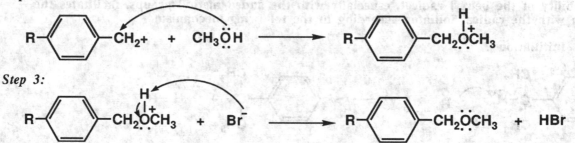

Rate of S_N1 reaction: $R = CH_3O- > CH_3- > H- > NO_2-$

A reasonable S_N1 mechanism is shown below:

Step 1: **Rate-limiting step**

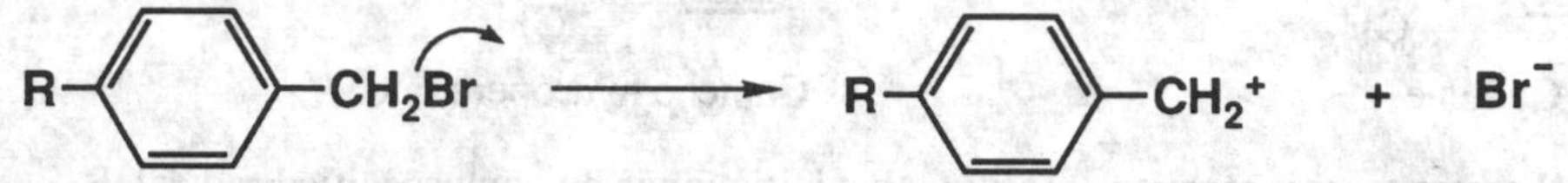

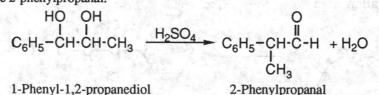

Step 2:

Step 3:

The rate-limiting step in this process is formation of the benzylic cation, so anything that stabilizes the benzylic cation will speed up the S_N1 reaction. Electron-donating groups such as the methoxy group are stabilizing to a benzylic cation. Electron-withdrawing groups such as the nitro group are destabilizing to a benzylic cation. Thus, the groups are listed in order from most electron-donating to most electron-withdrawing.

Problem 19.47 When warmed in dilute sulfuric acid, 1-phenyl-1,2-propanediol undergoes dehydration and rearrangement to give 2-phenylpropanal.

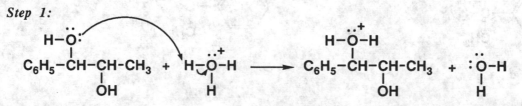

1-Phenyl-1,2-propanediol   2-Phenylpropanal

(a) Propose a mechanism for this example of a pinacol rearrangement (Section 9.8).

Step 1:

Step 2:

$$H-\overset{\overset{..+}{}}{\underset{|}{\ddot{O}}}-H$$

$$C_6H_5-\overset{|}{\underset{|}{CH}}-CH-CH_3 \longrightarrow C_6H_5-\overset{+}{CH}-CH-CH_3 + H_2\ddot{O}:$$
$$\quad\quad\quad\quad OH \quad\quad\quad\quad\quad\quad\quad\quad\quad\quad OH$$

Step 3: Note that the intermediate produced in this step is resonance stabilized

$$\overset{CH_3}{\underset{}{}}\quad\quad\quad\quad\overset{CH_3}{\underset{}{}}\quad\quad\quad\quad\overset{CH_3}{\underset{}{}}$$
$$C_6H_5-\overset{+}{CH}-\overset{|}{C}-H \longrightarrow C_6H_5-\overset{|}{CH}-\overset{+}{C}-H \longleftrightarrow C_6H_5-\overset{|}{CH}-C-H$$
$$\quad\quad\quad:\ddot{O}H \quad\quad\quad\quad\quad\quad :\overset{}{\underset{}{O}}-H \quad\quad\quad\quad\quad\quad +\ddot{O}-H$$

Step 4:

$$\overset{CH_3}{\underset{}{}}\quad\quad\quad\quad\quad\quad\quad\quad\quad\quad\overset{CH_3}{\underset{}{}}$$
$$C_6H_5-\overset{|}{CH}-C-H \; + \; :\ddot{O}-H \longrightarrow C_6H_5-\overset{|}{CH}-C-H \; + \; H-\overset{..+}{\ddot{O}}-H$$
$$\quad\quad\quad\; +\ddot{O}-H \quad\quad H \quad\quad\quad\quad\quad\quad\quad\quad O: \quad\quad\quad\quad H$$

(b) Account for the fact that 2-phenylpropanal is formed rather than its constitutional isomer, 1-phenyl-1-propanal.

$$\overset{O}{\underset{}{\|}}\quad\quad\quad\quad\quad\quad\quad\quad\overset{CH_3}{\underset{}{}}$$
$$C_6H_5-\overset{}{C}-CH_2-CH_3 \quad\quad\quad\quad C_6H_5-\overset{|}{CH}-C-H$$
$$\quad\quad\quad\quad\quad\quad\quad\quad\quad\quad\quad\quad\quad\quad\quad\quad\overset{\|}{O}$$

1-Phenyl-1-propanal   **2-Phenylpropanal**

In the observed reaction that leads to the aldehyde product, protonation of the benzylic hydroxyl followed by loss of H_2O gives a benzylic carbocation (Step 2). 1-Phenyl-1-propanal would result from protonation and loss of the other -OH group, but that would give a less stable secondary carbocation so it is not observed.

Problem 19.48 In the chemical synthesis of DNA and RNA, hydroxyl groups are normally converted to triphenylmethyl (trityl) ethers to protect the hydroxyl group from reaction with other reagents.

$$RCH_2OH + Ph_3CCl \xrightarrow{\text{tertiary amine}} RCH_2OCPh_3 + HCl \overset{\frown}{\quad} \text{neutralized by the tertiary amine}$$

triphenylmethyl chloride   a triphenylmethyl ether
(trityl chloride)   (a trityl ether)

Triphenylmethyl ethers are stable to aqueous base, but are rapidly cleaved in aqueous acid.

$$RCH_2OCPh_3 + H_2O \xrightarrow{H^+} RCH_2OH + Ph_3COH$$

(a) Why are triphenylmethyl ethers so readily hydrolyzed by aqueous acid?

The triphenylmethyl ethers are hydrolyzed according to the following mechanism.

Step 1:

Step 2:

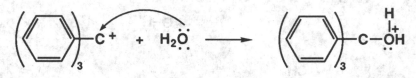

Note that the cation produced in this step is a highly resonance-stabilized benzylic cation due to the three adjacent phenyl rings.

Step 3:

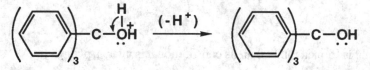

Step 4:

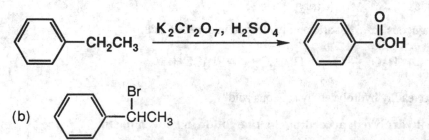

(b) How might the structure of the triphenylmethyl group be modified in order to increase or decrease its acid sensitivity?

Electron-releasing substituents like methoxy groups stabilize the triphenylmethyl cation and thereby increase the sensitivity of these triphenyl methyl ethers to acid. On the other hand, electron-withdrawing groups like nitro groups destabilize the triphenyl methyl cation and thereby decrease the sensitivity of these triphenylmethyl ethers to acid.

Synthesis

Problem 19.49 Using ethylbenzene as the only aromatic starting material, show how to synthesize the following compounds. In addition to ethylbenzene, use any other necessary organic or inorganic chemicals. Note that any compound already synthesized in one part of this problem may then be used to make any other compound in the problem.

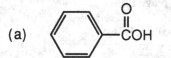

(a)

Oxidation of ethylbenzene using $K_2Cr_2O_7$ in H_2SO_4 gives benzoic acid.

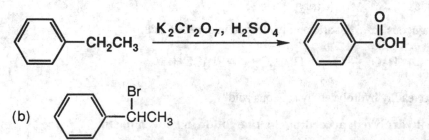

(b)

Bromination of the benzylic position using bromine or at elevated temperature or NBS in the presence of peroxide. The reaction involves a radical chain mechanism.

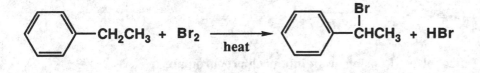

(c)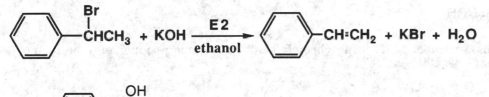

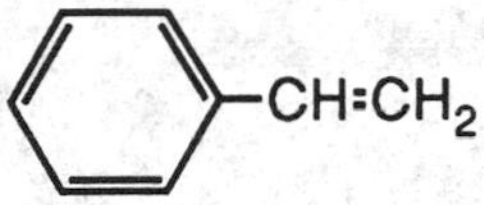

Dehydrohalogenation of the alkyl bromide from (b), brought about by a strong base such as KOH.

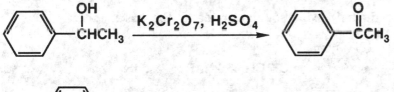

(d)

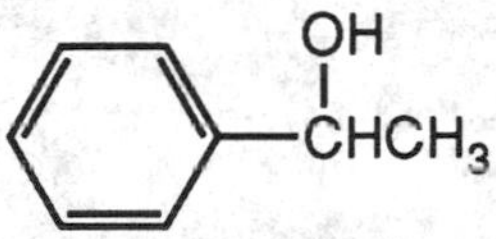

Acid-catalyzed hydration of a carbon-carbon double bond of (c). The same reaction may also be brought about by oxymercuration followed by reduction with NaBH$_4$.

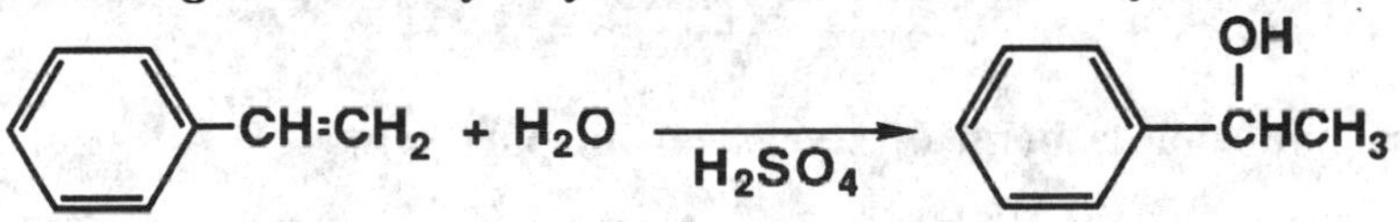

(e)

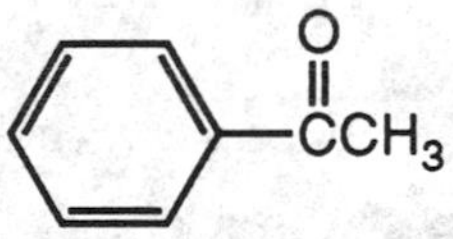

Oxidation of the secondary alcohol of (d) using chromic acid in aqueous sulfuric acid. The same oxidation may be brought about using the more selective oxidizing agents chromium trioxide in pyridine or pyridinium chlorochromate.

(f)

Hydroboration/oxidation of styrene from part (c).

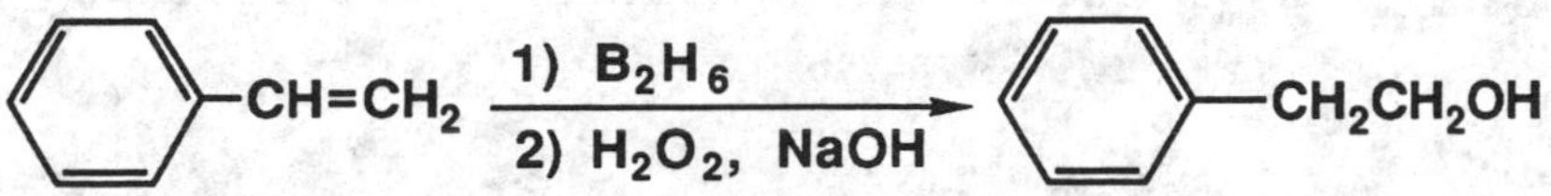

(g) 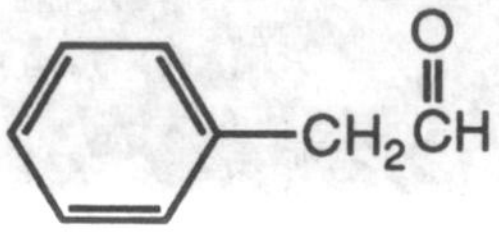 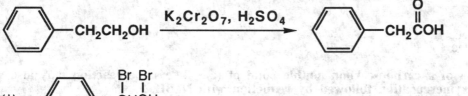 CH₂CH (with O double bond)

Oxidation of the primary alcohol of (f) using pyridinium chlorochromate.

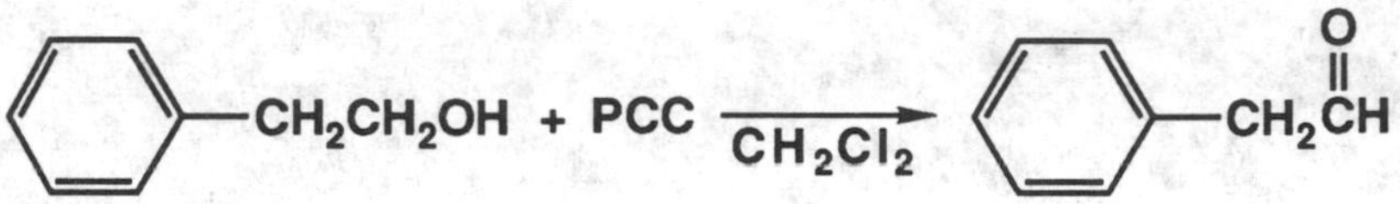

$$\text{—CH}_2\text{CH}_2\text{OH} + \text{PCC} \xrightarrow{\text{CH}_2\text{Cl}_2} \text{—CH}_2\overset{\text{O}}{\overset{\|}{\text{CH}}}$$

(h) 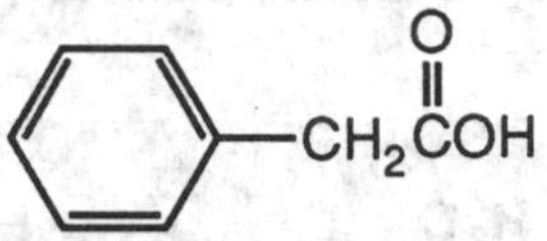 $$\text{—CH}_2\overset{\text{O}}{\overset{\|}{\text{COH}}}$$

Oxidation of the primary alcohol of (f) using chromic acid. The same product may be formed by similar oxidation of the aldehyde (g).

$$\text{—CH}_2\text{CH}_2\text{OH} \xrightarrow{\text{K}_2\text{Cr}_2\text{O}_7,\ \text{H}_2\text{SO}_4} \text{—CH}_2\overset{\text{O}}{\overset{\|}{\text{COH}}}$$

(i) 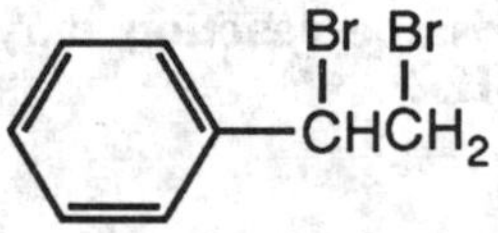 $$\overset{\text{Br}\ \text{Br}}{\underset{}{\text{—CHCH}_2}}$$

Addition of bromine to the carbon-carbon double bond of styrene from part (c).

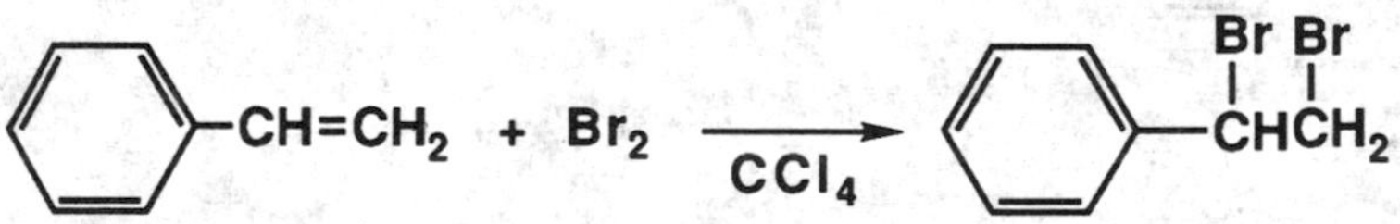

$$\text{—CH=CH}_2 + \text{Br}_2 \xrightarrow{\text{CCl}_4} \overset{\text{Br}\ \text{Br}}{\text{—CHCH}_2}$$

(j) 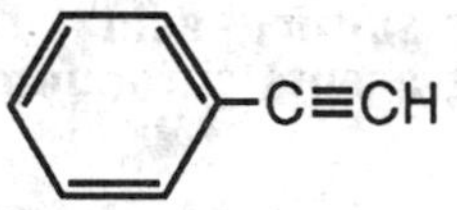 $$\text{—C}\equiv\text{CH}$$

A double dehydrohalogenation of product (i) using sodium amide as the base.

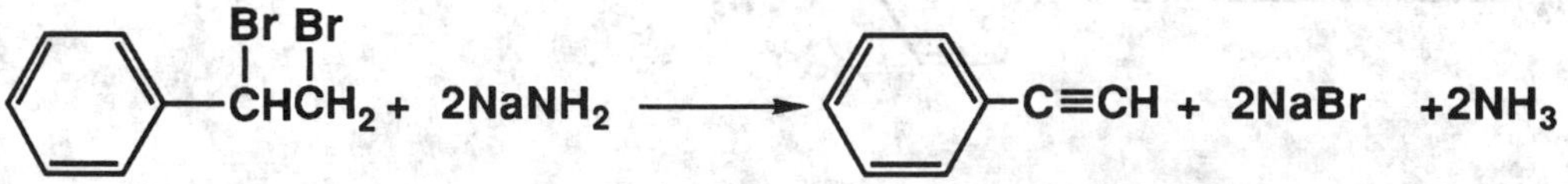

$$\overset{\text{Br}\ \text{Br}}{\text{—CHCH}_2} + 2\text{NaNH}_2 \longrightarrow \text{—C}\equiv\text{CH} + 2\text{NaBr} + 2\text{NH}_3$$

(k)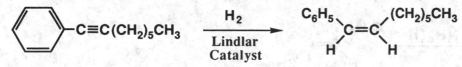

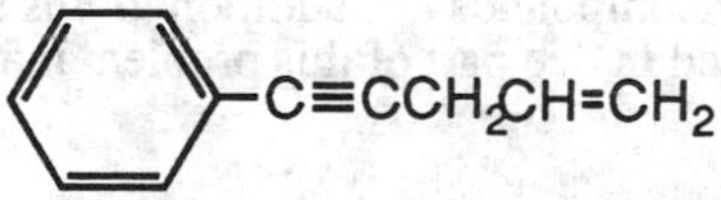

The terminal alkyne from (j) is deprotonated with sodium amide to produce the anionic species that reacts with allyl chloride to produce the desired product.

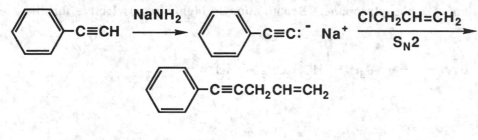

(l)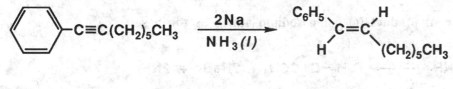

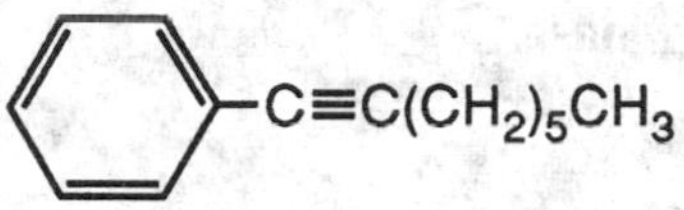

The deprotonated terminal alkyne from (k) reacts with hexyl chloride to produce the desired product.

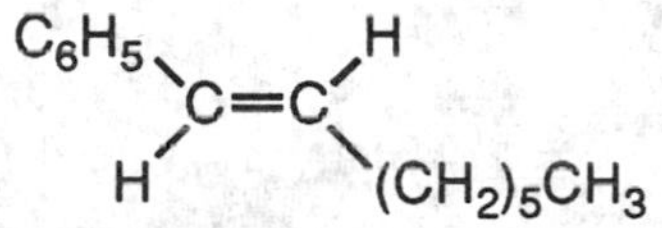

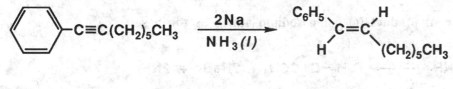

The alkyne produced in part (l) is reduced with sodium metal in liquid ammonia to produce the desired *trans* alkene.

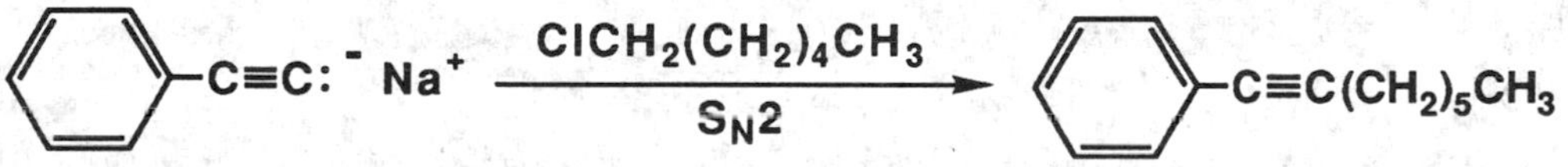

(m)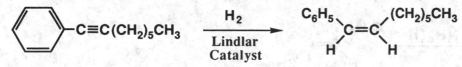

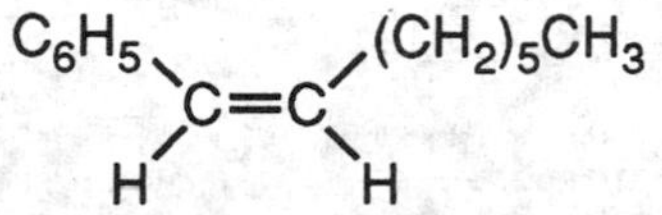

The alkyne produced in part (l) is reduced with hydrogen and the Lindlar catalyst to produce the desired *cis* alkene.

(n)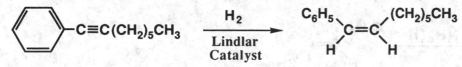

<u>Problem 19.50</u> Show how to convert 1-phenylpropane into the following compounds. In addition to this starting material, use any necessary inorganic reagents. Any compound synthesized in one part of this problem may then be used to make any other compound in the problem.

(a) $C_6H_5-\overset{\overset{\displaystyle Br}{|}}{C}HCH_2CH_3$

Radical chain bromination of 1-phenylpropane. Bromination is highly regioselective for the benzylic position.

$$C_6H_5-CH_2CH_2CH_3 \ + \ Br_2 \ \xrightarrow{\text{heat}} \ C_6H_5-\overset{\overset{\displaystyle Br}{|}}{C}HCH_2CH_3 \ + \ HBr$$

(b) $C_6H_5-CH=CHCH_3$

Dehydrohalogenation of product (a) using KOH or other strong base.

$$C_6H_5-\overset{\overset{\displaystyle Br}{|}}{C}HCH_2CH_3 \ + \ KOH \ \xrightarrow{\text{ethanol}} \ C_6H_5-CH=CHCH_3 \ + \ KBr \ + \ H_2O$$

(c) $C_6H_5-\overset{\overset{\displaystyle Cl}{|}}{C}H\text{-}\overset{\overset{\displaystyle Cl}{|}}{C}HCH_3$

Addition of chlorine to the double bond of product (b) by electrophilic addition.

$$C_6H_5-CH=CHCH_3 \ + \ Cl_2 \ \longrightarrow \ C_6H_5-\overset{\overset{\displaystyle Cl}{|}}{C}H\text{-}\overset{\overset{\displaystyle Cl}{|}}{C}HCH_3$$

(d) $C_6H_5-C\equiv CCH_3$

Double dehydrohalogenation of product (c) using sodium amide as the base.

$$C_6H_5-\overset{\overset{\displaystyle Cl}{|}}{C}H\text{-}\overset{\overset{\displaystyle Cl}{|}}{C}HCH_3 \ + \ 2NaNH_2 \ \longrightarrow \ C_6H_5-C\equiv CCH_3 \ + \ 2NaBr \ + \ 2NH_3$$

(e)

$$\underset{H}{\overset{C_6H_5}{}}C=C\underset{CH_3}{\overset{H}{}}$$

Chemical reduction of the alkyne from part (d) to a *cis* alkene using sodium metal in liquid ammonia.

$$C_6H_5-C\equiv CCH_3 \ + \ Na \ \xrightarrow{NH_3(l)} \ \underset{H}{\overset{C_6H_5}{}}C=C\underset{CH_3}{\overset{H}{}}$$

(f) 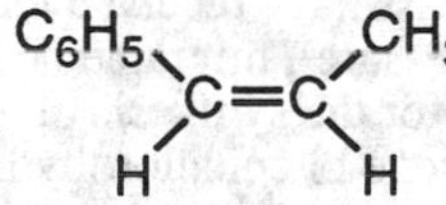 $\underset{H}{\overset{C_6H_5}{\diagdown}}C=C\underset{H}{\overset{CH_3}{\diagup}}$

Catalytic reduction of the alkyne using the Lindlar catalyst to reduce the alkyne to the alkene.

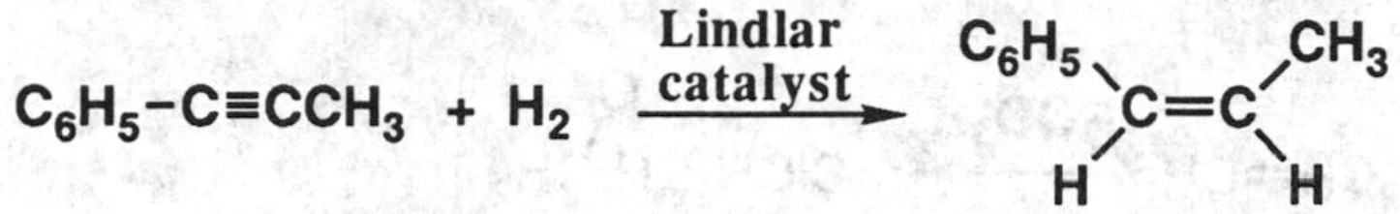

$$C_6H_5-C\equiv CCH_3 \;+\; H_2 \;\xrightarrow[\text{catalyst}]{\text{Lindlar}}\; \underset{H}{\overset{C_6H_5}{\diagdown}}C=C\underset{H}{\overset{CH_3}{\diagup}}$$

(g) 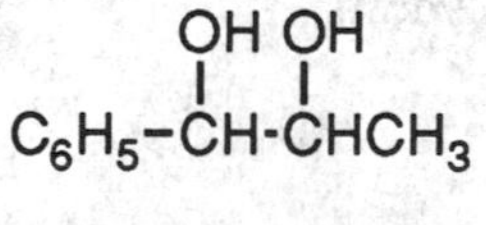 $C_6H_5-\overset{\overset{\displaystyle OH}{|}}{C}H-\overset{\overset{\displaystyle OH}{|}}{C}HCH_3$

Oxidation of the alkene from part (b) to a glycol with osmium tetroxide in the presence of hydrogen peroxide.

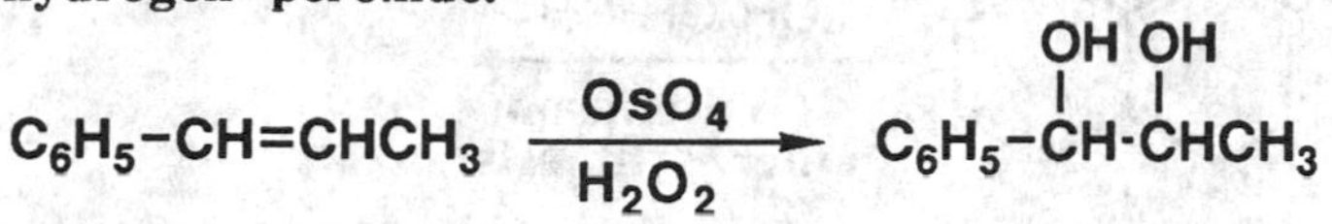

$$C_6H_5-CH=CHCH_3 \;\xrightarrow[\text{H}_2\text{O}_2]{\text{OsO}_4}\; C_6H_5-\overset{\overset{\displaystyle OH}{|}}{C}H-\overset{\overset{\displaystyle OH}{|}}{C}HCH_3$$

(h) $C_6H_5-\overset{\overset{\displaystyle OH}{|}}{C}HCH_2CH_3$

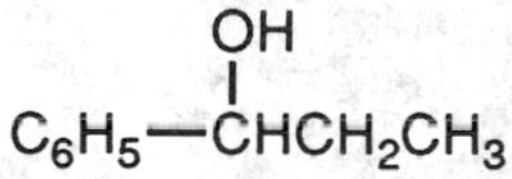

Acid-catalyzed hydration of the alkene from part (b). The reaction is highly regioselective because of the stability of the benzylic carbocation formed by protonation of the alkene. Alternatively, oxymercuration followed by reduction with $NaBH_4$ forms the same secondary alcohol.

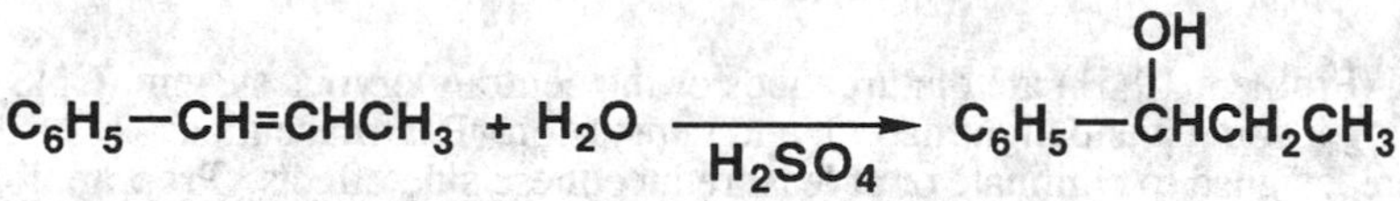

$$C_6H_5-CH=CHCH_3 \;+\; H_2O \;\xrightarrow[\text{H}_2\text{SO}_4]{}\; C_6H_5-\overset{\overset{\displaystyle OH}{|}}{C}HCH_2CH_3$$

(i) 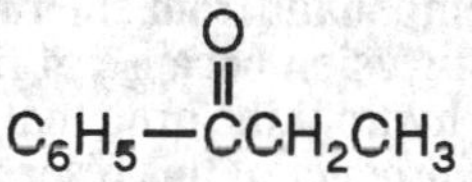 $C_6H_5-\overset{\overset{\displaystyle O}{\|}}{C}CH_2CH_3$

Oxidation of the secondary alcohol of (h) using chromic acid in aqueous sulfuric acid and the alcohol dissolved in acetone. Alternatively use chromium trioxide in pyridine or pyridinium chlorochromate as the oxidizing agent.

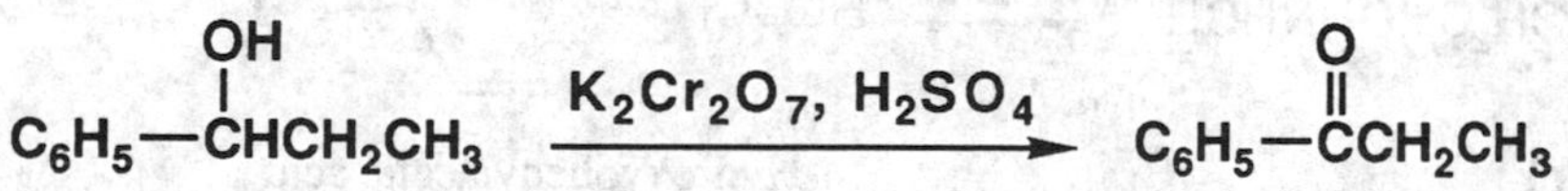

$$C_6H_5-\overset{\overset{\displaystyle OH}{|}}{C}HCH_2CH_3 \;\xrightarrow{K_2Cr_2O_7,\;H_2SO_4}\; C_6H_5-\overset{\overset{\displaystyle O}{\|}}{C}CH_2CH_3$$

Problem 19.51 Propranolol is a β-adrenergic receptor antagonist. Members of this class have received enormous clinical attention because of their effectiveness in treating hypertension (high blood pressure), migraine headaches, glaucoma, ischemic heart disease, and certain cardiac arrhythmias. Starting materials for the synthesis of propranolol are propene, 1-naphthol, and isopropylamine. Show how to convert propene to epichlorohydrin in stage 1, and then complete the synthesis of propranolol in stage 2.

Stage 1: Synthesis of epichlorohydrin

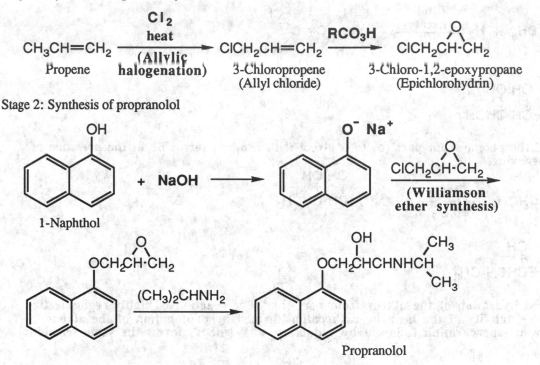

$$CH_3CH=CH_2 \xrightarrow[\text{(Allylic halogenation)}]{\overset{Cl_2}{\underset{\text{heat}}{}}} ClCH_2CH=CH_2 \xrightarrow{RCO_3H} ClCH_2CH\text{-}CH_2$$

Propene

3-Chloropropene
(Allyl chloride)

3-Chloro-1,2-epoxypropane
(Epichlorohydrin)

Stage 2: Synthesis of propranolol

Propranolol

Problem 19.52 Side effects of propranolol (Problem 19.51) are disturbances of the central nervous system (CNS) such as fatigue, sleep disturbances (including insomnia and nightmares), and depression. Pharmaceutical companies wondered if this drug could be redesigned to eliminate or at least reduce these side effects. Propranolol itself is highly lipophilic (hydrophobic) and readily passes through the blood-brain barrier, a lipid-like protective membrane that surrounds the capillary system in the brain and prevents hydrophilic compounds from entering the brain by passive diffusion. Propranolol, it was reasoned, enters the CNS by passive diffusion because of the lipid-like character of its naphthalene ring. The challenge, then, was to design a more hydrophilic drug that does not cross the blood-brain barrier but still retains a β-adrenergic antagonist property. A product of this research is atenolol, a potent β-adrenergic blocker that is hydrophilic enough that it crosses the blood-brain barrier to only a very limited extent.

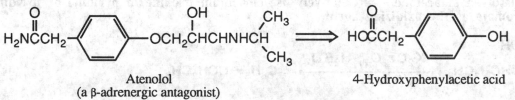

Atenolol
(a β-adrenergic antagonist)

4-Hydroxyphenylacetic acid

Propose a synthesis for atenolol starting with 4-hydroxyphenylacetic acid, epichlorohydrin (Problem 19.51), and isopropylamine.

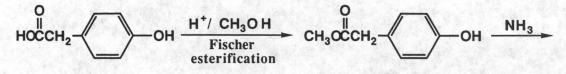

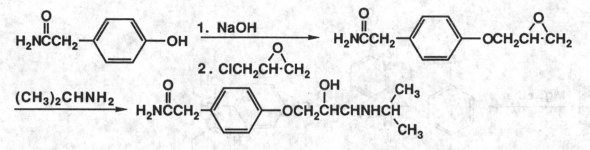

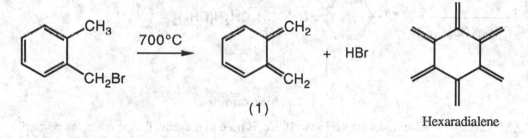

Problem 19.53 Benzylic bromination followed by loss of HBr by heating at high temperatures can be used to generate reactive intermediates such as (1). How do you take advantage of this observation to synthesize hexaradialene. (See L. G. Harruff, M. Brown, and V. Boekelheide, *J. Am. Chem. Soc.*, **100**, 2893, 1978).

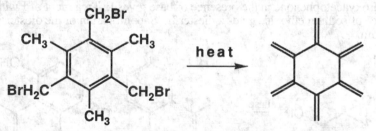

(1)

Hexaradialene

Using the above observation as a starting point, hexaradialene could be synthesized from the starting material shown below.

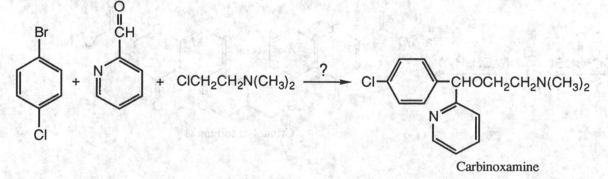

heat

Problem 19.54 Carbinoxamine is a histamine antagonist, specifically an H_1-antagonist. The maleic acid salt of the levorotatory isomer is sold as the prescription drug Rotoxamine. Propose a synthesis of carbinoxamine from the three compounds shown on the left of the reaction arrow. (Note: Aryl bromides form Grignard reagents much more readily than aryl chlorides.)

Carbinoxamine

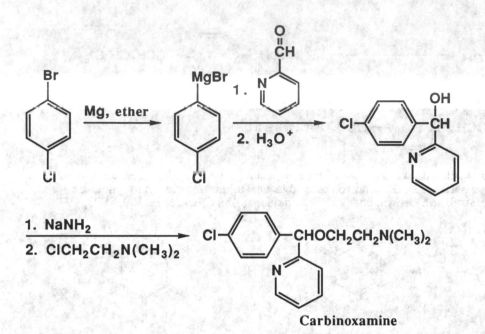

Carbinoxamine

<u>Problem 19.55</u> Cromolyn sodium, developed in the 1960s, is used to prevent allergic reactions primarily affecting the lungs, as for example exercise-induced emphysema. It is thought to block the release of histamine, which prevents the sequence of events leading to swelling, itching, and to constriction of bronchial tubes. Cromolyn sodium is synthesized in the following series of steps. Treatment of one mol of epichlorohydrin with two mol of 2,6-dihydroxyacetophenone in the presence of base gives I. Treatment of I with two mol of diethyl oxalate in the presence of sodium ethoxide gives a diester II. Saponification of the diester with aqueous NaOH gives cromolyn sodium.

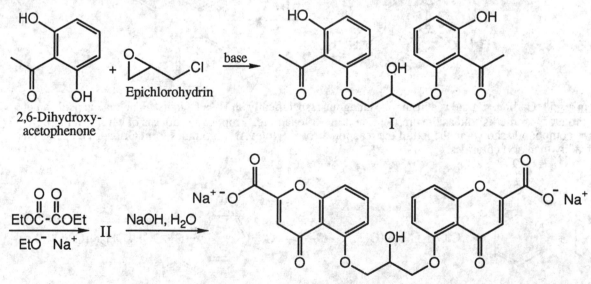

(a) Propose a mechanism for the formation of compound I.

Step 1:

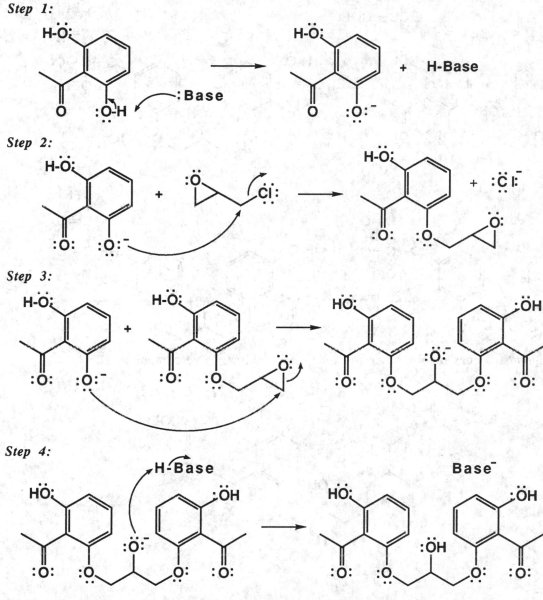

Step 2:

Step 3:

Step 4:

(b) Propose a structural formula for compound II and a mechanism for its formation.

Step 1:

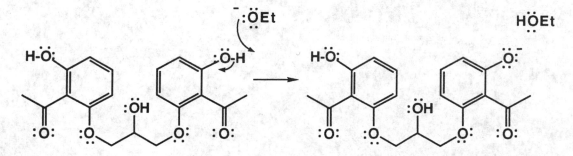

Step 2:

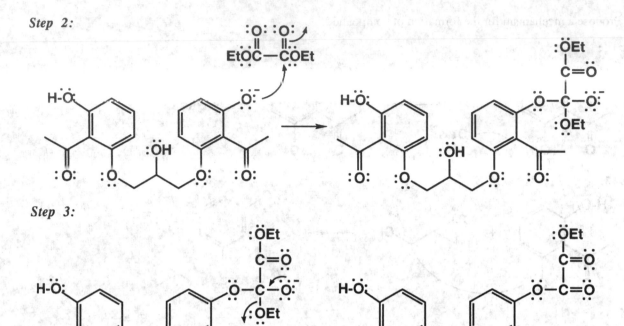

Step 3:

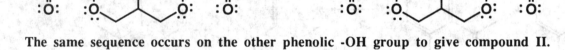

The same sequence occurs on the other phenolic -OH group to give compound II.

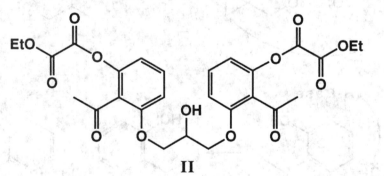

II

CHAPTER 20: AROMATICS II: REACTIONS OF BENZENE AND ITS DERIVATIVES

SUMMARY OF REACTIONS

Starting Material → Product ↓	Alkyl Benzenes	Anilines	Aryl Halides	Aryl Hydrazines	Aryl Ketones	Aryl Sulfonic Acids	Phenols		Nitro Aromatics
Aromatic Rings			20A 20.1A*			20B 20.1B			20C 20.1B
Aromatic Rings Acid Chlorides					20D 20.1C				
Aromatic Rings Alcohols	20E 20.1D								
Aromatic Rings Alkenes	20F 20.1D								
Aromatic Rings Alkyl Halides	20G 20.1C								
Aryl Halides	20H 20.3A		20I 20.3B			20J 20.3A	20K 20.3B		
Nitro Aromatics	20L 20.1B								

*Section in book that describes reaction.

REACTION 20A: BROMINATION AND CHLORINATION (Section 20.1A)

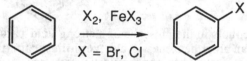

- **Aromatic rings** react with Br_2 in the presence of a **Lewis acid catalyst, such as $FeBr_3$**, to give an **aryl bromide**. This is an example of **electrophilic aromatic substitution.** ✻
- The mechanism involves an initial reaction between Br_2 and $FeBr_3$ to generate a molecular complex that can rearrange to give a Br^+ $FeBr_4^-$ ion pair. This reacts as a very strong electrophile with the weakly nucleophilic aromatic pi cloud to form a resonance-stabilized cation that loses a proton to give the final product.
- An analogous reaction can be carried out with **Cl_2 and $FeCl_3$** to give an **aryl chloride**.

REACTION 20B: SULFONATION (Section 20.1B)

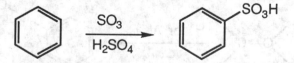

- **Aromatic rings** react with SO_3 in the presence of **sulfuric acid** to yield **aryl sulfonic acids** via **electrophilic aromatic substitution.** ✻
- The mechanism involves reaction of SO_3 as a very strong electrophile with the weakly nucleophilic aromatic pi cloud to form a resonance-stabilized cation that loses a proton to give the final product.
- In strongly acidic conditions, the electrophile can be the protonated form, SO_3H^+, instead of SO_3.

REACTION 20C: NITRATION (Section 20.1B)

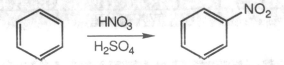

- **Aromatic rings** react with **nitric acid** in the presence of **sulfuric acid** to yield **nitro aromatic compounds** via **electrophilic aromatic substitution**. ✳
- The mechanism involves an initial reaction between nitric acid and sulfuric acid to yield the nitronium ion, NO_2^+. This reacts as a very strong electrophile with the weakly nucleophilic aromatic pi cloud to form a resonance-stabilized cation that loses a proton to give the final product.

REACTION 20D: FRIEDEL-CRAFTS ACYLATION (Section 20.1C)

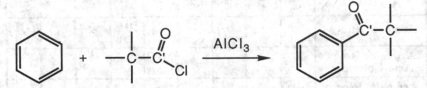

- **Aromatic rings** react with **acyl chlorides** in the presence of a Lewis acid catalyst, such as $AlCl_3$, to produce an aryl ketone via **electrophilic aromatic substitution**. ✳
- The mechanism involves an initial reaction between the acyl chloride and $AlCl_3$ to yield the acylium ion R-C^+=O. This reacts as a very strong electrophile with the weakly nucleophilic aromatic pi cloud to form a resonance stabilized cation that loses a proton to give the final product.
- Rearrangement is not a problem with acylium ions as it is with carbocations.

REACTION 20E: REACTIONS OF ALCOHOLS WITH AROMATIC RINGS IN THE PRESENCE OF STRONG ACID (Section 20.1D)

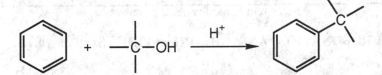

- **Aromatic rings** react with **alcohols** in the presence of a strong **acid catalyst**, such as H_3PO_4, H_2SO_4, and HF, to produce an alkyl benzene via **electrophilic aromatic substitution**. ✳
- The mechanism involves an initial reaction between the alcohol and strong acid to yield a carbocation. This reacts as a very strong electrophile with the weakly nucleophilic aromatic pi cloud to form a resonance stabilized cation that loses a proton to give the final product.
- Because carbocations are involved in the mechanism, rearrangements can be a problem, especially with primary or secondary alcohols.

REACTION 20F: REACTIONS OF ALKENES WITH AROMATIC RINGS IN THE PRESENCE OF STRONG ACID OR A LEWIS ACID (Section 20.1D)

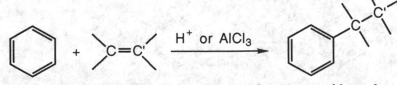

- **Aromatic rings** react with **alkenes** in the presence of a **strong acid catalyst, such as H_3PO_4, H_2SO_4, and HF**, or a **Lewis acid**, such as $AlCl_3$, to produce an **alkylbenzene** via **electrophilic aromatic substitution**. ✳

- The mechanism involves an initial reaction between the alkene and strong acid or Lewis acid to yield a carbocation. This reacts as a very strong electrophile with the weakly nucleophilic aromatic pi cloud to form a resonance-stabilized cation that loses a proton to give the final product.
- Because carbocations are involved in the mechanism, rearrangements can be a problem.

REACTION 20G: FRIEDEL-CRAFTS ALKYLATION (Section 20.1C)

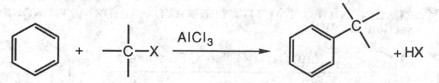

- **Aromatic rings** react with **alkyl halides** in the presence of a **Lewis acid catalyst**, such as $AlCl_3$, to produce an **alkylbenzene** via **electrophilic aromatic substitution**. ✳
- The mechanism involves an initial reaction between the alkyl halide and Lewis acid to yield an intermediate that can be thought of as a carbocation. This reacts as a very strong electrophile with the weakly nucleophilic aromatic pi cloud to form a resonance-stabilized cation that loses a proton to give the final product.
- Because carbocations are involved in the mechanism, rearrangements can be a problem, especially with primary or secondary alkyl halides.

REACTION 20H: REACTION OF AN ARYL HALIDE WITH SODIUM AMIDE (Section 20.3A)

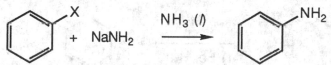

- **Aryl halides** react with **strongly basic nucleophiles,** such as **sodium amide**, to yield **anilines**. The $-NH_2$ group ends up on the ring carbon atom that was originally bonded to the halogen, as well as positions adjacent (ortho) to it. ✳
- The mechanism involves an initial reaction between the aryl halide and strong base to give a **benzyne intermediate**. This undergoes addition at either sp carbon atom to give the aniline products.

REACTION 20I: REACTION OF AN ARYL HALIDE WITH HYDRAZINE (Section 20.3B)

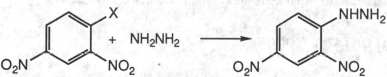

- **Activated aryl halides** react with **strong nucleophiles** such as **hydrazine** to give **aryl hydrazines** in a regioselective manner **via nucleophilic aromatic substitution**. The $-NHNH_2$ group ends up on the ring carbon atom that was originally bonded to the halogen. This reaction does not occur unless there are electron-withdrawing groups, such as nitro goups, ortho and/or para to the halogen. The electron-withdrawing groups activate the ring toward nucleophilic attack. ✳
- Unlike Reaction 20H that involves a benzyne intermediate, this reaction involves a nucleophilic attack of the ring carbon containing the halogen to give a negatively-charged **Meisenheimer complex**. Loss of halogen results in formation of the product.

REACTION 20J: REACTION OF AN ARYL HALIDE WITH SODIUM HYDROXIDE (Section 20.3A)

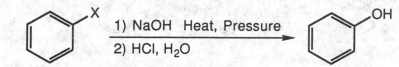

- **Aryl halides** react with **strongly basic nucleophiles** such as **sodium hydroxide** to yield **phenols via a benzyne intermediate**. The -OH group ends up on the ring carbon atom that was originally bonded to the halogen, as well as positions adjacent (ortho) to it. ✳
- The mechanism involves an initial reaction between the aryl halide and strong base to give a benzyne intermediate. This undergoes addition at either sp hybridized carbon atom to give the aniline products.
- This process requires extreme conditions and is generally **not useful as a laboratory method**.

REACTION 20K: REACTION OF AN ACTIVATED ARYL HALIDE WITH AQUEOUS BASE
(Section 20.3B)

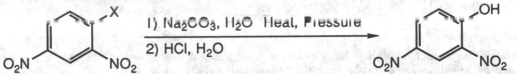

- **Activated aryl halides** react with **aqueous base** to give **phenols via nucleophilic aromatic substitution**. The -OH group ends up on the ring carbon atom that was originally bonded to the halogen. This reaction does not occur unless there are electron-withdrawing groups, such as nitro groups, ortho and/or para to the halogen. The electron-withdrawing groups activate the ring toward nucleophilic attack. ✳
- This reaction involves a nucleophilic attack of the ring carbon containing the halogen to give a negatively charged **Meisenheimer complex**. Loss of halogen results in formation of the product.

REACTION 20L: REDUCTION OF AROMATIC NITRO COMPOUNDS TO GIVE ANILINES
(Section 20.1B)

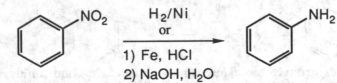

- **Aromatic nitro compounds** react with **H_2 in the presence of a transition metal catalyst such as Ni,** or alternatively using **Fe and HCl** followed by a **basic workup, to give aniline** derivatives. This is an important reaction because nitration of aromatic rings is easily accomplished and anilines are useful for a large number of other useful transformations. ✳
- Note that nitro groups are meta directing, while amino groups are ortho-para directing. This change in orientation preference can be exploited in the synthesis of complex aromatics.

SUMMARY OF IMPORTANT CONCEPTS

20.0 OVERVIEW
- **Aromatic rings,** such as benzene, **react with strong electrophiles** in a reaction that results in substitution of a ring hydrogen. **Electrophilic aromatic substitution is extremely useful** for preparing benzene derivatives. **Nucleophilic attack** on aromatic rings is **much less common**, but still possible when an aromatic ring has **electron-withdrawing substituents**. ✳

20.1 ELECTROPHILIC AROMATIC SUBSTITUTION
- A variety of electrophiles react with aromatic rings via **electrophilic aromatic substitution**. Examples include reactions 20A-20G. The general **mechanism involves attack on the electrophile** by the weakly nucleophilic aromatic pi cloud to form **a resonance-stabilized cation intermediate** that loses a proton to give the final product. ✳

20.2 DISUBSTITUTION AND POLYSUBSTITUTION

• **Substituents** other than hydrogen on an aromatic ring can have a **profound influence** on the **reaction rate** and **substitution pattern** of **electrophilic aromatic substitution reactions**. In particular, groups can either speed up or slow down the reaction, and can direct new groups meta or ortho-para. ✷

• These effects are the result of two types of interactions:

- **1)** A **resonance effect** in which the **positively-charged cation intermediate** in an electrophilic aromatic substitution is **stabilized**, or **destabilized**. Groups such as -NH$_2$ or -OH in which **different contributing structures reveal that lone pair electron density is added to the ring** help to **stabilize the cation intermediate**, thereby **lowering the activation energy** of the reaction, thereby speeding up the reaction. These groups are referred to as **activating**. Other groups like -NO$_2$ or

    -C≡N in which **different contributing structures reveal** that **electron density is removed from the ring destabilize the cation intermediate**, thereby **raising the activation energy** of the reaction, thereby **deactivating the ring**. ✷

- **2)** An **inductive effect** in which atoms that are **more electronegative** than the sp^2 ring carbons **pull electron density out** of the aromatic ring, or conversely, atoms that are **less electronegative** than the sp^2 ring carbons **add electron density** into the aromatic ring. Since the aromatic ring is acting as a weak nucleophile in the electrophilic aromatic substitution reactions, the **more electronegative atoms or groups (electron-withdrawing) reduce** electron density and thus the **nucleophilicity of the ring**, so they are **deactivating**. The **less electronegative atoms or groups (electron-releasing) increase** electron density and thus the **nucleophilicity of the ring**, so they are **activating**. **Alkyl groups are electron-releasing** and thus **activating** due to **hyperconjugation**.✷

• These inductive and resonance effects have **different levels of influence depending on their position** relative to the incoming electrophile. These effects can be seen by drawing all of the contributing structures for the positively-charged intermediate formed when an electrophile reacts with the weakly nucleophilic aromatic pi cloud. *[Being able to draw all of the important contributing structures of this positively-charged intermediate is the key to understanding substituent effects. Practice drawing the contributing structures with different types of substituents (activating, deactivating, halogens), and for substitution at the different positions, (ortho, meta, para.)]* ✷

- The effects of substituents that *activate* the ring are **most** *activating* ortho and para. For this reason, the **incoming electrophile reacts predominantly ortho and para** to the existing activating substituent. Such activating substituents are referred to as being **ortho-para directing**.

- The effects of most substituents that *deactivate* the ring are **maximally** *deactivating* ortho and para. For this reason, the **incoming electrophile reacts predominantly meta** to the existing activating substituent, that is, the less-deactivated positions. Such deactivating substituents are referred to as being **meta directing**. Halogens are the exception since they are deactivating, but ortho-para directing.

• **Substituents** can be divided into **three broad classes:**✷

- **1)** **Alkyl groups** and all **groups in which the atom bonded to the ring** has an **unshared pair of electrons** are **ortho-para directing**. With the exception of the halogens discussed below, these groups are **electron-releasing** and are thus **activating** toward electrophilic aromatic substitution. Examples include -R, -OH, -NH$_2$, -OR, etc.

- **2)** Groups in which the **atom bonded to the aromatic ring** bears a **partial or full positive charge** are **meta directing**. These groups often have **multiple bonds such as =O on the atom bonded to the aromatic ring**. These groups are **electron-withdrawing** and are thus **deactivating** toward electrophilic aromatic substitution. For example, -CN, -NO$_2$, -C(O)R, CO$_2$H, C(O)NH$_2$, -SO$_3$H, etc.

- **3)** **Halogens** are unique in that they are **ortho-para directing but deactivating**. This is not a contradiction, but rather the result of competing effects. Halogens show overall **inductive deactivation** of the **aromatic ring due to their electronegativity**, but **resonance stabilization** of the **cation intermediate due to their lone pair** and thus ortho-para direction.

• These effects have a large practical significance, since, in synthesizing polysubstituted aromatics, the **order of addition of the substituents must be taken into account**. For example, when making *m*-bromonitrobenzene from benzene, the nitro group (meta directing) must be added before the bromine atom (ortho-para directing). Adding the bromine atom first followed by the nitro group would result in the majority of product being the unwanted ortho and para isomers.✷

• When there are **more than one substituent already on the ring**, the following rules apply:

- **1)** The **more activating substituent dominates** when different groups are directing an incoming group to different positions. ✷

- **2)** **Non-bonded interactions prevent** any reasonable amount of **substitution** at the position that is **between two meta substituents**. ✷

CHAPTER 20
Solutions to the Problems

<u>Problem 20.1</u> Write the stepwise mechanism for sulfonation of benzene by hot, concentrated sulfuric acid. In this reaction, the electrophile is SO_3 formed as shown in the following equation. *Hint:* In thinking about a mechanism for this reaction, consider formal charges on sulfur and oxygen in the Lewis structure of sulfur trioxide.

$$H_2SO_4 \rightleftharpoons SO_3 + H_2O$$

In sulfonation of benzene, the electrophile is sulfur trioxide. In step 1, reaction of benzene with the electrophile yields a resonance-stabilized cation. In step 2, this intermediate loses a proton to complete the reaction: water is shown as the base accepting the proton. A second-proton transfer reaction in step 3 gives benzenesulfonic acid.

Step 1:

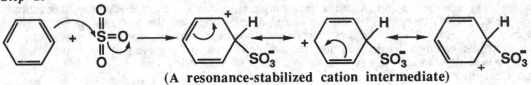

(A resonance-stabilized cation intermediate)

Step 2:

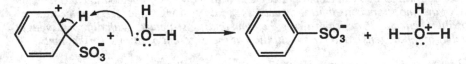

Step 3:

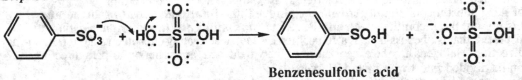

Benzenesulfonic acid

Note that in more strongly acidic conditions, the electrophile can be the protonated form, namely HSO_3^+, instead of SO_3.

<u>Problem 20.2</u> Write structural formulas for the products you expect from Friedel-Crafts alkylation or acylation of benzene with

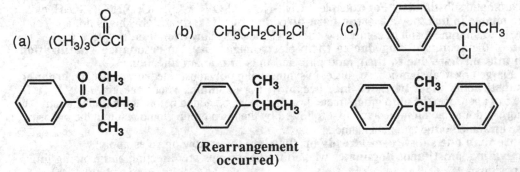

(a) $(CH_3)_3CCCl$ (with O double bond)

(b) $CH_3CH_2CH_2Cl$

(c) (benzene)—$CHCH_3$ with Cl

(Rearrangement occurred)

__Problem 20.3__ Write a mechanism for the formation of *tert*-butylbenzene from benzene and *tert*-butyl alcohol in the presence of phosphoric acid.

__A reasonable mechanism for this reaction involves formation of the *tert*-butyl carbocation, that then reacts with benzene via electrophilic aromatic substitution.__

Step 1:

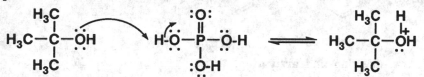

Step 2:

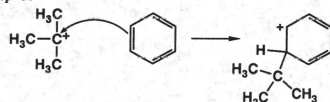

Step 3:

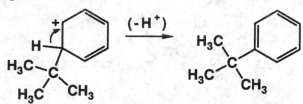

Step 4:

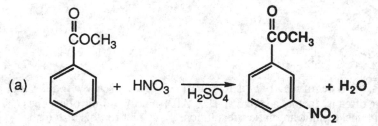

__Problem 20.4__ Complete the following electrophilic aromatic substitution reactions. Where you predict meta substitution, show only the meta product. Where you predict ortho-para substitution, show both products.

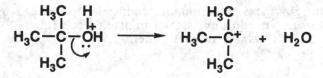

(a)   + HNO_3 $\xrightarrow{H_2SO_4}$ + H_2O

__The carboxymethyl group is meta directing and deactivating.__

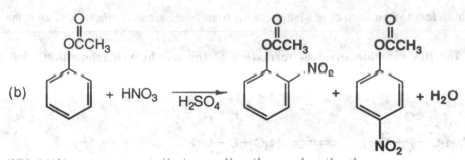

(b)

The acetoxy group is ortho/para directing and activating.

<u>Problem 20.5</u> Predict the major product(s) of each electrophilic aromatic substitution.

When there is more than one substituent on the ring, the predominant product is determined by the more activating (less deactivating) substituent. In addition, steric hindrance precludes reaction between two substituents that are meta with respect to each other.

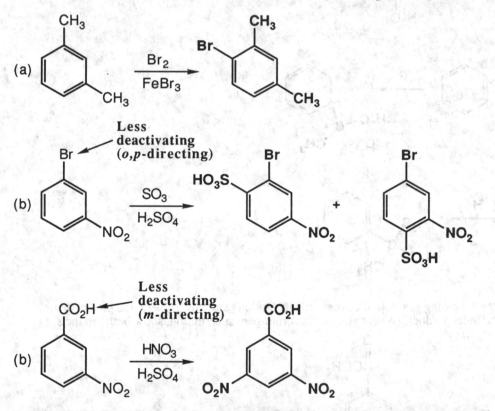

<u>Problem 20.6</u> In S_N2 reactions of alkyl halides, the order of reactivity is RI > RBr > RCl > RF. Alkyl iodides are considerably more reactive than alkyl fluorides, often by factors as great as 10^6. All 1-halo-2,4-dinitrobenzenes, however, react at approximately the same rate in nucleophilic aromatic substitutions. Account for this difference in relative reactivities.

Recall that the overall rate of a reaction is determined by the rate-limiting (slow) step. In an S_N2 reaction, departure of the leaving group is involved in the rate-limiting step, thus the nature of the leaving group influences the rate of the overall reaction. As shown in the text, the mechanism of nucleophilic substitution with 1-halo-2,4-dinitrobenzenes involves addition to the ring to form a Meisenheimer complex. Since this is the rate-limiting step and departure of the halide is not involved, the nature of the halogen has little influence on the overall rate of the process.

PROBLEMS
Electrophilic Aromatic Substitution: Monosubstitution
Problem 20.7 Write a stepwise mechanism for each of the following reactions. Use curved arrows to show the flow of electrons in each step.

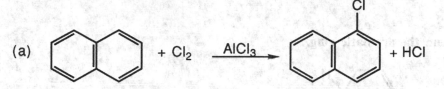

(a)

Chlorination of naphthalene in the presence of aluminum chloride is an example of electrophilic aromatic substitution. It is shown in three steps.

Step 1: Activation of chlorine to form an electrophile.

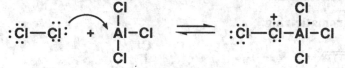

Step 2: Reaction of the electrophile with the aromatic ring, a nucleophile.

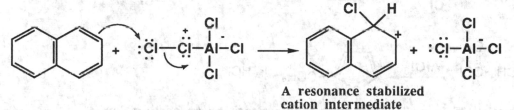

A resonance stabilized
cation intermediate

Step 3: Proton transfer and reformation of the aromatic ring.

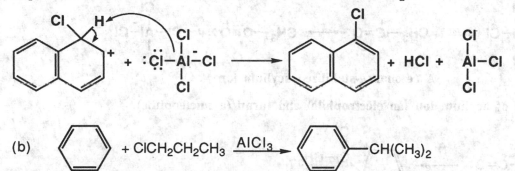

(b)

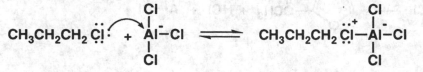

The reaction of benzene with 1-chloropropane in the presence of aluminum chloride involves initial formation of a complex between 1-chloropropane and aluminum chloride, and its rearrangement to an isopropyl cation. This cationic species is the electrophile that undergoes further reaction with benzene.

Step 1: Formation of a complex between 1-chloropropane (a Lewis base) and aluminum chloride (a Lewis acid).

$$CH_3CH_2CH_2 \ddot{C}l: \quad + \quad \underset{\underset{Cl}{|}}{\overset{\overset{Cl}{|}}{Al}} - Cl \quad \rightleftharpoons \quad CH_3CH_2CH_2 \overset{..+}{Cl} - \underset{\underset{Cl}{|}}{\overset{\overset{Cl}{|}}{Al}} - Cl$$

Step 2: **Rearrangement to form an isopropyl cation.**

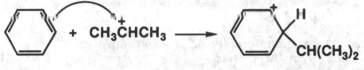

Step 3: **Electrophilic attack on the aromatic ring.**

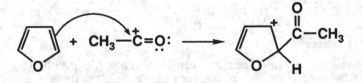

**Resonance-stabilized
cation intermediate**

Step 4: **Proton transfer to regenerate the aromatic ring.**

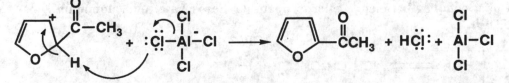

(c)

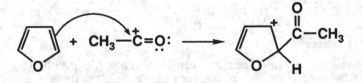

Friedel-Crafts acylation of furan involves electrophilic attack by an acylium ion.
Step 1: **Formation of resonance-stabilized acylium ion.**

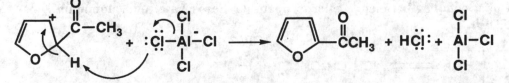

A resonance-stabilized acylium ion

Step 2: **Reaction of acylium ion (an electrophile) and furan (a nucleophile).**

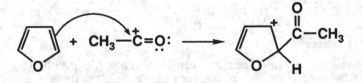

Step 3: **Proton transfer and regeneration of aromatic ring.**

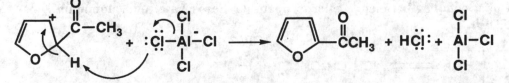

(d)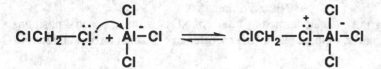

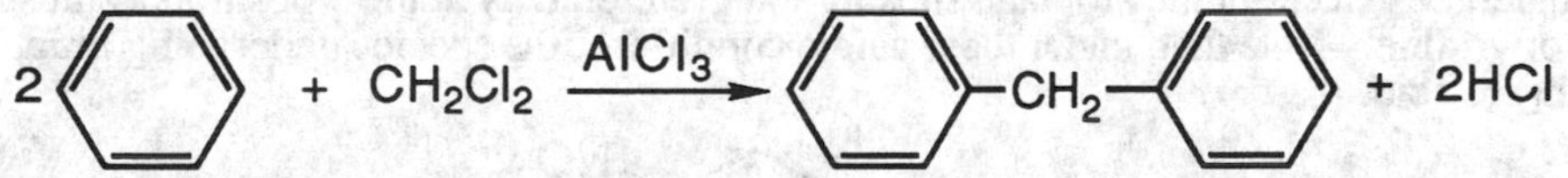

Formation of diphenylmethane involves two successive Friedel-Crafts alkylations.

Step 1:

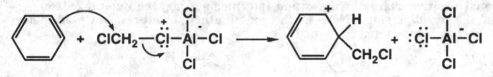

Step 2:

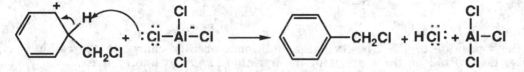

Step 3:

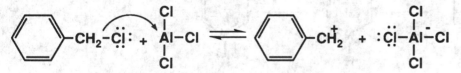

Formation of benzyl chloride completes the first Friedel-Crafts alkylation. This molecule then is a reactant for the second Friedel-Crafts alkylation.

Step 1:

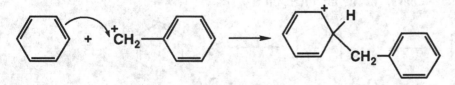

Step 2:

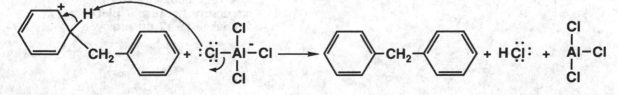

Step 3:

<u>Problem 20.8</u> Pyridine undergoes electrophilic aromatic substitution preferentially at the 3 position as illustrated by the synthesis of 3-nitropyridine. Note that, under these acidic conditions, the species undergoing nitration is not pyridine, but its conjugate acid.

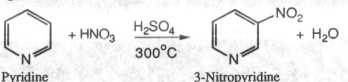

Pyridine   3-Nitropyridine

Write resonance contributing structures for the intermediate formed by attack of NO_2^+ at the 2, 3, and 4 positions of the conjugate acid of pyridine. From examination of these intermediates, offer an explanation for preferential nitration at the 3 position.

Pyridine is a base, and in the presence of a nitric acid-sulfuric acid mixture, it is protonated. It is the protonated form that must be attacked by the electrophile NO_2^+. For nitration at the 3-position, the additional positive charge in the cation intermediate may be delocalized on three carbon atoms of the pyridine ring. None of the contributing structures, however, places both positive charges on the same atom.

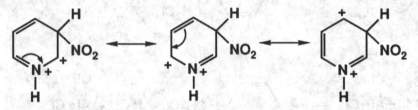

For nitration at the 4-position or the 2-position, the additional positive charge in the cation intermediate is also delocalized on three atoms of the pyridine ring, but one of these contributing structures has a charge of +2 on nitrogen. This situation is thus less stable than that which occurs for nitration at the 3-position.

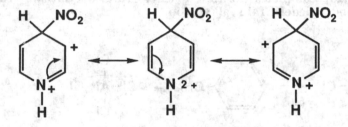

**A very poor contributing
structure because it places a
charge of +2 on nitrogen**

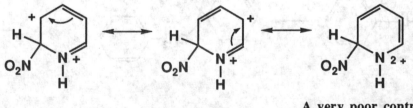

**A very poor contributing
structure because it places a
charge of +2 on nitrogen**

<u>Problem 20.9</u> Pyrrole undergoes electrophilic aromatic substitution preferentially at the 2 position as illustrated by the synthesis of 2 nitropyrrole.

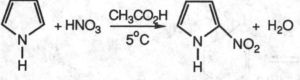

Pyrrole   2-Nitropyrrole

Write resonance contributing structures for the intermediate formed by attack of NO_2^+ at the 2 and 3 positions of pyrrole. From examination of these intermediates, offer an explanation for preferential nitration at the 2 position.

Pyrrole is nitrated under considerably milder conditions than pyridine. For nitration at the 2-position, the positive charge on the cation intermediate is delocalized over three atoms of the pyrrole ring whereas for nitration at the 3-position, it is delocalized over only two atoms. The intermediate with the greater degree of delocalization of charge has a lower activation energy for formation and hence it is formed at a faster rate.

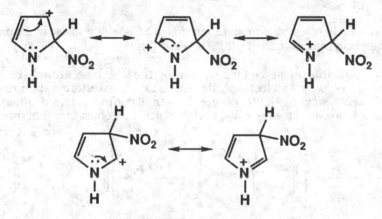

<u>Problem 20.10</u> Addition of *m*-xylene to the strongly acidic solvent HF/SbF₅ at -45° gives a new species, which shows ¹H-NMR resonances at δ 2.88 (3H), 3.00 (3H), 4.67 (2H), 7.93 (1 H), 7.83 (1H), and 8.68 (1H). Assign a structure to the species giving this spectrum.

The strong acid results in protonation of the aromatic ring to create a positively charged species as shown below.

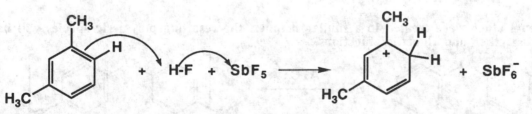

<u>Problem 20.11</u> Addition of *tert*-butylbenzene to the strongly acidic solvent HF/SbF₅ followed by aqueous work-up gives benzene. Propose a mechanism for this dealkylation reaction.

A reasonable mechanism for this reaction involves protonation of the aromatic ring, followed by heterolytic bond cleavage and release of the *tert*-butyl cation.

Step 1:

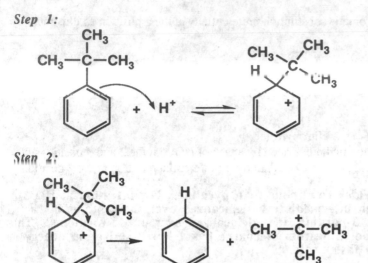

Step 2:

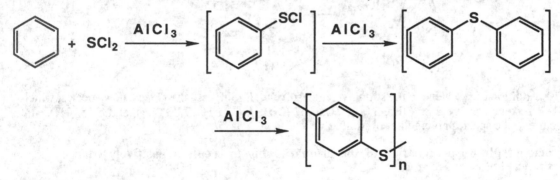

<u>Problem 20.12</u> What product do you predict from the reaction of SCl$_2$ with benzene in the presence of AlCl$_3$? What product results if diphenyl ether is treated with SCl$_2$ and AlCl$_3$?

The Lewis acid, AlCl$_3$, facilitates departure of one of the chlorine atoms from SCl$_2$, and the resulting electrophile takes part in electrophilic aromatic substitution reaction to create the -SCl derivative that then reacts again to create diphenyl sulfide. The diphenyl sulfide is activated compared to benzene, so this will react further to generate a polymeric product as shown.

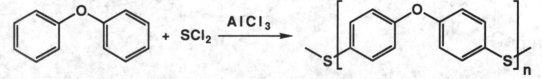

If diphenyl ether were treated in a similar manner, the resulting polymeric species will have alternating ether and thioether functions.

<u>Problem 20.13</u> Other groups besides H$^+$ can act as leaving groups in electrophilic aromatic substitution. One of the best leaving groups is the trimethylsilyl group (Me$_3$Si-). For example, treatment of Me$_3$SiC$_6$H$_5$ with CF$_3$CO$_2$D rapidly forms DC$_6$H$_5$. What are the properties of a silicon-carbon bond that allows you to predict this kind of reactivity?

Based on simple electronegativities, the C-Si bond is polarized such that a partial positive charge is on the Si atom. Furthermore, the heterolytic bond cleavage is facilitated because the (CH)$_3$Si$^+$ cation is so stable.

Disubstitution and Polysubstitution

__Problem 20.14__ The following groups are ortho-para directors. Draw a contributing structure for the resonance-stabilized aryl cation formed during electrophilic aromatic substitution that shows the role of each group in stabilizing the intermediate by further delocalizing its positive charge.

(a)   —OH

(b) $-O-\overset{\displaystyle O}{\overset{\|}{C}}CH_3$

(c)   —N(CH₃)₂

(d) $-N\overset{\displaystyle O}{\overset{\|}{H}C}CH_3$

(e)

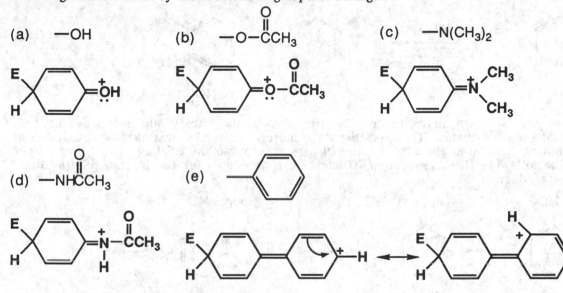

__Problem 20.15__   Predict the major product or products from treatment of each compound with HNO_3/H_2SO_4.

When there is more than one substituent on a ring, the predominant product is derived from the orientation preference of the most activating substituent.

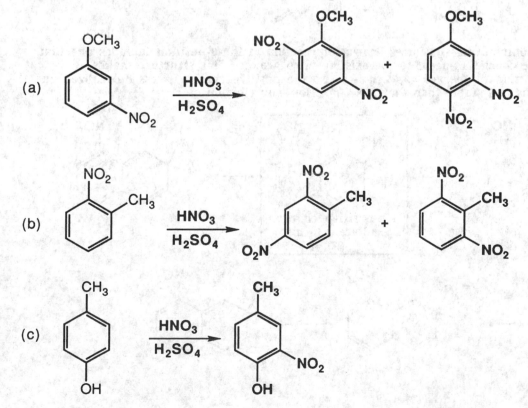

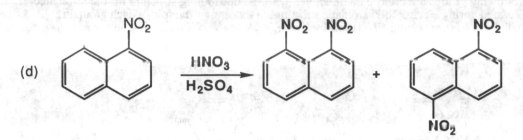

(d)

Nitration occurs as shown above, because the ring without the nitro group is less deactivated and reacts at the α-positions. The α-positions are more reactive, because in this case, none of the contributing structures place the positive charge adjacent to the existing nitro group. For example, the following five contributing structures can be drawn for the intermediate leading to the first product.

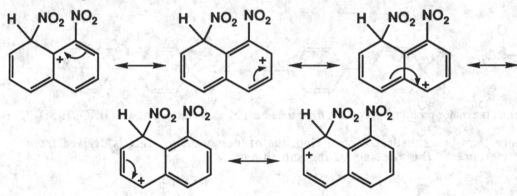

A similar set of contributing structures drawn for reaction at the β-position includes one that places the positive charge adjacent to the existing nitro group. This structure makes a negligible contribution to the resonance hybrid, so the positive charge is less delocalized and this intermediate is less stable than that for α-position substitution.

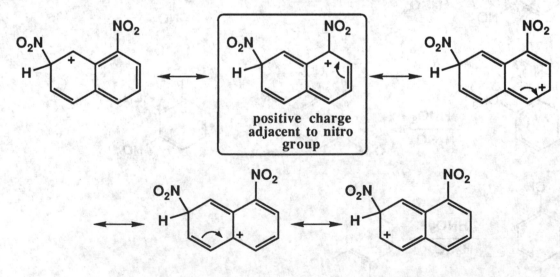

positive charge
adjacent to nitro
group

Problem 20.16 How do you account for the fact that *N*-phenylacetamide (acetanilide) is less reactive toward electrophilic aromatic substitution than aniline?

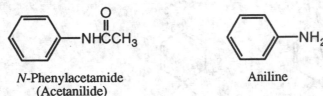

N-Phenylacetamide
(Acetanilide)   Aniline

The unshared pair of electrons on the nitrogen atom of acetanilide is involved in a resonance interaction with the carbonyl group of the amide, and, therefore, less available for stabilization of an aryl cation intermediate compared to aniline.

Problem 20.17 The trifluoromethyl group is almost exclusively meta directing as shown in the following example.

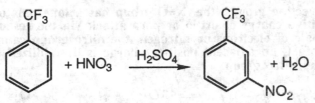

Draw contributing structures for the cation intermediate formed during nitration. First assume para attack and then meta attack. By reference to contributing structures you have drawn, account for the fact that nitration is essentially 100% in the meta position.

Following are contributing structures for meta and para attack of the electrophile. For meta attack, three contributing structures can be drawn and all make approximately equal contributions to the hybrid. Three contributing structures can also be drawn for ortho/para attack, one of which places a positive charge on carbon bearing the trifluoromethyl group; this structure makes only a negligible contribution to the hybrid. Thus, for meta attack, the positive charge on the aryl cation intermediate can be delocalized almost equally over three atoms of the ring giving this cation's formation a lower activation energy. For ortho/para attack, the positive charge on the aryl cation intermediate is delocalized over only two carbons of the ring, giving this cation's formation a higher activation energy.

meta **attack:**

ortho/para **attack:**

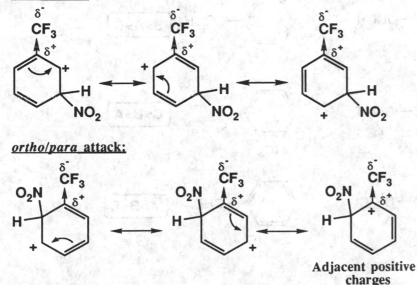

**Adjacent positive
charges**

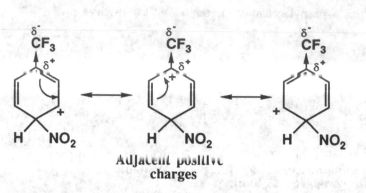

Adjacent positive charges

<u>Problem 20.18</u> Suggest a reason why the nitroso group, -N=O, is ortho-para directing although the nitro group, -NO$_2$, is meta directing.

Like other ortho-para directing groups, the -N=O group has a lone pair of electrons that can stabilize an adjacent positive charge of ortho or para attack via the resonance structures shown below. Without a lone pair of electrons on nitrogen, the nitro group cannot take part in similar stabilization. In fact, an adjacent positive charge is destabilized by the electron-withdrawing nature of the nitro group.

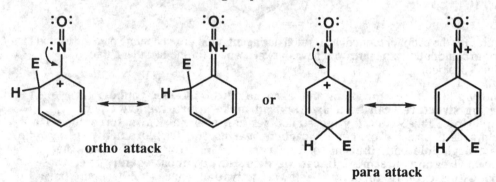

ortho attack   or   **para attack**

<u>Problem 20.19</u> Arrange the following in order of decreasing reactivity (fastest to slowest) toward electrophilic aromatic substitution.

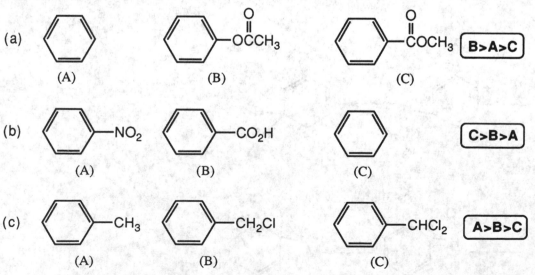

(a) (A)   (B)   (C) **B>A>C**

(b) (A)   (B)   (C) **C>B>A**

(c) (A)   (B)   (C) **A>B>C**

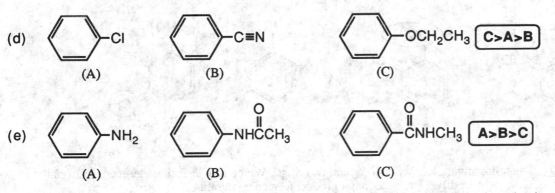

(d)

(A)   (B)   (C) [C>A>B]

-Cl   -C≡N   -OCH₂CH₃

(e)

(A)   (B)   (C) [A>B>C]

-NH₂   -NHCCH₃   -CNHCH₃

Problem 20.20 For each compound, indicate which group on the ring is the more strongly activating and then draw the structural formula of the major product formed by nitration of the compound.

In the following structures, the more strongly activating group is circled and arrows show the position(s) of nitration. Where both ortho and para nitration is possible, two arrows are shown. A broken arrow shows a product formed in only negligible amounts.

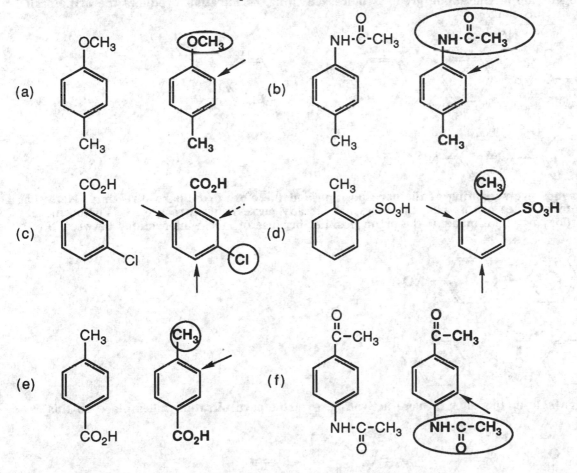

(a)   (b)

(c)   (d)

(e)   (f)

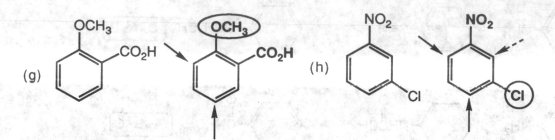

(g)   (h)

Problem 20.21 The following molecules each contain two rings. Which ring in each undergoes electrophilic aromatic substitution more readily? Draw the major product formed on nitration.

(a)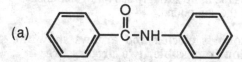

The nitrogen side of the amide group is more activating, so nitration produces the ortho/para products on this ring.

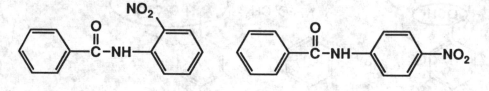

(b)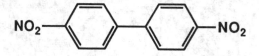

Phenyl groups are activating at the para position and therefore ortho/para directing. Note that nitration takes place on the ring that does not already possess the nitro group. Furthermore, nitration does not take place at the ortho position because of steric interactions between the rings.

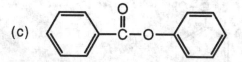

(c)

The oxygen side of the ester is more activating, so ortho/para nitration takes place on this ring.

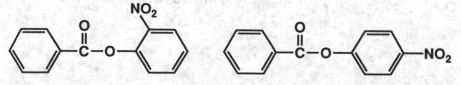

Problem 20.22 Reaction of phenol with acetone in the presence of an acid catalyst gives a compound known as bisphenol A. Bisphenol A is used in the production of epoxy and polycarbonate resins. Propose a mechanism for the formation of bisphenol A.

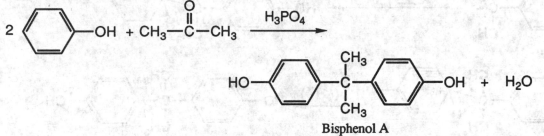

Bisphenol A

The reaction begins by proton transfer from phosphoric acid to acetone to form its conjugate acid which may be written as a hybrid of two contributing structures. The conjugate acid of acetone is an electrophile and reacts with phenol at the para position by electrophilic aromatic substitution to give 2-(4-hydroxy-phenyl)-2-propanol. Protonation of the tertiary alcohol in this molecule and departure of water gives a resonance-stabilized cation that reacts with a second molecule of phenol to give bisphenol A.

Step 1:

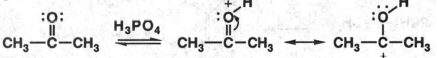

Step 2:

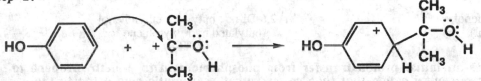

Step 3:

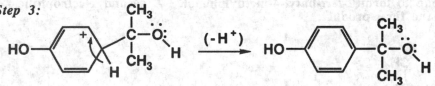

Step 4:

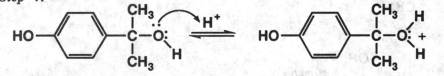

Step 5:

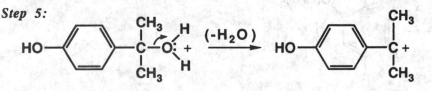

Step 6:

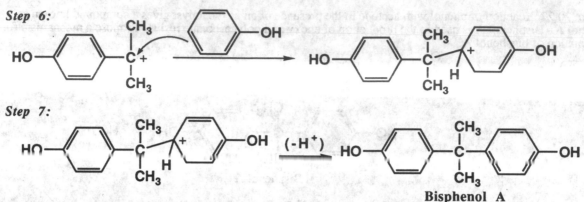

Step 7:

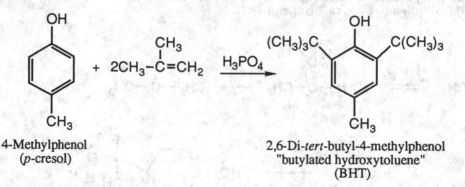

Bisphenol A

<u>Problem 20.23</u> 2,6-Di-*tert*-butyl-4-methylphenol, alternatively known as butylated hydroxytoluene (BHT), is used as an antioxidant in foods to "retard spoilage" (See the box Radical Autoxidation, pp. 261-262). BHT is synthesized industrially from 4-methylphenol (*p*-cresol) by reaction with 2-methylpropene in the presence of phosphoric acid. Propose a mechanism for this reaction.

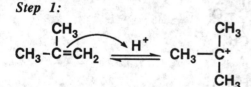

4-Methylphenol
(*p*-cresol)

2,6-Di-*tert*-butyl-4-methylphenol
"butylated hydroxytoluene"
(BHT)

The reaction involves an initial proton transfer from phosphoric acid to 2-methylpropene to give an electrophilic *tert*-butyl cation that then reacts with the aromatic ring ortho to the strongly activating -OH group to form 2-*tert*-butyl-4-methylphenol. A second electrophilic aromatic substitution gives the final product.

Step 1:

Step 2:

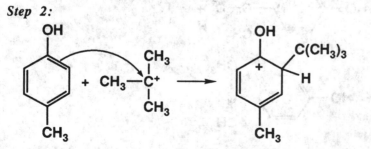

Step 3:

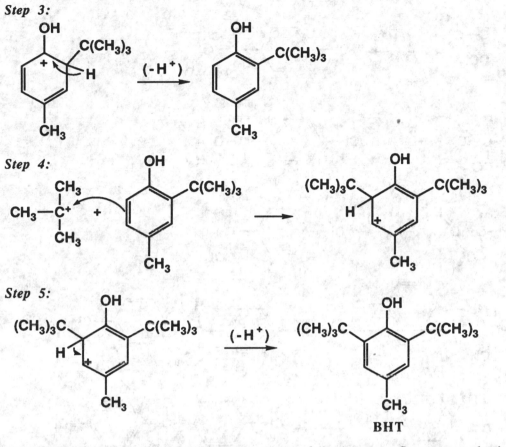

Step 4:

Step 5:

BHT

Problem 20.24 The insecticide DDT is prepared by the following route. Suggest a mechanism for this reaction. The abbreviation DDT is derived from the common name dichlorodiphenyltrichloroethane.

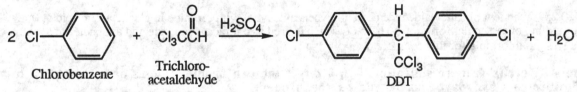

Chlorobenzene   Trichloro-   DDT
                     acetaldehyde

Propose a mechanism that is highly analogous to that proposed for the formation of Bisphenol A in Problem 20.22 above.

Step 1:

Step 2:

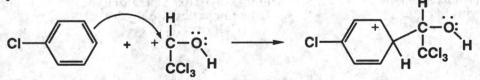

Step 3:

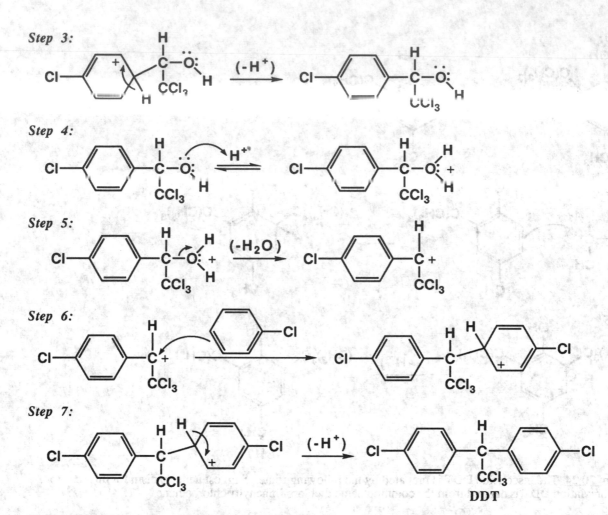

Step 4:

Step 5:

Step 6:

Step 7:

DDT

<u>Problem 20.25</u> Treatment of salicylaldehyde (2-hydroxybenzaldehyde) with bromine in glacial acetic acid at 0°C gives a compound of molecular formula $C_7H_4Br_2O_2$ which is used as a topical fungicide and antibacterial agent. Propose a structural formula for this compound.

Since the -OH group is more activating, it will direct the two bromine atoms ortho and para to give 3,5-dibromo-2-hydroxybenzaldehyde as the product.

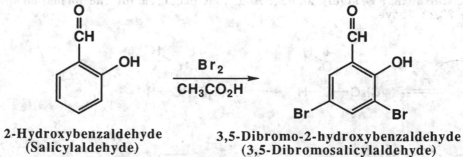

2-Hydroxybenzaldehyde
(Salicylaldehyde)

$\xrightarrow[\text{CH}_3\text{CO}_2\text{H}]{\text{Br}_2}$

3,5-Dibromo-2-hydroxybenzaldehyde
(3,5-Dibromosalicylaldehyde)
$(C_7H_4Br_2O_2)$

Problem 20.26 Propose a synthesis for 3,5-dibromo-2-hydroxybenzoic acid (3,5-dibromosalicylic acid) from phenol. [*Hint:* Recall the Kolbe carboxylation of phenol (Section 19.5E)].

A Kolbe reaction followed by bromination in glacial acetic acid leads to 3,5-dibromo-2-hydroxybenzoic acid.

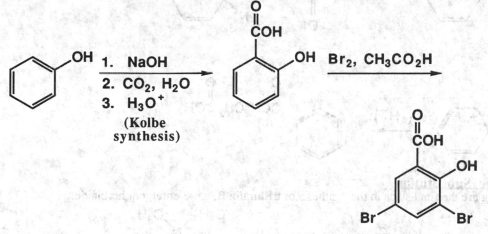

**3,5-Dibromo-2-hydroxybenzoic acid
(3,5-Dibromosalicylic acid)**

Problem 20.27 Treatment of benzene with succinic anhydride in the presence of polyphosphoric acid gives the following γ-ketoacid. Propose a mechanism for this reaction.

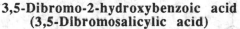

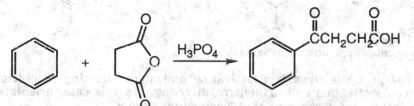

Succinic anhydride   4-Oxo-4-phenylbutanoic acid

Step 1:

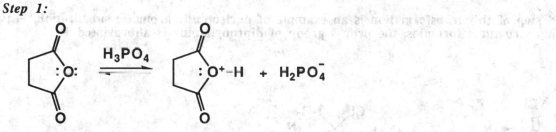

Step 2:

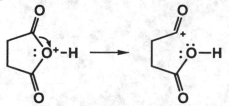

Step 3:

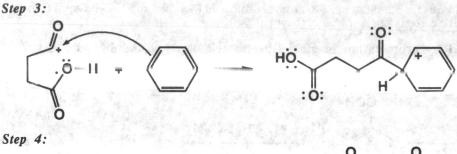

Step 4:

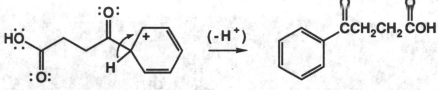

Nucleophilic Aromatic Substitution

Problem 20.28 Following are the final steps in the synthesis of trifluralin B, a pre-emergent herbicide.

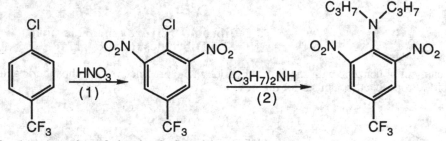

(a) Account for the orientation of nitration in Step (1).

The trifluoromethyl group is strongly deactivating and meta directing (Problem 20.17). Chlorine is weakly deactivating and ortho/para directing. In this case, orientation is determined by chlorine, the weaker of the deactivating groups.

(b) Propose a mechanism for the substitution reaction in Step (2).

The second step of this transformation is an example of nucleophilic aromatic substitution. In the following structural formulas, the propyl group of dipropylamine is abbreviated R.

Step 1:

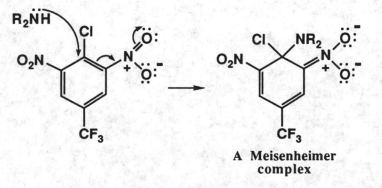

A Meisenheimer
complex

Step 2:

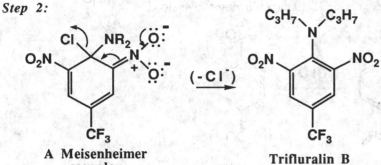

A Meisenheimer
complex   **Trifluralin B**

Syntheses
Problem 20.29 Show how to convert toluene to these compounds.

(a)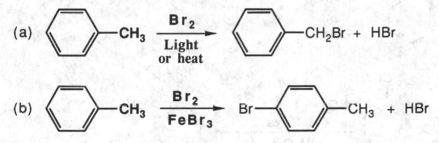

(b)

Problem 20.30 Show how to convert toluene to (a) 2,4-dinitrobenzoic acid and (b) 3,5-dinitrobenzoic acid.

Methyl is ortho/para directing. Therefore, toluene can be nitrated twice then oxidized with
chromic acid to convert the methyl group into the carboxyl group.

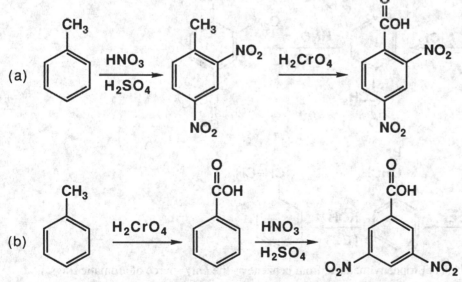

The reaction sequence is very similar to the last one, except now the order of the reactions is
reversed because the carboxylic acid group is a meta director.

Problem 20.31 Show reagents and conditions to bring about the following conversions.

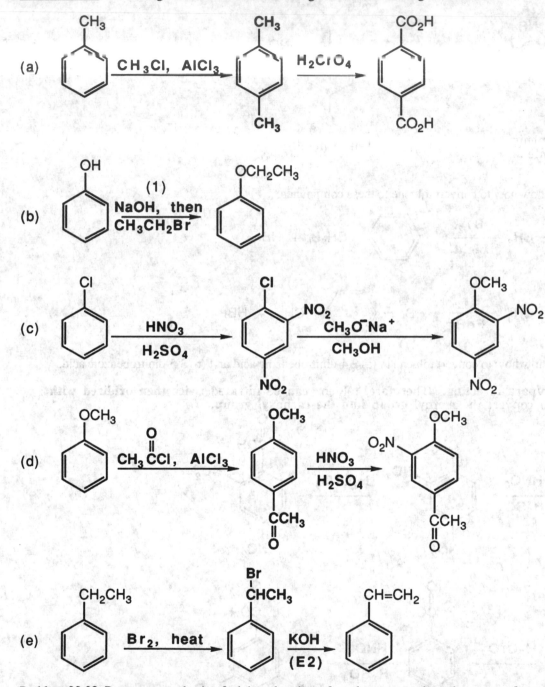

Problem 20.32 Propose a synthesis of triphenylmethane from benzene as the only source of aromatic rings, and any other necessary reagents.

Reaction of 3 mol benzene with 1 mol trichloromethane (chloroform) will give triphenylmethane.

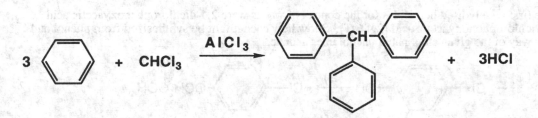

Problem 20.33 Propose a synthesis for each compound from the given starting material(s). Show all reagents and products in each step of your syntheses.

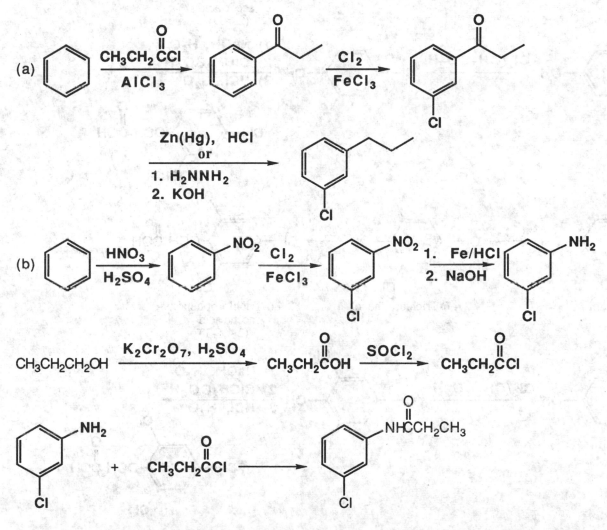

In the above synthesis scheme, the benzene ring is nitrated then chlorinated followed by reduction with Fe/HCl to give the 3-chloroaniline. The 1-propanol is converted to propanoic acid then the acid chloride to get ready for the final step in which the 3-chloroaniline is treated with propanoyl chloride to give the desired amide product.

<u>Problem 20.34</u> The first widely used herbicides for the control of weeds were 2,4-dichlorophenoxyacetic acid (2,4-D) and 2,4,6-trichlorophenoxyacetic acid (2,4,6-T). Show how each might be synthesized from phenol and chloroacetic acid by way of the given chlorinated phenol intermediate.

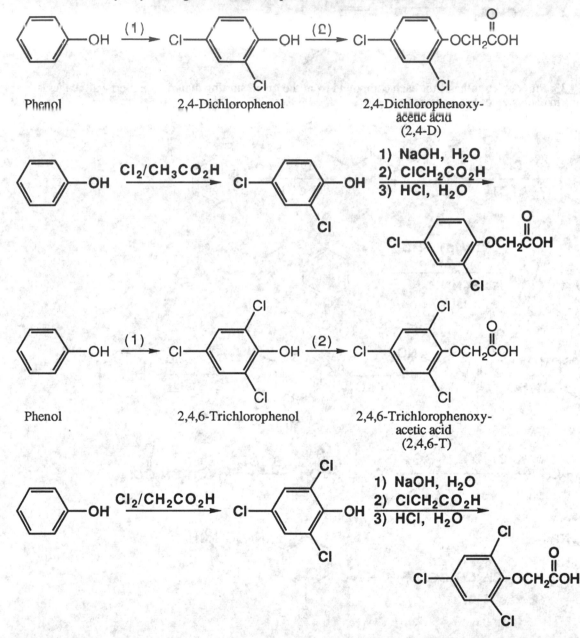

<u>Problem 20.35</u> Phenol is the starting material for the synthesis of 2,3,4,5,6-pentachlorophenol, known alternatively as pentachlorophenol or more simply as "penta." At one time, penta was widely used as a wood preservative for decks, siding, and outdoor wood furniture. Draw the structural formula for pentachlorophenol and describe its synthesis from phenol.

"Penta" is synthesized industrially from phenol. Because the -OH group is strongly activating, reaction of phenol with chlorine in a polar solvent such as acetic acid at room temperature gives 2,4,6-trichlorophenol. Further chlorination of this intermediate gives pentachlorophenol.

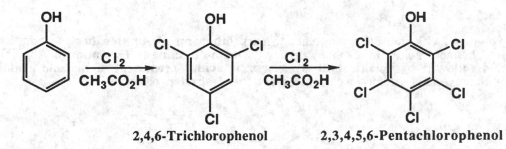

2,4,6-Trichlorophenol   2,3,4,5,6-Pentachlorophenol

<u>Problem 20.36</u> Starting with benzene, toluene, or phenol as the only sources of aromatic rings, show how to synthesize the following. Assume in all syntheses that mixtures of ortho-para products can be separated into the desired isomer.
(a) 1-Bromo-3-nitrobenzene

Nitro is meta directing; bromine is ortho/para directing. Therefore, to have the two substituents meta to each other, carry out nitration first followed by bromination.

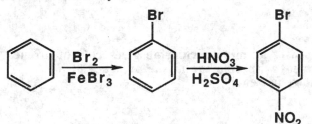

Benzene   Nitrobenzene   1-Bromo-3-nitrobenzene

(b) 1-Bromo-4-nitrobenzene

Reverse the order of steps from part (a). Nitro is meta directing; bromine is ortho-para directing. Therefore, to have the two substituents para to each other, carry out bromination first followed by nitration.

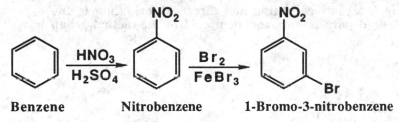

Benzene   Bromobenzene   1-Bromo-4-nitrobenzene

(c) 2,4,6-Trinitrotoluene (TNT)

The methyl group is ortho/para directing. Therefore, nitrate toluene three successive times.

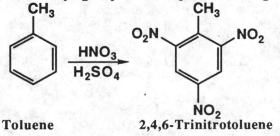

Toluene   2,4,6-Trinitrotoluene

(d) *m*-Chlorobenzoic acid

The carboxyl group and chlorine atom are meta to each other, an orientation best accomplished by chlorination of benzoic acid (the carboxyl group is meta-directing). Oxidation of toluene with chromic acid followed by acidification gives benzoic acid. Treatment of benzoic acid with chlorine in the presence of ferric chloride gives the desired product.

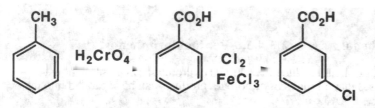

Toluene   **Benzoic acid**   **3-Chlorobenzoic acid**

(e) *p*-Chlorobenzoic acid

Start with toluene. The methyl group is weakly activating and directs chlorination to the ortho/para positions. Separate the desired para isomer and then oxidize the methyl group to a carboxyl group using chromic acid.

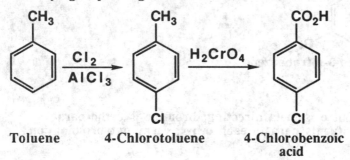

Toluene   **4-Chlorotoluene**   **4-Chlorobenzoic**
                                                    acid

(f) *p*-Dichlorobenzene

Treatment of benzene with chlorine in the presence of aluminum chloride gives chlorobenzene. The chlorine atom is ortho/para directing. Treatment with chlorine in the presence of aluminum chloride a second time gives 1,4-dichlorobenzene.

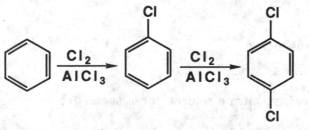

Benzene   **Chlorobenzene**   *p*-**Dichlorobenzene**

(g) *m*-Nitrobenzenesulfonic acid

Both the sulfonic acid group and the nitro group are meta directors. Therefore, the two electrophilic aromatic substitution reactions may be carried out in either order. The sequence shown is nitration followed by sulfonation.

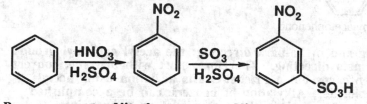

Benzene   **Nitrobenzene**   *m*-**Nitrobenzenesulfonic acid**

<u>Problem 20.37</u> 3,5-Dibromo-4-hydroxybenzenesulfonic acid is used as a disinfectant. Propose a synthesis of this compound from phenol.

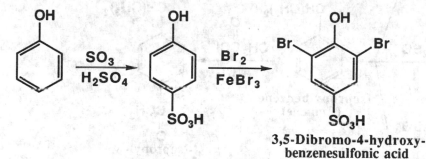

**3,5-Dibromo-4-hydroxy-
benzenesulfonic acid**

<u>Problem 20.38</u> 3,5-Dichloro-2-methoxybenzoic acid is used as a herbicide. Propose a synthesis of this compound from salicylic acid.

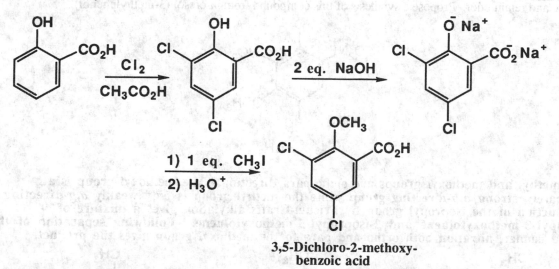

**3,5-Dichloro-2-methoxy-
benzoic acid**

Note that in the methylation step, the methyl iodide could have reacted with either the phenoxide or carboxylate groups. Recall that phenoxide is a stronger base, and thus, a stronger nucleophile, so the product shown is the predominant one.

<u>Problem 20.39</u> The following compound used in perfumery has a violet-like scent. Propose a synthesis of this compound from benzene.

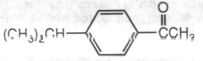

4-Isopropylacetophenone

The isopropyl group is weakly activating and ortho-para directing; the acetyl carbonyl group of the acetyl group is deactivating and meta directing. Therefore, start with benzene, convert it to isopropylbenzene (cumene) and then carry out a Friedel-Crafts acylation using acetyl chloride in the presence of aluminum chloride. Alkylation of benzene can be accomplished using a 2-halopropane, 2-propanol, or propene, each in the presence of an appropriate catalyst.

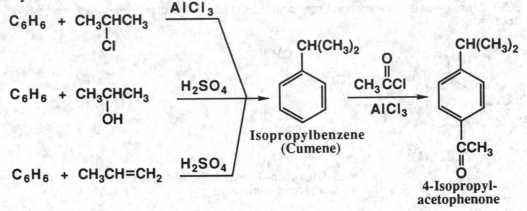

<u>Problem 20.40</u> Following is the structural formula of musk ambrette, a synthetic musk, essential in perfumes to enhance and retain odor. Propose a synthesis of this compound from *m*-cresol (3-methylphenol).

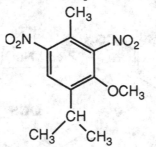

Both methyl and methoxyl groups are ortho/para directing. The methoxyl group is a moderately strong *o,p*-directing group while the methyl group is only weakly *o,p*-directing. Introduction of the isopropyl group by Friedel-Crafts alkylation gives a mixture of 4-isopropyl-3-methoxytoluene and 2-isopropyl-5-methoxytoluene. Following separation of the desired isomer, nitration both ortho and para to the methoxyl group gives the product.

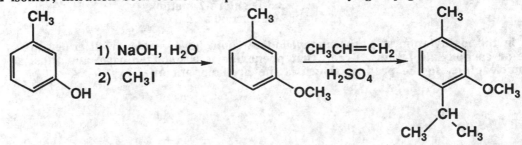

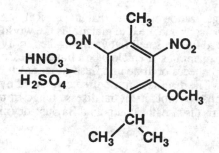

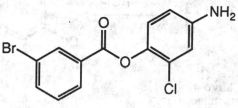

Problem 20.41 Propose a synthesis of this compound starting from toluene and phenol.

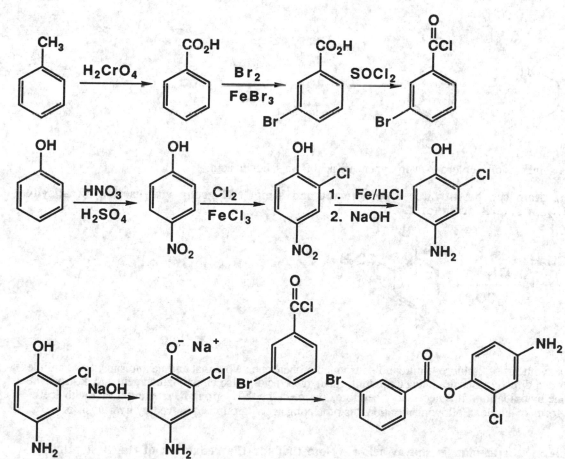

Problem 20.42 The World Health Organization estimates that the tropical disease schistosomiasis (bilharziasis) affects between 180 and 200 million persons and, next to malaria, is the world's most serious parasitic infection in humans. The disease is caused by blood flukes, small flatworms of the family Schistosomatidae, which live in the blood vessels of humans and other mammals. Female blood flukes release from 300 to 3000 eggs daily into the bloodstream. Those evacuated in the feces or urine into fresh water hatch to larvae, which find their way to host water snails, and develop further. The disease is most often contracted by humans from contaminated water populated by snails that carry the worms. Symptoms of the disease range from cough and fever to livor, lung, and brain damage. As one attack on this disease, the compound Bayluscide (niclosamide) has been developed to kill infected water snails.

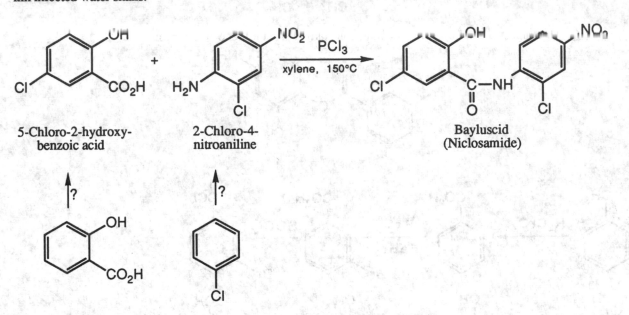

| 5-Chloro-2-hydroxy-benzoic acid | 2-Chloro-4-nitroaniline | Bayluscid (Niclosamide) |

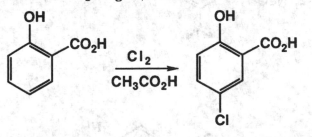

(a) Propose a synthesis of 5-chloro-2-hydroxybenzoic acid from salicylic acid.

The chlorine atom can be introduced *para* to the activating -OH group via reaction of salicylic acid with Cl₂ in glacial acetic acid.

(b) Propose a synthesis of 2-chloro-4-nitroaniline from chlorobenzene. Note that aniline and substituted anilines cannot be nitrated directly. Besides being a nitrating agent, nitric acid is also an oxidizing agent and destroys the aromatic amine by oxidation. It is necessary, therefore, to protect the -NH₂ group by reacting it first with acetic anhydride to form an amide. Following nitration, the protecting acetyl group is removed by hydrolysis of the amide.

A reasonable synthetic plan is shown below. Note that for the reduction of the first nitro compound, Fe/HCl could have been used in place of H₂/Ni.

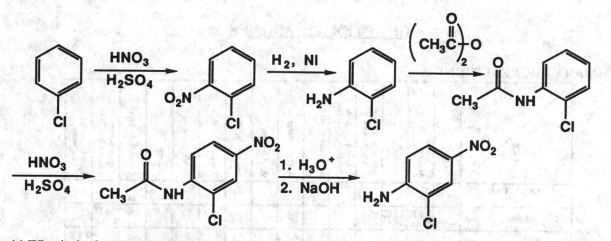

(c) What is the function of PCl₃ in the final stage of this synthesis?

The PCl₃ converts the carboxylic acid to an acid chloride, that then reacts with the amino group to create the desired amide bond.

CHAPTER 21: AMINES

SUMMARY OF REACTIONS

Starting Material ＼ Product	Alkenes	Amines	Ammonium Ions	Aromatic Nitroso Compounds	Acyl Bromides, Chlorides, Cyanides	Aryl Diazonium Salts	Aryl Fluorides	Aryl Iodides	Benzene Derivatives	Carbocations	N-Nitrosamines	Phenols	Primary Amines	Quaternary Ammonium Salts	Ring-Expanded Ketones
Alkyl Halides Azide													21A 21.8B*		
Amines		21B 21.8A	21C 21.6											21D 21.10	
Cyclic β-Aminoalcohols															21E 21.9D
Primary Aliphatic Amines										21F 21.9D					
Primary Aromatic Amines					21G 21.9E	21H 21.9E	21I 21.9E	21J 21.9E	21K 21.9E			21L 21.9E			
Quaternary Ammonium Hydroxides	21M 21.10														
Secondary Aliphatic Amines											21N 21.9C				
Tertiary Amines	21O 21.11														
Tertiary Aromatic Amines				21P 21.9B											

*Section in book that describes reaction.

REACTION 21A: ALKYLATION OF AZIDE ION TO PREPARE PRIMARY AMINES (Section 21.8B)

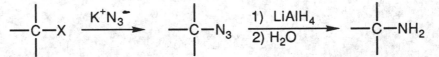

- **Primary amines** can be synthesized by **alkylation** of the **azide ion**, followed by **reduction** with **LiAlH₄**. ✳
- This strategy eliminates the problem of overalkylation that occurs with the alkylation of ammonia or amines.
- The azide ion can also be used to open epoxide rings, to produce β-**aminoalcohols** after reduction.

REACTION 21B: ALKYLATION OF AMMONIA AND AMINES (Section 21.8A)

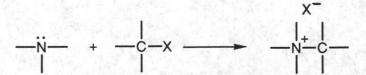

- **Ammonia** or **amines** can be **alkylated** with **primary** or **secondary alkyl halides**. Unfortunately, this approach is **not generally synthetically useful** because mixtures of **overalkylated products** usually result. The problem is that the initially alkylated product is also a good nucleophile and reacts further. ✳

REACTION 21C: PROTONATION OF AMINES (Section 21.6)

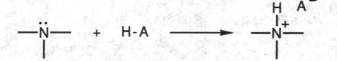

- **Amines are bases** due to the lone pair of electrons on nitrogen. ✳
- **Aliphatic amines** have **pK$_b$'s near 4**, which is slightly more basic than ammonia itself.
- **Aromatic amines** are much less basic, having **pK$_b$'s near 9**, because the nitrogen lone pair is conjugated with the adjacent aromatic ring. Electron-withdrawing groups on the aromatic ring decrease base strength of aromatic amines, and electron-donating groups increase base strength.
- **Aromatic heterocyclic amines** are less basic than non-aromatic heterocyclic amines, particularly if the lone pair of electrons is part of the aromatic ($4n + 2$) pi electrons.

REACTION 21D: EXHAUSTIVE METHYLATION (Section 21.10)

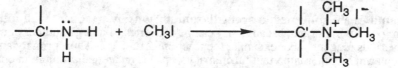

- **Quaternary ammonium** salts are produced when amines are treated with excess methyl iodide. ✳

REACTION 21E: β-AMINOALCOHOLS TREATED WITH NITROUS ACID UNDERGO REARRANGEMENT (Section 21.9D)

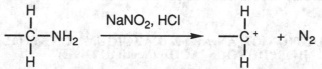

- **Cyclic β-aminoalcohols** react with **nitrous acid** to give **ring-expanded cyclic ketones**. This process is referred to as the **Tiffeneau-Demjanov** reaction. ✳
- The mechanism of the reaction involves initial formation of a diazonium ion, that then undergoes ring expansion through a concerted loss of N_2 and rearrangement via a 1,2 shift to give a cation, which then loses a proton to give the ring-expanded cyclic ketone.

REACTION 21F: REACTION OF PRIMARY ALIPHATIC AMINES WITH NITROUS ACID (Section 21.9D)

- Treatment of **primary amino groups** with nitrous acid produces a number of products that are derived from **carbocation intermediates**. ✳
- The mechanism of this reaction involves initial formation of an *N*-nitrosamine, which undergoes keto-enol tautomerization to give a diazotic acid. Protonation and loss of water produces an unstable aliphatic diazonium ion that loses N_2 to give the carbocation intermediate.
- The carbocation intermediate can take part in a variety of different reactions such as substitution, elimination, rearrangement, or a combination thereof, to give multiple products.

REACTION 21G: SANDMEYER REACTION (Section 21.9E)

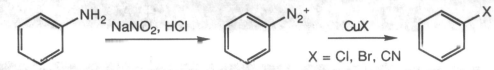

- A primary **aromatic amine** can be reacted with **nitrous acid**, followed by **CuBr, CuCl, or CuCN** to produce an **aryl bromide, chloride, or cyanide**, respectively. This reaction is referred to as the **Sandmeyer reaction.** *

REACTION 21H: FORMATION OF ARENEDIAZONIUM IONS (Section 21.9E)

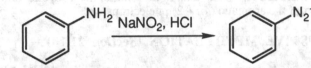

- **Primary aryl amines** are converted to **arenediazonium ions** by treatment with **nitrous acid.** *
- Arenediazonium ions are important, because they can be converted to a large number of other aryl species by treatment with various reagents. For example, see reactions 21G-21L. Thus, primary aromatic amino groups can be replaced by other important functional groups via an arenediazonium intermediate.

REACTION 21I: SCHIEMANN REACTION (Section 21.9E)

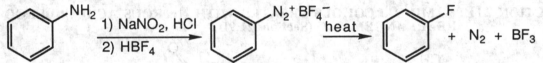

- A **primary aromatic amine** can be reacted with **nitrous acid**, followed by **HBF$_4$ or NaBF$_4$,** to produce an arenediazonium fluoroborate, which is converted to an **aryl fluoride** upon heating. This process is referred to as the **Schiemann reaction.** *
- The Schiemann reaction is thought to involve an aryl cation intermediate.

REACTION 21J: TREATMENT OF AN ARENEDIAZONIUM ION WITH KI (Section 21.9E)

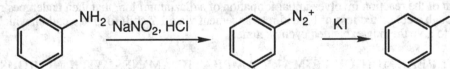

- A **primary aromatic amine** can be reacted with **nitrous acid**, followed by **KI**, to produce an **aryl iodide.** *

REACTION 21K: REDUCTION OF AN ARYL DIAZONIUM ION WITH HYPOPHOSPHOROUS ACID (Section 21.9E)

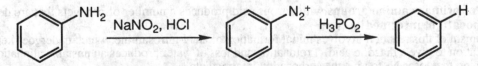

- A **primary aromatic amine** can be reacted with **nitrous acid**, followed by **H$_3$PO$_2$,** to produce a product in which the aryl amine function is replaced by hydrogen. *
- This reaction is useful for a variety of synthetic applications, especially when the amino group is desired to control temporarily orientation or reactivity during electrophilic aromatic substitution.

REACTION 21L: CONVERSION OF A PRIMARY ARYL AMINE TO A PHENOL (Section 21.9E)

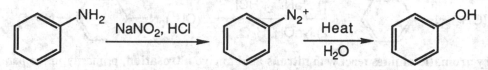

- **Heating an aryl diazonium salt in water** results in formation of **a phenol.** ✳
- An aryl cation intermediate is formed, that reacts with water to produce the phenol.

REACTION 21M: HOFMANN EXHAUSTIVE METHYLATION/ELIMINATION (Section 21.10)

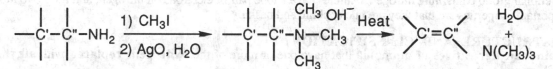

- Heating of a **quaternary ammonium hydroxide** results in formation of an **alkene**, water, and tertiary amine. This reaction is referred to as the **Hofmann elimination.** ✳
- The quaternary ammonium hydroxide can be produced by treating a quaternary ammonium halide with AgO.
- The reaction follows Hofmann's rule, in that elimination occurs predominantly in the direction of the least substituted carbon atom.

REACTION 21N: FORMATION OF *N*-NITROSAMINES FROM SECONDARY AMINES (Section 21.9C)

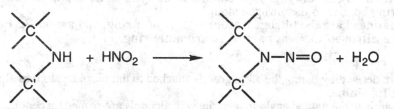

- **Aliphatic** and **aromatic secondary amines** react with **nitrous acid** to produce *N*-nitrosamines. ✳
- The reaction mechanism involves reaction of the lone pair of electrons on the amine nitrogen atom with the nitrosyl cation to form a new nitrogen-nitrogen bond. The resulting intermediate loses a proton to give the final product.
- Many *N*-nitrosamines have been found to be carcinogenic.

REACTION 21O: COPE ELIMINATION: PYROLYSIS OF A TERTIARY AMINE OXIDE (Section 21.11)

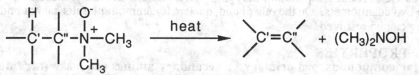

- **Heating an amine oxide** with at least one β-hydrogen results in formation of an **alkene** and an *N,N*-**dialkylhydroxylamine.** This reaction is referred to as the **Cope elimination.** ✳
- The reaction mechanism is considered to be a concerted syn elimination, and little selectivity is seen if more than one alkene is possible. Since the transition state involves the cyclic, planar flow of six electrons, this reaction is said to have **transition state aromaticity.**

REACTION 21P: NITROSATION OF TERTIARY AROMATIC AMINES (Section 21.9B)

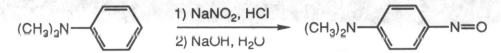

- **Tertiary aromatic amines** react with **nitrous acid** to give **nitrosation**, primarily in the para position.✳
- The reaction mechanism involves electrophilic aromatic substitution by the nitrosyl cation.

SUMMARY OF IMPORTANT CONCEPTS

21.0 OVERVIEW
• Nitrogen is the **fourth most common element** in organic molecules, and **amines** are the most common functional group containing nitrogen. Amines have a lone pair of electrons on nitrogen, and their two most important properties are that they are nucleophilic and basic. ✳

21.1 STRUCTURE AND CLASSIFICATION
• **Amines** are **derivatives of ammonia** that have one or more of the **hydrogens replaced** with **alkyl and/or aryl groups.** ✳
  - **Primary amines** have **one hydrogen** of ammonia **replaced by a carbon** in the form of an alkyl or aryl group.
  - **Secondary amines** have **two hydrogens** of ammonia **replaced by a carbon** in the form of alkyl and/or aryl groups.
  - **Tertiary amines** have **all three hydrogens** of ammonia **replaced by a carbon** in the form of alkyl and/or aryl groups.
  - **Quaternary ammonium ions** have four alkyl and/or aryl groups attached to nitrogen, resulting in a **positively–charged** species.
  - **Aliphatic amines** have only **alkyl groups** bonded to nitrogen, while **aromatic amines** have **at least one aromatic ring** bonded to the nitrogen atom.
  - **Heterocyclic amines** have the **nitrogen atom** as **part of a ring**, and **aromatic heterocyclic amines** have the **nitrogen atom** as **part of an aromatic ring.**

21.2 NOMENCLATURE
• **Common names** are derived by listing the alkyl groups attached to the nitrogen atom in alphabetical order, followed by the suffix **amine.**
• In **IUPAC names**, amines are named analogous to the way alcohols are named, except the suffix **e** of the parent chain is replaced by **amine.**
• IUPAC uses the common name **aniline** for derivatives of $C_6H_5NH_2$, although certain common names are retained for some substituted anilines such as **toluidine** and **anisidine.**
• Several common heterocycles retain their common names in IUPAC nomenclature as well, including **indole, purine, quinoline,** and **isoquinoline.**

21.3 CHIRALITY OF AMINES AND QUATERNARY AMMONIUM SALTS
• Secondary or tertiary **amines with three different groups attached are chiral,** but they **cannot** usually **be resolved, because** at room temperature, they undergo a process called **pyramidal inversion** that interconverts the two enantiomers. On the other hand, quaternary ammonium salts cannot undergo pyramidal inversion, so they can be resolved. ✳

21.4 PHYSICAL PROPERTIES
• **Amines are polar compounds,** and **primary or secondary amines** can make **intramolecular hydrogen bonds.** As a result, amines have **higher melting and boiling points than analogous hydrocarbons.** Since they can take part in hydrogen bonding, they are also **more soluble in water** than their hydrocarbon counterparts.

21.5 SPECTROSCOPIC PROPERTIES
• **Mass spectrometry** can be useful for identifying amines, because molecules with an **odd number of nitrogen atoms** have **an odd mass number.** This can be used to assign a molecular formula, since molecules containing C, H, and O will always have an even mass number. Furthermore, aliphatic amines undergo a **characteristic β-cleavage reaction,** usually so that the largest R group is lost. ✳

• In ^{1}H NMR spectra, an amine hydrogen usually is a **singlet** because of rapid intermolecular exchange. The signal appears in the region δ **0.5 to** δ **5.0**, depending on the conditions of the experiment. **Hydrogens** on the α**-carbon** usually occur in the region δ **2.2 to** δ **2.8**. In ^{13}C NMR spectra, **carbon atoms** bound to **amine nitrogen atoms** are **deshielded** by about **20 ppm** compared to where they would be in an analogous hydrocarbon. ✳

• **Primary and secondary amines** have **two and one N-H stretching vibrations**, respectively, in the region of **3300 to 3500 cm**$^{-1}$ of **infrared spectra**. Tertiary amines have no N-H stretching vibrations in infrared spectra. ✳

21.6 BASICITY

• **Amines are weak bases**, so aqueous solutions of amines are basic. The **lone pair of electrons on nitrogen** forms a new **bond** with a **hydrogen of water**, producing the **ammonium species and hydroxide**. This equilibrium is often described by K_b, that is defined by the following equation:

$$K_b = K_{eq}[H_2O] = \frac{[R_3NH^+][OH^-]}{[R_3N]}$$

where R is an alkyl group, aryl group, or hydrogen. ✳

- The **pK_b** of an amine is equal to the **-log K_b**.
- The **pK_a** of the conjugate acid of an amine is **related to** the **pK_b** by the following equation:

$$pK_a + pK_b = 14.00$$

• In general, any group attached to the nitrogen of an amine that **releases electron density increases basicity**, while any group that **withdraws electron density decreases basicity**.

• The **basicity of amines** and the **water solubility of ammonium salts** in water can be used to **separate amines from water-insoluble**, non-basic compounds.

CHAPTER 21
Solutions to the Problems

Problem 21.1 Identify all carbon stereocenters in coniine, nicotine, and cocaine.

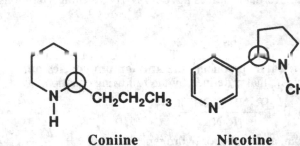

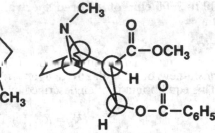

Coniine   **Nicotine**   **Cocaine**

Problem 21.2 Write structural formulas for these amines.
(a) 2-Methyl-1-propanamine   (b) Cyclohexanamine   (c) (R)-2-Butanamine

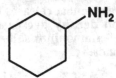

CH₃CHCH₂NH₂ (with CH₃ substituent)

CH₃CHCH₂NH₂

$CH_3CHCH_2NH_2$

Problem 21.3 Write structural formulas for these amines.
(a) Isobutylamine   (b) Triphenylamine   (c) Diisopropylamine

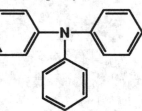

$CH_3CHCH_2NH_2$

$HN—CH(CH_3)_2$
$CH(CH_3)_2$

Problem 21.4 Write IUPAC and, where possible, common names for these amines.

(a)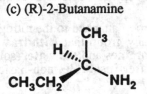

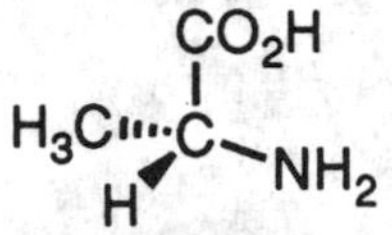

(R)-2-Aminopropanoic acid
(R-Alanine)

(b) $H_2NCH_2CH_2CH_2CO_2H$

4-Aminobutanoic acid
(γ-Aminobutyric acid)

(c) $(CH_3)_3CCH_2NH_2$

2,2-Dimethylpropanamine
(Neopentylamine)

Problem 21.5 Select the stronger acid from each pair of compounds.

(a)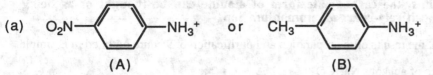

(A)   (B)

4-Nitroaniline (pK$_b$ 13.0) is a weaker base than 4-methylaniline (pK$_b$ 8.92). The decreased basicity of 4-nitroaniline is due to the electron-withdrawing effect of the *para* nitro group. Because 4-nitroaniline is the weaker base, its conjugate acid (A) is the stronger acid.

(b)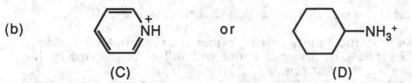

(C)   (D)

Pyridine (pK$_b$ 8.75) is a much weaker base than cyclohexanamine (pK$_b$ 3.34). The lone pair of electrons in the sp^2 orbital on the nitrogen atom of pyridine (C) has more s character, so these electrons are less available for binding to a proton. Because pyridine is a weaker base, its conjugate acid, (C), is the stronger acid

Problem 21.6 Complete each acid-base reaction and name the salt formed.

(a) $(CH_3CH_2)_3N$ + HCl ⟶ $(CH_3CH_2)_3NH^+ Cl^-$
Triethylammonium chloride

(b)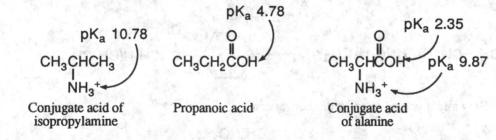

Piperidinium acetate

Problem 21.7 Following are structural formulas for propanoic acid and the conjugate acids of isopropylamine and alanine, along with pK$_a$ values for each functional group:

pK$_a$ 4.78

pK$_a$ 10.78   pK$_a$ 2.35

CH_3CHCH_3 CH_3CH_2COH CH_3CHCOH pK$_a$ 9.87
       | |
     NH_3^+ NH_3^+

Conjugate acid of   Propanoic acid   Conjugate acid
isopropylamine   of alanine

(a) How do you account for the fact that the -NH$_3^+$ group of the conjugate acid of alanine is a stronger acid than the -NH$_3^+$ group of the conjugate acid of isopropylamine?

The electron-withdrawing properties of the carboxyl group adjacent to the amine of alanine make the conjugate acid of alanine more acidic than the conjugate acid of isopropylamine.

(b) How do you account for the fact that the -CO$_2$H group of the conjugate acid of alanine is a stronger acid than the -CO$_2$H group of propanoic acid?

The -NH$_3^+$ group is electron-withdrawing, so the adjacent -CO$_2$H group is made more acidic by an inductive effect. This situation is analogous to the electron-withdrawing effects of halogens adjacent to carboxylic acids in molecules such as chloroacetic acid, which has a pK$_a$ of 2.86.

In addition, deprotonation of the carboxylic acid function of alanine results in formation of an overall neutral zwitterion. Thus, the carboxylate form of alanine can be thought of as being neutralized by the adjacent positively-charged ammonium ion.

Problem 21.8 In what way(s) might the results of the separation and purification procedure outlined in Example 21.8 be different if
(a) Aqueous NaOH is used in place of aqueous NaHCO$_3$?

If NaOH is used in place of aqueous NaHCO$_3$, then the phenol will be deprotonated along with the carboxylic acid, so they will be isolated together in fraction A.

(b) The starting mixture contains an aromatic amine, ArNH$_2$, rather than an aliphatic amine, RNH$_2$!

If the starting mixture contains an aromatic amine, ArNH$_2$, rather than an aliphatic amine, RNH$_2$, then the results will be the same. The aromatic amine will still be protonated by the HCl wash, and deprotonated by the NaOH treatment.

Problem 21.9 Show how to bring about each conversion in good yield. In addition to the given starting material, use any other reagents as necessary.

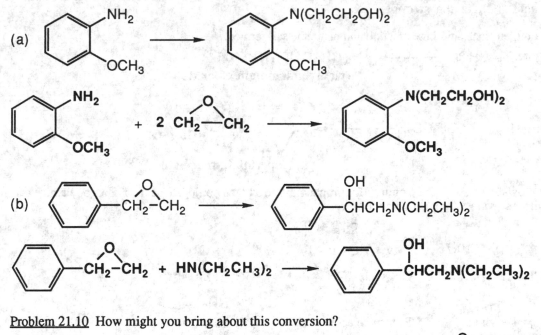

(a)

(b)

Problem 21.10 How might you bring about this conversion?

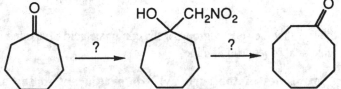

The synthesis begins with an aldol reaction using nitromethane, followed by reduction to give a β-aminoalcohol that then undergoes the ring expansion reaction.

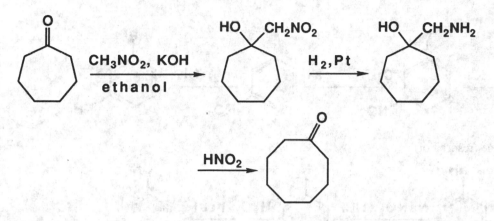

Problem 21.11 Show how to convert toluene to 3-hydroxybenzoic acid using the same set of reactions as in Example 21.11, but changing the order in which some of the steps are carried out.

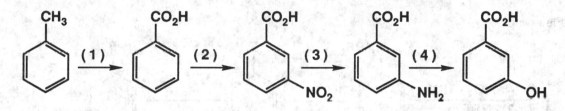

The key to this question is that the methyl group is converted to a meta-directing carboxyl group before the nitration reaction. This leads to the desired product with the hydroxy group in the 3 position.

Step 1: Oxidation at a benzylic carbon (Section 19.6A) can be brought about using chromic acid to give benzoic acid.
Step 2: Nitration of the aromatic ring using HNO_3 in H_2SO_4. The meta-directing carboxyl group gives predominantly the desired 3-nitrobenzoic acid product.
Step 3: Reduction of the nitro group to 3-aminobenzoic acid can be brought about using H_2 in the presence of Ni or other transition metal catalyst. Alternatively, it can be brought about using Zn, Sn, or Fe metal in aqueous HCl.
Step 4: Reaction of the aromatic amine with HNO_2 followed by heating gives 3-hydroxybenzoic acid.

Problem 21.12 Starting with 3-nitroaniline, show how to prepare the following compounds.
(a) 3-Nitrophenol

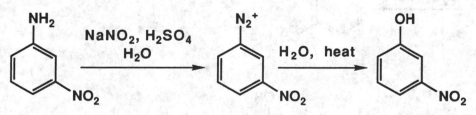

(b) 3-Bromoaniline

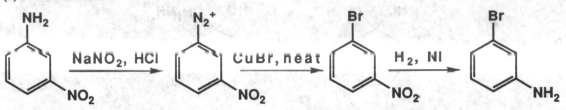

(c) 1,3-Dihydroxybenzene (resorcinol)

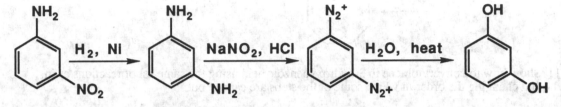

(d) 3-Fluoroaniline

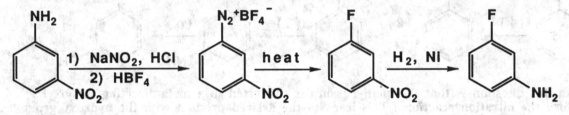

(e) 3-Fluorophenol

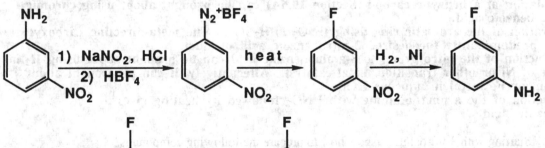

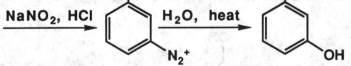

(f) 3-Hydroxybenzonitrile

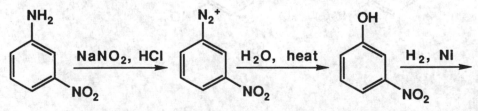

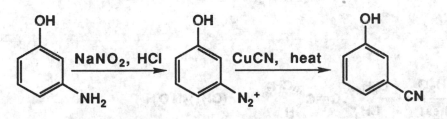

Problem 21.13 The procedure of methylation of an amine and thermal decomposition of quaternary ammonium hydroxides was first reported by Hofmann in 1851, but its value as a means of structure determination was not appreciated until 1881 when he published a report of its use in determining the structure of piperidine. Following are the results obtained by Hofmann:

$$C_5H_{11}N \xrightarrow[\substack{1.\ CH_3I\ (excess),\ K_2CO_3 \\ 2.\ Ag_2O,\ H_2O \\ 3.\ heat}]{} C_7H_{15}N \xrightarrow[\substack{4.\ CH_3I\ (excess),\ K_2CO_3 \\ 5.\ Ag_2O,\ H_2O \\ 6.\ heat}]{} CH_2=CHCH_2CH=CH_2$$

Piperidine   (A)   1,4-Pentadiene

(a) Show that these results are consistent with the structure of piperidine (Section 21.1).

As shown below, the structure of piperidine is consistent with the formulas given, as well as the final product.

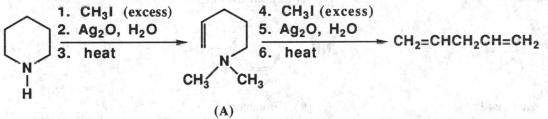

(A)

(b) Propose two additional structural formulas (excluding stereoisomers) for $C_5H_{11}N$ that are also consistent with the results obtained by Hofmann.

The following two molecules also have structures that are consistent with the formulas given, as well as the final product. Remember that in Hofmann eliminations, the least substituted alkene is formed predominantly.

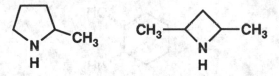

Problem 21.14 In Example 21.14, you considered the product of Cope elimination from the (2R, 3S) stereoisomer of 2-dimethylamino-3-phenylbutane. What is the product of Cope elimination from each of these stereoisomers? What is the product of Hofmann elimination from each of these stereoisomers?
(a) (2S, 3R) stereoisomer?

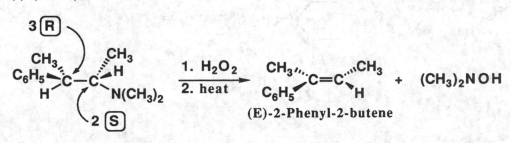

(b) (2S, 3S) stereoisomer?

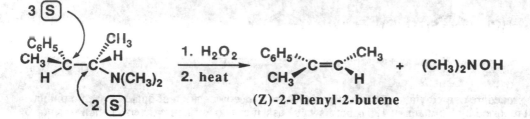

(Z)-2-Phenyl-2-butene

PROBLEMS
Structure and Nomenclature
Problem 21.15 Draw structural formulas for these amines and amine derivatives.

(a) *N,N*-Dimethylaniline

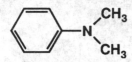

(b) Triethylamine

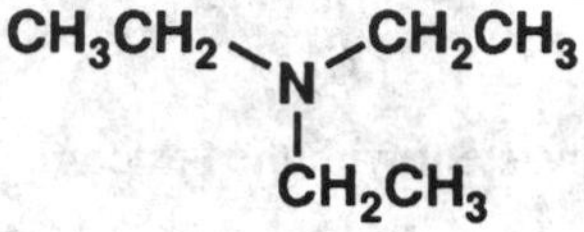

$$CH_3CH_2-N \begin{array}{c} CH_2CH_3 \\ \\ CH_2CH_3 \end{array}$$

(c) *tert*-Butylamine

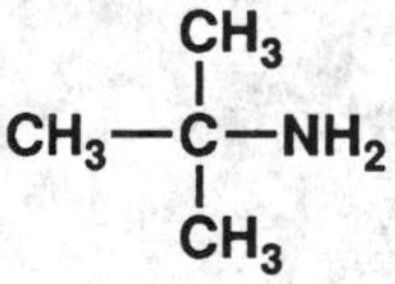

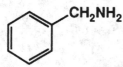

(d) 1,4-Benzenediamine

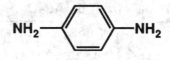

(e) 4-Aminobutanoic acid

$$H_2NCH_2CH_2CH_2CO_2H$$

(f) (R)-2-Butanamine

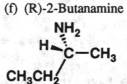

(g) Benzylamine

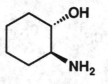

(h) *trans*-2-Aminocyclohexanol

(i) 1-Phenyl-2-propanamine (amphetamine)

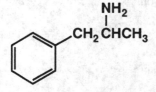

(j) Lithium diisopropylamide (LDA)

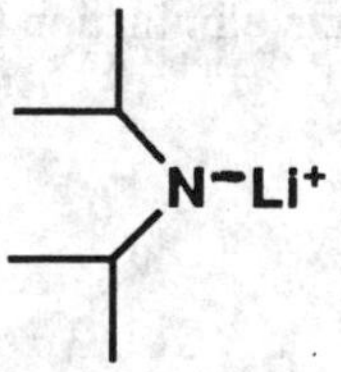

(k) Benzyltrimethylammonium hydroxide (Triton B)

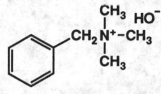

Problem 21.16 Give an acceptable name for these compounds.

(a)

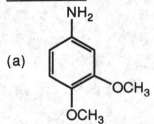

3,4-Dimethoxyaniline

(b)

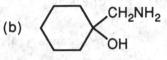

1-Aminomethylcyclohexanol

(c)

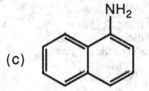

1-Aminonaphthalene

(d) $CH_3NH(CH_2)_2CH_3$

Methylpropylamine

(e)

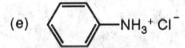

Aniline hydrochloride

(f)

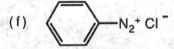

Benzenediazonium chloride

(g)

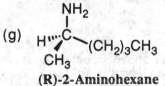

(R)-2-Aminohexane

(h)

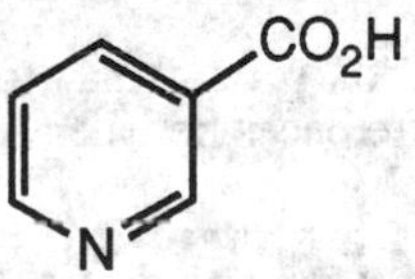

3-Pyridinecarboxylic acid

Problem 21.17 Classify each amine as primary, secondary, or tertiary; as aliphatic or aromatic.

(a)

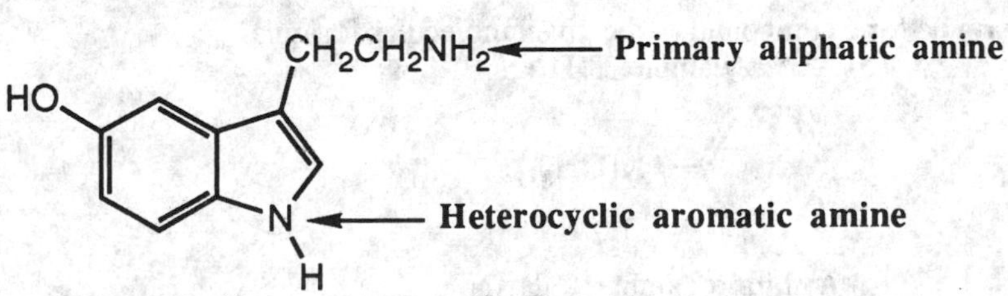

CH₂CH₂NH₂ ◄──── **Primary aliphatic amine**

HO

◄──── **Heterocyclic aromatic amine**

Serotonin
(a neurotransmitter)

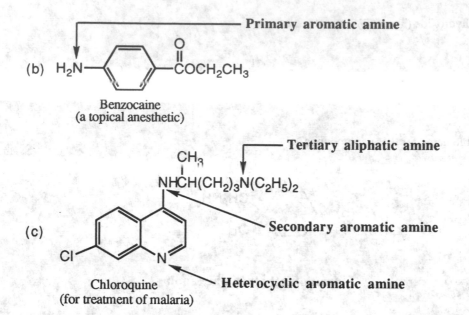

Primary aromatic amine

(b) H_2N—⬡—$\overset{\overset{O}{\|}}{C}OCH_2CH_3$

Benzocaine
(a topical anesthetic)

Tertiary aliphatic amine

(c)

Chloroquine
(for treatment of malaria)

Secondary aromatic amine

Heterocyclic aromatic amine

Problem 21.18 Epinephrine is a hormone secreted by the adrenal medulla. Among its actions, it is a bronchodilator. Salbutamol, sold as the R enantiomer, is one of the most effective and widely prescribed antiasthma drugs. The R enantiomer is 68 times more effective in the treatment of asthma than the S enantiomer.

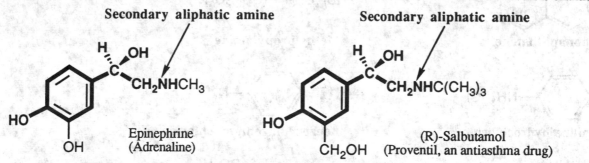

Secondary aliphatic amine

Epinephrine
(Adrenaline)

Secondary aliphatic amine

(R)-Salbutamol
(Proventil, an antiasthma drug)

(a) Classify each as a primary, secondary, or tertiary amine.
(b) Compare the similarities and differences between their structural formulas.

The parts of the molecules that are identical are indicated in bold on the above structures. As far as differences are concerned, epinephrine possesses a second hydroxyl group on the aromatic ring and a methyl group on the amine, while (R)-salbutamol has a hydroxymethyl on the ring and a *tert*-butyl group on the amine.

Problem 21.19 Draw the structural formula for a compound of the given molecular formula.
a) A 2° arylamine, C_7H_9N (b) A 3° arylamine, $C_8H_{11}N$

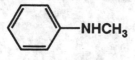

—$NHCH_3$

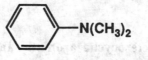

—$N(CH_3)_2$

(c) A 1° aliphatic amine, C_7H_9N (d) A chiral 1° amine, $C_4H_{11}N$

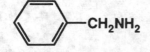

—CH_2NH_2

$\overset{\overset{CH_3}{|}}{CH_3CH_2CHNH_2}$

(e) A 3° heterocyclic amine, $C_6H_{11}N$

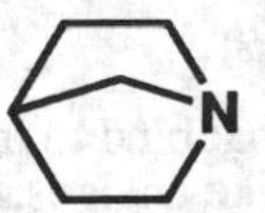

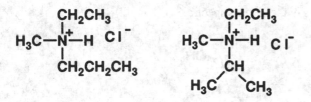

(f) A trisubstituted 1° arylamine, $C_9H_{13}N$

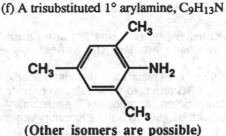

(Other isomers are possible)

(g) A chiral quaternary ammonium salt, $C_6H_{16}NCl$

There are two answers to this question.

$$H_3C-\overset{\overset{\displaystyle CH_2CH_3}{|}}{\underset{\underset{\displaystyle CH_2CH_2CH_3}{|}}{\overset{+}{N}}}-H \quad Cl^-$$

$$H_3C-\overset{\overset{\displaystyle CH_2CH_3}{|}}{\underset{\underset{\displaystyle \overset{\displaystyle CH}{\overset{}{H_3C}\quad CH_3}}{|}}{\overset{+}{N}}}-H \quad Cl^-$$

<u>Problem 21.20</u> Morphine and its *O*-methylated derivative, codeine, are among the most effective pain killers known. However, they possess two serious drawbacks: they are addictive, and repeated use induces a tolerance to the drug. Increasingly larger doses become necessary, which can lead to respiratory arrest. Many morphine analogs have been prepared in an effort to find drugs that are equally effective as pain killers but that have less risk of physical dependence and potential for abuse. Following are structural formulas for pentazocine (one-third the potency of codeine), meperidine (one-half the potency of morphine), and dextropropoxyphene (one-half the potency of codeine). Methadone, with a potency equal to that of morphine, is used to treat opiate withdrawal symptoms in heroin abusers.

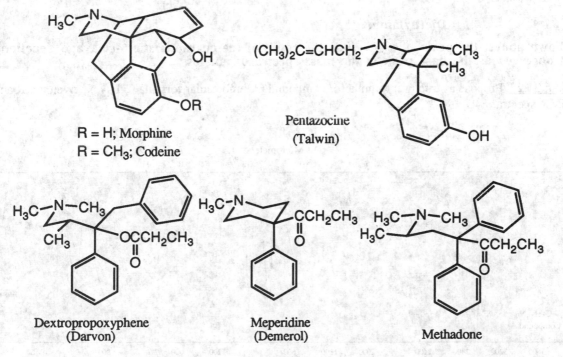

R = H; Morphine
R = CH$_3$; Codeine

Pentazocine
(Talwin)

Dextropropoxyphene
(Darvon)

Meperidine
(Demerol)

Methadone

(a) List the structural features common to each of these molecules.

Each of the above molecules contains a tertiary amine and a phenyl ring that is three sp^3 carbons away from the nitrogen atom of the amine.

(b) The Beckett-Casy rules are a set of empirical rules to predict the structure of molecules that bind to morphine receptors and act as analgesics. According to these rules, to provide an effective morphine-like analgesia, a molecule must have (1) an aromatic ring attached to (2) a quaternary carbon and (3) a nitrogen at a distance equal to two carbon-carbon single bond lengths from the quaternary center. Show that these structural requirements are present in the molecules given in this problem.

By inspection of the structures, it can be seen that all of these three structural requirements are present in the molecules mentioned in this problem.

Spectroscopy
Problem 21.21 Account for the formation of the base peaks in these mass spectra.
(a) Isobutylmethylamine, m/z 44

Isobutylmethylamine $m/z = 44$

As shown above, the peak at $m/z = 44$ is the result of the characteristic β-cleavage reaction often observed as the base peak in the mass spectra of amines.

(b) Diethylamine, m/z 58

Diethylamine $m/z = 58$

As shown above, the peak at $m/z = 58$ is the result of the characteristic β-cleavage reaction often observed as the base peak in the mass spectra of amines.

Problem 21.22 Propose a structural formula for compound (A), molecular formula $C_5H_{13}N$, given its IR and ^{1}H-NMR spectra.

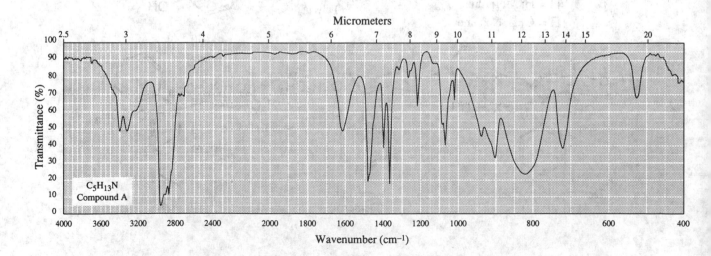

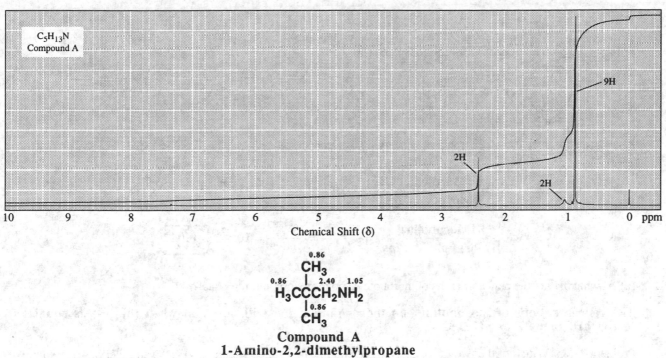

$$CH_3$$
0.86

$$H_3CCCH_2NH_2$$
0.86   2.40   1.05

$$CH_3$$
0.86

Compound A
1-Amino-2,2-dimethylpropane
(Neopentylamine)

The presence of two absorptions near 3300 to 3400 cm^{-1} in the IR spectrum indicate that compound A is a primary amine. Furthermore, in the ^{1}H-NMR, there are two sharp singlets at δ 0.86 and δ 2.40 integrating to 9H and 2H, respectively. The above structure is consistent with the index of hydrogen deficiency of zero, since there are no rings or pi bonds.

Basicity of Amines
Problem 21.23 Select the stronger base from each pair of compounds.

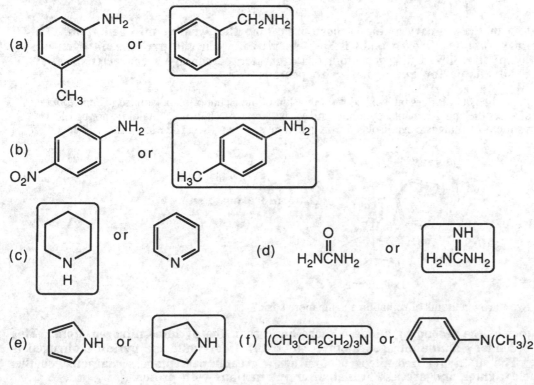

Problem 21.24 The pK$_a$ of morpholine is 8.33.

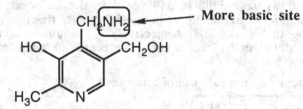

Morpholinium ion   Morpholine

(a) Calculate the ratio of morpholine to morpholinium ion in aqueous solution at pH 7.0.

$$K_a = \frac{[Morpholine][H^+]}{[Morpholinium\ Ion]} = 10^{-8.33} \quad At\ pH\ 7.0\ [H^+] = 10^{-7}$$

$$\frac{[Morpholine]}{[Morpholinium\ Ion]} = \frac{10^{-8.33}}{[H^+]} = \frac{10^{-8.33}}{10^{-7.0}} = 10^{-1.33} = \boxed{0.047}$$

(b) At what pH are the concentrations of morpholine and morpholinium ion equal?

The concentrations of morpholine and morpholinium ion will be equal when the pK$_a$ is equal to the pH, namely at pH 8.33.

Problem 21.25 Which of the two nitrogens in pyridoxamine (a form of vitamin B$_6$) is the stronger base? Explain your reasoning.

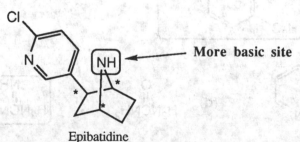

The nitrogen atom of the primary amine is more basic than the pyridine nitrogen atom. This is because the primary amine nitrogen atom is sp^3 hybridized, while the pyridine nitrogen is sp^2 hybridized. The sp^2 hybridized nitrogen atom has a greater percentage s character, so the electrons are less available for interactions with protons.

Problem 21.26 Epibatidine, a colorless oil isolated from the skin of the Ecuadorian poison frog *Epipedobates tricolor* has several times the analgesic potency of morphine.   It is the first chlorine-containing, nonopioid (nonopium-like in structure) analgesic ever isolated from a natural source. (See The Merck Index, 12th ed., # 3647.)

Epibatidine

(a) Which of the two nitrogen atoms of epibatidine is the more basic?

The nitrogen atom of the secondary amine is more basic than the pyridine nitrogen atom. This is because the secondary amine nitrogen atom is sp^3 hybridized, while the pyridine nitrogen is sp^2 hybridized. The sp^2 hybridized nitrogen atom has a greater percentage s character, so the electrons are held tighter and are less available for interactions with protons.

(b) Mark all stereocenters in this molecule.

The three stereocenters are marked with an asterisk (*).

<u>Problem 21.27</u> Aniline (pK_b 9.37) is a considerably stronger base than diphenylamine (pK_b 13.21).
Conversely, diphenylammonium ion is a considerably stronger acid (pK_a 0.79) than anilinium ion (pK_b 4.63).
Account for these marked differences in acidity/basicity.

$$C_6H_5NH_2 \qquad\qquad (C_6H_5)_2NH$$
$$\text{Aniline} \qquad\qquad\qquad \text{Diphenylamine}$$
$$(pK_a\ 4.63;\ pK_b\ 9.37) \qquad (pK_a\ 0.79;\ pK_b\ 13.21)$$

**The diphenylamine is a weaker base because the nitrogen lone pair is delocalized by
interaction with the pi system of both aromatic rings, as opposed to aniline in which the lone
pair on nitrogen is only delocalized by interaction with a single aromatic ring. This
delocalization and thus stabilization cannot take place when the amine is protonated. Thus,
the diphenylammonium ion is a stronger acid because loss of the proton results in greater
resonance stabilization (delocalization into two aromatic rings instead of one).**

<u>Problem 21.28</u> Complete the following acid-base reactions and predict the direction of equilibrium (to the right or
to the left) for each. Justify your prediction by citing values of pK_a for the stronger and weaker acid in each
equilibrium. For values of acid ionization constants, consult Table 3.1 (Strengths of Some Inorganic and Organic
Acids), Table 10.2 (Acidity of Alkanes, Alkenes, and Alkynes), Table 9.3 (Acidity of Alcohols), Section 19.5B
(Acidity of Phenols), Table 21.2 (Base Strengths of Amines), and Appendix 2 (Acid Ionization Constants for the
Major Classes of Organic Acids). Where no ionization constants are given, make the best estimate from the
information given in the reference tables and sections.

**In all cases, the equilibrium favors formation of the weaker acid (higher pK_a) and weaker base
(higher pK_b). Recall that $pK_a + pK_b = 14$ for any conjugate acid-base pair.**

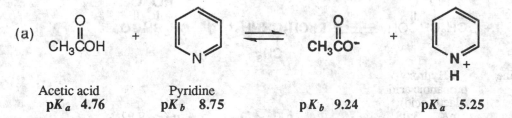

Acetic acid   Pyridine
pK_a 4.76 pK_b 8.75 pK_b 9.24 pK_a 5.25

**Equilibrium lies to the right, since the acetate anion and pyridinium ion are the weaker acid
and base, respectively.**

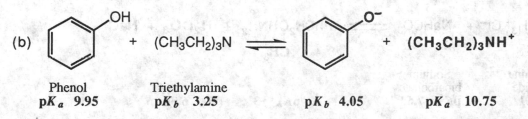

Phenol   Triethylamine
pK_a 9.95 pK_b 3.25 pK_b 4.05 pK_a 10.75

**Equilibrium lies to the right, since the phenoxide anion and the triethylammonium species are
the weaker base and acid, respectively.**

(c) $PhC\equiv CH$ + NH_3 $\rightleftharpoons$ $PhC\equiv C^-$ + NH_4^+

Phenylacetylene   Ammonia
pK_a ~25 pK_b 4.74 pK_b ~-11 pK_a 9.26

Equilibrium lies to the left, since the alkyne and the ammonia are the weaker acid and base, respectively.

(d) $PhC\equiv CH$ + $[(CH_3)_2CH]_2N^-Li^+$ $\rightleftharpoons$ $PhC\equiv C^-Li^+$ + $[(CH_3)_2CH]_2NH$

Phenylacetylene   Lithium
                      diisopropylamide
                          (LDA)
pK_a ~25 pK_b ~-19 pK_b ~-11 pK_a ~33

Equilibrium lies to the right, since the alkyne anion and the amine are the weaker base and acid, respectively.

(e) $PhCO^-Na^+$ + $(CH_3CH_2)_3NH^+Cl^-$ $\rightleftharpoons$ $PhCOH$ + $(CH_3CH_2)_3N$ + $NaCl$

Sodium   Triethylammonium
benzoate   chloride
pK_b 9.81 pK_a 10.75 pK_a 4.19 pK_b 3.25

Equilibrium lies to the left, since the carboxylate anion and the ammonium ion are the weaker base and acid, respectively.

(f) $PhCH_2\overset{\underset{\displaystyle CH_3}{|}}{C}HNH_2$ + $CH_3\overset{\underset{\displaystyle}{HO}}{\overset{|}{C}}H\overset{\underset{\displaystyle}{O}}{\overset{||}{C}}OH$ $\rightleftharpoons$ $PhCH_2\overset{\underset{\displaystyle CH_3}{|}}{C}HNH_3^+$ + $CH_3\overset{\underset{\displaystyle}{HO}}{\overset{|}{C}}H\overset{\underset{\displaystyle}{O}}{\overset{||}{C}}O^-$

1-Phenyl-2-propanamine   2-Hydroxy-
  (Amphetamine)   propanoic acid
                          (lactic acid)
pK_b ~3 pK_a 3.08 pK_a ~11 pK_b 9.92

Equilibrium lies to the right, since the ammonium ion and the carboxylate ion are the weaker acid and base, respectively.

(g) $PhCH_2\overset{\underset{\displaystyle CH_3}{|}}{C}HNH_3^+Cl^-$ + $NaHCO_3$ $\rightleftharpoons$ $PhCH_2\overset{\underset{\displaystyle CH_3}{|}}{C}HNH_2$ + H_2CO_3 + $NaCl$

Amphetamine   Sodium
hydrochloride   bicarbonate
pK_a ~11 pK_b 7.63 pK_b ~3 pK_a 6.36

Equilibrium lies to the left, since the ammonium ion and the bicarbonate ion are the weaker acid and base, respectively.

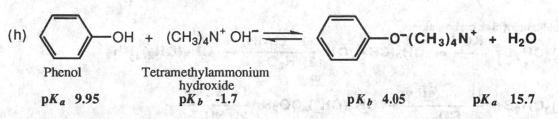

Phenol   Tetramethylammonium
                               hydroxide
pK_a 9.95   pK_b -1.7   pK_b 4.05   pK_a 15.7

Equilibrium lies to the right, since phenoxide and water are the weaker base and acid, respectively.

<u>Problem 21.29</u> Quinuclidine and triethylamine are both tertiary amines. Quinuclidine, however, is a considerably stronger base than triethylamine. Stated alternatively, the conjugate acid of quinuclidine is a considerably weaker acid than the conjugate acid of triethylamine. Propose an explanation for these differences in acidity/basicity. (*Hint:* The answer lies in the ease with which each ammonium ion can be solvated.)

Quinuclidine   Triethylamine
(pK_a 10.6)   (pK_a 8.6)

The ammonium ion of the protonated form of quinuclidine is more compact than that of triethylamine since the alkyl groups are "tied back"; thus the ammonium ion of protonated quinuclidine is solvated better. For this reason, the ammonium ion of quinuclidine is a weaker acid, so quinuclidine is a stronger base.

<u>Problem 21.30</u> Suppose you have a mixture of these three compounds. Devise a chemical procedure based on their relative acidity or basicity to separate and isolate each in pure form.

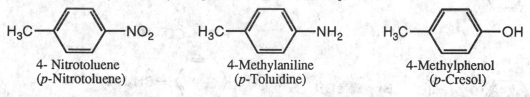

4- Nitrotoluene   4-Methylaniline   4-Methylphenol
(*p*-Nitrotoluene)   (*p*-Toluidine)   (*p*-Cresol)

These molecules can be separated by extraction into different aqueous solutions. First, the mixture is dissolved in an organic solvent such as ether in which all three compounds are soluble. Then, the ether solution is extracted with dilute aqueous HCl. Under these conditions, 4-methylaniline (a weak base) is converted to its protonated form and dissolves in the aqueous solution. The aqueous solution is separated, treated with dilute NaOH, and the water-insoluble 4-methylaniline separates, and is recovered. The ether solution containing the other two components is then treated with dilute aqueous NaOH. Under these conditions, 4-methylphenol (a weak acid) is converted to its phenoxide ion and dissolves in the aqueous solution. Acidification of this aqueous solution with dilute HCl forms water-insoluble 4-methylphenol that is then isolated. Evaporation of the remaining ether solution gives the 4-nitrotoluene, which is neither acidic or basic.

Preparation of Amines
<u>Problem 21.31</u> Propose a synthesis of 1-hexanamine from the following:
(a) A bromoalkane of six carbon atoms

$$CH_3(CH_2)_4CH_2Br + excess\ NH_3 \longrightarrow CH_3(CH_2)_5NH_2$$

$$or$$

$$CH_3(CH_2)_4CH_2Br \xrightarrow{K^+N_3^-} CH_3(CH_2)_5N_3 \xrightarrow[\text{2) } H_2O]{\text{1) } LIAlH_4} CH_3(CH_2)_5NH_2$$

(b) A bromoalkane of five carbon atoms

$$CH_3(CH_2)_3CH_2Br \xrightarrow{KCN} CH_3(CH_2)_4CN \xrightarrow[\text{2) } H_2O]{\text{1) } LIAlH_4} CH_3(CH_2)_5NH_2$$

or

$$CH_3(CH_2)_3CH_2Br \xrightarrow[\substack{\text{2) } CO_2 \\ \text{3) } H_3O^+}]{\text{1) Mg/ether}} CH_3(CH_2)_4CO_2H \xrightarrow[\text{2) } NH_3]{\text{1) } SOCl_2}$$

$$CH_3(CH_2)_4\overset{O}{\underset{}{C}}NH_2 \xrightarrow[\text{2) } H_2O]{\text{1) } LIAlH_4} CH_3(CH_2)_5NH_2$$

<u>Problem 21.32</u> Show how to convert each starting material into 4-methoxybenzylamine in good yield.

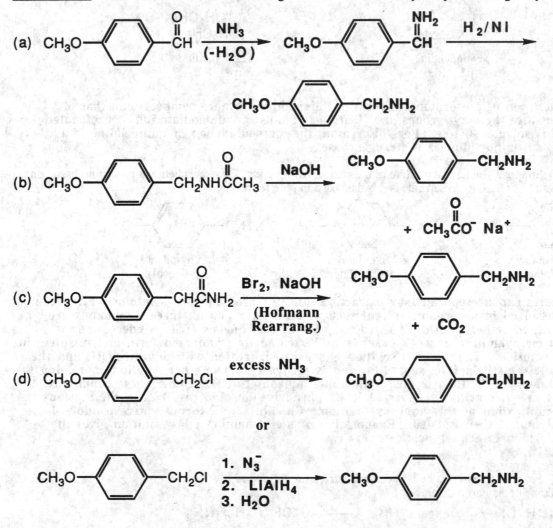

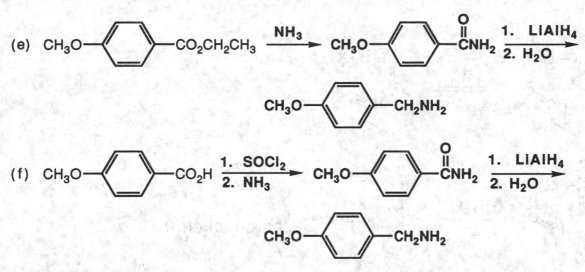

Reactions of Amines

Problem 21.33 *N*-Nitrosamines, by themselves, are not significant carcinogens. However, they are activated in the liver by a class of iron-containing enzymes (members of the cytochrome P-450 family). Activation involves the oxidation of a C-H bond next to the amine nitrogen to a C-OH group. Show how this hydroxylation product can be transformed into an alkyl diazonium ion, an active carcinogen, in the presence of an acid catalyst.

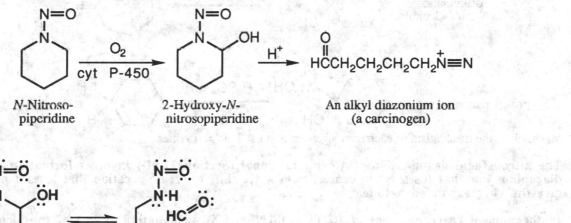

Step 1:

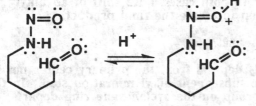

Step 2:

Step 3:

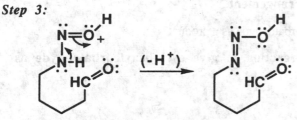

Step 4:

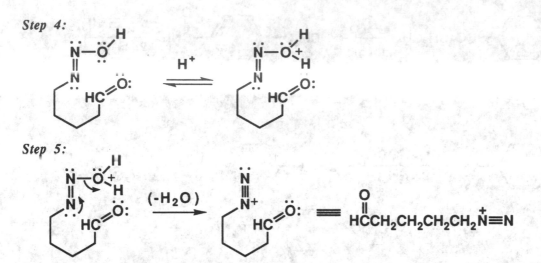

Step 5:

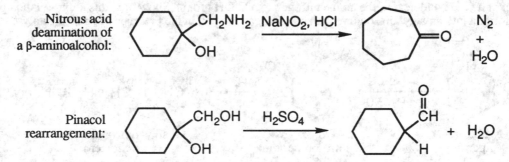

<u>Problem 21.34</u> Marked similarities exist between the mechanism of nitrous acid deamination of β-aminoalcohols and the pinacol rearrangement. Following are examples of each.

Nitrous acid
deamination of
a β-aminoalcohol:

Pinacol
rearrangement:

(a) Analyze the mechanism of each rearrangement and list their similarities.

The nitrous acid deamination of a β-aminoalcohol (Section 21.9D) involves formation of a diazonium ion that loses N$_2$ in concert with a 1,2 shift to create a cation that loses a proton to give the ring expanded ketone.

In the pinacol rearrangement of 1,2 diols (Section 9.8), protonation of one of the alcohol groups leads to departure of water to create a carbocation, followed by migration of an alkyl group to generate a resonance-stabilized cation that loses a proton to give the ketone product.

The similarities between the two mechanisms are that, in both cases, a 1,2 shift of an alkyl group or hydride produces a cation, that loses a proton to create the final product.

(b) Why does the first reaction give ring expansion but not the second?

In the case of the β-aminoalcohol shown above, the N$_2$ departs from the primary center not in the ring, requiring a concerted ring expansion for the subsequent alkyl migration step. In the case of the pinacol rearrangement, the tertiary -OH group on the cyclohexane ring departs, precluding the possibility of a ring expanding rearrangement.

(c) Suggest a β-aminoalcohol that would give cyclohexanecarbaldehyde as a product?

1-Hydroxymethylcyclohexanamine would undergo reaction to give cyclohexylcarbaldehyde as shown.

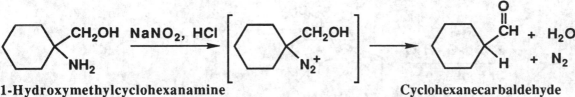

1-Hydroxymethylcyclohexanamine   **Cyclohexanecarbaldehyde**

<u>Problem 21.35</u> Propose a mechanism for this conversion. Your mechanism must account for the fact that there is retention of configuration at the stereocenter. (S)-Glutamic acid is one of the 20 amino acid building blocks of polypeptides and proteins (Chapter 27).

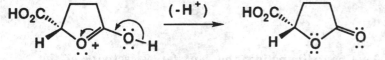

(S)-Glutamic acid

Several Steps:

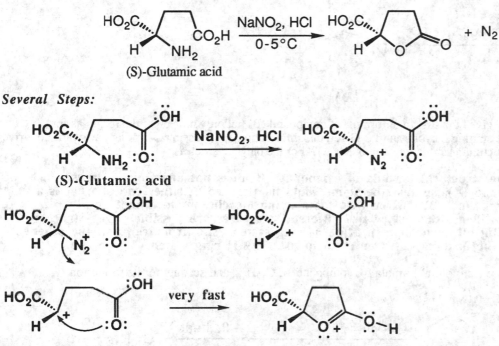

(S)-Glutamic acid

Note that the above ring closure must occur rapidly to avoid racemization of the stereocenter.

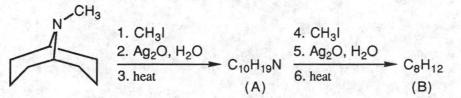

<u>Problem 21.36</u> The following sequence of methylation and Hofmann elimination was used in the determination of the structure of this bicyclic amine.

1. CH_3I
2. Ag_2O, H_2O
3. heat
$\longrightarrow C_{10}H_{19}N$
(A)

4. CH_3I
5. Ag_2O, H_2O
6. heat
$\longrightarrow C_8H_{12}$
(B)

(a) Propose structural formulas for compounds (A) and (B).

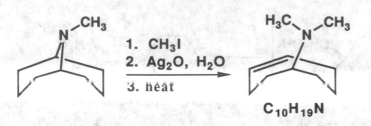

C10H19N

(A)

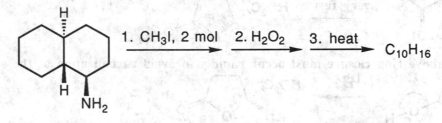

(b) Suppose you were given the structural formula of compound (B) but only molecular the formulas for compound (A) and the starting bicyclic amine. Given this information, is it possible, working backward, to arrive at an unambiguous structural formula for compound (A)? For the bicyclic amine?

Simply knowing the structural formula of compound (B) does not unambiguously establish the structure of compound (A), because the amine could be attached at either end of what is a double bond in compound (B). Given this, the starting bicyclic amine cannot be unambiguously identified either, since two different structures are possible. One structure contains the bridging nitrogen atom in a [3.3.1] ring system (shown above), and the other structure contains the bridging nitrogen atom in in a [4.2.1] ring system.

Problem 21.37 Propose a structural formula for compound A, $C_{10}H_{16}$, and account for its formation.

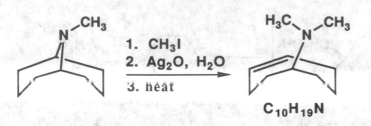

$$1. \text{ CH}_3\text{I, 2 mol} \quad 2. \text{ H}_2\text{O}_2 \quad 3. \text{ heat} \quad C_{10}H_{16}$$

Note how in the following example of a Cope elimination, the syn stereoselectivity of the transition state allows formation of the more substituted alkene, as shown.

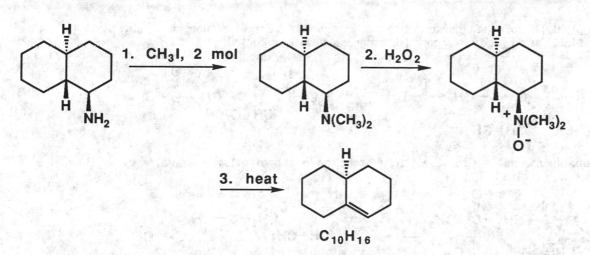

Problem 21.38
An amine of unknown structure contains one nitrogen and nine carbon atoms. The ^{13}C-NMR spectrum shows only five signals, all between 20 and 60 ppm. Three cycles of Hofmann elimination sequence (1. CH_3I; 2. Ag_2O, H_2O; 3. heat) give trimethylamine and 1,4,8-nonatriene. Propose a structural formula for the amine.

The bicyclic amine shown below is the only structure that would explain the five signals in the ^{13}C spectrum as well as the location of the double bonds in 1,4,8-nonatriene.

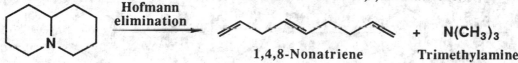

1,4,8-Nonatriene   Trimethylamine

Problem 21.39
The Cope elimination of tertiary amine N-oxides involves a planar transition state and cyclic redistribution of $(4n + 2)$ electrons. The pyrolysis of acetic esters to give an alkene and acetic acid is also thought to involve a planar transition state and cyclic redistribution of $(4n + 2)$ electrons. Propose a mechanism for pyrolysis of the following ester.

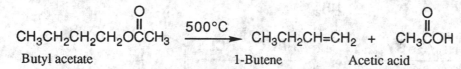

Butyl acetate   1-Butene   Acetic acid

A reasonable transition state structure including the likely flow of electrons is shown below:

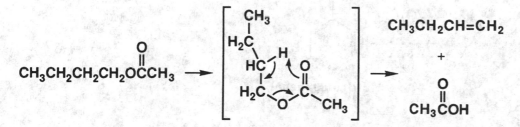

Problem 21.40 The following transformation is an example of the Carroll reaction, named after the English chemist, M. F. Carroll, who first reported it. Propose a mechanism for this reaction.

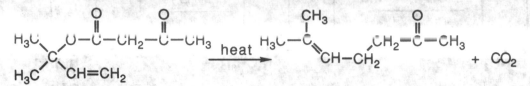

A reasonable mechanism for the Carroll reaction is shown below. Again, a cyclic flow of 6 electrons is involved.

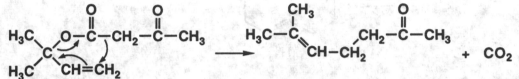

Synthesis
Problem 21.41 Propose steps for the following conversions using a reaction of a diazonium salt in at least one step of each conversion.
(a) Toluene to 4-methylphenol (*p*-cresol)

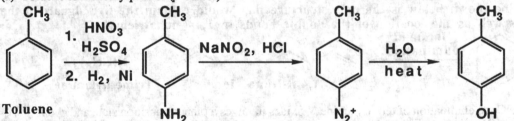

Toluene

(b) Nitrobenzene to 3-bromophenol

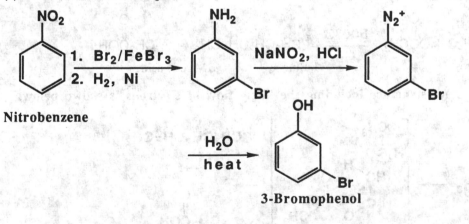

Nitrobenzene

3-Bromophenol

(c) Toluene to *p*-cyanobenzoic acid

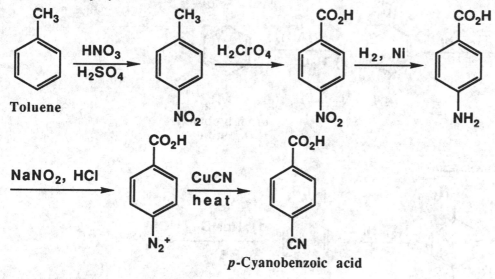

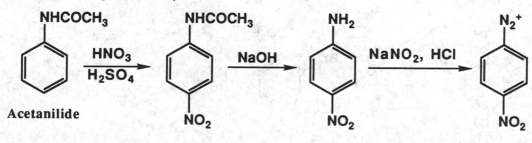

(d) Phenol to *p*-iodoanisole

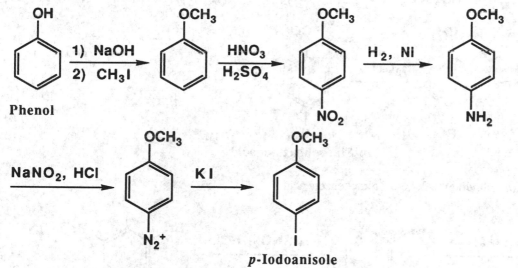

(e) Acetanilide to *p*-aminobenzylamine

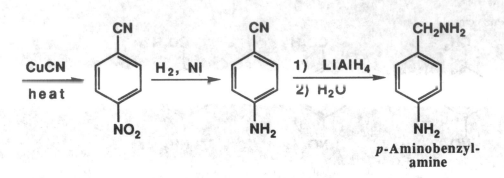

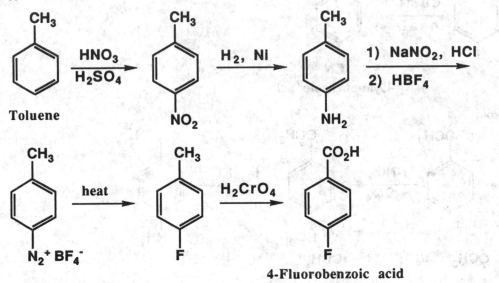

(f) Toluene to 4-fluorobenzoic acid

4-Fluorobenzoic acid

(g) 3-Methylaniline (*m*-toluidine) to 2,4,6-tribromobenzoic acid

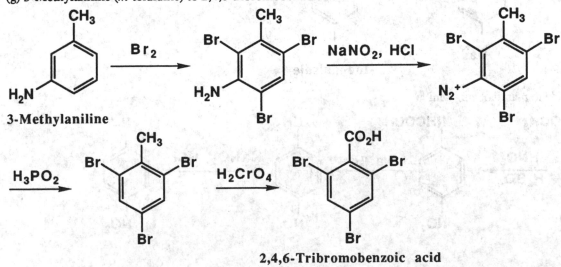

3-Methylaniline

2,4,6-Tribromobenzoic acid

Problem 21.42 Starting materials for the synthesis of the herbicide propranil are benzene and propanoic acid. Show how to bring about this synthesis.

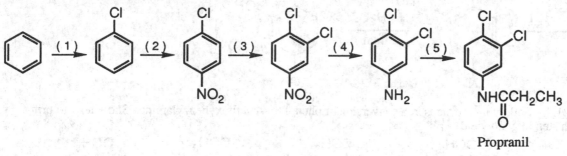

Propranil

The reagents that can be used in each step are listed below:
(1) Aromatic chlorination reaction using Cl_2 and $FeCl_3$.
(2) Aromatic nitration reaction using HNO_3 and H_2SO_4.
(3) Another aromatic chlorination reaction using Cl_2 and $FeCl_3$. The new Cl atom is directed to the correct position by the groups already present on the ring.
(4) This reduction can be carried out using H_2 and a transition metal catalyst.
(5) For this reaction, propanoic acid is first converted to the acid chloride by treatment with $SOCl_2$ and then reacted with the 3,4-dichloroaniline to give the desired propranil.

Problem 21.43 Show how to bring about each step in the following synthesis.

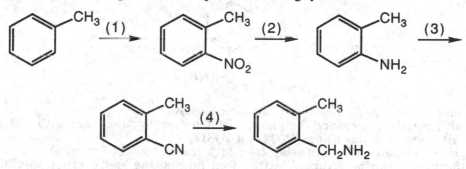

The reagents that can be used in each step are listed below:
(1) Aromatic nitration reaction using HNO_3 and H_2SO_4.
(2) This reduction can be carried out using H_2 and a transition metal catalyst.
(3) This transformation can be accomplished by first turning the amino group into a diazonium salt by reaction with $NaNO_2$ and HCl, followed by a Sandmeyer reaction using CuCN and heat.
(4) This reduction can be carried out using $LiAlH_4$ followed by H_2O.

Problem 21.44 Show how to bring about this synthesis.

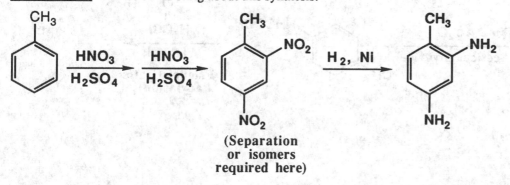

(Separation
or isomers
required here)

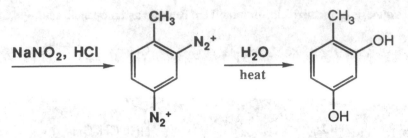

Problem 21.45 Following are steps in a conversion of phenol to 4-methoxybenzylamine. Show how to bring about each step in good yield.

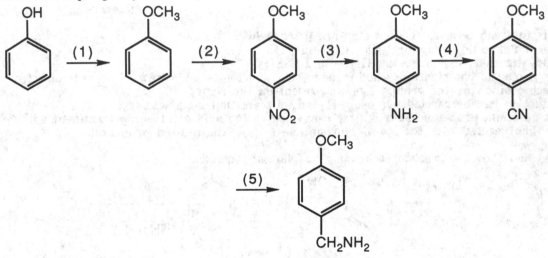

The reagents that can be used in each step are listed below:
(1) Reaction with NaOH to produce the phenoxide, followed by treatment with CH_3I.
(2) Aromatic nitration reaction using HNO_3 and H_2SO_4.
(3) This reduction can be carried out using H_2 and a transition metal catalyst.
(4) This transformation can be accomplished by first turning the amino group into a diazonium salt by reaction with $NaNO_2$ and HCl, followed by a Sandmeyer reaction using CuCN and heat.
(5) This reduction can be carried out using (1) H_2 and a transition metal or (2) $LiAlH_4$ followed by H_2O.

Problem 21.46 Following is a synthesis of diazepam, a prescription tranquilizer sold under several trade names, the most well known of which is Valium. This drug may well be the most commercially successful of all synthetic drugs.
Show reagents and conditions for the synthesis of diazepam from *p*-chloroaniline.

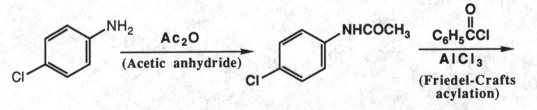

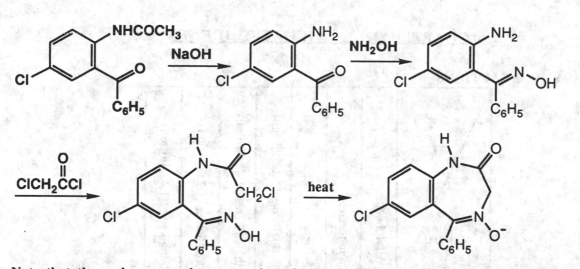

Note that the amino group is protected as the acetamide at the beginning of the synthesis. This is necessary for two reasons. First, without the acetamide protecting group, the benzoyl chloride used in the Friedel-Crafts acylation reaction would react with the amino group and make an unwanted amide. Second, the aluminum chloride catalyst used in the same Friedel-Crafts acylation reaction would become deactivated by reacting with the lone pair of electrons on the amino group. Similar catalyst deactivation does not occur with amide groups, because the nitrogen lone pair is involved with amide resonance.

CHAPTER 22: CONJUGATED DIENES

SUMMARY OF REACTIONS

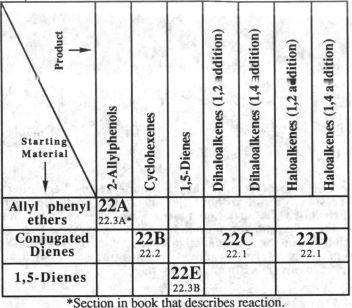

Starting Material ↓ / Product →	2-Allylphenols	Cyclohexenes	1,5-Dienes	Dihaloalkenes (1,2 addition)	Dihaloalkenes (1,4 addition)	Haloalkenes (1,2 addition)	Haloalkenes (1,4 addition)
Allyl phenyl ethers	22A 22.3A*						
Conjugated Dienes		22B 22.2		22C 22.1	22D 22.1		
1,5-Dienes			22E 22.3B				

*Section in book that describes reaction.

REACTION 22A: CLAISEN REARRANGEMENT OF ALLYL PHENYL ETHERS (Section 22.3A)

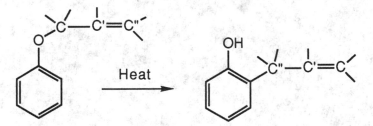

- The **Claisen rearrangement** transforms allyl phenyl ethers to 2-allylphenols, and involves an **intramolecular rearrangement**. The Claisen rearrangement is an example of a pericyclic reaction since bond-forming and bond-breaking are simultaneous in the cyclic transition state. The substituted cyclohexadienone intermediate that is produced undergoes keto-enol tautomerism to give the final product. ✳

REACTION 22B: THE DIELS-ALDER REACTION (Section 22.2)

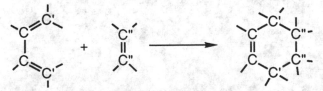

- Conjugated dienes undergo reaction with an alkene or alkyne called a **cycloaddition** reaction that results in production of a six-membered ring through the formation of two carbon-carbon bonds. This reaction is named for its discoverers, and hence is called the **Diels-Alder reaction**. ✳

- The Diels-Alder reaction is a very important and useful synthetic reaction because (1) it is one of the few reactions that creates six-membered rings (2), it is one of the few reactions that forms two carbon-carbon bonds simultaneously, and (3), it is stereoselective.

REACTION 22C: ADDITION OF X₂ (Section 22.1)

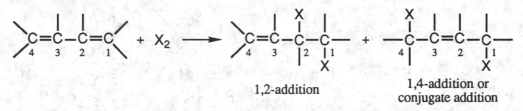

- Conjugated dienes undergo the same kind of addition reactions with X_2 as unconjugated alkenes, but the reaction usually gives mixtures of 1,2-addition and 1,4-addition products. The numbers do not refer to IUPAC nomenclature, but to the sp^2 atoms of the conjugated diene. Therefore, 1,2-addition indicates that addition takes place at adjacent atoms 1 and 2 of the conjugated diene, and 1,4-addition indicates that addition takes place at atoms 1 and 4 of the conjugated diene. ✻

REACTION 22D: ADDITION OF H-X (Section 22.1)

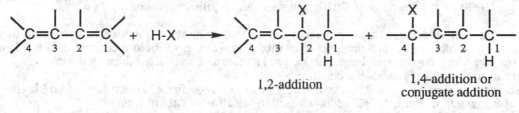

- Conjugated dienes undergo the same kind of addition reactions with H-X as unconjugated alkenes, but the reaction usually gives mixtures of 1,2-addition and 1,4-addition products.

REACTION 22E: COPE REARRANGEMENT (Section 22.3B)

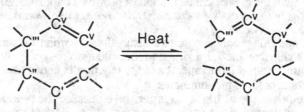

- Upon heating, **1,5-dienes undergo** an **isomerization** called the **Cope rearrangement.** This is an example of a pericyclic reaction since bond-forming and bond-breaking occur simultaneously in a cyclic transition state. ✻

SUMMARY OF IMPORTANT CONCEPTS

22.0 OVERVIEW
• This chapter involves the study of conjugated dienes, that is molecules having two carbon-carbon double bonds seaparted by one carbon-carbon single bond. Conjugated dienes undergo many of the same reactions characteristic of, unconjugated alkenes, but they also undergo their own unique set of reactions.

22.1 ELECTROPHILIC ADDITION TO CONJUGATED DIENES
• The two isomeric products (1,2 and 1,4 addition) observed with addition reactions to conjugated dienes can be explained by a two-step mechanism that involves rate limiting formation of an **allylic cation intermediate**, that then reacts with a nucleophile at the 2 or 4 position to yield products.

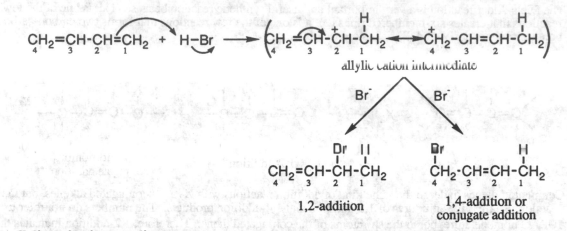

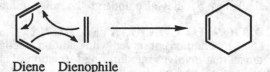

1,2-addition   1,4-addition or
                                conjugate addition

- The **allylic cation intermediate** is considerably more stable than a comparably substituted alkyl carbocation. This extra stability is the result of delocalization of the pi electrons and positive charge as described by the following contributing structures. ✳
- For conjugated dienes reacting with molecules such as HBr or Br_2, 1,2-addition products predominate over 1,4-addition products at low temperature, while at high temperature the 1,4-addition products predominate. These facts are explained by the concepts of **kinetic** and **thermodynamic control** of reaction product distributions.✳
- **Kinetic or rate control** occurs at low temperature when the distribution of products is determined by the relative **rates** of formation of each product. **Thermodynamic or equilibrium control** occurs when, at high temperature, the products equilibrate, thus the product distribution is determined by the relative **thermodynamic stabilities** of each product. ✳
  - The activation energy for 1,2-addition is lower than that for 1,4-addition. As a result, under **kinetic control** at low temperature, the 1,2-addition reaction has a **faster rate** of formation.
     As predicted by the two most important contributing structures, the allylic cation intermediate has the positive charge concentrated at carbon atoms number 2 and 4. However, the **greater concentration of positive charge** is at carbon number 2, because this is a **secondary carbon atom**. **Less positive charge** is concentrated at the **primary carbon atom 4**. As a result, there is a lower activation energy for reaction at the atom with the greater concentration of positive charge, namely carbon atom two. [*Recall that carbocations with more alkyl substituents are more stable, thus a secondary carbocation is more stable than a primary carbocation.*]
  - On the other hand, the 1,4-addition product contains a disubstituted carbon-carbon double bond, so it is thermodynamically more stable than the 1,2-addition product that has only one alkyl group attached to the double bond. Thus, at equilibrium (high temperature) the reaction is under **thermodynamic control**, and the **more stable** 1,4-addition product predominates. ✳
     The two products can **equilibrate at higher temperature** because the product molecules that have been formed **collide with enough energy to convert them back into the allylic cation intermediate** that can then form products again. In this way, the thermodynamically less stable 1,2-addition product can be converted into the thermodynamically more stable 1,4-addition product. At lower temperature, the product molecules do not collide with enough energy to convert them back into the allylic cation, so the 1,2-addition products that form the fastest remain as products, and kinetic control is observed.

22.5 THE DIELS-ALDER REACTION
- In the Diels-Alder reaction, the electrons in the pi bonds move to create the new bonds as shown below. The arrows are not meant to indicate a mechanism, but rather to help keep track of the electrons during the reaction.

Diene   Dienophile

- The alkene or alkyne that reacts with the diene is referred to as the **dienophile**.

- The Diels-Alder reaction is usually favorable (the product is more stable than the reactants) because the overall transformation involves breaking two carbon-carbon pi bonds in order to make two carbon-carbon sigma bonds. Recall that generally carbon-carbon sigma bonds are stronger than pi bonds, so Diels-Alder reactions generally have negative values for ΔH. *[It is always helpful to understand why a reaction occurs as well as simply how it occurs.]*

• Studies of the mechanism of Diels-Alder reactions have shown that the pi bonds are broken and the sigma bonds are made in a **single, concerted step**. In other words, the cyclic redistribution of bonding electrons occurs such that bond-forming and bond-breaking occur simultaneously. Reactions of this type involving a single step and a cyclic rearrangement of a Hückel number of bonding electrons are called **pericyclic reactions**.

- During the reaction, the diene and the dienophile approach each other in parallel planes, allowing interaction between the pi bonds of both molecules. The reaction is depicted in Figure 22.5.

• The above mechanism explains some important details of the Diels-Alder reaction.

- The diene must adopt the **s-*cis* conformation** in order to react with the dienophile. This conformation is required to allow both sigma bonds to be formed simultaneously with the dienophile.

As stated earlier, in order to allow for maximum overlap of the 2p orbitals, the four carbon atoms of conjugated dienes prefer to lie in the same plane. This means that there are two preferred conformations of conjugated dienes, the s-*cis* and s-*trans* conformations. The "s" indicates that it is the **single bond** between the two alkenes that is referred to with the *cis* or *trans* designation. ✳

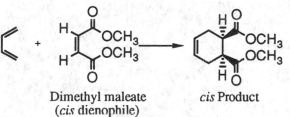

s-*trans*-1,3-Butadiene
cannot take part in
Diels-Alder reaction

s-*cis*-1,3-Butadiene
can take part in
Diels-Alder reaction

- Conjugated dienes that cannot adopt the s-*cis* conformation cannot take part in Diels-Alder reactions. On the other hand, conjugated dienes such as cyclopentadiene, which can only adopt the s-*cis* geometry, are very good reactants in Diels-Alder reactions.

- The stereochemistry of the dienophile is retained in the product of the Diels-Alder reaction, since the two sigma bonds are made simultaneously. For example, the *cis* stereochemistry of dimethyl maleate is retained in the product. ✳

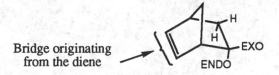

Dimethyl maleate
(*cis* dienophile)

cis Product

• When **cyclic conjugated dienes** are used for a Diels-Alder reaction, the product is a **bicyclic** structure. As a result, substituents on the dienophile can end up in the **endo** or **exo** positions. In practice, the substituents end up predominantly in the endo position.

- The **exo** position is the one that is on the opposite side as the bridge originating from the diene, while the **endo** position is the one that is on the same side as the bridge originating from the diene.✳

Bridge originating
from the diene

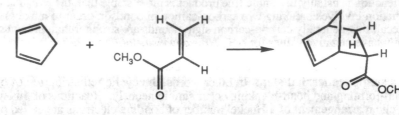

The endo product is formed
predominantly

- Substituents on the reactants can greatly alter the rates of Diels-Alder reactions. The reaction is **facilitated** by a combination of **electron-withdrawing substituents** on one of the reactants and **electron-releasing substituents** on the other. Usually, the electron-withdrawing groups are found on the dienophile. ✳
  - Examples of **electron-withdrawing groups** usually found on dienophiles include ketones, carboxylic esters, nitro groups, and nitriles. These groups pull electron density out of the pi bond, and make it easier for the reaction to take place.
- It is important to point out that cycloaddition reactions have specific electronic requirements, so alkenes do not dimerize to form a four-membered ring product, and dienes do not dimerize to form an eight-membered ring product.
- Certain Diels-Alder reactions can be catalyzed by Lewis acids such as $AlCl_3$. The Lewis acid is thought to operate by coordinating to the dienophile to help polarize it, thus facilitating reaction with the diene.

CHAPTER 22
Solutions to the Problems

<u>Problem 22.1</u> Predict the product(s) formed by addition of 1 mol of Br_2 to 2,4-hexadiene.

Predict both 1,2-addition and 1,4-addition.

$$CH_3-CH=CH-CH=CH\cdot CH_3 \xrightarrow{\textbf{Br}_2}$$

$$CH_3-\underset{\overset{|}{Br}}{CH}-\underset{\overset{|}{Br}}{CH}-CH=CH\cdot CH_3 \ + \ CH_3-\underset{\overset{|}{Br}}{CH}-CH=CH-\underset{\overset{|}{Br}}{CH}\cdot CH_3$$

<u>Problem 22.2</u> What combination of diene and dienophile undergoes Diels-Alder reaction to give each adduct.

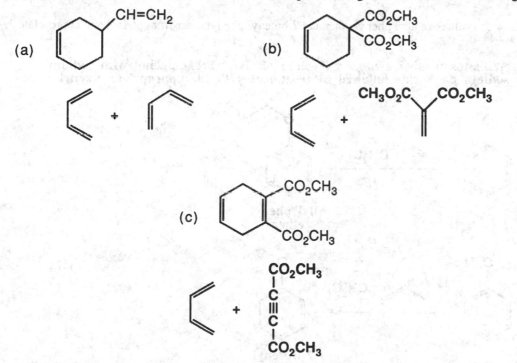

In each part of this problem, the diene is 1,3-butadiene. In (a), the dienophile is one of the double bonds of a second molecule of 1,3-butadiene. In part (b), the dienophile is a 1,1-disubstituted alkene and in part (c), it is a disubstituted alkyne.

<u>Problem 22.3</u> Which molecules can function as dienes in Diels-Alder reactions? Explain your reasoning.

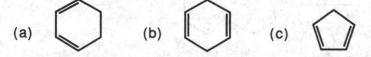

(a)   (b)   (c)

To function as a diene, the double bonds must be conjugated and able to assume an *s-cis* conformation. Both (a) 1,3-cyclohexadiene and (c) 1,3-cyclopentadiene can function as dienes. The double bonds in 1,4-cyclohexadiene are not conjugated and, therefore, this molecule cannot function as a diene.

<u>Problem 22.4</u> What diene and dienophile might you use to prepare the following Diels-Alder adduct?

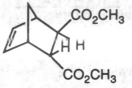

Use cyclopentadiene and the E-alkene.

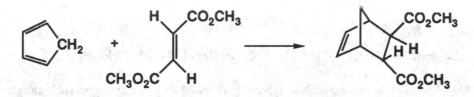

<u>Problem 22.5</u> Show how to synthesize allyl phenyl ether and 2-butenyl phenyl ether from phenol and appropriate alkenyl halides.

Prepare each by a Williamson ether synthesis (Section 11.4A). Treat phenol with sodium hydroxide to form sodium phenoxide followed by treatment with the appropriate alkenyl chloride.

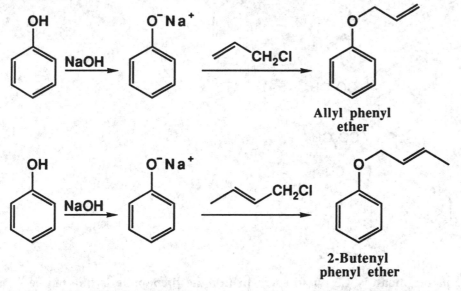

<u>Problem 22.6</u> Propose a mechanism for the following Cope rearrangement.

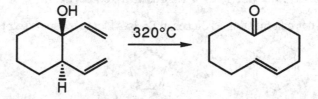

Cope rearrangement gives an enol that then undergoes keto-enol tautomerism (Section 15.11B) to give the observed product.

Step 1:

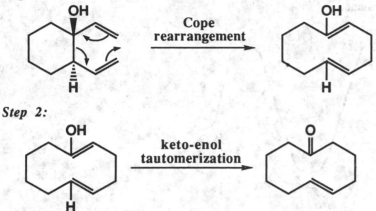

Step 2:

PROBLEMS
Structure and Stability
Problem 22.7 If an electron is added to 1,3-butadiene, into which molecular orbital does it go? If an electron is removed from 1,3-butadiene, from which molecular orbital is it taken?

When an electron is removed from a conjugated pi system, such as that in 1,3-butadiene, it is taken from the highest occupied molecular orbital (HOMO). When an electron is added to a conjugated pi system, is added to the lowest unoccupied molecular orbital (LUMO).

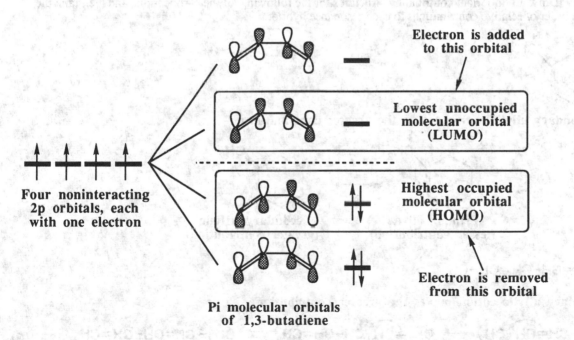

Problem 22.8 Draw a potential energy diagram (potential energy versus dihedral angle from 0° to 360°) for rotation about the 2,3 single bond in 1,3-butadiene.

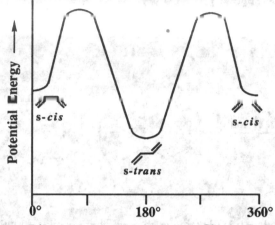

Dihedral Angle About 2,3 Single Bond

The s-*cis* and s-*trans* conformations are the most stable for 1,3-butadiene, because these provide for an overall planar geometry, which maximizes overlap of the 2p orbitals. The s-*trans* geometry is more stable than the s-*cis* geometry because of steric interactions.

Problem 22.9 Draw all important contributing structures for the following allylic carbocations and then rank the structures in order of relative contributions to each resonance hybrid.

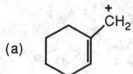

(a)

The secondary allylic cation makes the greater contribution.

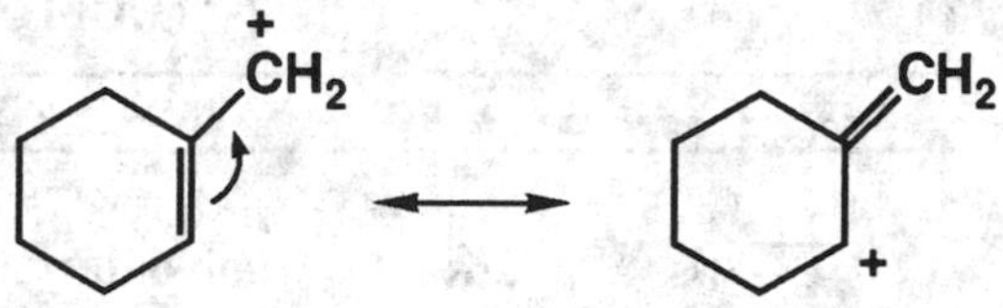

Primary allylic   Secondary allylic
(lesser contribution)   (greater contribution)

(b) $CH_2=CH-CH=CH-CH_2^+$

The secondary allylic cation makes the greater contribution.

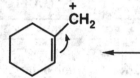

$CH_2=CH-CH=CH-CH_2^+ \longleftrightarrow CH_2=CH-\overset{+}{C}H-CH=CH_2 \longleftrightarrow \overset{+}{C}H_2-CH=CH-CH=CH_2$

Primary allylic   Secondary allylic   Primary allylic
(lesser contribution)   (greater contribution)   (lesser contribution)

(c)
$$CH_3 \overset{\overset{\displaystyle CH_3}{|}}{\underset{+}{C}} - CH = CH_2$$

The tertiary allylic cation makes the greater contribution.

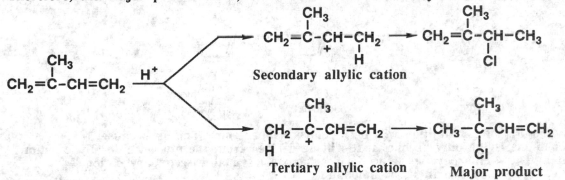

Tertiary allylic	Primary allylic
(greater contribution)	(lesser contribution)

Electrophilic Addition to Conjugated Dienes

Problem 22.10 Predict the structure of the major product formed by 1,2-addition of HCl to 2-methyl-1,3-butadiene (isoprene). To arrive at a prediction, first consider proton transfer to carbon 1 of this diene. Second, consider proton transfer to carbon 4 of this diene. Then compare the relative stabilities of the two allylic carbocation intermediates.

A tertiary allylic cation is more stable than a secondary allylic cation. Assume that because the tertiary allylic cation is the more stable of the two, the activation energy is lower for its formation, and accordingly, it is formed at a greater rate than the secondary allylic cation. Therefore, the major product of 1,2-addition is 3-chloro-3-methyl-1-butene.

Secondary allylic cation

Tertiary allylic cation   **Major product**

Problem 22.11 Predict the major product formed by 1,4-addition of HCl to isoprene. Follow the reasoning suggested in the previous problem.

The major product of 1,4-addition is 1-chloro-3-methyl-2-butene.

Less stable allylic cation

More stable allylic cation   **Major product**

<u>Problem 22.12</u> Predict the structure of the major 1,2-addition product formed by reaction of 1 mol of Br_2 with isoprene. Also predict the structure of the major 1,4-addition product formed under these conditions.

Following the reasoning developed in the previous problems, predict that the major product of 1,2-addition to isoprene is 3,4-dibromo-3-methyl-1-butene.

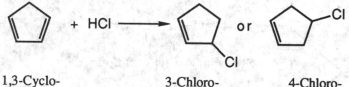

$$CH_2=\overset{\overset{\displaystyle CH_3}{|}}{C}-CH=CH_2 \;+\; Br_2 \;\;\xrightarrow{\text{1,2-addition}}\;\; CH_2-\overset{\overset{\displaystyle CH_3}{|}}{\underset{\underset{\displaystyle Br}{|}}{C}}-CH=CH_2$$

Major product

There is only one 1,4-addition product possible from addition of bromine.

$$CH_2=\overset{\overset{\displaystyle CH_3}{|}}{C}-CH=CH_2 \;+\; Br_2 \;\;\xrightarrow{\text{1,4-addition}}\;\; CH_2-\overset{\overset{\displaystyle CH_3}{|}}{C}=CH-CH_2$$

Br   Br

<u>Problem 22.13</u> Which of the two molecules shown do you expect to be the major product formed by 1,2-addition of HCl to cyclopentadiene? Explain.

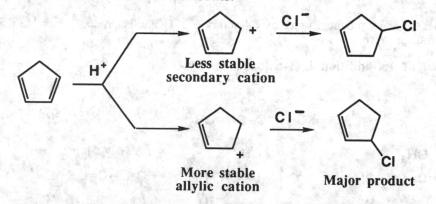

1,3-Cyclo-   3-Chloro-   4-Chloro-
pentadiene   cyclopentene   cyclopentene

The major product of 1,2-addition will be 3-chlorocyclopentene. This is because 3-chlorocyclopentene is the only product that will be derived from the more stable allylic cation intermediate. The 4-chlorocyclopentene derives from a less stable secondary cation intermediate, so it is formed in lesser amounts.

Less stable
secondary cation

More stable
allylic cation

Major product

<u>Problem 22.14</u> Predict the major product formed by 1,4-addition of HCl to cyclopentadiene.

The conjugate addition product would also be the 3-chlorocyclopentene. Because of symmetry in the ring, the 1,4-addition product is the same as the predominant 1,2-addition product, both being derived from the allylic cation.

Problem 22.15 Draw structural formulas for the two constitutional isomers of molecular formula $C_5H_6Br_2$ formed by adding 1 mol of Br_2 to cyclopentadiene.

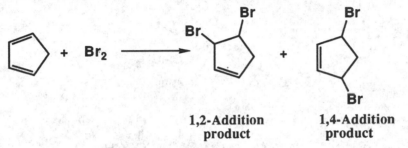

The two constitutional isomers are the 1,2 and 1,4-addition products.

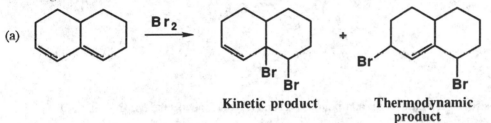

1,2-Addition product   1,4-Addition product

Problem 22.16 What are the expected kinetic and thermodynamic products from addition of one mol of Br_2 to the following dienes

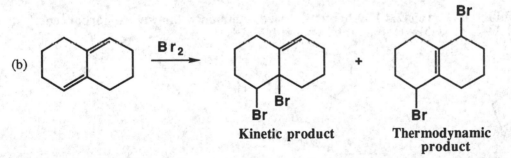

(a)

Kinetic product   Thermodynamic product

The kinetic product will be the 1,2-addition product that is derived from the more highly substituted and thus more stable bridged bromonium ion intermediate. The thermodynamic product is the most stable one, namely the 1,4-addition product, since this has the most highly substituted alkene as shown.

(b)

Kinetic product   Thermodynamic product

Because of symmetry in the molecule, both possible bromonium ion intermediates are the same. Thus, there is only one possible 1,2-addition product and this is the kinetic product. The thermodynamic product is the more stable one, namely the 1,4-addition product, because this has the most highly substituted alkene as shown.

Diels-Alder Reactions

Problem 22.17 Draw structural formulas for the products of reaction of cyclopentadiene with each dienophile.

Underneath each dienophile is the corresponding Diels-Alder adduct.

(a) $CH_2{=}CHCl$

(b) $CH_2{=}CHCOCH_3$ (with C=O shown above)

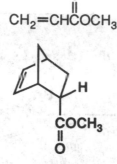

(c) $HC{\equiv}CH$

(d) $CH_3OCC{\equiv}CCOCH_3$ (with two C=O groups shown above)

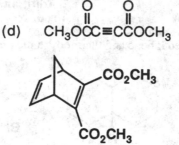

Problem 22.18 Propose structural formulas for compounds (A) and (B) and specify the configuration of compound (B).

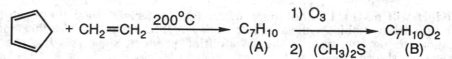

The Diels-Alder adduct is bicyclo[2.2.1]-2-heptene, more commonly known as norbornene. The oxidation product is cis-1,3-cyclopentanedicarbaldehyde.

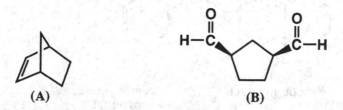

(A)   (B)

Problem 22.19 Under certain conditions, 1,3-butadiene can function both as a diene and a dienophile. Draw a structural formula for the Diels-Alder adduct formed by reaction of 1,3-butadiene with itself.

In the formulas below, one molecule of butadiene is shown as the diene and the other is shown as the dienophile.

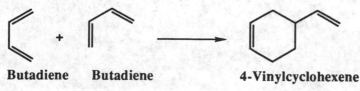

Butadiene   Butadiene   4-Vinylcyclohexene

Note that an electron pushing mechanism can be drawn for cyclization of two molecules of butadiene to 1,5-cyclooctadiene. This reaction does not take place. Rather, the Diels-Alder reaction shown above takes place instead.

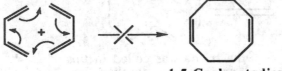

1,5-Cyclooctadiene

Problem 22.20 1,3-Butadiene is a gas at room temperature and a requires gas-handling apparatus to use in a Diels-Alder reaction. Butadiene sulfone is a convenient substitute for gaseous 1,3-butadiene. This sulfone is a solid at room temperature (mp 66°C) and, when heated above its boiling point of 110°C, decomposes by a reverse Diels-Alder reaction to give s-*cis*-1,3-butadiene and sulfur dioxide. Draw a Lewis structure for butadiene sulfone, and show by curved arrows the path of this reverse Diels-Alder reaction.

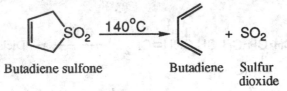

| Butadiene sulfone | | Butadiene | Sulfur dioxide |

The electron flow in this decomposition is the reverse of that observed in a Diels-Alder reaction.

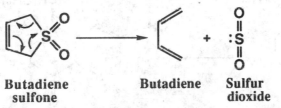

| **Butadiene sulfone** | **Butadiene** | **Sulfur dioxide** |

Problem 22.21 The following trienone undergoes an intramolecular Diels-Alder reaction to give the product shown. Show how the carbon skeleton of the triene must be coiled to give this product, and show by curved arrows the redistribution of electron pairs that takes place to give the product.

$$CH_2=CH-\underset{\underset{CH_3}{|}}{C}=CH-CH_2CH_2CH_2CH_2CH=CH_2 \xrightarrow{160°C}$$

Locate the position of the carbon-carbon double bond in the product and realize that the two carbons of this double bond were carbons 2 and 3 of the conjugated diene. The other two carbon atoms making up this six-membered ring were from the dienophile.

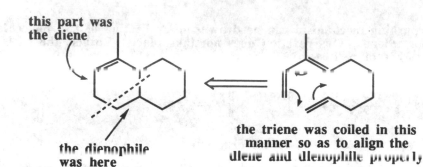

the triene was coiled in this
manner so as to align the
diene and dienophile properly

Problem 22.22 The following triene undergoes an intramolecular Diels-Alder reaction to give a bicyclic product. Propose a structural formula for the product. Account for the observation that the Diels-Alder reaction given in this problem takes place under milder conditions (at lower temperature) than the analogous Diels-Alder reaction shown in the previous problem?

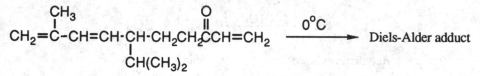

Follow the arrangement of diene and dienophile illustrated in the previous problem.

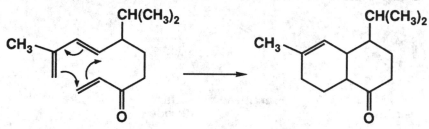

Problem 22.23 The following compound undergoes an intramolecular Diels-Alder reaction to give a bicyclic product. Propose a structural formula for the product.

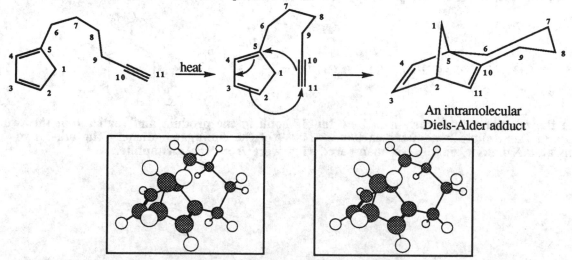

An intramolecular
Diels-Alder adduct

Problem 22.24 One of the published syntheses of warburganal (Problem 5.31) begins with the following Diels-Alder reaction. Propose a structure for Compound A.

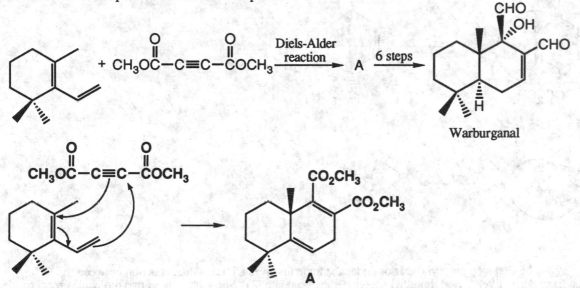

Warburganal

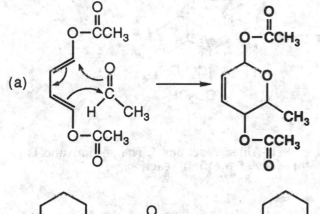

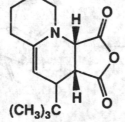

A

Problem 22.25 The Diels-Alder reaction is not limited to making six-membered rings with only carbon atoms. Predict the products of the following reactions that produce rings with atoms other than carbon in them.

(a)

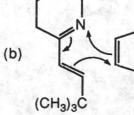

(b)

(c)

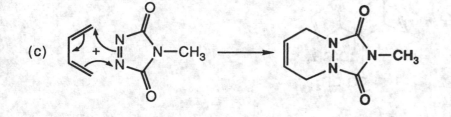

(d)

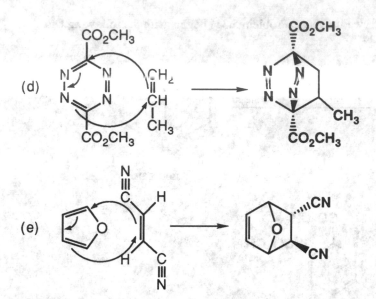

(e)

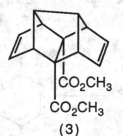

Problem 22.26 The first step in a synthesis of dodecahedrane involves a Diels-Alder reaction between the cyclopentadiene derivative (1) and dimethyl acetylenedicarboxylate (2). Show how these two molecules react to form the dodecahedrane synthetic intermediate (3). (See L. A. Paquette, R. J. Ternansky, D. W. Balogh, et al., *J. Am. Chem. Soc.,* **105**, 5446, 1983.)

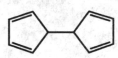   +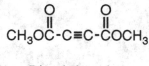

Cyclopentadienyl-   Dimethylacetylene-
cyclopentadiene   dicarboxylate

(1)   (2)   (3)

This reaction is best understood as two successive Diels-Alder reactions. The second one is an intramolecular reaction that takes place on the product of the first reaction.

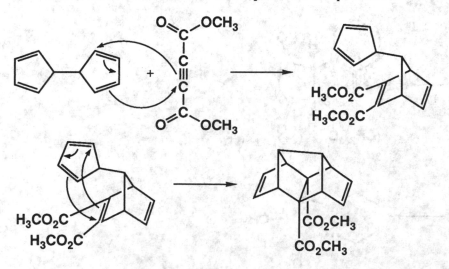

Problem 22.27 Bicyclo[2.2.1]-2,5-heptadiene can be prepared in two steps from cyclopentadiene and vinyl chloride. Provide a mechanism for each step.

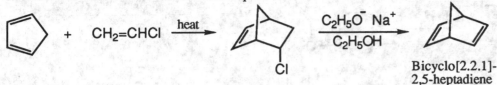

Bicyclo[2.2.1]-
2,5-heptadiene

Step 1:

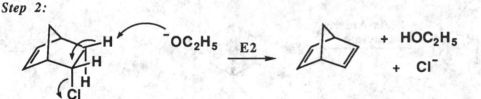

Step 2:

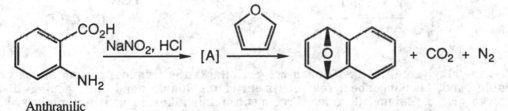

Problem 22.28 Treatment of anthranilic acid with nitrous acid gives an intermediate, A, that contains a diazonium ion and a carboxylate group. When this intermediate is heated in the presence of furan, a bicyclic compound is formed. Propose a structural formula for intermediate A and mechanism for formation of the bicyclic product.

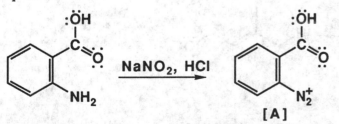

Step 1:

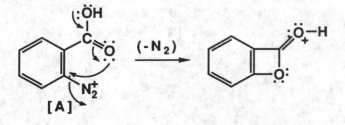

Step 2: **This step may or may not be concerted**

Step 3:

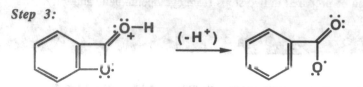

Step 4:

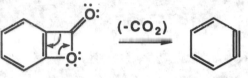

Benzyne

Step 5:

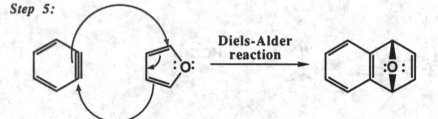

Diels-Alder reaction

Problem 22.29 Propose a mechanism for the following reaction, which is called an allylic rearrangement, or, alternatively, a conjugate addition.

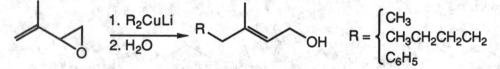

$$R = \begin{cases} CH_3 \\ CH_3CH_2CH_2CH_2 \\ C_6H_5 \end{cases}$$

Propose that the alkyl group of the Gilman reagent attacks the terminal carbon of the carbon-carbon double bond, bringing about rearrangement of the double bond and opening of the epoxide ring to give a lithium alkoxide. Treatment of this alkoxide with water gives the observed allylic alcohol.

Step 1:

Li(R)Cu—R

Step 2:

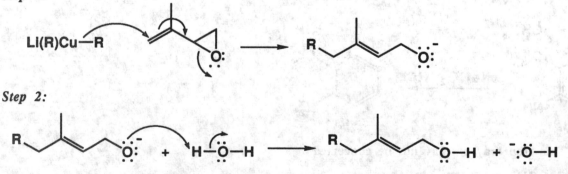

Problem 22.30 All attempts to synthesize cyclopentadienone yield only a Diels-Alder adduct. Cycloheptatrienone, however, has been prepared by several methods and is stable.

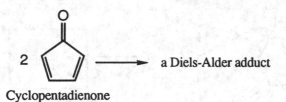

2   Cyclopentadienone   ⟶   a Diels-Alder adduct

Cycloheptatrienone

(a) Draw a structural formula for the Diels-Alder adduct formed by cyclopentadienone.

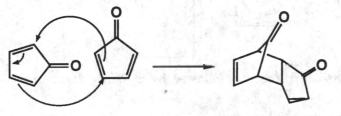

a Diels-Alder adduct

(b) How do you account for the marked difference in stability of these two ketones?

A major contributing structure for cyclopentadienone has only four pi electrons, an anti-aromatic number, thereby expaining the extreme reactivity of cyclopentadienone.

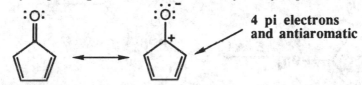

4 pi electrons
and antiaromatic

Cyclopentadienone contributing structures

On the other hand, a major contributing structure of cycloheptatrienone has six pi electrons, a Hückel number, thereby explaining the enhanced stability of cycloheptatrienone.

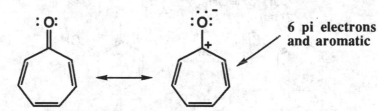

6 pi electrons
and aromatic

Cycloheptatrienone contributing structures

Problem 22.31 Following is a retrosynthetic scheme for the synthesis of the tricyclic diene on the left. Show how to accomplish this synthesis using the indicated carbon sources as starting materials.

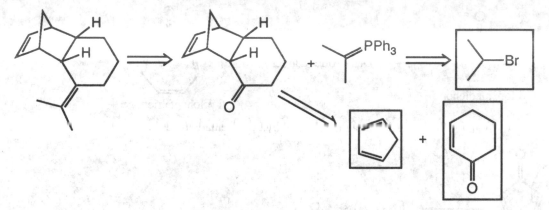

The key elements of this synthesis are a Diels-Alder reaction followed by a Wittig reaction.

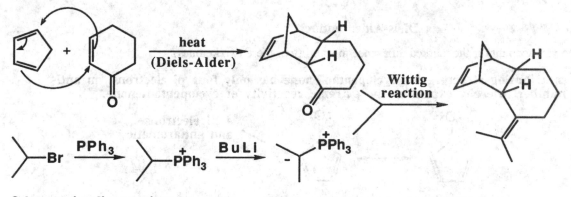

Other pericyclic reactions

Problem 22.32 Claisen rearrangement of an allyl phenyl ether with substituent groups in both ortho positions leads to formation of a para substituted product. Propose a mechanism for the following rearrangement.

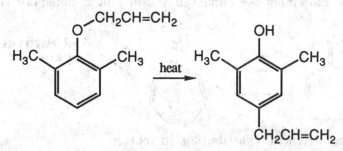

Propose rearrangement of the allyl group to the ortho position to form a substituted cyclohexadienone, as in the normal Claisen rearrangement. This intermediate, however, cannot undergo keto-enol tautomerism to become an aromatic compound. A second rearrangement, this time of the allyl group to the para position, gives a substituted cyclohexadienone that can undergo keto-enol tautomerism to give an aromatic compound.

Step 1:

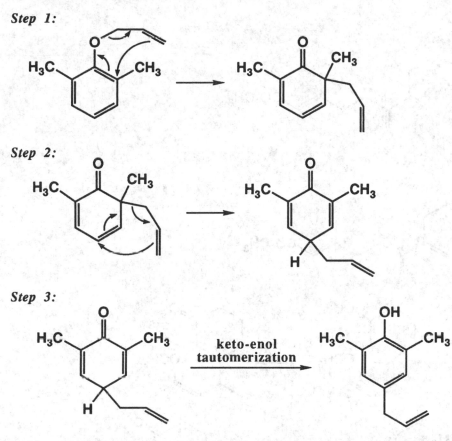

Step 2:

Step 3:

keto-enol
tautomerization

<u>Problem 22.33</u> Following are three examples of Cope rearrangements of 1,5-dienes. Show that each product can be formed in a single step by a mechanism involving redistribution of six electrons in a cyclic transition state.

Keep in mind as you do these problems that the Cope rearrangement involves flow of three pairs of electrons in a cyclic, six-membered transition state. Bonding in the six carbon atoms participating in the reaction is C=C-C-C-C=C (double-single-single-single-double). Find that combination of carbon atoms and you are well on your way. Numbers have been assigned to the atoms to keep track of the bonds made and broken.

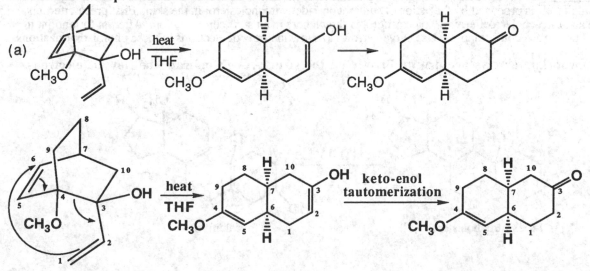

(a)

heat
THF

heat
THF

keto-enol
tautomerization

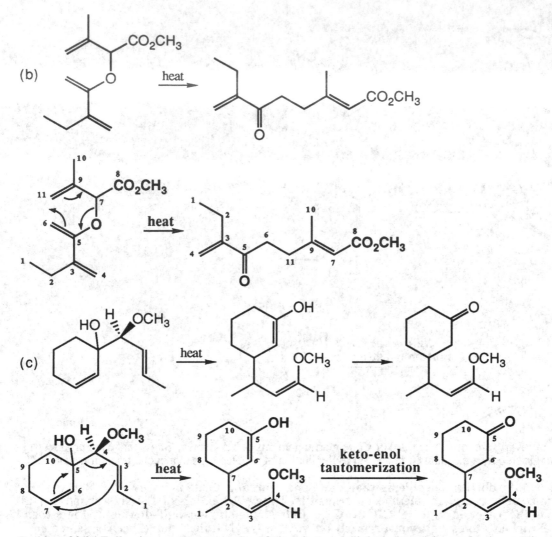

(b)

(c)

Problem 22.34 Following are two examples of photoinduced (light-induced) isomerizations. Vitamin D_3 (cholecalciferol) is produced by the action of sunlight on 7-dehydrocholesterol in the skin. First precalciferol is formed and then cholecalciferol. Cholecalciferol is shown here in an *s-cis* conformation. After its formation, it assumes an *s-trans* conformation. Use curved arrows to show the flow of electrons in these photoisomerizations.

The appropriate arrows are drawn directly on the structures to indicate the flow of electrons.

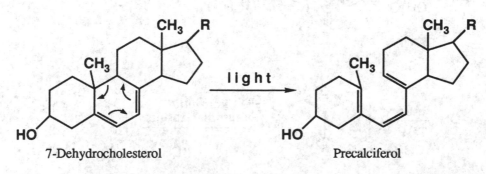

7-Dehydrocholesterol   Precalciferol

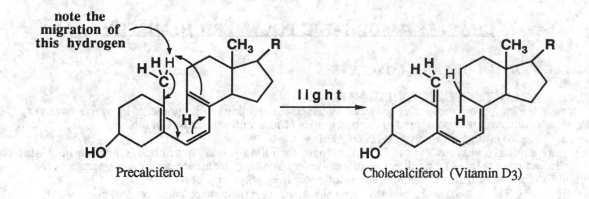

note the
migration of
this hydrogen

Precalciferol   Cholecalciferol (Vitamin D₃)

CHAPTER 23: ORGANIC POLYMER CHEMISTRY

SUMMARY OF IMPORTANT CONCEPTS

23.1 THE ARCHITECTURE OF POLYMERS
- **Polymers** are long chain molecules produced by the covalent linking of monomers. Polymers have **very high molecular weights** compared to the other types of molecules described in the book so far, typically ranging from 10,000 g/mol up to 1,000,000 g/mol. ✳
- Polymers can have **various types of architectures** including **linear, branched, comb, ladder, star** and **crosslinked networks.** The architecture of a polymer has a large influence over the properties of the polymer. Chain length also has a tremendous influence over polymer properties.
- **Plastics** are polymers that can be molded when hot, then retain their shape when cooled. ✳
  - **Thermoplastics** can be melted and become fluid so they can be molded when hot, but retain their shape when cooled.
  - **Thermoset plastics** can be molded when they are first prepared, but harden irreversibly when cooled.

23.2 POLYMER NOTATION AND NOMENCLATURE
- Polymer structures are denoted by drawing parentheses around the repeating unit. ✳
  - The **repeating unit** is the smallest fragment that contains all of the **nonredundant structure of the polymer chain.**
  - A **subscript n**, called the **average degree of polymerization,** indicates how many times the repeating unit is found in the polymer.
- An exception is polymers formed from symmetric monomers such as $(-CH_2CH_2)_n$ (polyethylene) and $(-CF_2CF_2-)_n$ (polytetrafluoroethylene), which are produced from $CH_2=CH_2$ and $CF_2=CF_2$, respectively.

22.3 MOLECULAR WEIGHTS OF POLYMERS
- Most polymers are mixtures of polymer molecules of variable molecular weights. Polymer molecular weights are defined as either **number average** and **weight molecular average weights.** ✳
  - The **number average molecular weight, M_n,** is calculated by counting the number of chains of a particular weight, then dividing this sum by the total number of chains.
  - The **weight average molecular weight, M_W,** is calculated by taking the total weight of each chain of a particular length, summing these weights and dividing by the total weight of the sample.
  - The ratio M_W/M_n is called the **polydispersity index,** and is used to characterize the distribution of different molecular weights in the polymer. A **monodisperse** polymer is one in which all of the polymer chains are the same length so that M_W/M_n is equal to one.

22.4 POLYMER MORPHOLOGY-Crystalline Versus Amorphous Materials
- Polymers that are solids usually have both **ordered crystalline domains** and **disordered amorphous domains.** ✳
  - High degrees of crystallinity occur in polymers having regular, compact structures, especially ones having strong intermolecular forces such as hydrogen bonding, dipolar interactions, etc.
  - **Increasing crystallinity increases** both the **melting temperature, T_m,** and **opacity** of a polymer.
  - **Amorphous polymers** have no long range order in the solid state. Because amorphous polymers lack crystalline domains to scatter light, they are **transparent** and thus are sometimes referred to as "glassy solids". Amorphous polymers turn soft and rubbery upon heating, and the temperature at which this occurs is called the glass transition temperature, T_g.
- **Elastomers** are polymers that return to their original form when stretched. If the temperature drops below the T_g for an elastomer, is it converted to a rigid glassy solid.

22.5 STEP-GROWTH POLYMERIZATIONS
- **Step-growth or condensation polymerizations** are polymerizations in which chain growth occurs in a stepwise manner. ✳
  - The monomers for step-growth polymerizations are generally formed from difunctional monomers, with the new bonds created in separate steps in a statistical fashion.
  - Because it is more likely that monomers react with growing chains, very long chains only show up when chains start reacting with each other. This only occurs very late in the polymerization process, usually after ~99% of the monomers are used up.

- The bonds formed during step-growth polymerizations are usually formed from **polar reactions**, and representative bond types include:

 Polyamides, formed from carboxyl and amino groups on monomers.

 Polyesters, formed from carboxyl and hydroxyl groups on monomers.

 Polycarbonates, formed from phosgene, ClC(O)Cl, and hydroxyl groups on monomers.

 Polyurethanes, formed from isocyanates and hydroxyl groups on monomers.

 Epoxy resins, formed from epoxide and nucleophilic groups such as amines on monomers.

22.6 CHAIN-GROWTH POLYMERIZATIONS

- **Chain-growth** polymerizations involve sequential addition reactions. In chain growth polymerizations, only the end of the growing chain possesses a reactive function that can react with another monomer. ✳
  - Typical monomers that undergo chain-growth polymerization include alkenes, alkynes, allenes, isocyanates and cyclic compounds such as lactones, lactams, ethers and epoxides.
- Chain-growth polymerizations can involve **radical chain reactions**, usually initiated by an initiator such as dibenzoyl peroxide.
  - The radical chain-growth reactions involve the normal radical reaction steps of **initiation, propagation** and **termination**.
  - In the case of polymerization, the **propagation** involves reaction with a monomer unit, so the chains successively add monomers.
  - The **termination step** can involve either **radical coupling**, two radicals reacting with each other, or **disproportionation**, abstraction of a hydrogen from one radical by another resulting in two dead chains.
  - **Chain transfer reactions** occur when hydrogen abstraction from solvent, a monomer or another chain by the endgroup of a growing chain. Chain transfer reactions can lead to branching of a polymer.
- Chain-growth polymerization can involve heterogeneous catalysts such as $TiCl_4$, $Al((CH_2CH_3)_2Cl$ on a $MgCl_2$ support, the so-called **Ziegler-Natta catalyst**. This type of catalysis is referred to as **heterogeneous** because the catalytic species is not soluble, but rather it exists on the surface of insoluble particles.
  - The **mechanism of Ziegler-Natta polymerization** of ethylene includes an initial formation of a titanium-ethylene bond, followed by insertion of ethylene into the titanium-carbon bond of the growing chain.
  - The Ziegler-Natta polymerization creates **high-density polyethylene (HDPE)** that is stronger, melts at a higher temperature, and is more opaque than the **low-density polyethylene (LDPE)** produced by radical polymerization.
- Soluble metal coordination complexes such as *bis*(cyclopentadienyl)dimethylzirconium $[Cp_2Zr(CH_3)_2]$ in the presence of methaluminoxane (MAO) also carry out a Ziegler-Natta type of catalysis. Since these new catalysts are soluble, there are referred to as **homogeneous catalysts** and the polymerizations are called **coordination polymerizations**. **Coordination polymerization** is a very active area of current research.
- Polymers can have stereocenters. Polymers with identical configurations at all stereocenters are called **isotactic** polymers, those with alternating configurations are called **syndiotactic** polymers, and those with random configurations are called **atactic** polymers. Generally, the more stereoregular the polymer, the more crystalline it is. ✳
- Chain-growth polymerizations can also be carried out using anionic or cationic species in the propagation steps, processes that can be referred to as **ionic polymerizations**.
  - The choice of ionic species for polymerization depends on the **electronics of the monomer**.
  - Vinyl monomers with **electron-withdrawing groups** (carbanion stabilizing) are used in **anionic polymerizations**.

 Anionic polymerizations are initiated by a strong nucleophile such as *sec*-butyllithium. The growing chain has an anionic and thus nucleophilic end group, that attacks other monomers so the chain grows. Anionic polymerizations can also be initiated by **electron transfer** from an electron rich species. Chain-transfer and chain-termination steps that disrupt radical polymerizations do not occur with anionic polymerizations, so initiated chains continue to grow until either all the monomer is consumed or the reaction is terminated by adding an external reagent. Such polymerizations are referred to as **living polymerizations**.

 Living polymerizations produce polymers with a **well-defined molecular weight size** distribution, determined by the initiator to monomer ratio.

 Electrophilic groups can be added to each end of a living polymer chain after all of the monomer has been consumed. Polymer chains with functional groups on both ends are called **telechelic polymers**.
  - Vinyl monomers with **electron-donating groups** (cation stabilizing) are used in **cationic polymerizations**.

 Cationic polymerization can be initiated by **strong protic acid** or a **Lewis acid**.

CHAPTER 23
Solutions to the Problems

<u>Problem 23.1</u> Given the following structure, determine the polymer's repeat unit, redraw the structure using the simplified parenthetical notation, and name the polymer.

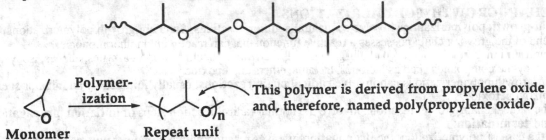

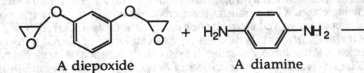

Polymer-
ization

This polymer is derived from propylene oxide and, therefore, named poly(propylene oxide)

Monomer   **Repeat unit**

<u>Problem 23.2</u> Write the repeating unit of the polymer formed from the following reaction and propose a mechanism for its formation.

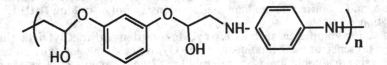

A diepoxide   **A diamine**

Following is the structural formula of the repeat unit of this polymer.

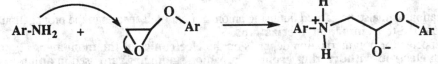

As a mechanism, propose nucleophilic attack of the amine on the less hindered carbon of the epoxide followed by proton transfer from nitrogen to oxygen.

Step 1: Nucleophilic ring opening of the epoxide

The diamine   **The diepoxide**

Step 2: Proton transfer from nitrogen to oxygen

<u>Problem 23.3</u> Show how to prepare polybutadiene that is terminated at both ends with primary alcohol groups.

Treat 1,3-butadiene with two mol of lithium metal to form a dianion followed by addition of monomer units as in Example 23.3 to form a living polymer. Cap the active end groups by treatment of the living polymer with ethylene oxide followed by aqueous acid.

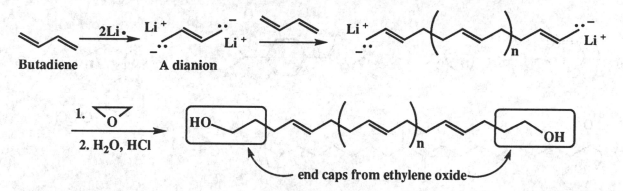

Butadiene   A dianion

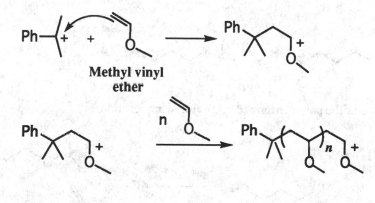

end caps from ethylene oxide

<u>Problem 23.4</u> Write a mechanism for the polymerization of methyl vinyl ether initiated by 2-chloro-2-phenylpropane and $SnCl_4$. Label the initiation, propagation, and termination steps.

The mechanism for this cationic polymerization is similar to that shown in Example 23.4 for the cationic polymerization of 2-methylpropene. Treatment of 2-chloro-2-phenylpropane with $SnCl_4$ forms the 2-phenyl-2-propyl cation, the initiating cation. The termination step shown here is loss of H⁺ from the end of the polymer chain to form a carbon-carbon double bond.

Initiation:

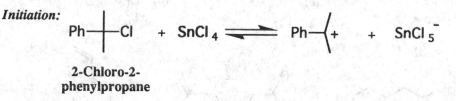

2-Chloro-2-
phenylpropane

Propagation:

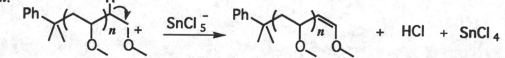

Methyl vinyl
ether

Termination:

Problem 23.5 Name the following polymers.

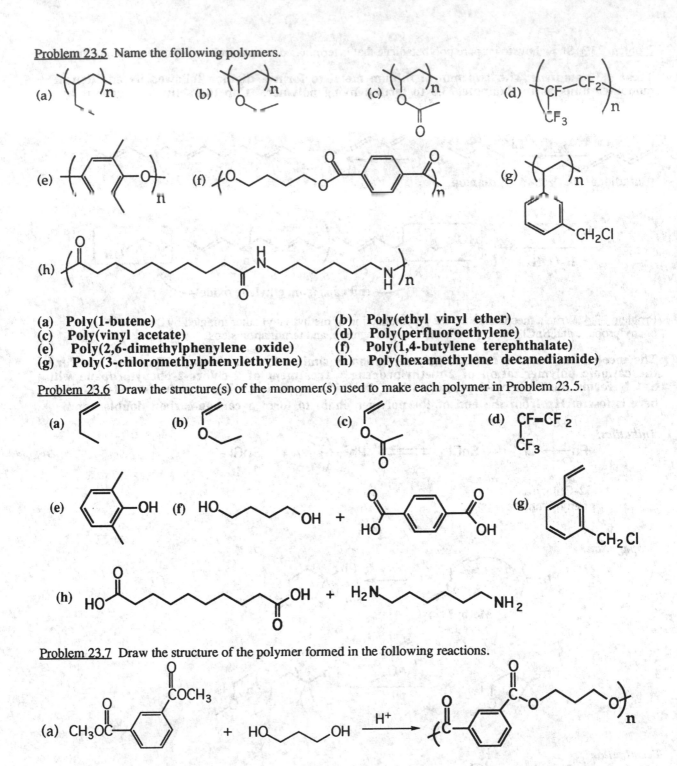

(a) **Poly(1-butene)**
(c) **Poly(vinyl acetate)**
(e) **Poly(2,6-dimethylphenylene oxide)**
(g) **Poly(3-chloromethylphenylethylene)**

(b) **Poly(ethyl vinyl ether)**
(d) **Poly(perfluoroethylene)**
(f) **Poly(1,4-butylene terephthalate)**
(h) **Poly(hexamethylene decanediamide)**

Problem 23.6 Draw the structure(s) of the monomer(s) used to make each polymer in Problem 23.5.

Problem 23.7 Draw the structure of the polymer formed in the following reactions.

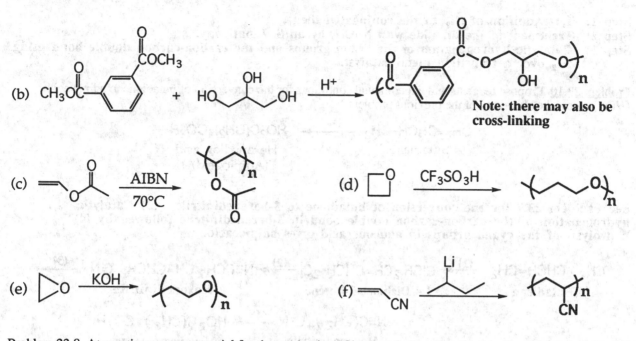

(b) ... + ... $\xrightarrow{H^+}$ ...

Note: there may also be cross-linking

(c) ... $\xrightarrow[70°C]{AIBN}$ ...

(d) ... $\xrightarrow{CF_3SO_3H}$ ...

(e) ... $\xrightarrow{KOH}$ ...

(f) ...

Problem 23.8 At one time, a raw material for the production of hexamethylenediamine was the pentose-based polysaccharides of agricultural wastes, such as oat hulls. Treatment of these wastes with sulfuric acid or hydrochloric acid gives furfural. Decarbonylation of furfural over a zinc-chromium-molybdenum catalyst gives furan. Propose reagents and experimental conditions for the conversion of furan to hexamethylenediamine.

oat hulls, corn cobs, sugar cane stalks, etc $\xrightarrow[H_2O]{H_2SO_4}$ **Furfural** $\xrightarrow[\text{catalyst}]{\text{Zn-Cr-Mo}}$ **Furan** $\xrightarrow{(1)}$ **Tetrahydrofuran (THF)** $\xrightarrow{(2)}$

$Cl(CH_2)_4Cl$ $\xrightarrow{(3)}$ $N\equiv C(CH_2)_4C\equiv N$ $\xrightarrow{(4)}$ $H_2N(CH_2)_6NH_2$

1,4-Dichloro-butane   **Hexanedinitrile (Adiponitrile)**   **1,6-Hexanediamine (Hexamethylenediamine)**

Step 1: Catalytic hydrogenation using H_2 over a transition metal catalyst.
Step 2: Cleavage of the ether using concentrated HCl at elevated temperature.
Step 3: Treatment of the dihalide with NaCN, by an S_N2 pathway.
Step 4: Catalytic hydrogenation of the cyano groups using H_2 over a transition metal catalyst.

Problem 23.9 Another raw material for the production of hexamethylenediamine is butadiene derived from thermal and catalytic cracking of petroleum. Propose reagents and experimental conditions for the conversion of butadiene to hexamethylenediamine.

$CH_2=CHCH=CH_2$ $\xrightarrow{(1)}$ $ClCH_2CH=CHCH_2Cl$ $\xrightarrow{(2)}$

Butadiene   **1,4-Dichloro-2-butene**

$N\equiv CCH_2CH=CHCH_2C\equiv N$ $\xrightarrow{(3)}$ $H_2N(CH_2)_6NH_2$

3-Hexenedinitrile   **1,6-Hexanediamine (Hexamethylenediamine)**

Step 1: 1,4- Addition of Cl$_2$ to the conjugated diene.
Step 2: Treatment of the dihalide with NaCN, by an S$_N$2 pathway.
Step 3: Catalytic hydrogenation of the cyano groups and the carbon-carbon double bond using H$_2$ over a transition metal catalyst.

Problem 23.10 Propose reagents and experimental conditions for the conversion of butadiene to adipic acid. (*Hint:* review the chemistry of the previous problem.)

$$CH_2=CHCH=CH_2 \longrightarrow HO_2C(CH_2)_4CO_2H$$

Butadiene   Hexanedioic acid
                                       (Adipic acid)

See Problem 23.9 for the conversion of butadiene to 3-hexenedinitrile. (3) Catalytic hydrogenation of the carbon-carbon double bond in 3-hexenedinitrile followed by (4) hydrolysis of the cyano groups in aqueous acid gives adipic acid.

$$CH_2=CHCH=CH_2 \xrightarrow{(1)} ClCH_2CH=CHCH_2Cl \xrightarrow{(2)} N\equiv CCH_2CH=CHCH_2C\equiv N \xrightarrow{(3)}$$

Butadiene   1,4-Dichloro-2-butene   3-Hexenedinitrile

$$N\equiv C(CH_2)_4C\equiv N \xrightarrow{(4)} HO_2C(CH_2)_4CO_2H$$

Hexanedinitrile   Hexanedioic acid
(Adiponitrile)   (Adipic acid)

Problem 23.11 Polymerization of 2-chloro-1,3-butadiene under Ziegler-Natta conditions gives a synthetic elastomer called neoprene. All carbon-carbon double bonds in the polymer chain have the trans configuration. Draw the repeat unit in neoprene.

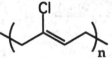

Problem 23.12 Poly(ethylene terephthalate) (PET) can be prepared by this reaction. Propose a mechanism for the step-growth reaction in this polymerization.

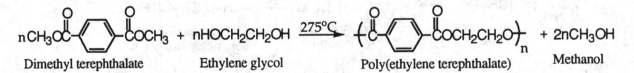

Dimethyl terephthalate   Ethylene glycol   Poly(ethylene terephthalate)   Methanol

Propose addition of a hydroxyl group to a carbonyl carbon of dimethyl terephthalate to form a tetrahedal carbonyl addition intermediate, followed by its collapse to give an ester bond of the polymer plus methanol etc. This is an example of transesterification.

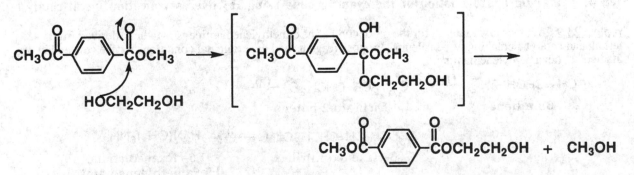

Problem 23.13 Identify the monomers required for the synthesis of these step-growth polymers.

(a)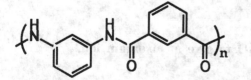

Kodel
(a polyester)

$$HO\overset{O}{\overset{||}{C}}-\underset{\text{}}{\bigcirc}-\overset{O}{\overset{||}{C}}OH \quad + \quad HOCH_2-\bigcirc-CH_2OH$$

(b)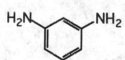

Quiana
(a polyamide)

$$HO\overset{O}{\overset{||}{C}}(CH_2)_6\overset{O}{\overset{||}{C}}OH \quad + \quad H_2N-\bigcirc-CH_2-\bigcirc-NH_2$$

Problem 23.14 Nomex, another aromatic polyamide (aramid) is prepared by polymerization of 1,3-benzenediamine and the diacid chloride of 1,3-benzenedicarboxylic acid. The physical properties of the polymer make it suitable for high strength, high temperature applications such as parachute cords and jet aircraft tires. Draw a structural formula for the repeating unit of Nomex.

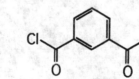

1,3-Benzenediamine   1,3-Benzene-
                          dicarbonyl chloride

polymerization ⟶ Nomex

Following is the repeat unit in Nomex.

Problem 23.15 Caprolactam, the monomer from which nylon 6 is synthesized, is prepared from cyclohexanone in two steps. In step 1, cyclohexanone is treated with hydroxylamine to form cyclohexanone oxime. Treatment of the oxime with concentrated sulfuric acid in step 2 gives caprolactam by a reaction called a Beckmann rearrangement. Propose a mechanism for the conversion of cyclohexanone oxime to caprolactam. (*Hint*: Proton transfer to oxygen of the oxime followed by loss of water gives a positively charged, electron deficient nitrogen atom, which then provides the driving force for the skeletal rearrangement.)

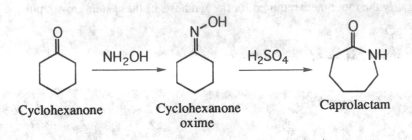

Cyclohexanone   Cyclohexanone   Caprolactam
                    oxime

The mechanism is shown divided into six steps.

Step 1: **Proton transfer from H_3O^+ to the oxygen atom of the oxime generates an oxonium ion, which converts OH, a poor leaving group, into OH_2, a better leaving group.**

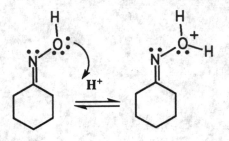

Step 2: **Departure of H_2O, which creates an electron-deficient nitrogen atom.**

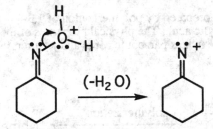

Step 3: **Migration of the electron pair of an adjacent carbon-carbon bond to nitrogen, which creates an electron-deficient carbon atom.**

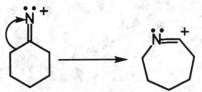

Step 4: **Reaction of the carbocation from step 2 with H_2O to give an oxonium ion.**

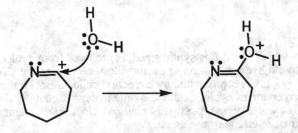

Step 5: **Proton transfer to solvent gives the enol of an amide.**

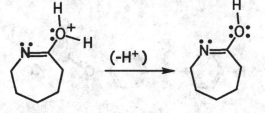

Step 6: **Tautomerization of the enol form of the amide gives caprolactam.**

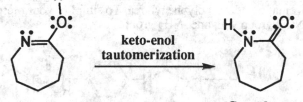

Caprolactam

<u>Problem 23.16</u> Nylon 6,10 is prepared by polymerization of a diamine and a diacid chloride. Draw the structural formula of each reactant and for the repeat unit in nylon 6,10.

The six-carbon diamine is 1,6-hexanediamine and the ten-carbon diacid chloride is decanedioyl chloride.

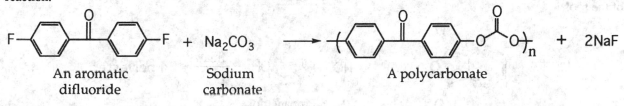

    1,6-Hexanediamine   **Decanedioyl**   **The repeat unit of nylon 6,10**
                                    chloride

<u>Problem 23.17</u> Polycarbonates (Section 23.5C) are also formed by using a nucleophilic aromatic substitution route (Section 20.3B) involving aromatic difluoro monomers and carbonate ion. Propose a mechanism for this reaction.

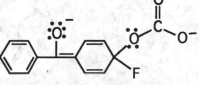

    An aromatic   **Sodium**   **A polycarbonate**
    difluoride   **carbonate**

Review Section 20.3B for the addition-elimination mechanism of nucleophilic aromatic substitution.

Step 1: **Nucleophilic addition of carbonate ion to the aromatic ring at the carbon bearing the fluorine atom forms a Meisenheimer complex.**

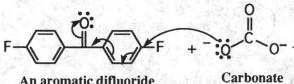

    An aromatic difluoride   **Carbonate**   **A Meisenheimer complex**
                                     ion

Step 2: **Collapse of the Meisenheimer complex with ejection for fluoride ion.**

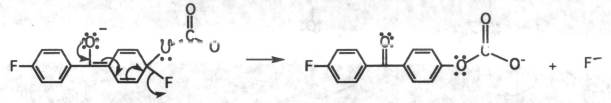

A Meisenheimer complex

<u>Problem 23.18</u> Propose a mechanism for the formation of this polyphenylurea. To simplify your presentation of the mechanism, consider the reaction of one -NCO group with one -NH₂ group.

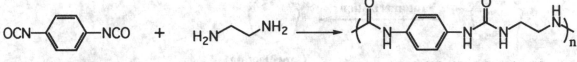

1,4-Benzenediisocyanate   1,2-Ethanediamine   Poly(ethylene phenylurea)

Propose (1) addition of an amino group to the C=O of the isocyanate group to give an enol, followed by (2) keto-enol tautomerism of the enol to give a disubstituted urea.

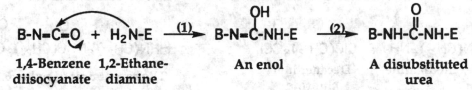

1,4-Benzene   1,2-Ethane-   An enol   A disubstituted
diisocyanate   diamine   urea

<u>Problem 23.19</u> When equal molar amounts of phthalic anhydride and 1,2,3-propanetriol are heated, they form an amorphous polyester. Under these conditions, polymerization is regioselective for the primary hydroxyl groups of the triol.

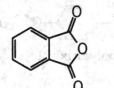   +   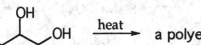   heat → a polyester

Phthalic anhydride   1,2,3-Propanetriol
(Glycerol)

(a) Draw a structural formula for the repeat unit of this polyester

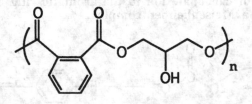

(b) Account for the regioselective reaction with the primary hydroxyl groups only.

The regioselectivity reflects the fact that the 1° hydroxyl groups are more accessible than the 2° hydroxyl group.

Problem 23.20 The polyester from Problem 23.19 can be mixed with additional phthalic anhydride (0.5 mol of phthalic anhydride for each mol of 1,2,3-propanetriol in the original polyester) to form a liquid resin. When this resin is heated, it forms a hard, insoluble, thermosetting polyester called glyptal.
(a) Propose a structure for the repeat unit in glyptal.

The polymer described in Problem 23.19 becomes cross linked as shown.

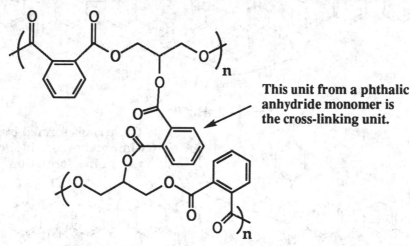

This unit from a phthalic anhydride monomer is the cross-linking unit.

(b) Account for the fact that glyptal is a thermosetting plastic.

Because of the extensive cross linking, the individual polymer chains can no longer be made to flow and, therefore, the polymer can not be made to assume a liquid state.

Problem 23.21 Propose a mechanism for the formation of the following polymer.

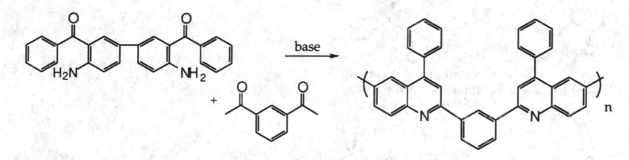

One way to attack this problem is to first determine which rings in the product are present in the original monomers, and which are formed during the polymerization. The monomer units are redrawn here to show that new rings are formed during polymerization by aldol reactions followed by dehydration and by imine formation.

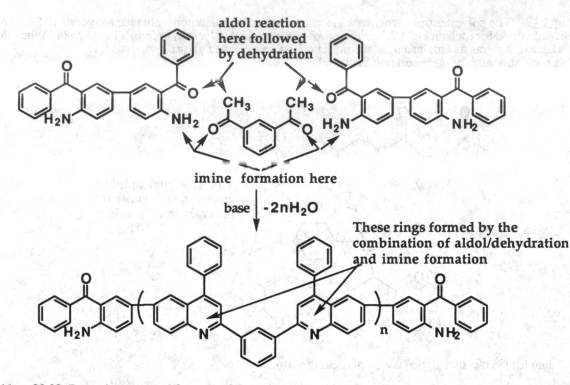

aldol reaction
here followed
by dehydration

imine formation here

base -2nH₂O

These rings formed by the
combination of aldol/dehydration
and imine formation

<u>Problem 23.22</u> Draw the structural formula of the polymer resulting from base-catalyzed polymerization of each
compound. Would you expect the polymers to be optically active? (S)-(+)-lactide is the dilactone formed from
two molecules of (S)-(+)-lactic acid.

(a)

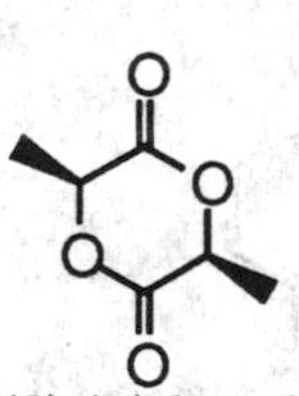

(S)-(+)-lactide

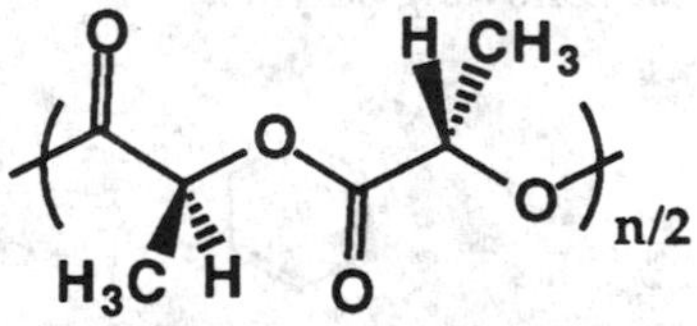

each stereocenter has the S
configuration; the polymer
is optically active

(b)

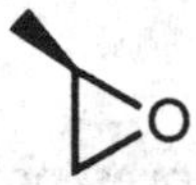

(R)-Propylene oxide.

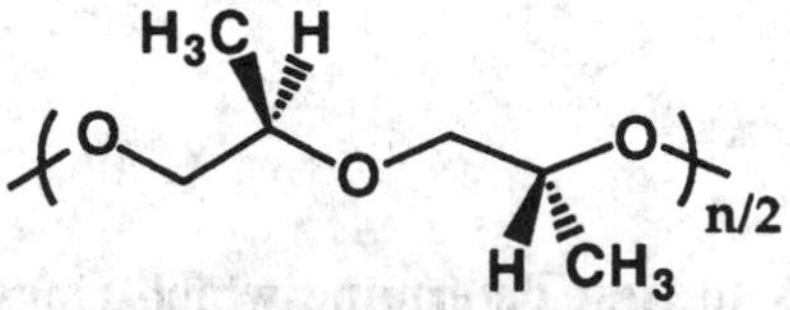

each stereocenter has the R
configuration; the polymer
is optically active

Problem 23.23 Poly(3-hydroxybutanoic acid), a biodegradable polyester, is an insoluble, opaque material that is difficult to process into shapes. In contrast, the copolymer of 3-hydroxybutanoic acid and 3-hydroxyoctanoic acid is a transparent polymer that shows good solubility in a number of organic solvents. Explain the difference in properties between these two polymers in terms of their structure.

Poly(3-hydroxybutanoic acid)   Poly(3-hydroxybutanoic acid -
3-hydroxyoctanoic acid) copolymer

The polymer chains of poly(3-hydroxybutanoic acid) can assume a highly ordered arrangement with a high degree of crystallinity, hence its insolubility and its opaque character. In contrast, the polymer chains of the copolymer of 3-hydroxybutanoic acid and 3-hydroxyoctanoic acid have bulky five-carbon chains which effectively prevent polymer chains from assuming any regular ordered structure. As a result, the polymer has little crystalline character, that is, it is an amorphous material with little crystalline character to reflect light.

Problem 23.24 How might you determine experimentally if a particular polymerization is propagating by a step-growth or a chain-growth mechanism?

Analyze the distribution of polymer molecular weights as a function of degree of polymerization. As discussed in the introduction to Section 23.5, high molecular weight polymers are not produced until very late in step-growth polymerization, typically past 99% conversion of monomers to polymers. Given the mechanism of chain-growth polymerization, high molecular weight polymer molecules are produced very early and continuously in the polymerization process.

Problem 23.25 Select the monomer in each pair that is more reactive toward cationic polymerization.

The more reactive monomer in each pair is the one forming the more stable carbocation. The first structure in each pair forms the more stable carbocation.

(a) $\equiv$ OCH$_3$ or $\diagdown\diagup$ (b) $\equiv$ OCH$_3$ or   OCCH$_3$
                                                               O

$^+$:OCH$_3$ $\longleftrightarrow$ $\nearrow$ $_+$OCH$_3$

More important contributing structure; carbon and oxygen have complete valence shells

$\delta+$
$_+$OCCH$_3$
$\cdot\cdot\parallel$
O
$\delta-$

This structure makes little contribution to the hybrid because of adjacent positive and partial positive charges.

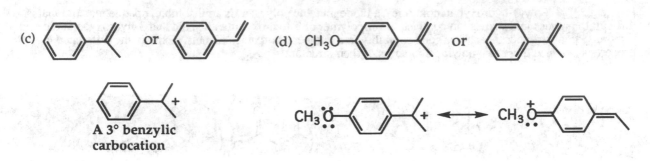

(c) or   (d) or

A 3° benzylic
carbocation

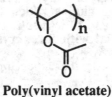

Problem 23.26 Polymerization of vinyl acetate gives poly(vinyl acetate). Hydrolysis of this polymer in aqueous sodium hydroxide gives poly(vinyl alcohol). Draw the repeat units of both poly(vinyl acetate) and poly(vinyl alcohol).

Following are structural formulas for the monomer and repeat unit of each polymer.

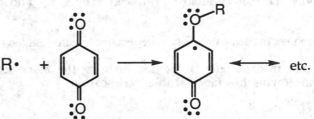

Vinyl acetate   Poly(vinyl acetate)   Poly(vinyl alcohol)

Problem 23.27 Benzoquinone can be used to inhibit radical polymerizations. This compound reacts with a radical intermediate, R•, to form a less reactive radical that does not participate in chain propagation steps and, thus, breaks the chain.

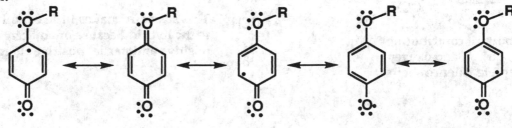

Draw a series of contributing structures for this less reactive radical and account for its stability.

This radical can be represented as a hybrid of five contributing structures; three place the single electron on carbon atoms of the ring, and two place it on the oxygen atoms bonded to the ring. This radical is stabilized by the significant degree of delocalization of the single electron.

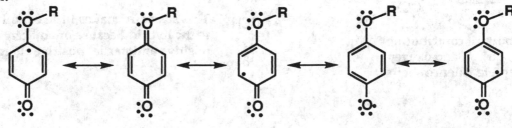

<u>Problem 23.28</u> Following is the structural formula of a section of polypropylene derived from three units of propylene monomer.

$$CH_3 \quad CH_3 \quad CH_3$$
$$-CH_2CH-CH_2CH-CH_2CH-$$

Polypropylene

Draw structural formulas for comparable sections of:
(a) poly(vinyl chloride)   (b) polytetrafluoroethylene

$$Cl \qquad Cl \qquad Cl$$
$$-CH_2CH-CH_2\ CH-CH_2CH-$$

$$-CF_2\ CF_2-CF_2\ CF_2-CF_2\ CF_2\cdot$$

(c) poly(methyl methacrylate) (Plexiglas)   (d) poly(1,1-dichloroethylene)

$$CH_3 \quad CH_3 \quad CH_3$$
$$O \qquad O \qquad O$$
$$O{=}C \quad O{=}C \quad O{=}C$$
$$-CH_2C-CH_2C-CH_2CH-$$
$$CH_3 \quad CH_3 \quad CH_3$$

$$-CH_2CCl_2-CH_2CCl_2-CH_2CCl_2$$

<u>Problem 23.29</u> Low-density polyethylene (LDPE) has a higher degree of chain branching than high-density polyethylene (HDPE). Explain the relationship between chain branching and density.

Unbranched polyethylene packs more efficiently into compact structures which have more mass per unit volume than structures formed by packing of branched polyethylene chains. Therefore, unbranched polyethylene has a higher density than branched-chain polyethylene.

<u>Problem 23.30</u> We saw how intramolecular chain transfer in radical polymerization of ethylene creates a four-carbon branch on a polyethylene chain. What branch is created by a comparable intramolecular chain transfer during radical polymerization of styrene.

A six-membered transition
state leading to
1,5-hydrogen abstraction

This four-carbon
branch is created

<u>Problem 23.31</u> Compare the densities of low-density polyethylene (LDPE) and high-density polyethylene (HDPE) with the densities of the liquid alkanes listed in Table 2.4. How might you account for the differences between them?

Given in the table are densities of several liquid alkanes reported in Table 2.4, plus densities for pentadecane, eicosane, and tricosane. As you can see, densities for these unbranched alkanes reach a maximum in the range 0.77-0.79 g/mL, which is significantly less than the density of both LDPE and HDPE. From this data, we conclude that both LDPE and HDPE pack more efficiently (have greater mass per unit volume) than their lower molecular weight counterparts.

Alkane	Formula	Density (g/mL)
Pentane	C_5H_{12}	0.626
Heptane	C_7H_{16}	0.684
Decane	$C_{10}H_{22}$	0.730
Pentadecane	$C_{15}H_{32}$	0.769
Eicosane	$C_{20}H_{42}$	0.789
Tricosane	$C_{30}H_{62}$	0.775
LDPE	$\text{(CH}_2)_n$	0.91-0.94
HDPE	$\text{(CH}_2)_n$	0.96

<u>Problem 23.32</u> Natural rubber is the all cis polymer of 2-methyl-1,3-butadiene (isoprene).

Poly(2-methyl-1,3-butadiene)
(Polyisoprene)

(a) Draw the structural formula for the repeat unit of natural rubber.

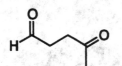

(b) Draw the structural formula of the product of oxidation of natural rubber by ozone followed by a workup in the presence of $(CH_3)_2S$. Name each functional group present in this product.

4-Oxopentanal
(a ketoaldehyde)

(c) The smog prevalent in many major metropolitan areas contains oxidizing agents, including ozone. Account for the fact that this type of smog attacks natural rubber (automobile tires, etc.) but does not attack polyethylene or polyvinyl chloride.

Polyethylene and poly(vinyl chloride) do not contain carbon-carbon double bonds.

Problem 23.33 Radical polymerization of styrene gives a linear polymer. Radical polymerization of a mixture of styrene and 1,4-divinylbenzene gives a cross linked network polymer of the type shown in Figure 23.1. Show by drawing structural formulas how incorporation of a few percent 1,4-divinylbenzene in the polymerization mixture gives a cross linked polymer.

Drawn here is a section of the copolymer showing cross linking by one molecule of 1,4-divinylbenzene. Benzene rings derived from, PhCH=CH$_2$, are shown as Ph.

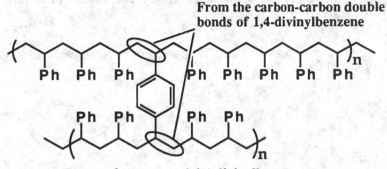

A copolymer of styrene and 1,4-divinylbenzene

Problem 23.34 One common type of cation exchange resin is prepared by polymerization of a mixture containing styrene and 1,4-divinylbenzene (Problem 23.33). The polymer is then treated with concentrated sulfuric acid to sulfonate a majority of the aromatic rings in the polymer.
(a) Show the product of sulfonation of each benzene ring.

The following is a structural formula for a section of the polymer. Structural formulas for only the sulfonated rings are written in full; unsulfonated benzene rings are shown as Ph.

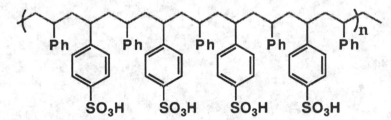

(b) Explain how this sulfonated polymer can act as a cation exchange resin.

The resin is shown in the acid or protonated form. When functioning as a cation exchange resin, cations displace H$^+$ and become bound to the negatively charged -SO$_3^-$ groups.

Problem 23.35 The most widely used synthetic rubber is a copolymer of styrene and butadiene called SB rubber. Ratios of butadiene to styrene used in polymerization vary depending on the end use of the polymer. The ratio used most commonly in the preparation of SB rubber for use in automobile tires is 1 mol styrene to 3 mol butadiene. Draw a structural formula of a section of the polymer formed from this ratio of reactants. Assume that all carbon-carbon double bonds in the polymer chain are in the cis configuration.

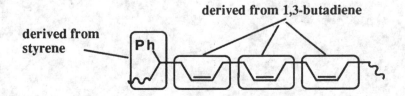

Problem 23.36 From what two monomer units is the following polymer made?

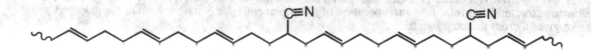

The section of polymer drawn here is derived six 1,3-butadiene monomer units and two acrylonitrile monomer units.

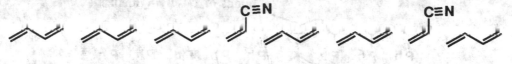

CHAPTER 24: CARBOHYDRATES

SUMMARY OF REACTIONS

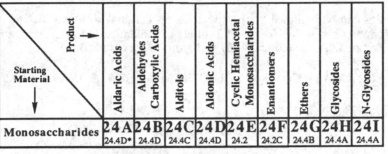

	Aldaric Acids	Aldehydes Carboxylic Acids	Alditols	Aldonic Acids	Cyclic Hemiacetal Monosaccharides	Enantiomers	Ethers	Glycosides	N-Glycosides
Monosaccharides	24A 24.4D*	24B 24.4D	24C 24.4C	24D 24.4D	24E 24.2	24F 24.2C	24G 24.4B	24H 24.4A	24I 24.4A

*Section in book that describes reaction.

REACTION 24A: OXIDATION TO ALDARIC ACIDS (Section 24.4D)

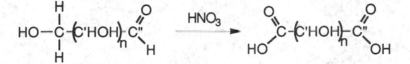

- Nitric acid oxidizes both the -CHO and terminal -CH$_2$OH groups of an aldose to carboxyl groups. ✶
- The product dicarboxylic acid is referred to as an aldaric acid.

REACTION 24B: OXIDATION BY PERIODIC ACID (Section 24.4D)

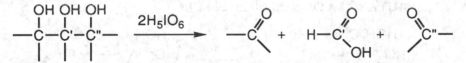

- Periodic acid cleaves carbon-carbon bonds of glycols (Reaction 9J, section 9.5H) to give either aldehydes or ketones, depending on the structure of the original glycol The reaction involves a cyclic periodate ester intermediate. ✶
- Periodic acid also cleaves α-hydroxy aldehydes and ketones to give an aldehyde, ketone or carboxylic acid. The exact product can be predicted by assuming that each C-C bond broken in the reaction is replaced by a carbon-oxygen bond.
- The reaction can be used for structure determination of carbohydrates by analyzing the amount of periodate used and the different fragments produced.

REACTION 24C: REDUCTION (Section 24.4C)

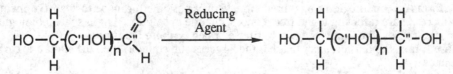

- The carbonyl group of a monosaccharide can be reduced to an alcohol to give an alditol.
- Different reducing agents can be used including NaBH$_4$ and metal-catalyzed hydrogenation with H$_2$. ✶
- The reduction actually occurs with the small amount of open-chain sugar that is present at any one time, until eventually, the reaction is complete. This reaction emphasizes how the cyclic form of a monosaccharide predominates in solution, but this is in equilibrium with a small amount of the open-chain form.

REACTION 24D: OXIDATION TO ALDONIC ACIDS (Section 24.4D)

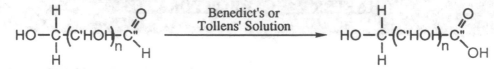

- **Monosaccharides** can be oxidized to **aldonic** acids by **Tollens' solution** (Silver ion in aqueous ammonia) or **Benedict's solution** (copper(II) sulfate in sodium carbonate in citrate buffer). ✳
- As opposed to reaction 24A, the terminal -CH$_2$OH group is not oxidized.

REACTION 24E: FORMATION OF CYCLIC HEMIACETALS (Section 24.2)

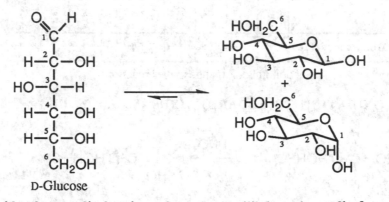

D-Glucose

- **Monosaccharides** form **cyclic hemiacetals**, and at equilibrium, the **cyclic forms predominate** to such an extent that only a small amounts of the open chain forms exist at any one time. Nevertheless, it is important to remember that this equilibration exists, because sometimes the small amount of open chain form can be important in a reaction mechanism. See for example, mutarotation (reaction 24E, section 24.2D). ✳

REACTION 24F: MUTAROTATION (Section 24.2C)

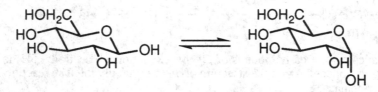

- The α- and **β-anomers** of cyclic monosaccharides slowly interconvert in aqueous solution, and the process is known as **mutarotation**. ✳
- The mechanism of mutarotation involves the open chain form as an intermediate, that quickly recloses to the different cyclic forms.
- A number of generalizations can be made about which forms of a given monosaccharide will predominate at equilibrium.
  Little free aldehyde or ketone is present.
  Pyranose forms (six-membered rings) predominate over possible furanose forms (five-membered rings). Note that here we are referring to the free monosaccharide in solution, but in special biological structures such as nucleic acids, the furanose form of ribose or deoxyribose is found.
  In pyranose forms, the diastereomer that has the larger group equatorial on the anomeric carbon atom predominates.

REACTION 24G: FORMATION OF ETHERS (Section 24.4B)

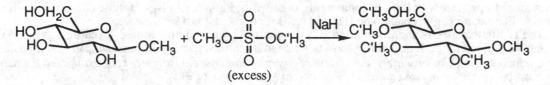

(excess)

- All of the **-OH** groups of a glycoside are converted to **methyl ethers** with **dimethyl sulfate** in the presence of sodium hydride. ✳
- The sodium hydride turns the weakly nucleophilic -OH groups into strong nucleophilic -O⁻ groups.

REACTION 24H: FORMATION OF GLYCOSIDES (Section 24.4A)

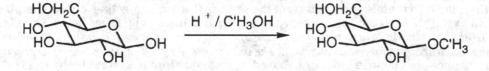

- Reaction of a **cyclic hemiacetal** with an **alcohol** like methanol in the presence of an **acid catalyst** produces a cyclic acetal called a **glycoside**. ✳
- The mechanism involves protonation of the glycosidic -OH group, loss of H₂O to generate a resonance stabilized cation, attack on this cation by the alcohol, and loss of a proton to generate the final product glycoside.
- The reaction is reversible, and treatment of a glycoside with aqueous acid regenerates the free monosaccharide.

REACTION 24I: FORMATION OF N-GLYCOSIDES (Section 24.4A)

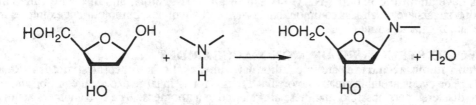

- *N*-glycosides can be formed from a **cyclic monosaccharide** with a compound containing an **N-H bond** to form *N*-glycosides. ✳
- The nucleic acids are based on D-ribose or 2-deoxy-D-ribose, in the furanose form, as the *N*-glycosides with the aromatic bases uracil, cytosine, thymine, adenine, or guanine. In DNA and RNA, these *N*-glycosides are found exclusively as the β-anomers.

SUMMARY OF IMPORTANT CONCEPTS

24.0 OVERVIEW
- **Carbohydrates** are **polyhydroxylated aldehydes** or **ketones**, or compounds that produce polyhydroxylated aldehydes or ketones upon hydrolysis. ✳
- **Carbohydrates** are the most **abundant organic molecules** in the world. They are essential to all forms of life, and perform such functions as energy storage (glucose, starch, glycogen), structural reinforcement (cellulose), and genetic information storage as components of nucleic acids (DNA, RNA).

24.1 MONOSACCHARIDES
- **Monosaccharides** usually have molecular formulas of $C_nH_{2n}O_n$ (3<n<8), and are the **monomers** from which **larger carbohydrates** are constructed. ✳

- Monosaccharides are named by using the suffix **ose**. The prefixes **tri, tetr, pent** are used to indicate 3,4, or 5 carbon atoms, respectively. An **aldehyde carbohydrate** is called an **aldose**, and is sometimes designated with an **aldo** prefix. A **ketone carbohydrate** is called an **ketose**, and is sometimes designated with a **keto** prefix. For example, glyceraldehyde is an aldotriose and fructose is a ketohexose.
  - The nomenclature of monosaccharides is dominated by common names. Even though IUPAC names can be derived for each different monosaccharide, the common names are much simpler and used almost exclusively. * *[Unfortunately, the system of common names is only slightly systematic, and the names must be simply memorized along with the corresponding structures.]*
- Monosaccharides usually have one or more stereocenters, so **stereochemistry** is of **major importance** with **monosaccharides**. A **Fischer projection** of a monosaccharide is used to show the structure of a monosaccharide and thus keep track of stereochemistry.
  - In a **Fischer projection**, the monosaccharide is usually drawn in the open chain form, and the carbonyl carbon atom is placed at the top of the structure. *[Even though it is understood that the cyclic hemiacetal predominates at equilibrium, the open chain Fischer projection is used for clarity]*
  - Since the sp³ carbon atoms are tetrahedral, a monosaccharide that is stretched out analogous to a Fisher projection will have some groups projecting forward, and some projecting backward. In a **Fischer projection, the horizontal lines represent the groups** that are **projecting above the plane of the paper**, and **vertical lines** are used to **represent groups** that are **projecting below the plane of the paper**. *[Although very difficult to master at first, Fisher projections are useful for understanding and describing the structures of monosaccharides. The only way to get good at using them is to practice.]* *
  - Monosaccharides are **classified** as **D** or **L, based** on a **comparison** to glyceraldehyde stereochemistry. In a monosaccharide, the **point of reference** is the **stereocenter** that is **farthest from** the **carbonyl group**. Since this is a carbon atom that is next to the last carbon atom in the chain (notice that the last carbon atom of the chain has two -H atoms, so it is not a stereocenter) it is referred to as the **penultimate carbon atom**. A monosaccharide that has the **same configuration** about the **penultimate carbon** as **D-glyceraldehyde** is classified as a **D monosaccharide**. In this case the -OH group will be on the right side of the carbon atom in the Fischer projection. Similarly, an **L monosaccharide** has a **configuration** about the **penultimate carbon atom** that is the same as the configuration of **L-glyceraldehyde**, with the -OH group being on the left in a Fischer projection.
  - The enantiomer of a given monosaccharide is not produced by simply changing the configuration of the penultimate carbon atom, but rather by reversing all of the stereocenters. *[This may seem obvious, but it is worth keeping in mind as the different sugars are examined.]*
- Some sugars have an **amino group (-NH₂) instead of all -OH groups**, and these are called **amino sugars**. Amino sugars are much less common that normal carbohydrates, but important examples include D-glucosamine and D-galactosamine.

24.2 THE CYCLIC STRUCTURE OF MONOSACCHARIDES

- The **open chain monosaccharides** are in equilibrium with a **cyclic hemiacetal structure** (Reaction 24E, Section 24.2). The cyclic acetal is greatly favored and thus is found in large excess at equilibrium.
- There are two diastereomers possible, and these are referred to a **anomers**. The two anomers are distinguished by the relative orientation of the **anomeric -OH** group (the -OH group on the so-called **anomeric carbon atom**, the one that was a carbonyl in the open chain form). *
- The two anomers are named as α– or β–. The α anomer is the one that has the anomeric **-OH group** on the **same side** of the **Fischer projection** as the **-OH group** on the **penultimate carbon** *[Remember that the -OH group on the penultimate carbon atom is the one that determines whether a monosaccharide is named as D or L]*. The β-anomer has the anomeric -OH group and the -OH group of the penultimate carbon atom on opposites sides in the Fischer projection. With D-glucose in the cyclic hemiacetal form, the α anomer is the one with the anomeric -OH group is axial, while for the β anomer the anomeric -OH group is equatorial. *
- **Cyclic monosaccharide structures** are usually drawn as **Haworth projections**, in which the five-membered or six-membered **cyclic hemiacetal** is drawn as **planar**, and **perpendicular** to the **plane of the paper**. They are usually drawn with the anomeric carbon to the right, and the hemiacetal oxygen atom in the back. A more accurate "chair" type of structure can be drawn for six-membered ring hemiacetals, showing which groups are axial and which are equatorial.

24.3 PHYSICAL PROPERTIES

- **Monosaccharides** are all **very soluble** in **water** due to all of the -OH groups that can take part in hydrogen bonding with the water molecules. Monosaccharides also taste sweet to differing degrees.

24.5 GLUCOSE ASSAYS: THE SEARCH FOR SPECIFICITY

• For various medical conditions, especially diabetes, an accurate determination of glucose levels in serum and other fluids is required. In order to carry out this determination quantitatively, some chemical and enzymatic tests have been developed.

24.6 L-ASCORBIC ACID (VITAMIN C)
• In nature, **L-ascorbic acid** is made through a series of enzyme-catalyzed reactions from D-glucose. Humans, primates and guinea pigs cannot carry out the complete synthesis of ascorbic acid, since not all the required enzymes are present.
• On an industrial level, ascorbic acid is synthesized through a combination of some chemical steps, as well as some microbiological fermentations, starting with inexpensive D-glucose.

24.7 DISACCHARIDES AND OLIGOSACCHARIDES
• The monosaccharides can be linked into **disaccharides**, **trisaccharides**, or higher **oligosaccharides**.
  - Important disaccharides include **maltose** (2 D-glucose molecules linked with a β-1,4-glycoside bond), **lactose** (D-glucose and D-galactose linked with a β–1,4-glycoside bond), and **sucrose** (D-glucose and D-fructose linked with an α-1,2-glycoside bond).
  - Larger oligosaccharides are also important in nature, including the **blood group substances** that are on the surfaces of red blood cells. These are used to differentiate the blood types.

24.8 POLYSACCHARIDES
• Monosaccharides can also be linked together into very large **polysaccharides**.
  - **Starch** is composed entirely of D-glucose units, as a straight chain polymer with α–1,4-glycoside linkages (called amylose) or as a branched polymer with both α-1,4-glycoside linkages and α-1,6-glycoside linkages. Starch is used for energy storage in plants.
  - **Glycogen** is the animal equivalent of starch, and is a highly branched structure with both α-1,4-glycoside linkages and α-1,6-glycoside linkages between D-glucose units. Glycogen is found mostly in the liver and muscles, and serves as the carbohydrate reserve.
  - **Cellulose** is a linear polymer of D-glucose linked by β-1,4 glycoside bonds. Cellulose is a structural component of plants, comprising almost half of the cell wall material of wood. Cotton is almost entirely composed of cellulose, and the synthetic textile fibers **rayon** and **acetate rayon** are chemically modified forms of cellulose.

Problem 24.1 (a) Draw Fischer projections for all 2-ketopentoses.
(b) Show which are D-ketopentoses, which are L-ketopentoses, and which are enantiomers.
(c) Refer to Table 24.2, and write names of the ketopentoses you have drawn.

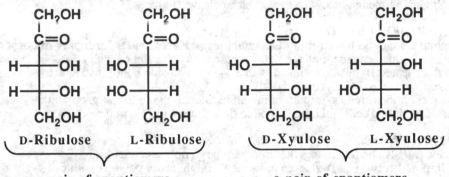

a pair of enantiomers   a pair of enantiomers

Problem 24.2 D-Mannose exists in aqueous solution as a mixture of α-D-mannopyranose and β-D-mannopyranose. Draw Haworth projections for these molecules.

D-Mannose differs in configuration from D-glucose at carbon-2. Therefore, the alpha and beta forms of D-mannopyranose differ from those of alpha and beta D-glucopyranoses only in the orientation of the -OH on carbon-2. Following are Haworth projections for these compounds.

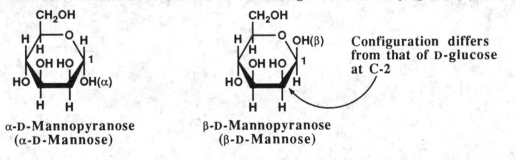

α-D-Mannopyranose   β-D-Mannopyranose
(α-D-Mannose)   (β-D-Mannose)

Problem 24.3 Draw chair conformations for α-D-mannopyranose and β-D-mannopyranose. Label the anomeric carbon atom in each.

D-Mannose differs in configuration from D-glucose at carbon-2. Draw chair conformations for the alpha and beta forms of D-glucopyranose and then invert the configuration of the -OH on carbon-2. For reference, the open chain form of D-mannose is also drawn.

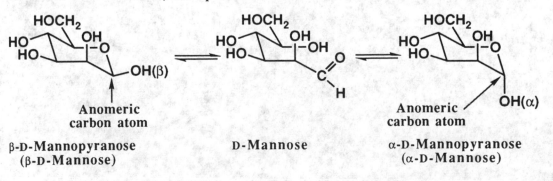

β-D-Mannopyranose   D-Mannose   α-D-Mannopyranose
(β-D-Mannose)   (α-D-Mannose)

Problem 24.4 Draw structural formulas for these glycosides. In each, label the anomeric carbon and the glycoside bond.
(a) Methyl β-D-fructofuranoside (methyl β-D-fructoside)

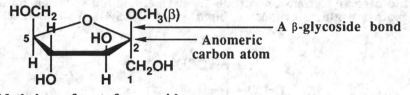

Methyl β-D-fructofuranoside

(b) Methyl α-D-mannopyranoside (methyl α-D-mannoside)

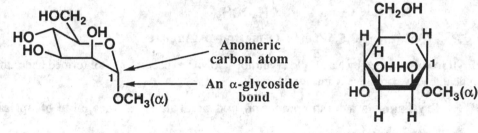

Methyl α-D-mannopyranoside   **Methyl α-D-mannopyranoside**
     (Chair conformation)   **(Haworth projection)**

Problem 24.5 Suppose that a β-D-glycopyranose is treated with methanol enriched in oxygen-18. Is the isotopic label found in the resulting methyl glycoside, in the water produced in the reaction, or in both the methyl glycoside and in the water?

Oxygen-18 will appear in the -OCH₃ of the glycoside according to the mechanism put forward in the solution to Example 24.5.

Problem 24.6 Draw a structural formula for the β-N-glycoside formed between 2-deoxy-D-ribofuranose and adenine.

Following are structural formulas for adenine, the monosaccharide hemiacetal, and the N-glycoside.

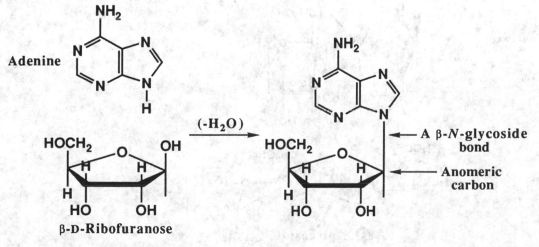

β-D-Ribofuranose

Problem 24.7 Suppose it were possible to convert D-glucose to a permethylated D-glucofuranoside. Draw an open-chain formula for the tetra-O-methyl-D-glucose that would be isolated after this permethylated derivative is treated with dilute aqueous HCl.

The furanoside form of D-glucose results from cyclic hemiacetal formation between the -OH on carbon-4 and the carbonyl group on carbon-1. Permethylation followed by acid-catalyzed hydrolysis of the methyl D-gluco-furanoside bond gives 2,3,5,6-tetra-O-methyl-D-glucose.

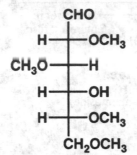

2,3,5,6-Tetra-O-methyl-D-glucose

Problem 24.8 D-Erythrose is reduced by NaBH$_4$ to erythritol. Do you expect the alditol formed under these conditions to be optically active or optically inactive?

Optically inactive. Erythritol is a meso compound and achiral. It is incapable of optical activity.

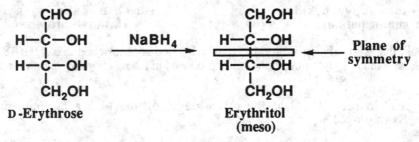

Problem 24.9 Draw Haworth and chair formulas for the α anomer of a disaccharide in which two units of D-glucopyranose are joined by a β-1,3-glycoside bond.

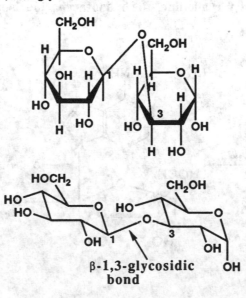

β-1,3-glycosidic bond

PROBLEMS
Monosaccharides
Problem 24.10 Explain the meaning of the designations D and L as used to specify the configuration of carbohydrates.

The designations D and L refer to the configuration of the chiral center farthest from the carbonyl group of the monosaccharide. When a monosaccharide is drawn in a Fischer projection, the reference -OH is on the right in a D-monosaccharide and on the left in an L-monosaccharide. Note that the conventions D and L specify the configuration at one and only one of however many chiral carbons there are in a particular monosaccharide.

Problem 24.11 Which compounds are D-monosaccharides and which are L-monosaccharides?

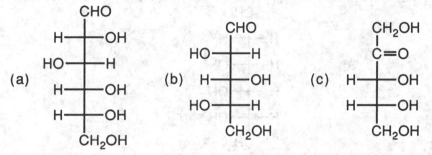

Compounds (a) and (c) are D-monosaccharides, and compound (b) is an L-monosaccharide.

Problem 24.12 Classify each monosaccharide in Problem 24.11 using the designations D/L and aldose/ketose and according to the number of carbon atoms it contains. For example, glucose is classified as a D-aldohexose.

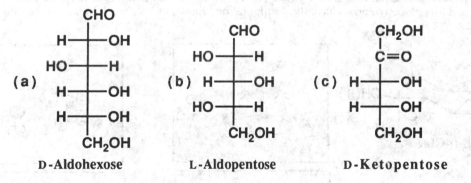

Problem 24.13 Write Fischer projections for L-ribose and L-arabinose.

L-Ribose and L-arabinose are the mirror images of D-ribose and D-arabinose, respectively. The most common error in answering this question is to start with the Fischer projection for the D sugar and then invert the configuration of carbon-4 only. While the monosaccharide thus drawn is an L-sugar, it is not the correct one. All of the stereocenters must be changed to draw the true enantiomers.

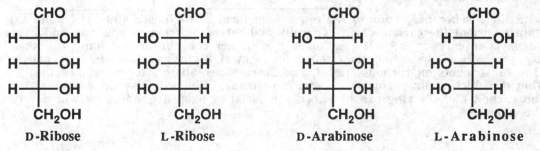

<u>Problem 24.14</u> What is the meaning of the prefix deoxy- as it is used in carbohydrate chemistry?

The prefix "deoxy-" means "without oxygen".

<u>Problem 24.15</u> Give L-fucose a name incorporating the prefix "deoxy-" that shows its relationship to galactose.

A systemic name for L-fucose is 6-deoxy-L-galactose.

<u>Problem 24.16</u> 2,6-Dideoxy-D-altrose, known alternatively as D-digitoxose, is a monosaccharide obtained on hydrolysis of digitoxin, a natural product extracted from foxglove (*Digitalis purpurea*). Digitoxin has found wide use in cardiology because it reduces pulse rate, regularizes heart rhythm, and strengthens heart beat. Draw the structural formula of 2,6-dideoxy-D-altrose.

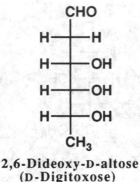

2,6-Dideoxy-D-altose
(D-Digitoxose)

The Cyclic Structure of Monosaccharides
<u>Problem 24.17</u> Build a molecular model of D-glucose and show that its six-membered hemiacetals (α-D-glucopyranose and β-D-glucopyranose) have the configurations shown in Figure 24.1 and not the mirror images of these structures.

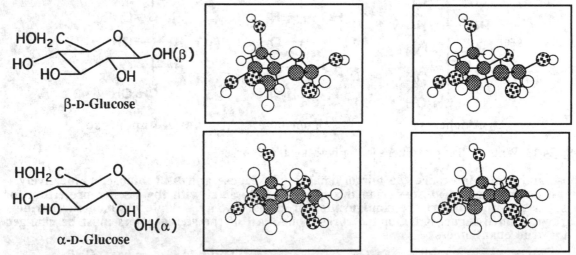

β-D-Glucose

α-D-Glucose

Start with the Fischer projection of the open-chain form of D-glucose and then build a model. Be certain to orient the carbon chain vertically and extending from you as you add hydrogens and hydroxyls to carbons 2-5. If you follow the Fischer convention and build the model correctly, you will discover that the open-chain form of D-glucose closes to a pyranose ring in which the substituents on carbons 2, 3, 4, and 5 are either all equatorial or all axial, depending on which chair conformation you have made. You will also discover that the -OH group on carbon-1 of the ring can be either equatorial or axial depending on which way you close the pyranose ring.

Problem 24.18 Draw α-D-glucopyranose (α-D-glucose) as a Haworth projection. Now, using only the information given here, draw Haworth projections for these monosaccharides.
(a) α-D-mannopyranose (α-D-mannose). The configuration of D-mannose differs from that of D-glucose only at carbon 2.
(b) α-D-gulopyranose (α-D-gulose). The configuration of D-gulose differs from that of D-glucose at carbons 3 and 4.

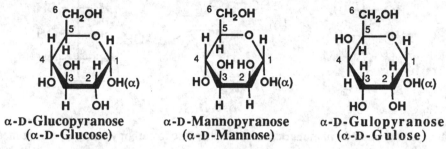

α-D-Glucopyranose   α-D-Mannopyranose   α-D-Gulopyranose
(α-D-Glucose)   (α-D-Mannose)   (α-D-Gulose)

Problem 24.19 Repeat Problem 24.18, using chair conformations instead of Haworth projections for the monosaccharides.

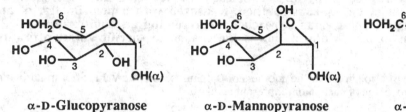
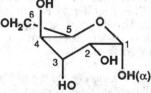

α-D-Glucopyranose   α-D-Mannopyranose   α-D-Gulopyranose
(α-D-Glucose)   (α-D-Mannose)   (α-D-Gulose)

Problem 24.20 Convert each Haworth projection to an open-chain Fischer projection and name the monosaccharide you have drawn.

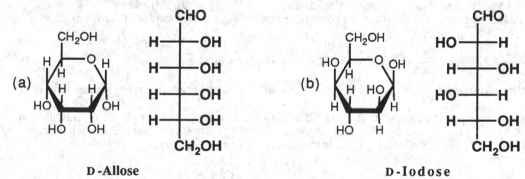
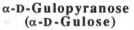

(a)   (b)

```
      CHO                                                  CHO
  H —— OH                                              HO —— H
  H —— OH                                               H —— OH
  H —— OH                                              HO —— H
  H —— OH                                               H —— OH
     CH₂OH                                                 CH₂OH
```

D-Allose   D-Iodose

<u>Problem 24.21</u> Convert each chair conformation to an open-chain Fischer projection and name the monosaccharide you have drawn.

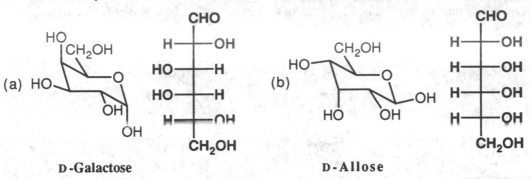

(a)   (b)

D-Galactose   D-Allose

<u>Problem 24.22</u> Explain the phenomenon of mutarotation with reference to carbohydrates. By what means is it detected?

Any monosaccharides of four or more carbons can exist in an open-chain form and two or more cyclic hemiacetal (i.e. furanose or pyranose) forms, each having a different specific rotation. The specific rotation of an aqueous solution, measured with a polarimeter, of any one form changes until an equilibrium value is reached, representing an equilibrium concentration of the different forms. Mutarotation is the change in specific rotation toward an equilibrium value.

<u>Problem 24.23</u> Following are specific rotations for the anomers of D-mannose and the value after mutarotation. Using these data, calculate the percentage of each anomer present at equilibrium.

$$\alpha\text{-D-mannose} \quad +29.3° \longrightarrow +14.5°$$
$$\beta\text{-D-mannose} \quad -16.3° \longrightarrow +14.5°$$

The difference in specific rotation between the anomers is 29.3° - (16.3°)=45.6°. The difference between the specific rotation of β-D-mannopyranose and the equilibrium value is 29.3°-14.5° = 14.8°. The percent β-anomer at equilibrium is

$$\frac{29.3 - 14.5}{29.3 - (-16.3)} \times 100 = \frac{14.8}{45.6} \times 100 = 32.4\% \text{ β-D-mannopyranose}$$

This means that there must be 100% - 32.4% or 67.6% of the α-anomer present at equilibrium.

<u>Problem 24.24</u> It has been proposed that structural characteristics of sweet-tasting compounds are (1) a hydrogen bond donating group, AH, (2) a hydrogen bond-accepting group, B, and (3) that atoms A and B are separated by between 2.5 Å and 4.0 Å. Identify the AH and B units in these nonnutritive sweeteners.

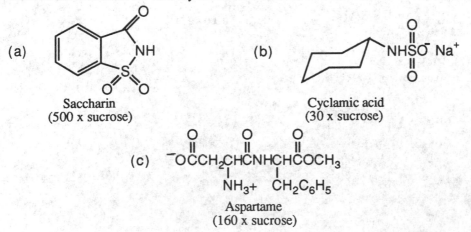

(a)

Saccharin
(500 x sucrose)

(b)

Cyclamic acid
(30 x sucrose)

(c)

Aspartame
(160 x sucrose)

The hydrogen bond donating groups are shown with a circle drawn around them, and the hydrogen bond accepting groups are shown with a square drawn around them.

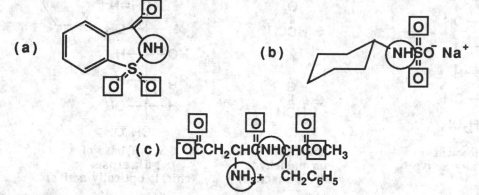

(a)

(b)

(c)

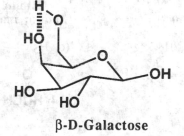

Of the groups identified above, the ones that satisfy the 0.25 and 0.4 nm conditional distance between hydrogen bond donor and acceptor groups would be the four atom systems of O=C-N-H or O=S-N-H in (a), O=S-N-H or O=S-O-H in (b), and the central O=C-N-H in (c). For (c), other possibilities exist in which the terminal -NH$_3^+$ group could come within the correct distance of a hydrogen bond acceptor group in certain conformations of the molecule.

<u>Problem 24.25</u> It has been observed that sugars that can form one or more strong intramolecular hydrogen bonds are less sweet than sugars which cannot form such hydrogen bonds. Draw chair conformations of β-D-galactose and β-D-mannose and identify one strong intramolecular hydrogen bond in each molecule.

The probable hydrogen bonds are shown below in the diagrams.

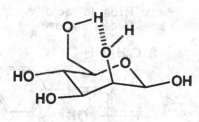

β-D-Mannose   β-D-Galactose

Reactions of Monosaccharides

<u>Problem 24.26</u> Draw Fischer projections for the product(s) formed by reaction of D-galactose with the following. In addition, state whether each product is optically active or inactive.
(a) NaBH$_4$ in H$_2$O   (b) H$_2$/Pt   (c) HNO$_3$, warm

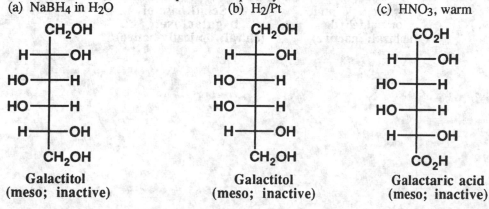

Galactitol   Galactitol   Galactaric acid
(meso; inactive)   (meso; inactive)   (meso; inactive)

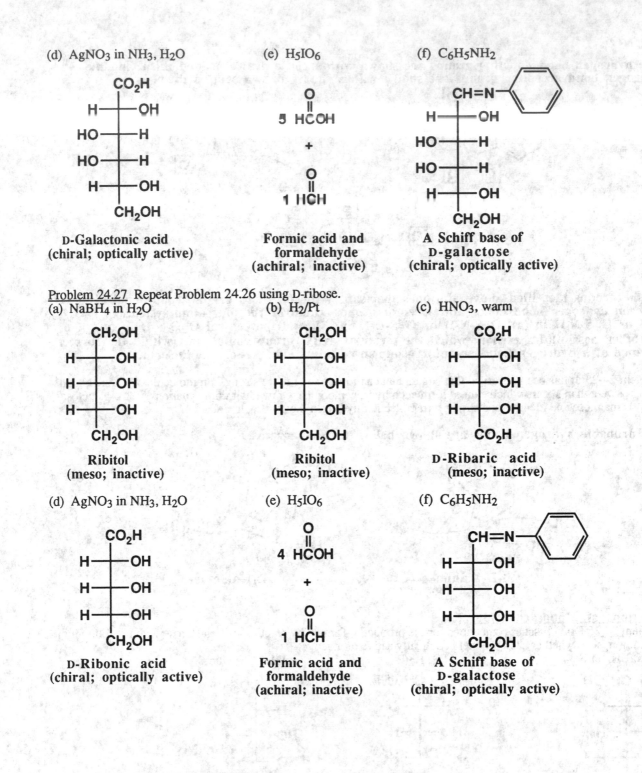

(d) AgNO₃ in NH₃, H₂O

CO₂H
H——OH
HO——H
HO——H
H——OH
CH₂OH

D-Galactonic acid
(chiral; optically active)

(e) H₅IO₆

O
‖
5 HCOH

+

O
‖
1 HCH

Formic acid and
formaldehyde
(achiral; inactive)

(f) C₆H₅NH₂

CH=N
H——OH
HO——H
HO——H
H——OH
CH₂OH

A Schiff base of
D-galactose
(chiral; optically active)

Problem 24.27 Repeat Problem 24.26 using D-ribose.

(a) NaBH₄ in H₂O

CH₂OH
H——OH
H——OH
H——OH
CH₂OH

Ribitol
(meso; inactive)

(b) H₂/Pt

CH₂OH
H——OH
H——OH
H——OH
CH₂OH

Ribitol
(meso; inactive)

(c) HNO₃, warm

CO₂H
H——OH
H——OH
H——OH
CO₂H

D-Ribaric acid
(meso; inactive)

(d) AgNO₃ in NH₃, H₂O

CO₂H
H——OH
H——OH
H——OH
CH₂OH

D-Ribonic acid
(chiral; optically active)

(e) H₅IO₆

O
‖
4 HCOH

+

O
‖
1 HCH

Formic acid and
formaldehyde
(achiral; inactive)

(f) C₆H₅NH₂

CH=N
H——OH
H——OH
H——OH
CH₂OH

A Schiff base of
D-galactose
(chiral; optically active)

Problem 24.28 An important technique for establishing relative configurations among isomeric aldoses and ketoses is to convert both terminal carbon atoms to the same functional group. This can be done either by selective oxidation or reduction. As a specific example, nitric acid oxidation of D-erythrose gives meso-tartaric acid (Table 4.1, Section 4.7). Similar oxidation of D-threose gives (S,S)-tartaric acid. Given this information and the fact that D-erythrose and D-threose are diastereomers, draw Fischer projections for D-erythrose and D-threose. Check your answers against Table 24.1.

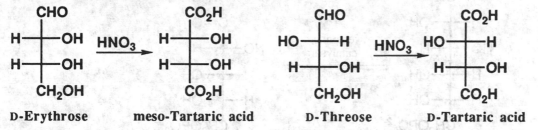

Problem 24.29 There are four D-aldopentoses (Table 24.1). If each is reduced with $NaBH_4$, which yield optically active alditols? Which yield optically inactive alditols?

D-Ribose and D-xylose yield different achiral (meso) alditols. D-Arabinose and D-lyxose yield the same chiral alditol.

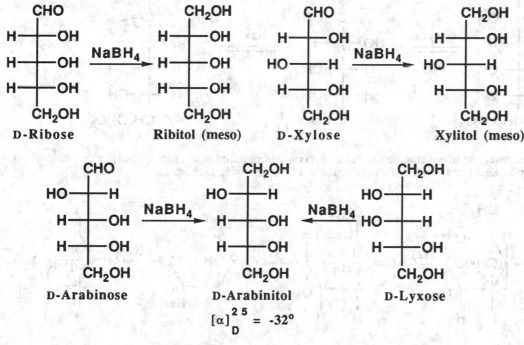

Problem 24.30 Name the two alditols formed by $NaBH_4$ reduction of D-fructose?

D-Glucitol and D-mannitol. Each differs in configuration only at carbon-2.

Problem 24.31 One pathway for the metabolism of glucose-6-phosphate is its enzyme-catalyzed conversion to fructose-6-phosphate. Show that this transformation can be regarded as two enzyme-catalyzed keto-enol tautomerisms.

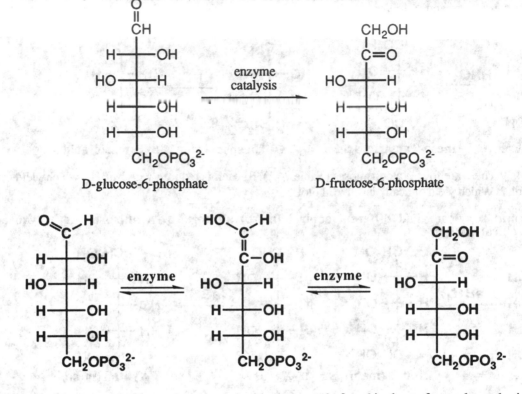

D-glucose-6-phosphate　　　　　　　　　　　D-fructose-6-phosphate

Problem 24.32 L-Fucose, one of several monosaccharides commonly found in the surface polysaccharides of animal cells, is synthesized biochemically from D-mannose in the following eight steps:

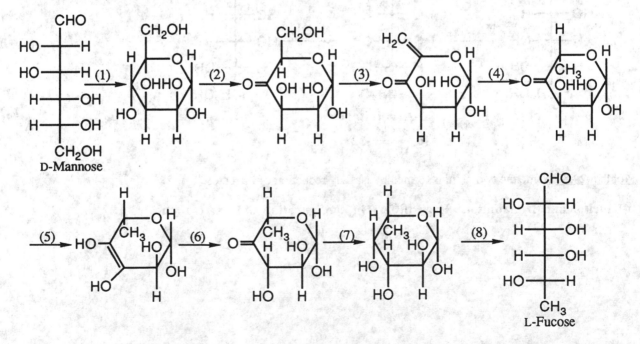

(a) Describe the type of reaction (i.e., oxidation, reduction, hydration, dehydration, etc.) involved in each step.

Following is the type of reaction in each step.
(1) Formation of a cyclic hemiacetal from a carbonyl group and a secondary alcohol.
(2) A two-electron oxidation of a secondary alcohol to a ketone.
(3) Dehydration of a β-hydroxyketone to an α,β-unsaturated ketone.
(4) A two-electron reduction of a carbon-carbon double bond to a carbon-carbon single bond.
(5) Keto-enol tautomerism of an α-hydroxyketone to form an enediol.
(6) Keto-enol tautomerism of an enediol to form an α-hydroxyketone.
(7) A two-electron reduction of a ketone to a secondary alcohol.
(8) Opening of a cyclic hemiacetal to form an aldehyde and an alcohol.

(b) Explain why it is that this monosaccharide derived from D-mannose now belongs to the L series.

It is the configuration at carbon-5 of this aldohexose that determines whether it is of the D-series or of the L-series. The result of steps 3 and 4 is inversion of configuration at carbon-5 and, therefore, conversion of a D-aldohexose to an L-aldohexose.

<u>Problem 24.33</u> Complete the following:

(a) Draw a structural formula for the compound formed when D-ribose is converted to methyl β-D-ribofuranoside and then permethylated with sodium hydride and dimethyl sulfate.

In the following equations, β-D-ribofuranose is first treated with methanol in the presence of an acid catalyst to give methyl β-D-ribofuranoside, which is in turn treated with dimethyl sulfate in the presence of sodium hydride to give the permethylated derivative.

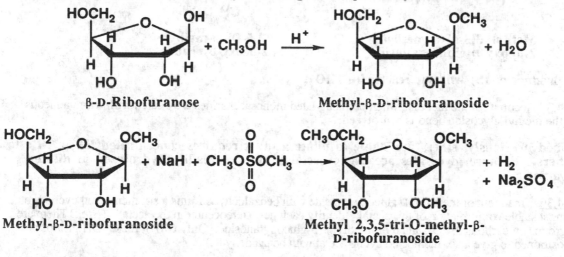

(b) Draw the structural formula for the product of hydrolysis of the permethylated compound formed in part (a).

Acid-catalyzed hydrolysis gives 2,3,5-tri-O-methyl-D-ribose.

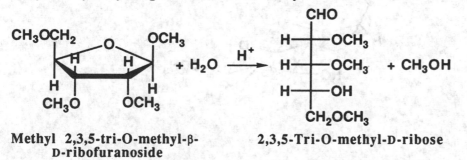

Methyl 2,3,5-tri-O-methyl-β-   2,3,5-Tri-O-methyl-D-ribose
     D-ribofuranoside

(c) With how many moles of periodic acid does the compound in part (b) react?

The product of (b) will not react with HIO₄.

<u>Problem 24.34</u> Repeat Problem 24.33 for 2-deoxy-D-ribose.

(a,b)   2-Deoxy-D-ribose is treated first with one mol of methanol and then with one mol of dimethyl sulfate in the presence of sodium hydride to give methyl 3,5-di-O-methyl-β-2-deoxy-D-ribofuranoside.

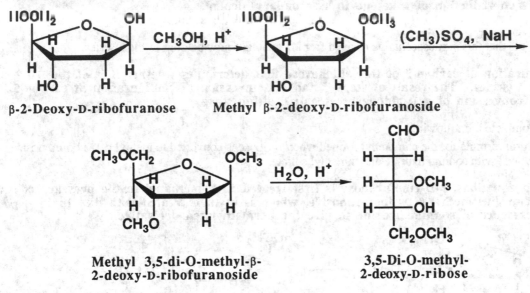

β-2-Deoxy-D-ribofuranose   Methyl β-2-deoxy-D-ribofuranoside

Methyl 3,5-di-O-methyl-β-   3,5-Di-O-methyl-
2-deoxy-D-ribofuranoside   2-deoxy-D-ribose

(c) **The product of (b) will not react with HIO₄.**

<u>Problem 24.35</u> Account for the fact that, when permethylated monosaccharides are treated with warm aqueous acid, only the methyl glycoside bond is hydrolyzed.

As described previously, ethers are stable to dilute acid; it requires concentrated HI or HBr to cleave ethers. Because glycosides are cyclic acetals, they are readily hydrolyzed in dilute aqueous acid.

<u>Problem 24.36</u> Treatment of methyl β-D-glucopyranoside with benzaldehyde forms a six-membered cyclic acetal. Draw the most stable conformation of this acetal. Identify each new stereocenter in the acetal? (*Hint*: There are free -OH groups on carbons 2, 3, 4, and 6 of methyl β-D-glucopyranoside. Only two of these -OH groups are properly positioned to give a six-membered cyclic acetal with benzaldehyde.)

Problem 24.37 Vanillin, the principal component of vanilla, occurs in vanilla beans and other natural sources as a β-D-glucopyranoside. Draw a structural formula for this glycoside, showing the D-glucose unit as a chair conformation. (See The Merck Index, 12th ed., #10069.)

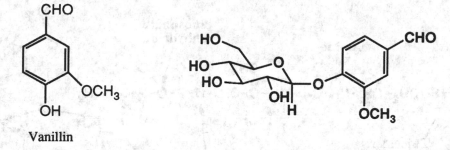

Vanillin

Problem 24.38 Hot water extracts of ground willow and poplar bark are an effective pain reliever. Unfortunately, the liquid is so bitter that most persons refuse it. The pain reliever in these infusions is salicin, a β-glycoside of D-glucopyranose and the phenolic -OH group of 2-(hydroxymethyl)phenol. Draw a structural formula for salicin, showing the glucose ring as a chain conformation. (See The Merck Index, 12th ed., 8476.)

Salicin can have the glycoside bond in the α or β conformation of the sugar, linked to the hydroxymethyl or phenol -OH group.

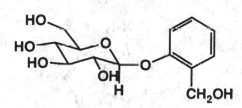

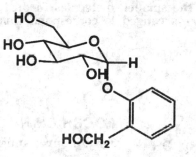

<u>Problem 24.39</u> Draw structural formulas for the products formed by hydrolysis at pH 7.4 (the pH of blood plasma) of all ester, thioester, amide, anhydride, and glycoside groups in acetyl coenzyme A. Name as many of these compounds as you can.

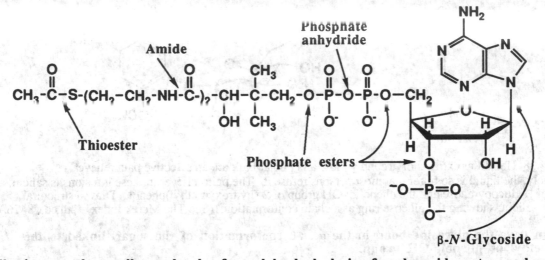

Following are the smaller molecules formed by hydrolysis of each amide, ester, and anhydride bond. They are arranged to correspond roughly to their location from left to right in acetyl CoA.

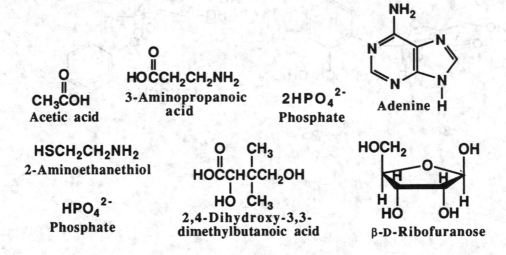

The molecule formed by amide formation between 3-aminopropanoic acid and 2,4-dihydroxy-3,3-dimethylbutanoic acid is given the special name pantothenic acid. Pantothenoic acid is a vitamin, most commonly contained in vitamin pills as calcium pantothenate. Its minimum daily requirement (MDR) has not yet been determined.

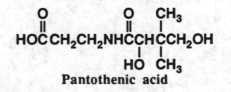

Pantothenic acid

Ascorbic Acid

Problem 24.40 Assign R or S configurations to each stereocenter in ascorbic acid.

There are two tetrahedral stereocenters in L-ascorbic acid, namely carbon-4 and carbon-5. The following stereo drawings show order of priority about each stereocenter and the assignment of R, S configuration.

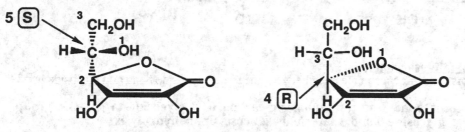

Problem 24.41 Write a balanced half-reaction to show that conversion of ascorbic acid to dehydroascorbic acid is an oxidation. How many electrons are involved in this oxidation?

The most direct way to see that this is a two-electron oxidation is to write a balanced half-reaction for the conversion of the enediol to a diketone.

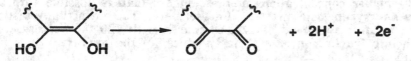

Problem 24.42 Given the fact that ascorbic acid and dehydroascorbic acid are the physiologically active forms of vitamin C, is ascorbic acid a biological oxidizing agent or a biological reducing agent? Explain.

Because L-ascorbic acid donates two electrons to another molecule or ion, it is a biological reducing agent. Conversely, L-dehydroascorbic acid is a biological oxidizing agent.

Problem 24.43 Ascorbic acid is a diprotic acid with the following acid ionization constants.
$$pK_{a1} = 4.10 \qquad pK_{a2} = 11.79$$
The two acidic hydrogens are those connected with the enediol part of the molecule. Which is the more acidic hydrogen? (*Hint:* Draw separately the anion derived by loss of each acidic hydrogen. Which anion is more stable, that is, which anion has the greater degree of delocalization of charge?)

Following are assignments of the two pK_a values.

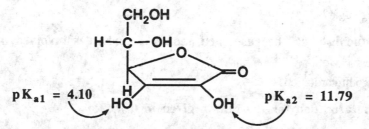

The anion derived from ionization of -OH on carbon-3 is stabilized by resonance interaction with the carbonyl oxygen. There is no comparable resonance stabilization of the anion derived from ionization of -OH on carbon-2.

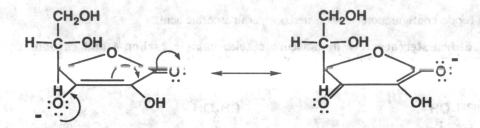

Disaccharides and Oligosaccharides

Problem 24.44 What is the difference in meaning between the terms glycoside bond and glucoside bond?

A glycoside bond is the bond from the anomeric carbon of a glycoside to an -OR group. A glucoside bond is a glycoside bond that yields glucose upon hydrolysis.

Problem 24.45 In making candy or sugar syrups, sucrose is boiled in water with a little acid, such as lemon juice. Why does the product mixture taste sweeter than the starting sucrose solution?

Sucrose is a disaccharide composed of the monosaccharides glucose and fructose linked through a glycoside bond. The acid catalyzes hydrolysis of the glycoside bond, and the monomeric glucose and fructose are more soluble than sucrose itself. Because of this, the syrups and candy have higher concentrations of these sugars than is possible with sucrose. As a result, the candy and syrup taste sweeter. Furthermore, fructose actually tastes sweeter than sucrose, having a relative sweetness of 174, compared with 100 for sucrose. Thus, converting the sucrose into fructose increases the sweetness of the mixture.

Problem 24.46 Trehalose, found in young mushrooms and the chief carbohydrate in the blood of certain insects, is a disaccharide consisting of two D-monosaccharide units joined by an α-1,1-glycoside bond.

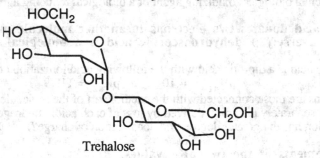

Trehalose

(a) Is trehalose a reducing sugar?

Trehalose is not a reducing sugar because each anomeric carbon is involved in formation of the glycoside bond.

(b) Does trehalose undergo mutarotation?

It will not undergo mutarotation for the reason given in (a).

(c) With how many moles of periodic acid does trehalose react? How many moles of formaldehyde are formed? How many moles of formic acid are formed?

It will react with four mol of HIO_4, two mol for each monosaccharide unit. Two mol of formic acid are formed; there is no formaldehyde formed.

(d) Draw structural formulas for the two O-methylated monosaccharides formed when trehalose is permethylated with dimethyl sulfate and sodium hydride and then warmed in dilute aqueous acid to hydrolyze the glycoside bond.

Two mol of 2,3,4,6-tetra-O-methyl-D-glucose are formed.

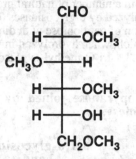

CHO
H
OCH3
CH3O
H
H
OCH3
H
OH
CH2OCH3

2,3,4,6-Tetra-O-methyl-D-glucose

Problem 24.47 The trisaccharide raffinose occurs principally in cottonseed meal.

HO
CH2OH
O
HO
OH
O
CH2
HO
O
HO
HO
OH
O
CH2OH
HO
HOCH2
O

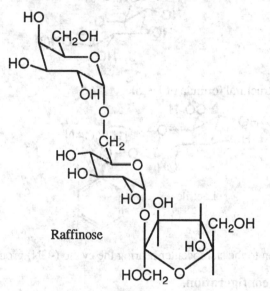

Raffinose

(a) Name the three monosaccharide units in raffinose.

The three monosaccharide units in raffinose, from top to bottom, are D-galactose, D-glucose, and D-fructose.

(b) Describe each glycoside bond in this trisaccharide.

Reading from left to right, they are D-galactopyranose joined by an α-1,6-glycoside bond to D-glucopyranose and then D-glucopyranose, in turn, joined by an α-1,2-glycoside bond to β-D-fructofuranose.

(c) Is raffinose a reducing sugar?

No, it is not a reducing sugar.

(d) With how many mol of periodic acid will raffinose react?

Raffinose will react with 5 mol of HIO₄; 2 mol for the D-galactopyranose ring, 2 mol for the D-glucopyranose ring, and 1 mol for the D-fructofuranose ring.

Problem 24.48 Gentiobiose is a disaccharide found in a number of natural products including gentian plants (Gentiana lutea). It is a reducing sugar and is hydrolyzed by β-glycosidases (enzymes with catalytic activity limited to hydrolysis of β-glycoside bonds). Reaction of gentiobiose with dimethyl sulfate in the presence of sodium hydride yields an octamethyl derivative, which, when hydrolyzed in warm aqueous acid, gives equimolar amounts of 2,3,4,6-tetra-O-methyl-D-glucose and 2,3,4-tri-O-methyl-D-glucose. Propose a structural formula for gentiobiose.

Gentiobiose consists of two units of D-glucopyranose joined by a β-1,6-glycoside bond. Drawn below is the beta anomer of gentiobiose.

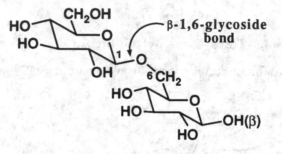

Problem 24.49 Following is the structural formula of laetrile.

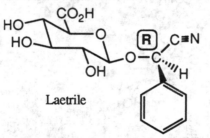

Laetrile

(a) Assign an R or S configuration to the stereocenter bearing the cyano (-CN) group.

The stereocenter has the R-configuration.

(b) Account for the fact that on hydrolysis in warm aqueous acid, laetrile liberates benzaldehyde and HCN.

Hydrolysis of the glycoside bond in aqueous acid gives D-glucuronic acid and the cyanohydrin of benzaldehyde. This cyanohydrin is in equilibrium with benzaldehyde and HCN.

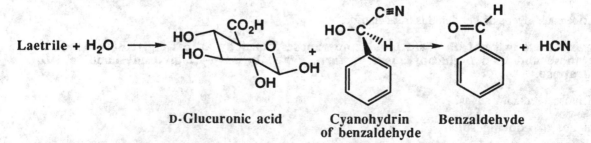

D-Glucuronic acid   Cyanohydrin   Benzaldehyde
                         of benzaldehyde

Polysaccharides
Problem 24.50 Following is the Fischer projection for N-acetyl-D-glucosamine:

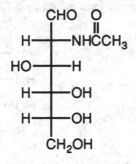

N-Acetyl-D-glucosamine

(a) Draw a chair conformation for the α- and β-pyranose forms of this monosaccharide.

Following are Haworth and chair formulas for the β-pyranose form of this monosaccharide. To draw the α-pyranose form, invert configuration at carbon-1.

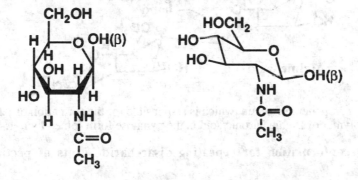

(b) Draw a chair conformation for the disaccharide formed by joining two units of the pyranose form of N-acetyl-D-glucosamine by a β-1,4-glycoside bond. If you drew this correctly, you have the structural formula for the repeating dimer of chitin, the structural polysaccharide component of the shell of lobster and other crustaceans.

Following are Haworth and chair formulas for the β-anomer of this disaccharide.

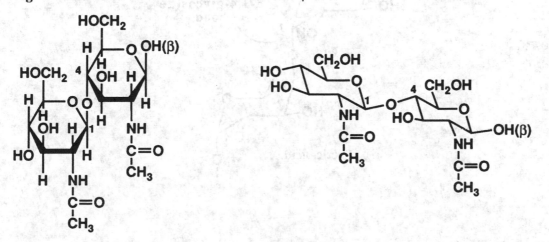

Problem 24.51 Propose structural formulas for the following polysaccharides:
(a) Alginic acid, isolated from seaweed, is used as a thickening agent in ice cream and other foods. Alginic acid is a polymer of D-mannuronic acid in the pyranose form joined by β-1,4-glycoside bonds.

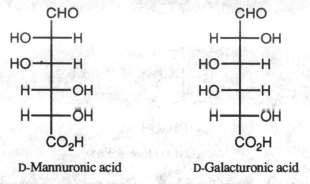

D-Mannuronic acid   D-Galacturonic acid

Following is the chair conformation for repeating disaccharide units of alginic acid.

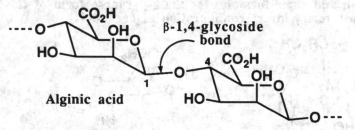

(b) Pectic acid is the main component of pectin, which is responsible for the formation of jellies from fruits and berries. Pectic acid is a polymer of D-galacturonic acid in the pyranose form joined by α-1,4-glycoside bonds.

Following is the chair conformation for repeating disaccharide units of pectic acid.

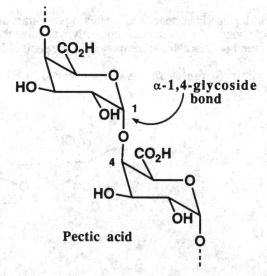

<u>Problem 24.52</u> Certain types of streptococci found in the mouth, especially *Streptococcus mutans*, have an enzyme system that uses sucrose as a starting material for the synthesis of high-molecular-weight polysaccharides known as dextrans. About 10% of the dry weight of dental plaque is composed of dextran. In one study of the dextran composition of dental plaque, dextran was methylated with methyl iodide in the presence of NaH and then the permethylated polysaccharide was hydrolyzed in dilute aqueous acid. The only monosaccharides obtained were the following four O-methyl derivatives of D-glucose.

Methylated D-glucose	Mole %
2,3,4,6-Tetra-O-methyl-D-glucose	14.6
2,4,6-Tri-O-methyl-D-glucose	50.5
2,3,4-Tri-O-methyl-D-glucose	20.9
2,4-Di-O-methyl-D-glucose	14.0

(a) Draw the structural formula of the open-chain form of each of these derivatives of D-glucose.

Following are Fischer projection formulas for each methylated derivative.

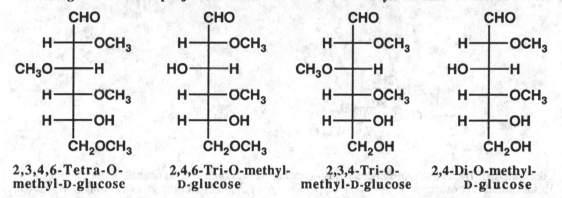

2,3,4,6-Tetra-O-methyl-D-glucose   2,4,6-Tri-O-methyl-D-glucose   2,3,4-Tri-O-methyl-D-glucose   2,4-Di-O-methyl-D-glucose

(b) The isolation of one of these derivatives is evidence that one group of glucose units in this dextran participates in only 1,3-glucoside bonds. Explain. What is the percentage of 1,3-glycoside bonds?

Following is a chair conformation of a unit of D-glucose bonded to two other monosaccharides, one by a glycoside bond to carbon-1, the other by a glycoside bond to carbon-3. After permethylation of the dextran followed by acid-catalyzed hydrolysis of glycoside bonds, this unit is isolated as 2,4,6-tri-O-methyl-D-glucose. Isolation of 50.5% of this derivative indicates that approximately half of the glycoside bonds are 1,3-glycoside bonds.

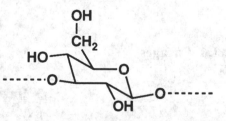

(c) The isolation of a second derivative of D-glucose is evidence that a second group of glucose units participates in only 1,6-glycoside bonds. Explain. What is the percentage of 1,6-glycoside bonds?

This glucose unit, after permethylation and acid-catalyzed hydrolysis of the fully methylated dextran, gives 2,3,4-tri-O-methyl-D-glucose, indicating that 20.9% of the glycoside bonds are 1,6-glycosides.

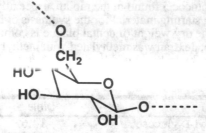

(d) The isolation of a third glucose derivative is evidence that a third group of glucose units participates in both 1,3- and 1,6-glycoside bonds and, therefore, serve as branch points in the polysaccharide chain. Explain. What is the percentage of chain branching?

This glucose unit participates in glycoside bonds at the anomeric carbon as well as at carbon-3 and carbon-6. After permethylation and acid-catalyzed hydrolysis it is isolated as 2,4-di-O-methyl-D-glucose indicating that approximately 14% of the glucose units are points of chain branching.

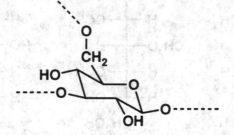

(e) The fourth derivative of D-glucose represents the monosaccharide end of branched chains. Compare the percentage of this terminal monosaccharide unit with the percentage of chain branching you determined in part (d).

Terminal glucose units are isolated as 2,3,4,6-tetra-O-methyl-D-glucose. The percent of terminal glucose units (14.6%) is almost identical to the degree of chain branching (14%).

(f) From all of this evidence, sketch the polysaccharide of dextran in the same manner as amylopectin is sketched in Figure 24.9.

Based on the information provided in the question, a reasonable cartoon for dextran is as follows. The main chain contains 1,3-linkages, with terminated branches that contain 1,6-linkages.

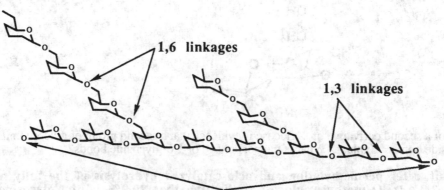

continuing polysaccharide chain

Problem 24.53 Digitalis is a preparation made from the dried seeds and leaves of the purple foxglove, *Digitalis purpurea* , a plant native to southern and central Europe and cultivated in the United States. The preparation is a mixture of several active components, including digitalin: Digitalis is used in medicine to increase the force of myocardial contraction and as a conduction depressant to decrease heart rate (the heart pumps more forcefully but less often).

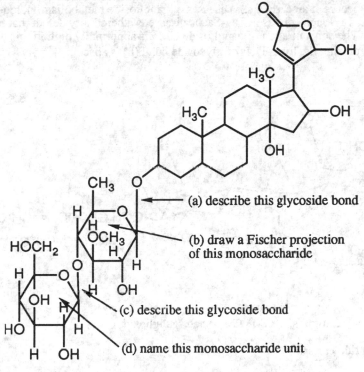

(a) describe this glycoside bond

(b) draw a Fischer projection of this monosaccharide

(c) describe this glycoside bond

(d) name this monosaccharide unit

Digitalin

a) **The indicated bond is a β-glycoside (the oxygen is equatorial).**
b) **The first monosaccharide corresponds to the following Fischer projection.**

$$
\begin{array}{c}
\text{O}\!\!=\!\!\text{C}\!\!-\!\!\text{H} \\
\text{H}\!-\!\text{C}\!-\!\text{OH} \\
\text{CH}_3\text{O}\!-\!\text{C}\!-\!\text{H} \\
\text{H}\!-\!\text{C}\!-\!\text{OH} \\
\text{H}\!-\!\text{C}\!-\!\text{OH} \\
\text{CH}_3
\end{array}
$$

c) **This bond is a β-1,4-glycoside bond.**
d) **This monosaccharide is glucose.**

<u>Problem 24.54</u> Following is the structural formula of ganglioside GM$_2$, a macromolecular glycolipid (meaning that it contains a lipid and monosaccharide units joined by glycoside bonds). In normal cells, this and other gangliosides are synthesized continuously and degraded by lysosomes, which are cell organelles containing digestive enzymes. If pathways for the degradation of gangliosides are inhibited, the gangliosides accumulate in the central nervous system causing all sorts of life-threatening consequences. In inherited diseases of ganglioside metabolism, death usually occurs at an early age. Diseases of ganglioside metabolism include Gaucher's disease, Niemann-Pick disease, and Tay-Sachs disease. Tay-Sachs disease is a hereditary defect that is transmitted as an autosomal recessive gene. The concentration of ganglioside GM$_2$ is abnormally high in this disease because the enzyme responsible for catalyzing the hydrolysis of glycoside bond (b) is absent.

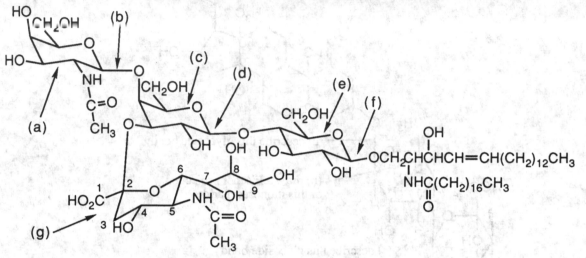

Ganglioside GM$_2$ (Tay-Sachs ganglioside)

(a) Name this monosaccharide unit.

This monosaccharide is *N*-acetyl-D-galactosamine

(b) Describe this glycoside bond (α or β, and between which carbons of each unit).

Because the group is equatorial, this is a β-1,4-glycoside bond.

(c) Name this monosaccharide unit.

This monosaccharide is D-galactose.

(d) Describe this glycoside bond.

This is also a β-1,4-glycoside bond.

(e) Name this monosaccharide unit.

This monosaccharide is D-glucose

(f) Describe this glycoside bond.

This is a β-glycoside bond.

(g) This unit is *N*-acetylneuraminic acid, the most abundant member of a family of amino sugars containing nine or more carbons and distributed widely throughout the animal kingdom. Draw the open-chain form of this amino sugar. Do not be concerned with the configuration of the five stereocenters in the open-chain form.

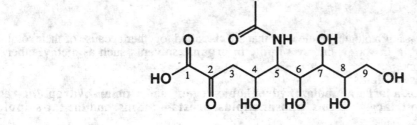

CHAPTER 25: LIPIDS
Outline of Important Concepts

25.0 OVERVIEW
• **Lipids** are a heterogeneous class of biological molecules that are classified together because of their solubility properties. They are **insoluble in water**, but are **soluble in organic solvents** such as diethyl ether, methylene chloride, and acetone. ✳
• There are **two main classes of lipids:**
  - The first class of lipids has a **large nonpolar hydrophobic region, and a polar hydrophilic region.** Members in this class are **triacylglycerols, phospholipids, prostaglandins,** and the **fat-soluble vitamins.**
  - Molecules in the second class, such as **cholesterol and compounds derived from it, contain a tetracyclic steroid ring nucleus.**

25.1 FATTY ACIDS
• **Fatty acids** are **long-chain monocarboxylic acids** produced by the **hydrolysis of fats** and oils. ✳
  - Nearly all fatty acids have an **even number of carbon atoms**, the most abundant are C_{16} (palmitic acid) and C_{18} (stearic and oleic acids). Different fatty acids can have **different numbers of double bonds**. The number of carbon atoms in the chain and the number of double bonds are separated by a colon when they are named. For example, a 16:2 fatty acid contains sixteen carbon atoms and two double bonds.
  - In most natural **unsaturated fatty acids**, the **Z (*cis*) isomer predominates**, and the **E (*trans*) isomer** is very **rare**. Because they are bent and cannot pack together well, the Z unsaturated fatty acids and molecules that contain them have lower melting points than analogous saturated fatty acids.

25.2 SOAPS, AND DETERGENTS
• **Natural soaps** are the sodium or potassium salts of fatty acids. They can be prepared from the base-promoted hydrolysis of the ester functions in triacylglycerols, a process called **saponification**. ✳
• **Soaps** act as **cleansing agents** because the long **hydrocarbon chains tend to cluster**, while the polar carboxylate groups remain in contact with the water. The so-called **micelle** structures that are formed "dissolve" nonpolar substances such as dirt and grease in the hydrophobic interior. **Natural soaps** form **insoluble salts** with the ions found in hard water such as **Ca(II), Mg(II),** or **Fe(III)**, leading to **soap scum**.
• **Synthetic detergents** are analogous to natural soaps except the polar group is a sulfonate, not a carboxylate. The synthetic detergents have the advantage that they do not readily form insoluble salts in hard water.
• **Triacylglycerols**, also called **triglycerides**, are triesters of glycerol and fatty acids.
  - Triacylglycerols rich in unsaturated fatty acids such as oleic and linoleic acids are generally liquids at room temperature, and are referred to as **oils**. Triacylglycerols rich in saturated fatty acids are generally solids or semisolids at room temperature, since the saturated fatty acid chains can pack together well. These triacylglycerols are called **fats**. ✳

25.3 PROSTAGLANDINS
• **Prostaglandins** are a class of compounds that have the **20-carbon skeleton** of **prostanoic acid**. Different prostaglandins have different biological activities, and they are usually very potent. As a result, much research has been invested in the understanding of the biological action of natural and synthetic prostaglandins. Prostaglandins are part of a larger family of biological molecules called **eicosanoids** that include **prostacyclins, thromboxanes,** and **leukotrienes**. ✳

25.4 STEROIDS
• **Steroids** are a group of lipids that have the characteristic tetracyclic steroid ring system. Cholesterol, a component of biological membranes, is a precursor to other important classes of steroids including the **androgens** (male sex hormones), **estrogens** (female sex hormones), **glucocorticoid hormones,** and **mineralocorticoid hormones**. ✳
• Steroids such as cholesterol are built up from two-carbon units derived from **acetyl CoA** via several intermediate steps. Various intermediate structures are produced along the way including **geranyl pyrophosphate, farnesyl pyrophosphate,** and **squalene**. The squalene is enzymatically oxidized to give an epoxide, that opens to give a tertiary carbocation. Formation of this carbocation sets into motion a very remarkable number of probably concerted chemical steps including four concerted cation-initiated cyclizations

and four 1,2 shifts. The product of this sequence, **lanosterol**, is converted to cholesterol though about 25 more enzyme-catalyzed steps.

25.5 PHOSPHOLIPIDS

- **Phospholipids** are derivatives of **phosphatidic acid**, having **glycerol esterified** with **two fatty acid molecules** and **one molecule of phosphoric acid**. Fatty acids such as palmitic, stearic, and oleic acid are the most common. The phosphoric acid group can be attached to other groups such as ethanolamine, choline, serine, or inositol. The phosphoric acid group is negatively charged at neutral pH. Thus, phospholipids have the long hydrophobic chains of the fatty acids, but also the very hydrophilic and charged phosphoric acid group.
- The unique structure of the **phospholipids** allows them to **self-assemble** in water to give a **bilayer** in which the polar headgroups lie on the surface exposed to the water molecules, and the hydrophobic fatty acid alkyl chains are buried within the bilayer. The hydrophobic interior of the bilayer can vary from rigid to fluid. Saturated fatty acid chains can pack together more easily, so they make more rigid bilayers than the kinked chains of unsaturated fatty acids.
- According to the **fluid mosaic model** of biological membranes, the membrane is composed of a phospholipid bilayer with membrane proteins associated on the inside and outside surfaces. Some proteins can also span the distance from the inside to the outside of the membrane bilayer.

25.6 FAT-SOLUBLE VITAMINS

- **Vitamins** are classified as either water-soluble or fat-soluble. The fat-soluble vitamins include **vitamin A** (important in the visual cycle of rod cells), **vitamin D** (important for the regulation of calcium and phosphorus metabolism), **vitamin E** (an antioxidant also important for red blood cell membranes), and **vitamin K** (important for blood clotting).

CHAPTER 25
Solutions to the Problems

<u>Problem 25.1</u> How many isomers are possible for a triglyceride containing one molecule each of palmitic, oleic, and stearic acids?

There are three constitutional isomers possible, the difference being which fatty acid is in the middle of the molecule:

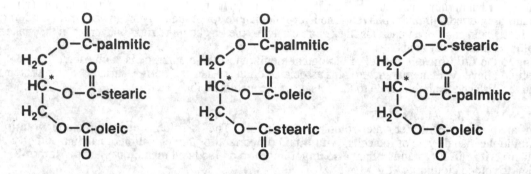

Each of these molecules has one stereocenter as indicated by the asterisk, so each constitutional isomer shown above can exist as a pair of enantiomers. Thus there are 2 x 3 = 6 total isomers possible. Note that for oleic acid, the carbon-carbon double bond is assumed to have the Z (*cis*) configuration only.

(b) Which of these constitutional isomers are chiral?

They are all chiral because they each have a stereocenter at the middle carbon of the glycerol moiety.

<u>Problem 25.2</u> Identify the hydrophobic region(s) and the hydrophilic region(s) of a triglyceride.

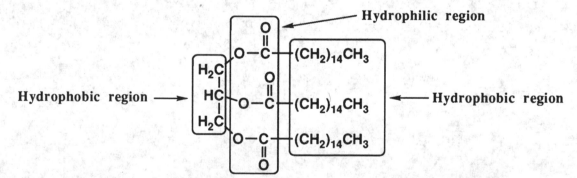

Note that the vast majority of the molecule is hydrophobic, thus explaining why triglycerides are so hydrophobic overall.

<u>Problem 25.3</u> Explain why the melting points of unsaturated fatty acids are lower than those of saturated fatty acids.

When fatty acids pack together better, the attractive dispersion forces between molecules are stronger, thereby increasing the melting point. Saturated fatty acids can adopt a much more compact structure compared to unsaturated fatty acids that have a kink induced by the *cis* double bond. The more compact saturated fatty acids can pack together better, so their melting points are higher.

Problem 25.4 Which would you expect to have the higher melting point, glyceryl trioleate or glyceryl trilinoleate?

The triglyceride with fewer *cis* double bonds will have the higher melting point. Each oleic acid unit has only one *cis* double bond, while each linoleic acid has two (Please see Table 25.1). Glycerol trioleate will have the higher melting point.

Problem 25.5 Explain why olive oil solidifies in the refrigerator, but corn oil does not.

Corn oil remains liquid because it has a melting temperature that is lower than the standard refrigerator temperature (4°C), while olive oil solidifies because it has a melting temperature that is above this temperature. The difference is due to the fact that corn oil has a significantly higher percentage of linoleic acid than olive oil (Table 25.2). Linoleic acid has 2 *cis* double bonds to disrupt chain packing and thereby raise the melting point.

Problem 25.6 Draw a structural formula for methyl lineolate. Be certain to show the correct stereochemistry about the carbon-carbon double bonds.

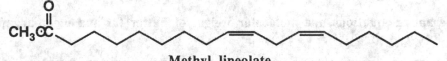

Methyl lineolate

Problem 25.7 Explain why coconut oil is a liquid triglyceride, even though most of its fatty acid components are saturated.

Triglycerides having fatty acids with shorter chains have lower melting points. As can be seen in Table 25.2, coconut oil is 45% lauric acid. Lauric acid is only a C12 fatty acid, so coconut oil has a melting point that is low enough to make it a liquid near room temperature.

Problem 25.8 What is meant by the term "hardening" as applied to fats and oils?

The term "hardening" refers to the process of catalytic hydrogenation using H_2 and a transition metal with polyunsaturated plant oils. By removing the (Z) double bonds, the reduction reaction allows the fatty acids to pack together better and thus the triacylglycerols become more solid.

Problem 25.9 How many mol of H_2 are used in the catalytic hydrogenation of 1 mol of a triglyceride derived from glycerol, stearic acid, linoleic acid, and arachidonic acid?

One molecule of H_2 is used per double bond in the triglyceride. Stearic acid does not have any double bonds, linoleic acid has 2 and arachidonic acid has 4 double bonds, respectively. Thus, 2 + 4 = 6 mol of H_2 will be used per mol of the triglyceride.

Problem 25.10 Saponification number is defined as the number of milligrams of potassium hydroxide required for saponification of 1.00 g of fat or oil
(a) Write a balanced equation for the saponification of tristearin.

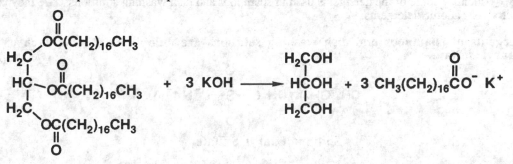

(b) The molecular weight of tristearin is 890 g/mol. Calculate the saponification number of tristearin.

The molecular weight of potassium hydroxide is 56 g/mol.

$$(3 \times 56 \text{ g/mol})\left(\frac{1 \text{ g}}{890 \text{ g/mol}}\right) = 0.189 \text{ g}$$

Therefore, as determined in the above equation, it would take 189 milligrams of KOH to saponify 1 g of tristearin, so it has a saponification number of 189.

<u>Problem 25.11</u> The saponification number of butter fat is approximately 230; that of oleomargarine is approximately 195. Calculate the average molecular weight of butter fat and of oleomargarine.

$$\frac{(3 \times 56 \text{ g/mol})(1 \text{ g})}{0.230 \text{ g}} = 730 \text{ g/mol} \qquad \frac{(3 \times 56 \text{ g/mol})(1 \text{ g})}{0.195 \text{ g}} = 862 \text{ g/mol}$$

As shown in the above equations, the molecular weight of butter fat and oleomargarine are 730 g/mol and 862 g/mol, respectively.

<u>Problem 25.12</u> Characterize the structural features necessary to make a good synthetic detergent.

A good synthetic detergent should have a long hydrocarbon tail and a very polar group at one end. This combination will allow for the production of micelle structures in aqueous solution that will dissolve hydrophobic dirt such as grease and oil. The very polar group should not form insoluble salts with the ions normally found in hard water such as Ca(II), Mg(II), and Fe(III).

<u>Problem 25.13</u> Following are structural formulas for a cationic detergent and a neutral detergent. Account for the detergent properties of each.

$$CH_3(CH_2)_6CH_2\overset{\displaystyle CH_3}{\underset{\displaystyle CH_2C_6H_5}{\overset{+}{N}}}CH_3 \quad Cl^- \qquad\qquad HOCH_2\overset{\displaystyle HOCH_2}{\underset{\displaystyle HOCH_2}{C}}CH_2O\overset{O}{\overset{\|}{C}}(CH_2)_{14}CH_3$$

Benzyldimethyloctylammonium chloride   Pentaerythrityl palmitate
(a cationic detergent)   (a neutral detergent)

In each case there is a long hydrocarbon tail attached to a very polar group. This combination will allow for the production of micelle structures in aqueous solution that will dissolve nonpolar, hydrophobic dirt such as grease and oil. In the case of benzyldimethyloctylammonium chloride, the polar group is the positively-charged ammonium group, while for the pentaerythrityl palmitate the polar group is composed of the triol functions.

<u>Problem 25.14</u> Identify some of the detergents used in shampoos and dish washing solutions. Are they primarily anionic, neutral, or cationic detergents.

Most detergents in shampoos and dish washing solutions are anionic detergents such as sodium lauryl sulfate.

$$CH_3(CH_2)_{10}CH_2O-\overset{O}{\underset{O}{\overset{\|}{\underset{\|}{S}}}}-O^- \ Na^+$$

Sodium Lauryl Sulfate

Problem 25.15 Show how to convert palmitic acid (hexadecanoic acid) into the following:
(a) Ethyl palmitate

$$CH_3(CH_2)_{14}\overset{\overset{O}{\|}}{C}OH \quad + \quad CH_3CH_2OH \quad \xrightarrow{\;H^+\;} \quad CH_3(CH_2)_{14}\overset{\overset{O}{\|}}{C}OCH_2CH_3$$

Ethyl palmitate

(b) Palmitoyl chloride

$$CH_3(CH_2)_{14}\overset{\overset{O}{\|}}{C}OH \quad + \quad SOCl_2 \quad \longrightarrow \quad CH_3(CH_2)_{14}\overset{\overset{O}{\|}}{C}Cl$$

Palmitoyl chloride

(c) 1-Hexadecanol (cetyl alcohol)

$$CH_3(CH_2)_{14}\overset{\overset{O}{\|}}{C}OH \quad \xrightarrow[\text{2) } H_2O]{\text{1) } LiAlH_4,\ \text{ether or THF}} \quad CH_3(CH_2)_{14}CH_2OH$$

**1-Hexadecanol
(Cetyl alcohol)**

(d) 1-Hexadecamine

$$CH_3(CH_2)_{14}\overset{\overset{O}{\|}}{C}OH \quad + \quad SOCl_2 \quad \longrightarrow \quad CH_3(CH_2)_{14}\overset{\overset{O}{\|}}{C}Cl \quad \xrightarrow{\;NH_3\;}$$

$$CH_3(CH_2)_{14}\overset{\overset{O}{\|}}{C}NH_2 \quad \xrightarrow[\text{2) } H_2O]{\text{1) } LiAlH_4,\ \text{ether or THF}} \quad CH_3(CH_2)_{14}CH_2NH_2$$

1-Hexadecanamine

(e) *N,N*-Dimethylhexadecanamide

$$CH_3(CH_2)_{14}\overset{\overset{O}{\|}}{C}OH \quad + \quad SOCl_2 \quad \longrightarrow \quad CH_3(CH_2)_{14}\overset{\overset{O}{\|}}{C}Cl$$

$$CH_3(CH_2)_{14}\overset{\overset{O}{\|}}{C}Cl \quad + \quad HN(CH_3)_2 \quad \longrightarrow \quad CH_3(CH_2)_{14}\overset{\overset{O}{\|}}{C}N(CH_3)_2$$

***N,N*-Dimethylhexadecanamide**

Problem 25.16 Palmitic acid (hexadecanoic acid) is the source of the hexadecyl (cetyl) group in the following compounds. Each is a mild surface-acting germicide and fungicide and is used as a topical antiseptic and disinfectant.

Cetylpyridinium chloride   Benzylcetyldimethylammonium chloride

(a) Cetylpyridinium chloride is prepared by treating pyridine with 1-chlorohexadecane (cetyl chloride). Show how to convert palmitic acid to cetyl chloride.

$$CH_3(CH_2)_{14}\overset{\overset{O}{\|}}{C}OH \xrightarrow[\text{2) } H_2O]{\text{1) } \textbf{LiAlH}_4, \textbf{ ether or THF}} CH_3(CH_2)_{11}CH_2OH \xrightarrow{\textbf{SOCl}_2}$$

$$CH_3(CH_2)_{14}CH_2Cl$$

1-Chlorohexadecane
(Cetyl chloride)

(b) Benzylcetyldimethylammonium chloride is prepared by treating benzyl chloride with N,N-dimethyl-1-hexadecanamine. Show how this tertiary amine can be prepared from palmitic acid.

$$CH_3(CH_2)_{14}\overset{\overset{O}{\|}}{C}OH + SOCl_2 \longrightarrow CH_3(CH_2)_{14}\overset{\overset{O}{\|}}{C}Cl \xrightarrow{HN(CH_3)_2}$$

$$CH_3(CH_2)_{14}\overset{\overset{O}{\|}}{C}N(CH_3)_2 \xrightarrow[\text{2) } H_2O]{\text{1) } \textbf{LiAlH}_4, \textbf{ ether or THF}} CH_3(CH_2)_{14}CH_2N(CH_3)_2$$

N,N-**Dimethyl-1-**
hexadecanamine

Problem 25.17 Lipases are enzymes that catalyze the hydrolysis of esters, especially esters of glycerol. Because enzymes are chiral catalysts, they catalyze the hydrolysis of only one enantiomer of a racemic mixture. For example, porcine pancreatic lipase catalyzes the hydrolysis of only one enantiomer of the following racemic epoxyester. Calculate the number of grams of epoxyalcohol that can be obtained from 100 g of racemic epoxy ester by this method.

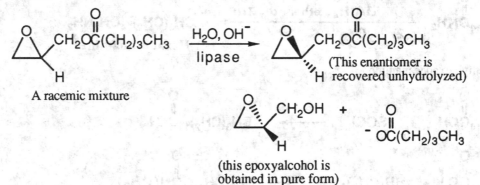

The molecular weight for the epoxyester starting material ($C_8H_{14}O_3$) is 158 g/mol, while the molecular weight of the epoxyalcohol product ($C_3H_6O_2$) is 74 g/mol. Because the starting material is racemic, there is actually only 100 g/2 or 50 g of the starting material epoxyester enantiomer that will be converted to product. As a result, the number of grams of the epoxyalcohol product is calculated as:

$$\frac{(50 \text{ g})(74 \text{ g}/\text{mol})}{(158 \text{ g}/\text{mol})} = 23.4 \text{ g}$$

Thus, 23.4 g of epoxyalcohol will be produced in this enzyme-catalyzed reaction.

Prostaglandins
Problem 25.18 Examine the structure of PGF$_{2\alpha}$ and
(a) Identify all stereocenters

The stereocenters are indicated with an asterisk.

(b) Identify all double bonds about which *cis-trans* isomerism occurs

These double bonds are indicated by the arrows.

(c) State the number of stereoisomers possible for a molecule of this structure.

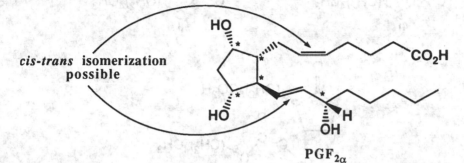

cis-trans **isomerization possible**

PGF$_{2\alpha}$

There are 2^5 stereoisomers possible and 2^2 *cis-trans* isomers possible for a grand total of 32 × 4 = 128 possible stereoisomers.

<u>Problem 25.19</u> Doxaprost, an orally active bronchodilator patterned after the natural prostaglandins (Section 25.3), is synthesized in the following series of reactions, starting with ethyl 2-oxycyclopentanecarboxylate. Except for the Nef reaction in Step 8, we have seen examples of all other types of reactions involved in this synthesis.

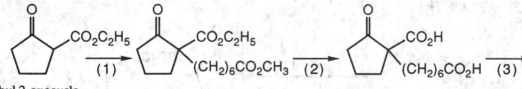

Ethyl 2-oxocyclo-
pentanecarboxylate

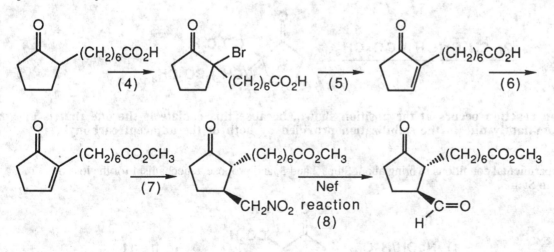

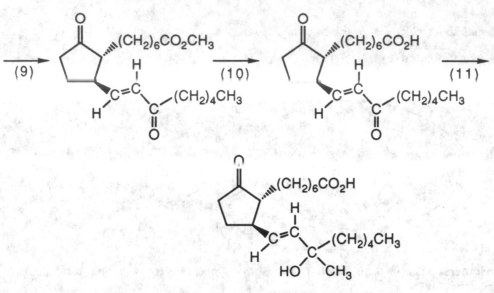

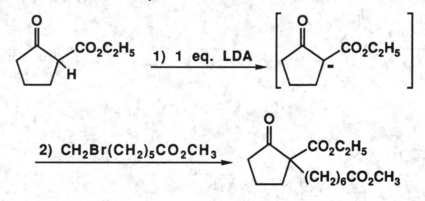

Doxaprost
(an orally active bronchodilator)

(a) Propose a set of experimental conditions to bring about the alkylation in Step 1. Account for the regioselectivity of the alkylation, that is, that it takes place on the carbon between the two carbonyl groups rather than on the other side of the ketone carbonyl.

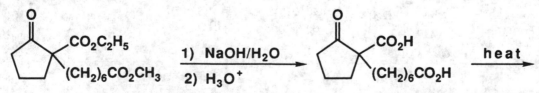

The alkylation reaction occurs at the position shown, because this enolate is the one that is formed predominantly due to the stabilization provided by both of the adjacent carbonyl functions.

(b) Propose experimental conditions to bring about Steps 2 and 3, and propose a mechanism for the loss of carbon dioxide in Step 3.

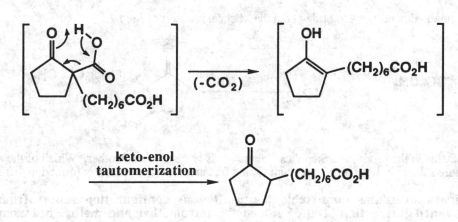

(c) Propose experimental conditions for bromination of the ring in Step 4 and dehydrobromination in Step 5.

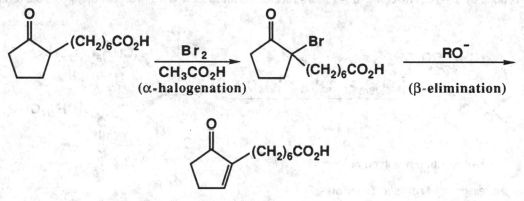

(d) Write equations to show that Step 6 can be brought about using either methanol or diazomethane (CH_2N_2) as a source of the -CH_3 in the methyl ester.

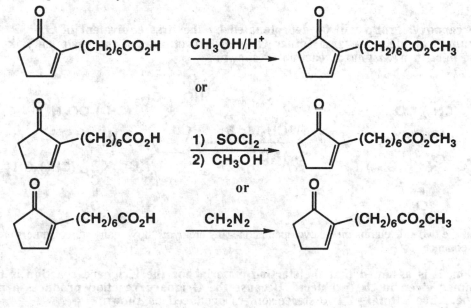

(e) Describe experimental conditions to bring about the Michael reaction of Step 7.

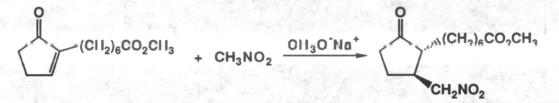

(f) The two side chains in the product of Step 7 can be either *cis* or *trans* to each other. Which of the two do you expect to be the more stable configuration? Account for the fact that the *trans* isomer is formed in this step.

The *trans* configuration is the more stable, because the *cis* configuration suffers from increased non-bonded interaction strain between the nitromethyl and methyl hexanoate groups.

(g) Step 9 is done by a Wittig reaction. Suggest a structural formula for a Wittig reagent that gives the product shown.

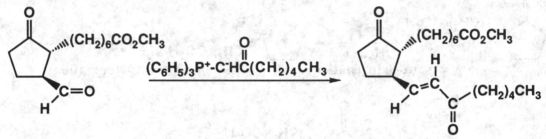

(h) Name the type of reaction involved in Step 10.

Step 10 is an ester hydrolysis reaction.

(i) Step 11 can best be described as a Grignard reaction with methylmagnesium bromide under very carefully controlled conditions. In addition to the observed reaction, what other Grignard reactions might take place in Step 11?

Of course, the carboxyl group will be deprotonated by the first equivalent of methylmagnesium bromide. As far as other Grignard reactions go, the cyclopentyl ketone might also take part in a Grignard reaction as shown:

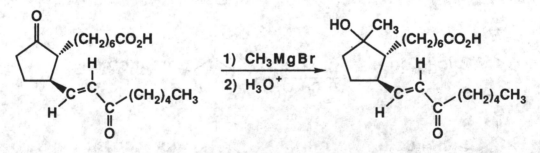

(j) Assuming that the two side chains on the cyclopentanone ring are *trans*, how many stereoisomers are produced in this synthetic sequence?

For this question, it is assumed that the starting material for the Grignard reaction is the single stereoisomer given in the problem. Because the Grignard reaction produces a new stereocenter, there are a total of two stereoisomers produced as shown:

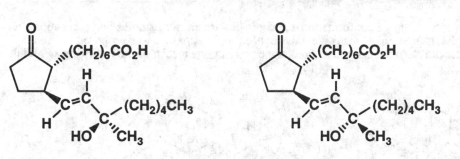

Steroids

Problem 25.20 Examine the structural formulas of testosterone (a male sex hormone) and progesterone (a female sex hormone). What are the similarities in structure between the two? What are the differences?

Overall, these structures are remarkably similar. Both of these steroids contain the standard four ring steroid structure with the axial methyl groups at C10 and C13. In addition, both structures contain an ene-one group in the A ring. On the other hand, the two structures differ in the nature of the D ring substituent at C17. In testosterone, the substituent is a hydroxyl group and in progesterone it is a ketomethyl group.

Problem 25.21 Examine the structural formula of cholic acid and account for the ability of this and other bile acids to emulsify fats and oils and thus aid in their digestion.

Cholic acid has the characteristic structure of a soap, so it can emulsify hydrophobic substances. In particular, cholic acid has a large hydrophobic steroid nucleus, and a highly polar carboxylate group.

Problem 25.22 Following is a structural formula for cortisol (hydrocortisone). Draw a conformational representation of this molecule.

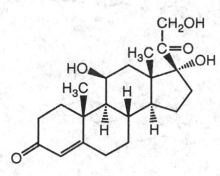

Cortisol
(Hydrocortisone)
structural formula

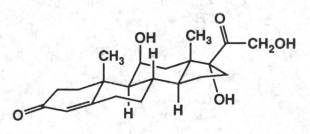

Cortisol
(Hydrocortisone)
conformational formula

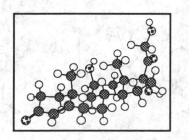

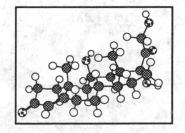

Problem 25.23 Much of our understanding of conformational analysis has arisen from studies on the reactions of rigid steroid nuclei. For example, the concept of *trans*-diaxial ring opening of epoxides was proposed to explain the stereospecific reactions seen with steroidal epoxides. Predict the product when each of the following steroidal epoxides is treated with LiAlH₄;

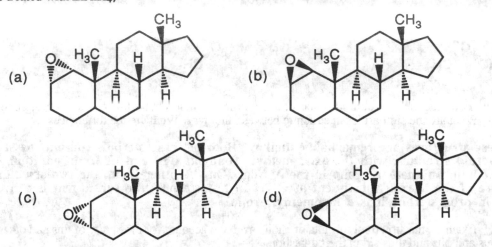

The idea here is that the hydride reagent attacks such that the epoxide opens to give a diaxial product. This dictates that only one product is observed for each reaction, the one that is shown below:

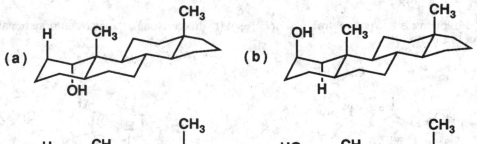

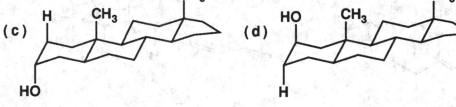

Problem 25.24 Addition of the HOCl in the following reaction is both regioselective and stereoselective. Only one stereoisomer is formed. Alternative regioisomers, as for example the regioisomer with -OH on carbon 5 and -Cl on carbon 6, are not formed.

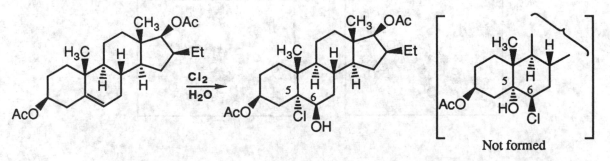

(a) Show the four stereoisomers that place -Cl and -OH *trans* to each other on carbons 5 and 6. (In two of these, -Cl is on carbon 5 and in the other two, it is on carbon 6).

Following are the four stereoisomers that place the -Cl and -OH *trans* to each other on carbons 5 and 6. Note that in structures C and D the A-B ring fusion is *cis*.

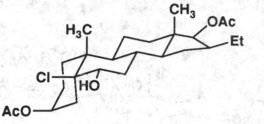

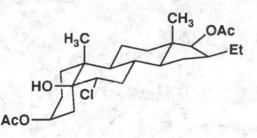

A (OH and Cl are *trans* diaxial)   **B (OH and Cl are *trans* diaxial)**

C (OH and Cl are *trans* diequatorial)   **D (OH and Cl are *trans* diequatorial)**

(b) Draw a conformational representation for the product formed in this reaction.

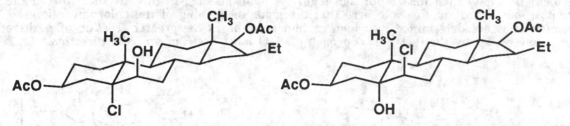

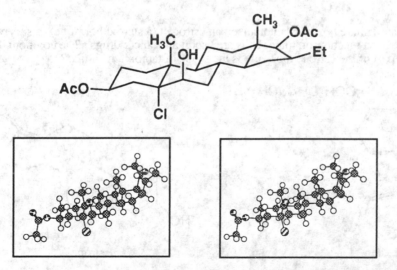

(c) According to the mechanism proposed in Section 5.3F, addition of HOCl is initiated by interaction of the alkene and chlorine to form a bridged chloronium ion intermediate. From which face of this steroid , top or bottom, is it more likely for chlorine to approach?

The chlorine atom is more likely to approach from the bottom, because approach from the top is partially blocked by the axial methyl group on C10.

(d) Show that both the regioselectivity and the stereoselectivity of this addition are consistent with the mechanism proposed in Section 5.3F for addition of HOCl to an alkene.

The mechanism proposed in Section 5.3F is fully consistent with formation of this product. Chlorine approaches from the bottom (the less–hindered side) to form a chloronium ion intermediate. H$_2$O then must approach from the opposite side, in this instance toward the axial position on carbon 6. Note here that the mode of opening of the chloronium ion intermediate is not determined by which carbon atom bears the greater fraction of positive charge, but rather by the fact that approach of H$_2$O must be from an axial direction.

Step 1:

Step 2:

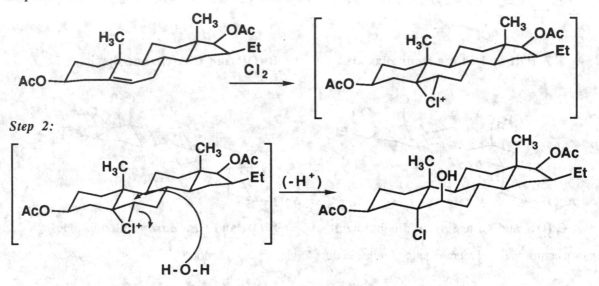

Problem 25.25 Because some types of tumors need an estrogen (a steroid hormone) to survive, compounds that compete with the estrogen receptor on tumor cells are useful anticancer drugs. The compound tamoxifen is one such drug. To what part of the estrone molecule is the shape of tamoxifen similar?

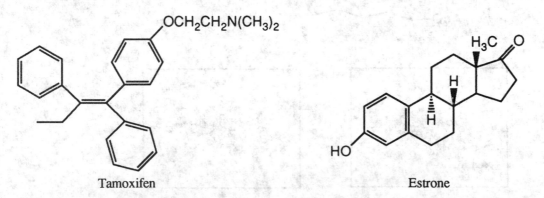

Tamoxifen   Estrone

Both tamoxifen and estrone are very hydrophobic. Drawn below are highlighted regions of tamoxifen and estrone that emphasize structural similarity. It should be pointed out that some liberties are taken in the following structures when it comes to some bond angles in tetrahedral and trigonal carbon atoms.

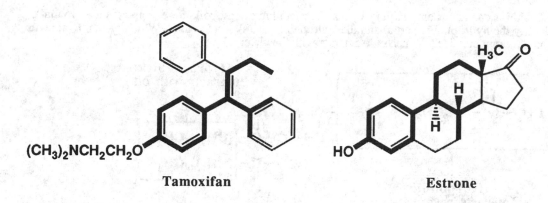

Tamoxifan   Estrone

Phospholipids

Problem 25.26 The hydrophobic effect is one of the most important noncovalent forces directing the self-assembly of biomolecules in aqueous solution. The hydrophobic effect arises from tendencies (1) to arrange polar groups so that they interact with the aqueous environment by hydrogen bonding and (2) to arrange nonpolar groups so that they are shielded from the aqueous environment. Show how the hydrophobic effect is involved in directing:

(a) Formation of micelles by soaps and detergents.

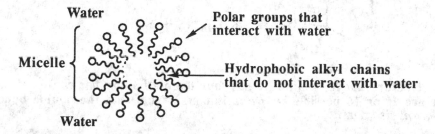

In micelles, the hydrophobic hydrocarbon tails are associated with each other to form the hydrophobic interior, while the polar groups are associated with each other on the outside surface where they interact with water.

(b) Formation of lipid bilayers by phospholipids.

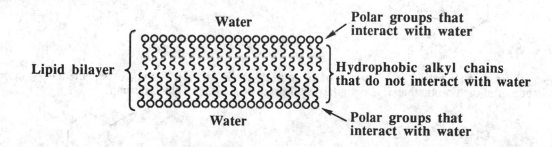

In lipid bilayers, the hydrophobic hydrocarbon tails are associated with each other to form the hydrophobic inner layer, while the polar head groups are associated with each other on both outside surfaces where they interact with water.

Problem 25.27 Lecithins can act as emulsifying agents. The lecithin of egg yolk, for example, is used to make mayonnaise. Identify the hydrophobic part(s) and the hydrophilic part(s) of a lecithin. Which parts interact with the oils used in making mayonnaise? Which parts interact with the water?

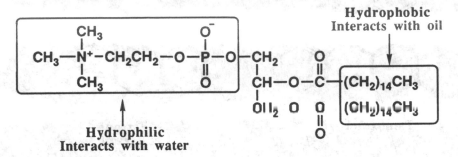

Fat-soluble Vitamins

Problem 25.28 Examine the structural formula of vitamin A and state the number of *cis-trans* isomers possible for this molecule.

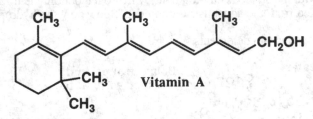

As shown in the structure above, vitamin A has four double bonds that can be either *cis* or *trans*, thus there are 2^4 or 16 possible *cis-trans* isomers. Note that the double bond in the ring cannot have *cis-trans* isomers.

Problem 25.29 The form of vitamin A present in many food supplements is vitamin A palmitate. Draw the structural formula of this molecule.

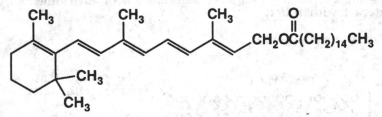

Problem 25.30 Examine the structural formulas of vitamins A, D_3, E, and K_1. Do you expect them to be more soluble in water or in dichloromethane? Do you expect them to be soluble in blood plasma?

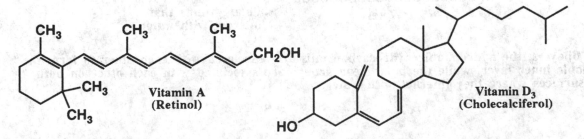

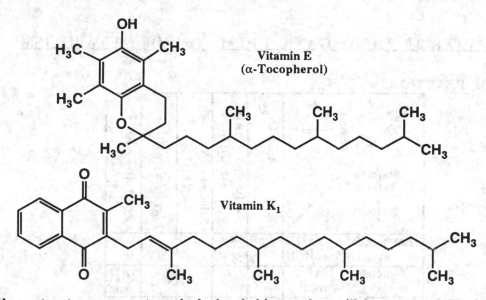

Vitamin E
(α-Tocopherol)

Vitamin K₁

All of these structures are extremely hydrophobic, so they will be more soluble in organic solvents such as dichloromethane than polar solvents such as water. Since blood plasma is an aqueous solution, these vitamins will only be sparingly soluble in blood plasma.

CHAPTER 26: THE ORGANIC CHEMISTRY OF METABOLISM

SUMMARY OF REACTIONS

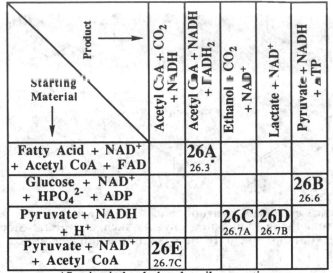

Starting Material \ Product →	Acetyl CoA + CO₂ + NADH	Acetyl CoA + NADH + FADH₂	Ethanol + CO₂ + NAD⁺	Lactate + NAD⁺	Pyruvate + NADH + ATP
Fatty Acid + NAD⁺ + Acetyl CoA + FAD		**26A** 26.3*			
Glucose + NAD⁺ + HPO₄²⁻ + ADP					**26B** 26.6
Pyruvate + NADH + H⁺			**26C** 26.7A	**26D** 26.7B	
Pyruvate + NAD⁺ + Acetyl CoA	**26E** 26.7C				

*Section in book that describes reaction.

REACTION 26A: β-OXIDATION OF FATTY ACIDS (Section 26.3)

$$CH_3(CH_2)_{14}\overset{O}{\overset{\|}{C}}OH \; + \; 8 \; CoA\text{-}SH \;\; \xrightarrow[\text{ATP} \quad \text{AMP} \; + \; P_2O_7^{4-}]{} \;\; 8 \; CH_3\overset{O}{\overset{\|}{C}}SCoA \; + \; 7 \; NADH$$

$$+ \; 7 \; NAD^+ \; + \; 7 \; FAD \qquad\qquad\qquad\qquad\qquad\qquad + \; 7 \; FADH_2$$

- The overall process of **fatty acid β-oxidation** involves **conversion of a fatty acid** in the presence of coenzyme A (CoA-SH) to molecules of the thioester species **acetyl coenzyme A** along with **reduction of NAD⁺ and FAD.** ✳
- The first step in the process is the conversion of the fatty acid into an activated form as the thioester derivative of CoA-SH in the cytoplasm. This reaction requires the hydrolysis of ATP to AMP and the pyrophosphate ion. An acyl-AMP mixed anhydride is an intermediate in the process.
- The activated fatty acid is transported into mitochondria where the following four enzyme-catalyzed reactions take place. *[It is very helpful to follow the chemical "logic" of the steps in fatty acid β-oxidation. Notice how the acyl fatty acid is first oxidized to a create a new double bond that is then hydrated and oxidized to set up the reverse Claisen type of cleavage reaction.]*
  1) The alpha- and beta- carbons of the fatty acid chain are oxidized to a double bond while FAD is reduced to FADH₂. Enzyme: fatty acyl-CoA dehydrogenase.
  2) The double bond is hydrated to give a β-hydroxyacyl-CoA. The hydroxyl group is added stereoselectively to carbon 3 producing exclusively the R stereoisomer. Enzyme: Enoyl-CoA hydrase.
  3) The β-hydroxy group is oxidized to a ketone while NAD⁺ is reduced to NADH. Enzyme: (R)-β-hydroxyacyl-CoA dehydrogenase.
  4) The carbon chain is cleaved between carbons 2 and 3 by an enzymatic reaction that is the functional equivalent of a reverse Claisen condensation using thioesters. The reaction produces the two-carbon fragment acetyl-CoA and an acyl-CoA that is now two carbons shorter than the original fatty acid. Enzyme: thiolase.
- This new shorter acyl-CoA undergoes additional cycles of reactions 1) - 4) until the entire fatty acid is converted to acetyl-CoA.

REACTION 26B: GLYCOLYSIS (Section 26.6)

$$C_6H_{12}O_6 + 2\,NAD^+ + 2\,HPO_4^{2-} \xrightarrow{\text{Glycolysis}} 2\,CH_3\overset{\displaystyle O}{\overset{\|}{C}}CO_2^- + 2\,NADH + 2\,ATP$$

Glucose   + 2 ADP   Pyruvate

- **Glycolysis** is a metabolic pathway for converting the monosaccharide **glucose** into **2 molecules of pyruvate**, along with the **reduction of two molecules of NAD$^+$ to NADH** and the **synthesis of two molecules of ATP** from 2 molecules of ADP and phosphoric acid. It is the synthesis of the **high energy phosphoric anhydride bonds of ATP** that accounts for the energy harvesting of glycolysis. From an evolutionary standpoint, glycolysis is a very old anaerobic metabolic process for producing energy from nutrient molecules and for providing precursors to aerobic pathways such as the tricarboxylic acid cycle. *
- The ten enzyme-catalyzed reactions of glycolysis are listed below. The first five steps are preparing the molecules for the last five steps, especially steps 7 and 10, which are the energy harvesting steps. *[It is very helpful to follow the chemical "logic" of the steps in glycolysis. Notice how the six carbon sugar is set up for cleavage via a reverse aldol type of reaction. Also notice how high energy bonds are formed then used to drive production of ATP. Finally, notice how keto-enol tautomerization is used several times during the process.]*
    1) Transfer of a phosphoric acid group from ATP to the -OH on C6 of glucose. This step is exothermic since a high energy phosphoric anhydride bond is broken and a lower energy phosphoric ester bond to glucose is created. Enzyme: hexokinase.
    2) The α-D-glucose-6-phosphate is isomerized to α-D-fructose-6-phosphate via a keto-enol tautomerization to form an enediol intermediate. Enzyme: phosphoglucoisomerase.
    3) α-D-Fructose-6-phosphate is phosphorylated at the -OH group on C1 to form α-D-fructose-1,6-diphosphate. Enzyme: phosphofructokinase.
    4) α-D-Fructose-1,6-diphosphate is cleaved into dihydroxyacetone phosphate and glyceraldehyde 3-phosphate. There is an imine intermediate in the reaction, which is the functional equivalent of a reverse aldol reaction. Enzyme: aldolase.
    5) Dihydroxyacetone phosphate is isomerized to glyceraldehyde 3-phosphate via a keto-enol tautomerization to form an enediol intermediate. Enzyme: triose phosphate isomerase.
    6) The aldehyde group of glyceraldehyde 3-phosphate is first oxidized to a carboxylic acid derivative in the form of a thioester with the thiol group of coenzyme A. NAD$^+$ is reduced to NADH in the process. The activated thioester is then converted to a high energy phosphoric anhydride, 1,3-bisphosphoglycerate. Enzyme: glyceraldehyde-3-phosphate dehydrogenase.
    7) The high energy acyl-phosphoric mixed anhydride bond of 1,3-bisphosphoglycerate is converted to a phosphoric anhydride bond of ATP to create 3-phosphoglycerate. Enzyme: phosphoglycerate kinase.
    8) 3-Phosphoglycerate is isomerized to 2-phosphoglycerate. Enzyme: phosphoglycerate mutase.
    9) The 3-OH group of 2-Phosphoglycerate is lost in a dehydration to give a 2-3 double bond in the product phosphoenolpyruvate. Note that the phosphate group keeps the molecule in the relatively high energy enol form. Enzyme: enolase.
    10) The phosphate group of phosphoenolpyruvate is transferred to ADP to generate a high energy phosphoric anhydride in ATP and the ketone pyruvate. Notice how pyruvate is just the more stable keto form of enol pyruvate. Enzyme: pyruvate kinase.

REACTION 26C: REDUCTION OF PYRUVATE TO ETHANOL - ALCOHOL FERMENTATION (Section 26.7B)

$$CH_3\overset{\displaystyle O}{\overset{\|}{C}}CO_2^- + 2\,H^+ + NADH \xrightarrow[\text{Fermentation}]{\text{Alcoholic}} CH_3CH_2OH + CO_2 + NAD^+$$

Pyruvate   Ethanol

- **Yeast and other organisms** can **convert pyruvate to ethanol and CO_2** in the **absence of oxygen**. That is why fermentation of beer and wine must be carried out in sealed vessels that exclude air. The CO_2 is responsible for the natural carbonation of beverages such as beer and champagne. There are two steps involved; a decarboxylation to give CO_2 and acetaldehyde followed by reduction of the acetaldehyde to ethanol. Enzymes: pyruvate decarboxylase, alcohol dehydrogenase. *

REACTION 26D: REDUCTION OF PYRUVATE TO LACTATE (Section 26.7A)

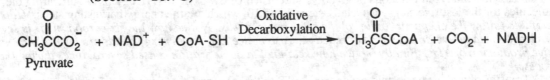

$$\underset{\text{Pyruvate}}{CH_3\overset{O}{\overset{\|}{C}}CO_2^-} + H^+ + NADH \underset{\text{Lactate}}{\overset{\text{Lactate}}{\overset{\text{Dehydrogenase}}{\rightleftharpoons}}} \underset{\text{Lactate}}{CH_3\overset{OH}{\overset{|}{C}H}CO_2^-} + NAD^+$$

- In vertebrates, **pyruvate is converted to lactate,** an **important anaerobic process** for **regenerating NAD+ from NADH.** In the overall process of lactate fermentation, glucose is converted all the way to two molecules of lactic acid, a relatively strong acid that is fully dissociated at the usual pH of blood. Thus, this process generates lactate and protons. The buildup of lactate is associated with muscle fatigue. Enzyme: lactate dehydrogenase. ✳

REACTION 26E: OXIDATIVE DECARBOXYLATION OF PYRUVATE TO ACETYL-CoA (Section 26.7C)

$$\underset{\text{Pyruvate}}{CH_3\overset{O}{\overset{\|}{C}}CO_2^-} + NAD^+ + CoA\text{-}SH \xrightarrow{\overset{\text{Oxidative}}{\text{Decarboxylation}}} CH_3\overset{O}{\overset{\|}{C}}SCoA + CO_2 + NADH$$

- **Under aerobic conditions, pyruvate is oxidized and decarboxylated to give Acetyl CoA** and CO_2. The acetyl-CoA then becomes fuel for the tricarboxylic acid cycle. Enzyme: pyruvate dehydrogenase complex.

SUMMARY OF IMPORTANT CONCEPTS

26.0 OVERVIEW
• The biochemical pathways of **metabolism** such as **β-oxidation of fatty acids and glycolysis** involve numerous enzyme-catalyzed reactions that operate in sequential fashion to effect complex overall processes. **These pathways are actually the biochemical equivalents of organic functional group reactions that have been covered in the previous chapters of the book.** ✳

26.1 FIVE KEY PARTICIPANTS IN GLYCOLYSIS AND β-OXIDATION
• **ATP, ADP, and AMP** are **phosphorylated** derivatives of the nucleoside **adenosine.** AMP has a single phosphoric acid group attached to the 5'-OH group of adenosine. ADP has an addition phosphoric acid group attached through a phosphoric anhydride bond. ATP has a third phosphoric acid group attached to the other two through a second phosphoric anhydride bond. ✳
  - **ATP, ADP, and AMP** are involved with **transfer and storage of phosphoric acid groups** within the cell. ✳
• **NAD+, nicotinamide adenine dinucleotide,** is composed of a unit of **ADP joined by a phosphoric ester bond** to the terminal -CH$_2$OH group of **β-D-ribofuranose** that is linked to **nicotinamide via a β-N-glycosyl bond. NAD+ is a two electron oxidizing agent,** since it is reduced by two electrons and a proton to give NADH. In NADH, the nicotinamide is reduced. Notice that in the following structure, the proton and electrons are shown as being independent of one another. In actual enzyme reactions, they can be considered to be combined in the form of a hydride, H-. ✳

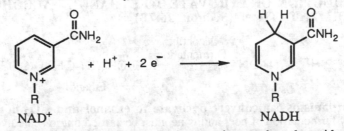

NAD+   NADH

  - NAD+ is a cofactor in enzymatic two-electron oxidation reactions such as the oxidation of a secondary alcohol to a ketone or the oxidation of an aldehyde to a carboxylic acid.
• **FAD, flavin adenine dinucleotide,** is composed of a **flavin group** attached to the five carbon sugar **ribotol,** that is in turn attached to the terminal phosphoric acid group of **ADP.** The flavin group of FAD is

reduced by the equivalent of two electrons and two protons to give $FADH_2$. In actual enzyme reactions, the 2 protons and 2 electrons can be thought of as a hydride, H^-, and a proton.

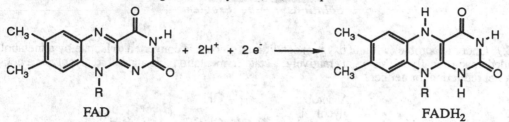

FAD $FADH_2$

- FAD is a cofactor in two-electron oxidation reactions such as the oxidation of a carbon-carbon single bond to a carbon-carbon double bond.

• **Every time a substrate is oxidized in the metabolic pathways, either FAD or NAD+ must be reduced, and *vice versa*. ✻**

26.2 FATTY ACIDS AS A SOURCE OF ENERGY

• **Fatty acids**, as triglycerides, are the **main storage form of energy** in most organisms. The -CH_2- groups of the fatty acid alkyl chains can be oxidized further than oxygenated species such as carbohydrates, so fatty acids are very potent sources of energy.

26.4 DIGESTION AND ABSORPTION OF CARBOHYDRATES

• **Carbohydrates** provide about **50 - 60% of daily energy needs**. The **carbohydrates** are consumed in the form of **disaccharides or polysaccharides**, which are first **hydrolyzed** to **monosaccharides** by enzymes called **glycosidases** in the mouth or small intestine. ✻

CHAPTER 26
Solutions to the Problems

Problem 26.1 Under anaerobic (without oxygen) conditions, glucose is converted to lactate by a metabolic pathway called anaerobic glycolysis or, alternatively, lactate fermentation. Is anaerobic glycolysis a net oxidation, a net reduction, or neither?

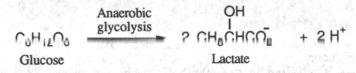

$$C_6H_{12}O_6 \xrightarrow[\text{glycolysis}]{\text{Anaerobic}} 2 \ CH_3CHCO_2^- + 2 \ H^+$$

Glucose   Lactate

The overall process of anaerobic glycolysis that converts glucose to lactate is neither an oxidation or a reduction. No electrons are involved in the balanced half-reaction.

Problem 26.2 Does lactate fermentation result in an increase or decrease in blood pH?

Lactate fermentation leads to an increase of the H+ concentration in the bloodstream, therefore the bloodstream pH must decrease.

Problem 26.3 Write structural formulas for palmitic, oleic, and stearic acids, the three most abundant fatty acids.

Palmitic and stearic acids are fully saturated, having 16 and 18 carbons in their chains, respectively. Oleic acid has 18 carbons and a single *cis* double bond.

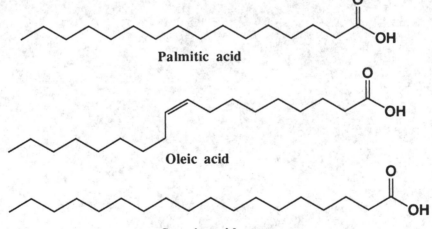

Palmitic acid

Oleic acid

Stearic acid

Problem 26.4 A fatty acid must be activated before it can be metabolized in cells. Write a balanced equation for the activation of palmitic acid.

Activation of a fatty acid involves formation of a thioester with coenzyme A. The proton is derived from the thiol group of CoA-SH.

$$CH_3(CH_2)_{14}\overset{O}{\overset{\|}{C}}O^- + CoA\text{-}SH + ATP \longrightarrow CH_3(CH_2)_{14}\overset{O}{\overset{\|}{C}}SCoA$$

Palmitic acid

$$+ \ AMP + P_2O_7^{4-} + H^+$$

Problem 26.5 Name three coenzymes necessary for β-oxidation of fatty acids. From what vitamin is each derived?

The three coenzymes needed for β-oxidation are:
  1) Coenzyme A (CoA-SH) derived from the vitamin pantothenic acid.
  2) Nicotine adenine dinucleotide (NAD⁺) derived from the vitamin niacin.
  3) Flavin adenine dinucleotide (FAD) derived from the vitamin riboflavin (vitamin B₂).
All three coenzymes contain a molecule of adenosine

Problem 26.6 We have examined β-oxidation of saturated fatty acids, such as palmitic acid and stearic acid. Oleic acid, an unsaturated fatty acid, is also a common component of dietary fats and oils. This unsaturated fatty acid is degraded by β-oxidation but, at one stage in its degradation, requires an additional enzyme named enoyl-CoA isomerase. Why is this enzyme necessary, and what isomerization does it catalyze? (*Hint:* Consider both the configuration of the carbon-carbon double bond in oleic acid and its position in the carbon chain.)

If you count the carbon atoms in oleic acid carefully, you will see that after three rounds of β-oxidation you are left with the following fragment that is then isomerized by enoyl-CoA isomerase to the *trans*-enoyl-CoA derivative needed for the next step of β-oxidation.

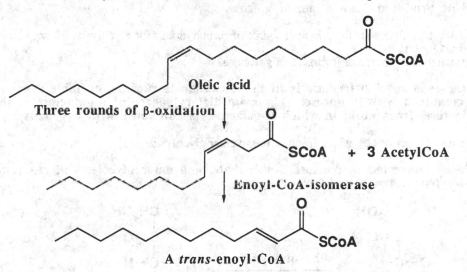

GLYCOLYSIS
Problem 26.7 Name one coenzyme required for glycolysis. From what vitamin is it derived?

The one coenzyme required for glycolysis is NAD⁺, which is derived from the vitamin niacin.

Problem 26.8 Number the carbon atoms of glucose 1 through 6 and show from which carbon atom the carboxyl group of each molecule of pyruvate is derived.

By numbering the carbon atoms of glucose and following the different atoms through the pathway it can be seen that the carboxyl group carbon atoms are derived from carbon atoms 3 and 4 of glucose.

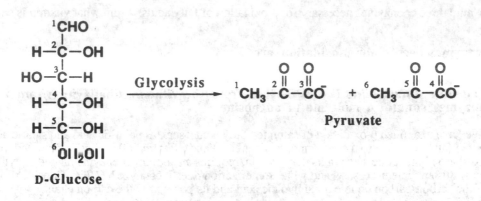

D-Glucose

Problem 26.9 How many mole of lactate are produced from three moles of glucose?

During anaerobic glycolysis, 2 mol of lactate are produce for each mol of glucose used. 6 mol of lactate will be produced from 3 mol of glucose.

Problem 26.10 Although glucose is the principal source of carbohydrates for glycolysis, fructose and galactose are also metabolized for energy.
(a) What is the main dietary source of fructose? of galactose?

The main dietary source of D-fructose is in the disaccharide sucrose, or table sugar, in which D-fructose is combined with D-glucose. The main dietary source of D-galactose is the disaccharide lactose, from milk, in which D-galactose is combined with D-glucose.

(b) Propose a series of reactions by which fructose might enter glycolysis.

Fructose could be converted to fructose 6-phosphate, and enter glycolysis at reaction 3, where it will be converted to fructose 1,6-bisphosphate.

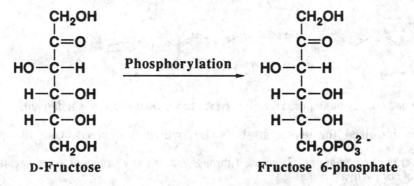

(c) Propose a series of reactions by which galactose might enter glycolysis.

D-Galactose can be isomerized at C-4 to produce D-glucose and thereby enter glycolysis at the beginning.

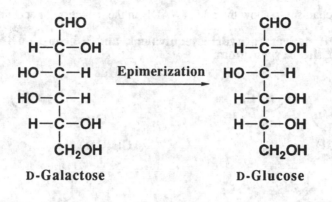

D-Galactose   D-Glucose

Problem 26.11 How many mol of ethanol are produced per mole of sucrose through the reactions of glycolysis and alcoholic fermentation? How many mol of CO_2 are produced?

A total of 4 mol of ethanol and 4 mol of carbon dioxide are produced from 1 mol of sucrose. This can be seen be remembering that 1 mol of the disaccharide sucrose is first hydrolyzed to 1 mol of glucose and 1 mol of fructose. Each of these 6-carbon monosaccharides enter glycolysis to give 2 mol of pyruvate, so a total of 4 mol of pyruvate are produced for each mol of sucrose used. Each mol of pyruvate is converted to 1 mol of ethanol and 1 mol of carbon dioxide, so a total of 4 mol of ethanol and 4 mol of carbon dioxide are produced for each mol of sucrose.

Problem 26.12 Glycerol is derived from hydrolysis of triglycerides and phospholipids. Propose a series of reactions by which the carbon skeleton of glycerol might enter glycolysis and be oxidized to pyruvate.

Glycerol enters glycolysis through the following enzyme catalyzed steps that lead to glyceraldehyde-3-phosphate, which is converted into pyruvate according to the normal glycolysis pathway.

Glycerol   Glycerol-3-   Dihydroxyacetone   Glyceraldehyde-3-
                phosphate   phosphate   phosphate

Problem 26.13 Ethanol is oxidized in the liver to acetate ion by NAD^+.
(a) Write a balanced equation for this oxidation.

The production of acetate ion from ethanol is an overall 4 electron process, so two mol of NAD^+ are required for every mole of ethanol. In addition, 2 protons are produced along with the proton that will dissociate from acetic acid to give acetate.

$$CH_3CH_2OH + 2\ NAD^+ \longrightarrow CH_3\overset{\overset{\displaystyle O}{\|}}{C}O^- + 2\ NADH + 3\ H^+$$

(b) Do you expect the pH of blood plasma to increase, decrease, or remain the same as a result of metabolism of a significant amount of ethanol?

The pH of blood plasma will drop due to the protons produced as the result of metabolism of a significant amount of ethanol.

<u>Problem 26.14</u> Write a mechanism to show the role of NADH in the reduction of acetaldehyde to ethanol.

For this reaction, NADH delivers a hydride equivalent, and a group on the enzyme (denoted as A) delivers a proton to the oxygen atom.

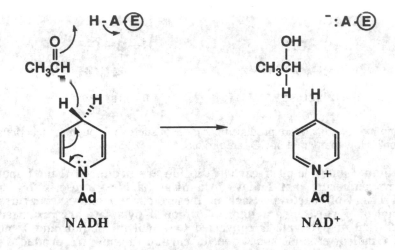

<u>Problem 26.15</u> When pyruvate is reduced to lactate by NADH, two hydrogens are added to pyruvate; one to the carbonyl carbon, the other to the carbonyl oxygen. Which of these hydrogens is derived from NADH?

As can be seen in the mechanism given in the answer to Problem 26.14, the NADH delivers a hydride equivalent, H-. This species is highly nucleophilic and reacts with the electrophilic carbonyl carbon atom.

<u>Problem 26.16</u> Review the oxidation reactions of glycolysis and β-oxidation and compare the types of functional groups oxidized by NAD^+ with those oxidized by FAD.

NAD^+ oxidizes a secondary alcohol to a ketone (reaction 3 of β-oxidation) as well as an aldehyde to a carboxylic acid derivative (reaction 6 of glycolysis). FAD oxidizes a carbon-carbon single bond to a carbon-carbon double bond (reaction 1 of β-oxidation).

<u>Problem 26.17</u> Why is glycolysis called an anaerobic pathway?

Glycolysis is called an anaerobic pathway because no oxygen is used. Glycolysis probably first evolved in organisms that appeared before there was oxygen in the environment.

<u>Problem 26.18</u> Which carbons of glucose appear as CO_2 as a result of alcoholic fermentation?

As shown in the answer to Problem 26.8, it is carbons 3 and 4 of D-glucose that end up as the carboxylic acid carbons of pyruvate. These same two carbon atoms, carbons 3 and 4, end up as CO_2 as a result of alcoholic fermentation.

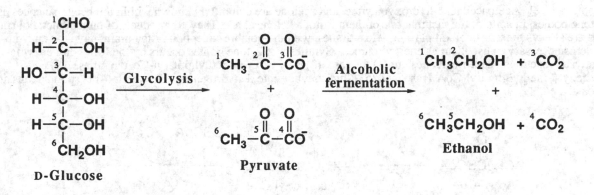

D-Glucose → Glycolysis → Pyruvate → Alcoholic fermentation → Ethanol

Problem 26.19 Which steps in glycolysis require ATP? Which steps produce ATP?

Reactions 1 and 3 of glycolysis require ATP, while reactions 7 and 10 produce ATP.

Problem 26.20 The respiratory quotient (RQ) is used in studies of energy metabolism and exercise physiology. It is defined as the ratio of the volume of carbon dioxide produced to the volume of oxygen used:

$$RQ = \frac{\text{Volume } CO_2}{\text{Volume } O_2}$$

(a) Show that RQ for glucose is 1.00. (*Hint:* Look at the balanced equation for complete oxidation of glucose to carbon dioxide and water.)

In the balanced reaction for the complete oxidation of glucose into CO_2 and H_2O, 6 mol of O_2 are used and 6 mol of CO_2 are produced, so the RQ is 6/6 = 1.00.

$$C_6H_{12}O_6 + 6\ O_2 \longrightarrow 6\ CO_2 + 6\ H_2O$$
D-Glucose

(b) Calculate RQ for triolein, a triglyceride of molecular formula $C_{57}H_{104}O_6$.

In the balanced equation for the complete oxidation of triolein, 80 mol of O_2 are used and 57 mol of CO_2 are produced for each mol of triolein consumed. The RQ = 57/80 = 0.71

$$C_{57}H_{104}O_6 + 80\ O_2 \longrightarrow 57\ CO_2 + 52\ H_2O$$
Triolein

(c) For an individual on a normal diet, RQ is approximately 0.85. Would this value increase or decrease if ethanol were to supply an appreciable portion of caloric needs?

In the balanced equation for the complete oxidation of ethanol, C_2H_6O, 3 mol of O_2 are used and 2 mol of CO_2 are produced for each mole of ethanol consumed. The RQ = 2/3 = 0.67, so the individual's RQ would decrease if ethanol were to supply an appreciable portion of caloric needs.

$$C_2H_6O + 3\ O_2 \longrightarrow 2\ CO_2 + 3\ H_2O$$
Ethanol

<u>Problem 26.21</u>   Acetoacetate, β-hydroxybutyrate, and acetone are commonly known within the health sciences as ketone bodies, in spite of the fact that one of them is not a ketone at all. They are products of human metabolism and are always present in blood plasma. Most tissues, with the notable exception of the brain, have the enzyme systems necessary to use them as energy sources. Synthesis of ketone bodies occurs by the following enzyme-catalyzed reactions. Enzyme names are (1) thiolase, (2) β-hydroxy-β-methylglutaryl-CoA synthase, (3) β-hydroxy-β-methylglutaryl-CoA lyase, and (5) β-hydroxybutyrate dehydrogenase. Reaction (4) is spontaneous and uncatalyzed.

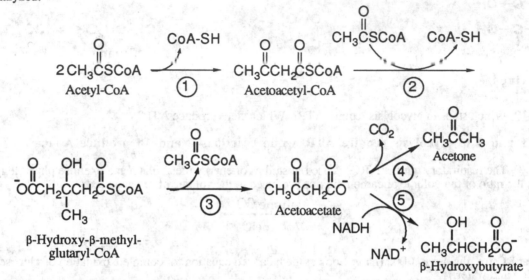

Describe the type of reaction involved in each step and the type of mechanism by which each occurs.

Reaction 1 is a Claisen condensation (Section 18.3) between two molecules of acetyl-CoA.
Reaction 2 is an aldol reaction (Section 18.1) that can be thought of as taking place between the enolate of acetyl-CoA and the ketone carbonyl of acetoacetyl-CoA.
Reaction 3 is a reverse aldol reaction (Section 18.1) that generates acetyl-CoA and acetoacetate.
Reaction 4 is a decarboxylation of a β-ketoacid (Section 16.10A) that generates CO_2 and acetone from acetoacetate.
Reaction 5 is a reduction of the ketone group of acetoacetate to a secondary alcohol (Section 15.14).

<u>Problem 26.22</u> Show that (S)-3-hydroxy-3-methylglutaryl-CoA is a branch point connecting the synthesis of ketone bodies, terpenes, and cholesterol and the steroid hormones. (Hint: review Section 18.4 (Claisen condensations in the Biological World) and Section 25.4B (The Biosynthesis of Steroids).

(S)-3-Hydroxy-3-methylglutaryl-CoA, often referred to as HMG-CoA, is produced from the condensation of acetyl-CoA as described above. HMG-CoA is then decomposed to give ketone bodies, condensed with more acetyl- CoA on the way to terpenes, or converted to 3R-mevalonate on the way to lanosterol and eventually cholesterol.

$$3 \ \text{Acetyl-CoA} \longrightarrow \longrightarrow \ ^{-}\text{OOCH}_2\text{CCH}_2\text{CSCoA}$$

β-**Hydroxy-β-methyl-**
glutaryl-CoA
(HMG-CoA)

Ketone
bodies

Terpenes

Cholesterol

Problem 26.23 A connecting point between anaerobic glycolysis and β-oxidation is formation of acetyl-CoA. Which carbon atoms of glucose appear as methyl groups of acetyl-CoA? Which carbon atoms of palmitic acid appear as methyl groups of acetyl-CoA?

As shown in the answer to Problem 26.8, it is carbons 3 and 4 of D-glucose that end up as the carboxylic acid carbons of pyruvate. These same two carbon atoms, carbons 3 and 4, end up as CO_2 as a result of oxidation and decarboxylation to acetyl CoA. This means that it is carbons 1 and 6 that end up as the methyl groups of acetyl CoA.

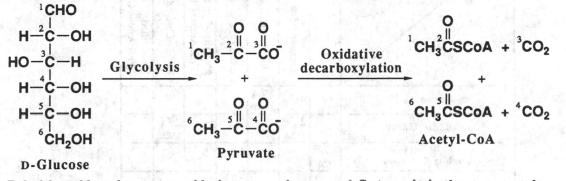

Palmitic acid undergoes β-oxidation to produce acetyl-CoA, so it is the even number carbon atoms (2,4,6,8,10,12,14,16) that end up being the methyl groups.

CHAPTER 27: AMINO ACIDS AND PROTEINS

SUMMARY OF REACTIONS

Starting Material ↓ / Product →	Amide	Ammonium Ion	N-Benzylcarbamate (Z-)	Carboxylate	Phenylthiohydantoin Free Amino Group	Purple-Colored Anion	Substituted γ-Lactone Free Amino Group
α-Amino Acid			27A 27.5C*				
α-Amino Group		27B 27.2A				27C 27.2E	
Amino Group Carboxyl Group	27D 27.5E						
α-Carboxyl Group				27E 27.2A			
Peptide Bond Carboxyl Side of Methionine							27F 27.4B
Peptide Bond N-Terminal Amino Acid						27G 27.4B	

*Section in book that describes reaction.

REACTION 27A: THE BENZYLOXYCARBONYL (Z-) PROTECTING GROUPS FOR AMINES (Section 27.5C)

$$ ^-OCC'HRNH_3^+ \ + \ ClC''OC'''H_2Ph \ \xrightarrow[\text{2) HCl, } H_2O]{\text{1) NaOH}} \ HOCC'HRNHC''OC'''H_2Ph $$

- The **amino groups** of amino acids can be **reversibly blocked** as a **benzylcarbamate**, also known as the **benzyloxycarbonyl** or **Z** group. This protecting group can be used during peptide synthesis, and is stable to dilute base. ✳
- The Z group is put on via the chloroformate as shown above, and is removed by treatment with HBr in anhydrous acetic acid or hydrogenolysis using H_2 and Pd.

REACTION 27B: PROTONATION OF THE α-AMINO GROUP (Section 27.2A)

$$ ^-OCC'HRNH_2 \ + \ H^+ \ \rightleftharpoons \ ^-OCC'HRNH_3^+ $$

- The α-**amino group** of amino acids are **relatively basic with a pK$_a$ near 10.** ✳

REACTION 27C: THE NINHYDRIN REACTION (Section 27.2E)

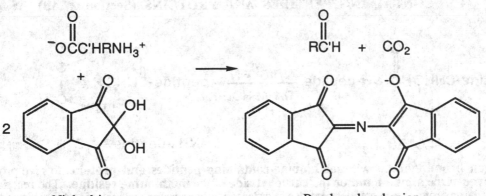

Ninhydrin   Purple-colored anion

- **Ninhydrin** reacts with **primary amino** groups such as the α-amino group of amino acids to give a **deep purple-colored anion**. This reaction can be used to detect even very small amounts of primary amino groups. Ninhydrin has been used extensively to detect unreacted α–amino groups during solid-phase peptide synthesis. Secondary amines like those in proline react with ninhydrin to form an orange-colored compound. ✶

REACTION 27D: PEPTIDE BOND FORMATION USING DICYCLOHEXYL-CARBODIIMIDE (DCC) (Section 27.5E)

$$\underset{RO\overset{\displaystyle O}{\overset{\|}{C}}C'HR'NH_2}{} + \underset{HO\overset{\displaystyle O}{\overset{\|}{C}}C''HR''NH-Z}{} \xrightarrow{\text{DCC}} RO\overset{\displaystyle O}{\overset{\|}{C}}C'HR'NH\overset{\displaystyle O}{\overset{\|}{C}}C'''HR''NH-Z$$

- **Carboxyl** and **amino groups** react to form an **amide bond** in the presence of **dicyclohexylcarbodiimide (DCC)**. DCC can be used to form peptide bonds between an amino acid with an appropriate amino protecting group, and an amino acid with an appropriate carboxyl protecting group. The protecting groups are necessary to prevent unwanted side reactions. The DCC reaction is used extensively in solid-phase peptide synthesis, including automated solid-phase peptide synthesis. ✶
- The mechanism involves initial reaction of the carboxyl group with DCC to create an activated ester that is attacked by the nucleophilic α–amino group to produce the amide bond. The DCC is converted into the relatively insoluble dicyclohexylurea (DCU) in the reaction.

REACTION 27E: ACIDITY OF THE α-CARBOXYL GROUP (Section 27.2A)

$$HO\overset{\displaystyle O}{\overset{\|}{C}}C'HRNH_3^+ \rightleftharpoons {}^-O\overset{\displaystyle O}{\overset{\|}{C}}C'HRNH_3^+ + H^+$$

- The **α-carboxyl group** is **relatively acidic**, having a pK_a value near **2.0**, due to the inductive effect of the adjacent electron-withdrawing $-NH_3^+$ group. ✶

REACTION 27F: THE CYANOGEN BROMIDE CLEAVAGE REACTION OF METHIONINE CONTAINING PEPTIDES AND PROTEINS (Section 27.4B)

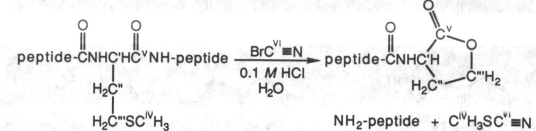

- **Cyanogen bromide** reacts with **methionine-containing peptides and proteins** to give products that are **cleaved** at the **amide bond** on the **carboxyl side** of the **methionine residue**. The reaction is used to cleave large proteins into smaller peptides in order to facilitate sequence analysis. A substituted γ-lactone is the other product of the reaction. ✳

REACTION 27G: THE EDMAN DEGRADATION (Section 27.4B)

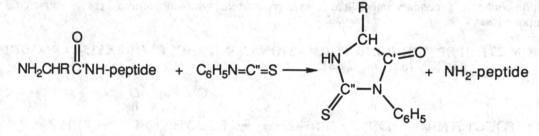

A Phenylthiohydantoin

- Reacting peptides or proteins with **phenyl isothiocyanate** causes **removal of the *N*-terminal amino acid** as a **phenylthiohydantoin** that can be **isolated and identified**. This reaction is referred to as the **Edman degradation** after its inventor Pehr Edman. ✳
- The Edman degradation can be run sequentially on a peptide of unknown sequence, so that the exact amino acid sequence can be determined. Since usually only twenty to thirty amino acids can be sequenced using the Edman degradation, a large protein must first be fragmented into smaller pieces using the CNBr reaction and/or limited hydrolysis with a protease such as trypsin or chymotrypsin. The Edman degradation can be used for automated sequencing.

SUMMARY OF IMPORTANT CONCEPTS

27.0 OVERVIEW

- **Proteins** are composed of chains of **amino acids** linked together by amide (**peptide**) bonds. Two triumphs of chemistry are that the sequence of amino acids in proteins can be determined chemically, and that amino acids can be joined together synthetically to produce functional proteins. ✳

27.1 AMINO ACIDS

- **Amino acids** are compounds that contain a carboxyl group and an amino group. The α-**amino acids** (H_2N-CHR-CO_2H) are the most important class of amino acids in biochemistry. ✳
  - The **amino group** is **basic**, while the **carboxyl** group is relatively **acidic**, thus a **proton is transferred** from the carboxyl group to the amine to create an **internal salt** called a **zwitterion** at neutral pH. ✳
  - **Except for glycine** (H_2N-CH_2-CO_2H), the α-**carbon atoms** of amino acids are **stereocenters**. According to the D and L designations used with carbohydrates, the vast majority of amino acids in living systems are of the **L-series**. This corresponds to the **S configuration** in the R-S convention except for cysteine in which the L designation corresponds to the R configuration because of the priorities assigned with the R-S system. Isoleucine and threonine have a second stereocenter on their side chains.

- The **20 protein-derived amino acids** are usually grouped according to the chemical properties of the **side chains** as either **nonpolar, polar but non-ionized, acidic,** and **basic side chains**.
- Besides these 20 amino acids, small amounts of other amino acids are found in nature. For example, L-ornithine and L-citrulline are components of the urea cycle. In addition, D-amino acids are found as structural components of lower forms of life.

27.2 ACID-BASE PROPERTIES OF AMINO ACIDS
- The **Henderson-Hasselbalch equation** can be used to calculate the ratio of a conjugate base to weak acid at any pH. This can be used to calculate the amount of protonated carboxylic acid or protonated amine functions in amino acids at a given pH. ✳

$$pH = pK_a + \log \frac{[conjugate\ base]}{[weak\ acid]}$$

Henderson-Hasselbalch equation

- The **isoelectric point, pI**, for an amino acid is the pH at which the majority of molecules in solution have no charge. At the isoelectric point, the molecules are more likely to aggregate and thus precipitate. For this reason, amino acids can be precipitated from solution at the isoelectric point, a process known as **isoelectric precipitation**. ✳
- Charged molecules move in an applied electric field toward the electrode carrying the charge opposite their own. This process is called **electrophoresis**, and it can be used to separate amino acids on the basis of charge.
- Of the twenty amino acids needed by humans to synthesize proteins, ten amino acids must be supplied in the diet. These ten amino acids are referred to as **essential amino acids**. ✳

27.3 POLYPEPTIDES AND PROTEINS
- **Proteins** are long chains of amino acids **linked** together by **amide bonds** between the α-**amino group** of one amino acid and the α-**carboxyl group** of another. These amides bonds are given the special name of **peptide bonds**. ✳

27.4 PRIMARY STRUCTURE OF POLYPEPTIDES AND PROTEINS
- The **primary (1°) structure** of a protein or polypeptide is the **sequence of amino acids** in the polypeptide chain. The sequence of amino acids that make up a protein is determined in several steps. ✳
  - First, the **proportion of different amino acids** is determined by **amino acid analysis**. The **polypeptide is hydrolyzed** into individual amino acids by heating in 6 M HCl or 4 M NaOH, then techniques such as ion-exchange chromatography are used to **separate, identify, and quantitate** the **different amino acids** present.
  - Next the **polypeptide is selectively cleaved into fragments** using the **CNBr** reaction (Reaction 27F, Section 27.4B) and/or limited proteolysis. The fragments are subjected to sequence analysis using the **Edman degradation** (Reaction 27G, Section 27.4B). ✳

27.5 SYNTHESIS OF POLYPEPTIDES
- The **synthesis of polypeptides** is **complicated** by the fact that certain α-**amino groups**, α-**carboxyl groups**, and some **side chains** must be **blocked** with carefully chosen **protecting groups** to avoid unwanted side reactions. The α-amino protecting groups are usually carbamates such as the Z group (Reaction 27A, Section 27.5C), while α-carboxyl protecting groups are usually esters such as methyl or benzyl. The *tert*-butoxycarbonyl (BOC) group is another carbamate protecting group commonly used to block the α-amino group during peptide synthesis. The BOC group is usually removed using anhydrous acid such as trifluoroacetic acid (TFA).
  - The amide bond is produced through the use of coupling reagents such as DCC (Reaction 27D, Section 27.5E). ✳
- Synthesis of long peptides is greatly facilitated by the use of the **solid-phase strategy**. The first amino acid of the **peptide chain** is **attached to a solid support** usually at the carboxyl terminus, and sequential reactions are carried out to add amino acids to the growing chain in the desired order. The ninhydrin reaction (Reaction 27C, Section 27.2E) is used to check for complete coupling at each step. Following synthesis, the completed peptide chain is cleaved from the solid support and all protecting groups removed from the side chains to yield the desired product. ✳
  - The solid support usually consists of very small polystyrene beads.
  - For the **addition** of **each new amino acid residue**, the following steps are carried out:

The α-**amino protecting group** on the amino acid attached most recently to the growing chain on the solid support is **removed**.

The **next α-amino-protected amino acid** is **coupled using DCC**.
- The α-amino protecting group of this newest amino acid is removed, and the process is repeated in the proper order with the appropriate α-amino-protected amino acids until the entire chain is assembled.

27.6 THREE-DIMENSIONAL SHAPES OF POLYPEPTIDES AND PROTEINS
• The **amide bond is planar**. In other words, the carbonyl carbon atom, the carbonyl oxygen atom, the amide nitrogen atom, the amide hydrogen atom, and both α-carbon atoms are all in the same plane. ✳
- The planarity is explained by considering that an amide is accurately represented as the **resonance hybrid of two contributing structures,** one with a carbon-oxygen double bond and **one with a carbon-nitrogen double bond**.
- The partial double bond character of the amide means that **two configurations are possible** for an amide, an *s-trans* or *s-cis* configuration. **Almost all peptide bonds** in proteins are in the *s-trans* **configuration** in which the two α-carbon atoms are *trans* to each other.
• Due to the planarity and rigidity of peptide bonds, polypeptide chains form **secondary (2°) structures** such as α-**helixes** and β-**sheets**. These structures are reinforced by hydrogen bonds between the oxygen atoms and hydrogen atoms of the peptide bonds. ✳
• A polypeptide chain exhibits even higher order structure, referred to as **tertiary (3°) structure**, that describes the way in which the secondary structural units of the chain are oriented in three-dimensions. The 3° structure can be held together by a combination of forces including disulfide bonds between two cysteine residues. ✳
• More than one folded polypeptide chain can come together to form a functional complex, and the association of more than one chain is referred to as **quaternary (4°) structure**. ✳

CHAPTER 27
Solutions to the Problems

<u>Problem 27.1</u> Of the 20 protein-derived amino acids shown in Table 27.1, which contain (a) no stereocenter, (b) two stereocenters.

The only amino acid with no stereocenters is glycine (Gly, G). Both isoleucine (Ile, I) and threonine (Thr, T) have two stereocenters as shown with asterisks in the structures below.

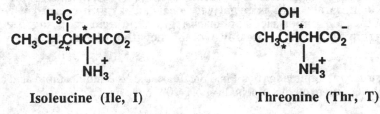

Isoleucine (Ile, I)   Threonine (Thr, T)

<u>Problem 27.2</u> Draw a structural formula for lysine, and estimate the net charge on each functional group at pH values of 3.0, 7.0, and 10.0.

The net charge on the functional groups is calculated as described in Example 27.2. The results of the calculations are shown on the structures at the indicated pH:

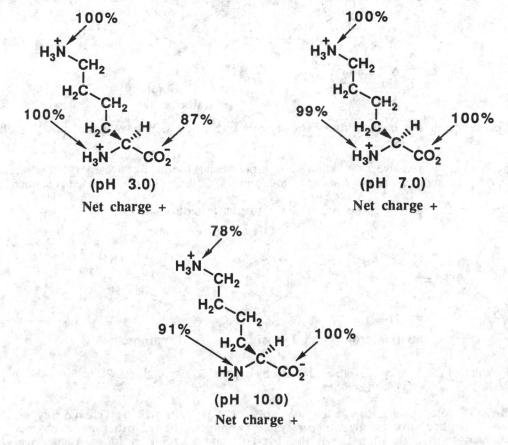

(pH 3.0)   (pH 7.0)
Net charge +   Net charge +

(pH 10.0)
Net charge +

<u>Problem 27.3</u> The isoelectric point of histidine is 7.64. Toward which electrode does histidine migrate on paper electrophoresis at pH 7.0?

An amino acid will have at least a partial positive charge at any pH that is below its isoelectric point. A pH of 7.0 is below the isoelectric point of histidine (7.64), so it will have a partial positive charge. Therefore, at this pH histidine migrates toward the negative electrode.

<u>Problem 27.4</u> Describe the behavior of a mixture of glutamic acid, arginine, and valine on paper electrophoresis at pH 6.0.

The pI's for glutamic acid, arginine, and valine are 3.08, 10.76, and 6.00, respectively. Therefore, at pH 6.0 glutamic acid is negatively charged, arginine is positively charged, and valine is neutral. Thus, on paper electrophoresis, glutamic acid will migrate toward the positive electrode, arginine will migrate toward the negative electrode, and valine will not move.

<u>Problem 27.5</u> Draw a structural formula for Lys-Phe-Ala. Label the *N*-terminal amino acid and the *C*-terminal amino acid. What is the net charge on this tripeptide at pH 6.0?

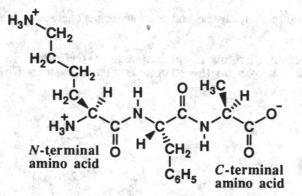

Due to the presence of the basic lysine residue, this tripeptide will have a net positive charge at pH 6.0

<u>Problem 27.6</u> Chymotrypsin catalyzes the hydrolysis of peptide bonds formed by the carboxyl groups of phenylalanine, tyrosine, and tryptophan. What structural feature(s) do these side chains have in common?

Phenylalanine, tyrosine and tryptophan all have aromatic side chains.

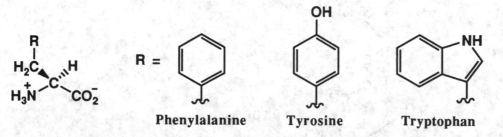

<u>Problem 27.7</u> Which of these tripeptides are hydrolyzed by trypsin? By chymotrypsin?
(a) Tyr-Gln-Val   (b) Thr-Phe-Ser

Based on the substrate specificities listed in Table 27.3, trypsin will not cleave any of these tripeptides because there are no arginine or lysine residues. On the other hand chymotrypsin will cleave peptides (a) and (b) between the Tyr-Gln and Phe-Ser residues, respectively.

Problem 27.8 Deduce the amino acid sequence of an undecapeptide (11 amino acids) from the experimental results shown in the accompanying table.

Experimental Procedure	Amino Acid Composition
Undecapeptide	Ala,Arg,Glu,Lys$_2$,Met,Phe,Ser,Thr,Trp,Val
Edman degradation	Ala
Trypsin-Catalyzed Hydrolysis	
Fragment E	Ala,Glu,Arg
Fragment F	Thr,Phe,Lys
Fragment G	Lys
Fragment H	Met,Ser,Trp,Val
Chymotrypsin-Catalyzed Hydrolysis	
Fragment I	Ala,Arg,Glu,Phe,Thr
Fragment J	Lys$_2$,Met,Ser,Trp,Val
Reaction with Cyanogen Bromide	
Fragment K	Ala,Arg,Glu,Lys$_2$,Met,Phe,Thr,Val
Fragment L	Trp,Ser

Based on the Edman degradation result, alanine (Ala) is the *N*-terminal residue of the peptide. Fragment E must have Arg on the *C*-terminal end because it is a peptide produced by trypsin cleavage. Since we know Ala is the *N*-terminal residue, this means fragment E must be of the sequence Ala-Glu-Arg.
There must be two lysine residues or an arginine and a lysine residue adjacent to each other based on the appearance of a single lysine residue as Fragment G. Since Fragment J has two lysines and no arginine residues, the two lysine residues must be adjacent to each other.
Methionine must be the third to the last residue, because CNBr treatment created fragment L that is only Ser and Trp. In addition, Trp and Ser must be the last two residues. Combining this information with the knowledge that there are two lysine residues adjacent to each other indicates the Fragment J is of the sequence Lys-Lys-Val-Met-Ser-Trp. Note that Val must come after the two Lys residues because of Fragment H. In addition, Trp has to be on the *C*-terminus or a residue would have been cleaved off by chymotrypsin.
Phenylalanine must be on the *C* terminus of Fragment I since it results from chymotrypsin cleavage. We already know that Fragment I must start with Ala-Glu-Arg, so the entire sequence of Fragment I must be Ala-Glu-Arg-Thr-Phe.
Putting Fragments I and J together gives the following sequence for the entire peptide:

Ala-Glu-Arg-Thr-Phe-Lys-Lys-Val-Met-Ser-Trp

Problem 27.9 At pH 7.4, with what amino acid side chains can the side chain of lysine form salt linkages.

At pH 7.4, the only negatively charged side chains are the carboxylates of glutamic acid and aspartic acid. Therefore, these are the amino acid side chains with which the side chain of lysine can form a salt linkage.

Amino Acids
Problem 27.10 What amino acids do these abbreviations stand for?
(a) Phe   **Phenylalanine**   (b) Ser   **Serine**   (c) Asp   **Aspartic acid**
(d) Gln   **Glutamine**   (e) His   **Histidine**   (f) Gly   **Glycine**
(g) Tyr   **Tyrosine**

Problem 27.11 Why are Glu and Asp often referred to as acidic amino acids.

The side chains of glutamic acid (Glu) and aspartic acid (Asp) have carboxyl groups, so these amino acids are referred to as acidic amino acids.

Problem 27.12 Why is Arg often referred to as a basic amino acid? Which two other amino acids are also basic amino acids?

The guanidino group of arginine (Arg) is strongly basic, so this amino acid is referred to as being a basic amino acid. Note that this means arginine is positively charged at neutral pH.

Lysine (Lys) and histidine (His) are also referred to as basic amino acids because their side chains contain a basic primary amine and imidazole functions, respectively.

Problem 27.13 Referring to Tables 27.1 and Table 27.2, identify the
(a) One achiral amino acid   Glycine
(b) Two amino acids that have diastereomers   Threonine, Isoleucine
(c) The two sulfur-containing amino acids   Methionine, Cysteine
(d) Four amino acids with aromatic side chains   Tyrosine, Histidine,
                                                        Tryptophan, Phenylalanine
(e) Amino acid with the most basic side chain   Arginine
(f) Amino acid with the most acidic side chain   Aspartic acid

Problem 27.14   As discussed in the Chemistry in Action box "Vitamin K, Blood Clotting, and Basicity," (Chapter 25), vitamin K participates in carboxylation of glutamate residues of the blood-clotting protein prothrombin.
(a) Write a structural formula for γ-carboxyglutamate.

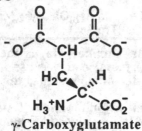

γ-Carboxyglutamate

(b) Account for the fact that the presence of γ-carboxyglutamate escaped detection for many years; on routine amino acid analyses, only glutamate was detected.

This amino acid was not detected because it is a β-dicarboxylic acid and therefore easily decarboxylated under conditions of routine amino acid analysis.

Problem 27.15 Isoleucine has two tetrahedral stereocenters, and four stereoisomers are possible. The protein-derived stereoisomer, L-isoleucine, is named (2S,3S)-(+)-2-amino-3-methylpentanoic acid.
(a) What is the meaning of the designation (+) in this name?

The (+) designation indicates that a sample of this compound will rotate plane polarized light in the clockwise direction.

(b) Draw a stereorepresentation showing the configuration of each stereocenter in L-isoleucine.

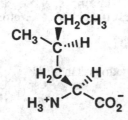

(2S,3S)-(+)-2-Amino-3-methylpentanoic acid
L-Isoleucine

Problem 27.16 The amino acid threonine has two stereocenters. The stereoisomer found in proteins has the configuration 2S,3R. Draw a Fischer projection of this stereoisomer and also a three-dimensional representation using solid, wedged, and dashed lines.

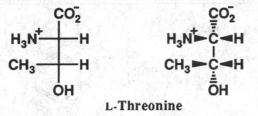

L-Threonine

Problem 27.17 Histamine is biosynthesized from one of the 20 protein-derived amino acids. Suggest which amino acid is its biochemical precursor, and the type of organic reaction(s) involved in its biosynthesis (e.g., oxidation, reduction, decarboxylation, nucleophilic substitution).

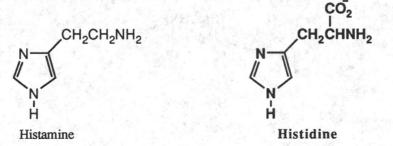

Histamine   Histidine

Histamine is derived from the amino acid histidine and is the result of a biosynthetic decarboxylation reaction. Note how both histamine and histidine are drawn in the form present at basic pH.

Problem 27.18 Both norepinephrine and epinephrine are biosynthesized from the same protein-derived amino acid. From which amino acid are they synthesized and what types of reactions are involved in their biosynthesis?

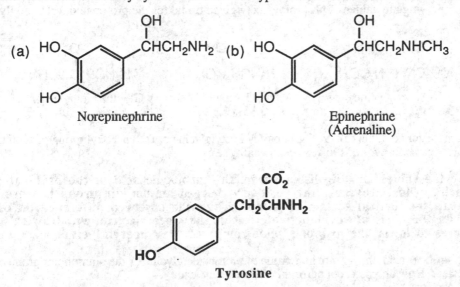

(a) Norepinephrine

(b) Epinephrine
(Adrenaline)

Tyrosine

Norepinephrine and epinephrine are derived from the amino acid tyrosine. In both cases, biosynthesis of these molecules involves decarboxylation, aromatic hydroxylation (a two-electron oxidation) ortho to the original aromatic -OH group, and hydroxylation (a second two-electron oxidation) of the benzylic methylene group. Epinephrine is also methylated on the α-amino group. Note how all of the molecules in the problem are drawn in the form present at basic pH.

Problem 27.19 From which amino acid are serotonin and melatonin biosynthesized and what types of reactions are involved in their biosynthesis?

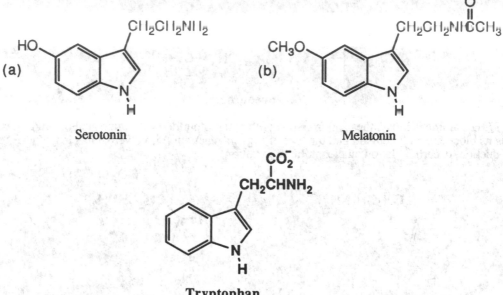

(a) Serotonin

(b) Melatonin

Tryptophan

Serotonin and melatonin are derived from the amino acid tryptophan. In both cases, biosynthesis of these molecules involves decarboxylation. In the case of serotonin there is also an aromatic hydroxylation (a two-electron oxidation). For melatonin the phenolic -OH group of seratonin is methylated and the primary amino group is acetylated. Note how all of the molecules in the problem are drawn in the form present at basic pH.

Problem 27.20 Following are values of pK$_a$ for N-acetylglycine, and for the protonated forms of glycine and glycine methyl ester.

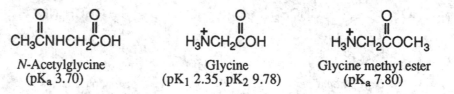

N-Acetylglycine (pK$_a$ 3.70)

Glycine (pK$_1$ 2.35, pK$_2$ 9.78)

Glycine methyl ester (pK$_a$ 7.80)

(a) Which is the stronger acid, the carboxyl group of N-acetylglycine or the carboxyl group of protonated glycine? How do you account for this difference in acidity?

As revealed by the values for pK$_a$ listed above, the carboxylic acid of the protonated glycine is the stronger acid. This is because the positively-charged ammonium group is more electron withdrawing than the neutral acetamido group. Thus, the observed differences in pK$_a$ can be accounted for on the basis of an inductive effect of the more electron-withdrawing ammonium group that serves to lower the pK$_a$ of glycine compared to N-acetylglycine.

(b) Which is the stronger acid, the ammonium group of protonated glycine or the ammonium group of protonated glycine methyl ester? How do you account for this difference in acidity?

For this problem, it is helpful to think in terms of inductive effects operating in conjugate acid-base pairs. The ammonium ion of the glycine methyl ester is the stronger acid, and therefore the weaker conjugate base. The ammonium ion of glycine is the weaker acid, and therefore the stronger conjugate base. The glycine is the stronger conjugate base because above a pH of about 3, glycine contains a deprotonated α-carboxylate group that is negatively-

charged and thus substantially more electron releasing than the neutral methyl ester group of glycine methyl ester.

<u>Problem 27.21</u> For lysine and arginine, the isoelectric point, pI, occurs at a pH where the net charge on the nitrogen-containing groups is +1 and balances the charge of -1 on the α–carboxyl group. Calculate pI for these amino acids.

The pI will occur when the nitrogen-containing groups have a total charge of +1, and this occurs halfway between their respective pK$_a$ values.
For lysine:

$$pI = \frac{1}{2}\ (pK_a\alpha\text{-NH}_3^+ + pK_a\text{side chain-NH}_3^+) = \frac{8.95\ +\ 10.53}{2} = \boxed{9.74}$$

For arginine:

$$pI = \frac{1}{2}\ (pK_a\alpha\text{-NH}_3^+ + pK_a\text{side chain guanidium}) = \frac{9.04\ +\ 12.48}{2} = \boxed{10.76}$$

<u>Problem 27.22</u> For aspartic and glutamic acids, the isoelectric point occurs at a pH where the net charge on the two carboxyl groups is -1 and balances the charge of +1 on the α–amino group. Calculate pI for these amino acids.

The pI will occur when the two acid groups have a total charge of -1, and this occurs halfway between their respective pK$_a$ values.
For aspartic acid:

$$pI = \frac{1}{2}\ (pK_a\alpha\text{-CO}_2\text{H} + pK_a\text{side chain-CO}_2\text{H}) = \frac{2.10\ +\ 3.86}{2} = \boxed{2.98}$$

For glutamic acid:

$$pI = \frac{1}{2}\ (pK_a\alpha\text{-CO}_2\text{H} + pK_a\text{side chain-CO}_2\text{H}) = \frac{2.10\ +\ 4.07}{2} = \boxed{3.08}$$

<u>Problem 27.23</u> Draw the structural formula for the form of each amino acid most prevalent at pH 1.0.
(a) Threonine   (b) Arginine

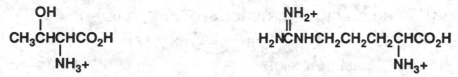

(c) Methionine   (d) Tyrosine

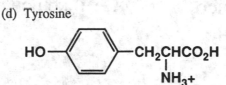

<u>Problem 27.24</u> Draw the structural formula for the form of each amino most prevalent at pH 10.0.
(a) Leucine   (b) Valine

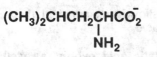
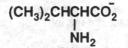

(c) Proline

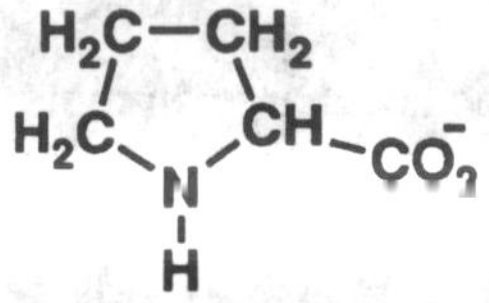

$$H_2C-CH_2$$
$$H_2C \quad CH-CO_2^-$$
$$\underset{\underset{H}{|}}{N}$$

(d) Aspartic acid

$$^-O_2CCH_2CHCO_2^-$$
$$\underset{NH_2}{|}$$

Problem 27.25 At pH 7.4, the pH of blood plasma, do the majority of protein-derived amino acids bear a net negative charge or a net positive charge?

The majority of amino acids have a pI near 5 or 6, so they will bear a net negative charge at pH 7.4.

Problem 27.26 Write the zwitterion form of alanine and show its reaction with:
(a) 1 mol NaOH

$$CH_3CHCO_2^- \atop {\underset{NH_3+}{|}} \quad + \text{ 1 mol NaOH} \longrightarrow \quad CH_3CHCO_2^- \atop {\underset{NH_2}{|}}$$

(b) 1 mol HCl

$$CH_3CHCO_2^- \atop {\underset{NH_3+}{|}} \quad + \text{ 1 mol HCl} \longrightarrow \quad CH_3CHCO_2H \atop {\underset{NH_3+}{|}}$$

Problem 27.27 Write the form of lysine most prevalent at pH 1.0 and then show its reaction with the following. Consult Table 27.2 for pK_a values of the ionizable groups in lysine.

At pH 1.0, the most prevalent form of lysine has both amino groups as well as the carboxylic acid group protonated and a total charge of +2 as shown in the following structure.

$$pK_a \text{ 9.95} \frown \overset{+}{H_3N}CH_2CH_2CH_2CH_2CHCO_2H \frown pK_a \text{ 2.10}$$
$$\underset{NH_3^+}{|} \frown pK_a \text{ 9.82}$$

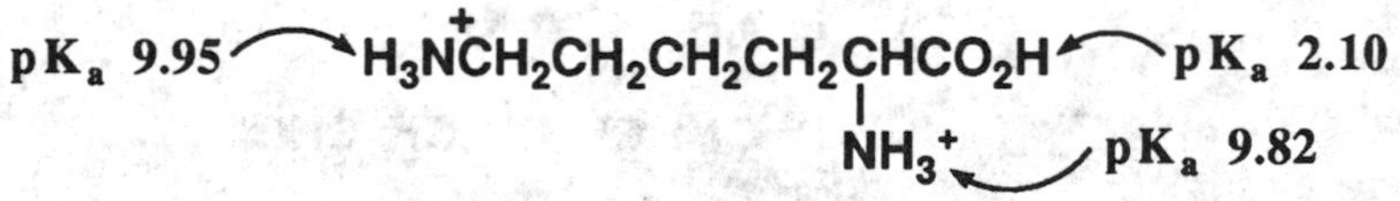

(a) 1 mol NaOH

$$\overset{+}{H_3N}CH_2CH_2CH_2CH_2CHCO_2^- \atop {\underset{NH_3^+}{|}}$$

(b) 2 mol NaOH

$$\overset{+}{H_3N}CH_2CH_2CH_2CH_2CHCO_2^- \atop {\underset{NH_2}{|}}$$

(c) 3 mol NaOH

$$H_2NCH_2CH_2CH_2CH_2CHCO_2^- \atop {\underset{NH_2}{|}}$$

Problem 27.28 Write the form of aspartic acid most prevalent at pH 1.0 and then show its reaction with the following. Consult Table 27.2 for pK_a values of the ionizable groups in aspartic acid.

At pH 1.0, the most prevalent form of aspartic acid has both carboxylic acid groups as well as the amino group protonated and a total charge of +1 as shown in the following structure.

$$pK_a \text{ 3.86} \frown HO_2CCH_2CHCO_2H \frown pK_a \text{ 2.10}$$
$$\underset{NH_3^+}{|} \frown pK_a \text{ 9.82}$$

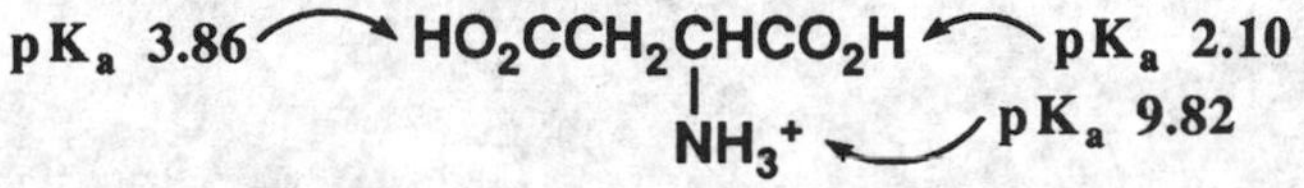

(a) 1 mol NaOH   (b) 2 mol NaOH   (c) 3 mol NaOH

$$HO_2CCH_2\underset{\underset{NH_3^+}{|}}{CH}CO_2^-$$ $$^-O_2CCH_2\underset{\underset{NH_3^+}{|}}{CH}CO_2^-$$ $$^-O_2CCH_2\underset{\underset{NH_2}{|}}{CH}CO_2^-$$

Problem 27.29 Account for the fact that the isoelectric point of glutamine (pI 5.65) is higher than the isoelectric point of glutamic acid (pI 3.08).

Amino acids have no net charge at their pI. For this to happen with glutamic acid, the net charge on the α-carboxyl and side chain carboxyl groups must be -1 to balance the +1 charge of the α-amino group. This will occur at a pI = $(1/2)(2.10 + 4.07)$ = 3.08. The amide side chain of glutamine is already neutral near neutral pH, so the pI of the amino acid is determined by the values for the only ionizable groups, namely the α-carboxyl group and α-amino groups, according to the equation pI = $(1/2)(2.17 + 9.03)$ = 5.6. This value is near that of the other amino acids with non-ionizable functional groups on their side chains.

Problem 27.30 Enzyme-catalyzed decarboxylation of glutamic acid gives 4-aminobutanoic acid (Section 27.1D). Estimate the pI of 4-aminobutanoic acid.

There is little if any inductive effect operating between the amino and carboxyl groups of 4-aminobutanoic acid because there are three methylene groups between them. Thus, the pK_a of the amino group of 4-aminobutanoic acid is like that of a simple amino group, near 10.0. Similarly, the pK_a of the carboxyl group is like that of a simple carboxyl group, near 4.5. Given these estimates for the pK_a values, the pI would be:

$$pI = \frac{1}{2}(pK_a \, \alpha - CO_2H + pK_a \, \alpha - NH_3^+) = \frac{1}{2}(4.5 + 10.0) = 7.25$$

Problem 27.31 Given pK_a values for ionizable groups in Table 27.2, sketch curves for the titration of (a) glutamic acid with NaOH, and (b) histidine with NaOH.

Glutamic acid has pK_a values of 2.10, 4.07, and 9.47 so the titration curve would look something like the following:

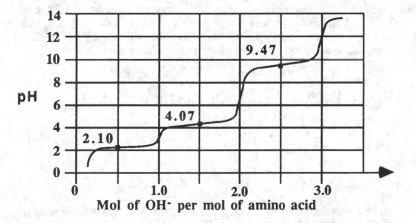

Histidine has pK$_a$ values of 1.77, 6.10, and 9.18 so the titration curve would look something like the following:

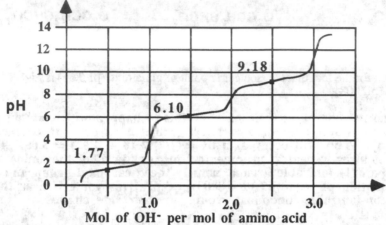

Mol of OH⁻ per mol of amino acid

<u>Problem 27.32</u> Guanidine and the guanidino group present in arginine are two of the strongest organic bases known. Account for this basicity.

The guanidino group is strongly basic because of resonance stabilization of the protonated guanidinium ion as shown below:

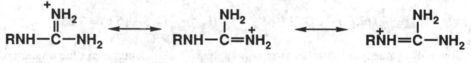

R = H or alkyl group

<u>Problem 27.33</u> A chemically modified guanidino group is present in cimetidine (Tagamet) , a widely prescribed drug for the control of gastric acidity and peptic ulcers. Cimetidine reduces gastric acid secretion by inhibiting the interaction of histamine with gastric H2 receptors. In the development of this drug, a cyano group was added to the substituted guanidino group to significantly alter its basicity. Do you expect this modified guanidino group to be more basic or less basic than the guanidino group of arginine? Explain.

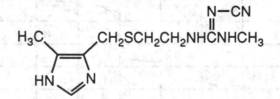

Cimetidine
(Tagamet)

A cyano group is electron-withdrawing. As a result, the guanidino function will have less electron density available to interact with a proton, and will be less basic than a similar guanidino group without the cyano group.

<u>Problem 27.34</u> Only three amino acids have appreciable absorption in the ultraviolet spectrum. Which three amino acids contribute to the commonly quoted λ_{max} of 2800 Å for proteins?

The three amino acids that have appreciable absorption near 280 nm are the aromatic amino acids phenylalanine, tyrosine, and tryptophan.

Problem 27.35 Draw a structural formula for the product formed when alanine is treated with the following reagents.
(a) Aqueous NaOH

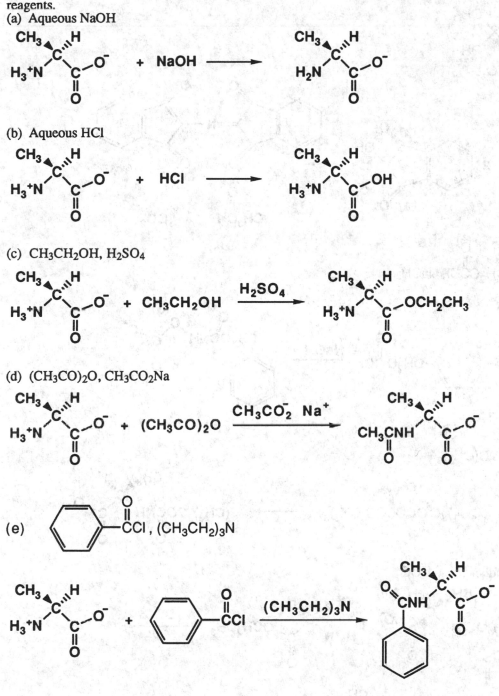

(b) Aqueous HCl

(c) CH₃CH₂OH, H₂SO₄

(d) (CH₃CO)₂O, CH₃CO₂Na

(e)

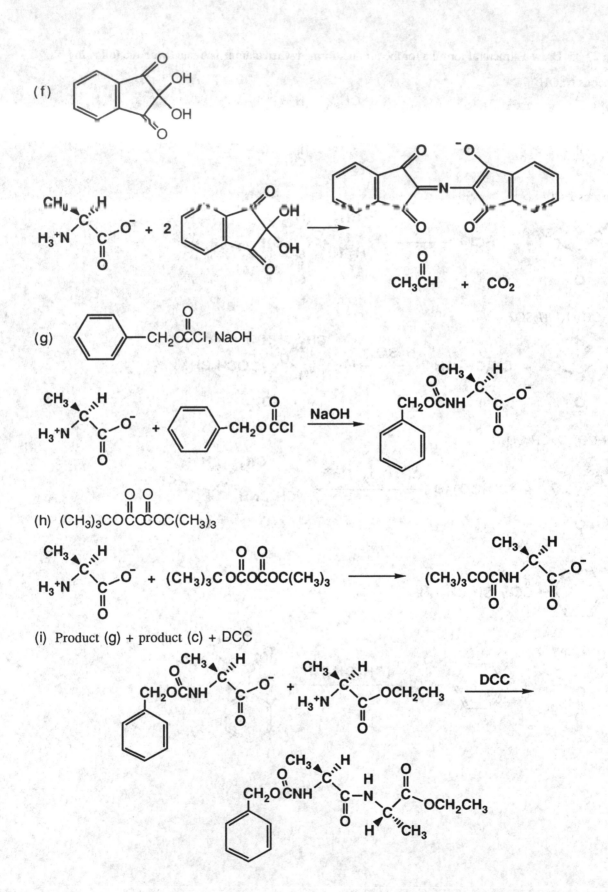

(f)

(g)

(h) $(CH_3)_3COCOCOC(CH_3)_3$

(i) Product (g) + product (c) + DCC

Problem 27.36 At what pH would you carry out an electrophoresis to separate the amino acids in each mixture of amino acids?

Recall that an amino acid below its isoelectric point will have some degree of positive charge, an amino acid above its isoelectric point will have some degree of negative charge, and an amino acid at its isoelectric point will have no net charge.

(a) Ala, His, Lys

Electrophoresis could be carried out at pH 7.64, the isoelectric point of histidine (His). At this pH, the histidine is neutral and would not move, the lysine (Lys) will be positively charged and will move toward the negative electrode, and the alanine (Ala) will be slightly negatively charged and will move toward the positive electrode.

(b) Glu, Gln, Asp

Electrophoresis could be carried out at pH 3.08, the isoelectric point of glutamic acid (Glu). At this pH, the glutamic acid is neutral and would not move, the glutamine (Gln) will be positively charged and will move toward the negative electrode, and the aspartic acid (Asp) will be slightly negatively charged and will move toward the positive electrode.

(c) Lys, Leu, Tyr

Electrophoresis could be carried out at pH 6.04, the isoelectric point of leucine (Leu). At this pH, the leucine is neutral and would not move, the lysine (Lys) will be positively charged and will move toward the negative electrode, and the tyrosine (Tyr) will be slightly negatively charged and will move toward the positive electrode.

Problem 27.37 Do the following molecules migrate to the cathode or to the anode on electrophoresis at the specified pH?

The key to determining which way the molecules migrate is to estimate the net charge on the molecules at the given pH. Molecules with a net positive charge will migrate toward the negative electrode and molecules with a net negative charge will migrate toward the positive electrode. Molecules at a pH below their isoelectric point (Table 27.2) have a net positive charge, molecules at a pH above their isoelectric point have a net negative charge, and molecules at a pH that equals their isoelectric point have no net charge.

(a) Histidine at pH 6.8

pI = 7.64, so at pH 6.8 histidine has a net positive charge and migrates toward the negative electrode (cathode).

(b) Lysine at pH 6.8

pI = 9.74, so at pH 6.8 lysine has a net positive charge and migrates toward the negative electrode (cathode).

(c) Glutamic acid at pH 4.0

pI = 3.08, so at pH 4.0 glutamic acid has a net negative charge and migrates toward the positive electrode (anode).

(d) Glutamine at pH 4.0

pI = 5.65, so at pH 4.0 glutamine has a net positive charge and migrates toward the negative electrode (cathode).

(e) Glu-Ile-Val at pH 6.0

The glutamic acid residue has a carboxyl group that will be largely deprotonated at pH 6.0, so the overall molecule will have a net negative charge and will migrate toward the positive electrode (anode).

(f) Lys-Gln-Tyr at pH 6.0

The lysine residue has an amino group on its side chain that will be protonated at pH 6.0, so the molecule will have a net positive charge and will migrate toward the negative electrode (cathode).

Problem 27.38 Examine the amino acid sequence of human insulin (Figure 27.17). Do you expect human insulin to have an isoelectric point nearer that of the acidic amino acids (pI 2.0 - 3.0), the neutral amino acids (pI 5.5 - 6.5), or the basic amino acids (pI 9.5 - 11.0)?

A listing of the amino acids present are shown below:

aspartic acid (Asp)	0
glutamic acid (Glu)	4
histidine (His)	2
lysine (Lys)	1
arginine (Arg)	1

The charge will only be neutral when there are four positively charged residues to neutralize the four negative charges of the carboxylates from the four Glu residues. For this to happen, the Lys, Arg, and both His residues must be positively charged. Since the imidazole of His is not protonated until the pH is below 6 or so, the entire molecule will only be neutral around this pH. Thus, insulin is expected to have an isoelectric point nearer to that of the neutral amino acids. Its isoelectric point is 5.30 to 5.35.

Primary Structure of Polypeptides and Proteins

Problem 27.39 If a protein contains four different SH groups, how many different disulfide bonds are possible if only a single disulfide bond is formed? How many different disulfides are possible if two disulfide bonds are formed?

If only one disulfide bond were to be formed from the four different cysteine residues, then there are a total of 6 different disulfide bonds that can be formed. There are three possibilities if two disulfide bonds are to be formed.

Problem 27.40 How many different tetrapeptides can be made if:
(a) The tetrapeptide contains one unit each of Asp, Glu, Pro, and Phe?

There could be any of the four residues in the first position, any of the remaining three amino acids in the second position and so on. Thus, there are 4 x 3 x 2 x 1 = 24 possible tetrapeptides.

(b) All 20 amino acids can be used, but each only once?

Using the same logic as in (a), there are 20 x 19 x 18 x 17 = 116,280 possible tetrapeptides.

Problem 27.41 A decapeptide has the following amino acid composition:

$$Ala_2, Arg, Cys, Glu, Gly, Leu, Lys, Phe, Val$$

Partial hydrolysis yields the following tripeptides:

Cys-Glu-Leu + Gly-Arg-Cys + Leu-Ala-Ala + Lys-Val-Phe + Val-Phe-Gly

One round of Edman degradation yields a lysine phenylthiohydantoin. From this information, deduce the primary structure of this decapeptide.

Due to the Edman degradation result, the Lys residue must be at the *N*-terminus. Given this information, the rest of the peptide sequence is deduced because of overlap among the tripeptide sequences as shown below.

The complete peptide is:
          Lys-Val-Phe-Gly-Arg-Cys-Glu-Leu-Ala-Ala

The peptides fit as follows:
          Lys-Val-Phe
              Val-Phe-Gly
                  Gly-Arg-Cys
                      Cys-Glu-Leu
                          Leu-Ala-Ala

<u>Problem 27.42</u> A tetradecapeptide (14 amino acid residues) gives the following peptide fragments on partial hydrolysis. From this information, deduce the primary structure of this polypeptide. Fragments are grouped according to size.

Pentapeptide Fragments	Tetrapeptide Fragments
Phe-Val-Asn-Gln-His	Gln-His-Leu-Cys
His-Leu-Cys-Gly-Ser	His-Leu-Val-Glu
Gly-Ser-His-Leu-Val	Leu-Val-Glu-Ala

The complete peptide is:
      Phe-Val-Asn-Gln-His-Leu-Cys-Gly-Ser-His-Leu-Val-Glu-Ala

The peptides fit as follows:
        Phe-Val-Asn-Gln-His
                Gln-His-Leu-Cys
                      His-Leu-Cys-Gly-Ser
                                Gly-Ser-His-Leu-Val
                                        His-Leu-Val-Glu
                                                Leu-Val-Glu-Ala

<u>Problem 27.43</u> 2,4-Dinitrofluorobenzene, very often known as Sanger's reagent after the English chemist Frederick Sanger who popularized its use, reacts selectively with the *N*-terminal amino group of a polypeptide chain. Sanger was awarded the 1958 Nobel Prize for chemistry for his work in determining the primary structure of bovine insulin. One of the few persons to be awarded two Nobel Prizes, he also shared the 1980 award in chemistry with American chemists, Paul Berg and Walter Gilbert, for the development of chemical and biological analyses of DNAs.

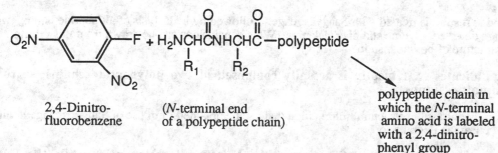

2,4-Dinitro-
fluorobenzene

(*N*-terminal end
of a polypeptide chain)

polypeptide chain in
which the *N*-terminal
amino acid is labeled
with a 2,4-dinitro-
phenyl group

Following reaction with 2,4-dinitrofluorobenzene, all amide bonds of the polypeptide chain are hydrolyzed, and the amino acid labeled with a 2,4-dinitrophenyl group is separated by either paper or column chromatography and identified.

(a) Write the structural formula for the product formed by treatment of the *N*-terminal amino group with Sanger's reagent and propose a mechanism for its formation. (*Hint:* Review nucleophilic aromatic substitution, Section 20.3B).

Step 1:

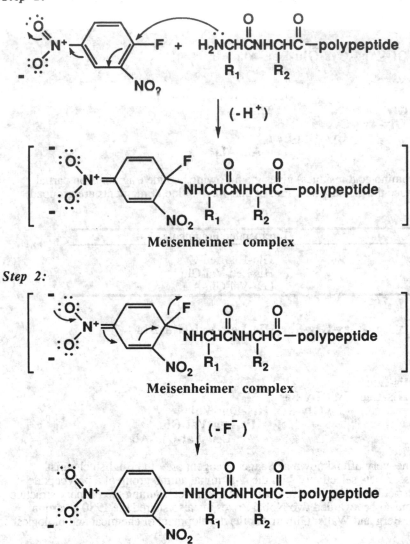

Meisenheimer complex

Step 2:

Meisenheimer complex

(b) When bovine insulin is treated with Sanger's reagent followed by hydrolysis of all peptide bonds, two labeled amino acids are detected: glycine and phenylalanine. What conclusions can be drawn from this information about the primary structure of bovine insulin?

This result indicates that insulin is actually composed of two polypeptide chains, so there are two *N*-terminal residues.

(c) Compare and contrast the structural information that can be obtained from use of Sanger's reagent with that from use of the Edman degradation.

Sanger's reagent only allows identification of the *N*-terminal amino acid, while the Edman degradation can sequentially determine the sequence of amino acids at the *N*-terminus.

<u>Problem 27.44</u> Write structural formulas for the products formed after one cycle of Edman degradation on the tripeptide Ser-Leu-Phe.

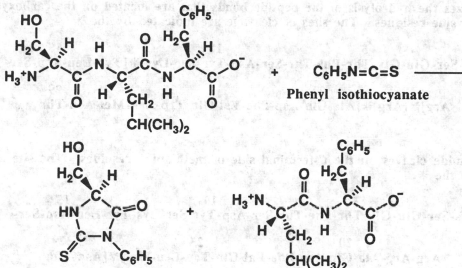

<u>Problem 27.45</u> Following is the primary structure of glucagon a polypeptide hormone of 29 amino acids. Glucagon is produced in the α-cells of the pancreas and helps maintain blood glucose levels in a normal concentration range.

```
   1               5                 10                15
His-Ser-Glu-Gly-Thr-Phe-Thr-Ser-Asp-Tyr-Ser-Lys-Tyr-Leu-Asp-Ser-Arg-Arg-

       20              25              29
   Ala-Gln-Asp-Phe-Val-Gln-Trp-Leu-Met-Asn-Thr
```

Glucagon

Which peptide bonds are hydrolyzed when this polypeptide is treated with
(a) Phenyl isothiocyanate

This reagent only hydrolyzes the *N*-terminal amino acid, so the His-Ser bond would be hydrolyzed. The site of cleavage is indicated by the ✳.

```
    1                5                   10                15
 His]✳[Ser-Glu-Gly-Thr-Phe-Thr-Ser-Asp-Tyr-Ser-Lys-Tyr-Leu-Asp-Ser-Arg-

         20              25              29
     Arg-Ala-Gln-Asp-Phe-Val-Gln-Trp-Leu-Met-Asn-Thr
```

(b) Chymotrypsin

Chymotrypsin catalyzes the hydrolysis of the peptide bonds that are located on the carboxyl side of phenylalanine, tyrosine, and tryptophan residues. The sites of cleavage are indicated by the ✳.

```
    1               5                      10                  15
 His-Ser-Glu-Gly-Thr-Phe]✳[Thr-Ser-Asp-Tyr]✳[Ser-Lys-Tyr]✳[Leu-Asp-Ser-

         20                  25                  29
     Arg-Arg-Ala-Gln-Asp-Phe]✳[Val-Gln-Trp]✳[Leu-Met-Asn-Thr
```

(c) Trypsin

Trypsin catalyzes the hydrolysis of the peptide bonds that are located on the carboxyl side of arginine and lysine residues. The sites of cleavage are indicated by the *.

1   5   10   15
His-Ser-Glu-Gly-Thr-Phe-Thr-Ser-Asp-Tyr-Ser-Lys]*[Tyr-Leu-Asp-Ser-

              20   25   29
Arg]*[Arg]*[Ala-Gln-Asp-Phe-Val-Gln-Trp-Leu-Met-Asn-Thr

(d) Br-CN

Cyanogen bromide cleaves on the *C*-terminal side of methionine residues. The site of cleavage is indicated by the *.

1   5   10   15
His-Ser-Glu-Gly-Thr-Phe-Thr-Ser-Asp-Tyr-Ser-Lys-Tyr-Leu-Asp-Ser-

              20   25   29
Arg-Arg-Ala-Gln-Asp-Phe-Val-Gln-Trp-Leu-Met]*[Asn-Thr

Problem 27.46 Glutathione (G-SH), one of the most common tripeptides in animals, plants, and bacteria, is a scavenger of oxidizing agents. In reacting with oxidizing agents, glutathione is converted to G-S-S-G.

$$
\underset{\underset{CO_2^-}{|}}{\overset{+}{H_3N}CHCH_2CH_2}\overset{\overset{O}{||}}{C}NH\underset{\underset{CH_2SH}{|}}{CH}\overset{\overset{O}{||}}{C}NHCH_2CO_2^-
$$

Glutathione

(a) Name the amino acids in this tripeptide.

The amino acids in glutathione are glutamic acid (Glu), cysteine (Cys), and glycine (Gly).

(b) What is unusual about the peptide bond formed by the *N*-terminal amino acid?

The *N*-terminal glutamic acid is linked to the next residue by an amide bond between the carboxyl group of the side chain, not the α-carboxyl group.

(c) Write a balanced half-reaction for the reaction of two molecules of glutathione to form a disulfide bond. Is glutathione a biological oxidizing agent or a biological reducing agent?

$$2G\text{-}SH \longrightarrow G\text{-}S\text{-}S\text{-}G + 2H^+ + 2e^-$$

The glutathione is oxidized in this process, so it is a biological reducing agent.

(d) Write a balanced equation for reaction of glutathione with molecular oxygen, O_2, to form G-S-S-G and H_2O. Is molecular oxygen oxidized or reduced in this process?

$$2G\text{-}SH + 1/2\ O_2 \longrightarrow G\text{-}S\text{-}S\text{-}G + H_2O$$

The molecular oxygen is reduced in this process.

Synthesis of Polypeptides

Problem 27.47 In a variation of the Merrifield solid-phase peptide synthesis, the amino group is protected as by a fluorenylmethoxycarbonyl (FMOC) group. This protecting group is removed by treatment with a weak base such as the secondary amine, piperidine. Write a balanced equation and propose a mechanism for this deprotection.

Fluorenylmethoxycarbonyl
(FMOC) group

The key to the mechanism is that the FMOC group is unusually acidic for a hydrocarbon, since deprotonation generates a relatively stable dibenzocyclopentadienyl anion. The stability of this anion is analogous to cyclopentadiene itself, as described in Section 19.2E. The following is a reasonable mechanism for the removal of FMOC protecting groups in weak base:

Step 1:

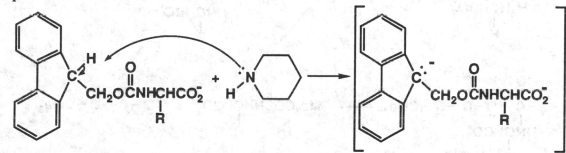

Step 2:

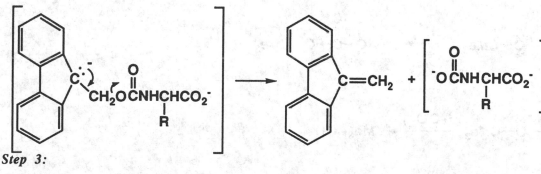

Step 3:

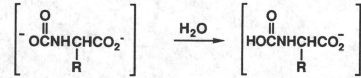

Step 4:

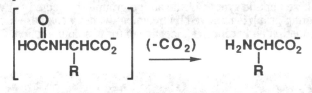

Problem 27.48 The BOC-protecting group may be added by treatment of an amino acid with di-*tert*-butyl dicarbonate as shown in the following reaction sequence. Propose a mechanism to account for formation of these products.

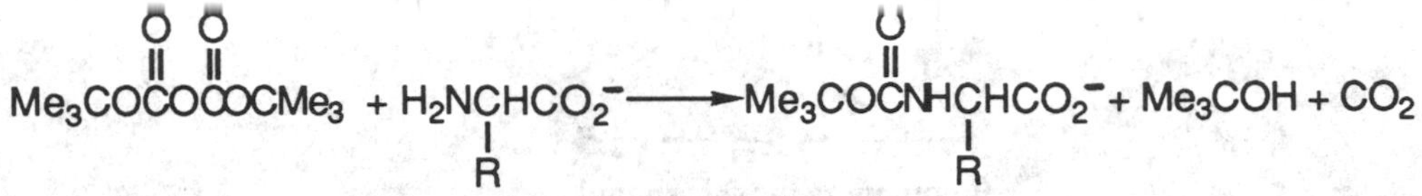

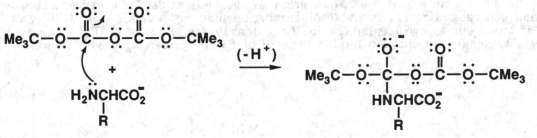

Di-*tert*-butyl dicarbonate   BOC-Amino acid

Step 1:

:O:⟋   :O:
Me₃C—Ö—C—Ö—C—Ö—CMe₃   (-H⁺)   :Ö:⁻   :O:
            ⟍   ────→   Me₃C—Ö—C—Ö—C—Ö—CMe₃
            + |
        H₂NCHCO₂⁻   HNCHCO₂⁻
            | |
            R   R

Step 2:

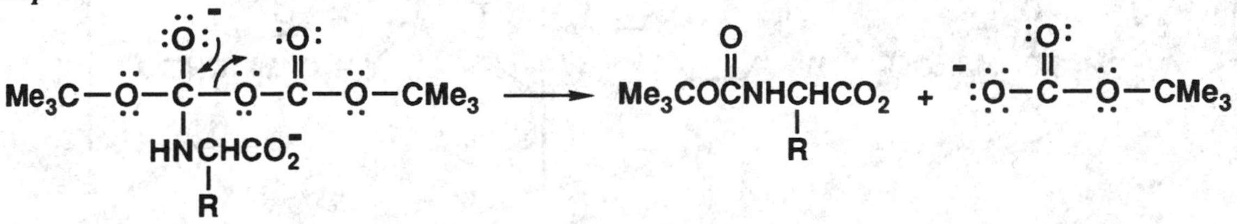

:Ö:⁻⟍   :O: O   :O:
Me₃C—Ö—C—Ö—C—Ö—CMe₃   ────→   Me₃COCNHCHCO₂   +   ⁻:Ö—C—Ö—CMe₃
        | |
    HNCHCO₂⁻   R
        |
        R

Step 3:

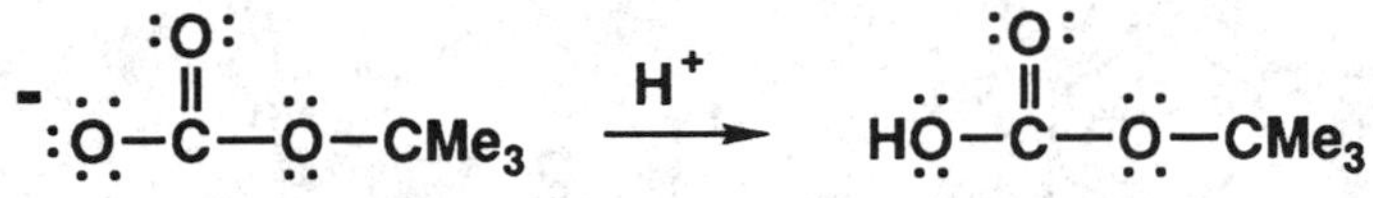

    :O: :O:
⁻:Ö—C—Ö—CMe₃   H⁺   HÖ—C—Ö—CMe₃
                  ────→

Step 4:

    :O:
HÖ—C—Ö—CMe₃   (-CO₂)   Me₃COH
              ──────→

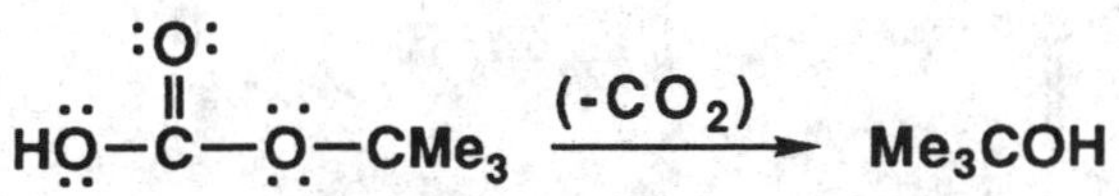

Problem 27.49 In peptide synthesis with BOC protecting groups, acid is used for deprotection. What is the initial fate of the *tert*-butyl group during acid deprotection? Why is a nucleophile such as anisole often added to the peptide deprotection mixture?

The *tert*-butyl group initially is converted to a carbocation.

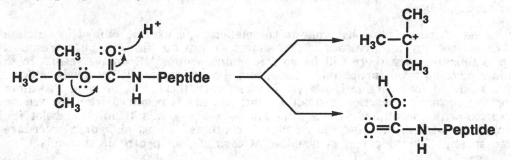

The anisole is added to react with the *tert*-butyl carbocations before they can take part in unwanted side reactions with the peptide.

Problem 27.50 The side chain carboxyl groups of aspartic acid and glutamic acid are often protected as benzyl esters.

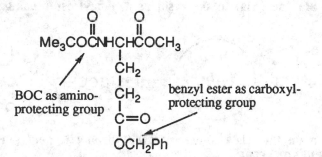

(a) Show how to convert the side-chain carboxyl group to a benzyl ester using benzyl chloride as a source of the benzyl group.

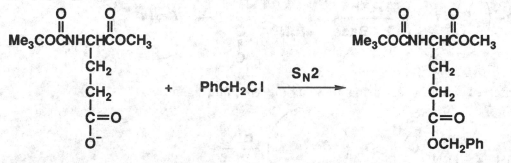

(b) How do you deprotect the side-chain carboxyl under mild conditions without removing the BOC protecting group at the same time?

The benzyl esters can be removed under very mild conditions using hydrogenolysis using hydrogen in the presence of a transition metal catalyst.

Problem 27.51 In solid-phase peptide synthesis, it is important that the coupling reaction give as close to a 100% yield as possible, otherwise shorter peptides (failure sequences) accumulate. Using a large excess of the BOC-protected amino acid and long reaction times maximize the coupling yield but can also lead to wasteful, expensive, and slow peptide syntheses. Suppose you remove a small amount of the Merrifield resin from the reaction vessel after a coupling reaction. How could you tell, based on the chemistry covered in this chapter, if the coupling reaction has gone to completion?

A great way to see if the coupling has gone to completion is to use the ninhydrin reaction (Section 27.2E) to look for the presence of free amino groups on the solid-phase resin. If the coupling step is finished, then there will be no free amino groups left on the resin. In this case, the amines have all been turned into amides and the chains end in the newly added N-blocked amino acid. If the reaction is not yet finished, then there will be some free amino groups left on the resin. In practice, a small amount of resin is removed from the reaction vessel and reacted with the ninhydrin reagent. The presence of significant blue color indicates that the coupling is not complete and the reaction is continued. This ninhydrin procedure can be quantitated to establish the extent of coupling at each step of a peptide synthesis.

Problem 27.52 Outline a synthesis of the tripeptide Phe-Val-Ala from its constituent amino acids using the Merrifield solid-support synthesis.

For the solid phase synthesis of this molecule, the following steps will be required. There are several brief washing steps with pure solvents between each step mentioned below.
1) Attach BOC-protected alanine (Ala) to the resin as a benzyl ester using the carboxylate cesium salt.

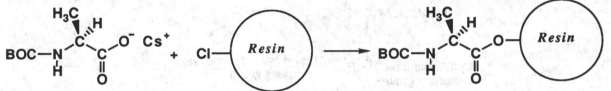

2) Remove the BOC group on the Ala residue with trifluoroacetic acid (TFA), then wash with base to deprotonate ammonium group.

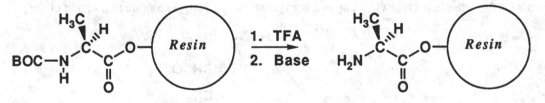

3) Couple BOC-protected valine (Val) using DCC as the coupling agent.

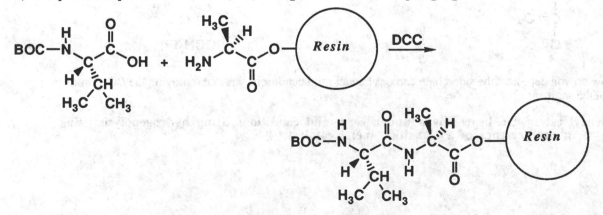

4) Remove the BOC group on the Val residue with trifluoroacetic acid (TFA), then wash with base to deprotonate ammonium group.

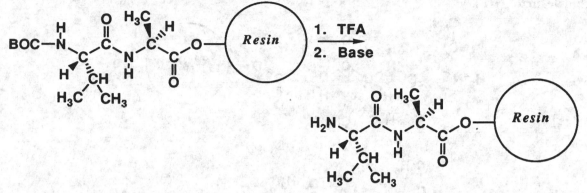

5) Couple BOC-protected phenylalanine (Phe) using DCC as the coupling agent.

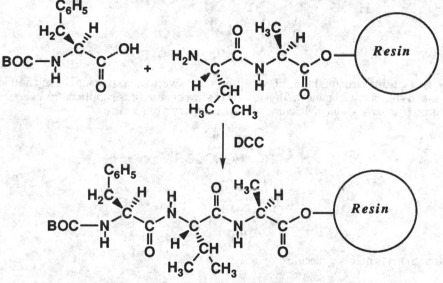

6) Remove the BOC group on the Phe residue with trifluoroacetic acid (TFA).

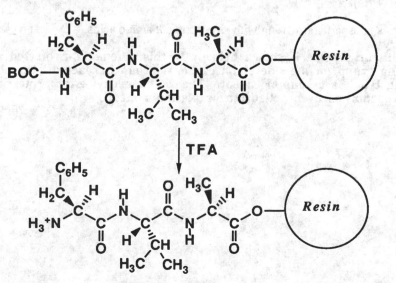

7) Remove the completed peptide from the solid phase resin, usually by treatment with hydrofluoric acid (HF).

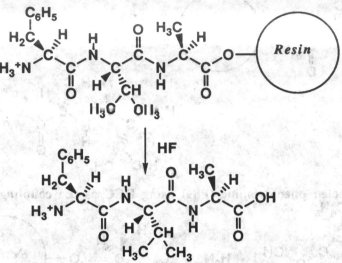

Problem 27.53 Following is the structural formula of the artificial sweetener aspartame. Each amino acid has the L configuration. L-aspartyl-L-phenylalanine methyl ester has a sweet taste (it is significantly sweeter than sugar) whereas its enantiomer, D-aspartyl-D-phenylalanine methyl ester, has a bitter taste.

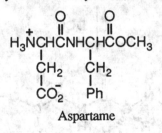

Aspartame

(a) Name the two amino acids in this molecule.

Aspartame is composed of aspartic acid (Asp) attached via a peptide bond to the methyl ester of phenylalanine (Phe).

(b) Propose a synthesis of aspartame starting with its constituent amino acids.

A reasonable synthetic scheme is shown below. In this scheme, Z-protected aspartic acid with a benzyl protecting group on the side chain carboxyl group is reacted with phenylalanine methyl ester using DCC as a coupling agent. The Z and benzyl ester protecting groups are removed by hydrogenolysis over palladium or platinum metal.

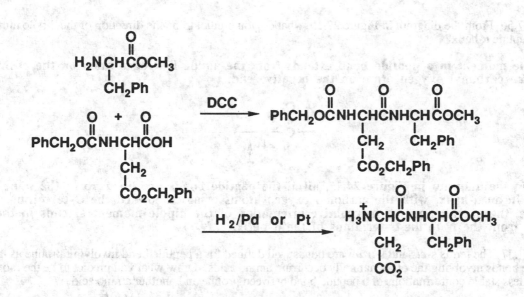

Three-Dimensional Shapes of Polypeptides and Proteins

Problem 27.54 Following is a diagram of the groups that make up the backbone of a polypeptide chain:

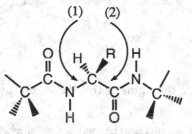

Draw a Newman projection looking down bond (1) with the nitrogen atom toward the front. Also, draw a Newman projection down bond (2) with the tetrahedral carbon in the front. What favorable conformations can you identify on the basis of these projections?

Bond 1:   Bond 2:

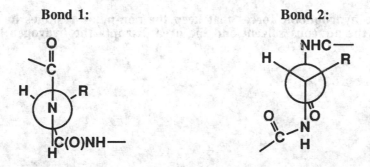

The favorable conformations are ones that place the peptide chains anti or gauche with respect to each other and the side chain R group.

Problem 27.55 Examine the α-helix conformation. Are amino acid side chains arranged all inside the helix, all outside the helix, or is their arrangement random?

All of the amino acid side chains extend outside the helix.

<u>Problem 27.56</u> From the diagram in Figure 27.15, what do you predict to be the direction of the dipole moment of a polypeptide α-helix?

The dipole moment in a peptide bond extends from the amide hydrogen atom on the positive end, to the carbonyl oxygen atom on the negative end.

As seen in the diagram in Figure 24.15, all of the peptide bonds are lined up in the same direction in an α-helix, with the carbonyl oxygen atoms oriented toward the C-terminus. Therefore, the overall positive to negative orientation of the dipole moment extends in the direction from the N to the C-terminus of the α-helix.

<u>Problem 27.57</u> The terms s-*cis* and s-*trans* are not as well defined for a peptide bond involving proline as they are for peptide bonds involving the other naturally occurring amino acids. Draw what you predict to be the more stable and less stable conformations of a peptide bond between proline and another amino acid.

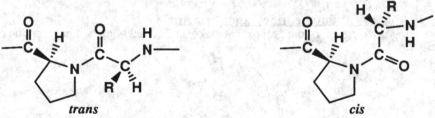

Both of these structures have about the same stability, so *cis* peptide bonds are often found in proteins at proline residues.

<u>Problem 27.58</u> Denaturation of a protein is a physical change, the most readily observable result of which is loss of biological activity. Denaturation stems from changes in secondary, tertiary, and quaternary structure through disruption of noncovalent interactions including hydrogen bonding and hydrophobic interactions. Two common denaturing agents are sodium dodecyl sulfate (SDS) and urea. What kinds of noncovalent interactions might each reagent disrupt?

The SDS disrupts the hydrophobic forces that keep the non-polar residues in the interior of the structure away from the aqueous solvent and the urea disrupts the hydrogen bonds of the protein.

CHAPTER 28: NUCLEIC ACIDS

28.0 OVERVIEW
• The **genetic information** inside living cells is stored and transmitted in the form of long stretches of deoxyribonucleic acid (DNA) often referred to as **genes**. The genetic information is relayed to the cell in **two stages, transcription** of the **DNA to ribonucleic acids (RNA)** and then **translation** of the **RNA** to give **proteins.** ✳

28.1 NUCLEOSIDES AND NUCLEOTIDES
• A **nucleoside** is a compound containing a **heterocyclic aromatic amine (base) bonded** to a monosaccharide, D-**ribose** or 2-**deoxy-D-ribose**, via a β-N-**glycoside bond.** ✳
  - The **heterocyclic bases** most common to the nucleic acids are **adenine (A)** and **guanine (G)**, both **purines**, as well as **cytosine (C)**, a **pyrimidine**. DNA also has the pyrimidine **thymidine (T)**, while RNA contains the pyrimidine **uracil (U)**. Uracil has a methyl group at the 5 position on the pyrimidine ring that is not found in thymine, otherwise they are the same.
  - The β-N-**glycoside bond** of nucleosides is between the **anomeric (C-1') carbon atom of the monosaccharide** and the N-**1 of the pyrimidine base** or N-**9 of the purine base**, respectively.
  - Nucleosides are named after the heterocyclic base as well as the type of monosaccharide attached. If the monosaccharide is D-ribose, then the common ribonucleosides are named **adenosine, guanosine, cytidine, and uridine**. If the monosaccharide is 2-deoxy-D-ribose then the common deoxyribonucleosides are named **2'-deoxyadenosine, 2'-deoxyguanosine, 2'-deoxycytidine, and 2'-deoxythymidine**.
• A **nucleotide** is a **nucleoside** that has one or more **phosphate groups** attached to one of the hydroxyl groups, usually at the **5' and/or 3' positions**. Nucleotides usually have between one and three phosphate groups attached, and are **named according** to the **nucleoside** present, followed by the position and number of phosphate groups attached. For example, **adenosine 5'-triphosphate (ATP)** is the name given the molecule that has adenosine with three phosphates attached at the 5'-position of the D-ribose ring. The three phosphates groups are held together via phosphate anhydride linkages. The phosphate groups are deprotonated and thus negatively-charged at neutral pH. ✳

28.2 THE STRUCTURE OF DNA
• There are **three levels of structural complexity of DNA.** ✳
• The **primary structure** of a nucleic acid refers to the **sequence of nucleotides** that are **linked via single phosphate** units between the **5' position of one nucleotide** and the **3' position** of the **adjacent nucleotide**. These chains of nucleotides linked 5' to 3' with phosphates can be extremely long. Because of this structure, the backbone is often referred to as being a **sugar-phosphate backbone**. Because the phosphodiester units have a negative charge, the backbone is polyanionic. ✳
• The **secondary structure of DNA** is **best thought of as a double helix** as described by Watson and Crick. ✳
  - In the **DNA double helix**, the two strands of nucleic acids are **antiparallel**. In other words, one strand is in the **5' to 3' direction**, while the other is in the **3' to 5' direction**.
  - The **bases project inward toward the axis** of the double helix, and they pair in a specific manner with the bases of the opposite strand. **Guanine makes three specific hydrogen bonds with cytosine, and adenine makes two specific hydrogen bonds with thymine**. The specific pattern of hydrogen bonding ensures that these are the only two sets of base pairs normally observed. This means that the two strands are held together by these hydrogen bonds, so the sequences must be complementary for them to be paired in a double helix. ✳
  - Since each of these base pairs is composed of a purine and a pyrimidine, they are the **same general size** and the double helix is relatively regular in structure.
• There are **several forms** of the **double helix**.
  - The most common type of DNA helix is called the **B-form helix**. It is a **"right-handed helix"** with so-called minor and major grooves of similar depths but different widths.
  - An **A-form helix** is also known. It is also a right-handed helix, but it has a **different conformation** of the **2'-deoxy-D-ribose ring** leading to different grooves and a slightly different number of bases per turn of the helix.
  - A so-called **Z-form helix** has also been found that is a **left-handed helix**.

- The **different types of helices** can be **interconverted** depending on parameters such as **temperature, ionic strength, polarity of the solvent,** and the nature of the **cations** associated with the negatively-charged helix backbone.
- The **long pieces of DNA** are **flexible** and can exhibit **tertiary structure** called **supercoiling.** Supercoiling involves a **different number of helical turns** than normal along the DNA helix. Supercoiling is observed in circular pieces of DNA, as well as in long pieces of linear DNA wound around histone proteins.

28.3 RIBONUCLEIC ACIDS (RNA)
- RNA is different from DNA in three important ways:
  - β-**D-Ribose** is found in **RNA** instead of the β-D-2'-deoxyribose found in DNA.
  - The pyrimidine base uracil is found in RNA instead of the thymine found in DNA.
  - **RNA is single-stranded,** rather than double-stranded like DNA.
- RNA is found in three major forms within the cell, listed in decreasing order of abundance:
  - **Ribosomal RNA (rRNA)** is present in **ribosomes, the particles responsible for protein synthesis.** Ribosomes contain 60% RNA and 40% protein.
  - **Transfer RNA (tRNA)** are relatively small nucleic acid molecules, 73-94 bases in length, that **carry amino acids to the appropriate sites of protein synthesis** on the ribosome.
  - **Messenger RNA (mRNA)** are short-lived, single-stranded pieces of RNA that result from transcription of DNA. The mRNA serves as the actual template for protein synthesis on the ribosome.

28.4 THE GENETIC CODE
- The **genetic code is a triplet code** since three nucleic acid bases code for a single amino acid or a "stop" signal. There are 4^3 **or 64 possible sequences** of three nucleic acid bases, and these code for 20 different amino acids and stop signals. Therefore, the **genetic code is degenerate** in that **different sequences can code for the same amino acid,** and there are three different "stop" sequences.

28.5 SEQUENCING DNA
- **Sequencing of nucleic acids** is accomplished by **selectively cleaving** a strand of nucleic acids at a given base or bases, then **separating** the cleaved fragments by **electrophoresis on a polyacrylamide gel.** The individual steps in this process are as follows:
  - Double stranded DNA is cleaved at specific sites using enzymes called **restriction enzymes** that selectively cleave a 4-8 base sequence of DNA. The resulting fragments, referred to as **restriction fragments,** are purified. A variety of restriction enzymes are commercially available.
  - Both strands of the restriction fragment are **labeled with radioactive** ^{32}P in the form of phosphate that is added to the **5'-OH group** *via* **an enzyme-catalyzed reaction.** This produces a restriction fragment that has two ^{32}P labels; one ^{32}P on the 5' end of each strand.
  - In order for sequencing to take place in an unambiguous manner, only one strand can have a ^{32}P label. This is because trying to carry out sequencing reactions with two different ^{32}P labels on the same sequence of DNA will lead to double ^{32}P signals on the polyacrylamide gel.
    In theory, the two labeled single strands could be separated by heating, then isolated as single-stranded pieces of DNA to be sequenced individually. In practice, this is rarely done because of the difficulty associated with the isolation of single-stranded fragments of the same length.
    More often, a new restriction enzyme is used to cut the doubly ^{32}P-labeled fragment in the middle, thereby creating two new shorter fragments, each with only a single ^{32}P label at the 5' end of one of the strands. These singly-labeled fragments are easily separated and isolated because they are generally of different lengths. The singly-labeled fragments are sequenced individually.
  - The singly-labeled DNA fragments are placed into separate equivalent samples, that are then subjected to **limited base-specific cleavage reactions.** Each sample undergoes a reaction with different base specificity, and conditions are adjusted so that, on average, **each strand is cleaved only once.**
    There are chemical, base-specific cleavage reactions that cleave the DNA at **G, G or A, C, and T residues.** These reactions are run separately.
  - The cleaved fragments are subjected to **electrophoresis on a polyacrylamide gel,** where they separate according to size. The **radioactivity from the** ^{32}P **label is used to visualize** the different fragments by exposing the finished gel to **photographic film.**
  - Shorter pieces of DNA run faster than longer pieces of DNA on the polyacrylamide gels. Therefore, the **lengths of the fragments** in the different lanes correspond to **locations of the different bases.** For example, the G cleavage reaction will generate fragments of DNA corresponding in length to the locations of the G residues, and so on. When the lengths of the fragments for each of the different cleavage reaction are

compared, an entire sequence can be determined. Up to 400 or more bases can be sequenced on a single polyacrylamide gel.

- A different strategy based on an enzyme called a polymerase and chain terminating 2',3'-dideoxy nucleotides has been developed to generate labeled sequences of different lengths for use in sequencing. It is this latter method that is generally used in automated DNA sequencing.

CHAPTER 28
Solutions to the Problems

Problem 28.1 Draw structural formulas for these compounds.
(a) 2'-Deoxythymidine 5'-monophosphate

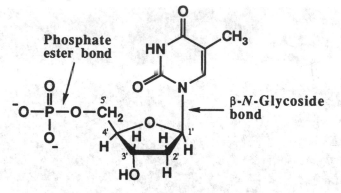

(b) 2'-Deoxythymidine 3'-monophosphate

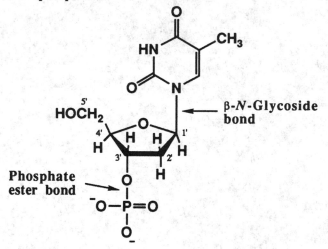

Problem 28.2 Write the structural formula for the section of DNA that contains the base sequence CTG and is phosphorylated on the 3' end only.

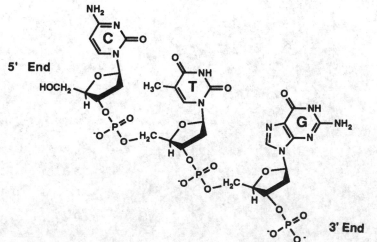

Problem 28.3 Write the complementary DNA base sequence for 5'-CCGTACGA-3'.

The complementary sequence would be 3'-GGCATGCT-5'

Problem 28.4 Here is a portion of the nucleotide sequence in phenylalanine tRNA.

3'-ACCACCUGCUCAGGCCUU-5'

Write the nucleotide sequence of its DNA complement.

Remember that the base uracil (U) in RNA is complementary to adenine (A) in DNA. The complement DNA sequence of the above RNA sequence would be:

5'-TGGTGGACGAGTCCGGAA-3'.

Problem 28.5 The following section of DNA codes for oxytocin, a polypeptide hormone.

3'-ACG-ATA-TAA-GTT-TTA-ACG-GGA-GAA-CCA-ACT-5'

(a) Write the base sequence of the mRNA synthesized from this section of DNA.

The base sequence of the mRNA synthesized from this section of DNA would be:

5'-UGC-UAU-AUU-CAA-AAU-UGC-CCU-CUU-GGU-UGA-3'

(b) Given the sequence of bases in part (a), write the primary structure of oxytocin.

The primary sequence of oxytocin would be:

Amino terminus- Cys-Tyr-Ile-Gln-Asn-Cys-Pro-Leu-Gly -Carboxyl terminus

Note how the last codon, UGA, does not code for an amino acid, but rather is the stop signal.

Problem 28.6 The following is another section of the bovine rhodopsin gene. Which of the endonucleases given in Example 28.6 will catalyze cleavage of this section.

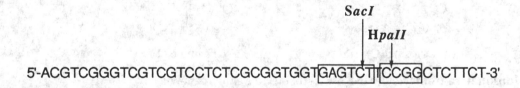

The SacI and HpaII cleavage sites are shown on the sequence above.

Problem 28.7 In what order will the excision fragments in Example 28.7 appear on the developed photographic plate? Remember that only the 5' end of the original restriction fragment is labeled with phosphorus-32.

On the polyacrylamide gel, shorter fragments migrate faster. Thus, fragment (i) will be closest to the bottom, fragment (ii) will be in the middle, and fragment (iii) will be closest to the top of the gel.

PROBLEMS
Nucleosides and Nucleotides
<u>Problem 28.8</u> Two important drugs in the treatment of acute leukemia are 6-mercaptopurine and 6-thioguanine. In each of these drugs, the oxygen at carbon 6 of the parent molecule is replaced by divalent sulfur. Draw structural formulas for the enethiol forms of 6-mercaptopurine and 6-thioguanine.

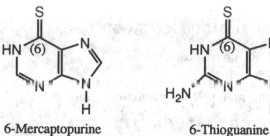

6-Mercaptopurine   6-Thioguanine

The enethiol forms are shown below:

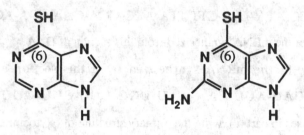

<u>Problem 28.9</u> Following are structural formulas for cytosine and thymine. Draw two additional tautomeric forms for cytosine and three additional tautomeric forms for thymine.

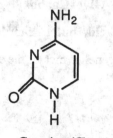   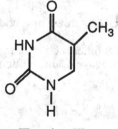

Cytosine (C)   Thymine (T)

Three additional tautomeric forms for cytosine are shown here:

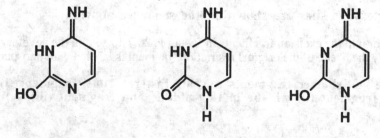

Four additional tautomeric forms for thymine are shown here:

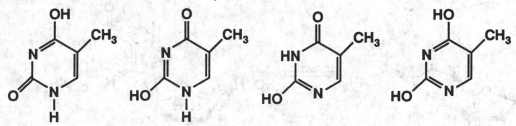

<u>Problem 28.10</u> Draw structural formulas for a nucleoside composed of
(a) β-D-Ribose and adenine

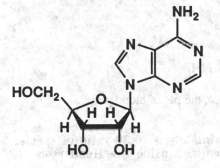

(b) β-D-Deoxyribose and cytosine

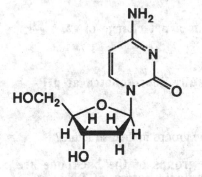

<u>Problem 28.11</u> Nucleosides are stable in water and in dilute base. In dilute acid, however, the glycoside bond of a nucleoside undergoes hydrolysis to give a pentose and a heterocyclic aromatic amine base. Propose a mechanism for this acid-catalyzed hydrolysis.

Acid-catalyzed glycoside bond hydrolysis is most pronounced for purine nucleosides. A reasonable mechanism involves protonation of the heterocyclic base to create a good leaving group that is displaced by water to produce the product pentose and free base. The reaction of guanosine in acid is shown below.

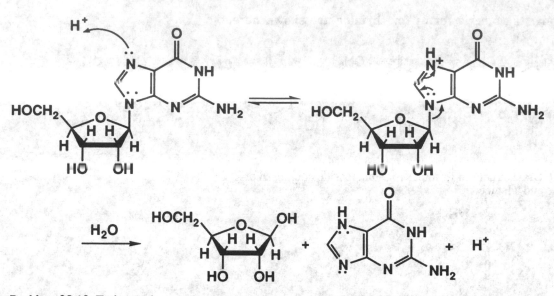

<u>Problem 28.12</u> Estimate the net charge on the following nucleotides at pH 7.4, the pH of blood plasma:
(a) ATP

The fourth pK_a of ATP is 7.0, so that at pH 7.4 there is approximately a 70:30 ratio of species with a net charge of -4 or -3, respectively. This ratio was determined using the Henderson-Hasselbalch equation (Section 27.2B).

(b) GMP

The phosphate groups are fully deprotonated at pH 7.4, so GMP has a net charge of -2.

(c) dGMP

Regardless of the type of sugar residue, the phosphate groups are fully deprotonated at pH 7.4, so dGMP also has a net charge of -2.

The Structure of DNA
<u>Problem 28.13</u> Why are deoxyribonucleic acids called acids? What are the acidic groups in their structure?

Deoxyribonucleic acids are called acids because the phosphodiester groups of the backbone are acidic. At neutral pH, they are fully deprotonated, leading to the anionic nature of DNA.

<u>Problem 28.14</u> Human DNA contains approximately 30.4% A. Estimate the percentages of G, C, and T and compare them with the values presented in Table 28.1.

The A residues must be paired with T residues, so estimate that there is also 30.4% T. A and T must therefore account for 30.4% + 30.4% = 60.8% of the bases. That leaves (100% - 60.8%) / 2 = 39.2% / 2 = 19.6% each for G and C. In Table 28.1, there is actually slightly less T than expected, so there is also slightly more G and C than expected.

Problem 28.15 Draw the structural formula of the DNA tetranucleotide 5'-A-G-C-T-3'. Estimate the net charge on this tetranucleotide at pH = 7.0. What is the complementary tetranucleotide to this sequence?

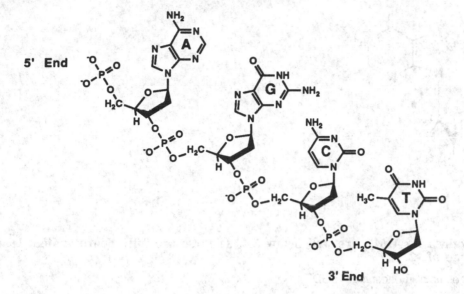

As shown in the preceding structure, there is a net charge of -5 on this tetranucleotide at pH 7.0. This oligonucleotide is self-complementary, that is the complementary oligonucleotide also has the sequence 5'-A-G-C-T-3'.

Problem 28.16 Write the DNA complement for 5'-ACCGTTAAT-3'. Be certain to label which is the 5' end and which is the 3' end of the complement strand.

The complementary sequence is 3'-TGGCAATTA-5'

Problem 28.17 Write the DNA complement for 5'-TCAACGAT-3'.

The complementary sequence is 3'-AGTTGCTA-5'

Problem 28.18 Write the structural formula for each nucleotide and estimate its net charge at pH 7.4, the pH of blood plasma.
(a) 2'-Deoxyadenosine 5'-triphosphate (dATP)

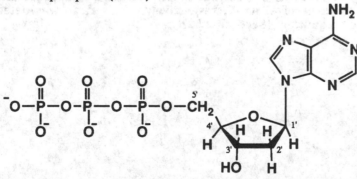

The values for the first three pK_a's of dATP are all below 5.0, so these are fully deprotonated at pH 7.4. The fourth pK_a of dATP is 7.0, so that at pH 7.4 there is approximately a 70:30 ratio of species with a net charge of -4 or -3, respectively. This ratio was determined using the Henderson-Hasselbalch equation (Section 27.2B).

(b) Guanosine 3'-monophosphate (GMP)

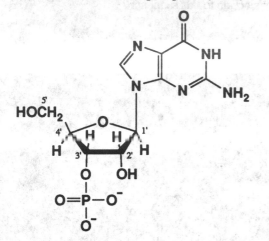

The two pK_a values for GMP are well below 7.4, so these are fully deprotonated, leading to an overall charge of -2.

(c) 2'-Deoxyguanosine 5'-diphosphate (dGDP)

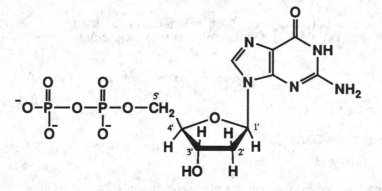

The values for the first two pK_a's of dGDP are all below 5.0, so these are fully deprotonated at pH 7.4. The third pK_a of dATP is 6.7, so that at pH 7.4 there is approximately a 83:17 ratio of species with a net charge of -3 or -2, respectively. This ratio was determined using the Henderson-Hasselbalch equation (Section 27.2B).

Problem 28.19 Cyclic-AMP, first isolated in 1959, is involved in many diverse biological processes as a regulator of metabolic and physiological activity. In it, a single phosphate group is esterified with both the 3' and 5' hydroxyls of adenosine. Draw the structural formula of cyclic-AMP.

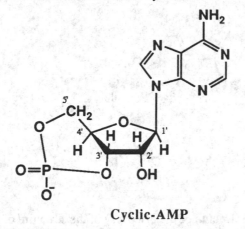

Cyclic-AMP

Problem 28.20 Discuss the role of the hydrophobic interactions in stabilizing:
(a) Double-stranded DNA

In the DNA double helix, the relatively hydrophobic bases are stacked on the inside, surrounded by the relatively hydrophilic sugar-phosphate backbone that is on the outside of the structure. The stacking of the hydrophobic bases minimizes contact with water.

(b) Lipid bilayers

In lipid bilayers, the hydrophobic hydrocarbon tails are associated with each other to form the hydrophobic inner layer, while the polar head groups are associated with each other on both outside surfaces.

(c) Soap micelles

In micelles, the hydrophobic hydrocarbon tails are associated with each other to form the hydrophobic interior, while the polar groups are associated with each other on the outside surface.

Problem 28.21 At elevated temperatures, nucleic acids become denatured, that is, they unwind into single-stranded DNA. Account for the observation that the higher the G-C content of a nucleic acid, the higher the temperature required for its thermal denaturation.

G-C base pairs have three hydrogen bonds between them, while A-T base pairs have only two. Thus, the G-C base pairs are held together with stronger overall attractive forces and require higher temperatures to denature.

Problem 28.22 The Watson-Crick pattern of hydrogen bonding is not the only type of interaction possible for nucleic acids. Draw the structure of an A-T base pair in which the purine uses N-7 instead of N-1 as a hydrogen bond acceptor.

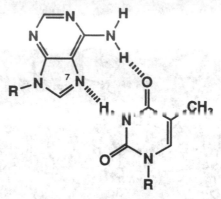

Problem 28.23 Reading J.D. Watson's account of the discovery of the structure of DNA, *The Double Helix*, you will find that for a time in their model building studies, he and Crick were using alternative (and incorrect, at least in terms of their final model of the double helix) tautomeric structures for some of the heterocyclic bases.
(a) Write at least one alternative tautomeric structure for adenine.

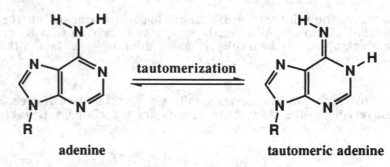

adenine   tautomeric adenine

(b) Would this structure still base-pair with thymine, or would it now base-pair more efficiently with a different base and if so, with what base?

This tautomeric form of adenine would not be able to base pair with thymine, but it would be able to form a reasonable base pair with cytosine, as shown below:

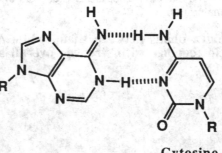

Cytosine

Problem 28.24 The following questions deal with the chemistry and physical properties of the sulfur and nitrogen mustard compounds discussed in the box Chemistry in Action: Mustard Gases and the Treatment of Neoplastic Diseases.
(a) Account for the fact that sulfur mustards undergo more rapid reaction with nucleophiles than do nitrogen mustards.

The sulfur mustard cyclizes to form a positively-charged sulfonium species, which is therefore more reactive with nucleophiles than the neutral aziridine formed by the nitrogen mustard. In addition, a sulfide is a better nucleophile than an amine, so the sulfur mustard will cyclize more readily.

(b) Account for the fact that substitution of phenyl for methyl in a nitrogen mustard decreases the nucleophilicity of nitrogen.

The lone pair of electrons on nitrogen are partially delocalized into the phenyl ring, thereby reducing nucleophilicity compared with the methyl derivative.

(c) Account for the fact that substitution of carboxyl in the para position of the aromatic ring further decreases the nucleophilicity of nitrogen.

The carboxyl group is electron withdrawing, thus increasing the delocalization of the nitrogen lone pair into the ring. This further reduces nucleophilicity.

(d) Account for the fact that N-7 of guanine is more nucleophilic than -NH$_2$ at C-2; than N-1; than N-9.

The N-7 position of guanine is highly nucleophilic, because the lone pair of electrons on N-7 is in an sp^2 hybrid orbital that is pointing out and away from the aromatic ring, preventing any delocalization. The lone pair of electrons on N-9 is in a 2p orbital and is part of the aromatic π system, so it is not nucleophilic. The lone pair of electrons on N-1 and on the exocyclic -NH$_2$ group of C-2 are partially delocalized by resonance, thus limiting their nucleophilicity.

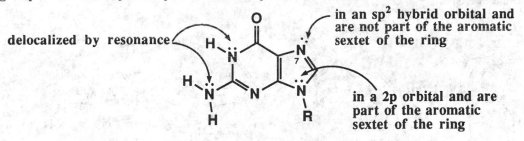

in an sp^2 hybrid orbital and are not part of the aromatic sextet of the ring

delocalized by resonance

in a 2p orbital and are part of the aromatic sextet of the ring

(e) Draw the structural formula for the product of reaction of two molecules of guanine, each at N-7, with one molecule of nitrogen mustard, that is, show how a nitrogen mustard cross-links DNA.

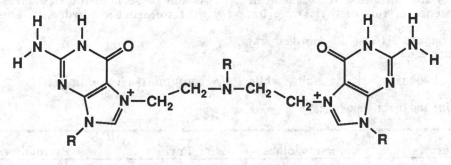

(f) Consider a guanine alkylated at N-7 by a nitrogen mustard. Draw the enol form of this guanine.

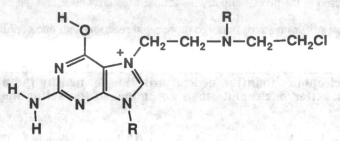

(g) Account for the fact that the enol form of guanine hydrogen bonds with thymine rather than with cytosine.

As shown below, the enol form of guanine could potentially make three hydrogen bonds with thymine.

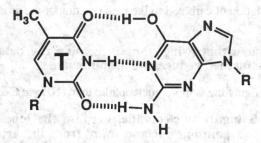

Ribonucleic Acids (RNA)
Problem 28.25 Compare the degree of hydrogen bonding in the base pair A-T found in DNA with that in the base pair A-U found in RNA.

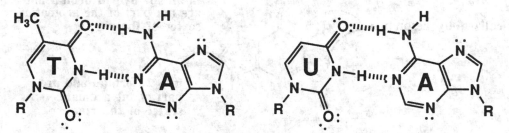

The only difference between uracil (U) and thymine (T) is the presence of a methyl group at the 5 position of thymine, that is absent in uracil. As can be seen in the structures, the presence or absence of this methyl group has very little influence on hydrogen bonding.

Problem 28.26 Compare DNA and RNA in these ways:
(a) Monosaccharide units

DNA contains 2'-deoxy-D-ribose units, while RNA contains D-ribose units.

(b) Principal purine and pyrimidine bases

DNA		RNA	
Purines	Pyrimidines	Purines	Pyrimidines
Adenine	Thymine	Adenine	Uracil
Guanine	Cytosine	Guanine	Cytosine

(c) Primary structure

The monosaccharide unit in DNA is 2'-deoxy-D-ribose, the monosaccharide in RNA is D-ribose. The bases are the same between the two types of nucleic acids, except thymine is found in DNA while uracil is found in RNA. DNA is usually double stranded and RNA is primarily single stranded. In both DNA and RNA, the primary sequence consists of linear chains of the nucleic acids linked by phosphodiester bonds involving the 3' and 5' hydroxyl groups of the monosaccharide units.

(d) Location in the cell

DNA is found in cell nuclei, while the bulk of RNA occurs as ribosome particles in the cytoplasm.

(e) Function in the cell

DNA serves to store and transmit genetic information, and RNA is primarily involved with the transcription and translation of that genetic information during the production of proteins.

Problem 28.27 What type of RNA has the shortest lifetime in cells?

Messenger RNA has the shortest lifetime in cells, usually on the order of a few minutes or less. This short lifetime is thought to allow for very tight control over how much protein is produced in the cell at any one time.

Problem 28.28 Write the mRNA complement for 5'-ACCGTTAAT-3'. Be certain to label which is the 5' end and which is the 3' end of the mRNA strand.

The mRNA complement would be 3'-UGGCAAUUA-5'

Problem 28.29 Write the mRNA complement for 5'-TCAACGAT-3'.

The mRNA complement would be 3'-AGUUGCUA-5'

The Genetic Code
Problem 28.30 What does it mean to say that the genetic code is degenerate?

The genetic code is referred to as degenerate because more than one codon can code for the same amino acid. This is because there are 64 different codons, but only twenty amino acids and a stop signal for which coding is needed.

Problem 28.31 Write the mRNA codons for
(a) Valine GUU, GUC, GUA, GUG   (b) Histidine CAU, CAC
(c) Glycine   GGU, GGC, GGA, GGG

Problem 28.32 Aspartic acid and glutamic acid have carboxyl groups on their side chains and are called acidic amino acids. Compare the codons for these two amino acids.

All of the codons for these two acidic amino acids begin with GA. The codons for aspartic acid are GAU and GAC, while the codons for glutamic acid are GAA and GAG.

Problem 28.33 Compare the structural formulas of the amino acids phenylalanine and tyrosine. Compare also the codons for these two amino acids.

Phenylalanine   Tyrosine

Phenylalanine has a phenyl group while tyrosine has a phenol group. The mRNA codons for phenylalanine are UUU and UUC, while the mRNA codons for tyrosine are UAU and UAC.

<u>Problem 28.34</u> Glycine, alanine, and valine are classified as nonpolar amino acids. Compare the codons for these three amino acids. What similarities do you find? What differences do you find?

Glycine	Alanine	Valine
GGU	GCU	GUU
GGC	GCC	GUC
GGA	GCA	GUA
GGG	GCG	GUG

All of these amino acids have four mRNA codons, all codons start with G, and in each case, the first two bases of the codon are identical for a given amino acid. This makes the last base irrelevant.

<u>Problem 28.35</u> Codons in the set CUU, CUC, CUA, and CUG all code for the amino acid leucine. In this set, the first and second bases are identical, and the identity of the third base is irrelevant. For what other sets of codons is the third base also irrelevant, and for what amino acid(s) does each set code?

The third base is also irrelevant for GUX (valine), GCX (alanine), GGX (glycine), ACX (threonine), CCX (proline), CGX (arginine), and UCX (serine). In the preceding codons, X stands for any of the bases.

<u>Problem 28.36</u> Compare the codons with a pyrimidine, either U or C, as the second base. Do the majority of the amino acids specified by these codons have hydrophobic or hydrophilic side chains?

The majority of amino acids with a pyrimidine in the second position of their codons are hydrophobic. This set contains phenylalanine, leucine, isoleucine, methionine, valine, proline, and alanine. Only serine and threonine have a pyrimidine in the second position and also have a somewhat hydrophilic side chain.

<u>Problem 28.37</u> Compare the codons with a purine, either A or G, as the second base. Do the majority of the amino acids specified by these codons have hydrophilic or hydrophobic side chains?

The majority of amino acids with a purine in the second position of their codons are hydrophilic. This set contains histidine, glutamine, asparagine, lysine, aspartic acid, glutamic acid, arginine, cysteine, and serine. Only glycine and tryptophan are not hydrophilic, while tyrosine is a special case that is aromatic with a polar group.

<u>Problem 28.38</u> What polypeptide is coded for by this mRNA sequence?

5'-GCU-GAA-GUC-GAG-GUG-UGG-3'

This mRNA codes for the following polypeptide:

Amino terminus- Ala-Glu-Val-Glu-Val-Trp -Carboxyl terminus.

<u>Problem 28.39</u> The alpha chain of human hemoglobin has 141 amino acids in a single polypeptide chain. Calculate the minimum number of bases on DNA necessary to code for the alpha chain. Include in your calculation the bases necessary for specifying termination of polypeptide synthesis.

The minimum number of bases needed for the alpha chain of human hemoglobin must code for the 141 amino acids as well as three extra bases for the stop codon. Therefore, the minimum number of bases that will be required is (3 x 141) + (1 x 3) = 426 bases.

Problem 28.40 In HbS, the human hemoglobin found in individuals with sickle-cell anemia, glutamic acid at position 6 in the beta chain is replaced by valine.
(a) List the two codons for glutamic acid and the four codons for valine.

The two mRNA codons for glutamic acid are GAA and GAG, while the four mRNA codons for valine are GUU, GUC, GUA, and GUG.

(b) Show that one of the glutamic acid codons can be converted to a valine codon by a single substitution mutation, that is, by changing one letter in the codon.

Both of the glutamic acid codons can be converted to valine by replacing the central A with a U residue.